STUDENT'S SOLUTIONS MANUAL

ROCKSWOLD • KRIEGER • SCHNEIDER
BLOCK • PURCELL

coordinated by
TERRY KRIEGER

ALGEBRA AND TRIGONOMETRY WITH MODELING AND VISUALIZATION

THIRD EDITION

AND

PRECALCULUS WITH MODELING AND VISUALIZATION

THIRD EDITION

Gary Rockswold

Minnesota State University, Mankato

PEARSON

Addison
Wesley

Boston San Francisco New York
London Toronto Sydney Tokyo Singapore Madrid
Mexico City Munich Paris Cape Town Hong Kong Montreal

Reproduced by Pearson Addison-Wesley from electronic files supplied by the author.

Copyright © 2006 Pearson Education, Inc.
Publishing as Pearson Addison-Wesley, 75 Arlington Street, Boston, MA 02116.

ISBN   0-321-28074-1

    6 7 8 BRR 08 07

Preface

This solutions manual is written as an aid for students studying the text *College Algebra and Trigonometry with Modeling and Visualization 3/e* by Gary Rockswold. It contains answers and solutions to the odd-numbered exercises.

This manual is consistent with the text and gives solutions that are accessible to college algebra students. The solutions are written in a clear format with hundreds of graphs and tables to help with the learning process. Solutions frequently include not only symbolic solutions but also graphical and numerical solutions to help students understand the problem completely. It is recommended that you make a genuine attempt to solve a problem before looking in this solutions manual for help. Regular class attendance is also recommended. Following these suggestions will promote insight and understanding of college algebra.

It is the author's hope that you will find this manual helpful when studying the text *College Algebra and Trigonometry with Modeling and Visualization 3/e*. Please feel free to send comments to the email address below. Your opinion is important. Best wishes for an enjoyable and successful college algebra and trigonometry course.

Terry A. Krieger

tkrieger@charter.net

Table of Contents

Chapter 1: Introduction to Functions and Graphs

1.1: Numbers, Data and Problem Solving

Classifying Numbers

1. 45,000 is a real, rational, and natural number. It is also an integer.

3. -3 is an integer, real number, and rational number.

5. 7.5 is a real and rational number.

7. $90\sqrt{2}$ is a real number.

9. Natural number: $\sqrt{9} = 3$; integers: -3, $\sqrt{9}$; rational numbers: $-3, \frac{2}{9}, \sqrt{9}, 1.\overline{3}$; irrational numbers: $\pi, -\sqrt{2}$

11. Natural number: None; integer: $-\sqrt{4} = -2$; rational numbers: $\frac{1}{3}, 5.1 \times 10^{-6}, -2.33, 0.\overline{7}, -\sqrt{4}$; irrational number: $\sqrt{13}$

13. Shoe sizes are normally measured to within half sizes. Rational numbers are most appropriate.

15. Gasoline is usually measured to a fraction of a gallon using rational numbers.

17. Temperature is typically measured to the nearest degree in a weather forecast. Since temperature can include negative numbers, the integers would be most appropriate.

Real Number Computation

19. Percent change $= \dfrac{13 - 8}{8} \times 100 = \dfrac{5}{8} \times 100 = 0.625 \times 100 = 62.5\%$

21. Percent change $= \dfrac{0.85 - 1.4}{1.4} \times 100 = -0.393 \times 100 = -39.3\%$

23. $185,800 = 1.858 \times 10^5$

25. $0.03892 = 3.892 \times 10^{-2}$

27. $2450 = 2.45 \times 10^3$

29. $0.56 = 5.6 \times 10^{-1}$

31. $-0.0087 = -8.7 \times 10^{-3}$

33. $206.8 = 2.068 \times 10^2$

35. $854,000 = 8.54 \times 10^5$

37. $10^{-6} = 0.000001$

39. $2 \times 10^8 = 200,000,000$

41. $1.567 \times 10^2 = 156.7$

43. $-5.68 \times 10^{-1} = -0.568$

45. $5 \times 10^5 = 500,000$

47. $0.045 \times 10^5 = 4500$

49. $67 \times 10^3 = 67{,}000$

51. $(4 \times 10^3)(2 \times 10^5) = 4 \cdot 2 \times 10^{3+5} = 8 \times 10^8;\ 800{,}000{,}000$

53. $(5 \times 10^2)(7 \times 10^{-4}) = 5 \cdot 7 \times 10^{2-4} = 35 \times 10^{-2} = 3.5 \times 10^{-1};\ 0.35$

55. $(1.2 \times 10^{-3})(1.2 \times 10^{-3}) = (1.2 \times 1.2) \times 10^{-3+(-3)} = 1.44 \times 10^{-6};\ 0.00000144$

57. $\dfrac{4 \times 10^4}{2 \times 10^3} = \dfrac{4}{2} \times 10^{4-3} = 2 \times 10^1;\ 20$

59. $\dfrac{6.3 \times 10^{-2}}{3 \times 10^1} = \dfrac{6.3}{3} \times 10^{-2-1} = 2.1 \times 10^{-3};\ 0.0021$

61. $\dfrac{4 \times 10^{-3}}{8 \times 10^{-1}} = \dfrac{4}{8} \times 10^{-3-(-1)} = 0.5 \times 10^{-2} = 5 \times 10^{-3};\ 0.005$

63. $(9.87 \times 10^6)(34 \times 10^{11}) = 3.3558 \times 10^{19} \approx 3.36 \times 10^{19}$

65. $\left(\dfrac{101 + 23}{0.42}\right)^2 + \sqrt{3.4 \times 10^{-2}} \approx 87166 + 0.2 \approx 87166.2 \approx 8.72 \times 10^4$

67. $(8.5 \times 10^{-5})(-9.5 \times 10^7)^2 \approx (8.5 \times 10^{-5})(9.025 \times 10^{15}) \approx 76.7 \times 10^{10} \approx 7.67 \times 10^{11}$

69. $\sqrt[3]{192} \approx 5.769$

71. $|\pi - 3.2| \approx 0.058$

73. $\dfrac{(0.3 + 1.5)}{(5.5 - 1.2)} \approx 0.419$

75. $\dfrac{1.5^3}{\sqrt{2} + \pi - 5} \approx \dfrac{3.375}{-2.732} \approx -1.235$

77. $15 + \dfrac{4 + \sqrt{3}}{7} \approx 15.819$

Problem Solving

79. From 1985 to 2002 tuition and fees increased by $\dfrac{18{,}273 - 6{,}121}{6{,}121} \approx 198.5\%$ while the CPI increased by

    $\dfrac{179.9 - 107.6}{107.6} \approx 67.2\%$. The percent increase in the cost of tuition and fees is considerably more than the

    percent increase in the CPI.

81. The time-and-a-half hourly rate for the seven hours worked over the 40 hours would be $12 + \dfrac{1}{2}(12) = \$18$.

    The hourly rate for the 8 hours worked on Sunday would be $2(12) = \$24$. The employee's pay is

    $40(12) + 7(18) + 8(24) = \$798$.

83. The distance Mars travels around the sun is $2\pi r = 2\pi(141{,}000{,}000) \approx 885{,}929{,}128$ miles.

    The number of hours in 1.88 years is $365 \times 1.88 \times 24 \approx 16{,}469$ hours. So Mars' speed is

    $\dfrac{885{,}929{,}128}{16{,}469} \approx 53{,}794$ miles per hour.

85. (a) $\dfrac{3.7 \times 10^{11}}{2.03 \times 10^{8}} \approx \$1{,}820$ per person.

(b) $\dfrac{5.54 \times 10^{12}}{2.81 \times 10^{8}} \approx \$19{,}715$ per person.

(c) There are different ways to estimate this.

1) During the last 30 years the debt per person has increased by $19{,}715 - 1{,}820 = 17{,}895$.

Thus, during the next 30 years one might predict the debt per person to be $19{,}715 + 17{,}895 = \$37{,}610$.

2) During the past 30 years the debt per person has increased by a factor of $\dfrac{19{,}715}{1{,}820} \approx 10.8$.

During the next 30 years one might predict the debt to increase by a factor of 10.8 or

$10.8 \times \$19{,}715 = \$212{,}922$. *Answers may vary.*

87. The area of the film is πr^2 or $\pi(11.5)^2 \approx 415.5 \text{ cm}^2$ and the volume of the drop is 0.12 cm^3. The thickness of

the film is equal to the volume divided by the area. This is, $\dfrac{0.12}{415.5} \approx 2.9 \times 10^{-4}$ cm.

89. (a) It would take $\dfrac{5.54 \times 10^{12}}{100} \approx 5.54 \times 10^{10}$ or 55.4 billion $100-dollar bills to equal the federal debt. The

height of the stacked bills would be $\dfrac{5.54 \times 10^{10}}{250} \approx 2.216 \times 10^{8}$ inches or $\dfrac{2.216 \times 10^{8}}{12} \approx 18{,}466{,}667$ feet.

(b) There are 5280 feet in one mile so $\dfrac{18{,}466{,}667}{5280} \approx 3497$ miles. It would reach farther than the distance

between Los Angles and New York $\approx$ Yes

91. $\dfrac{4.5 + 4.2 + 4.0 + 4.7 + 5.8}{5} = \dfrac{23.2}{5} \approx 4.64\%$; this is not an integer but is a rational number $\Rightarrow$ No, Yes

93. (a) $V = \pi r^2 h \Rightarrow V = \pi(1.3)^2(4.4) \Rightarrow V = 7.436\pi \approx 23.4 \text{ in}^3$

(b) $1 \text{ in}^3 = 0.55$ fluid ounces $\Rightarrow 23.4 \cdot 0.55 = 12.87$ fluid ounces; Yes it can hold 12 fluid ounces.

95. (a) Alcohol consumption on a per capita basis increased each ten-year period between 1940 and 1980. Between

1980 and 1998 alcohol consumption declined.

(b) There were 22,513,000 people who consumed 2.19 gallons of alcohol on the average. This amounts to

$225{,}130{,}000 \times 2.19 \approx 4.9 \times 10^{8}$ gallons of alcohol.

(c) $\dfrac{2.19 - 2.76}{2.76} \times 100 \approx -20.7\%$

1.2: Visualization of Data

Data Involving One Variable

1. (a)

(b) Maximum: 6; minimum -2

(c) $\dfrac{3 + (-2) + 5 + 0 + 6 + (-1)}{6} \approx 1.8\overline{3}$

3. (a)

(b) Maximum: 30; minimum -20

(c) $\dfrac{(-10) + 20 + 30 + (-20) + 0 + 10}{6} = 5$

5.

-30	-30	-10	5	15	25	45	55	61

(a) The maximum is 61 and the minimum is -30.

(b) The average is $\dfrac{-30 - 30 - 10 + 5 + 15 + 25 + 45 + 55 + 61}{9} \approx 15.11$ and the median is 15. The range is $61 - (-30) = 91$.

7. $\sqrt{15} \approx 3.87$, $2^{2.3} \approx 4.92$, $\sqrt[3]{69} \approx 4.102$, $\pi^2 \approx 9.87$, $2^{\pi} \approx 8.82$, 4.1

$\sqrt{15}$	4.1	$\sqrt[3]{69}$	$2^{2.3}$	2^{π}	π^2

(a) The maximum is π^2 and the minimum is $\sqrt{15}$.

(b) The average is $\dfrac{\sqrt{15} + 4.1 + \sqrt[3]{69} + 2^{2.3} + 2^{\pi} + \pi^2}{6} \approx 5.95$ and the median is $\dfrac{\sqrt[3]{69} + 2^{2.3}}{2} \approx 4.51$. The range is $\pi^2 - \sqrt{15} \approx 6.00$.

9. (a) $S = \{(-1, 5), (2, 2), (3, -1), (5, -4), (9, -5)\}$

(b) $D = \{-1, 2, 3, 5, 9\}$

$R = \{-5, -4, -1, 2, 5\}$

11. (a) $S = \{(1, 5), (4, 5), (5, 6), (4, 6), (1, 5)\}$

(b) $D = \{1, 4, 5\}$

$R = \{5, 6\}$

13. (a)

(b) Mean: $\dfrac{840 + 227 + 280 + 196 + 306 + 165}{6} \approx 336$; Median $= \dfrac{227 + 280}{2} = 254$;

Range: $840 - 165 = 675$; The average area of the six largest islands in the world is 336,000 square miles. Half of the islands have areas of less than 250,000 square miles and half have more. The largest difference in area between any two islands is 675,000 square miles.

(c) The largest island is Greenland with 840,000 square miles.

15. (a)

(b) Average $= \dfrac{19.3 + 18.5 + 29.0 + 7.31 + 16.1 + 22.8 + 20.3}{7} \approx 19.0$; Median $= 19.3$;

Range $= 29.0 - 7.31 \approx 21.7$. The average of the maximum elevations of the seven continents is 19,000 feet. About half of these continents have maximum elevations below 19,300 feet and about half are above. The largest difference between these elevations is 21,700 feet.

(c) The mountain with the highest elevation is Mount Everest in Asia.

17. (a) $\dfrac{26{,}518{,}000 + 367{,}000(5)}{6} = 4{,}725{,}500$

(b) 367,000

(c) One large salary can raise the average considerably, while the majority of the players have considerably lower salaries.

19. *Answers may vary.* 16, 18, 26; No

Distance Formula

21. $d = \sqrt{(5-2)^2 + (2-(-2))^2} = \sqrt{3^2 + 4^2} = \sqrt{25} = 5$

23. $d = \sqrt{(9-7)^2 + (1-(-4))^2} = \sqrt{2^2 + 5^2} = \sqrt{29} \approx 5.39$

25. $d = \sqrt{(3.6-(-6.5))^2 + (-2.9-2.7)^2} = \sqrt{10.1^2 + (-5.6)^2} = \sqrt{133.37} \approx 11.55$

27. $d = \sqrt{(-3-(-3))^2 + (10-2)^2} = \sqrt{0^2 + 8^2} = \sqrt{64} = 8$

29. $d = \sqrt{\left(\dfrac{3}{4}-\dfrac{1}{2}\right)^2 + \left(\dfrac{1}{2}-\left(-\dfrac{1}{2}\right)\right)^2} = \sqrt{\left(\dfrac{1}{4}\right)^2 + 1^2} = \sqrt{\dfrac{1}{16}+1} = \sqrt{\dfrac{17}{16}} = \dfrac{\sqrt{17}}{4} \approx 1.03$

31. $d = \sqrt{\left(-\dfrac{1}{10}-\dfrac{2}{5}\right)^2 + \left(\dfrac{4}{5}-\dfrac{3}{10}\right)^2} = \sqrt{\left(-\dfrac{1}{2}\right)^2 + \left(\dfrac{1}{2}\right)^2} = \sqrt{\dfrac{1}{4}+\dfrac{1}{4}} = \sqrt{\dfrac{1}{2}} = 0.71$

33. $d = \sqrt{(-30-20)^2 + (-90-30)^2} = \sqrt{(-50)^2 + (-120)^2} = \sqrt{2500 + 14{,}400} = \sqrt{16{,}900} = 130$

35. $d = \sqrt{(0-a)^2 + (-b-0)^2} = \sqrt{(-a)^2 + (-b)^2} = \sqrt{a^2 + b^2}$

37. $d = \sqrt{(a-a)^2 + (a-b)^2} = \sqrt{0 + (a-b)^2} = \sqrt{(a-b)^2} = |a-b|$

39. $d = \sqrt{(3-0)^2 + (4-0)^2} = \sqrt{3^2 + 4^2} = \sqrt{9 + 16} = \sqrt{25} = 5$

$d = \sqrt{(7-3)^2 + (1-4)^2} = \sqrt{4^2 + (-3)^2} = \sqrt{16 + 9} = \sqrt{25} = 5$

The side between (0,0) and (3,4) and the side between (3,4) and (7,1) have equal length, so the triangle is isosceles.

41. (a) See Figure 41.

(b) $d = \sqrt{(0-(-40))^2 + (50-0)^2} = \sqrt{40^2 + 50^2} = \sqrt{1600 + 2500} = \sqrt{4100} \approx 64.0$ miles.

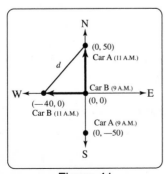

Figure 41

Midpoint Formula

43. Use the midpoint formula: $M = \left(\dfrac{1980 + 2000}{2}, \dfrac{77.4 + 79.5}{2} \right) = (1990, 78.45)$. According to the midpoint

    estimate the life expectancy was 78.45. This compares favorably to the actual life expectancy of 78.8.

45. Since 1986 is not between 1980 and 1983, we can think of the point (1983, 436,885) as the midpoint and the

    1986 data as an endpoint. Let y represent the number of inmates in 1986. The midpoint between the data

    points (1980, 329,821) and (1986, y) is equal to (1983, 436,885).

    $M = \left(\dfrac{1980 + 1986}{2}, \dfrac{329{,}821 + y}{2} \right) = (1983, 436{,}885)$. According to the midpoint estimate

    $\dfrac{329{,}821 + y}{2} = 436{,}885$ or $y = 543{,}949$. *Alternate solution:* A faster way to solve this exercise is to realize

    that the increase between 1980 and 1983 was $436{,}885 - 329{,}821 = 107{,}064$. Since the midpoint is midway

    between, the estimate for 1986 would be $436{,}885 + 107{,}064 = 543{,}949$.

47. Assuming the distance of 0 meters requires 0 second to run, the midpoint between the data points (0, 0) and

    (200, 20) is equal to $M = \left(\dfrac{0 + 200}{2}, \dfrac{0 + 20}{2} \right) = (100, 10)$. According to the midpoint estimate the time

    required to run 100 meters is 10 seconds or half the time required to run 200 meters.

49. $M = \left(\dfrac{1 + 5}{2}, \dfrac{2 + (-3)}{2} \right) = (3, -0.5)$

51. $M = \left(\dfrac{-30 + 50}{2}, \dfrac{50 - 30}{2} \right) = (10, 10)$

53. $M = \left(\dfrac{1.5 + (-5.7)}{2}, \dfrac{2.9 + (-3.6)}{2} \right) = (-2.1, -0.35)$

55. $M = \left(\dfrac{\sqrt{2} + \sqrt{2}}{2}, \dfrac{\sqrt{5} + (-\sqrt{5})}{2} \right) = (\sqrt{2}, 0)$

57. $M = \left(\dfrac{a + (-a)}{2}, \dfrac{b + 3b}{2} \right) = (0, 2b)$

59. $M = \left(\dfrac{(a + b) + (a - b)}{2}, \dfrac{(a - b) + (b - a)}{2} \right) = \left(\dfrac{2a}{2}, \dfrac{0}{2} \right) = (a, 0)$

Data Involving Two Variables

61. (a) The domain is $D = \{0, -3, -2, 7\}$ and the range is $R = \{5, 4, -5, -3, 0\}$

    (b) The minimum x-value is -3, and the maximum x-value is 7. The minimum y-value is -5, and the maximum

    y-value is 5.

    (c) The axes must include at least $-3 \le x \le 7$ and $-5 \le y \le 5$. It would be appropriate to extend each axis

    slightly beyond these intervals and let each tick mark represent 1 unit. Plot the points (0, 5), (-3, 4),

    (-2, -5), (7, -3), and (0, 0).

    (d) See Figure 61.

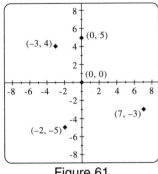

Figure 61

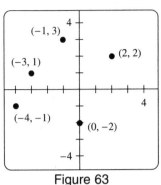

Figure 63

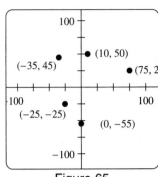

Figure 65

63. (a) The domain is $D = \{2, -3, -4, -1, 0\}$ and the range is $R = \{2, 1, -1, 3, -2\}$

    (b) The minimum x-value is -4, and the maximum x-value is 2. The minimum y-value is -2, and the maximum y-value is 3.

    (c) The axes must include at least $-4 \le x \le 2$ and $-2 \le y \le 3$. It would be appropriate to extend each axis slightly beyond these intervals. Plot the points $(2, 2)$, $(-3, 1)$, $(-4, -1)$, $(-1, 3)$, and $(0, -2)$.

    (d) See Figure 63.

65. (a) The domain is $D = \{10, -35, 0, 75, -25\}$ and the range is $R = \{50, 45, -55, 25, -25\}$

    (b) The minimum x-value is -35, and the maximum x-value is 75. The minimum y-value is -55, and the maximum y-value is 50.

    (c) The axes must include at least $-35 \le x \le 75$ and $-55 \le y \le 50$ (beyond these intervals)... and let each tick mark represent 5 units. It would be appropriate to extend each axis slightly beyond these intervals. Plot the points $(10, 50)$, $(-35, 45)$, $(0, -55)$, $(75, 25)$, and $(-25, -25)$.

    (d) See Figure 65.

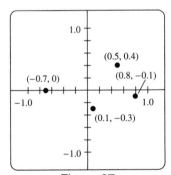

Figure 67

67. (a) The domain is $D = \{0.1, 0.5, -0.7, 0.8\}$ and the range is $R = \{-0.3, 0.4, 0, -0.1\}$

    (b) The minimum x-value is -0.7, and the maximum x-value is 0.8. The minimum y-value is -0.3, and the maximum y-value is 0.4.

    (c) The axes must include at least $-0.7 \le x \le 0.8$ and $-0.3 \le y \le 0.4$ (beyond these intervals)... and let each tick mark represent 0.1 units. It would be appropriate to extend each axis slightly beyond these intervals. Plot the points $(0.1, -0.3)$, $(0.5, 0.4)$, $(-0.7, 0)$, and $(0.8, -0.1)$.

    (d) See Figure 67.

69. *x*-axis: 10 tick marks; *y*-axis: 10 tick marks. See Figure 69.

71. *x*-axis: 10 tick marks; *y*-axis: 5 tick marks. See Figure 71.

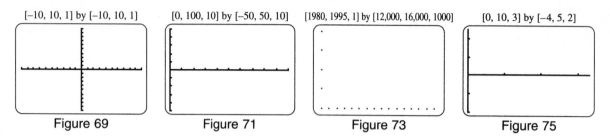

Figure 69   Figure 71   Figure 73   Figure 75

73. *x*-axis: 16 tick marks; *y*-axis: 5 tick marks. See Figure 73.

75. *x*-axis: 3 tick marks; *y*-axis: 2 tick marks. See Figure 75.

77. Graph (b)

79. Graph (a)

81. Plot the points $(1, 3)$, $(-2, 2)$, $(-4, 1)$, $(-2, -4)$ and $(0, 2)$ in $[-5, 5, 1]$ by $[-5, 5, 1]$, See Figure 81.

83. Plot the points $(10, -20)$, $(-40, 50)$, $(30, 60)$, $(-50, -80)$, $(70, 0)$, and $(0, -30)$

    in $[-100, 100, 10]$ by $[-100, 100, 10]$, See Figure 83.

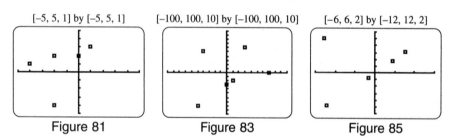

Figure 81   Figure 83   Figure 85

85. Plot the points $(3.1, 6.2)$, $(-5.1, 10.1)$, $(-0.7, -1.4)$, $(1.8, 3.6)$ and $(-4.9, -9.8)$ in $[-6, 6, 2]$ by $[-12, 12, 2]$,

    See Figure 85.

87. (a) *x*-min: 1979; *x*-max: 2001; *y*-min: 15; *y*-max: 37

    (b) [1978, 2002, 2] by [10, 40, 5]. *Answers may vary.*

    (c) Plot the points $(1979, 37)$, $(1985, 25)$, $(1988, 18)$, $(1993, 15)$ and $(2001, 22)$ in $[1978, 2002, 2]$ by $[10, 40, 5]$

        See Figure 87c.

    (d) See Figure 87d.

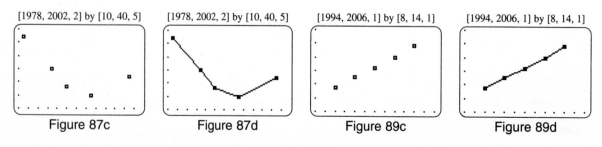

Figure 87c   Figure 87d   Figure 89c   Figure 89d

89. (a) *x*-min: 1996; *x*-max: 2004; *y*-min: 9.7; *y*-max: 12.8

    (b) [1994, 2006, 1] by [8, 14, 1]. *Answers may vary.*

    (c) Plot the points (1996, 9.7), (1998, 10.5), (2000, 11.2), (2002, 12.0) and (2004, 12.8) in

        [1994, 2006, 1] by [8, 14, 1], See Figure 89c.

    (d) See Figure 89d.

91. Number of doctorate degrees conferred in the United States. *Answers may vary.*

1.3: Functions and Their Representations

Evaluating and Representing Functions

1. If $f(-2) = 3$, then the point (–2, 3) is on the graph of f.

3. If (7, 8) is on the graph of f, then $f(7) = 8$.

5. See Figure 5.

7. See Figure 7.

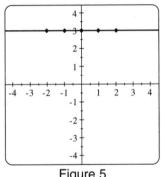

Figure 5

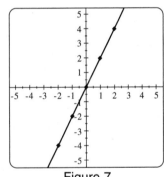

Figure 7

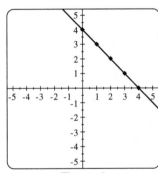

Figure 9

9. See Figure 9.

11. See Figure 11.

13. See Figure 13.

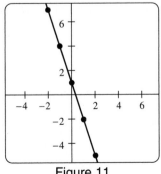

Figure 11

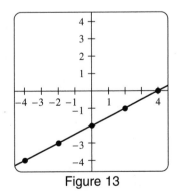

Figure 13

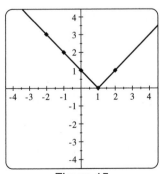

Figure 15

15. See Figure 15.

17. See Figure 17.

19. See Figure 19.

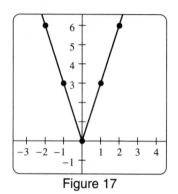

Figure 17

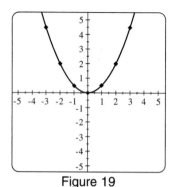

Figure 19

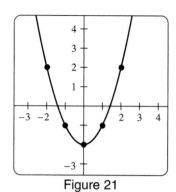

Figure 21

21. See Figure 21.

23. (a) $f(x) = x^3 \Rightarrow f(-2) = (-2)^3 = -8$ and $f(5) = 5^3 = 125$.

   (b) The domain of f includes all real numbers.

25. (a) $f(x) = \sqrt{x} \Rightarrow f(-1) = \sqrt{-1}$ which is not a real number, and $f(a+1) = \sqrt{a+1}$.

   (b) The domain of f includes all non-negative real numbers.

27. (a) $f(x) = \dfrac{1}{x-1} \Rightarrow f(-1) = \dfrac{1}{-1-1} = -\dfrac{1}{2}$ and $f(a+1) = \dfrac{1}{(a+1)-1} = \dfrac{1}{a}$.

   (b) The domain of f includes all real numbers not equal to 1 $(x \neq 1)$.

29. (a) $f(x) = -7 \Rightarrow f(6) = -7$ and $f(a-1) = -7$. (f is a constant function.)

   (b) The domain of f includes all real numbers.

31. (a) $f(x) = \dfrac{1}{x^2} \Rightarrow f(4) = \dfrac{1}{4^2} = \dfrac{1}{16}$ and $f(-7) = \dfrac{1}{(-7)^2} = \dfrac{1}{49}$.

   (b) The domain of f includes all real numbers not equal to 0 $(x \neq 0)$.

33. (a) $f(x) = \dfrac{1}{x^2 - 9} \Rightarrow f(4) = \dfrac{1}{4^2 - 9} = \dfrac{1}{16 - 9} = \dfrac{1}{7}$ and

   $$f(a-5) = \dfrac{1}{(a-5)^2 - 9} = \dfrac{1}{a^2 - 10a + 25 - 9} = \dfrac{1}{a^2 - 10a + 16}.$$

   (b) The domain of f includes all real numbers not equal to 3 or -3 $(x \neq 3, x \neq -3)$.

35. (a) $f(x) = \dfrac{1}{\sqrt{2-x}} \Rightarrow f(1) = \dfrac{1}{\sqrt{2-1}} = \dfrac{1}{\sqrt{1}} = \dfrac{1}{1} = 1$ and $f(a+2) = \dfrac{1}{\sqrt{2-(a+2)}} = \dfrac{1}{\sqrt{-a}}$

   (b) $2 - x > 0 \Rightarrow -x > -2 \Rightarrow x < 2$

37. (a) $D = \{$all real numbers$\}$

   (b) $g(x) = 2x - 1 \Rightarrow g(-1) = 2(-1) - 1 = -3$ and $g(2) = 2(2) - 1 = 3$

   (c) $g(-1) = -3$ and $g(2) = 3$

39. (a) $D = \{\text{all real numbers}\}$

    (b) $g(x) = 3 - 2|x| \Rightarrow g(-1) = 3 - 2|-1| = 3 - 2 = 1$ and $g(2) = 3 - 2|2| = 3 - 4 = -1$

    (c) $g(-1) = 1$ and $g(2) = -1$

41. (a) $D = \{\text{all real numbers}\}$

    (b) $g(x) = x^2 - 3 \Rightarrow g(-1) = (-1)^2 - 3 = 1 - 3 = -2$ and $g(2) = (2)^2 - 3 = 4 - 3 = 1$

    (c) $g(-1) = -2$ and $g(2) = 1$

43. The domain of f includes all real numbers x such that $-3 \le x \le 3$.

    The range of f includes all real numbers y such that $0 \le y \le 3$.

    $f(0) = 3$.

45. The domain of f includes all real numbers x such that $-2 \le x \le 4$.

    The range of f includes all real numbers y such that $-2 \le y \le 2$.

    $f(0) = -2$.

47. The domain of f includes all real numbers x.

    The range of f includes all real numbers y such that $y \le 2$.

    $f(0) = 2$.

49. The domain of f includes all real numbers x such that $x \ge -1$.

    The range of f includes all real numbers y such that $y \le 2$.

    $f(0) = 0$.

51. (a) $f(2) = 7$

    (b) $f = \{(1, 7), (2, 7), (3, 8)\}$

    (c) $D = \{1, 2, 3\}; \ R = \{7, 8\}$

53. (a) Since the point $(0, -2)$ lies on the graph of f, $f(0) = -2$. Similarly $f(2) = 2$.

    (b) We must find all points (x, y) on the graph of f where $y = 0$. There is only one point where this occurs: $(1, 0)$. Thus, when $x = 1, f(x) = 0$.

55. (a) Since the point $(0, 0)$ lies on the graph of f, $f(0) = 0$. Similarly $f(2) = 4$.

    (b) We must find all points (x, y) on the graph of f where $y = 0$. There is only one point where this occurs: $(0, 0)$. Thus, when $x = 0, f(x) = 0$.

57. (a) Since the point $(0, 0)$ lies on the graph of f, $f(0) = 0$. Similarly $f(2) = 0$.

    (b) We must find all points (x, y) on the graph of f where $y = 0$. There are three points where this occurs: $(-2, 0), (0, 0)$ and $(2, 0)$. Thus, when $x = -2, 0,$ or $2, f(x) = 0$.

59. Graph $f(x) = 0.25x^2$ in $[-4.7, 4.7, 1]$ by $[-3.1, 3.1, 1]$ by letting $Y_1 = 0.25X\char`^2$. See Figure 59.

    (a) From the graph it appears that $f(2) \approx 1$.

    (b) Evaluating $f(2) = 0.25(2)^2 = 0.25(4) = 1$.

    (c) See Figure 59c.

$[-4.7, 4.7, 1]$ by $[-3.1, 3.1, 1]$ $[-4.7, 4.7, 1]$ by $[-3.1, 3.1, 1]$

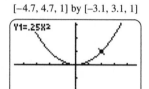   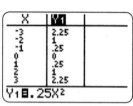   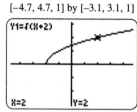   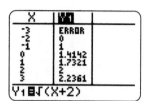

Figure 59   Figure 59c   Figure 61   Figure 61c

61. Graph $f(x) = \sqrt{x + 2}$ in $[-4.7, 4.7, 1]$ by $[-3.1, 3.1, 1]$ by letting $Y_1 = \sqrt{(X + 2)}$. See Figure 61.

    (a) From the graph it appears that $f(2) = 2$.

    (b) Evaluating $f(2) = \sqrt{2 + 2} = \sqrt{4} = 2$.

    (c) See Figure 61c.

63. Verbal: Square the input x.

    Graphical: Graph $Y_1 = X\char`^2$. See Figure 63.

    Numerical:

-2	-1	0	1	2
4	1	0	1	4

$[-10, 10, 1]$ by $[-10, 10, 1]$ $[-6, 6, 1]$ by $[-4, 4, 1]$ $[-6, 6, 1]$ by $[-4, 4, 1]$

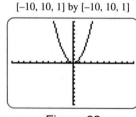   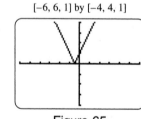   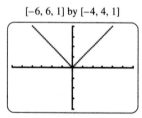

Figure 63   Figure 65   Figure 67

65. Verbal: Multiply the input x by 2, add 1, and then take the absolute value.

    Graphical: Graph $Y_1 = \text{abs}(2X + 1)$. See Figure 65.

    Numerical:

-2	-1	0	1	2
3	1	1	3	5

67. Verbal: Compute the absolute value of x.

    Graphical: Graph $Y_1 = \text{abs}(X)$. See Figure 67.

    Numerical:

-2	-1	0	1	2
2	1	0	1	2

69. Verbal: Add 1 to the input x and then take the square root of the result.

Graphical: Graph $Y_1 = \sqrt{(X + 1)}$ See Figure 69.

Numerical:

−2	−1	0	1	2
Error	0	1	$\sqrt{2}$	$\sqrt{3}$

[−6, 6, 1] by [−4, 4, 1]

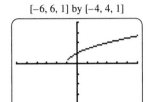

Figure 69

Figure 75

71. (a) $g = \{(-1, 2), (0, 4), (1, -3), \text{ and } (2, 2)\}$

(b) $D = \{-1, 0, 1, 2\};\ R = \{2, 4, -3\}$

73.

Bills (millions)	0	1	2	3	4	5	6
Counterfeit Bills	0	9	18	27	36	45	54

75. (a) See Figure 75

(b) $f(1975) = 7.7$; in 1975 there were 7700 radio stations on the air.

(c) $D = \{1950, 1975, 2001\}$ and $R = \{2.8, 7.7, 13.1\}$.

Identifying Functions

77. This is a graph of a function because every vertical line intersects the graph at most once. Both the domain and the range are all real numbers.

79. This is not a graph of a function because some vertical lines can intersect the graph twice. Because a vertical line can intersect the graph twice, two functions are necessary to create this graph.

81. This is a graph of a function because every vertical line intersects the graph at most once.

The domain is $-4 \leq x \leq 4$. The range is $0 \leq y \leq 4$.

83. Yes. The calculation of the cube root requires the input of a real number. It produces a single output. A number has only one cube root.

85. No. On most English exams, more than one person passes. Therefore, for a given English exam as input, there is more than one output. A listing does not typically compute an output.

87. No. The ordered pairs $(1, 2)$ and $(1, 3)$ belong to the set S. The domain element 1 has more than one range element associated with it.

89. Yes. Each element in its domain is associated with exactly one range element.

91. No. The ordered pair $(1, 10.5)$ and $(1, -0.5)$ belong to the set S. The domain element 1 has more than one range element associated with it.

93. No. The ordered pair (1, –1) and (1, 1) belong to the relation. The domain element 1 has more than one range element associated with it.

95. Yes. Each element in the domain of f is associated with exactly one range element.

97. $g(x) = 12x \Rightarrow g(10) = 12(10) = 120$ inches; There are 120 inches in 10 feet.

99. $g(x) = 0.25x \Rightarrow g(10) = 0.25(10) = \$2.50$; There are 2.5 dollars in 10 quarters.

101. $g(x) = 60 \cdot 60 \cdot 24 \cdot x \Rightarrow g(x) = 86{,}400x \Rightarrow g(10) = 86{,}400(10) = 864{,}000$ seconds; There are 864,000 seconds in 10 days.

Applications

103. Let $x =$ time in seconds that elapsed for the observer to hear the thunder from the lightning bolt. The distance is found by multiplying the speed by the time. Since 5280 ft. = 1 mi., the distance in miles is given by

$$f(x) = \frac{1150}{5280}x = \frac{115}{528}x; \ f(15) = \frac{115}{528}(15) \approx 3.3; \ \text{with a 15-second delay, the lightning bolt was about 3.3}$$

miles away.

[0, 3, 1] by [–20, 20, 5]

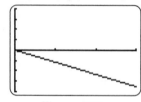

Figure 105a   Figure 105b

105. Verbal: Multiply the input x by –5.8 to obtain the change in temperature.

Symbolic: $f(x) = -5.8x$.

Graphical: $Y_1 = -5.8X$. See Figure 105a.

Numerical: Table $Y_1 = -5.8X$. See Figure 105b.

1.4: Types of Functions and Their Rates of Change

Slope

1. $m = \dfrac{5-6}{2-4} = 0.5$

3. $\dfrac{-2-4}{5-(-1)} = \dfrac{-6}{6} = -1$

5. $\dfrac{-8-(-8)}{7-12} = \dfrac{0}{-5} = 0$

7. $\dfrac{0.4-(-0.1)}{-0.3-0.2} = \dfrac{0.5}{-0.5} = -1$

9. $m = \dfrac{7.6-9.2}{-0.3-(-0.5)} = -8$

11. $m = \dfrac{7.9 - 5.6}{1994 - 1997} = -\dfrac{23}{30} \approx -0.7667$

13. $m = \dfrac{8 - 6}{-5 - (-5)} =$ undefined

15. $m = \dfrac{\frac{7}{10} - (-\frac{3}{5})}{-\frac{5}{6} - \frac{1}{3}} = \dfrac{\frac{13}{10}}{-\frac{7}{6}} = \dfrac{13}{10} \cdot \left(-\dfrac{6}{7}\right) = -\dfrac{39}{35}$

17. Slope $= 2$; the graph rises 2 units for every unit increase in x.

19. Slope $= -\dfrac{3}{4}$; the graph falls $\dfrac{3}{4}$ unit for every unit increase in x; or equivalently, the graph falls 3 units for every 4-unit increase in x.

21. Slope $= 0$; the graph neither falls nor rises for every unit increase in x since the y-value is always -5.

23. Slope $= -1$; the graph falls 1 unit for every unit increase in x.

25. (a) $D(2) = 75(2) = 150$ miles

    (b) Slope $= 75$; the car is traveling away from the rest stop at 75 miles per hour.

27. (a) Buying no carpet should and does cost $\$0$.

    (b) Slope $= \dfrac{100}{5} = 20$

    (c) The carpet costs $\$20$ per square yard.

29. (a) To find the median age in 1980 and 2000, we must evaluate $A(1980)$ and $A(2000)$.

    $A(1980) = 0.243(1980) - 450.8 = 30.34$ and $A(2000) = 0.243(2000) - 450.8 = 35.2$

    In 1980, the median age was 30.34 years and in 2000, it increased to 35.2 years.

    (b) Since $A(t) = 0.243t - 450.8$, the slope of its graph is $m = 0.243$. The value of 0.243 means that the median age in the United States has increased by approximately 0.243 for each year from 1980 to 2000.

31. Between each pair of points, the y-values increase 4 units for each unit increase in x. Therefore, the data is linear. The slope of the line passing through the data points is 4.

33. The y-values do not increase by a constant amount for each 2-unit increase in x. The data is nonlinear.

35. The y-values decrease 0.5 units for each unit increase in x. Therefore, the data is linear. The slope of the line passing through the data points is -0.5.

Linear and Nonlinear Functions

37. $f(x) = -2x + 5$ is a linear function with a slope of $m = -2$. See Figure 37.

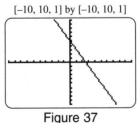
[−10, 10, 1] by [−10, 10, 1]
Figure 37

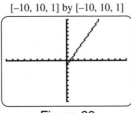
[−10, 10, 1] by [−10, 10, 1]
Figure 39

39. $f(x) = 2x^{1.01} - 1$ is a nonlinear function because the exponent on x is not equal to one. See Figure 39.

41. $f(x) = 1$ is a constant (and linear) function. See Figure 41.

43. From its graph, we see that $f(x) = 0.2x^3 - 2x + 1$ represents a nonlinear function. See Figure 43.

[–10, 10, 1] by [–10, 10, 1]   [–10, 10, 1] by [–10, 10, 1]   [–10, 10, 1] by [–10, 10, 1]   [–10, 10, 1] by [–10, 10, 1]

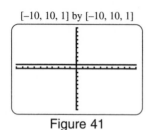
Figure 41

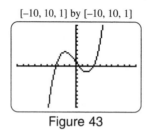
Figure 43

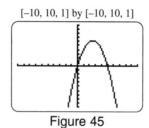
Figure 45

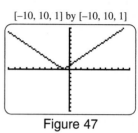
Figure 47

45. From its graph, we see that $f(x) = 5x - x^2$ represents a nonlinear function. See Figure 45.

47. From its graph, we see that $f(x) = |x + 1|$ represents a nonlinear function. See Figure 47.

49. From its graph, we see that $f(x) = x^2 - 1$ represents a nonlinear function. See Figure 49.

51. $f(x) = 2\sqrt{x}$ is a nonlinear function since $2\sqrt{x} = 2x^{1/2}$ has an exponent different than 1. See Figure 51.

[–10, 10, 1] by [–10, 10, 1]   [–10, 10, 1] by [–10, 10, 1]   [0, 4000, 500] by [0, 280, 35]   [1965, 2000, 5] by [0, 35,000, 5000]

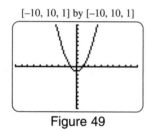
Figure 49

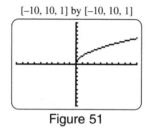
Figure 51

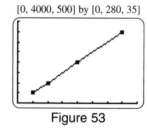
Figure 53

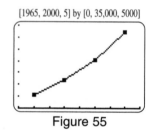
Figure 55

53. (a) See Figure 53.

    (b) From the line graph it can be seen that f is linear. The slopes of the line segments are all equal.

55. (a) See Figure 55.

    (b) From the line graph it can be seen that the data is not linear. The function f is nonlinear.

57. (a) Since the wind speed is not constant, it cannot be represented exactly by a constant function.

    (b) The average of the twelve wind speeds is 7 miles per our. It would be reasonable to approximate the wind speed at Myrtle Beach by $f(x) = 7$.

    (c) A scatterplot of the data and the function $f(x) = 7$ are shown in Figure 57.

[0, 15, 3] by [0, 10, 1]

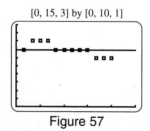
Figure 57

59. If the distance is constant, then the car is not moving. The car's velocity is zero.

61. $f(x) = \dfrac{x}{16}$

63. $f(x) = 50x$ (miles)

65. $f(x) = 500$

67. $f(x) = 6x + 1$

Curve Sketching

69. Sketch a line that is sloping upward from left to right. The line must pass through the point $(3, -1)$.

   See Figure 69. *Answers may vary.*

71. See Figure 71. *Answers may vary.*

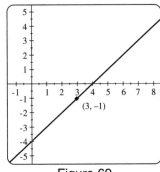
Figure 69

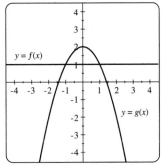
Figure 71

73. See Figure 73. *Answers may vary.*

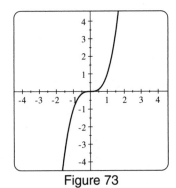
Figure 73

Average Rates of Change

75. The average rate of change from -3 to -1 is $\dfrac{f(-1) - f(-3)}{-1 - (-3)} = \dfrac{4 - 4}{2} = 0.$

   The average rate of change from 1 to 3 is $\dfrac{f(3) - f(1)}{3 - 1} = \dfrac{4 - 4}{2} = 0.$

77. The average rate of change from –3 to –1 is $\dfrac{f(-1) - f(-3)}{-1 - (-3)} = \dfrac{3.7 - 1.3}{2} = \dfrac{2.4}{2} = 1.2.$

The average rate of change from 1 to 3 is $\dfrac{f(3) - f(1)}{3 - 1} = \dfrac{1.3 - 3.7}{2} = -\dfrac{2.4}{2} = -1.2.$

79. (a) $f(x) = x^2 \Rightarrow f(1) = 1^2 = 1$ and $f(2) = 2^2 = 4 \Rightarrow (1, 1), (2, 4);$ using the slope formula for rate of

change we get $\dfrac{4 - 1}{2 - 1} = \dfrac{3}{1} = 3.$ So $f(x)$ rises 3 units for each 1 unit increase of x.

(b) See Figure 79.

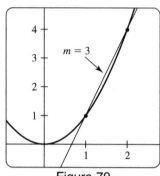

Figure 79

81. If $f(x) = 7x - 2$ then $\dfrac{f(4) - f(1)}{4 - 1} = 7.$ The slope of the graph is 7.

83. If $f(x) = x^3 - 2x$ then $\dfrac{f(4) - f(2)}{4 - 2} = \dfrac{56 - 4}{2} = 26..$ The slope of the line passing through the points

$(2, f(2))$ and $(4, f(4))$ is 26.

85. If $f(x) = \sqrt{2x - 1}$ then $\dfrac{f(3) - f(1)}{3 - 1} \approx 0.6180.$ The slope of the line passing through the points

$(1, f(1))$ and $(3, f(3))$ is approximately 0.6180.

87. (a) Make a line graph using the ordered pairs (x, E) for the data points. See Figure 87.

(b) From 0 to 10 the average rate of change is $\dfrac{f(10) - f(0)}{10 - 0} = \dfrac{69.9 - 72.3}{10} = -0.24.$ Similarly, the average

rate of change is -0.98 from 10 to 20, -0.97 from 20 to 30, -0.95 from 30 to 40, -0.93 from 40 to 50,

-0.85 from 50 to 60, -0.76 from 60 to 70, and -0.63 from 70 to 80. The average rate of change is always

negative. As a woman becomes older, her remaining life expectancy decreases. For example, if the rate of

change is -0.63, then for each year a woman lives, her remaining life expectancy decreases by 0.63 years.

(c) A person 20 years old can expect to live an additional 60.1 years. Therefore, their life expectancy is 80.1

years. A person 70 years old has a life expectancy of $70 + 15.5 = 85.5$ years. The person 70 years old has

a longer life expectancy because that person has already survived the risk that a 20-year-old must endure

between the ages of 20 and 70.

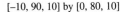

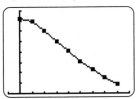

Figure 87

The Difference Quotient

89. (a) $f(x + h) = 3$

    (b) $\dfrac{f(x + h) - f(x)}{h} = \dfrac{3 - 3}{h} = \dfrac{0}{h} = 0$

91. (a) $f(x + h) = -2(x + h) = -2x - 2h$

    (b) $\dfrac{f(x + h) - f(x)}{h} = \dfrac{-2x - 2h - (-2x)}{h} = \dfrac{-2h}{h} = -2$

93. (a) $f(x + h) = 2(x + h) + 1 = 2x + 2h + 1$

    (b) $\dfrac{f(x + h) - f(x)}{h} = \dfrac{2x + 2h + 1 - (2x + 1)}{h} = \dfrac{2h}{h} = 2$

95. (a) $f(x + h) = 3(x + h)^2 + 1 = 3(x^2 + 2xh + h^2) + 1 = 3x^2 + 6xh + 3h^2 + 1$

    (b) $\dfrac{f(x + h) - f(x)}{h} = \dfrac{3x^2 + 6xh + 3h^2 + 1 - (3x^2 + 1)}{h} = \dfrac{6xh + 3h^2}{h} = 6x + 3h$

97. (a) $f(x + h) = -(x + h)^2 + 2(x + h) = -(x^2 + 2xh + h^2) + 2x + 2h = -x^2 - 2xh - h^2 + 2x + 2h$

    (b) $\dfrac{f(x + h) - f(x)}{h} = \dfrac{-x^2 - 2xh - h^2 + 2x + 2h - (-x^2 + 2x)}{h} = \dfrac{-2xh - h^2 + 2h}{h} = -2x - h + 2$

99. (a) $f(x + h) = 2(x + h)^2 - (x + h) + 1 = 2(x^2 + 2xh + h^2) - (x + h) + 1 =$

    $2x^2 + 4xh + 2h^2 - x - h + 1$

    (b) $\dfrac{f(x + h) - f(x)}{h} = \dfrac{2x^2 + 4xh + 2h^2 - x - h + 1 - (2x^2 - x + 1)}{h} = \dfrac{4xh + 2h^2 - h}{h} = 4x + 2h - 1$

101. (a) $d(t + h) = 8(t + h)^2 = 8(t^2 + 2th + h^2) = 8t^2 + 16th + 8h^2$

    (b) $\dfrac{d(t + h) - d(t)}{h} = \dfrac{8t^2 + 16th + 8h^2 - (8t^2)}{h} = \dfrac{16th + 8h^2}{h} = 16t + 8h$

    (c) If $t = 4$ and $h = 0.05$, then the difference quotient becomes: $16t + 8h = 16(4) + 8(0.05) =$

    $64 + 0.04 = 64.4$. The average rate of change, or average velocity, of the car during the time interval

    from 4 to 4.05 seconds is 64.4 feet per second.

Chapter 1 Review Exercises

1. -2 is an integer, rational number, and real number. $\dfrac{1}{2}$ is both a rational and a real number. 0 is an integer, rational number, and real number. 1.23 is both a rational and a real number. $\sqrt{7}$ is a real number. $\sqrt{16} = 4$ is a natural number, integer, rational number, and real number.

3. $\dfrac{P_2 - P_1}{P_1} = \dfrac{1.75 - 1.25}{1.25} = \dfrac{0.50}{1.25} = 0.40 = 40\%$ increase in price

5. $1{,}891{,}000 = 1.891 \times 10^6$

7. $0.0000439 = 4.39 \times 10^{-5}$

9. $1.52 \times 10^4 = 15{,}200$

11. (a) $\sqrt[3]{1.2} + \pi^3 \approx 32.07$

    (b) $\dfrac{3.2 + 5.7}{7.9 - 4.5} \approx 2.62$

    (c) $\sqrt{5^2 + 2.1} \approx 5.21$

    (d) $1.2(6.3)^2 + \dfrac{3.2}{\pi - 1} \approx 49.12$

13. (a) $(9.3 \times 10^3)(2.1 \times 10^{-4}) = 19.53 \times 10^{-1} = 1.953 \times 10^0 = 1.953$

    (b) $\dfrac{2.46 \times 10^6}{4.1 \times 10^2} = 0.6 \times 10^4 = 6 \times 10^3$

15.

-23	-5	8	19	24

    (a) Maximum $= 24$; Minimum $= -23$

    (b) Average $= \dfrac{-23 + (-5) + 8 + 19 + 24}{5} = 4.6$; Median $= 8$; Range $= 24 - (-23) = 47$

17. (a) $S = \{(-15, -3), (-10, -1), (0, 1), (5, 3), (20, 5)\}$

    (b) $D = \{-15, -10, 0, 5, 20\}$ and $R = \{-3, -1, 1, 3, 5\}$

19. (a) $D = \{-1, 4, 0, -5, 1\}$ and $R = \{-2, 6, -5, 3, 0\}$

    (b) Xmax $= 4$, Xmin $= -5$, Ymax $= 6$, Ymin $= -5$

    (c) & (d) The data is plotted in Figure 19.

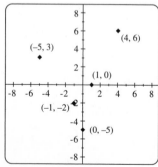

Figure 19

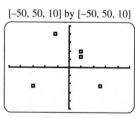

Figure 21

21. The relation {(10, 13), (−12, 40), (−30, −23), (25, −22), (10, 20)} is plotted in Figure 21. It is not a function since both (10, 13) and (10, 20) are contained in the set. Notice that these points are lined up vertically.

23. $d = \sqrt{(2 - (-4))^2 + (-3 - 5)^2} = \sqrt{6^2 + (-8)^2} = \sqrt{36 + 64} = \sqrt{100} = 10$

25. $d = \sqrt{(5 - (-3))^2 + (-4 - (-4))^2} = \sqrt{8^2 + 0^2} = \sqrt{64} = 8$

27. $m = \left(\dfrac{24 + (-20)}{2}, \dfrac{-16 + 13}{2}\right) = \left(\dfrac{4}{2}, \dfrac{-3}{2}\right) = \left(2, \dfrac{-3}{2}\right)$

29. Using the distance formula for the 3 sides:

$d = \sqrt{(1 - (-3))^2 + (2 - 5)^2} = \sqrt{4^2 + (-3)^2} = \sqrt{16 + 9} = \sqrt{25} = 5$

$d = \sqrt{((-3) - 0)^2 + (5 - 9)^2} = \sqrt{(-3)^2 + (-4)^2} = \sqrt{9 + 16} = \sqrt{25} = 5$

$d = \sqrt{(1 - 0)^2 + (2 - 9)^2} = \sqrt{1^2 + (-7)^2} = \sqrt{1 + 49} = \sqrt{50} = 7.07$

Two sides are 5 units long and the other side is 7.07 units long; therefore the triangle is isosceles ⇒ Yes.

31. See Figure 31.

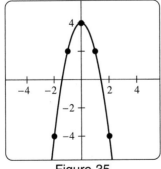

Figure 31

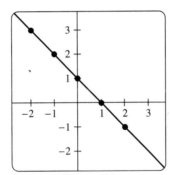

Figure 33

33. See Figure 33.

35. See Figure 35.

37. See Figure 37.

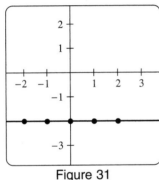

Figure 35

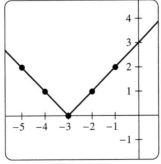

Figure 37

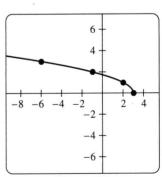

Figure 39

39. See Figure 39.

41. Symbolic: $f(x) = 16x$.

   Numerical: Table f starting at $x = 0$, incrementing by 25. See Figure 41a.

   Graphical: Graph $Y_1 = 16X$ in [0, 100, 10] by [0, 1800, 300]. See Figure 41b.

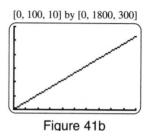

[0, 100, 10] by [0, 1800, 300]

x	0	25	50	75	100
$f(x)$	0	400	800	1200	1600

Figure 41a   Figure 41b

43. (a) $f(-3) = 2$ and $f(1) = -2$

   (b) $f(x) = 0$ when $x = -1$ and when $x = 3$

45. $f(x) = \sqrt[3]{x} \Rightarrow f(-8) = -2$ and $f(1) = 1$. The domain of f is all real numbers.

47. (a) $f(x) = 5 \Rightarrow f(-3) = 5$ and $f(1.5) = 5$

   (b) $D = \{$all real numbers$\}$

49. (a) $f(x) = x^2 - 3 \Rightarrow f(-10) = (-10)^2 - 3 = 97$ and $f(a + 2) = (a + 2)^2 - 3 = a^2 + 4a + 4 - 3 =$
   $a^2 + 4a + 1$

   (b) $D = \{$all real numbers$\}$

51. $f(x) = \dfrac{1}{x^2 - 4} \Rightarrow f(-3) = \dfrac{1}{(-3)^2 - 4} = \dfrac{1}{5}$ and $f(a + 1) = \dfrac{1}{(a^2 + 2a + 1) - 4} = \dfrac{1}{a^2 + 2a - 3}$

   The domain of f is $D = \{x \mid x \neq \pm 2\}$.

53. No, since $x = y^2 + 5$ is not a function. An input $x = 6$ produces outputs of $y = \pm 1$.

55. Since any vertical line intersects the graph of f at most once, it is a function.

57. Yes, it is a function. All points produce a single output.

59. $f(x) = 60 \cdot 60 \cdot 24 \cdot t \Rightarrow f(x) = 86{,}400t$

61. $f(0) = 7$ and $f(1) = 7 \Rightarrow (0, 7)$ and $(1, 7)$; $m = \dfrac{7 - 7}{1 - 0} = \dfrac{0}{1} = 0$

63. $f(0) = 4$ and $f(1) = -2 \Rightarrow (0, 4)$ and $(1, -2)$; $m = \dfrac{-2 - 4}{1 - 0} = \dfrac{-6}{1} = -6$

65. $m = \dfrac{4 - 7}{3 - (-1)} = -\dfrac{3}{4}$

67. $m = \dfrac{4 - 4}{-2 - 8} = \dfrac{0}{-10} = 0$

69. $f(x) = 8 - 3x$ represents a linear function.

71. $f(x) = |x + 2|$ represents a nonlinear function.

73. $f(x) = 6$ represents a constant (and linear) function.

75. See Figure 75.

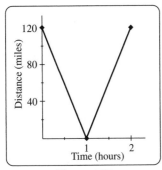

Figure 75

77. Linear, but not constant since the y-values decrease 8 units for every 2-unit increase in x.

79. $f(x + h) = 5(x + h) + 1 = 5x + 5h + 1$

$$\frac{f(x + h) - f(x)}{h} = \frac{5x + 5h + 1 - (5x + 1)}{h} = \frac{5h}{h} = 5$$

Applications

81. $\dfrac{2.28 \times 10^8}{3 \times 10^5} = 760$ seconds $= 12\dfrac{2}{3}$ minutes

83. The paint on the circular piece of paper can be thought of as a thin cylinder.

The volume of a cylinder is: $V = \pi r^2 h$. Substitute 0.25 in^3 in for V and 10 in. for r:

$0.25 = \pi(10)^2 h \Rightarrow 0.25 = 100\pi h \Rightarrow h = \dfrac{0.25}{100\pi} \approx 7.96 \times 10^{-4}$

The thickness of the paper is about 0.000796 inches.

85 (a) $D(2) = 280 - 70(2) = 280 - 140 = 140$ miles

(b) $D(1) = 280 - 70(1) = 280 - 70 = 210$. Now using the slope formula for $(1, 210)$ and $(2, 140)$ we get

$\dfrac{140 - 210}{2 - 1} = \dfrac{-70}{1} = -70$; the driver is moving toward the rest stop at 70 miles per hour.

87. (a) Plot the points $(0, 100)$, $(1, 10)$, $(2, 6)$, $(3, 3)$, and $(4, 2)$ and make a line graph. See Figure 87. The data

decreases rapidly at first. This means that a large number of eggs never develop into mature adults.

(b) Since any vertical line could intersect the graph at most once, this graph could represent a function.

(c) From 0 to 1, $\dfrac{10 - 100}{1 - 0} = -90$; from 1 to 2, $\dfrac{6 - 10}{2 - 1} = -4$; from 3 to 2, $\dfrac{3 - 6}{3 - 2} = -3$;

from 3 to 4, $\dfrac{2 - 3}{4 - 3} = -1$; during the first year, the population of sparrows decreased, on average, by 90

birds. The other average rates of change can be interpreted similarly.

[–1, 5, 1] by [0, 110, 10]

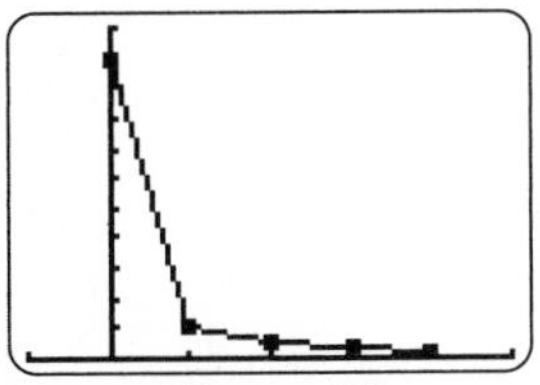

Figure 87

89. (a) See Figure 89a. f is nonlinear.

(b) $f(x) = 0.5x^2 + 50 \Rightarrow \dfrac{f(4) - f(1)}{4 - 1} = \dfrac{58 - 50.5}{3} = 2.5$

(c) The average rate of change in outside temperature from 1 P.M. to 4 P.M. was 2.5° F per hour. The slope of the line segment from (1, 50.5) to (4, 48) is 2.5. See Figure 89c.

[1, 5, 1] by [40, 70, 5]

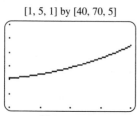

Figure 89a

[1, 5, 1] by [50, 63, 1]

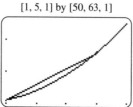

Figure 89c

Chapter 2: Linear Functions and Equations

2.1: Linear Functions and Models

Exact and Approximate Models

1. Evaluating f for $x = 1, 2, 3, 4$ we get: $f(1) = 5(1) - 2 = 3$, value agrees with table.

    $f(2) = 5(2) - 2 = 8$, value agrees with table. $f(3) = 5(3) - 2 = 13$, value agrees with table.

    $f(4) = 5(4) - 2 = 18$, value agrees with table. Since all agree, it models the data exactly.

3. Evaluating f for $x = -6, 0, 6, 12$ we get: $f(-6) = 3.7 - 1.5(-6) = 12.7$, value agrees with table.

    $f(0) = 3.7 - 1.5(0) = 3.7$, value agrees with table. $f(6) = 3.7 - 1.5(6) = -5.3$, value does not agree with

    table. $f(12) = 3.7 - 1.5(12) = -14.3$, value does not agree with table. Since $f(6)$ and $f(12)$ do not agree but

    are close and the others agree, it models the data approximately.

Graphs of Linear Functions

5. (a) Slope $= \dfrac{\text{rise}}{\text{run}} = \dfrac{2}{1} = 2$; y-intercept: -1; x-intercept: 0.5

    (b) $f(x) = mx + b \Rightarrow f(x) = 2x - 1$

    (c) 0.5

7. (a) Slope $= \dfrac{\text{rise}}{\text{run}} = \dfrac{3}{4}$; y-intercept: -3; x-intercept: 4

    (b) $f(x) = mx + b \Rightarrow f(x) = \dfrac{3}{4}x - 3$

    (c) 4

9. (a) Slope $= \dfrac{\text{rise}}{\text{run}} = \dfrac{100}{5} = 20$; y-intercept: -50; x-intercept: 2.5

    (b) $f(x) = mx + b \Rightarrow f(x) = 20x - 50$

    (c) 2.5

11. $f(x) = 3x + 2$. See Figure 11. $m = 3$; y-intercept $= 2$

13. $f(x) = \dfrac{1}{2}x - 2$. See Figure 13. $m = \dfrac{1}{2}$; y-intercept $= -2$

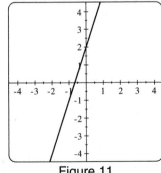

Figure 11

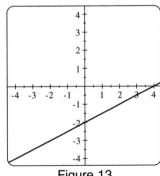

Figure 13

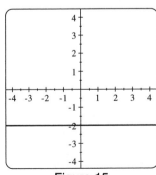

Figure 15

15. $g(x) = -2$. See Figure 15. $m = 0$; y-intercept $= -2$

17. $f(x) = 4 - \dfrac{1}{2}x$. See Figure 17. Slope $= -\dfrac{1}{2}$; y-intercept $= 4$

19. $g(x) = \dfrac{1}{2}x$. See Figure 19. Slope $= \dfrac{1}{2}$; y-intercept $= 0$

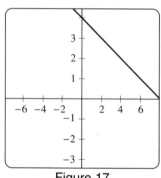
Figure 17

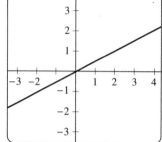
Figure 19

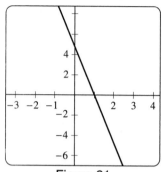
Figure 21

21. $g(x) = 5 - 5x$. See Figure 21. Slope $= -5$; y-intercept $= 5$

23. $f(x) = 20x - 10$. See Figure 23. Slope $= 20$; y-intercept $= -10$

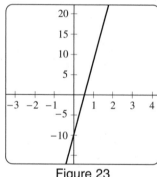
Figure 23

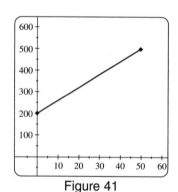
Figure 41

25. $f(x) = mx + b \Rightarrow f(x) = -\dfrac{3}{4}x + \dfrac{1}{3}$

27. $f(x) = mx + b \Rightarrow f(x) = 15x + 0$, or $f(x) = 15x$

29. Since the slope $= 0.5 = \dfrac{0.5}{1} = \dfrac{-0.5}{-1}$, to go from $(1, 4.5)$ to another point on the line, you can move 0.5 unit

down and 1 unit to the left. This gives the point $(1 - 1, 4.5 - 0.5) = (0, 4)$, so the y-intercept is 4;

$f(x) = mx + b \Rightarrow f(x) = 0.5x + 4$

31. $f(-2) = 10$ and $f(2) = 10 \Rightarrow$ we have points $(-2, 10)$ and $(2, 10)$; the slope of this line is $m = \dfrac{10 - 10}{2 - (-2)} =$

$\dfrac{0}{4} = 0$. The average rate of change is 0. For two distinct real numbers a and b the points are $(a, 10)$ and

$(b, 10)$; the slope of this line is $m = \dfrac{10 - 10}{b - a} = \dfrac{0}{b - a} = 0$. The average rate of change is 0.

33. $f(-2) = -\frac{1}{4}(-2) = \frac{1}{2}$ and $f(2) = -\frac{1}{4}(2) = -\frac{1}{2}$ $\Rightarrow$ we have points $\left(-2, \frac{1}{2}\right)$ and $\left(2, -\frac{1}{2}\right)$; the slope of this

line is $m = \frac{-\frac{1}{2} - \frac{1}{2}}{2 - (-2)} = \frac{-1}{4} = -\frac{1}{4}$. The average rate of change is $-\frac{1}{4}$. For two distinct real numbers a and b

the points are $f(a) = -\frac{1}{4}a$ and $f(b) = -\frac{1}{4}b$ or $\left(a, -\frac{1}{4}a\right)$ and $\left(b, -\frac{1}{4}b\right)$; the slope of this line is

$m = \frac{-\frac{1}{4}b - (-\frac{1}{4}a)}{b - a} = \frac{-\frac{1}{4}(b - a)}{b - a} = -\frac{1}{4}$. The average rate of change is $-\frac{1}{4}$.

35. $f(-2) = 4 - 3(-2) = 4 + 6 = 10$ and $f(2) = 4 - 3(2) = 4 - 6 = -2$ $\Rightarrow$ we have points $(-2, 10)$ and

$(2, -2)$; the slope of this line is $m = \frac{-2 - 10}{2 - (-2)} = \frac{-12}{4} = -3$. The average rate of change is -3. For two

distinct real numbers a and b the points are $f(a) = 4 - 3a$ and $f(b) = 4 - 3b$ or $(a, 4 - 3a)$ and $(b, 4 - 3b)$;

the slope of this line is $m = \frac{(4 - 3b) - (4 - 3a)}{b - a} = \frac{-3b + 3a}{b - a} = \frac{-3(b - a)}{b - a} = \frac{-3}{1} = -3$. The average rate of

change is -3.

Modeling with Linear Functions

37. Since the height of the Empire State Building is constant; the graph that has no rate of change is D.

39. As time increases the distance to the finish line decreases; the graph that shows this decline in distance as time

increases is C.

41. (a) Slope: 6 and y-intercept: 200; $f(x) = 6x + 200$.

(b) An appropriate domain is $D = \{x \mid 0 \le x \le 50\}$. See Figure 41.

(c) The y-intercept is 200, which indicates that the tank initially contains 200 gallons of fuel oil.

(d) No, the x-intercept of $-\frac{100}{3}$ corresponds to negative time, which does not apply to the situation.

43. (a) Slope: -0.26 and y-intercept: 16.7; $f(x) = 16.7 - 0.26x$.

(b) Since 1990 corresponds to $x = 0$, 2002 corresponds to $x = 12$; $f(x) = 16.7 - 0.26(12) = 13.58$, which

means that in 2002 there were about 13.58 births per 1000 people in the United States. This is slightly

lower than the actual value of 13.9.

45. $V(t) = 32t$; t represents time in seconds; $D = \{t \mid 0 \le t \le 3\}$

47. $P(t) = 21.5 + 0.581t$; t represents years after 1900; $D = \{t \mid 0 \le t \le 100\}$

49. Slope $= \frac{3}{1} = 3$; y-intercept $= -7$; $f(x) = 3x - 7$.

51. Slope $= \frac{-12.5}{5} = -\frac{5}{2} = -2.5$; $(1, -2.5)$ $\Rightarrow$ $(1 - 1, -2.5 + 2.5) = (0, 0)$, so y-intercept $= 0$;

$f(x) = -2.5x + 0 = -2.5x$

53. (a) No, the data cannot be modeled exactly by a linear function since rate of change is not constant, $\left(\dfrac{0.6}{10} \neq \dfrac{0.9}{10}\right)$.

   (b) A possible typical rate of change is $\dfrac{0.8}{10} = 0.08$, so let slope be 0.08 and since (0, 2.2) is a data point, the

   y-intercept is 2.2; $f(x) = 0.08x + 2.2$ models the data. *Answers may vary slightly.*

   (c) See Figure 53. The number of miles traveled has been increasing by about 0.08 trillion miles per year.

   (d) 1970 corresponds to $x = 0 \Rightarrow$ 2003 corresponds to $x = 33$; $f(33) = 0.08(33) + 2.2 = 4.84$, so the

   number of passenger miles will be about 4.8 trillion miles in 2003.

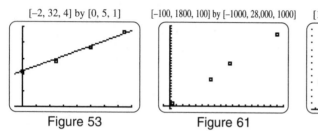

[–2, 32, 4] by [0, 5, 1]   [–100, 1800, 100] by [–1000, 28,000, 1000]   [1973, 2000, 2] by [0, 24, 2]

Figure 53   Figure 61   Figure 63

55. No, since $f(x) = ax \Rightarrow f(2x) = a(2x) = 2 \cdot ax$, if the speed doubles, then the braking distance only doubles,

   not quadruples.

Linear Regression

57. (a) Enter the x-values into the list L_1 and the y-values into the list L_2 in the statistical feature of your graphing

   calculator; the scatterplot of the data indicates that the correlation coefficient will be positive (and very

   close to +1).

   (b) $y = ax + b$, where $a \approx 3.0929$ and $b \approx -2.2143$; $r \approx 0.9989$

   (c) $y = 3.0929(2.4) - 2.2143 \approx 5.209$

59. (a) Enter the x-values into the list L_1 and the y-values into the list L_2 in the statistical feature of your graphing

   calculator; the scatterplot of the data indicates that the correlation coefficient will be negative (and very

   close to –1).

   (b) $y = ax + b$, where $a \approx -3.8857$ and $b \approx 9.3254$; $r \approx -0.9996$

   (c) $y = -3.8857(2.4) + 9.3254 \approx -0.00028$. *Due to rounding answers may very slightly.*

61. (a) The data points (50, 990), (650, 9300), (950, 15,000) and (1700, 25,000) are plotted in Figure 61. The data

   appears to have a linear relationship.

   (b) Use the linear regression feature on your graphing calculator to find the values of a and b in the equation

   $y = ax + b$. In this instance, $a \approx 14.680$ and $b \approx 277.82$.

   (c) We must find the x-value when $y = 37,000$. This can be done by solving the equation

   $37,000 = 14.680x + 277.82 \Rightarrow 14.680x = 36,722.18 \Rightarrow x \approx 2500$ light years away. One could also

   solve the equation graphically to obtain the same approximation.

63. (a) The data points (1975, 8.0), (1977, 9.1), (1979, 10.3), (1981, 12.1), (1983, 13.3), (1985, 14.8), (1987, 15.7), (1989, 17.0), (1991, 18.3), (1993, 20.5), (1995, 20.7) and (1998, 21.6) are plotted in Figure 63. The data appears to have a linear relationship.

(b) Use the statistical package on the graphing calculator to find the values of $a \approx 0.6301$ and $b \approx -1236.4$. Therefore the regression line equation is $y = 0.6301x - 1236.4$.

(c) The slope of 0.6301 means that the percentage of women in state legislatures has increased by approximately 0.63% each year from 1975 to 1998.

(d) Let $x = 2003$, then $y = 0.6301(2003) - 1236.45 \approx 25.6$. In 2003 the regression model finds that about 25.8% of the people serving in state legislatures will be women. This is higher than the actual percent, which is 22.3%.

2.2: Equations of Lines

Equations of Lines

1. Find Slope: $m = \dfrac{(-2) - 2}{3 - 1} = \dfrac{-4}{2} = -2$, using $(x_1, y_1) = (1, 2)$ and point slope form: $y = m(x - x_1) + y_1$

   we get $y = -2(x - 1) + 2$. See Figure 1.

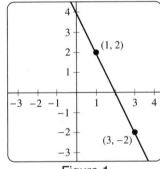

Figure 1

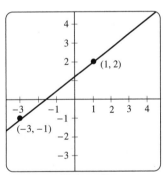

Figure 3

3. Find Slope: $m = \dfrac{2 - (-1)}{1 - (-3)} = \dfrac{3}{4}$, using $(x_1, y_1) = (-3, -1)$ and point slope form: $y = m(x - x_1) + y_1$ we get

   $y = \dfrac{3}{4}(x + 3) - 1$. See Figure 3.

5. The point-slope form is given by $y = m(x - h) + k$. Thus, $m = -2.4$ and $(h, k) = (4, 5) \Rightarrow$

   $y = -2.4(x - 4) + 5 \Rightarrow y = -2.4x + 9.6 + 5 \Rightarrow y = -2.4x + 14.6$.

7. First find the slope between the points $(1, -2)$ and $(-9, 3)$. $m = \dfrac{3 - (-2)}{-9 - 1} = -\dfrac{1}{2}$

   $y = -\dfrac{1}{2}(x - 1) - 2$ or $y = -\dfrac{1}{2}(x + 9) + 3 \Rightarrow y = -\dfrac{1}{2}x - \dfrac{9}{2} + 3 \Rightarrow y = -\dfrac{1}{2}x - \dfrac{3}{2}$.

9. $m = \dfrac{25 - 5}{1990 - 1980} = \dfrac{20}{10} = 2.$ Thus, $y = 2(x - 1980) + 5$ or $y = 2(x - 1990) + 25 \Rightarrow$

$y = 2x - 3980 + 25 \Rightarrow y = 2x - 3955.$

11. Using the points $(0, -1)$ and $(3, 1)$, we get $m = \dfrac{1 - (-1)}{3 - 0} = \dfrac{2}{3}$ and $b = -1;\ y = mx + b \Rightarrow y = \dfrac{2}{3}x - 1.$

13. Using the points $(-2, 1.8)$ and $(1, 0)$, we get $m = \dfrac{0 - 1.8}{1 - (-2)} = \dfrac{-1.8}{3} = -\dfrac{18}{30} = -\dfrac{3}{5};$ to find b, we use $(1, 0)$ in

$y = mx + b$ and solve for b: $0 = -\dfrac{3}{5}(1) + b \Rightarrow b = \dfrac{3}{5};\ y = -\dfrac{3}{5}x + \dfrac{3}{5}.$

15. Since the line has a y-intercept of 5, $b = 5$. Since the slope is -7.8, $m = -7.8$. Thus $y = -7.8x + 5.$

17. The line passes through the points $(0, 45)$ and $(90, 0)$.

$m = \dfrac{0 - 45}{90 - 0} = -\dfrac{1}{2};\ b = 45$ and $m = -\dfrac{1}{2} \Rightarrow y = -\dfrac{1}{2}x + 45$

19. The line has a slope of 4 and passes through the point $(-4, -7);\ y = 4(x + 4) - 7 \Rightarrow y = 4x + 9.$

21. The perpendicular slope is equal to $\dfrac{3}{2}$ and the line passes through the point $(1980, 10);$

$y = \dfrac{3}{2}(x - 1980) + 10 \Rightarrow y = \dfrac{3}{2}x - 2960$

23. $m = 3$ and $b = 5 \Rightarrow y = -3x + 5$

25. $m = \dfrac{0 - (-6)}{4 - 0} = \dfrac{6}{4} = \dfrac{3}{2}$ and $b = -6;\ y = mx + b \Rightarrow y = \dfrac{3}{2}x - 6$

27. $m = \dfrac{\frac{2}{3} - \frac{3}{4}}{\frac{1}{5} - \frac{1}{2}} = \dfrac{-\frac{1}{12}}{-\frac{3}{10}} = \dfrac{5}{18};$ using the point-slope form with $m = \dfrac{5}{18}$ and $\left(\dfrac{1}{2}, \dfrac{3}{4}\right)$ we get

$y = \dfrac{5}{18}\left(x - \dfrac{1}{2}\right) + \dfrac{3}{4} \Rightarrow y = \dfrac{5}{18}x - \dfrac{5}{36} + \dfrac{3}{4} \Rightarrow y = \dfrac{5}{18}x + \dfrac{11}{18}.$

29. $y = \dfrac{2}{3}x + 3 \Rightarrow m = \dfrac{2}{3};$ the parallel line has a slope $\dfrac{2}{3};$ since it passes through $(0, -2.1),$

the y-intercept $= -2.1;\ y = mx + b \Rightarrow y = \dfrac{2}{3}x - 2.1.$

31. $y = -2x \Rightarrow m = -2;$ the perpendicular line has slope $\dfrac{1}{2};$ since it passes through $(-2, 5)$, the equation is

$y = \dfrac{1}{2}(x + 2) + 5 = \dfrac{1}{2}x + 1 + 5 = \dfrac{1}{2}x + 6.$

33. $y = -x + 4 \Rightarrow m = -1;$ the perpendicular line has slope 1; since it passes through $(15, -5)$, the equation is

$y = 1(x - 15) - 5 = x - 15 - 5 = x - 20.$

35. Find slope, $m = \dfrac{1 - 3}{-3 - 1} = \dfrac{-2}{-4} = \dfrac{1}{2},$ a line parallel to this line has the same slope $m = \dfrac{1}{2},$ now using

$(x_1, y_1) = (5, 7),\ m = \dfrac{1}{2}$ and point-slope form $y = m(x - x_1) + y$ we get $y = \dfrac{1}{2}(x - 5) + 7 \Rightarrow y = \dfrac{1}{2}x + \dfrac{9}{2}.$

37. Find slope, $m = \dfrac{\frac{2}{3} - \frac{1}{2}}{-3 - (-5)} = \dfrac{\frac{1}{6}}{2} = \dfrac{1}{12}$, a line perpendicular to this line will have a slope $m = -\dfrac{12}{1} \Rightarrow -12$,

 now using $(x_1, y_1) = (-2, 4)$, $m = -12$ and point-slope form $y = m(x - x_1) + y$ we get

 $y = -12(x + 2) + 4 \Rightarrow y = -12x - 24 + 4 \Rightarrow y = -12x - 20$.

39. $x = -5$

41. $y = 6$

43. Since the line $y = 15$ is horizontal, the perpendicular line is vertical with the equation $x = 4$.

45. The line parallel to $x = 4.5$ is also vertical and has an equation $x = 19$.

47. c

49. b

51. e

Interpolation and Extrapolation

53. (a) Since the point $(0, -3.2)$ is on the graph, the y-intercept is -3.2 The data is exactly linear, so one can use

 any two points to determine the slope. Using the points $(0, -3.2)$ and $(1, -1.7)$, $m = \dfrac{-1.7 - (-3.2)}{1 - 0} = 1.5$.

 The slope-intercept form of the line is $y = 1.5x - 3.2$.

 (b) When $x = -2.7$, $y = 1.5(-2.7) - 3.2 = -7.25$. This calculation involves interpolation.

 When $x = 6.3$, $y = 1.5(6.3) - 3.2 = 6.25$. This calculation involves extrapolation.

55. (a) Since the data is exactly linear, one can use any two points to determine the slope. Using the points

 $(5, 94.7)$ and $(23, 56.9)$, $m = \dfrac{56.9 - 94.7}{23 - 5} = -2.1$. The point-slope form of the line is

 $y = -2.1(x - 5) + 94.7$ and the slope-intercept form of the line is $y = -2.1x + 105.2$.

 (b) When $x = -2.7$, $y = -2.1(-2.7) + 105.2 = 110.87$. This calculation involves extrapolation.

 When $x = 6.3$, $y = -2.1(6.3) + 105.2 = 91.97$. This calculation involves interpolation.

57. (a) The slope between $(1998, 3305)$ and $(1999, 3185)$ is -120, and the slope between $(1999, 3185)$ and

 $(2000, 3089)$ is -96. Using the average of -120 and -96, we will let $m = -108$,

 $f(x) = -108(x - 1998) + 3305$, or $f(x) = -108x + 219{,}089$ approximately models the data.

 Answers may vary.

 (b) $f(1996) = -108(1996) + 219{,}089 = 3521$; this estimated value is too high (compared to the actual value

 of 2776); this estimate involved extrapolation.

Determining Intercepts

59. Let $4x - 5y = 20$.

   x-intercept: Substitute $y = 0$ and solve for x. $4x - 5(0) = 20 \Rightarrow 4x = 20 \Rightarrow x = 5$

   y-intercept: Substitute $x = 0$ and determine y. $4(0) - 5y = 20 \Rightarrow -5y = 20 \Rightarrow y = -4$

   See Figure 59.

61. Let $x - y = 7$.

   x-intercept: Substitute $y = 0$ and solve for x. $x - 0 = 7 \Rightarrow x = 7$

   y-intercept: Substitute $x = 0$ and determine y. $0 - y = 7 \Rightarrow -y = 7 \Rightarrow y = -7$

   See Figure 61.

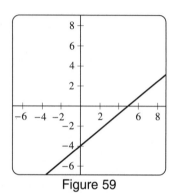
Figure 59

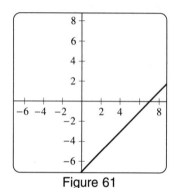
Figure 61

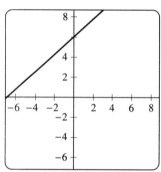
Figure 63

63. Let $6x - 7y = -42$.

   x-intercept: Substitute $y = 0$ and solve for x. $6x - 7(0) = -42 \Rightarrow 6x = -42 \Rightarrow x = -7$

   y-intercept: Substitute $x = 0$ and determine y. $6(0) - 7y = -42 \Rightarrow -7y = -42 \Rightarrow y = 6$

   See Figure 63.

65. Let $y = 8x - 5$.

   x-intercept: Substitute $y = 0$ and solve for x. $0 = 8x - 5 \Rightarrow 5 = 8x \Rightarrow x = \dfrac{5}{8}$

   y-intercept: Substitute $x = 0$ and determine y. $y = 8(0) - 5 \Rightarrow y = -5$

   See Figure 65.

67. Let $y = 3(x - 2) - 5$.

   x-intercept: Substitute $y = 0$ and solve for x. $0 = 3(x - 2) - 5 \Rightarrow 0 = 3x - 6 - 5 \Rightarrow x = \dfrac{11}{3}$

   y-intercept: Substitute $x = 0$ and determine y. $y = 3(0 - 2) - 5 \Rightarrow y = -11$

   See Figure 67.

69. Let $y - 3x = 7$.

   x-intercept: Substitute $y = 0$ and solve for x. $0 - 3x = 7 \Rightarrow -3x = 7 \Rightarrow x = -\dfrac{7}{3}$

   y-intercept: Substitute $x = 0$ and determine y. $y - 3(0) = 7 \Rightarrow y - 0 = 7 \Rightarrow y = 7$

   See Figure 69.

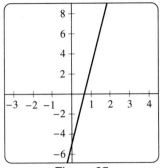

Figure 65

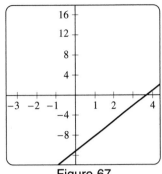

Figure 67

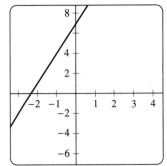

Figure 69

71. Let $0.2x + 0.4y = 0.8$.

   x-intercept: Substitute $y = 0$ and solve for *x*. $0.2x + 0.4(0) = 0.8 \Rightarrow 0.2x = 0.8 \Rightarrow x = 4$

   y-intercept: Substitute $x = 0$ and determine *y*. $0.2(0) + 0.4y = 0.8 \Rightarrow 0.4y = 0.8 \Rightarrow y = 2$

   See Figure 71.

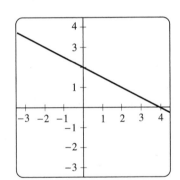

Figure 71

73. Let $\dfrac{x}{5} + \dfrac{y}{7} = 1$.

   x-intercept: Substitute $y = 0$ and solve for *x*. $\dfrac{x}{5} + \dfrac{0}{7} = 1 \Rightarrow \dfrac{x}{5} = 1 \Rightarrow x = 5$

   y-intercept: Substitute $x = 0$ and determine *y*. $\dfrac{0}{5} + \dfrac{y}{7} = 1 \Rightarrow \dfrac{y}{7} = 1 \Rightarrow y = 7$

   a and *b* represent the *x* and *y* intercepts respectively.

75. Let $\dfrac{x}{4} + \dfrac{y}{-3} = 1$.

   x-intercept: Substitute $y = 0$ and solve for *x*. $\dfrac{x}{4} + \dfrac{0}{-3} = 1 \Rightarrow \dfrac{x}{4} = 1 \Rightarrow x = 4$

   y-intercept: Substitute $x = 0$ and determine *y*. $\dfrac{0}{4} + \dfrac{y}{-3} = 1 \Rightarrow \dfrac{y}{-3} = 1 \Rightarrow y = -3$

   a and *b* represent the *x* and *y* intercepts respectively.

77. Let $\dfrac{2x}{3} + \dfrac{4y}{5} = 1$.

x-intercept: Substitute $y = 0$ and solve for x. $\dfrac{2x}{3} + \dfrac{4(0)}{5} = 1 \Rightarrow \dfrac{2x}{3} = 1 \Rightarrow x = \dfrac{3}{2}$

y-intercept: Substitute $x = 0$ and determine y. $\dfrac{2(0)}{3} + \dfrac{4y}{5} = 1 \Rightarrow \dfrac{4y}{5} = 1 \Rightarrow y = \dfrac{5}{4}$

a and b represent the x and y intercepts respectively

Applications

79. (a) Find the slope: $m = \dfrac{37{,}000 - 25{,}000}{2010 - 2003} = \dfrac{12{,}000}{7}$. Using the first point $(2003, 25000)$ for (x_1, y_1) and

$m = \dfrac{12{,}000}{7}$ we get: $y = \dfrac{12{,}000}{7}(x - 2003) + 25{,}000;$ the cost of attending a private college or university

is increasing by $\dfrac{12{,}000}{7} \approx \$1714$ per year on average.

(b) $y = \dfrac{12{,}000}{7}(2007 - 2003) + 25{,}000 \Rightarrow y = \dfrac{12{,}000}{7}(4) + 25{,}000 \Rightarrow y = 6857 + 25{,}000 \Rightarrow$

$y = \$31{,}857$

(c) $y = \dfrac{12{,}000}{7}(x - 2003) + 25{,}000 \Rightarrow y = \dfrac{12{,}000}{7}x - 3{,}433{,}714 + 25{,}000 \Rightarrow$

$y = 1714x - 3{,}408{,}714$ (approximate)

81. (a) The average rate of change $= \dfrac{161 - 128}{4 - 1} = \dfrac{33}{3} = 11 \Rightarrow$ the biker is traveling 11 mile per hour.

(b) Using $m = 11$ and the point $(1, 128)$. we get $y = 11(x - 1) + 128 = 11x - 11 + 128 \Rightarrow y = 11x + 117$.

(c) Find the y-intercept in $y = 11x + 117 \Rightarrow y = 117;$ the biker is initially 117 miles from the interstate highway.

(d) 1 hour and 15 minutes $= 1.25$ hours; $y = 11(1.25) + 117 = 13.75 + 117 = 130.75;$ the biker is 130.75

miles from the interstate highway after 1 hour and 15 minutes.

83. (a) See Figure 83.

(b) Using the first and last points find slope. $m = \dfrac{8.8 - 1.0}{2004 - 1999} = \dfrac{7.8}{5} = 1.56$. Now using the first point

$(1999, 1.0)$ for (x_1, y_1) and $m = 1.56$ we get: $y = 1.56(x - 1999) + 1.0$. The daily world wide spam

message numbers increased 1.56 billion per year on average. *Answers may vary.*

(c) $y = 1.56(2007 - 1999) + 1.0 \Rightarrow y = 1.56(8) + 1.0 \Rightarrow y = 12.48 + 1 \Rightarrow y \approx 13.5$ billion.

Answers may vary.

85. (a) See Figure 85.

(b) Using the first and last points find slope. $m = \dfrac{1.8 - 1.4}{2002 - 1998} = \dfrac{0.4}{4} = 0.1$. Now using the first point

and slope $m = 0.1$ we get: $y = 0.1(x - 1998) + 1.4$. The U.S. sales of Toyota vehicles has increased

by 0.1 million per year.

(c) Each one year increase has 0.1 increase $\Rightarrow$ Exact.

[1998, 2005, 1] by [0, 10, 1]   [1997, 2003, 1] by [0, 2, 0.2]

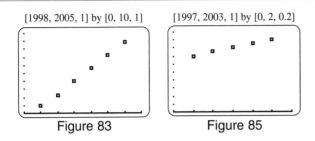

Figure 83   Figure 85

87. (a) The annual fixed cost would be $350 \times 12 = \$4{,}200$. The variable cost of driving x miles is $0.29x$. Thus,

   $f(x) = 0.29 + 4200$.

   (b) The y-intercept is 4200, which represents the annual fixed costs. This means that even if the car is not driven, it will still cost $\$4{,}200$ each year to own it.

89. (a) Scatterplot the data in the table.

   (b) Start by picking a data point for the line to pass through. If we choose (1996, 9.7), f is represented by

   $f(x) = m(x - 1996) + 9.7$. Using trial and error, the slope m is between 0 and 1. Let $m = 0.38$.

   Answers may vary. Thus, $f(x) = 0.38(x - 1996) + 9.7$. The graph of f together with the scatterplot is shown in Figure 89.

   (c) A slope of $m = 0.38$ means that Asian-American population is predicted to increase by approximately 0.38 million (380,000) people each year.

   (d) To predict the population in the year 2005, evaluate $f(2005) = 0.38(2005 - 1996) + 9.7 = 13.12$ million people.

[1995, 2003, 1] by [9, 13, 1]   [0, 3, 1] by [-2, 2, 1]

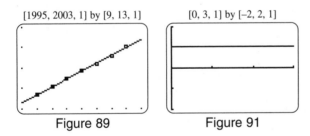

Figure 89   Figure 91

Perspectives and Viewing Rectangles

91. (a) Graph $Y_1 = X/1024 + 1$ in $[0, 3, 1]$ by $[-2, 2, 1]$ as in Figure 91. The line appears to be horizontal in this viewing rectangle. However, we know that the graph of the line is not horizontal because it's slope is $\dfrac{1}{1024} \neq 0$.

   (b) The resolution of most graphing calculator screens is not good enough to show the slight increase in the y-values. Since the x-axis is 3 units long, this increase in y-values amounts to only $\dfrac{1}{1024} \times 3 \approx 0.003$ units, which does not show up on the screen.

93. (a) From Figure 93a, one can see that the lines do not appear to be perpendicular.

(b) The lines are graphed in the specified viewing rectangles and shown in Figures 93b-d, respectively. In the windows [−15, 15, 1] by [−10, 10, 1] and [−3, 3, 1] by [−2, 2, 1] the lines appear to be perpendicular.

(c) The lines appear perpendicular when the distance shown along the *x*-axis is approximately 1.5 times farther than the distance along the *y*-axis. For example, in window [−12, 12,1] by [−8, 8, 1], the lines will appear perpendicular. The distance along the *x*-axis is 24 while the distance along the y-axis is 16. Notice that 1.5 × 16 = 24. This is called a "square window" and can be set automatically on some graphing calculators.

[−10, 10, 1] by [−10, 10, 1]   [−15, 15, 1] by [−10, 10, 1]   [−10, 10, 1] by [−3, 3, 1]   [−3, 3, 1] by [−2, 2, 1]

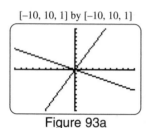
Figure 93a

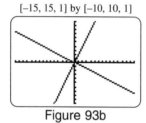
Figure 93b

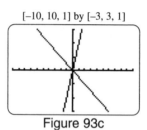
Figure 93c

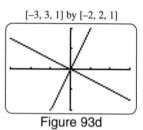
Figure 93d

Graphing a Rectangle

95. (i)   The slope of the line connecting $(0, 0)$ and $(2, 2)$ is 1. Let $y_1 = x$.

(ii) A second line passing through $(0, 0)$ has a slope of −1. Let $y_2 = -x$.

(iii) A third line passing through $(1, 3)$ has a slope of 1. Let $y_3 = (x - 1) + 3 = x + 2$.

(iv) A fourth line passing through $(2, 2)$ has a slope of −1. Let $y_4 = -(x - 2) + 2 = -x + 4$.

97. (i)   The slope of the line connecting $(-4, 0)$ and $(0, 4)$ is given by $y_1 = x + 4$.

(ii) A second line passing through $(4, 0)$ and $(0, -4)$ has a slope of 1. Let $y_2 = x - 4$.

(iii) A third line passing through $(0, -4)$ and $(-4, 0)$ has a slope of −1. Let $y_3 = -x - 4$.

(iv) A fourth line passing through $(0, 4)$ and $(4, 0)$ is −1. Let $y_4 = -x + 4$.

Direct Variation

99. Since *y* is directly proportional to *x*, the variation equation $y = kx$ must hold. To find the value of *k* use the value $y = 7.5$ when $x = 3$ from the table. Solve the equation $7.5 = k(3) \Rightarrow k = 2.5$. The variation equations is $y = 2.5x$ and hence $y = 2.5(8) = 20$ when $x = 8$. A graph of $Y_1 = 2.5X$ together with the data points is shown in Figure 99.

[0, 10, 1] by [0, 24, 2]   [0, 100, 10] by [0, 6, 1]

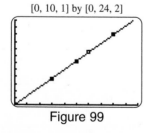
Figure 99

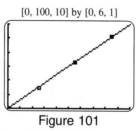
Figure 101

101. Since y is directly proportional to x, the variation equation $y = kx$ must hold. To find the value of k use the value $y = 1.50$ when $x = 25$ from the table. Solve the equation $1.50 = k(25) \Rightarrow k = 0.06$. The variation equations is $y = 0.06x$ and hence when $y = 5.10$, $x = \dfrac{5.1}{0.06} = 85$. A graph of $Y_1 = 0.06X$ together with the data points is shown in Figure 101.

103. Let y represent the cost of tuition and x represent the number of credits taken. Since the cost of tuition is directly proportional to the number of credits taken, the variation equation $y = kx$ must hold. If cost is $y = \$720.50$ when the number of credits is $x = 11$, we may find the constant of proportionality k by solving $720.50 = k(11) \Rightarrow k = 65.50$. The variation equation is $y = 65.50x$. Therefore, the cost of taking 16 credits is $y = 65.50(16) = \$1048$.

105. (a) Since the points $(0, 0)$ and $(300, 3)$ lie on the graph of $y = kx$, the slope of the graph is $\dfrac{3 - 0}{300 - 0} = 0.01$ and $y = 0.01x$, so $k = 0.01$.

  (b) $y = 0.01(110) = 1.1$ millimeters.

  (c) No, ozone was still present. There actually was a substantial thinning of the ozone layer to about one-third of the normal amount.

107. (a) Using $F = kx \Rightarrow 8 = k(15) \Rightarrow k = \dfrac{8}{15}$

  (b) The variation equation is $y = \dfrac{8}{15}x$.

  Therefore, a 25 pound weight will stretch the spring $y = \dfrac{8}{15}(25) = 13\dfrac{1}{3}$ inches.

109. Start by calculating a constant of proportionality for data points $(20, 118)$. $118 = k(20) \Rightarrow k = 5.9$. Similarly, calculate a constant of proportionality for data point $(40, 324)$. $324 = k(40) \Rightarrow k = 8.1$. Since the constant of proportionalities are not the same, we may conclude that stopping distance is not directly proportional to speed. Note that doubling the speed more than doubles the stopping distance.

111. (a) For $(150, 26)$, $\dfrac{F}{x} \approx 0.173$; for $(180, 31)$, $\dfrac{F}{x} \approx 0.172$; for $(210, 36)$, $\dfrac{F}{x} \approx 0.171$; for $(320, 54)$, $\dfrac{F}{x} \approx 0.169$; the ratios give the constant of proportionality or the coefficient of friction.

  (b) From the table it appears that approximately 0.17 lbs of force is needed to push a 1 lb cargo box $\Rightarrow$ $k = 0.17x$.

  (c) See Figure 111.

  (d) $F = 0.17(275) \Rightarrow F = 46.75$ lbs of force.

[125, 350, 25] by [0, 75, 5]

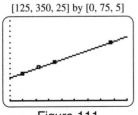

Figure 111

2.3: Linear Equations

Identifying Linear and Nonlinear Equations

1. $3x - 1.5 = 7 \Rightarrow 3x - 1.5 - 7 = 0 \Rightarrow 3x - 8.5 = 0$; the equation is linear.

3. $2\sqrt{x} + 2 = 1 \Rightarrow 2\sqrt{x} = -1 \Rightarrow \sqrt{x} = -\dfrac{1}{2}$, this has no real solution $\Rightarrow$ it is not linear.

5. $7x - 55 = 3(x - 8) + x \Rightarrow 7x - 55 = 3x - 24 + x \Rightarrow 7x - 4x - 55 + 24 = 0 \Rightarrow 3x - 31 = 0$; the equation is linear.

Solving Linear Equations Symbolically

7. $2x - 8 = 0 \Rightarrow 2x = 8 \Rightarrow x = 4$ Checking: $2(4) - 8 = 0 \Rightarrow 8 - 8 = 0 \Rightarrow 0 = 0$

9. $-5x + 3 = 23 \Rightarrow -5x = 20 \Rightarrow x = -4$ Checking: $-5(-4) + 3 = 23 \Rightarrow 20 + 3 = 23 \Rightarrow 23 = 23$

11. $4(z - 8) = z \Rightarrow 4z - 32 = z \Rightarrow 3z = 32 \Rightarrow z = \dfrac{32}{3}$ Checking: $4\left(\dfrac{32}{3} - 8\right) = \dfrac{32}{3} \Rightarrow 4\left(\dfrac{8}{3}\right) = \dfrac{32}{3} \Rightarrow$
$\dfrac{32}{32} = \dfrac{32}{32}$

13. $-5(3 - 4t) = 65 \Rightarrow -15 + 20t = 65 \Rightarrow 20t = 80 \Rightarrow t = 4$ Checking: $-5[3 - 4(4)] = 65 \Rightarrow$
$-5(3 - 16) = 65 \Rightarrow -5(-13) = 65 \Rightarrow 65 = 65$

15. $k + 8 = 5k - 4 \Rightarrow -4k = -12 \Rightarrow k = 3$ Checking: $3 + 8 = 5(3) - 4 \Rightarrow 11 = 15 - 4 \Rightarrow 11 = 11$

17. $2(1 - 3x) + 1 = 3x \Rightarrow 2 - 6x + 1 = 3x \Rightarrow -6x + 3 = 3x \Rightarrow -9x = -3 \Rightarrow x = \dfrac{1}{3}$

    Checking: $2\left[1 - 3\left(\dfrac{1}{3}\right)\right] + 1 = 3\left(\dfrac{1}{3}\right) \Rightarrow 2(1 - 1) + 1 = 1 \Rightarrow 0 + 1 = 1 \Rightarrow 1 = 1$

19. $-5(3 - 2x) - (1 - x) = 4(x - 3) \Rightarrow -15 + 10x - 1 + x = 4x - 12 \Rightarrow 11x - 16 = 4x - 12 \Rightarrow$
$7x = 4 \Rightarrow x = \dfrac{4}{7}$ Checking: $-5\left[3 - 2\left(\dfrac{4}{7}\right)\right] - \left(1 - \dfrac{4}{7}\right) = 4\left(\dfrac{4}{7} - 3\right) \Rightarrow -5\left(\dfrac{13}{7}\right) - \dfrac{3}{7} = 4\left(-\dfrac{17}{7}\right) \Rightarrow$
$\dfrac{65}{7} - \dfrac{3}{7} = -\dfrac{68}{7} \Rightarrow -\dfrac{68}{7} = -\dfrac{68}{7}$

21. $\dfrac{2}{7}n + \dfrac{1}{5} = \dfrac{4}{7} \Rightarrow \dfrac{2}{7}n = \dfrac{13}{35} \Rightarrow n = \dfrac{13}{10}$ Checking: $\dfrac{2}{7}\left(\dfrac{13}{10}\right) + \dfrac{1}{5} = \dfrac{4}{7} \Rightarrow \dfrac{26}{70} + \dfrac{1}{5} = \dfrac{4}{7} \Rightarrow \dfrac{40}{70} = \dfrac{4}{7} \Rightarrow$
$\dfrac{4}{7} = \dfrac{4}{7}$

23. $\dfrac{1}{2}(d - 3) - \dfrac{2}{3}(2d - 5) = \dfrac{5}{12} \Rightarrow \dfrac{1}{2}d - \dfrac{3}{2} - \dfrac{4}{3}d + \dfrac{10}{3} = \dfrac{5}{12} \Rightarrow -\dfrac{5}{6}d + \dfrac{11}{6} = \dfrac{5}{12} \Rightarrow -\dfrac{5}{6}d = -\dfrac{17}{12} \Rightarrow$
$d = \dfrac{17}{10}$ Checking: $\dfrac{1}{2}\left(\dfrac{17}{10} - 3\right) - \dfrac{2}{3}\left[2\left(\dfrac{17}{10}\right) - 5\right] = \dfrac{5}{12} \Rightarrow \dfrac{1}{2}\left(-\dfrac{13}{10}\right) - \dfrac{2}{3}\left(\dfrac{34}{10} - 5\right) = \dfrac{5}{12} \Rightarrow$
$\dfrac{1}{2}\left(-\dfrac{13}{10}\right) - \dfrac{2}{3}\left(-\dfrac{16}{10}\right) = \dfrac{5}{12} \Rightarrow -\dfrac{13}{20} + \dfrac{32}{30} = \dfrac{5}{12} \Rightarrow \dfrac{25}{60} = \dfrac{5}{12} \Rightarrow \dfrac{5}{12} = \dfrac{5}{12}$

25. $\dfrac{x-5}{3} + \dfrac{3-2x}{2} = \dfrac{5}{4} \Rightarrow 12\left(\dfrac{x-5}{3} + \dfrac{3-2x}{2} = \dfrac{5}{4}\right) \Rightarrow 4x - 20 + 18 - 12x = 15 \Rightarrow -8x - 2 = 15 \Rightarrow$

$-8x = 17 \Rightarrow x = -\dfrac{17}{8}$ Checking: $\dfrac{-\frac{17}{8} - 5}{3} + \dfrac{3 - 2(-\frac{17}{8})}{2} = \dfrac{5}{4} \Rightarrow \dfrac{-\frac{57}{8}}{3} + \dfrac{3 + \frac{34}{8}}{2} = \dfrac{5}{4} \Rightarrow -\dfrac{19}{8} + \dfrac{\frac{58}{8}}{2} = \dfrac{5}{4} \Rightarrow$

$-\dfrac{19}{8} + \dfrac{29}{8} = \dfrac{5}{4} \Rightarrow \dfrac{10}{8} = \dfrac{5}{4} \Rightarrow \dfrac{5}{4} = \dfrac{5}{4}$

27. $0.1z - 0.05 = -0.07z \Rightarrow 0.17z = 0.05 \Rightarrow z = \dfrac{0.05}{0.17} \Rightarrow z = \dfrac{5}{17}$ Checking:

$0.1\left(\dfrac{5}{17}\right) - 0.05 = -0.07\left(\dfrac{5}{17}\right) \Rightarrow \dfrac{5}{170} - \dfrac{5}{100} = -\dfrac{35}{1700} \Rightarrow \dfrac{50}{1700} - \dfrac{85}{1700} = -\dfrac{35}{1700} \Rightarrow -\dfrac{35}{1700} = -\dfrac{35}{1700}$

29. $0.15t + 0.85(100 - t) = 0.45(100) \Rightarrow 0.15t + 85 - 0.85t = 45 \Rightarrow -0.7t = -40 \Rightarrow t = \dfrac{40}{0.7} \Rightarrow t = \dfrac{400}{7}$

Checking: $0.15\left(\dfrac{400}{7}\right) + 0.85\left(100 - \dfrac{400}{7}\right) = 0.45(100) \Rightarrow \dfrac{6000}{700} + 85 - \dfrac{34{,}000}{700} = 45 \Rightarrow$

$\dfrac{60}{7} + 85 - \dfrac{340}{7} = 45 \Rightarrow 85 - \dfrac{280}{7} = 45 \Rightarrow 85 - 40 = 45 \Rightarrow 45 = 45$

31. (a) $5x - 1 = 5x + 4 \Rightarrow -1 = 4 \Rightarrow$ there is no solution.

   (b) Since no x-value satisfies the equation, it is contradiction.

33 . (a) $3(x - 1) = 5 \Rightarrow 3x - 3 = 5 \Rightarrow 3x = 8 \Rightarrow x = \dfrac{8}{3}$

   (b) Since one x-value is a solution and other x-values are not, the equation is conditional.

35. (a) $0.5(x - 2) + 5 = 0.5x + 4 \Rightarrow 0.5x - 1 + 5 = 0.5x + 4 \Rightarrow 0.5x + 4 = 0.5x + 4 \Rightarrow$ every x-value

   satisfies this equation.

   (b) Since every x-value satisfies the equation, it is an identity.

37. (a) $\dfrac{t + 1}{2} = \dfrac{3t - 2}{6} \Rightarrow 6\left(\dfrac{t + 1}{2} = \dfrac{3t - 2}{6}\right) \Rightarrow 3t + 3 = 3t - 2 \Rightarrow 3 = -2 \Rightarrow$ there is no solution

   (b) Since no x-value satisfies the equation, it is contradiction.

39. (a) $\dfrac{1 - 2x}{4} = \dfrac{3x - 1.5}{-6} \Rightarrow -6(1 - 2x) = 4(3x - 1.5) \Rightarrow -6 + 12x = 12x - 6 \Rightarrow 0 = 0 \Rightarrow$ every

   x-value satisfies this equation.

   (b) Since every x-value satisfies the equation, it is an identity.

41. (a) $0.5(x - 1980) + 5 = 10 \Rightarrow 0.5x - 990 + 5 = 10 \Rightarrow 0.5x = 995 \Rightarrow x = \dfrac{995}{0.5} = 1990$

   (b) Since one x-value is a solution and other x-values are not, the equation is conditional.

43. (a) $\dfrac{1}{2}(x - 3) = \sqrt{2}(4 + x) \Rightarrow \dfrac{1}{2}x - \dfrac{3}{2} = 4\sqrt{2} + \sqrt{2}x \Rightarrow \dfrac{1}{2}x - \sqrt{2}x = 4\sqrt{2} + \dfrac{3}{2} \Rightarrow$

   $(0.5 - \sqrt{2})x = 4\sqrt{2} + 1.5 \Rightarrow x = \dfrac{4\sqrt{2} + 1.5}{0.5 - \sqrt{2}} \approx -7.828$

   (b) Since one x-value is a solution and other x-values are not, the equation is conditional.

45. (a) $\pi(2x - 2.1) = 16 + 2\pi x \Rightarrow 2\pi x - 2.1\pi = 16 + 2\pi x \Rightarrow -2.1\pi = 16.$ There is no solution.

   (b) Since no x-value satisfies the equation, it is contradiction.

Solving Linear Equations Graphically

47. In the graph the lines intersect at $(3, -1)$. The solution is the x-value of $x = 3$.

49. (a) From the graph, when $f(x)$ or $y = -1, x = 4$

    (b) From the graph, when $f(x)$ or $y = 0, x = 2$

    (c) From the graph, when $f(x)$ or $y = 2, x = -2$

51. Let $y_1 = x + 4$ and $y_2 = 1 - 2x$ and graph each. See Figure 51. The lines intersect at $x = -1$.

53. Let $y_1 = -x + 4$ and $y_2 = 3x$ and graph each. See Figure 53. The lines intersect at $x = 1$.

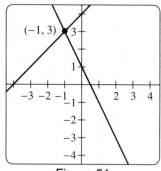
Figure 51

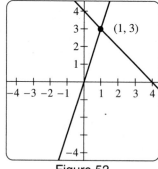
Figure 53

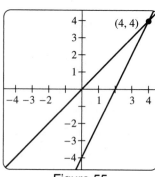
Figure 55

55. Let $y_1 = 2(x - 1) - 2$ and $y_2 = x$ and graph each. See Figure 55. The lines intersect at $x = 4$.

57. Graph $Y_1 = 5X - 1.5$ and $Y_2 = 5$. Their graphs intersect at $(1.3, 5)$. The solution is $x = 1.3$. See Figure 57.

59. Graph $Y_1 = 3X - 1.7$ and $Y_2 = 1 - X$. Their graphs intersect at $(0.675, 0.325)$. The solution is $x = 0.675$.

    See Figure 59.

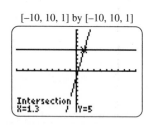
$[-10, 10, 1]$ by $[-10, 10, 1]$
Figure 57

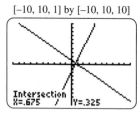
$[-10, 10, 1]$ by $[-10, 10, 10]$
Figure 59

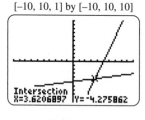
$[-10, 10, 1]$ by $[-10, 10, 10]$
Figure 61

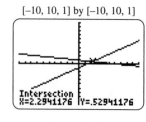
$[-10, 10, 1]$ by $[-10, 10, 1]$
Figure 63

61. Graph $Y_1 = 3.1(X - 5)$ and $Y_2 = 0.2X - 5$. Their graphs intersect near $(3.6207, -4.2759)$. The solution is

    $x \approx 3.621$. See Figure 61.

63. Graph $Y_1 = (6 - X)/7$ and $Y_2 = (2X - 3)/3$. Their graphs intersect near $(2.294, 0.5294)$. The solution is

    $x \approx 2.294$. See Figure 63.

Solving Linear Equations Numerically

65. $Y_1 = Y_2 = 4$ when $x = -2$.

67. One way to solve this equation is to table $Y_1 = 2X - 7$ and determine the x-value where $Y_1 = -1$. The results are shown in Figure 67. It can be seen that the solution occurs when $x = 3$.

X	Y₁	
0	-7	
1	-5	
2	-3	
3	-1	
4	1	
5	3	
6	5	

X=3

Figure 67

X	Y₁	
8.3	9.4	
8.4	9.6	
8.5	9.8	
8.6	10	
8.7	10.2	
8.8	10.4	
8.9	10.6	

X=8.6

Figure 69

69. Table $Y_1 = 2X - 7.2$ and determine the x-value where $Y_1 = 10$. The results are shown in Figure 69. It can be seen that the solution occurs when $x = 8.6$.

71. One way to solve this equation is to table $Y_1 = (5X - 16) - (5 - X)$ and determine the x-value where $Y_1 = 0$. This corresponds to the x-intercept method. The results are shown in Figure 71. $Y_1 = 0$ when $x = 3.5$.

X	Y₁	
3.3	-1.2	
3.4	-.6	
3.5	0	
3.6	.6	
3.7	1.2	
3.8	1.8	
3.9	2.4	

X=3.5

Figure 71

Solving Linear Equations by More than One Method

73. (a) $x - 3 = 2x + 1 \Rightarrow -x = 4 \Rightarrow x = -4$

   (b) Using the intersection-of-graphs method, graph $Y_1 = X - 3$ and $Y_2 = 2X + 1$. Their point of intersection is shown in Figure 73b as $(-4, -7)$. The solution is the x-value of $x = -4$.

   (c) Table $Y_1 = X - 3$ and $Y_2 = 2X + 1$, starting at $x = -5$, incrementing by 1. Figure 73c shows a table where $Y_1 = Y_2$ at $x = -4$.

[–12, 8, 1] by [–12, 8, 1]

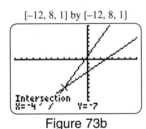

Figure 73b

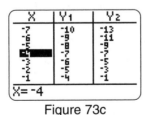

Figure 73c

[–10, 10, 1] by [–10, 10, 1]

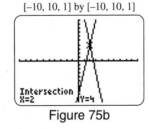

Figure 75b

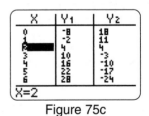

Figure 75c

75. (a) $6x - 8 = -7x + 18 \Rightarrow 13x = 26 \Rightarrow x = 2$

   (b) Using the intersection-of-graphs method, graph $Y_1 = 6X - 8$ and $Y_2 = -7X + 18$. Their point of intersection is shown in Figure 75b as $(2, 4)$. The solution is the x-value of $x = 2$.

   (c) Table $Y_1 = 6X - 8$ and $Y_2 = -7X + 18$, starting at $x = 0$, incrementing by 1. Figure 75c shows a table where $Y_1 = Y_2$ at $x = 2$.

77. (a) $5.5x - 16 = 2.3x + 8 \Rightarrow 3.2x = 24 \Rightarrow x = \dfrac{24}{3.2} = 7.5$

    (b) Using the *x*-intercept method, graph $Y_1 = (5.5X - 16) - (2.3X + 8)$. The *x*-intercept is 7.5 as shown in Figure 77b. The solution is $x = 7.5$.

    (c) Table $Y_1 = (5.5X - 16) - (2.3X + 8)$, starting at $x = 5$, incrementing by 0.5. Figure 77c shows a table where $Y_1 = 0$ at $x = 7.5$.

[−20, 20, 5] by [−20, 20, 5]

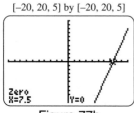

Figure 77b   Figure 77c

Applications

79. $f(x) = 16{,}190$ and $f(x) = 604.8(x - 1970) + 2280 \Rightarrow 604.8(x - 1970) + 2280 = 16{,}190 \Rightarrow$

    $604.8(x - 1970) = 16{,}190 - 2280 \Rightarrow x - 1970 = \dfrac{16{,}190 - 2280}{604.8} \Rightarrow x = 1970 + \dfrac{16{,}190 - 2280}{604.8} \Rightarrow$

    $x = 1992.9993$. The per capita income was \$16,190 in 1993.

81. (a) Since each student needs 900 cubic feet each hour, the ventilation required by *x* students per hour would be

    $V(x) = 900x$.

    (b) Fifty students would require a ventilation of $V(50) = 900(50) = 45{,}000$ cubic feet per hour.

    (c) $\dfrac{45{,}000}{10{,}000} = 4.5$; The air in the classroom should be replaced 4.5 times per hour.

    (d) It should be increased by $\dfrac{3000}{900} = 3\dfrac{1}{3}$ times. Smoking areas require more than triple the ventilation.

83. The graph of *f* must pass through the points (1980, 64) and (1990, 70). It's slope is $m = \dfrac{70 - 64}{1990 - 1980} = 0.6$.

    Thus $f(x) = 0.6(x - 1980) + 64$. Find the year when $f(x) = 72.4$:

    $0.6(x - 1980) + 64 = 72.4 \Rightarrow 0.6(x - 1980) = 72.4 - 64 \Rightarrow x - 1980 = \dfrac{72.4 - 64}{0.6} \Rightarrow$

    $x = 1980 + \dfrac{72.4 - 64}{0.6} \Rightarrow x = 1994$

85. The follow sketch describes the situation: Let $x =$ height of streetlight. See Figure 85.

    Using similar triangle, we get $\dfrac{x}{15 + 7} = \dfrac{5.5}{7} \Rightarrow x = \dfrac{(5.5)(22)}{7} \Rightarrow x \approx 17.29$. The streetlight is about

    17.29 feet. high.

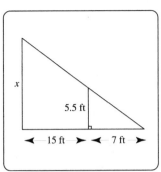

Figure 85

87. (a) It is reasonable to expect that f is linear because if the number of gallons of gas doubles so should the amount of oil. Five gallons of gasoline requires five times the oil that one gallon of gasoline would. The increase in oil is always equal to 0.16 pint for each additional gallon of gasoline. Oil is mixed at a constant rate, so a linear function describes this amount.

   (b) $f(3) = 0.16(3) = 0.48$; 0.48 pint of oil should be added to 3 gallons of gasoline to get the correct mixture.

   (c) 1 quart = 2 pints; $0.16x = 2 \Rightarrow x = 12.5$; 12.5 gallons of gasoline should be mixed with 1 quart of oil.

89. Use similar triangles to find the radius of the cone when the water is 7 feet deep: $\dfrac{r}{3.5} = \dfrac{7}{11} \Rightarrow r \approx 2.23$ ft. Use $V = \dfrac{1}{3}\pi r^2 h$ to find the volume of the water in the cone at $h = 7$ ft.: $V = \dfrac{1}{3}\pi(2.23)^2(7) \approx 36.4$ cu. ft.

91. The rate of deaths is $\dfrac{8.6}{1000} = 0.86\%$. That means that 0.86% of the population died in 1997. Let P be the population of the United States. Then $0.0086P = 2.31 \Rightarrow P = \dfrac{2.31}{0.0086} \approx 268.6$. There were approximately 268.6 million people in the United States in 1997.

93. To calculate the sale price subtract 25% of the regular price from the regular price.

   $f(x) = x - 0.25x \Rightarrow f(x) = 0.75x$. An item which normally costs $\$56.24$ will be on sale for

   $f(56.24) = 0.75(56.24) = \$42.18$.

95. Let $t =$ time spent traveling at 55 mph and $6 - t =$ time spent traveling at 70 mph. Using $d = rt$, we get

   $d = 55t + 70(6 - t) \Rightarrow 372 = 55t + 420 - 70t \Rightarrow -48 = -15t \Rightarrow t = 3.2$ and $6 - t = 2.8$; the car traveled 3.2 hours at 55 mph and 2.8 hours at 70 mph.

97. (a) It would take a little less time than the faster gardener, who can rake the lawn in 3 hours, since the second gardener would take 5 hours to rake the same lawn; It would take both gardeners about 2 hours working together. *Answers may vary.*

   (b) Let $x =$ time to rake the lawn working together; in 1 hour one gardener can rake $\dfrac{1}{3}$ of the lawn, whereas the second gardener can rake $\dfrac{1}{5}$ of the lawn; in x hours both gardeners working together can rake $\dfrac{x}{3} + \dfrac{x}{5}$ of the lawn; $\dfrac{x}{3} + \dfrac{x}{5} = 1 \Rightarrow 5x + 3x = 15 \Rightarrow 8x = 15 \Rightarrow x = \dfrac{15}{8} = 1.875$ hours.

99. Let x = amount of pure water to be added and $x + 5$ = final amount of the 15% solution; since pure water has 0% sulfuric acid, we get the equation $0\%x + 40\%(5) = 15\%(x + 5) \Rightarrow .40(5) = .15(x + 5) \Rightarrow$ $40(5) = 15(x + 5) \Rightarrow 200 = 15x + 75 \Rightarrow 15x = 125 \Rightarrow x = \dfrac{125}{15} \approx 8.333$; about 8.33 liters of pure water should be added.

101. Let x = amount of the $2.50 per pound candy and $5 - x$ = amount of the $4.00 per pound candy; we get the equation $2.50x + 4.00(5 - x) = 17.60 \Rightarrow 250x + 400(5 - x) = 1760 \Rightarrow$ $250x + 2000 - 400x = 1760 \Rightarrow -150x = -240 \Rightarrow x = 1.6$ and $5 - x = 3.4$; add 1.6 pounds of $2.50 candy to 3.4 pounds of $4.00 candy.

103. $P = 2w + 2l \Rightarrow 180 = 2w + 2(w + 18) \Rightarrow 180 = 4w + 36 \Rightarrow 4w = 144 \Rightarrow w = 36$ and $w + 18 = 54$; the window is 36 inches by 54 inches.

105. $2(x + 6) + 2(2x + 6) = 174 \Rightarrow 2x + 12 + 4x + 12 = 174 \Rightarrow 6x + 24 = 174 \Rightarrow 6x = 150 \Rightarrow$ $x = 25 \Rightarrow 2x = 50$. The swimming pool is 25 ft. by 50 ft. See Figure 105.

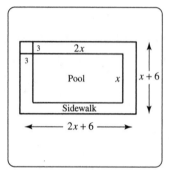

Figure 105

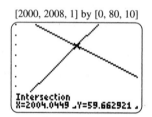

[2000, 2008, 1] by [0, 80, 10]

Figure 111c

X	Y1	Y2
2000	90	0
2001	82.5	14.75
2002	75	29.5
2003	67.5	44.25
2004	60	59
2005	52.5	73.75
2006	45	88.5

X=2004

Figure 111e

107. Let x = amount invested at 5% and $5000 - x$ = amount invested at 7%; $0.05x + 0.07(5000 - x) = 325 \Rightarrow$ $5x + 7(5000 - x) = 32,500 \Rightarrow 5x + 35,000 - 7x = 32,500 \Rightarrow -2x = -2500 \Rightarrow x = 1250$ and $5000 - x = 3750$; $1250 is invested at 5% and $3750 is invested at 7%.

109. $C = \dfrac{5}{9}(F - 32)$ and $F = C \Rightarrow F = \dfrac{5}{9}(F - 32) \Rightarrow F = \dfrac{5}{9}F - \dfrac{160}{9} \Rightarrow \dfrac{4}{9}F = -\dfrac{160}{9} \Rightarrow F = -40$; $-40°\,F$ is equivalent to $-40°\,C$.

111. (a) Using (2002, 75) and (2006, 45), find slope: $m = \dfrac{45 - 75}{2006 - 2002} \Rightarrow \dfrac{-30}{4} \Rightarrow -7.5$. Therefore, $C(x) = -7.5(x - 2002) + 75 \Rightarrow C(x) = -7.5x + 15,090$. Now using (2002, 29) and (2006, 88), find slope: $m = \dfrac{88 - 29}{2006 - 2002} = \dfrac{59}{4} = 14.75$. Therefore, $L(x) = 14.75(x - 2002) + 29 \Rightarrow$ $L(x) = 14.75x - 29,500.5$.

(b) The slope or sales of CRT monitors is decreasing 7.5 million per year, on average. The slope or sales of LCD monitors is increasing 14.75 million per year, on average.

(c) Graph $C(x) = -7.5x + 15,090$ and $L(x) = 14.75x - 29,500$. See Figure 111c. The lines of the functions intersect at $x = 2004$ or year 2004.

(d) Set $L(x) = C(x)$ and solve: $14.75x - 29,500 = -7.5x + 15,090 \Rightarrow 22.25x = 44,590 \Rightarrow x \approx 2004$

(e) Table $Y_1 = -7.5X + 15,090$ and $Y_2 = 14.75X - 29,500$, starting at 2000, incrementing by 1. Figure 111e shows a table where $Y_1 = Y_2$ and $x = 2004$.

Intermediate Value Property

113. To find the point $(x, 3.5)$ that lies on the graph of f solve the equation $3.5 = 2.5x - 6.5$.

$3.5 = 2.5x - 6.5 \Rightarrow 10 = 2.5x \Rightarrow x = 4$. The point is $(4, 3.5)$. Since the graph f is continuous and $-4 \le 3.5 \le 6$, there must be a number x such that $2.5x - 6.5 = 3.5$ and $1 \le x \le 5$. There must be a point where the graph of f crosses the line $y = 3.5$. See Figure 113.

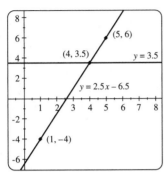

Figure 113

115. $f(x) = x^2 - 5 \Rightarrow f(2) = 2^2 - 5 = -1$ and $f(3) = 3^2 - 5 = 4$; because $f(2) < 0$ and $f(3) > 0$, by the intermediate value property, there exists an x-value between 2 and 3 such that $f(x) = 0$.

117. $f(x) = 2x^3 - 1 \Rightarrow f(0) = 2(0)^3 - 1 = -1$ and $f(1) = 2(1)^3 - 1 = 1$; because $f(0) < 0$ and $f(1) > 0$, by the intermediate value property, there exists an x-value between 0 and 1 such that $f(x) = 0$.

2.4: Linear Inequalities

Interval Notation

1. $[5, \infty)$

3. $[4, 19)$

5. $(-\infty, -37]$

7. $[-1, \infty)$

9. $(-3, 5]$

11. $(-\infty, -2)$

Solving Linear Inequalities Symbolically

13. $2x + 6 \ge 10 \Rightarrow 2x \ge 4 \Rightarrow x \ge 2;\ [2, \infty)$

15. $-2(x - 10) + 1 > 0 \Rightarrow -2x + 21 > 0 \Rightarrow -2x > -21 \Rightarrow x < 10.5;\ (-\infty, 10.5)$

17. $\dfrac{t + 2}{3} \ge 5 \Rightarrow t + 2 \ge 15 \Rightarrow t \ge 13;\ [13, \infty)$

19. $-3(z - 4) \ge 2(1 - 2z) \Rightarrow -3z + 12 \ge 2 - 4z \Rightarrow z \ge -10;\ [-10, \infty)$

21. $\dfrac{1 - x}{4} < \dfrac{2x - 2}{3} \Rightarrow 3(1 - x) < 4(2x - 2) \Rightarrow 3 - 3x < 8x - 8 \Rightarrow -11x < -11 \Rightarrow x > 1;\ (1, \infty)$

23. $2x - 3 > \dfrac{1}{2}(x + 1) \Rightarrow 2x - 3 > \dfrac{1}{2}x + \dfrac{1}{2} \Rightarrow \dfrac{3}{2}x > \dfrac{7}{2} \Rightarrow x > \dfrac{7}{3};\ \left(\dfrac{7}{3}, \infty\right)$

25. $-1 \le 2t \le 4 \Rightarrow -\dfrac{1}{2} \le t \le 2;\ \left[-\dfrac{1}{2}, 2\right]$

27. $3 \le 4 - x \le 20 \Rightarrow -1 \le -x \le 16 \Rightarrow 1 \ge x \ge -16;\ [-16, 1]$

29. $0 < \dfrac{7x - 5}{3} \le 4 \Rightarrow 0 < 7x - 5 \le 12 \Rightarrow 5 < 7x \le 17 \Rightarrow \dfrac{5}{7} < x \le \dfrac{17}{7};\ \left(\dfrac{5}{7}, \dfrac{17}{7}\right]$

31. $5 > 2(x + 4) - 5 > -5 \Rightarrow 5 > 2x + 8 - 5 > -5 \Rightarrow 2 > 2x > -8 \Rightarrow 1 > x > -4;\ (-4, 1)$

33. $3 \le \dfrac{1}{2}x + \dfrac{3}{4} \le 6 \Rightarrow 12 \le 2x + 3 \le 24 \Rightarrow 9 \le 2x \le 21 \Rightarrow \dfrac{9}{2} \le x \le \dfrac{21}{2};\ \left[\dfrac{9}{2}, \dfrac{21}{2}\right]$

35. $3x - 1 < 2(x - 3) + 1 \Rightarrow 3x - 1 < 2x - 6 + 1 \Rightarrow x < -4;\ (-\infty, -4)$

37. $\dfrac{1}{2} \le \dfrac{1 - 2t}{3} < \dfrac{2}{3} \Rightarrow \dfrac{3}{2} \le 1 - 2t < 2 \Rightarrow \dfrac{1}{2} \le -2t < 1 \Rightarrow -\dfrac{1}{4} \ge t > -\dfrac{1}{2} \Rightarrow -\dfrac{1}{2} < t \le -\dfrac{1}{4};\ \left(-\dfrac{1}{2}, -\dfrac{1}{4}\right]$

39. $\dfrac{1}{2}z + \dfrac{2}{3}(3 - z) - \dfrac{5}{4}z \ge \dfrac{3}{4}(z - 2) + z \Rightarrow \dfrac{1}{2}z + 2 - \dfrac{2}{3}z - \dfrac{5}{4}z \ge \dfrac{3}{4}z - \dfrac{3}{2} + z \Rightarrow -\dfrac{17}{12}z + 2 \ge \dfrac{7}{4}z - \dfrac{3}{2}$

$-\dfrac{38}{12}z \ge -\dfrac{7}{2} \Rightarrow z \le \dfrac{21}{19};\ \left(-\infty, \dfrac{21}{19}\right]$

Solving Linear Inequalities Graphically

41. Graph $y_1 = x + 2$ and $y_2 = 2x$. See Figure 41. $y_1 > y_2$ when the line of y_1 is above the line of y_2, which by the graph is left of the intersection point $(2, 4)$ and includes point $(2, 4) \Rightarrow \{x \mid x \le 2\}$.

43. Graph $y_1 = \dfrac{2}{3}x - 2$ and $y_2 = -\dfrac{4}{3}x + 4$. See Figure 43. $y_1 > y_2$ when the line of y_1 is above the line of y_2, which by the graph is right of the intersection point $(3, 0)$ and does not include point $(3, 0) \Rightarrow \{x \mid x > 3\}$.

[–10, 10, 1] by [–10, 10, 1]   [–10, 10, 1] by [–10, 10, 1]   [–10, 10, 1] by [–10, 10, 1]   [–10, 10, 1] by [–10, 10, 1]

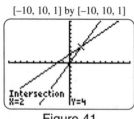
Figure 41

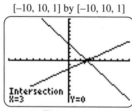
Figure 43

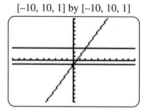
Figure 45

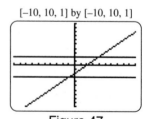
Figure 47

45. Graph $y_1 = -1$, $y_2 = 2x - 1$, and $y_3 = 3$. See Figure 45. $y_1 \le y_2 \le y_3$ when the line of y_2 is in between the lines of y_1 and y_3, which by the graph is in between the intersection points $(0, -1)$ and $(2, 3)$ and it does include each point $\Rightarrow \{x \mid 0 \le x \le 2\}$.

47. Graph $y_1 = -3$, $y_2 = x - 2$, and $y_3 = 2$. See Figure 47. $y_1 < y_2 \le 2$ when the line of y_2 is in between the lines of y_1 and y_3, which by the graph is in between the intersection points $(-1, -3)$ and $(4, 2)$, it does not include the point $(-1, -3)$ but does include $(4, 2) \Rightarrow \{x \mid -1 < x \le 4\}$.

49. (a) $y = \dfrac{3}{2}x - 3$, then $ax + b = 0$ gives us $\dfrac{3}{2}x - 3 = 0 \Rightarrow \dfrac{3}{2}x = 3 \Rightarrow x = 2$

  (b) $ax + b < 0$ gives us $\dfrac{3}{2}x - 3 < 0 \Rightarrow x < 2 \Rightarrow (-\infty, 2)$

  (c) $ax + b \geq 0$ gives us $\dfrac{3}{2}x - 3 \geq 0 \Rightarrow x \geq 2 \Rightarrow [2, \infty)$

51. (a) $y = -x - 2$, then $ax + b = 0$ gives us $-x - 2 = 0 \Rightarrow -x = 2 \Rightarrow x = -2$

  (b) $ax + b < 0$ gives us $-x - 2 < 0 \Rightarrow x > -2 \Rightarrow (-2, \infty)$

  (c) $ax + b \geq 0$ gives us $-x - 2 \geq 0 \Rightarrow x \leq -2 \Rightarrow (-\infty, -2]$

53. $x - 3 \leq \dfrac{1}{2}x - 2$, subtract $-\dfrac{1}{2}x - 2$ from each side, $y_1 = x - 3 - \left(\dfrac{1}{2}x - 2\right) \Rightarrow y_1 = \dfrac{1}{2}x - 1$, graph this

  equation. See Figure 53. The solution set for $y_1 \leq 0$ occurs where the graph of the line is below the x-axis, the

  x-intercept is 2, $\Rightarrow x \leq 2 \Rightarrow (-\infty, 2]$; solving symbolically, $x - 3 \leq \dfrac{1}{2}x - 2 \Rightarrow \dfrac{1}{2}x \leq 1 \Rightarrow x \leq 2 \Rightarrow$

  $(-\infty, 2]$.

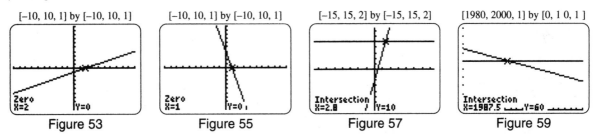

| [−10, 10, 1] by [−10, 10, 1] | [−10, 10, 1] by [−10, 10, 1] | [−15, 15, 2] by [−15, 15, 2] | [1980, 2000, 1] by [0, 1 0, 1] |

Figure 53   Figure 55   Figure 57   Figure 59

55. $2 - x < 3x - 2$, subtract $3x - 2$ from each side, $y_1 = 2 - x - (3x - 2) \Rightarrow y_1 = -4x + 4$, graph this

  equation. See Figure 55. The solution set for $y_1 < 0$ occurs where the graph of the line is below the x-axis, the

  x-intercept is 1, $\Rightarrow x > 1 \Rightarrow (1, \infty)$; solving symbolically, $2 - x < 3x - 2 \Rightarrow -4x < -4 \Rightarrow x > 1 \Rightarrow$

  $(1, \infty)$.

57. Graph $Y_1 = 5X - 4$ and $Y_2 = 10$. Their graphs intersect at the point (2.8, 10). The graph of Y_1 is above

  the graph of Y_2 for x-values to the right of this intersection point or where $x > 2.8$, $\{x \mid x > 2.8\}$.

  See Figure 57.

59. Graph $Y_1 = -2(X - 1990) + 55$ and $Y_2 = 60$. Their graphs intersect at the point (1987.5, 60). The graph

  of Y_1 is above the graph of Y_2 for x-values to the left of this intersection point or where $x \leq 1987.5$,

  $\{x \mid x \leq 1987.5\}$. See Figure 59.

61. Graph $Y_1 = \sqrt{5}(X - 1.2) - \sqrt{3}X$ and $Y_2 = 5(X + 1.1)$. Their graphs intersect near the point

  $(-1.820, -3.601)$. The graph of Y_1 is below the graph of Y_2 for x-values to the right of this intersection

  point or when $x > k$, where $k \approx -1.82$, $\{x \mid x > -1.82\}$. See Figure 61.

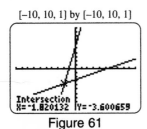

[−10, 10, 1] by [−10, 10, 1]

Figure 61

63. Graph $Y_1 = 3$, $Y_2 = 5X - 17$ and $Y_3 = 15$, as shown in Figure 63. The graphs intersect at the points $(4, 3)$ and $(6.4, 15)$. The solutions to $Y_1 \le Y_2 < Y_3$ are the x-values between these two points or $4 \le x < 6.4 \Rightarrow$ $[4, 6.4)$.

65. Graph $Y_1 = 1.5$, $Y_2 = 9.1 - 0.5X$ and $Y_3 = 6.8$, as shown in Figure 65. The graphs intersect at the points $(4.6, 6.8)$ and $(15.2, 1.5)$. The solutions to $Y_1 \le Y_2 \le Y_3$ are the x-values between these two points or $4.6 \le x \le 15.2 \Rightarrow [4.6, 15.2]$.

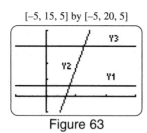
[–5, 15, 5] by [–5, 20, 5]
Figure 63

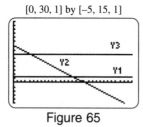
[0, 30, 1] by [–5, 15, 1]
Figure 65

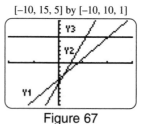
[–10, 15, 5] by [–10, 10, 1]
Figure 67

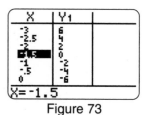
Figure 73

67. Graph $Y_1 = X - 4$, $Y_2 = 2X - 5$ and $Y_3 = 6$, as shown in Figure 67. The graphs intersect at the points $(1, -3)$ and $(5.5, 6)$. The solutions to $Y_1 < Y_2 < Y_3$ are the x-values between these two points or $1 < x < 5.5 \Rightarrow (1, 5.5)$.

69. (a) The graphs intersect at the point $(8, 7)$. Therefore, $g(x) = f(x)$ is satisfied when $x = 8$.

   (b) $g(x) > f(x)$ whenever the y-values on the graph of g are above the y-values on the graph of f. This occurs to the left of the point of intersection. Therefore the x-values that satisfy this inequality are $x < 8$.

Solving Linear Inequalities Numerically

71. From the table, $Y_1 = 0$ when $x = 4$. $Y_1 > 0$ when $x < 4$ and $Y_1 \le 0$ when $x \ge 4 \Rightarrow \{x \mid x < 4\}; \{x \mid x \ge 4\}$.

73. From the table shown in Figure 73, $Y_1 = 0$ when $x = -1.5$ or $-\dfrac{3}{2}$. $Y_1 > 0$ when $x < -\dfrac{3}{2}$.

75. From the table shown in Figure 75, Y_1 is between 1 and 4 for x values between 1 and 4, $\Rightarrow [1, 4]$.

77. From the table shown in Figure 77, Y_1 is between -0.75 and 0.75 for x values between -0.05 and 0.85, $\Rightarrow$ $\left[-\dfrac{1}{20}, \dfrac{17}{20}\right)$.

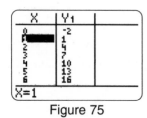
Figure 75

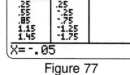
Figure 77

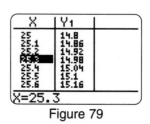
Figure 79

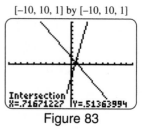
[–10, 10, 1] by [–10, 10, 1]
Figure 83

79. From the table shown in Figure 79, $Y_1 \approx 15$ when $x \approx 25.3$. $Y_1 < 15$ when $x < 25.3$; $(-\infty, 25.3)$.

You Decide the Method

81. Symbolically: $2x - 8 > 5 \Rightarrow 2x > 13 \Rightarrow x > \dfrac{13}{2}$. The solution interval is $\left(\dfrac{13}{2}, \infty\right)$.

83. Graphically: Let $Y_1 = 5.1X - \pi$ and $Y_2 = \sqrt{(3)} - 1.7X$. Graph Y_1 and Y_2 as shown in Figure 83.

    Their graphs intersect near $(0.717, 0.514)$. The graph of Y_1 is above Y_2 for $x \geq 0.717$.

    The solution interval is $[0.717, \infty)$.

Applications

85. (a) The graphs intersect at the point $(1000, 100)$. A loan amount of $1000 results in annual interest of $100.

    (b) A loan amount greater than $1000 results in annual interest of more than $100.

    (c) A loan amount less than $1000 results in annual interest of less than $100.

    (d) A loan amount greater than or equal to $1000 results in annual interest of $100 or more.

87. (a) Car A is traveling the fastest since it passes Car B. Its graph has the greater slope.

    (b) The cars are the same distance from St. Louis when their graphs intersect. This point of intersection occurs

    at $(2.5, 225)$. The cars are both 225 miles from St. Louis after 2.5 hours.

    (c) Car B is ahead of Car A when $0 \leq x < 2.5$.

89. (a) Graph $Y_1 = 85 - 29X$ and $Y_2 = 32$. These graphs intersect near the point $(1.8, 32)$ as shown in Figure 89.

    At an altitude of approximately 1.8 miles the temperature is 32°F. The temperature is below 32°F above

    this altitude. Since the domain is limited to an altitude of 6 miles, the region where the temperature is below

    freezing is above 1.8 miles and up to 6 miles. The solution is $1.8 < x \leq 6$ where 1.8 is approximate.

    (b) The x-intercept represents the altitude where the temperature is 0°F.

    (c) $T(x) = 32 \Rightarrow 85 - 29x = 32 \Rightarrow -29x = -53 \Rightarrow x = \dfrac{53}{29} \approx 1.8$. Thus, $\dfrac{53}{29} < x \leq 6$.

[−10, 10, 1] by [−10, 10, 1]   [0, 10, 1] by [50,000, 100,000, 110,000]

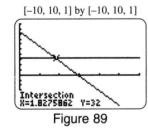
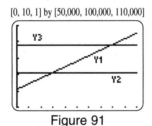

Figure 89   Figure 91

91. (a) The slope of the graph of P is 3421. This means that the median price of a single-family home has

    increased by approximately $3421 per year.

    (b) Graph $Y_1 = 3421X + 61000$, $Y_2 = 71,000$ and $Y_3 = 88,000$. These graphs are shown in Figure 91.

    The points of intersection are located near $(2.92, 71,000)$ and $(7.89, 88,000)$. For approximately

    $2.92 \leq x \leq 7.89$ or, rounded to the nearest year, between 1983 and 1988, the median price was between

    $71,000 and $88,000. (Note that $x = 0$ corresponds to 1980.)

93. (a) Using $(1997, 100)$ and $(2002, 400)$, find slope: $m = \dfrac{400 - 100}{2002 - 1997} = \dfrac{300}{5} = 60 \Rightarrow$

    $T(x) = 60(x - 1997) + 100.$

    (b) $160 \leq 60x - 119,720 \leq 280 \Rightarrow 119,880 \leq 60x \leq 120,000 \Rightarrow 1998 \leq x \leq 2000$, from 1998 to 2000.

95. (a) Using (2000, 6) and (2004, 30), find slope: $m = \dfrac{30 - 6}{2004 - 2000} = \dfrac{24}{4} = 6 \Rightarrow T(x) = 6(x - 2000) + 6.$

   (b) $24 \le 6x - 11{,}994 \Rightarrow 12{,}018 \le 6x \Rightarrow 2003 \le x$, from 2003 to 2006.

97. (a) Increasing the ventilation increases the percentage of pollutants removed. The slope is 1.06. This means that for each additional liter per second of air removed, the amount of pollutants decreases by 1.06%.

   (b) Graph $Y_1 = 1.06X + 7.18$, $Y_2 = 50$ and $Y_3 = 70$. See Figure 97. The points of intersection occur near (40.4, 50) and (59.3, 70). If the range hood removes between 40.4 and 59.3 liters of air per second, it will remove between 50% and 70% of the pollutants.

[10, 75, 5] by [0, 100, 5]

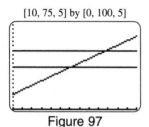

Figure 97

99. $r = \dfrac{C}{2\pi}$ and $1.99 \le r \le 2.01 \Rightarrow 1.99 \le \dfrac{C}{2\pi} \le 2.01 \Rightarrow 3.98\pi \le C \le 4.02\pi$

101. (a) $m = \dfrac{4.5 - (-1.5)}{2 - 0} = \dfrac{6}{2} = 3$ and y-intercept $= -1.5$; $f(x) = 3x - 1.5$ models the data.

   (b) $f(x) > 2.25 \Rightarrow 3x - 1.5 > 2.25 \Rightarrow 3x > 3.75 \Rightarrow x > 1.25$

103. (a) $m = \dfrac{48{,}650 - 47{,}975}{2003 - 2002} = 675$ and let the vertical intercept be at the point (2002, 47,975);

   The function $f(x) = 675(x - 2002) + 47{,}975$ or $f(x) = 675x - 1{,}303{,}375$ models the data.

   (b) $f(x) \le 52{,}700 \Rightarrow 675x - 1{,}303{,}375 \le 52{,}700 \Rightarrow 675x \le 1{,}356{,}075 \Rightarrow x \le 2009 \Rightarrow 2002, 2003,$ ... 2009; from 2002 to 2009 the person's salary was less than or equal to $52,700.

2.5: Piecewise-Defined Functions

Evaluating and Graphing Piecewise-Defined Functions

1. (a) The maximum speed limit is 55 mph and the minimum is 30 mph.

   (b) The speed limit is 55 for $0 \le x < 4$, $8 \le x < 12$, and $16 < x \le 20$. This is $4 + 4 + 4 = 12$ miles.

   (c) $f(4) = 40, f(12) = 30$, and $f(18) = 55$.

   (d) The graph has breaks in it or is discontinuous when $x = 4, 6, 8, 12$, and 16. The speed limit changes at each of these discontinuities.

3. (a) The initial amount in the pool occurs when $x = 0$. $f(0) = 50$ or 50,000 gallons. The final amount of water in the pool occurs when $x = 5$. Then, $f(5) = 30$ or 30,000 gallons.

   (b) The water level remained constant during the first day and the fourth day, when $0 \le x \le 1$ or $3 \le x \le 4$.

   (c) $f(2) = 45$ thousand and $f(4) = 40$ thousand

   (d) During the second and third days, the amount of water changed from 50,000 gallons to 40,000 gallons. This represents 10,000 gallons in 2 days or 5000 gallons per day were being pumped out of the pool.

5. (a) $f(1.5) = 30; f(4) = 10$

    (b) $m_1 = 20$ indicates that the car is moving away from home at 20 mph; $m_2 = -30$ indicates that the car is moving toward home at 30 mph; $m_3 = 0$ indicates that the car is not moving; $m_4 = -10$ indicates that the car is moving toward home at 10 mph.

    (c) The driver starts at home and drives away from home at 20 mph for 2 hours until the car is 40 miles from home. The driver then travels toward home at 30 mph for 1 hour until the car is 10 miles from home. Then the car does not move for 1 hour. Finally, the driver returns home in 1 hour at 10 mph.

7. (a) $D = \{x \mid -5 \le x \le 5\}$

    (b) $f(-2) = 2, f(0) = 0 + 3 = 3, f(3) = 3 + 3 = 6$

    (c) See Figure 7.

    (d) f is continuous.

9. (a) $D = \{x \mid -1 \le x \le 2\}$

    (b) $f(-2) =$ undefined, $f(0) = 3(0) = 0, f(3) =$ undefined

    (c) See Figure 9.

    (d) f is not continuous.

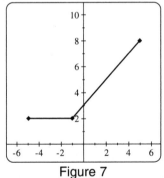

Figure 7

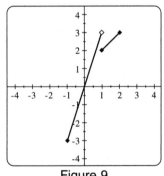

Figure 9

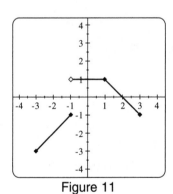

Figure 11

11. (a) $D = \{x \mid -3 \le x \le 3\}$

    (b) $f(-2) = -2, f(0) = 1, f(3) = 2 - 3 = -1$

    (c) See Figure 11.

    (d) f is not continuous.

13   $f(-4) = -\frac{1}{2}(-4) + 1 = 3, (-4, 3); f(-2) = -\frac{1}{2}(-2) + 1 = 2, (-2, 2);$ graph a segment from $(-4, 3)$ to

$(-2, 2)$, use a closed dot for each point. See Figure 13.

$f(-2) = 1 - 2(-2) = 5, (-2, 5); f(1) = 1 - 2(1) = -1, (1, -1);$ graph a segment from $(-2, 5)$ to $(1, -1)$, use

an open dot at $(-2, 5)$ and a closed dot for $(1, -1)$. See Figure 13.

$f(1) = \frac{2}{3}(1) + \frac{4}{3} = 2, (1, 2); f(4) = \frac{2}{3}(4) + \frac{4}{3} = 4, (4, 4);$ graph a segment from $(1, 2)$ to $(4, 4)$, use an open

dot at $(1, 2)$ and a closed dot for $(4, 4)$. See Figure 13.

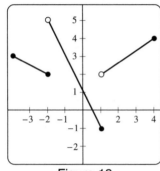
Figure 13

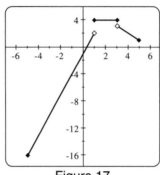
Figure 17

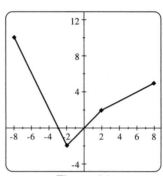
Figure 21

15. $f(-3) = 3(-3) - 1 = -10, f(1) = 4, f(2) = 4,$ and $f(5) = 6 - 5 = 1$

17. See Figure 17. f is not continuous.

19. $g(-8) = -2(-8) - 6 = 10, g(-2) = -2(-2) - 6 = -2, g(2) = 0.5(2) + 1 = 2$ and $g(8) = 0.5(8) + 1 = 5$

21. See Figure 21. g is continuous.

Greatest Integer Function

23. (a) Graph $Y_1 = \text{int}(2X - 1)$ as shown in Figure 23.

   (b) $f(-3.1) = [2(-3.1) - 1] = [-7.2] = -8$ and $f(1.7) = [2(1.7) - 1] = [2.4] = 2$

[-10, 10, 1] by [-10, 10, 1]
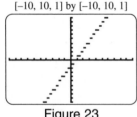
Figure 23

[-10, 10, 1] by [-10, 10, 1]
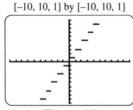
Figure 25

[6, 18, 1] by [0, 8, 1]
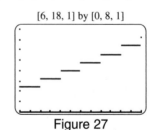
Figure 27

25. (a) Graph $Y_1 = 2(\text{int}(-X)) + 1$ as shown in Figure 25.

   (b) $f(-3.1) = 2[-3.1] + 1 = 2(-4) + 1 = -7$ and $f(1.7) = 2[1.7] + 1 = 2(1) + 1 = 3$

27. (a) Let $f(x) = 0.8\left[\dfrac{x}{2}\right]$ for $6 \le x \le 18$

   (b) Graph $Y_1 = 0.8(\text{int}(X/2))$ as shown in Figure 27.

   (c) $f(8.5) = 0.8\left[\dfrac{8.5}{2}\right] = 0.8[4.25] = 0.8(4) = \$3.20$ and $f(15.2) = 0.8\left[\dfrac{15.2}{2}\right] = 0.8[7.6] = 0.8(7) = \$5.60$

Absolute Value Equations and Inequalities

29. The vertex is $x + 1 = 0 \Rightarrow x = -1 \Rightarrow (-1, 0)$, find any other point, $x = 0 \Rightarrow (0, 1)$, graph the absolute value

    V graph with the vertex $(-1, 0)$, point $(0, 1)$ and its reflection through $x = -1$. See Figure 29.

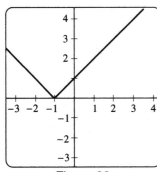

Figure 29

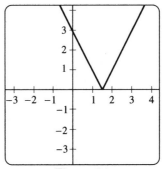

Figure 31

31. The vertex is $2x - 3 = 0 \Rightarrow 2x = 3 \Rightarrow x = \dfrac{3}{2} \Rightarrow \left(\dfrac{3}{2}, 0\right)$, find another point $(0, 3)$, graph the absolute

    value. See Figure 31.

33. (a) The graph of $Y_1 = 2X$ is shown in Figure 33a.

    (b) The graph of $y = |2x|$ is similar to the graph of $y = 2x$ except that it is reflected across x-axis whenever
        $2x < 0$. The graph of $Y_1 = |2X|$ is shown in Figure 33b.

    (c) The x-intercept occurs when $2x = 0$ or when $x = 0$. The x-intercept is located at $(0, 0)$.

[−10, 10, 1] by [−10, 10, 1]

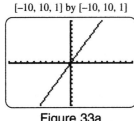

Figure 33a

[−10, 10, 1] by [−10, 10, 1]

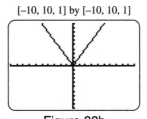

Figure 33b

35. (a) The graph of $Y_1 = 3X - 3$ is shown in Figure 35a.

    (b) The graph of $y = |3x - 3|$ is similar to the graph of $y = 3x - 3$ except that it is reflected across x-axis
        whenever $3x - 3 < 0$ or $x < 1$. The graph of $Y_1 = |3X - 3|$ is shown in Figure 35b.

    (c) The x-intercept occurs when $3x - 3 = 0$ or when $x = 1$. The x-intercept is located at $(1, 0)$.

[−10, 10, 1] by [−10, 10, 1]

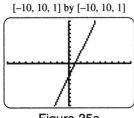

Figure 35a

[−10, 10, 1] by [−10, 10, 1]

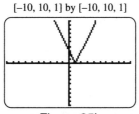

Figure 35b

37. (a) The graph of $Y_1 = 6 - 2X$ is shown in Figure 37a.

(b) The graph of $y = |6 - 2x|$ is similar to the graph of $y = 6 - 2x$ except that it is reflected across x-axis whenever $6 - 2x < 0$ or $x > 3$. The graph of $Y_1 = |6 - 2X|$ is shown in Figure 37b.

(c) The x-intercept occurs when $6 - 2x = 0$ or when $x = 3$. The x-intercept is located at $(3, 0)$.

[-10, 10, 1] by [-10, 10, 1]   [-10, 10, 1] by [-10, 10, 1]

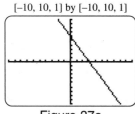   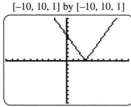

Figure 37a   Figure 37b

39. $|-2x| = 4 \Rightarrow |2x| = 4 \Rightarrow 2x = -4$ or $2x = 4 \Rightarrow x = -2$ or 2

41. $|5x - 7| = 2$, then $5x - 7 = -2$ or $5x - 7 = 2$. If $5x - 7 = -2$ then $5x = 5 \Rightarrow x = 1$.

Or if $5x - 7 = 2$ then $5x = 9 \Rightarrow x = \dfrac{9}{5}$; 1 or $\dfrac{9}{5}$.

43. $|3 - 4x| = 5$, then $3 - 4x = -5$ or $3 - 4x = 5$. If $3 - 4x = -5$ then $-4x = -8 \Rightarrow x = 2$.

Or if $3 - 4x = 5$ then $-4x = 2 \Rightarrow x = -\dfrac{1}{2}$; $-\dfrac{1}{2}$ or 2.

45. $|-6x - 2| = 0$, then $-6x - 2 = 0 \Rightarrow -6x = 2 \Rightarrow x = -\dfrac{1}{3}$.

47. $|17x - 6| = -3$ has no solution since the absolute value of any quantity is always greater than or equal to 0.

49. $|1.2x - 1.7| - 1 = 3 \Rightarrow |1.2x - 1.7| = 4$, then $1.2x - 1.7 = -4$ or $1.2x - 1.7 = 4$.

If $1.2x - 1.7 = -4$ then $1.2x = -2.3 \Rightarrow x = -\dfrac{2.3}{1.2} \Rightarrow x = -\dfrac{23}{12}$. Or if $1.2x - 1.7 = 4$ then $1.2x = 5.7 \Rightarrow$

$x = \dfrac{5.7}{1.2} \Rightarrow x = \dfrac{19}{4}$; $\Rightarrow -\dfrac{23}{12}$ or $\dfrac{19}{4}$.

51. $|4x - 5| + 3 = 2 \Rightarrow |4x - 5| = -1$ has no solution since the absolute value of any quantity is always greater than or equal to 0.

53. $|2x - 9| = |8 - 3x| \Rightarrow 2x - 9 = 8 - 3x$ or $2x - 9 = -(8 - 3x) \Rightarrow$

$2x + 3x = 8 + 9$ or $2x - 3x = -8 + 9 \Rightarrow 5x = 17$ or $-x = 1 \Rightarrow x = \dfrac{17}{5}$ or -1

55. $\left|\dfrac{3}{4}x - \dfrac{1}{4}\right| = \left|\dfrac{3}{4} - \dfrac{1}{4}x\right| \Rightarrow \dfrac{3}{4}x - \dfrac{1}{4} = \dfrac{3}{4} - \dfrac{1}{4}x$ or $\dfrac{3}{4}x - \dfrac{1}{4} = -\left(\dfrac{3}{4} - \dfrac{1}{4}x\right) \Rightarrow$

$\dfrac{3}{4}x + \dfrac{1}{4}x = \dfrac{3}{4} + \dfrac{1}{4}$ or $\dfrac{3}{4}x - \dfrac{1}{4}x = -\dfrac{3}{4} + \dfrac{1}{4} \Rightarrow x = 1$ or $\dfrac{1}{2}x = -\dfrac{1}{2} \Rightarrow x = -1$ or 1

57. $\left|15x - 5\right| = \left|35 - 5x\right| \Rightarrow 15x - 5 = 35 - 5x$ or $15x - 5 = -(35 - 5x) \Rightarrow$

$15x - 5 = 35 - 5x \Rightarrow 20x = 40 \Rightarrow x = 2$ or $15x - 5 = -(35 - 5x) \Rightarrow 15x - 5 = -35 + 5x \Rightarrow$

$10x = -30 \Rightarrow x = -3 \Rightarrow x = -3$ or 2

59. (a) $f(x) = g(x)$ when $x = -1$ or 7.

(b) $f(x) < g(x)$ between these x-values or when $-1 < x < 7$.

(c) $f(x) > g(x)$ outside of these x-values or when $x < -1$ or $x > 7$.

61. (a) Graph $Y_1 = abs(2X - 5)$ and $Y_2 = 10$. See Figures 61a and 61b.

    (b) Table $Y_1 = abs(2X - 5)$ starting at –5, incrementing by 2.5. See Figure 61c.

    (c) $|2x - 5| = 10 \Rightarrow 2x - 5 = -10$ or $2x - 5 = 10 \Rightarrow x = -2.5$ or 7.5

    From each method, the solution to $|2x - 5| < 10$ lies between –2.5 and 7.5, exclusively: $-2.5 < x < 7.5$.

[–10, 10, 1] by [–5, 15, 1]   [–10, 10, 1] by [–5, 15, 1]

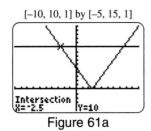   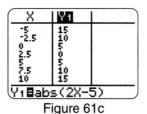   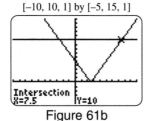

Figure 61a   Figure 61b   Figure 61c

63. (a) Graph $Y_1 = abs(5 - 3X)$ and $Y_2 = 2$. See Figures 63a and 63b.

    (b) Table $Y_1 = abs(5 - 3X)$ starting at $-\dfrac{1}{3}$, incrementing by $\dfrac{2}{3}$. See Figure 63c.

    (c) $|5 - 3x| = 2 \Rightarrow 5 - 3x = -2$ or $5 - 3x = 2 \Rightarrow x = \dfrac{7}{3}$ or 1

    From each method, the solution to $|5 - 3x| > 2$ lies outside of 1 and $\dfrac{7}{3}$, exclusively: $x < 1$ or $x > \dfrac{7}{3}$.

[–1, 4, 1] by [–1, 3, 1]   [–1, 4, 1] by [–1, 3, 1]

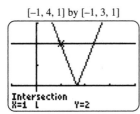   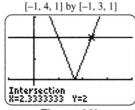   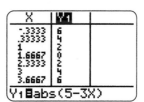

Figure 63a   Figure 63b   Figure 63c

65. $|2.1x - 0.7| = 2.4 \Rightarrow 2.1x - 0.7 = -2.4$ or $2.1x - 0.7 = 2.4 \Rightarrow x = -\dfrac{17}{21}$ or $\dfrac{31}{21}$

    The solution to $|2.1x - 0.7| \geq 2.4$ lies outside of $-\dfrac{17}{21}$ and $\dfrac{31}{21}$, inclusively: $x \leq -\dfrac{17}{21}$ or $x \geq \dfrac{31}{21}$.

    This can be verified by graphing $Y_1 = abs(2.1X - 0.7)$ and $Y_2 = 2.4$ and locating the points of intersection.

67. $|3x| + 5 = 6 \Rightarrow |3x| = 1 \Rightarrow 3x = -1$ or $3x = 1 \Rightarrow x = -\dfrac{1}{3}$ or $\dfrac{1}{3}$.

    The solution to $|3x| + 5 > 6$ lies outside of $-\dfrac{1}{3}$ and $\dfrac{1}{3}$, exclusively: $x < -\dfrac{1}{3}$ or $x > \dfrac{1}{3}$.

    This can be verified by graphing $Y_1 = abs(3X) + 5$ and $Y_2 = 6$ and locating the points of intersection.

69. $\left|\dfrac{2}{3}x - \dfrac{1}{2}\right| = -\dfrac{1}{4}$ has no solution since the absolute value of any quantity is always greater than or equal to 0.

    There are no solutions to $\left|\dfrac{2}{3}x - \dfrac{1}{2}\right| \leq -\dfrac{1}{4}$.

    This can be verified by graphing $Y_1 = abs(2X/3 - 0.5)$ and $Y_2 = -0.25$. There are no points of intersection.

You Decide the Method

71. The solutions to $|3x - 1| < 8$ satisfy $s_1 < x < s_2$ where s_1 and s_2 are the solutions to $|3x - 1| = 8$.

$|3x - 1| = 8$ is equivalent to $3x - 1 = -8 \Rightarrow x = -\dfrac{7}{3}$ and $3x - 1 = 8 \Rightarrow x = 3$.

The interval is $\left(-\dfrac{7}{3}, 3\right)$.

73. The solutions to $|7 - 4x| \leq 11$ satisfy $s_1 \leq x \leq s_2$ where s_1 and s_2 are the solutions to $|7 - 4x| = 11$.

$|7 - 4x| = 11$ is equivalent to $7 - 4x = -11 \Rightarrow x = \dfrac{9}{2}$ and $7 - 4x = 11 \Rightarrow x = -1$.

The interval is $\left[-1, \dfrac{9}{2}\right]$.

75. The solutions to $|0.5x - 0.75| < 2$ satisfy $s_1 < x < s_2$ where s_1 and s_2 are the solutions to $|0.5x - 0.75| = 2$.

$|0.5x - 0.75| = 2$ is equivalent to $0.5x - 0.75 = -2 \Rightarrow x = -\dfrac{5}{2}$ and $0.5x - 0.75 = 2 \Rightarrow x = \dfrac{11}{2}$.

The interval is $\left(-\dfrac{5}{2}, \dfrac{11}{2}\right)$.

77. The solutions to $|2x - 3| > 1$ satisfy $x < s_1$ or $x > s_2$ where s_1 and s_2 are the solutions to $|2x - 3| = 1$.

$|2x - 3| = 1$ is equivalent to $2x - 3 = -1 \Rightarrow x = 1$ and $2x - 3 = 1 \Rightarrow x = 2$.

The interval is $(-\infty, 1) \cup (2, \infty)$.

79. The solutions to $|-3x + 8| \geq 3$ satisfy $x \leq s_1$ or $x \geq s_2$ where s_1 and s_2 are the solutions to $|-3x + 8| = 3$.

$|-3x + 8| = 3$ is equivalent to $-3x + 8 = -3 \Rightarrow x = \dfrac{11}{3}$ and $-3x + 8 = 3 \Rightarrow x = \dfrac{5}{3}$.

The interval is $\left(-\infty, \dfrac{5}{3}\right] \cup \left[\dfrac{11}{3}, \infty\right)$.

81. The solutions to $|0.25x - 1| > 3$ satisfy $x < s_1$ or $x > s_2$ where s_1 and s_2 are the solutions to $|0.25x - 1| = 3$.

$|0.25x - 1| = 3$ is equivalent to $0.25x - 1 = -3 \Rightarrow x = -8$ and $0.25x - 1 = 3 \Rightarrow x = 16$.

The interval is $(-\infty, -8) \cup (16, \infty)$.

Sketching Graphs

83. Since the tank is either being filled or emptied at a constant rate, the graph will be piecewise linear. The graph will initially start at the origin and increase with a slope of 5 until the y-values reach a height of 100. This will occur at the point (20, 100), since after 20 minutes there are $5 \times 20 = 100$ gallons in the tank. It will take 50 minutes to empty a 100-gallon tank at 2 gallons per minute. The graph must pass through the point (70, 0). The resulting graph is shown in Figure 83.

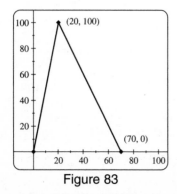

Figure 83

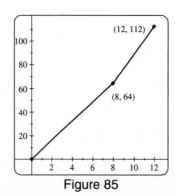

Figure 85

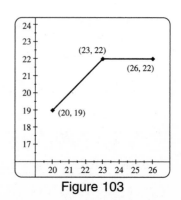

Figure 103

85. The graph will be piecewise linear because there are two different rates of pay. During the first 8 hours the person earns a constant rate of $8.00 per hour. In 8 hours the individual will earn $64. Graph a line segment connecting the points (0, 0) and (8, 64). After working 8 hours in one day, the person earns $1.5 \times 8 = \$12.00$ for each additional hour. In 12 hours the person will earn $\$64 + 4 \times \$12 = \$112$. Graph a line segment between (8, 64) and (12, 112). The resulting graph is shown in Figure 85.

Concepts

87. $|-6| = 6$

89. Since the inputs or domain of absolute values can be positive or negative the domain of $|f(x)|$ is also $(-2, 4]$.

91. Since all solutions or the range of absolute values must be positive, all negative solutions will change to reciprocal positive solutions, therefore if the range of $f(x)$ is $(-\infty, 0]$, the range of $|f(x)|$ is $[0, \infty)$

Applications

93. (a) $|T - 43| = 24 \Rightarrow T - 43 = -24$ or $T - 43 = 24 \Rightarrow T = 19$ or 67. The average monthly temperature range is $19°F \leq T \leq 67°F$.

    (b) The monthly average temperatures in Marquette vary between a low of 19°F and a high of 67°F. The monthly averages are always within 24 degrees of 43°F.

95. (a) $|T - 50| = 22 \Rightarrow T - 50 = -22$ or $T - 50 = 22 \Rightarrow T = 28$ or 72. The average monthly temperature range is $28°F \leq T \leq 72°F$.

    (b) The monthly average temperatures in Boston vary between a low of 28°F and a high of 72°F. The monthly averages are always within 22 degrees of 50°F.

97. (a) $|T - 61.5| = 12.5 \Rightarrow T - 61.5 = -12.5$ or $T - 61.5 = 12.5 \Rightarrow T = 49$ or 74. The average monthly temperature range is $49°F \leq T \leq 74°F$.

    (b) The monthly average temperatures in Buenos Aires vary between a low of 49°F and a high of 74°F. The monthly averages are always within 12.5 degrees of 61.5°F.

99. The solutions to $|d - 3| \leq 0.004$ satisfy $s_1 \leq d \leq s_2$ where s_1 and s_2 are the solutions to $|d - 3| = 0.004$. $|d - 3| = 0.004$ is equivalent to $d - 3 = -0.004 \Rightarrow d = 2.996$ and $d - 3 = 0.004 \Rightarrow d = 3.004$. The solution set is $\{d \,|\, 2.996 \leq d \leq 3.004\}$. The acceptable diameters are from 2.994 and 3.004 inches.

101. $\left|\dfrac{Q - A}{A}\right| \leq 0.02$ or $\left|\dfrac{Q - 35}{35}\right| \leq 0.02$, then $-0.02 \leq \dfrac{Q - 35}{35} \leq 0.02 \Rightarrow -0.70 \leq Q - 35 \leq 0.7 \Rightarrow$

     $34.3 \leq Q \leq 35.7$

103. (a) Let x represent Shaquille's age. Since the size is always one less than the age, let $f(x) = x - 1$.

     (b) On [20, 23], graph $y = x - 1$ and on (23, 26] graph $y = 22$. This is shown in Figure 103.

105. (a) A sketch of a piecewise-constant function is shown in Figure 105.

(b) The likelihood of being a victim of crime peaks from age 16 up to age 19, and then it decreases sharply as the person becomes older..

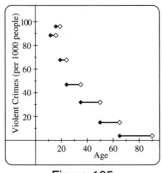

Figure 105

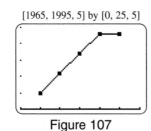

[1965, 1995, 5] by [0, 25, 5]

Figure 107

107. (a) Make a line graph of the points (1970, 5), (1975, 11), (1980, 17), (1985, 23), and (1990, 23) as shown in Figure 107. Cesarean births increased at a constant rate of 6% every five-year period between 1970 and 1985. They leveled off at 23%.

(b) $a = m = \dfrac{23 - 5}{1985 - 1970} = \dfrac{18}{15} = 1.2$; $b = 5$; $c = 23$

(c) $f(1978) = 1.2(1978 - 1970) + 5 = 14.6$; $f(1988) = 23$

In 1978, 14.6% of births were Cesarean births, and in 1988 the percentage was 23%.

Chapter 2 Review Exercises

1. (a) Using the points $(0, 6)$ and $(2, 2)$, $m = \dfrac{2 - 6}{2 - 0} = \dfrac{-4}{2} = -2$; y-intercept: 6; x-intercept: 3.

(b) $f(x) = -2x + 6$

(c) $f(x) = 0 \Rightarrow 0 = -2x + 6 \Rightarrow x = 3$.

3. $m = \dfrac{0 - 2.5}{2 - 1} = \dfrac{-2.5}{1} = -2.5$; $(1, 2.5) \Rightarrow (1 - 1, 2.5 - (-2.5)) = (0, 5)$, so $b = 5$; $f(x) = -2.5x + 5$

5. See Figure 5.

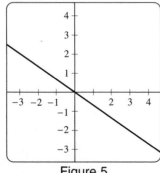

Figure 5

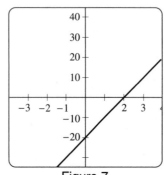

Figure 7

7. See Figure 7.

9. Using point-slope form $y = m(x - x_1) + y_1$, we get $y = -2(x + 2) + 3 \Rightarrow f(x) = -2x - 1$

11. $y = 7(x + 3) + 9 \Rightarrow y = 7x + 21 + 9 \Rightarrow y = 7x + 30$

13. Let $m = -3$. Then, $y = -3(x - 1) - 1 \Rightarrow y = -3x + 2$.

15. The line segment has slope $m = \dfrac{0 - 3.1}{5.7 - 0} = -\dfrac{31}{57}$; the parallel line has slope $m = -\dfrac{31}{57}$;

$y = -\dfrac{31}{57}(x - 1) - 7 \Rightarrow y = -\dfrac{31}{57}x - \dfrac{368}{57}$

17. The line is vertical passing through $(6, -7)$, so the equation is $x = 6$.

19. The line is horizontal passing through $(1, 3)$, so the equation is $y = 3$.

21. The equation of the horizontal line with y-intercept -8 is $y = -8$.

23. For x-intercept: $y = 0 \Rightarrow 5x - 4(0) = 20 \Rightarrow 5x = 20 \Rightarrow x = 4$; for y-intercept: $x = 0 \Rightarrow$

$5(0) - 4y = 20 \Rightarrow -4y = 20 \Rightarrow y = -5$; using $(4, 0)$ and $(0, -5)$ graph the equation. See Figure 23.

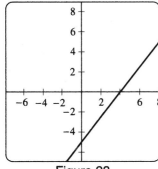

Figure 23

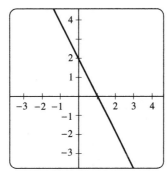

Figure 25

25. For x-intercept: $y = 0 \Rightarrow 0 = -2x + 2 \Rightarrow -2 = -2x \Rightarrow x = 1$; for y-intercept: $x = 0 \Rightarrow$

$y = -2(0) + 2 \Rightarrow y = 2$; using $(1, 0)$ and $(0, 2)$ graph the equation. See Figure 25.

27. Graphical: Graph $Y_1 = 5X - 22$ and $Y_2 = 10$. Their graphs intersect at $(6.4, 10)$ as shown in Figure 27. The

solution is $x = 6.4$.

Symbolic: $5x - 22 = 10 \Rightarrow 5x = 32 \Rightarrow x = \dfrac{32}{5} = 6.4$

[-15, 15, 5] by [-15, 15, 5]

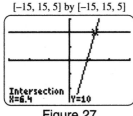

Figure 27

[-10, 10, 1] by [-10, 10, 1]

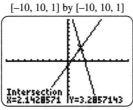

Figure 29

29. Graphical: Graph $Y_1 = -2(3X - 7) + X$ and $Y_2 = 2X - 1$. Their graphs intersect near $(2.143, 3.286)$ as

shown in Figure 29. The solution is $x \approx 2.143$.

Symbolic: $-2(3x - 7) + x = 2x - 1 \Rightarrow -6x + 14 + x = 2x - 1 \Rightarrow -7x = -15 \Rightarrow x = \dfrac{15}{7} \approx 2.143$

31. Graphical: Graph $Y_1 = 5X - 0.5(4 - 3X)$ and $Y_2 = 1.5 - (2X + 3)$. Their graphs intersect near

    $(0.059, -1.618)$ as shown in Figure 31. The solution is $x \approx 0.059$.

    Symbolic: $5x - \dfrac{1}{2}(4 - 3x) = \dfrac{3}{2} - (2x + 3) \Rightarrow 10x - (4 - 3x) = 3 - 2(2x + 3) \Rightarrow$

    $10x - 4 + 3x = 3 - 4x - 6 \Rightarrow 17x = 1 \Rightarrow x = \dfrac{1}{17} \approx 0.059$

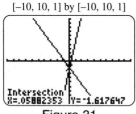

[–10, 10, 1] by [–10, 10, 1]

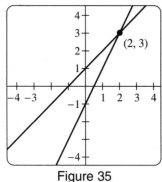

Figure 31   Figure 33

33. Let $Y_1 = 3.1X - 0.2 - 2(X - 1.7)$ and approximate where $Y_1 = 0$. From Figure 33 this occurs when $x \approx -2.9$.

35. Graph $y_1 = x + 1$ and $y_2 = 2x - 1$. See Figure 35. The lines intersect at $(2, 3) \Rightarrow x = 2$.

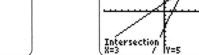

Figure 35   Figure 47

37. (a) $4(6 - x) = -4x + 24 \Rightarrow 24 - 4x = -4x + 24 \Rightarrow 0 = 0 \Rightarrow$ all real numbers are solutions.

    (b) When all real solutions work then Identity.

39. (a) $5 - 2(4 - 3x) + x = 4(x - 3) \Rightarrow 5 - 8 + 6x + x = 4x - 12 \Rightarrow 7x - 3 = 4x - 12 \Rightarrow 3x = -9 \Rightarrow$

    $x = -3$

    (b) If has an answer then Conditional.

41. $(-3, \infty)$

43. $\left[-2, \dfrac{3}{4}\right)$

45. $(-\infty, 0) \cup (2, \infty)$

47. Graphical: Graph $Y_1 = 3X - 4$ and $Y_2 = 2 + X$. Their graphs intersect at $(3, 5)$. The graph of Y_1 is below

    the graph of Y_2 to the left of the point of intersection. Thus, $3x - 4 \leq 2 + x$ holds when $x \leq 3$ or $(-\infty, 3]$.

    See Figure 47.

    Symbolic: $3x - 4 \leq 2 + x \Rightarrow 2x \leq 6 \Rightarrow x \leq 3$ or $(-\infty, 3]$

49. Graphical: Graph $Y_1 = (2X - 5)/2$ and $Y_2 = (5X + 1)/5$. Their graphs are parallel and never intersect. The

graph of Y_2 is always above the graph of Y_1, so $Y_1 < Y_2$ for al values of x; the inequality $\dfrac{2x - 5}{2} < \dfrac{5x + 1}{5}$

holds when $-\infty < x < \infty$, or $(-\infty, \infty)$. See Figure 49.

51. Graphical: Graph $Y_1 = -2$, $Y_2 = 5 - 2X$ and $Y_3 = 7$. See Figure 51. Their graphs intersect at the points

$(-1, 7)$ and $(3.5, -2)$. The graph of Y_2 is between the graphs of Y_1 and Y_3 when $-1 < x \le 3.5$. In interval

notation the solution is $(-1, 3.5]$.

Symbolic: $-2 \le 5 - 2x < 7 \Rightarrow -7 \le -2x < 2 \Rightarrow \dfrac{7}{2} \ge x > -1 \Rightarrow -1 < x \le \dfrac{7}{2}$ or $\left(-1, \dfrac{7}{2}\right]$ or $\left(-1, \dfrac{7}{2}\right]$

[−10, 10, 1] by [−10, 10, 1]   [−10, 10, 1] by [−10, 10, 1]   [−10, 10, 1] by [−10, 10, 1]   [−10, 10, 1] by [−10, 10, 1]

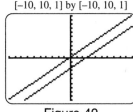   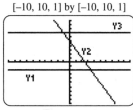   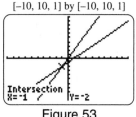   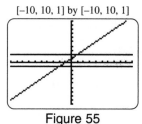

Figure 49   Figure 51   Figure 53   Figure 55

53. Graph $Y_1 = 2x$ and $Y_2 = x - 1$. See Figure 53. The lines intersect at $(-1, -2)$. $Y_1 > Y_2$ when the graph of

Y_1 is above the graph of Y_2, this happen when $x > -1 \Rightarrow (-1, \infty)$.

55. Graph $Y_1 = -1$, $Y_2 = 1 + x$ and $Y_3 = 2$. See Figure 55. The two points of intersection are $(-2, -1)$ and $(1, 2)$.

$Y_1 \le Y_2 \le Y_3$ when the graph of $1 + x$ is inbetween these two intersection points $\Rightarrow -2 \le x \le 1 \Rightarrow [-2, 1]$.

57. (a) The graphs intersect at $(2, 1)$. The solution on $f(x) = g(x)$ is $x = 2$.

(b) the graph of g is above the graph of f to the right of point $(2, 1)$. Thus, $g(x) > f(x)$ when $x > 2$ or $(2, \infty)$.

(c) The graph of g is below the graph of f to the left of the point $(2, 1)$. Thus, $g(x) < f(x)$ when

$x < 2$ or $(-\infty, 2)$.

59. (a) $f(-2) = 8 + 2(-2) = 4$; $f(-1) = 8 + 2(-1) = 6$; $f(2) = 5 - 2 = 3$; $f(3) = 3 + 1 = 4$.

(b) A graph of f is shown in Figure 59. It is essentially a piecewise line graph with the points $(-3, 2)$, $(-1, 6)$,

$(2, 3)$, and $(5, 6)$. Since there are no breaks in the graph, f is continuous.

(c) From the graph we can see that there are two x-values where $f(x) = 3$. They occur when

$8 + 2x = 3 \Rightarrow x = -2.5$ and when $5 - x = 3 \Rightarrow x = 2$. The solutions are $x = -2.5$ or 2.

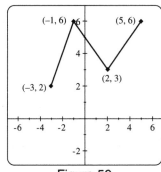   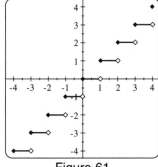

Figure 59   Figure 61

61. When $0 \le x < 1$, $[x] = 0$. When $1 \le x < 2$, $[x] = 1$. In general, when $n \le x < n + 1$, $[x] = n$.

A graph of $f(x) = [x]$ is shown in Figure 61.

63. $|2x - 5| - 1 = 8 \Rightarrow |2x - 5| = 9 \Rightarrow 2x - 5 = -9$ or $2x - 5 = 9$, if $2x - 5 = -9 \Rightarrow 2x = -4 \Rightarrow$
    $x = -2$ or if $2x - 5 = 9 \Rightarrow 2x = 14 \Rightarrow x = 7, \Rightarrow x = -2$ or 7

65. $|6 - 4x| = -2$ has no solution since the absolute value of any quantity is always greater than or equal to 0.

67. $|x| = 3 \Rightarrow x = \pm 3$. The solutions to $|x| > 3$ lie left of $x = -3$ and right of $x = 3$. That is,
    $x < -3$ or $x > 3$. This can be supported by graphing $Y_1 = |x|$ and $Y_2 = 3$ and determining where the
    V-shaped graph of Y_1 is above the graph of Y_2. To support this result numerically, table $Y_1 = \text{abs}(X)$ starting at
    -9 and incrementing by 3.

69. $|3x - 7| = 10 \Rightarrow 3x - 7 = 10$ or $3x - 7 = -10 \Rightarrow x = \dfrac{17}{3}$ or $x = -1$. The solutions to $|3x - 7| > 10$
    lie left of $x = -1$ or right of $x = \dfrac{17}{3}$. That is, $x < -1$ or $x > \dfrac{17}{3}$. This can be supported by graphing
    $Y_1 = |3x - 7|$ and $Y_2 = 10$ and determining where the graph of Y_1 is above the graph of Y_2. To support this
    result numerically, table $Y_1 = \text{abs}(3X - 7)$ starting at -3 and incrementing by $\dfrac{1}{3}$.

71. $|x + 1| = |3x + 2| \Rightarrow x + 1 = 3x + 2$ or $x + 1 = -(3x + 2) \Rightarrow x - 3x = 2 - 1$ or $x + 3x = -2 - 1 \Rightarrow$
    $-2x = 1$ or $4x = -3 \Rightarrow x = -\dfrac{1}{2}$ or $x = -\dfrac{3}{4}$, or $x = -0.5$ or $x = -0.75$.

73. The solutions to $|3 - 2x| < 9$ satisfy $s_1 < x < s_2$ where s_1 and s_2 are the solutions to $|3 - 2x| = 9$.
    $|3 - 2x| = 9$ is equivalent to $3 - 2x = -9 \Rightarrow x = 6$ and $3 - 2x = 9 \Rightarrow x = -3$.
    The solution is is $-3 < x < 6$.

75. The solutions to $\left|\dfrac{1}{3}x - \dfrac{1}{6}\right| \geq 1$ satisfy $x \leq s_1$ or $x \geq s_2$ where s_1 and s_2 are the solutions to $\left|\dfrac{1}{3}x + \dfrac{1}{6}\right| = 1$.
    $\left|\dfrac{1}{3}x - \dfrac{1}{6}\right| = 1$ is equivalent to $\dfrac{1}{3}x - \dfrac{1}{6} = -1 \Rightarrow x = -\dfrac{5}{2}$ and $\dfrac{1}{3}x - \dfrac{1}{6} = 1 \Rightarrow x = \dfrac{7}{2}$.
    The solution is $x \leq -\dfrac{5}{2}$ or $x \geq \dfrac{7}{2}$.

77. (a) Graph $Y_1 = 1550(x - 1980) + 21{,}000$ and $Y_2 = 36{,}500$. Their graphs intersect at the point
        $(1990, 36{,}500)$. See Figure 77. This means that in 1990 the median U.S. family income was $\$36{,}500$.

    (b) $1550(x - 1980) + 21{,}000 = 36{,}500 \Rightarrow 1550(x - 1980) = 15{,}500 \Rightarrow x = \dfrac{15{,}500}{1550} + 1980 = 1990$

[1980, 1993, 1] by [20,000, 40,000, 1000]   [1995, 2007, 1] by [200, 400, 50]   [1995, 2007, 1] by [200, 400, 50]

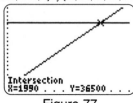
Figure 77

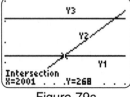
Figure 79a

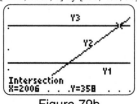
Figure 79b

79. (a) $268 \leq 18x - 35{,}750 \leq 358 \Rightarrow 36{,}018 \leq 18x \leq 36{,}108 \Rightarrow 2001 \leq x \leq 2006$. Medicare costs will be
        between 268 and 358 billion dollars from 2001 to 2006.

    (b) Graph $Y_1 = 268$, $Y_2 = 18X - 35{,}750$, and $Y_3 = 358$. Their graphs intersect at the points $(2001, 268)$ and
        $(2006, 358)$. See Figures 79a & 79b. Medicare costs will be between 268 and 358 billion dollars from 2001
        to 2006. This supports the answer in part (a).

81. Since the graph is piecewise linear, each line segment represents a constant speed. Initially, the car is at home. After 1 hour it is 30 miles form home and has traveled at a constant speed of 30 mph. After 2 hours it is 50 miles away. During the second hour the car travels 20 mph. During the third hour the car travels toward home at 30 mph until it is 20 miles away. During the fourth hour the car travels away from home at 40 mph until it is 60 miles away from home. The last hour the car travels 60 miles at 60 mph until it arrives back at home.

83. Since the tank is initially full, plot the point (0, 500). Since the tank is drained at a constant rate of 50 gallons per minute, the tank will be empty in 10 minutes. Draw a line segment between (0, 500) and (10, 0). At 25 gallons per minute, it will require 20 minutes to fill a 500-gallon tank. Draw a second line segment connecting (10, 0) and 30, 500). See Figure 83.

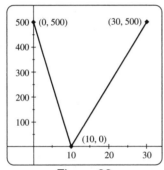

Figure 83

85. The midpoint is computed by $\left(\dfrac{2000 + 2004}{2}, \dfrac{143{,}247 + 167{,}933}{2} \right) = (2002, 155{,}590).$

The population was about 155,590.

87. Let $x =$ time it takes for both working together; the first worker can shovel $\dfrac{1}{50}$ of the sidewalk in 1 minute, and the second worker can shovel $\dfrac{1}{30}$ of the sidewalk in 1 minute; for the entire job, we get the equation $\dfrac{x}{50} + \dfrac{x}{30} = 1 \Rightarrow 30x + 50x = 1500 \Rightarrow 80x = 1500 \Rightarrow x = 18.75$; it takes the two workers 18.75 minutes to shovel the sidewalk together.

89. Let $t =$ time spent jogging at 7 mph, then $1.8 - t =$ time spent jogging at 8 mph; since $d = rt$ and the total distance jogged is 13.5 miles, we get the equation $7t + 8(1.8 - t) = 13.5 \Rightarrow 7t + 14.4 - 8t = 13.5 \Rightarrow -t = -0.9 \Rightarrow t = 0.9$ and $1.8 - t = 0.9$; the runner jogged 0.9 hours at 7 mph and 0.9 hours at 8 mph.

91. (a) Calculating the slope for each data pair in the table gives 2, 2, 2, and 1; using the average of these four slope values gives a slope of 1.75 for the linear function f; $f(x) = 1.75(x - 1994) + 504$, or $f(x) = 1.75x - 2985.5$ is a linear function that models the data. *Answers may vary.*

(b) $f(1997) = 1.75(1997) - 2985.5 = 509.25 \approx 509$. This estimate involves interpolation. *Answers may vary.*

93. The tank initially contains 25 gallons.

On the interval [0, 4] the slope is 5. The 5 gal./min. inlet pipe is open. The other pipes are closed.

On the interval (4, 8] the slope is –3 . Both inlet pipes are closed. The outlet pipe is open.

On the interval (8, 12] the slope is 7. Both inlet pipes are open. The outlet pipe is closed.

On the interval (12, 16] the slope is 4. All pipes are open.

On the interval (16, 24] the slope is –1. The 2 gal./min. inlet pipe and the outlet pipe are open.

On the interval (24, 28] the slope is 0. All pipes are closed.

95. (a) $f(x) = 6.15x - 12,059 \Rightarrow 70 = 6.15x - 12,059 \Rightarrow 12,129 = 6.15x \Rightarrow x \approx 1972.20$; the number of

species first exceeded 70 in 1973.

(b) The function f will not continue to be a good model far into the future. Once the water is no longer

polluted, the number of species will reach its natural level.

[1955, 2005, 10] by [1000, 7000, 1000]

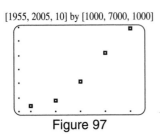

Figure 97

97. (a) The scatterplot in Figure 97 of the points (1960, $1394), (1970, $1763), (1980, $3176), (1990, $5136), and

(2000, $6880) indicates that the correlation coefficient should be positive, and somewhat close to +1.

(b) $y = ax + b$, where $a = 143.45$ and $b \approx -280,361.2 \Rightarrow y = 143.45x - 280,361.2$; $r \approx 0.978$

(c) $y = 143.45(1995) - 280,361.2 \approx 5821.55$; the estimated cost of driving a mid-size car in 1995 was

$5821.55.

Chapters 1-2 Cumulative Review Exercises

1. $\dfrac{1.69 - 1.89}{1.89} = \dfrac{-0.20}{1.89} = -10.6\%$

3. Move the decimal point six places to the right; $6.7 \times 10^6 \Rightarrow 6,700,000$

Move the decimal point four places to the left; $1.45 \times 10^{-4} \Rightarrow 0.000145$

5. (a) Yes, each input has only one output.

(b) $D = \{-1, 0, 1, 2, 3\}$; $R = \{0, 3, 4, 6\}$

7. $D = \sqrt{[2 - (-3)]^2 + ((-3) - 5)^2} = \sqrt{25 + 64} = \sqrt{89}$

9. (a) $D =$ all real numbers $\Rightarrow \{x \mid -\infty < x < \infty\}$; $R = \{y \mid y \geq -2\}$; $f(-1) = -1$

(b) $D = \{x \mid -3 \leq x \leq 3\}$; $R = \{y \mid -3 \leq y \leq 2\}$; $f(-1) = -\dfrac{1}{2}$

11. (a) $f(2) = 5(2) - 3 = 7$; $f(a - 1) = 5(a - 1) - 3 = 5a - 5 - 3 = 5a - 8$

(b) The domain of f includes all real numbers. $D = \{x \mid -\infty \le x \le \infty\}$

13. (a) $f(2) = \sqrt{2(2) - 1} = \sqrt{3}$; $f(a - 1) = \sqrt{2(a - 1) - 1} = \sqrt{2a - 2 - 1} = \sqrt{2a - 3}$

(b) The domain of f includes all real numbers greater than or equal to $\dfrac{1}{2} \Rightarrow D = \{x \mid x \ge \dfrac{1}{2}\}$.

15. No, this is not a graph of a function because some vertical lines can intersect the graph twice.

17. $f(1) = (1)^2 - 2(1) + 1 = 1 - 2 + 1 = 0 \Rightarrow (1, 0)$; $f(2) = (2)^2 - 2(2) + 1 = 4 - 4 + 1 = 1 \Rightarrow (2, 1)$

Find slope $m = \dfrac{1 - 0}{2 - 1} = \dfrac{1}{1} = 1$, the average rate of change is 1.

19. (a) $m = \dfrac{2}{3}$; y-intercept: -2, x-intercept: 3

(b) $f(x) = mx + b \Rightarrow f(x) = \dfrac{2}{3}x - 2$

(c) $y = 0$ at $x = 3$

21. Using point-slope form $y = m(x - x_1) + y_1 \Rightarrow y = -3\left(x - \dfrac{2}{3}\right) - \dfrac{2}{3} \Rightarrow y = -3x + \dfrac{4}{3} \Rightarrow f(x) = -3x + \dfrac{4}{3}$

23. $m = \dfrac{26 - 52}{2004 - 2000} = \dfrac{-26}{4} = \dfrac{-13}{2}$

25. Find slope $m = \dfrac{\frac{1}{2} - (-5)}{-3 - 1} = \dfrac{\frac{11}{2}}{-4} = -\dfrac{11}{8}$; using $(1, 5)$ and point-slope form: $y = -\dfrac{11}{8}(x - 1) - 5 \Rightarrow$

$y = -\dfrac{11}{8}x + \dfrac{11}{8} - 5 \Rightarrow y = -\dfrac{11}{8}x - \dfrac{29}{8}$

27. All the lines parallel to the y-axis will have an undefined slope, $\Rightarrow y$ changes but x remains constant

$\Rightarrow x = -1$.

29. Finding the slope of points $(2.4, 5.6)$ and $(3.9, 8.6)$ we get $m = \dfrac{8.6 - 5.6}{3.9 - 2.4} = \dfrac{3}{1.5} = 2$, a line parallel to this has

the same slope. Using point-slope form: $y = 2(x + 3) + 5 \Rightarrow y = 2x + 11$.

31. For $-2x + 3y = 6$: x-intercept, then $y = 0 \Rightarrow -2x + 3(0) = 6 \Rightarrow -2x = 6 \Rightarrow x = -3$,

y-intercept, then $x = 0 \Rightarrow -2(0) + 3y = 6 \Rightarrow 3y = 6 \Rightarrow y = 2$. See Figure 31.

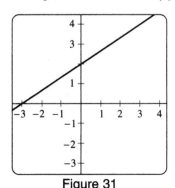

Figure 31

33. $4x - 5 = 1 - 2x \Rightarrow 6x = 6 \Rightarrow x = 1$

35. $\frac{2}{3}(x - 2) - \frac{4}{5}x = \frac{4}{15} + x \Rightarrow \frac{2}{3}x - \frac{4}{3} - \frac{4}{5}x = \frac{4}{15} + x \Rightarrow 15\left(\frac{2}{3}x - \frac{4}{3} - \frac{4}{5}x = \frac{4}{15} + x\right) \Rightarrow$

$10x - 20 - 12x = 4 + 15x \Rightarrow -17x = 24 \Rightarrow x = -\frac{24}{17}$

37. Graph $y_1 = x + 1$ and $y_2 = 2x - 2$. See Figure 37a. The lines intersect at point $(3, 4) \Rightarrow x = 3$. Make a table

of $y_1 = x + 1$ and $y_2 = 2x - 2$ for x values from 0 to 5. See Figure 37b. Both equations equal 4 at $x = 3$.

[−10, 10, 1] by [−10, 10, 1]

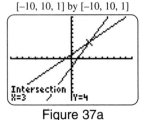

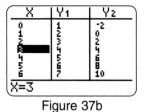

Figure 37a   Figure 37b

39. $(-\infty, 5)$

41. $(-\infty, -2) \cup (2, \infty)$

43. $-5x + 3 < 3 - x \Rightarrow -4x < 0 \Rightarrow x > 0 \Rightarrow (0, \infty)$

45. $-3(1 - 2x) + x \le 4 - (x + 2) \Rightarrow -3 + 6x + x \le 4 - x - 2 \Rightarrow 7x - 3 \le -x + 2 \Rightarrow 8x \le 5 \Rightarrow$

$x \le \frac{5}{8} \Rightarrow \left(-\infty, \frac{5}{8}\right]$

47. (a) $(2, 1)$

(b) The graph of the line $f(x)$ is above the line $g(x)$ left of $x = 2 \Rightarrow x < 2$

(c) The graph of the line $f(x)$ is below the line $g(x)$ right of $x = 2 \Rightarrow x \ge 2$

49. $|d + 1| = 5$ then $d + 1 = 5$ or $d + 1 = -5$. If $d + 1 = -5 \Rightarrow d = -6$, or if $d + 1 = 5 \Rightarrow d = 4, \Rightarrow$

$d = -6$ or 4.

51. $|2t| - 4 = 10 \Rightarrow |2t| = 14$ then $2t = 14$ or $2t = -14$. If $2t = -14 \Rightarrow t = -\frac{14}{2} \Rightarrow t = -7$, or if

$2t = 14 \Rightarrow t = 7, \Rightarrow t = -7$ or 7.

53. The solutions to $|2t - 5| \le 5$ satisfy $s_1 \le t \le s_2$ where s_1 and s_2 are the solutions to $|2t - 5| = 5$.

$|2t - 5| = 5$ is equivalent to $2t - 5 = -5 \Rightarrow t = 0$ and $2t - 5 = 5 \Rightarrow t = 5$.

The interval is $[0, 5]$.

55. The solutions to $\left|4 - \frac{1}{2}k\right| \ge 10$ satisfy $s_1 \le k \le s_2$ where s_1 and s_2 are the solutions to $\left|4 - \frac{1}{2}k\right| = 10$.

$\left|4 - \frac{1}{2}k\right| = 10$ is equivalent to $4 - \frac{1}{2}k = -10 \Rightarrow k = 28$ and $4 - \frac{1}{2}k = 10 \Rightarrow k = -12$.

The interval is $(-\infty, -12] \cup [28, \infty)$.

Applications

57. $V = \pi r^2 h \Rightarrow 24 = \pi(1.5)^2 h \Rightarrow 24 = \pi(2.25)h \Rightarrow h \approx 3.40$ inches

59. Area of a circle $= \pi r^2, d = 40$ then $r = 20, A = \pi(20)^2 \Rightarrow A = 400\pi \Rightarrow A = 1256.6, \dfrac{V}{A} = \dfrac{2}{1256.6} =$

    0.0016 inch.

61. If car B is at the origin on a coordinate plane then car A travels $60\left(\dfrac{5}{4}\right) = 75$ miles and ends up 35 miles north

    $(0, 35)$ of where car B started. car B travels $70\left(\dfrac{5}{4}\right) = 87.5$ west or $(-87.5, 0)$. Using the distance formula:

    $$D = \sqrt{(-87.5 - 0)^2 + (0 - 35)^2} \Rightarrow D = \sqrt{7656.25 + 1225} \Rightarrow D = \sqrt{8881.25} \Rightarrow D \approx 94.2$$

63. (a) $D(x) = 270 - 72x$

    (b) Since $72(x) = 270 \Rightarrow x = 3.75$, the driver arrives home in 3.75 hours so times after that are

    unnecessary, therefore $\{x \mid 0 \le x \le 3.75\}$. See figure 63.

    (c) x-intercept: when $y = 0 \Rightarrow 0 = 270 - 72(x) \Rightarrow 72x = 270 \Rightarrow x = 3.75$; the driver arrives home after

    3.75 hours. y-intercept: when $x = 0 \Rightarrow y = 270 - 72(0) \Rightarrow y = 270$; the driver is initially 270 miles

    from home.

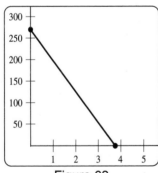

Figure 63

65. Let $x =$ time run at 8 mph and $\left(\dfrac{7}{4} - x\right) =$ time run at 10 mph. Then $8x + 10\left(\dfrac{7}{4} - x\right) = 15 \Rightarrow$

    $8x + \dfrac{70}{4} - 10x = 15 \Rightarrow -2x = -\dfrac{10}{4} \Rightarrow x = \dfrac{5}{4} \Rightarrow x = 1.25$ hours at 8mph and then 0.5 hours at 10 mph.

67. (a) Using $(2001, 9.30)$ and $(2006, 14.25)$ the slope; $m = \dfrac{14.25 - 9.30}{2006 - 2001} = \dfrac{4.95}{5} = 0.99$;

    $f(x) = 0.99(x - 2001) + 9.3$

    (b) $12.25 = 0.99(x - 2001) + 9.30 \Rightarrow 12.25 = 0.99x - 1971.69 \Rightarrow 1983.94 = 0.99x \Rightarrow x \approx 2004$

69. (a) The slope of the line passing through $(58, 91)$ and $(64, 111)$ is $m = \dfrac{111 - 91}{64 - 58} = \dfrac{10}{3}$. Thus, let

$f(x) = \dfrac{10}{3}(x - 58) + 91.$

(b) $f(61) = \dfrac{10}{3}(61 - 58) + 91 = 101$. The recommended minimum weight is 101 pounds for a person 61 inches tall. Since 61 inches is the midpoint between 58 and 64 inches, we can use a midpoint approximation. The midpoint between $(58, 91)$ and $(64, 111)$ is $\left(\dfrac{58 + 64}{2}, \dfrac{91 + 111}{2}\right) = (61, 101)$. The recommended minimum weight is again 101. The midpoint formula gives midpoint on the graph of f between the two given points. Therefore, the answers are the same.

Chapter 3: Quadratic Functions and Equations

3.1: Quadratic Functions and Models

Basics of Quadratic Functions

1. $f(x) = 1 - 2x + 3x^2$ is quadratic; leading coefficient: 3; $f(-2) = 1 - 2(-2) + 3(-2)^2 = 17$.

3. $f(x) = \dfrac{1}{x^2 - 1}$ is neither linear nor quadratic.

5. $f(x) = \dfrac{1}{2} - \dfrac{3}{10}x$ is linear.

7. $f(x) = -3x^2 + 9$ is quadratic; leading coefficient: -3; $f(-2) = -3(-2)^2 + 9 = -3$

9. Leading coefficient: positive; vertex: $(1, 0)$; axis of symmetry: $x = 1$

11. Leading coefficient: negative; vertex: $(-3, -2)$; axis of symmetry: $x = -3$

13. The graph of g is narrower than the graph of f.

15. The graph of g is wider than the graph of f and opens downward rather than upward.

Vertex Formula

17. $f(x) = -3(x - 1)^2 + 2 \Rightarrow$ vertex: $(1, 2)$; leading coefficient: -3; $f(x) = -3x^2 + 6x - 1$

19. $f(x) = 5 - 2(x - 4)^2 \Rightarrow$ vertex: $(4, 5)$; leading coefficient: -2; $f(x) = -2x^2 + 16x - 27$

21. $f(x) = \dfrac{3}{4}(x + 5)^2 - \dfrac{7}{4} \Rightarrow$ vertex: $\left(-5, -\dfrac{7}{4}\right)$; leading coefficient: $\dfrac{3}{4}$; $f(x) = \dfrac{3}{4}x^2 + \dfrac{15}{2}x + 17$

23. $f(x) = \dfrac{1}{2}\left(x + \dfrac{3}{4}\right)^2 \Rightarrow$ vertex: $\left(-\dfrac{3}{4}, 0\right)$; leading coefficient: $\dfrac{1}{2}$; $f(x) = \dfrac{1}{2}x^2 + \dfrac{3}{4}x + \dfrac{9}{32}$

25. It may be helpful to write f in standard form as follows: $f(x) = -x^2 + 0x + 6$.

    To find the vertex symbolically, use the vertex formula with $a = -1$ and $b = 0$.

    $x = -\dfrac{b}{2a} = -\dfrac{0}{2(-1)} = \dfrac{0}{2} = 0$. The x-coordinate of the vertex is 0. $y = f\left(-\dfrac{b}{2a}\right) = f(0) = 6 - (0)^2 = 6$.

    The y-coordinate of the vertex is 6. Thus, the vertex is $(0, 6)$. The graph of $Y_1 = 6 - X^2$ shows graphical

    support for this vertex. See Figure 25.

    [−10, 10, 1] by [−10, 10, 1]

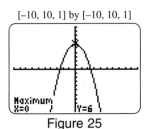

    Figure 25

27. It may be helpful to write f in standard form as follows: $f(x) = x^2 - 6x + 0$.

To find the vertex symbolically, use the vertex formula with $a = 1$ and $b = -6$.

$x = -\dfrac{b}{2a} = -\dfrac{(-6)}{2(1)} = \dfrac{6}{2} = 3$. The x-coordinate of the vertex is 3.

$y = f\left(-\dfrac{b}{2a}\right) = f(3) = (3)^2 - 6(3) = -9$. The y-coordinate of the vertex is –9. Thus, the vertex is (3, –9).

The graph of $Y_1 = X^2 - 6X$ shows graphical support for this vertex. See Figure 27.

29. The function is already in standard form: $f(x) = 2x^2 - 4x + 1$.

To find the vertex symbolically, use the vertex formula with $a = 2$ and $b = -4$.

$x = -\dfrac{b}{2a} = -\dfrac{(-4)}{2(2)} = \dfrac{4}{4} = 1$. The x-coordinate of the vertex is 1.

$y = f\left(-\dfrac{b}{2a}\right) = f(1) = 2(1)^2 - 4(1) + 1 = -1$. The y-coordinate of the vertex is −1. Thus, the vertex

is (1, −1). The graph of $Y_1 = 2X^2 - 4X + 1$ shows graphical support for this vertex. See Figure 29.

[−10, 10, 1] by [−15, 5, 1]   [−10, 10, 1] by [−10, 10, 1]   [−40, 40, 5] by [−40, 40, 5]   [−10, 10, 1] by [−10, 10, 1]

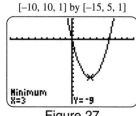
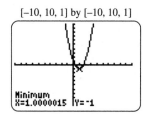
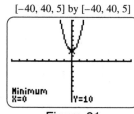
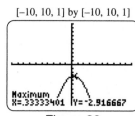

Figure 27   Figure 29   Figure 31   Figure 33

31. It may be helpful to write f in standard form as follows: $f(x) = \dfrac{1}{2}x^2 + 0x + 10$.

To find the vertex symbolically, use the vertex formula with $a = \dfrac{1}{2}$ and $b = 0$.

$x = -\dfrac{b}{2a} = -\dfrac{0}{2\left(\frac{1}{2}\right)} = 0$. The x-coordinate of the vertex is 0.

$y = f\left(-\dfrac{b}{2a}\right) = f(0) = \dfrac{1}{2}(0)^2 + 0(0) + 10 = 10$. The y-coordinate of the vertex is 10. Thus, the vertex

is (0, 10). The graph of $Y_1 = \dfrac{1}{2}X^2 + 10$ shows graphical support for this vertex. See Figure 31.

33. The function is already in standard form: $f(x) = -\dfrac{3}{4}x^2 + \dfrac{1}{2}x - 3$.

To find the vertex symbolically, use the vertex formula with $a = -\dfrac{3}{4}$ and $b = \dfrac{1}{2}$.

$x = -\dfrac{b}{2a} = -\dfrac{\frac{1}{2}}{2\left(-\frac{3}{4}\right)} = \dfrac{1}{3}$. The x-coordinate of the vertex is $\dfrac{1}{3}$.

$y = f\left(-\dfrac{b}{2a}\right) = f\left(\dfrac{1}{3}\right) = -\dfrac{3}{4}\left(\dfrac{1}{3}\right)^2 + \dfrac{1}{2}\left(\dfrac{1}{3}\right) - 3 = -\dfrac{35}{12}$. The y-coordinate of the vertex is $-\dfrac{35}{12}$. Thus,

the vertex is $\left(\dfrac{1}{3}, -\dfrac{35}{12}\right)$. The graph of $Y_1 = -\dfrac{3}{4}X^2 + \dfrac{1}{2}X - 3$ shows graphical support for this vertex.

See Figure 33.

35. The function is already in standard form: $f(x) = x^2 - 3.8x - 2$.

    To find the vertex symbolically, use the vertex formula with $a = 1$ and $b = -3.8$.

    $x = -\dfrac{b}{2a} = -\dfrac{(-3.8)}{2(1)} = \dfrac{3.8}{2} = 1.9$. The x-coordinate of the vertex is 1.9.

    $y = f\left(-\dfrac{b}{2a}\right) = f(1.9) = (1.9)^2 - 3.8(1.9) - 2 = -5.61$. The y-coordinate of the vertex is -5.61. Thus,

    the vertex is $(1.9, -5.61)$. The graph of $Y_1 = X^2 - 3.8X - 2$ shows graphical support for this vertex.

    See Figure 35.

37. It may be helpful to write f in standard form as follows: $f(x) = -6x^2 - 3x + 1.5$.

    To find the vertex symbolically, use the vertex formula with $a = -6$ and $b = -3$.

    $x = -\dfrac{b}{2a} = -\dfrac{(-3)}{2(-6)} = -\dfrac{3}{12} = -0.25$. The x-coordinate of the vertex is -0.25.

    $y = f\left(-\dfrac{b}{2a}\right) = f(-0.25) = 1.5 - 3(-0.25) - 6(-0.25)^2 = 1.875$. The y-coordinate of the vertex

    is 1.875. Thus, the vertex is $(-0.25, 1.875)$. The graph of $Y_1 = 1.5 - 3X - 6X^2$ shows graphical support

    for this vertex. See Figure 37.

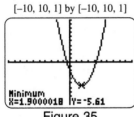

[–10, 10, 1] by [–10, 10, 1]

Figure 35

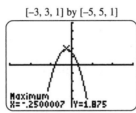

[–3, 3, 1] by [–5, 5, 1]

Figure 37

39. $f(x) = x^2 + 4x - 5 \Rightarrow$ leading coefficient: 1; $x = -\dfrac{b}{2a} = -\dfrac{4}{2(1)} = -2$ and

    $f(-2) = (-2)^2 + 4(-2) - 5 = -9 \Rightarrow$ vertex: $(-2, -9)$; since $a = 1$, $f(x) = 1(x - (-2))^2 - 9$, or

    $f(x) = (x + 2)^2 - 9$

41. $f(x) = x^2 - 3x \Rightarrow$ leading coefficient: 1; $x = -\dfrac{b}{2a} = -\dfrac{(-3)}{2(1)} = \dfrac{3}{2}$ and

    $f\left(\dfrac{3}{2}\right) = \left(\dfrac{3}{2}\right)^2 - 3\left(\dfrac{3}{2}\right) = -\dfrac{9}{4} \Rightarrow$ vertex: $\left(\dfrac{3}{2}, -\dfrac{9}{4}\right)$; since $a = 1$, $f(x) = 1\left(x - \dfrac{3}{2}\right)^2 - \dfrac{9}{4}$, or

    $f(x) = \left(x - \dfrac{3}{2}\right)^2 - \dfrac{9}{4}$

43. $f(x) = 2x^2 - 5x + 3 \Rightarrow$ leading coefficient: 2; $x = -\dfrac{b}{2a} = -\dfrac{(-5)}{2(2)} = \dfrac{5}{4}$ and

    $f\left(\dfrac{5}{4}\right) = 2\left(\dfrac{5}{4}\right)^2 - 5\left(\dfrac{5}{4}\right) + 3 = -\dfrac{1}{8} \Rightarrow$ vertex: $\left(\dfrac{5}{4}, -\dfrac{1}{8}\right)$; since $a = 2$, $f(x) = 2\left(x - \dfrac{5}{4}\right)^2 - \dfrac{1}{8}$

45. $f(x) = -\dfrac{1}{2}x^2 - \dfrac{3}{2}x + 1 \Rightarrow$ leading coefficient: $-\dfrac{1}{2}$; $x = -\dfrac{b}{2a} = -\dfrac{(-\frac{3}{2})}{2(-\frac{1}{2})} = -\dfrac{3}{2}$ and

    $f\left(-\dfrac{3}{2}\right) = -\dfrac{1}{2}\left(-\dfrac{3}{2}\right)^2 - \dfrac{3}{2}\left(-\dfrac{3}{2}\right) + 1 = \dfrac{17}{8} \Rightarrow$ vertex: $\left(-\dfrac{3}{2}, \dfrac{17}{8}\right)$; since $a = -\dfrac{1}{2}$, $f(x) = -\dfrac{1}{2}\left(x + \dfrac{3}{2}\right)^2 + \dfrac{17}{8}$

47. $f(x) = 2x^2 - 8x - 1 \Rightarrow$ leading coefficient: 2; $x = -\dfrac{b}{2a} = -\dfrac{(-8)}{2(2)} = 2$ and

$f(2) = 2(2)^2 - 8(2) - 1 = -9 \Rightarrow$ vertex: $(2, -9)$; since $a = 2$, $f(x) = 2(x - 2)^2 - 9$

49. $f(x) = 2 - 9x - 3x^2 \Rightarrow$ leading coefficient: -3; $x = -\dfrac{b}{2a} = -\dfrac{(-9)}{2(-3)} = -1.5$ and

$f(-1.5) = 2 - 9(-1.5) - 3(-1.5)^2 = 8.75 \Rightarrow$ vertex: $(-1.5, 8.75)$; since $a = -3$,

$f(x) = -3(x - (-1.5))^2 + 8.75$, or $f(x) = -3(x + 1.5)^2 + 8.75$

51. The vertex of the parabola is $(2, -2)$, so $f(x) = a(x - h)^2 + k \Rightarrow f(x) = a(x - 2)^2 - 2$; since $(0, 2)$ is a

point on the parabola, $f(0) = a(0 - 2)^2 - 2 \Rightarrow 2 = a(-2)^2 - 2 \Rightarrow 4 = 4a \Rightarrow a = 1$; $f(x) = (x - 2)^2 - 2$

53. The vertex of the parabola is $(2, -3)$, so $f(x) = a(x - h)^2 + k \Rightarrow f(x) = a(x - 2)^2 - 3$; since $(0, -1)$ is a

point on the parabola, $f(0) = a(0 - 2)^2 - 3 \Rightarrow -1 = a(4) - 3 \Rightarrow -1 = 4a - 3 \Rightarrow 2 = 4a \Rightarrow a = \dfrac{1}{2}$;

$f(x) = \dfrac{1}{2}(x - 2)^2 - 3$.

55. The vertex of the parabola is $(-1, 3)$, so $f(x) = a(x - h)^2 + k \Rightarrow f(x) = a(x + 1)^2 + 3$; since $(0, 1)$ is a

point on the parabola, $f(0) = a(0 + 1)^2 + 3 \Rightarrow 1 = a(1) + 3 \Rightarrow 1 = a + 3 \Rightarrow a = -2$;

$f(x) = -2(x + 1)^2 + 3$.

57. The vertex of the parabola is $(2, 6)$, so $f(x) = a(x - h)^2 + k \Rightarrow f(x) = a(x - 2)^2 + 6$; since $(0, -6)$ is a

point on the parabola, $f(0) = a(0 - 2)^2 + 6 \Rightarrow -6 = 4a + 6 \Rightarrow -12 = 4a \Rightarrow a = -3$;

$f(x) = -3(x - 2)^2 + 6$.

59. The graph of $f(x) = x^2$ is shown in Figure 59.

61. The graph of $f(x) = -\dfrac{1}{2}x^2$ is shown in Figure 61.

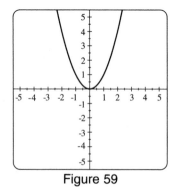
Figure 59

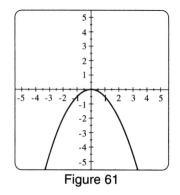
Figure 61

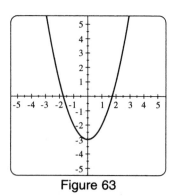
Figure 63

63. The graph of $f(x) = x^2 - 3$ is shown in Figure 63.

65. The graph of $f(x) = (x - 2)^2 + 1$ is shown in Figure 65.

67. The graph of $f(x) = -3(x + 1)^2 + 3$ is shown in Figure 67.

69. The graph of $f(x) = x^2 - 2x - 2$ is shown in Figure 69.

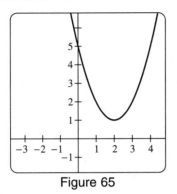

Figure 65

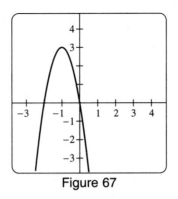

Figure 67

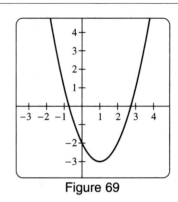

Figure 69

71. The graph of $f(x) = -x^2 + 4x - 2$ is shown in Figure 71.

73. The graph of $f(x) = 2x^2 - 4x - 1$ is shown in Figure 73.

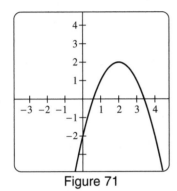

Figure 71

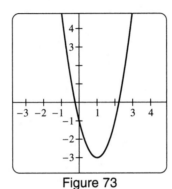

Figure 73

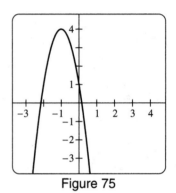

Figure 75

75. The graph of $f(x) = -3x^2 - 6x + 1$ is shown in Figure 75.

77. The graph of $f(x) = \dfrac{1}{2}x^2 - 2x + 2$ is shown in Figure 77.

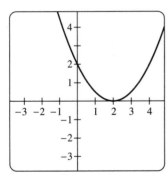

Figure 77

Applications and Models

79. Perimeter of fence $= 2l + 2w = 1000 \Rightarrow l = \dfrac{1000 - 2w}{2} \Rightarrow l = 500 - w$.

If $A = lw$ then $A = (500 - w)w \Rightarrow A = 500w - w^2 \Rightarrow A = -w^2 + 500w$, this is a parabola opening

downward and by the vertex formula, maximum area occurs when $w = -\dfrac{b}{2a} \Rightarrow w = -\dfrac{500}{2(-1)} = 250$.

The dimensions that maximize area are 250 ft. by 250 ft.

81. (a) $R(x) = x(40 - 2x) \Rightarrow R(2) = 2(40 - 2(2)) = 2(36) = 72$; the company receives \$72,000 for producing 2000 CD players.

 (b) $R(x) = 40x - 2x^2$ is a quadratic function; to find the value at which the maximum value occurs, we need to find the vertex: $x = -\dfrac{b}{2a} = -\dfrac{40}{2(-2)} = 10$ and $R(10) = 40(10) - 2(10)^2 = 200 \Rightarrow (10, 200)$ is the vertex; thus, the company needs to produce 10,000 CD players to maximize it's revenue.

 (c) Since $R(10) = 200$, the maximum revenue for the company is \$200,000.

83. (a) $s(1) = -16(1)^2 + 44(1) + 4 = 32$; the baseball was 32 feet high after 1 second.

 (b) For $s(t) = -16t^2 + 44t + 4$, the vertex formula gives $t = -\dfrac{b}{2a} = -\dfrac{44}{2(-16)} = 1.375$ and $f(1.375) = -16(1.375)^2 + 44(1.375) + 4 = 34.25$; the maximum height of the baseball was 34.25 feet.

85. A stone thrown from the ground level would first rise to some maximum height and then fall back to the ground. This is represented in figure d.

87. When the furnace first fails to work the temperature will begin to drop. Then, after the furnace is repaired, the temperature will begin to rise. This is represented in figure a.

89. (a) When $g = 32$, $v_0 = 88$, and $h_0 = 25$ the function becomes $f(x) = -\dfrac{1}{2}(32)x^2 + 88x + 25$.

 Graph $Y_1 = -16X^2 + 88X + 25$. By tracing along the graph the maximum height is found to be approximately 146 feet. This height occurs when $x \approx 2.75$ seconds. See Figure 89.

 (b) To find the maximum height symbolically, use the vertex formula with $a = -16$ and $b = 88$.

 $x = -\dfrac{b}{2a} = -\dfrac{88}{2(-16)} = \dfrac{88}{32} = 2.75$. The maximum height occurs at $x = 2.75$ seconds.

 $y = f\left(-\dfrac{b}{2a}\right) = f(2.75) = -16(2.75)^2 + 88(2.75) + 25 = 146$. The maximum height is 146 feet.

[0, 6, 1] by [0, 160, 10]   [0, 15, 5] by [0, 400, 100]   [1980, 1996, 2] by [−50,000, 500,000, 10,000]

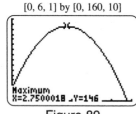   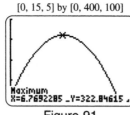   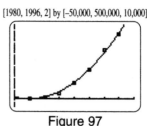

Figure 89   Figure 91   Figure 97

91. (a) When $g = 13$, $v_0 = 88$, and $h_0 = 25$ the function becomes $f(x) = -\dfrac{1}{2}(13)x^2 + 88x + 25$.

 Graph $Y_1 = -6.5X^2 + 88X + 25$. By tracing along the graph the maximum height is found to be approximately 323 feet. This height occurs when $x \approx 6.77$ seconds. See Figure 91.

 (b) To find the maximum height symbolically, use the vertex formula with $a = -6.5$ and $b = 88$.

 $x = -\dfrac{b}{2a} = -\dfrac{88}{2(-6.5)} = \dfrac{88}{13} \approx 6.77$. The maximum height occurs at $x \approx 6.77$ seconds.

 $y = f\left(-\dfrac{b}{2a}\right) = f(6.77) = -6.5(6.77)^2 + 88(6.77) + 25 \approx 322.8$. The maximum height is 322.8 feet.

93. The smallest y-value is -3 when $x = 1$; the symmetry in the y-values about $x = 1$ indicates the axis of symmetry is $x = 1$ and the vertex is $(1, -3)$, so $f(x) = a(x - 1)^2 - 3$; since $(0, -1)$ is a data point, $f(0) = -1$; $f(0) = a(0 - 1)^2 - 3 \Rightarrow -1 = a - 3 \Rightarrow a = 2$; the function $f(x) = 2(x - 1)^2 - 3$ models the data exactly.

95. The largest y-value is 4 when $x = -1$; the symmetry in the y-values about $x = -1$ indicates the axis of symmetry is $x = -1$ and the vertex is $(-1, 4)$, so $f(x) = a(x + 1)^2 + 4$; since $(0, 2)$ is a data point, $f(0) = 2$; $f(0) = a(0 + 1)^2 + 4 \Rightarrow 2 = a + 4 \Rightarrow a = -2$; the function $f(x) = -2(x + 1)^2 + 4$ models the data exactly.

97. (a) Scatterplot the data in $[1980, 1996, 2]$ by $[-50,000, 500,000, 10,000]$.

   (b) Start by letting the vertex be $(1982, 1586)$. With $h = 1982$ and $k = 1586$, f is represented by $f(x) = a(x - 1982)^2 + 1586$. Since the data is increasing to the right of the vertex, a is positive. Using trial and error, a is approximately 3100. *Answers may vary.* Thus, $f(x) = 3100(x - 1982)^2 + 1586$. The graph of f together with the data is shown in Figure 97.

   (c) Table $f(x) = 3100(x - 1982)^2 + 1586$ starting at $x = 1994$, incrementing by 1. There will be approximately 795,186 AIDS cases in the year 1998, if the trend in the table continues. That is, $f(1998) = 795,186$ cases of AIDS are predicted by f in the year 1998.

99. (a) Since the minimum occurs at $t = 4$, let $(4, 90)$ be the vertex and write $H(t) = a(t - 4)^2 + 90$. Using the first data point $(0, 122)$ we find a. $H(0) = a(0 - 4)^2 + 90 \Rightarrow 122 = a(0 - 4)^2 + 90 \Rightarrow$ $122 = 16a + 90 \Rightarrow 16a = 32 \Rightarrow a = 2$; $H(t) = 2(t - 4)^2 + 90$; Domain of $H = \{0 \le t \le 4\}$

   (b) $H(1.5) = 2(1.5 - 4)^2 + 90 \Rightarrow H(1.5) = 2(-2.5)^2 + 90 \Rightarrow H(1.5) = 12.5 + 90 \Rightarrow H(1.5) = 102.5$ or 102.5 beats per minute.

101. (a) Using $(1900, 5.3)$ and the vertex formula $f(x) = a(x - h)^2 + k \Rightarrow f(x) = a(x - 1900)^2 + 5.3$. Now using the last point $(2000, 65.6) \Rightarrow f(2000) = a(2000 - 1900)^2 + 5.3 \Rightarrow 65.6 = 10,000a + 5.3 \Rightarrow$ $60.3 = 10,000a \Rightarrow a = 0.006$; $f(x) = 0.006(x - 1900)^2 + 5.3$

   (b) $f(2010) = 0.006(2010 - 1900)^2 + 5.3 \Rightarrow (f)2010 = 0.006(12,100) + 5.3 = 77.9$ million.

Quadratic Regression

103. Enter the data into your calculator. See Figure 103a. Select quadratic regression from the STAT menu. See Figure 103b. In figure 103c, the modeling function is given (approximately) by $f(x) = 3.125x^2 + 2.05x - 0.9$. $f(3.5) = 3.125(3.5)^2 + 2.05(3.5) - 0.9 \approx 44.56$.

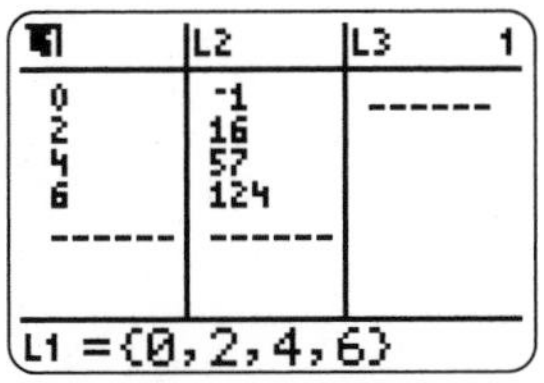

Figure 103a

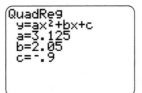
Figure 103b

Figure 103c

105. Enter the data into your calculator. See Figure 105a. Select quadratic regression from the STAT menu. See Figure 105b. In figure 105c, the modeling function is given (approximately) by

$f(x) = 1.415x^2 + 1.731x + 0.995$. $f(3.5) = 1.415(3.5)^2 + 1.731(3.5) + 0.995 \approx 24.39$.

[1935, 2005, 5] by [1.9, 2.6, 0.1]

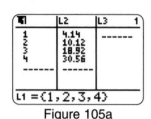   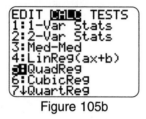   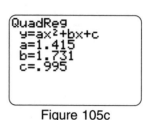   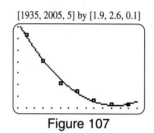

Figure 105a   Figure 105b   Figure 105c   Figure 107

107. (a) Entering the data and using quadratic regression gives the function:

$f(x) = 0.000217857x^2 - 0.8680357x + 866.58$. See figure 107.

(b) $f(1975) = 0.000217857(1975)^2 - 0.8680357(1975) + 866.58 \approx 1.99$, which compares favorably with the actual value of 2.01.

3.2: Quadratic Equations and Problem Solving

Quadratic Equations

1. $x^2 + x - 11 = 1 \Rightarrow x^2 + x - 12 = 0$. Factoring this we get: $(x + 4)(x - 3) = 0$. Then $x + 4 = 0 \Rightarrow x = -4$ or $x - 3 = 0 \Rightarrow x = 3$. Therefore $x = -4, 3$.

3. $t^2 = 2t \Rightarrow t^2 - 2t = 0$. Factoring this we get: $t(t - 2) = 0$. Then $t = 0$ or $t - 2 = 0 \Rightarrow t = 2$. Therefore $t = 0, 2$.

5. $2z^2 + 13z + 15 = 0$. Factoring this we get: $(2z + 3)(z + 5) = 0$. Then $2z + 3 = 0 \Rightarrow 2z = -3 \Rightarrow z = \dfrac{-3}{2}$ or $z + 5 = 0 \Rightarrow z = -5$. Therefore $z = -5, \dfrac{-3}{2}$.

7. $x(3x + 14) = 5 \Rightarrow 3x^2 + 14x - 5 = 0$. Factoring this we get: $(3x - 1)(x + 5) = 0$. Then $3x - 1 = 0 \Rightarrow 3x = 1 \Rightarrow x = \dfrac{1}{3}$ or $x + 5 = 0 \Rightarrow x = -5$. Therefore $x = -5, \dfrac{1}{3}$.

9. $6x^2 + \dfrac{5}{2} = 8x \Rightarrow 6x^2 - 8x + \dfrac{5}{2} = 0 \Rightarrow 12x^2 - 16x + 5 = 0$. Factoring this we get: $(2x - 1)(6x - 5) = 0$. Then $2x - 1 = 0 \Rightarrow 2x = 1 \Rightarrow x = \dfrac{1}{2}$ or $6x - 5 = 0 \Rightarrow 6x = 5 \Rightarrow x = \dfrac{5}{6}$. Therefore $x = \dfrac{1}{2}, \dfrac{5}{6}$.

11. $(t + 3)^2 = 5 \Rightarrow t^2 + 6t + 4 = 0$. Using the quadratic formula: $x = \dfrac{-b \pm \sqrt{b^2 - 4ac}}{2a} \Rightarrow$

$x = \dfrac{-6 \pm \sqrt{6^2 - 4(1)(4)}}{2(1)} \Rightarrow x = \dfrac{-6 \pm \sqrt{20}}{2} \Rightarrow x = \dfrac{-6 \pm 2\sqrt{5}}{2} \Rightarrow x = -3 \pm \sqrt{5}$

13. $4x^2 - 13 = 0 \Rightarrow 4x^2 + 0x - 13 = 0$. Using the quadratic formula: $x = \dfrac{-b \pm \sqrt{b^2 - 4ac}}{2a} \Rightarrow$

$x = \dfrac{0 \pm \sqrt{0^2 - 4(4)(-13)}}{2(4)} \Rightarrow x = \dfrac{\pm \sqrt{208}}{8} \Rightarrow x = \dfrac{\pm 4\sqrt{13}}{8} \Rightarrow x = \dfrac{\pm \sqrt{13}}{2}$

15. $2(x - 1)^2 + 4 = 0 \Rightarrow 2(x^2 - 2x + 1) + 4 = 0 \Rightarrow 2x^2 - 4x + 6 = 0$. Using the quadratic formula:

$$x = \frac{-b \pm \sqrt{b^2 - 4ac}}{2a} \Rightarrow x = \frac{-(-4) \pm \sqrt{(-4)^2 - 4(2)(6)}}{2(2)} \Rightarrow x = \frac{4 \pm \sqrt{-32}}{4} = \frac{4 \pm 4\sqrt{-2}}{4} =$$

$1 \pm \sqrt{-2}$. Since there is no real solution to $\sqrt{-2}$ there is no real solution to the equation.

17. $\frac{1}{2}x^2 - 3x + \frac{1}{2} = 0 \Rightarrow x^2 - 6x + 1 = 0$. Using the quadratic formula: $x = \frac{-b \pm \sqrt{b^2 - 4ac}}{2a} \Rightarrow$

$$x = \frac{-(-6) \pm \sqrt{(-6)^2 - 4(1)(1)}}{2(1)} \Rightarrow x = \frac{6 \pm \sqrt{32}}{2} \Rightarrow x = \frac{6 \pm 4\sqrt{2}}{2} \Rightarrow x = 3 \pm 2\sqrt{2}$$

19. $-3z^2 - 2z + 4 = 0$. Using the quadratic formula: $x = \frac{-b \pm \sqrt{b^2 - 4ac}}{2a} \Rightarrow$

$$x = \frac{-(-2) \pm \sqrt{(-2)^2 - 4(-3)(4)}}{2(-3)} \Rightarrow x = \frac{2 \pm \sqrt{52}}{-6} \Rightarrow x = \frac{-2 \pm 2\sqrt{13}}{6} \Rightarrow x = \frac{-1 \pm \sqrt{13}}{3}$$

21. $25k^2 + 1 = 10k \Rightarrow 25k^2 - 10k + 1 = 0$. Using the quadratic formula: $x = \frac{-b \pm \sqrt{b^2 - 4ac}}{2a} \Rightarrow$

$$x = \frac{-(-10) \pm \sqrt{(-10)^2 - 4(25)(1)}}{2(25)} \Rightarrow x = \frac{10 \pm \sqrt{0}}{50} \Rightarrow x = \frac{1}{5}$$

23. $-0.3x^2 + 0.1x = -0.02 \Rightarrow -30x^2 + 10x + 2 = 0$. Using the quadratic formula: $x = \frac{-b \pm \sqrt{b^2 - 4ac}}{2a} \Rightarrow$

$$x = \frac{-10 \pm \sqrt{10^2 - 4(-30)(2)}}{2(-30)} \Rightarrow x = \frac{-10 \pm \sqrt{340}}{-60} \Rightarrow x = \frac{10 \pm 2\sqrt{85}}{60} \Rightarrow x = \frac{5 \pm \sqrt{85}}{30}$$

25. For the x-intercept, $y = 0 \Rightarrow 6x^2 + 13x - 5 = 0$. Factoring this we get $(2x + 5)(3x - 1) = 0$. Then

$$2x + 5 = 0 \Rightarrow 2x = -5 \Rightarrow x = \frac{-5}{2} \text{ or } 3x - 1 = 0 \Rightarrow 3x = 1 \Rightarrow x = \frac{1}{3}. \text{ Therefore } x = \frac{-5}{2}, \frac{1}{3}.$$

27. For the x-intercept, $y = 0 \Rightarrow -4x^2 + 12x - 9 = 0$. Using quadratic formula:

$$x = \frac{-12 \pm \sqrt{12^2 - 4(-4)(-9)}}{2(-4)} \Rightarrow x = \frac{-12 \pm \sqrt{144 - 144}}{-8} \Rightarrow x = \frac{-12 \pm \sqrt{0}}{-8} \Rightarrow x = \frac{3}{2}$$

29. For the x-intercept, $y = 0 \Rightarrow -3x^2 + 11x - 6 = 0$. Using quadratic formula:

$$x = \frac{-11 \pm \sqrt{11^2 - 4(-3)(-6)}}{2(-3)} \Rightarrow x = \frac{-11 \pm \sqrt{121 - 72}}{-6} \Rightarrow x = \frac{-11 \pm \sqrt{49}}{-6} \Rightarrow x = \frac{11 \pm 7}{6} \Rightarrow$$

$$x = \frac{2}{3}, 3$$

31. (a) Graphical: Graph $Y_1 = X^2 + 2X$ and locate the x-intercepts. From Figure 31a and Figure 31b the

    solutions to the equation are $x = -2, 0$.

    (b) Numerical: Table $Y_1 = X^2 - X - 6$ starting at $x = -3$, incrementing by 1. $Y_1 = 0$ when $x = -2, 0$.

    See Figure 31c.

    (c) Symbolic: $x^2 + 2x = 0 \Rightarrow x(x + 2) = 0$. Then $x = 0$ or $x + 2 = 0 \Rightarrow x = -2$. Therefore $x = -2, 0$.

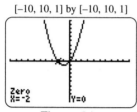
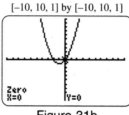

$[-10, 10, 1]$ by $[-10, 10, 1]$ $[-10, 10, 1]$ by $[-10, 10, 1]$

Figure 31a   Figure 31b   Figure 31c

33. (a) Graphical: Graph $Y_1 = X^2 - X - 6$ and locate the x-intercepts. From Figures 33a & 33b the solutions to

    the equation are $x = -2$ or 3.

    (b) Numerical: Table $Y_1 = X^2 - X - 6$ starting at $x = -3$, incrementing by 1. $Y_1 = 0$ when $x = -2$ or 3.

    See Figure 33c.

    (c) Symbolic: $x^2 - x - 6 = 0 \Rightarrow (x - 3)(x + 2) = 0 \Rightarrow x = 3$ or -2

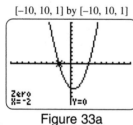
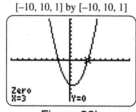
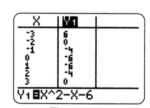

$[-10, 10, 1]$ by $[-10, 10, 1]$ $[-10, 10, 1]$ by $[-10, 10, 1]$

Figure 33a   Figure 33b   Figure 33c

35. (a) Graphical: Graph $Y_1 = 2X^2 - 6$ and locate the x-intercepts. From Figures 35a & 35b the solutions to the

    equation are $x \approx \pm 1.7$.

    (b) Numerical: Table $Y_1 = 2X^2 - 6$ starting at $x = -2\sqrt{3}$, incrementing by $\sqrt{3}$. $Y_1 \approx 0$ when $x \approx \pm 1.7$.

    See Figure 35c.

    (c) Symbolic: $2x^2 = 6 \Rightarrow x^2 = 3 \Rightarrow x = \pm\sqrt{3} \approx \pm 1.7$

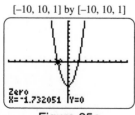
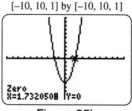

$[-10, 10, 1]$ by $[-10, 10, 1]$ $[-10, 10, 1]$ by $[-10, 10, 1]$

Figure 35a   Figure 35b   Figure 35c

37. (a) Graphical: Graph $Y_1 = 4X^2 - 12X + 9$ and locate the x-intercept(s). From Figure 37a the solution to

the equation is $x = 1.5$.

(b) Numerical: Table $Y_1 = 4X^2 - 12X + 9$ starting at $x = 0$, incrementing by 0.5. $Y_1 = 0$ when $x = 1.5$.

See Figure 37b.

(c) Symbolic: $4x^2 - 12x + 9 = 0 \Rightarrow (2x - 3)(2x - 3) = 0 \Rightarrow x = \dfrac{3}{2}$

[−5, 5, 1] by [−5, 5, 1]

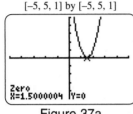

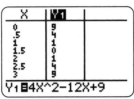

[−1, 1, 0.1] by [−10, 10, 1]

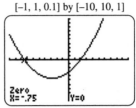

[−1, 1, 0.1] by [−10, 10, 1]

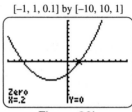

Figure 37a   Figure 37b   Figure 39a   Figure 39b

39. $20x^2 + 11x = 3 \Rightarrow 20x^2 + 11x - 3 = 0$. Graph $Y_1 = 20X^2 + 11X - 3$ and locate the x-intercepts.

From Figures 39a & 39b, the solutions to the equation are $x = -0.75$ or 0.2.

41. $2.5x^2 = 4.75x - 2.1 \Rightarrow 2.5x^2 - 4.75x + 2.1 = 0$. Graph $Y_1 = 2.5X^2 - 4.75X + 2.1$ and locate the

x-intercepts. From Figures 41a & 41b, the solutions to the equation are $x = 0.7$ or 1.2.

[0, 2, 0.1] by [−1, 1, 0.1]

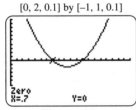

[0, 2, 0.1] by [−1, 1, 0.1]

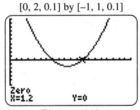

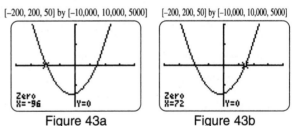

[−200, 200, 50] by [−10,000, 10,000, 5000]   [−200, 200, 50] by [−10,000, 10,000, 5000]

Figure 41a   Figure 41b   Figure 43a   Figure 43b

43. $x(x + 24) = 6912 \Rightarrow x^2 + 24x - 6912 = 0$. Graph $Y_1 = X^2 + 24X - 6912$ and locate the x-intercepts.

From Figures 43a & 43b, the solutions to the equation are $x = -96$ or 72.

45. (a) $3x^2 = 12 \Rightarrow 3x^2 - 12 = 0$

(b) $b^2 - 4ac = 0^2 - 4(3)(-12) = 144 > 0$. There are two real solutions.

(c) $3x^2 = 12 \Rightarrow x^2 = 4 \Rightarrow x = \pm 2$

47. (a) $x^2 - 2x = -1 \Rightarrow x^2 - 2x + 1 = 0$

(b) $b^2 - 4ac = (-2)^2 - 4(1)(1) = 0$. There is one real solution.

(c) $x^2 - 2x + 1 = 0 \Rightarrow (x - 1)(x - 1) = 0 \Rightarrow x = 1$

49. (a) $4x - 2 = x^2 \Rightarrow x^2 - 4x + 2 = 0$

(b) $b^2 - 4ac = (-4)^2 - 4(1)(2) = 8 > 0$. There are two real solutions.

(c) $x = \dfrac{-(-4) \pm \sqrt{(-4)^2 - 4(1)(2)}}{2(1)} = \dfrac{4 \pm \sqrt{8}}{2} = 2 \pm \sqrt{2} \approx 0.6$ or 3.4

51. (a) $x^2 + 1 = x \Rightarrow x^2 - x + 1 = 0$.

(b) $b^2 - 4ac = (-1)^2 - 4(1)(1) = -3 < 0$. There are no real solutions.

(c) There are no real solutions.

53. (a) $2x^2 + 3x = 12 - 2x \Rightarrow 2x^2 + 5x - 12 = 0$.

    (b) $b^2 - 4ac = (5)^2 - 4(2)(-12) = 121 > 0$. There are two real solutions.

    (c) $2x^2 + 5x - 12 = 0 \Rightarrow (2x - 3)(x + 4) = 0 \Rightarrow x = 1.5 \text{ or } -4$

55. (a) $\frac{1}{4}x^2 + 3x = x - 4 \Rightarrow \frac{1}{4}x^2 + 2x + 4 = 0$

    (b) $b^2 - 4ac = (2)^2 - 4\left(\frac{1}{4}\right)(4) = 0$. There is one real solution.

    (c) $\frac{1}{4}x^2 + 2x + 4 = 0 \Rightarrow \left(\frac{1}{2}x + 2\right)\left(\frac{1}{2}x + 2\right) = 0 \Rightarrow x = -4$

57. (a) $x\left(\frac{1}{2}x + 1\right) = -\frac{13}{2} \Rightarrow \frac{1}{2}x^2 + x + \frac{13}{2} = 0$

    (b) $b^2 - 4ac = (1)^2 - 4\left(\frac{1}{2}\right)\left(\frac{13}{2}\right) = -12 < 0$. There are no real solutions.

    (c) There are no real solutions.

59. (a) $3x^2 = 1 - x \Rightarrow 3x^2 + x - 1 = 0$

    (b) $b^2 - 4ac = 1^2 - 4(3)(-1) = 13 > 0$. There are two real solutions.

    (c) $x = \dfrac{-1 \pm \sqrt{1^2 - 4(3)(-1)}}{2(3)} = \dfrac{-1 \pm \sqrt{13}}{2(3)} = -\dfrac{1}{6} \pm \dfrac{1}{6}\sqrt{13} \approx 0.4343 \text{ or } -0.7676$

61. (a) Since the parabola opens upward, $a > 0$.

    (b) Since the zeros of f are -6 and 2, the solutions to $ax^2 + bx + c = 0$ are also -6 and 2.

    (c) Since there are two real solutions, the discriminant is positive.

63. (a) Since the parabola opens upward, $a > 0$.

    (b) Since the only zero of f is -4, the solutions is also -4.

    (c) Since there is one real solutions, the discriminant is equal to zero.

65. $x^2 + 4x = 6 \Rightarrow x^2 + 4x + 4 = 6 + 4 \Rightarrow (x + 2)^2 = 10 \Rightarrow x + 2 = \pm\sqrt{10} \Rightarrow x = -2 \pm \sqrt{10}$

67. $x^2 + 5x - 4 = 0 \Rightarrow x^2 + 5x = 4 \Rightarrow x^2 + 5x + \dfrac{25}{4} = 4 + \dfrac{25}{4} \Rightarrow \left(x + \dfrac{5}{2}\right)^2 = \dfrac{41}{4} \Rightarrow$

    $x + \dfrac{5}{2} = \pm\sqrt{\dfrac{41}{4}} \Rightarrow x = -\dfrac{5}{2} \pm \dfrac{\sqrt{41}}{2} \Rightarrow x = -\dfrac{5}{2} \pm \dfrac{1}{2}\sqrt{41}$

69. $3x^2 - 6x - 2 = 0 \Rightarrow 3x^2 - 6x = 2 \Rightarrow x^2 - 2x = \dfrac{2}{3} \Rightarrow x^2 - 2x + 1 = \dfrac{2}{3} + 1 \Rightarrow (x - 1)^2 = \dfrac{5}{3} \Rightarrow$

    $x - 1 = \pm\sqrt{\dfrac{5}{3}} \Rightarrow x = 1 \pm \sqrt{\dfrac{5}{3}}$

71. $x^2 - 8x = 10 \Rightarrow x^2 - 8x + 16 = 10 + 16 \Rightarrow (x - 4)^2 = 26 \Rightarrow x - 4 = \pm\sqrt{26} \Rightarrow x = 4 \pm \sqrt{26}$

73. $\dfrac{1}{2}t^2 - \dfrac{3}{2}t = 1 \Rightarrow t^2 - 3t = 2 \Rightarrow t^2 - 3t + \dfrac{9}{4} = 2 + \dfrac{9}{4} \Rightarrow \left(t - \dfrac{3}{2}\right)^2 = \dfrac{17}{4} \Rightarrow t - \dfrac{3}{2} = \pm\sqrt{\dfrac{17}{4}} \Rightarrow$

    $t = \dfrac{3}{2} \pm \sqrt{\dfrac{17}{4}} \Rightarrow t = \dfrac{3 \pm \sqrt{17}}{2}$

75. $-2z^2 + 3z + 1 = 0 \Rightarrow z^2 - \dfrac{3}{2}z = \dfrac{1}{2} \Rightarrow z^2 - \dfrac{3}{2}z + \dfrac{9}{16} = \dfrac{1}{2} + \dfrac{9}{16} \Rightarrow \left(z - \dfrac{3}{4}\right)^2 = \dfrac{17}{16} \Rightarrow$

$z - \dfrac{3}{4} = \pm\sqrt{\dfrac{17}{16}} \Rightarrow z = \dfrac{3}{4} \pm \sqrt{\dfrac{17}{16}} \Rightarrow z = \dfrac{3 \pm \sqrt{17}}{4}$

77. $-5x^2 + 7x + 2 = 0 \Rightarrow x^2 - \dfrac{7}{5}x = \dfrac{2}{5} \Rightarrow x^2 - \dfrac{7}{5}x + \dfrac{49}{100} = \dfrac{2}{5} + \dfrac{49}{100} \Rightarrow \left(x - \dfrac{7}{10}\right)^2 = \dfrac{89}{100} \Rightarrow$

$x - \dfrac{7}{10} = \pm\sqrt{\dfrac{89}{100}} \Rightarrow x = \dfrac{7}{10} \pm \sqrt{\dfrac{89}{100}} \Rightarrow x = \dfrac{7 \pm \sqrt{89}}{10}$

79. $-\dfrac{3}{2}z^2 - \dfrac{1}{4}z + 1 = 0 \Rightarrow z^2 + \dfrac{1}{6}z = \dfrac{2}{3} \Rightarrow z^2 + \dfrac{1}{6}z + \dfrac{1}{144} = \dfrac{2}{3} + \dfrac{1}{144} \Rightarrow \left(z + \dfrac{1}{12}\right)^2 = \dfrac{97}{144} \Rightarrow$

$z + \dfrac{1}{12} = \pm\sqrt{\dfrac{97}{144}} \Rightarrow z = \dfrac{-1}{12} \pm \sqrt{\dfrac{97}{144}} \Rightarrow z = \dfrac{-1 \pm \sqrt{97}}{12}$

81. D = all real numbers except when the denominator $x^2 - 5 = 0 \Rightarrow x^2 = 5 \Rightarrow x = \pm\sqrt{5} \Rightarrow$

$D = \{x \,|\, x \ne \sqrt{5}, x \ne -\sqrt{5}\}$

83. D = all real numbers except when the denominator $t^2 - t - 2 = 0 \Rightarrow (t - 2)(t + 1) = 0 \Rightarrow t = -1, 2 \Rightarrow$

$D = \{t \,|\, t \ne -1, x \ne 2\}$

85. $4x^2 + 3y = \dfrac{y + 1}{3} \Rightarrow 3y - \dfrac{y + 1}{3} = -4x^2 \Rightarrow 9y - y - 1 = -12x^2 \Rightarrow 8y - 1 = -12x^2 \Rightarrow$

$8y = -12x^2 + 1 \Rightarrow y = \dfrac{-12x^2 + 1}{8}$; yes y is a function of x since one x-input produces only one y-output.

87. $3x^2 + 4y^2 = 12 \Rightarrow 4y^2 = 12 - 3x^2 \Rightarrow 2y = \pm\sqrt{12 - 3x^2} \Rightarrow y = \pm\dfrac{\sqrt{12 - 3x^2}}{2}$;

no y is not a function of x because one x-input produces two y-outputs.

89. $x - 25y^2 = 50 \Rightarrow -25y^2 = -x + 50 \Rightarrow 25y^2 = x - 50 \Rightarrow 5y = \pm\sqrt{x - 50} \Rightarrow y = \pm\dfrac{\sqrt{x - 50}}{5}$;

no y is not a function of x because one x-input produces two y-outputs.

91. $V = \dfrac{1}{3}\pi r^2 h$ for r, $\dfrac{3V}{\pi h} = r^2 \Rightarrow r = \pm\sqrt{\dfrac{3V}{\pi h}}$

93. $K = \dfrac{1}{2}mv^2$ for v, $\dfrac{2K}{m} = v^2 \Rightarrow v = \pm\sqrt{\dfrac{2K}{m}}$

95. $s = -16t^2 + 100t$ for t, $-s = 16t^2 - 100t \Rightarrow \dfrac{-s}{16} = t^2 - \dfrac{25}{4}t \Rightarrow \dfrac{-s}{16} + \dfrac{625}{64} = t^2 - \dfrac{25}{4}t + \dfrac{625}{64} \Rightarrow$

$\left(t - \dfrac{25}{8}\right)^2 = \dfrac{625 - 4s}{64} \Rightarrow t - \dfrac{25}{8} = \pm\sqrt{\dfrac{625 - 4s}{64}} \Rightarrow t = \dfrac{25}{8} \pm \sqrt{\dfrac{625 - 4s}{64}} \Rightarrow t = \dfrac{25 \pm \sqrt{625 - 4s}}{8}$

Applications and Models

97. The height when it hits the ground will $= 0 \Rightarrow$ let $75 - 16t^2 = 0 \Rightarrow -16t^2 = -75 \Rightarrow t^2 = \dfrac{75}{16} \Rightarrow$

$t = \pm\sqrt{\dfrac{75}{16}} \Rightarrow t = \pm\dfrac{\sqrt{75}}{4} \Rightarrow t \approx 2.2$ seconds.

99. $90,000 = 2375x^2 + 5134x + 5020 \Rightarrow 2375x^2 + 5134 - 84,980 = 0$. Using the quadratic formula we get:

$$x = \frac{-5134 \pm \sqrt{(5134)^2 - 4(2375)(-84,980)}}{2(2375)} \Rightarrow x \approx 5.$$ Therefore $1984 + 5 = 1989$.

101. (a) If the speed doubles, the radius of the curve quadruples, or increases by a factor of 4.

    (b) Any ordered pair in the table may be used to determine the constant a. $R(x) = ax^2$ and

$$R(10) = 50 \Rightarrow a(10)^2 = 50 \Rightarrow a = \frac{50}{10^2} = 0.5.$$ Thus, let $R(x) = 0.5x^2$.

    (c) If $R = 500$, $500 = 0.5x^2 \Rightarrow 1000 = x^2 \Rightarrow x = \sqrt{1000} \approx 31.6$. 31.6 mph is the maximum safe speed on a curve with a 500 foot radius.

103. (a) $E(15) = 1.4$, $1987 + 15 = 2002$; in 2002 there were 1.4 million Wal-Mart employees.

    (b) Let $(0, 0.2)$ be the vertex $\Rightarrow E(x) = a(x - 0)^2 + 0.2 \Rightarrow E(x) = ax^2 + 0.2$, using the last point we get

$2.2 = a(20)^2 + 0.2 \Rightarrow 2.2 = 400a + 0.2 \Rightarrow 400a = 2 \Rightarrow a = 0.005 \Rightarrow E(x) = 0.005x^2 + 0.2$;

Answers may vary.

    (c) See Figure 103.

    (d) $3 = 0.005x^2 + 0.2 \Rightarrow 2.8 = 0.005x^2 \Rightarrow 560 = x^2 \Rightarrow x = 24 \Rightarrow 1987 + 24 = 2011$;

Answers may vary.

[−5, 25, 5] by [0, 2.5, 0.5]   [−15, 15, 1] by [−100, 100, 10]

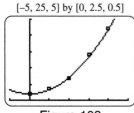 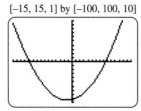

Figure 103   Figure 107

105. (a) Since $s(t) = 32$ when $t = 0$, $s(t) = -16t^2 + v_0t + h_0 \Rightarrow 32 = -16(0)^2 + v_0(0) + h_0 \Rightarrow h_0 = 32$ and so $s(t) = -16t^2 + v_0t + 32$; since $s(1) = 176$, $176 = -16(1)^2 + v_0(1) + 32 \Rightarrow v_0 = 160$; thus, $s(t) = -16t^2 + 160t + 32$ models the data.

    (b) When the projectile strikes the ground, $s(t) = 0$; $0 = -16t^2 + 160t + 32 \Rightarrow 0 = t^2 - 10t - 2 \Rightarrow$

$$t = \frac{10 \pm \sqrt{10^2 - 4(1)(-2)}}{2} \approx 10.2 \text{ or } -0.2;$$ since t cannot be negative, $t \approx 10.2$; the projectile strikes the ground after about 10.2 seconds.

107. Graphical: First, we must determine a formula for the area. Let x represent the height of the computer screen. Then, $x + 2.5$ is the width. The area of the screen is height times width, computed $A(x) = x(x + 2.5)$. we must solve the quadratic equation $x(x + 2.5) = 93.5$ or $x^2 + 2.5x - 93.5 = 0$. Graph

$Y_1 = X^2 + 2.5X - 93.5$ and determine any zeros. Figure 107 shows that the equation has two zeros, one negative and one positive. The positive zero is located at $x = 8.5$. Therefore, the height is 8.5 inches and the width is $8.5 + 2.5 = 11$ inches.

Symbolic: The quadratic equation $x^2 + 2.5x - 93.5 = 0$ can be solved by the quadratic formula.

$$x = \frac{-b \pm \sqrt{b^2 - 4ac}}{2a} = \frac{-2.5 \pm \sqrt{2.5^2 - 4(1)(-93.5)}}{2(1)} = \frac{-2.5 \pm 19.5}{2} = 8.5, -11.$$ The positive answer

gives a height of 8.5 inches. It follows that the width is 11 inches.

109. Let x = width of the metal sheet in inches and $x + 10$ = length of the metal sheet in inches. Make a sketch to find expressions for the dimensions of the box. See Figure 109. The width of the box is $x - 8$ inches, the length of the box is $x + 2$ inches and the height of the box is 4 inches; the volume of the box, which is given as 476 cubic inches, is determined by the length times the width times the height: $4(x - 8)(x + 2) = 476 \Rightarrow$ $(x - 8)(x + 2) = 119 \Rightarrow x^2 - 6x - 16 = 119 \Rightarrow x^2 - 6x - 135 = 0 \Rightarrow (x + 9)(x - 15) = 0 \Rightarrow$ $x = -9$ or $x = 15$. Since x cannot be negative, $x = 15$ inches; thus, the dimensions of the metal sheets are 15 inches by 25 inches, and the dimensions of the box are 4 inches by 7 inches by 17 inches.

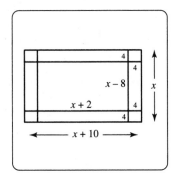

Figure 109

111. The dimensions of the picture are x inches by $x + 4$ inches; since the area of the picture is 320 square inches, $A = lw \Rightarrow 320 = x(x + 4) \Rightarrow x^2 + 4x = 320 \Rightarrow x^2 + 4x - 320 = 0 \Rightarrow (x - 16)(x + 20) = 0 \Rightarrow$ $x = 16$ or $x = -20$; since $x = -20$ has no physical meaning, $x = 16$ inches; thus, the picture is 16 inches by 20 inches, and so the frame is $16 + 4 = 20$ inches by $20 + 4 = 24$ inches.

113. Since the diameter of the semicircle is x, the radius is $\dfrac{x}{2}$; the area of the semicircle

$$= \frac{1}{2}\pi\left(\frac{x}{2}\right)^2 = \frac{1}{2}\pi\left(\frac{x^2}{4}\right) = \frac{1}{8}\pi x^2;$$ the area of the square $= x^2$; thus, the total area of the window, which is

463 square inches, is $x^2 + \dfrac{1}{8}\pi x^2$; $463 = x^2 + \dfrac{1}{8}\pi x^2 \Rightarrow 463 = \left(1 + \dfrac{\pi}{8}\right)x^2 \Rightarrow x^2 = \dfrac{463}{\left(1 + \frac{\pi}{8}\right)} \Rightarrow$

$x = \sqrt{\dfrac{463}{1 + \frac{\pi}{8}}} \approx 18.23$ inches.

3.3: Quadratic Inequalities

Quadratic Inequalities

1. (a) The inequality $f(x) < 0$ is satisfied when the graph of f is below the x-axis. This occurs when

    $-3 < x < 2$.

   (b) The inequality $f(x) \geq 0$ is satisfied when the graph of f is above the x-axis or intersects it. This occurs

    when $x \leq -3$ or $x \geq 2$.

3. (a) The inequality $f(x) \leq 0$ is satisfied when the graph of f is below the x-axis or intersects it. This only occurs when $x = -2$.

   (b) The inequality $f(x) > 0$ is satisfied when the graph of f is above the x-axis. This occurs when $x \neq -2$.

5. (a) The inequality $f(x) > 0$ is satisfied when the graph of f is above the x-axis. Since the graph of f is always below the x-axis, this inequality has no solution.

   (b) The inequality $f(x) < 0$ is satisfied when the graph of f is below the x-axis. This occurs when $-\infty < x < \infty$.

7. (a) $0 = 2x^2 + 6x + \dfrac{5}{2} \Rightarrow 4x^2 + 12x + 5 = 0$, using the quadratic formula we get:

   $$x = \frac{-12 \pm \sqrt{(12)^2 - 4(4)(5)}}{2(4)} \Rightarrow x = \frac{-12 \pm \sqrt{64}}{8} \Rightarrow x = \frac{-12 \pm 8}{8} \Rightarrow x = \frac{-4}{8} \Rightarrow$$

   $$x = -\frac{1}{2} \text{ or } x = -\frac{20}{8} \Rightarrow x = -2\frac{1}{2}, x = -2\frac{1}{2}, -\frac{1}{2}$$

   (b) $\left(-\dfrac{5}{2}, -\dfrac{1}{2}\right)$

   (c) $\left(-\infty, -\dfrac{5}{2}\right) \cup \left(-\dfrac{1}{2}, \infty\right)$

9. (a) $0 = -5x^2 + 2x + 7$, using the quadratic formula we get:

   $$x = \frac{-2 \pm \sqrt{(2)^2 - 4(-5)(7)}}{2(-5)} \Rightarrow x = \frac{-2 \pm \sqrt{144}}{-10} \Rightarrow x = \frac{-2 \pm 12}{-10} \Rightarrow x = -1, \frac{7}{5}$$

   (b) $(-\infty, -1) \cup \left(\dfrac{7}{5}, \infty\right)$

   (c) $\left(-1, \dfrac{7}{5}\right)$

11. (a) $x^2 - x - 12 = 0$, then $(x - 4)(x + 3) = 0 \Rightarrow x = -3, 4$. Using $x = 0$, which is inbetween -3 and 4 produces $0^2 - 0 - 12 = -12 \Rightarrow$ less than 0.

    (b) $x^2 - x - 12 < 0$ is $(-3, 4)$

    (c) $x^2 - x - 12 > 0$ is $(-\infty, -3) \cup (4, \infty)$

13. (a) $k^2 - 5 = 0 \Rightarrow k^2 = 5 \Rightarrow k = \pm\sqrt{5}$. Using $x = 0$, which is inbetween $-\sqrt{5}$ and $\sqrt{5}$ produces $0^2 - 5 = -5 \Rightarrow$ less than 0.

    (b) $k^2 - 5 < 0$ is $[-\sqrt{5}, \sqrt{5}]$

    (c) $k^2 - 5 > 0$ is $(-\infty, -\sqrt{5}] \cup [\sqrt{5}, \infty)$

15. (a) $3x^2 + 8x = 0 \Rightarrow x(3x + 8) = 0 \Rightarrow x = -\dfrac{8}{3}, 0$. Using $x = -1$, which is inbetween $-\dfrac{8}{3}$ and 0 produces $3(-1)^2 + 8(-1) = -5 \Rightarrow$ less than 0.

    (b) $3x^2 + 8x \leq 0$ is $\left[-\dfrac{8}{3}, 0\right]$

    (c) $3x^2 + 8x \geq 0$ is $\left(-\infty, -\dfrac{8}{3}\right] \cup [0, \infty)$

17. (a) $-4x^2 + 12x - 9 = 0 \Rightarrow (2x - 3)(-2x + 3) = 0 \Rightarrow x = \dfrac{3}{2}$. Using $x = 0$, which is less than $\dfrac{3}{2}$ produces

$-4(0)^2 + 12(0) - 9 = -9 \Rightarrow$ less than 0. Also using $x = 2$, which is more than $\dfrac{3}{2}$ produces

$-4(2)^2 + 12(2) - 9 = -1$ also less than 0.

(b) $-4x^2 + 12x - 9 < 0$ is $x \neq \dfrac{3}{2}$ or $\left(-\infty, \dfrac{3}{2}\right) \cup \left(\dfrac{3}{2}, \infty\right)$

(c) $-4x^2 + 12x - 9 > 0$ has no solution

19. (a) $12z^2 - 23z + 10 = 0 \Rightarrow (3z - 2)(4z - 5) = 0 \Rightarrow x = \dfrac{2}{3}, \dfrac{5}{4}$. Using $x = 1$, which is inbetween $\dfrac{2}{3}$ and $\dfrac{5}{4}$

produces $12(1)^2 - 23(1) + 10 = -1 \Rightarrow$ less than 0.

(b) $12z^2 - 23z + 10 \leq 0$ is $\left[\dfrac{2}{3}, \dfrac{5}{4}\right]$

(c) $12z^2 - 23z + 10 \geq 0$ is $\left(-\infty, \dfrac{2}{3}\right] \cup \left[\dfrac{5}{4}, \infty\right)$

21. (a) $x^2 + 2x - 1 = 0$, using the quadratic formula we get: $x = \dfrac{-2 \pm \sqrt{(2)^2 - 4(1)(-1)}}{2(1)} \Rightarrow$

$x = \dfrac{-2 \pm \sqrt{8}}{2} \Rightarrow x = \dfrac{-2 \pm 2\sqrt{2}}{2} \Rightarrow x = -1 \pm \sqrt{2}$. Using $x = 0$, which is inbetween

$-1 - \sqrt{2}$ and $-1 + \sqrt{2}$ produces $0^2 + 2(0) - 1 = -1 \Rightarrow$ less than 0.

(b) $x^2 + 2x - 1 < 0$ is $(-1 - \sqrt{2}, -1 + \sqrt{2})$

(c) $x^2 + 2x - 1 > 0$ is $(-\infty, -1 - \sqrt{2}) \cup (-1 + \sqrt{2}, \infty)$

23. (a) $f(x) > 0$ when $x < -1$ or $x > 1$

(b) $f(x) \leq 0$ when $-1 \leq x \leq 1$

25. (a) $f(x) > 0$ when $-6 < x < -2$

(b) $f(x) \leq 0$ when $x \leq -6$ or $x \geq -2$

27. Start by solving $x^2 - 3x - 4 = 0 \Rightarrow (x - 4)(x + 1) = 0 \Rightarrow x = -1$ or 4; thus,

$x^2 - 3x - 4 < 0$ when $-1 < x < 4$.

29. Start by solving $x^2 + x - 6 = 0 \Rightarrow (x + 3)(x - 2) = 0 \Rightarrow x = -3$ or 2; thus,

$x^2 + x - 6 > 0$ when $x < -3$ or $x > 2$.

31. Start by solving the equation $x^2 = 4 \Rightarrow x = \pm 2$. Next write the quadratic inequality with $a > 0$.

$x^2 \leq 4 \Rightarrow x^2 - 4 \leq 0$. The graph of $y = x^2 - 4$ is a parabola opening up. It will be less than or equal to 0

between (and including) the solutions to equality. The solutions are $-2 \leq x \leq 2$.

33. Since $x^2 - 4x + 4 = (x - 2)(x - 2) = (x - 2)^2 \geq 0$, the solutions are all real numbers.

35. Start by solving the equation $-x^2 + x + 6 = 0 \Rightarrow x^2 - x - 6 = 0 \Rightarrow (x + 2)(x - 3) = 0 \Rightarrow x = -2, 3$.

$-x^2 + x + 6 \leq 0$ is equivelent to $x^2 - x - 6 \geq 0$. The graph of $y = -x^2 + x + 6$ is a parabola opening

downward. It will be below the x-axis or intersect it outside of the solutions of the equality. The solutions are

$x \leq -2$ or $x \geq 3$.

37. Start by solving the equation $6x^2 - x = 1 \Rightarrow 6x^2 - x - 1 = 0 \Rightarrow (3x + 1)(2x - 1) = 0 \Rightarrow x = -\dfrac{1}{3}, \dfrac{1}{2}$. The graph of $y = 6x^2 - x - 1$ is a parabola opening up. It will be below the x-axis between the solutions of equality. The solutions to $6x^2 - x - 1 < 0$ are $-\dfrac{1}{3} < x < \dfrac{1}{2}$.

39. Start by solving the equation $(x + 4)(x - 10) = 0 \Rightarrow x = -4, 10$. The graph of $y = (x + 4)(x - 10) = x^2 - 6x - 40$ is a parabola opening up. It will intersect or be below the x-axis between (and including) the solutions of equality. The solutions to $(x + 4)(x - 10) \le 0$ are $-4 \le x \le 10$.

41. Start by solving the equation $x^2 + 4x + 3 = 0 \Rightarrow (x + 1)(x + 3) = 0 \Rightarrow x = -1, -3$. The graph of $y = x^2 + 4x + 3$ is a parabola opening up. It will be below the x-axis between the solutions of equality. The solutions to $x^2 + 4x + 3 < 0$ are $-3 < x < -1$.

43. Start by solving the equation $x^2 + 2x - 35 = 0 \Rightarrow (x + 7)(x - 5) = 0 \Rightarrow x = -7, 5$. The graph of $y = x^2 + 2x - 35$ is a parabola opening up. It will be above or intersect the x-axis outside (and including) the solutions of equality. The solutions to $x^2 + 2x - 35 \ge 0$ are $x \le -7$ or $x \ge 5$.

45. Start by solving the equation $x^2 = x \Rightarrow x^2 - x = 0 \Rightarrow x(x - 1) = 0 \Rightarrow x = 0, 1$. The graph of $y = x^2 - x$ is a parabola opening up. It will be above or intersect the x-axis outside (and including) the solutions of equality. The solutions to $x^2 - x \ge 0$ are $x \le 0$ or $x \ge 1$.

47. Start by solving the equation $x(x - 1) - 6 = 0 \Rightarrow x^2 - x - 6 = 0 \Rightarrow (x + 2)(x - 3) = 0 \Rightarrow x = -2, 3$. The graph of $y = x^2 - x - 6$ is a parabola opening up. It will be above or intersect the x-axis outside (and including) the solutions of equality. The solutions to $x^2 - x - 6 \ge 0$ are $x \le -2$ or $x \ge 3$.

49. Start by solving the equation $x^2 - 5 = 0 \Rightarrow x^2 = 5 \Rightarrow x = \pm\sqrt{5}$. The graph of $y = x^2 - 5$ is a parabola opening up. It will intersect or be below the x-axis between (and including) the solutions of equality. The solutions are $-\sqrt{5} \le x \le \sqrt{5}$.

51. Start by graphing $Y_1 = 7X^2 - 179.8X + 515.2$ in $[-4, 30, 2]$ by $[-750, 250, 50]$. The inequality given by $7x^2 - 179.8x + 515.2 \ge 0$ is satisfied when the graph Y_1 is above or intersects the x-axis. This occurs when $x \le k$ or $x \ge 22.4$, where $k \approx 3.29$.

53. For $x^2 - 9x + 14 \le 0$, first solve $x^2 - 9x + 14 = 0 \Rightarrow (x - 7)(x - 2) = 0 \Rightarrow x = 2, 7$. These boundary numbers give us the disjoint intervals: $(-\infty, 2), (2, 7), (7, \infty)$. See Figure 53. For $x^2 - 9x + 14 \le 0$, it is $2 \le x \le 7$.

Interval	Test Value x	$x^2 - 9x + 14$	Positive or Negative?
$(-\infty, 2)$	0	14	Positive
$(2, 7)$	4	-6	Negative
$(7, \infty)$	10	24	Positive

Interval	Test Value x	$x^2 - 3x - 10$	Positive or Negative?
$(-\infty, -2)$	-3	8	Positive
$(-2, 5)$	0	-10	Negative
$(5, \infty)$	6	8	Positive

Figure 53   Figure 55

55. For $x^2 \ge 3x + 10 \Rightarrow x^2 - 3x - 10 \ge 0$, first solve $x^2 - 3x - 10 = 0 \Rightarrow (x - 5)(x + 2) = 0 \Rightarrow x = -2, 5$. These boundary numbers give us the disjoint intervals: $(-\infty, -2), (-2, 5), (5, \infty)$. See Figure 55. For $x^2 - 3x - 10 \ge 0$, it is $x \le -2$ or $x \ge 5$.

57. For $x^2 - \dfrac{1}{2}x - 5 \not< 0$, first solve $x^2 - \dfrac{1}{2}x - 5 = 0 \Rightarrow 2x^2 - x - 10 = 0 \Rightarrow (2x - 5)(x + 2) = 0 \Rightarrow$

$x = -2, \dfrac{5}{2}$. These boundary numbers give us the disjoint intervals: $(-\infty, -2), \left(-2, \dfrac{5}{2}\right), \left(\dfrac{5}{2}, \infty\right)$. See Figure 57.

For $x^2 - \dfrac{1}{2}x - 5 < 0$, it is $-2 < x < \dfrac{5}{2}$.

Interval	Test Value x	$x^2 - 1/2x - 5$	Positive or Negative?
$(-\infty, -2)$	-4	13	Positive
$(-2, 5/2)$	0	-5	Negative
$(5/2, \infty)$	4	9	Positive

Interval	Test Value x	$x^2 + 4x - 3$	Positive or Negative?
$(-\infty, -4.6)$	-5	2	Positive
$(-4.6, 0.6)$	0	-3	Negative
$(0.6, \infty)$	1	2	Positive

Figure 57   Figure 59

59. For $x^2 > 3 - 4x \Rightarrow x^2 + 4x - 3 > 0$, first solve $x^2 + 4x - 3 = 0$, using the quadratic formula we get:

$x = \dfrac{-4 \pm \sqrt{(4)^2 - 4(1)(-3)}}{2(1)} \Rightarrow x = \dfrac{-4 \pm \sqrt{28}}{2} \Rightarrow x = -2 + \sqrt{7}$ or $x = -2 - \sqrt{7}$. These boundary

numbers give us the disjoint intervals: $(-\infty, -2 - \sqrt{7}), (-2 - \sqrt{7}, -2 + \sqrt{7}), (-2 + \sqrt{7}, \infty)$.

See Figure 59. For $x^2 + 4x - 3 > 0$, it is $x < -2 - \sqrt{7}$ or $x > -2 + \sqrt{7}$.

Applications

61. Graph $Y_1 = 2375X^2 + 5134X - 5020$, $Y_2 = 90{,}000$ and $Y_3 = 200{,}000$ in $[-1, 11, 1]$ by

$[5000, 210{,}000, 10{,}000]$. The intersection points are approximately $(4.9977492, 90{,}000)$ and

$(8.044127, 200{,}000)$. From the graphs, we see that $90{,}000 \le Y_1 \le 200{,}000$ when $4.998 \le X \le 8.044$. Since

1984 corresponds to $x = 0$, the number of AIDS deaths was from 90,000 to 200,000 for the years from 1989 to

1992.

63. (a) Since the units are in thousands, the number of subscribers exceeded 2 million when

$163x^2 - 146x + 205 > 2000 \Rightarrow 163x^2 - 146x - 1795 > 0$.

(b) Graph $Y_1 = 163X^2 - 146X - 1795$. Estimate the x-values where Y_1 is positive. From Figure 63, it can

be seen that $Y_1 > 0$ when $x > k$, where $k \approx 3.8$. This corresponds to 1989 or later.

$[0, 6, 1]$ by $[-2000, 2000, 1000]$

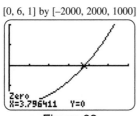

Figure 63

65. (a) The data cannot be modeled by a linear function because the rate at which the height of the water decreases will not be constant. When the tank first starts to drain, the large quantity of water in the tank will cause the water to drain rapidly. As the tank empties, there will be less water in the tank and the draining rate will decrease. This is evident in the data.

(b) From the table, it appears that the height of the water was between 5 and 10 centimeters from approximately 43 to 95 seconds after draining began.

(c) The graph of $Y_1 = 0.0003636X^2 - 0.1511X + 15.92$ with the data shown in Figure 65c.

(d) To find the times when the water height was between 5 and 10 centimeters, we will graph the function $Y_1 = 5$, $Y_2 = 0.0003636X^2 - 0.1511X + 15.92$, and $Y_3 = 10$. in the same window. The intersection points are approximately (93.1, 5) and (43.8, 10). See Figure 65d. The water level was between 5 an 10 centimeters from approximately 43.8 to 93.1 seconds after draining started.

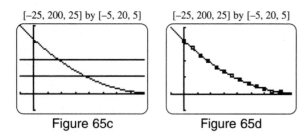

[−25, 200, 25] by [−5, 20, 5]   [−25, 200, 25] by [−5, 20, 5]

Figure 65c   Figure 65d

3.4: Transformations of Graphs

Vertical and Horizontal Translations

1. The vertex is (0, –2). The parabola $y = x^2$ has been shifted 2 units to the left. The equation is $y = (x + 2)^2$.

3. The endpoint is (–3, 0). The curve $y = \sqrt{x}$ has been shifted 3 units to the left. The equation is $y = \sqrt{x + 3}$.

5. The corner point is (–2, –1). The graph of $y = |x|$ has been shifted 2 units to the left and 1 unit down.

   The equation is $y = |x + 2| - 1$.

7. The vertex is (2, 1). The parabola $y = x^2$ has been shifted 2 units to the right and 1 unit up.

   The equation is $y = (x - 2)^2 + 1$.

9. The corner point is (0, –3). The graph of $y = |x|$ has been shifted 3 units down. The equation is $y = |x| - 3$.

Transforming Graphical Representations

11. (a) See Figure 11a.

    (b) See Figure 11b.

    (c) See Figure 11c.

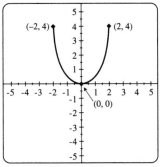

Figure 11a

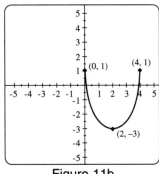

Figure 11b

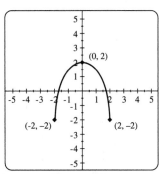

Figure 11c

13. (a) See Figure 13a.

    (b) See Figure 13b.

    (c) See Figure 13c.

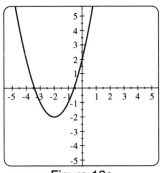

Figure 13a

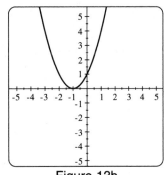

Figure 13b

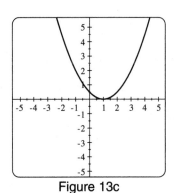

Figure 13c

15. (a) See Figure 15a.

    (b) See Figure 15b.

    (c) See Figure 15c.

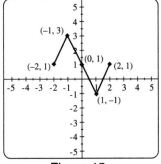

Figure 15a

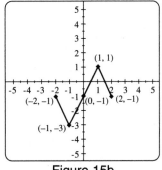

Figure 15b

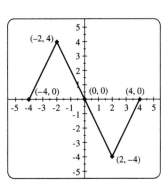

Figure 15c

17. (a) See Figure 17a.

   (b) See Figure 17b.

   (c) See Figure 17c.

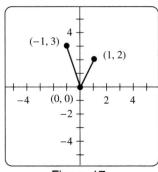

Figure 17a

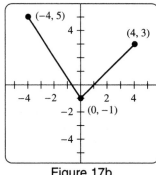

Figure 17b

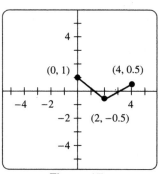

Figure 17c

19. Shift the graph of $y = x^2$ right 3 units and upward 1 unit.

21. Shift the graph of $y = x^2$ left 1 unit and vertically shrink it with a factor $\dfrac{1}{4}$.

23. Reflect the graph of $y = \sqrt{x}$ across the x-axis and shift it left 5 units.

25. Reflect the graph of $y = \sqrt{x}$ across the y-axis and vertically stretch it with factor 2.

27. Reflect the graph of $y = |x|$ across the y-axis and shift it left 1 unit.

29. To shift the graph of $f(x) = x^2$ to the right 2 units, replace x with $(x - 2)$ in the formula for $f(x)$.

   This results in $y = f(x - 2) = (x - 2)^2$. To shift the graph of this new equation down 3 units, subtract 3 from

   the formula to obtain $y = f(x - 2) - 3 = (x - 2)^2 - 3$. See Figure 29.

[–10, 10, 1] by [–10, 10, 1]   [–10, 10, 1] by [–10, 10, 1]

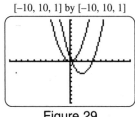

Figure 29

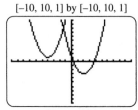

Figure 31

31. To shift the graph of $f(x) = x^2 - 4x + 1$ to the left 6 units, replace x with $(x + 6)$ in the formula for $f(x)$.

   This results in $y = f(x + 6) = (x + 6)^2 - 4(x + 6) + 1$. To shift the graph of this new equation up 4 units,

   add 4 to the formula to obtain $y = f(x + 6) + 4 = (x + 6)^2 - 4(x + 6) + 1 + 4 = (x + 6)^2 - 4(x + 6) + 5$.

   See Figure 31.

33. (a) $g(x) = 3(x + 3)^2 + 2(x + 3) - 5$

   (b) $g(x) = 3x^2 + 2x - 9$

35. (a) $g(x) = -(x - 2)^3 + 2(x - 2)^2 - 4(x - 2) + 5$

   (b) $g(x) = -(x + 8)^3 + 2(x + 8)^2 - 4(x + 8) - 4$

37. (a) $g(x) = -(2(x - 2)^2 - 3(x - 2) + 1)$

   (b) $g(x) = 2(-(x + 4))^2 - 3(-(x + 4)) + 1$

39. (a) $g(x) = 10(x - 2000)^2 + 80$

    (b) $g(x) = (-x + 20)^2$

41. Shift the graph of $y = x^2$ down 3 units. The vertex is located at $(0, -3)$. See Figure 41.

43. Shift the graph of $y = x^2$ to the left 4 units. The vertex is located at $(-4, 0)$. See Figure 43.

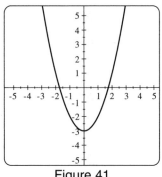

Figure 41

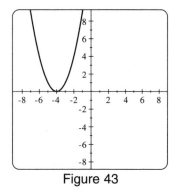

Figure 43

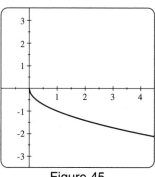

Figure 45

45. Reflect the graph of $y = \sqrt{x}$ across the *x*-axis. See Figure 45.

47. Reflect the graph of $y = x^2$ across the *x*-axis and shift the graph 4 units up. See Figure 47.

49. Shift the graph of $y = \sqrt{x}$ to the left 1 unit. See Figure 49.

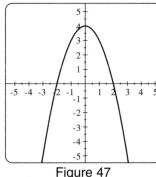

Figure 47

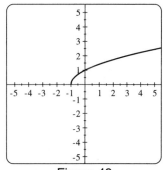

Figure 49

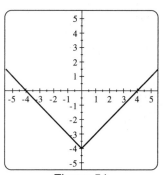

Figure 51

51. Reflect the graph of $y = |x|$ down 4 units. See Figure 51.

53. Reflect the graph of $y = \sqrt{x}$ to the right 3 units and up 2 units. See Figure 53.

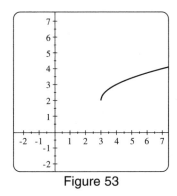

Figure 53

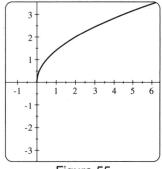

Figure 55

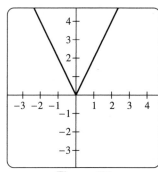

Figure 57

55. Stretch the curve of $y = \sqrt{x}$ vertically since $y = \sqrt{2x} \Rightarrow y = \sqrt{2} \cdot \sqrt{x} \Rightarrow y \approx 1.41\sqrt{x}$. See Figure 55.

57. Like $f(x) = |x|$ but narrower by a factor of 2, vertex $(0, 0)$. See Figure 57.

59. Reflect the graph of $f(x) = \sqrt{x}$ across the x-axis and then shift up 1 unit, the vertex is $(0, 1)$.

   See Figure 59.

61. Shift the graph of $f(x) = \sqrt{x}$ left 1 unit, then reflect this graph across the x-axis and then reflect this graph

   across the y-axis. See Figure 61.

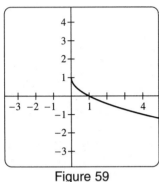

Figure 59

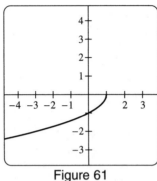

Figure 61

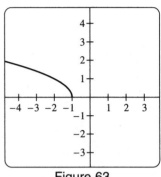

Figure 63

63. Shift the graph of $f(x) = \sqrt{x}$ right 1 unit, then reflect this graph across the y-axis. See Figure 63.

65. Shift the graph of $f(x) = x^3$ right 1 unit. See Figure 65.

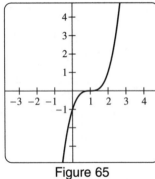

Figure 65

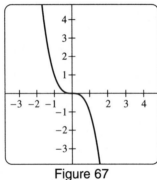

Figure 67

67. Reflect the graph of $f(x) = x^3$ across the x-axis. See Figure 67.

69. The graph of $f(x)$ is shifted up 2 units $\Rightarrow$ $(-12, 8)$, $(0, 10)$, and $(8, -2)$.

71. The graph of $f(x)$ is shifted right 2 units and up 1 unit $\Rightarrow$ $(-10, 7)$, $(2, 9)$, and $(10, -3)$.

73. The graph is reflected across the x-axis $\Rightarrow$ the y-coordinate is $-\dfrac{1}{2}$ times the given y-coordinate $\Rightarrow$

   $(-12, -3)$, $(0, -4)$, and $(8, 2)$.

75. The graph is reflected across the y-axis $\Rightarrow$ the x-coordinate is the reciprocal of -2 or $-\dfrac{1}{2}$ times the given

   x-coordinate $\Rightarrow$ $(6, 6)$, $(0, 8)$, and $(-4, -4)$.

77. To reflect this function across the *x*-axis, graph $y = -f(x) = -(x^2 - 2x - 3) = -x^2 + 2x + 3$. See Figure 77a.

To reflect this function across the *y*-axis, graph $y = f(-x) = (-x)^2 - 2(-x) - 3 = x^2 + 2x - 3$. See Figure 77b.

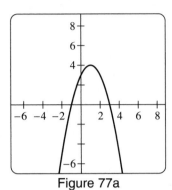
Figure 77a

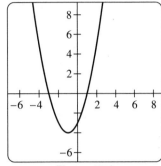
Figure 77b

79. To reflect this function across the *x*-axis, graph $y = -f(x) = -(2x^2 + x - 1) = -2x^2 - x + 1$. See Figure 79a.

To reflect this function across the *y*-axis, graph $y = f(-x) = 2(-x)^2 + (-x) - 1 = 2x^2 - x - 1$. See Figure 79b.

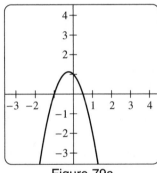
Figure 79a

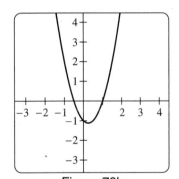
Figure 79b

81. To reflect this function across the *x*-axis, graph $y = -f(x) = -(|x + 1| - 1) = -|x + 1| + 1$. See Figure 81a.

To reflect this function across the *y*-axis, graph $y = f(-x) = |-x + 1| - 1 = |-x + 1| - 1$. See Figure 81b.

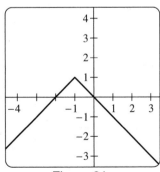
Figure 81a

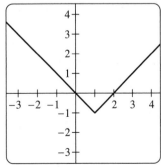
Figure 81b

83. (a) To reflect this line graph across the *x*-axis, make a table of values for $y = -f(x)$, that is, change each (x, y) point to $(x, -y)$. Plot the points $(-3, -2)$, $(-1, -3)$, $(1, 1)$, and $(2, 2)$. See Figure 83a.

    (b) To reflect this line graph across the *y*-axis, make a table of values for $y = f(-x)$, that is, change each (x, y) point to $(-x, y)$. Plot the points $(3, 2)$, $(1, 3)$, $(-1, -1)$, and $(-2, -2)$. See Figure 83b.

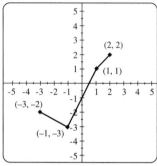
Figure 83a

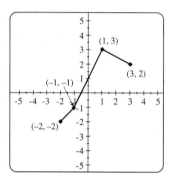
Figure 83b

Transforming Numerical Representations

85. Since $g(x) = f(x) + 7$, the output for $g(x)$ can be determined by adding 7 to the output of $f(x)$. See Figure 85.

x	1	2	3	4	5
$f(x)$	5	1	6	2	7
$g(x)$	12	8	13	9	14

Figure 85

87. Since $g(x) = f(x - 2)$, the output for $g(x)$ can be determined by translating the output of $f(x)$ right two units with respect to the *x*-values. For example, $g(4) = f(4 - 2) = f(2) = -5$ and $g(2) = f(2 - 2) = f(0) = -3$. However, the value of $g(-4) = f(-4 - 2) = f(-6)$ is undefined because -6 is not in the domain of f. See Figure 87.

x	-2	0	2	4	6
$f(x)$	2	-3	-5	-9	—
$g(x)$	5	2	-3	-5	-9

Figure 87

89. Since $g(x) = f(x + 1) - 2$, the output for $g(x)$ can be determined by translating the output of $f(x)$ left one unit with respect to the *x*-values and subtracting 2. For example,
$g(1) = f(1 + 1) - 2 = f(2) - 2 = 4 - 2 = 2$ and $g(2) = f(2 + 1) - 2 = f(3) - 2 = 3 - 2 = 1$.
However, the value of $g(5) = f(5 + 1) - 2 = f(6) - 2$ is undefined because 6 is not in the domain of f. See Figure 89.

x	0	1	2	3	4
$f(x)$	—	2	4	3	7
$g(x)$	0	2	1	5	6

Figure 89

91. Since $g(x) = f(-x) + 1$, the output for $g(x)$ can be determined by translating the output of $f(x)$ reflected across

the y-axis and adding 1. For example, $g(-2) = f(-(-2)) + 1 = f(2) + 1 = -1 + 1 = 0$ and

$g(-1) = f(-(-1)) + 1 = f(1) + 1 = 2 + 1 = 3$. See Figure 91.

x	-2	-1	0	1	2
$f(x)$	11	8	5	2	-1
$g(x)$	0	2	6	9	12

Figure 91

Applications

93. Let $(1997, 3.8)$ be the vertex (h, k). Translate the graph of $y = x^2$ right 1997 units and upward 3.8 units. To

determine a value for a, graph the data and $y = a(x - 1997)^2 + 3.8$ for different values of a. By trial and error

a value of $a = \dfrac{1}{4}$ appears to fit the data. Thus $f(x) = \dfrac{1}{4}(x - 1997)^2 + 3.8$.

95. For $g(x)$ change x to $(x - 1984) \Rightarrow g(x) = 2375(x - 1984)^2 + 5134(x - 1984) + 5020$

97. (a) When $x = 2$, the y-value is about 9. There are 9 hours of daylight on February 21 at 60°N latitude.

(b) Since February 21 is 2 months after the shortest day and October 21 is 2 months before the shortest day,

the days should have about the same number of daylight hours. A reasonable conjecture would be 9 hours.

(c) The amount of daylight would increase to the left or right of $x = 0$. There are approximately the same

number of daylight hours either x months before or after December 21. the graph should be symmetric

about the y-axis, as shown in Figure 97.

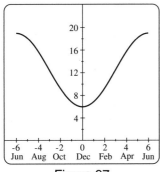

Figure 97

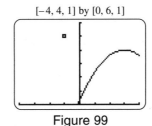

$[-4, 4, 1]$ by $[0, 6, 1]$

Figure 99

Using Transformations to Represent Data Dynamically

99. In 15 seconds, the plane has moved $15(0.2) = 3$ kilometers to the left. To show this movement, translate the

graph of the mountain 3 kilometers (units) to the right; $y = -0.4(x - 3)^2 + 4$. See Figure 99.

101. (a) In 4 hours, the cold front has moved $4(40) = 160$ miles, which is $\dfrac{160}{100} = 1.6$ units on the graph. To show

this movement, shift the graph 1.6 units down; $y = \dfrac{1}{20}x^2 - 1.6$. See Figure 101a.

(b) The graph of the cold front has shifted $\dfrac{250}{100} = 2.5$ units down and $\dfrac{210}{100} = 2.1$ units to the right. To show

this movement, the new equation should be $y = \dfrac{1}{20}(x - 2.1)^2 - 2.5$. The location of Columbus, Ohio is

at $(5.5, -0.8)$ relative to the location of Des Moines at $(0, 0)$. Since $(5.5, -0.8)$ is within the parabola

representing the cold front at midnight, the front reaches Columbus by midnight. See Figure 101b.

$[-15, 15, 1]$ by $[-10, 10, 1]$ $[-15, 15, 1]$ by $[-10, 10, 1]$

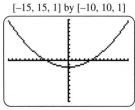   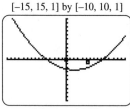

Figure 101a   Figure 101b

Chapter 3 Review Exercises

1. Leading coefficient: negative; vertex: $(2, 4)$; axis of symmetry: $x = 2$

3. $f(x) = -2(x - 5)^2 + 1 \Rightarrow f(x) = -2(x^2 - 10x + 25) + 1 \Rightarrow f(x) = -2x^2 + 20x - 49$

The leading coefficient is -2.

5. $f(x) = x^2 + 6x - 1 \Rightarrow f(x) = x^2 + 6x + 9 - 9 - 1 \Rightarrow f(x) = (x^2 + 6x + 9) - 10 \Rightarrow$

$f(x) = (x + 3)^2 - 10$; the vertex is $(-3, -10)$.

7. $-\dfrac{b}{2a} = -\dfrac{2}{2(-3)} = \dfrac{1}{3}$ and $f\left(\dfrac{1}{3}\right) = -3\left(\dfrac{1}{3}\right)^2 + 2\left(\dfrac{1}{3}\right) - 4 = -\dfrac{11}{3}$; the vertex is $\left(\dfrac{1}{3}, -\dfrac{11}{3}\right)$.

9. The graph $y = x^2$ has been shifted 1 unit left, 2 units up, and reflected through the x-axis $\Rightarrow$

$f(x) = -(x + 1)^2 + 2$.

11. Shift the graph of $f(x) = x^2$ up 3 units, reflect it across the x-axis, and make it narrower by the scale factor 3.

The vertex is located at $(0, 3)$. See Figure 11.

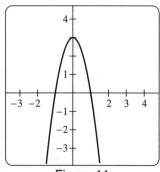   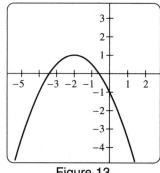   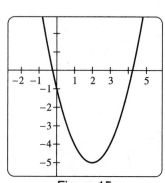

Figure 11   Figure 13   Figure 15

13. Shift the graph of $g(x) = x^2$ left 2 units, up 1 unit, reflect it across the x-axis, and make it 2 times wider $\left(\text{the reciprocal of } \dfrac{1}{2}\right)$. The vertex is located at $(-2, 1)$. See Figure 13.

15. $h(t) = t^2 - 4t - 1 \Rightarrow h(t) = t^2 - 4t + 4 - 5 \Rightarrow h(t) = (t - 2)^2 - 5$. Shift the graph of $h(t) = t^2$ right 2 units, and down 5 units. The vertex is $(2, -5)$. See Figure 15.

17. Shift the graph of $f(x) = |x|$ left 3 units, and reflect it across the x-axis. The vertex is located at $(-3, 0)$. See Figure 17.

19. $g(t) = t^2 - 2t \Rightarrow g(t) = t^2 - 2t + 1 - 1 \Rightarrow g(t) = (t - 1)^2 - 1$. Shift the graph of $g(t) = t^2$ right 1 unit, and down 1 unit. The vertex is $(1, -1)$. See Figure 19.

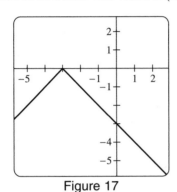

Figure 17

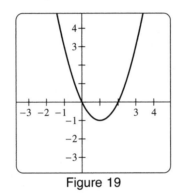

Figure 19

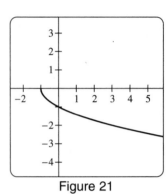

Figure 21

21 Shift the graph of $f(x) = \sqrt{x}$ left 1 unit and reflect across the x-axis. The vertex is $(-1, 0)$. See Figure 21.

23. (a) Since the parabola is opening upward, $a > 0$.

(b) The real solutions to $ax^2 + bx + c = 0$ appear to be $x = -3$ or $x = 2$, the x-intercepts.

(c) The discriminant must be positive since $ax^2 + bx + c = 0$ has two real solutions.

25. $x^2 - x - 20 = 0 \Rightarrow (x - 5)(x + 4) = 0 \Rightarrow x = -4, 5$

27. $x^2 = 4x \Rightarrow x^2 - 4x = 0 \Rightarrow x(x - 4) = 0 \Rightarrow x = 0, 4$

29. $4z^2 - 7 = 0 \Rightarrow 4z^2 = 7 \Rightarrow z^2 = \dfrac{7}{4} \Rightarrow z = \pm\sqrt{\dfrac{7}{4}} \Rightarrow z = \pm\dfrac{\sqrt{7}}{2}$

31. $-2t^2 - 3t + 14 = 0 \Rightarrow (2t + 7)(-t + 2) = 0 \Rightarrow 2t + 7 = 0 \Rightarrow 2t = -7 \Rightarrow t = \dfrac{-7}{2}$ or $-t + 2 = 0 \Rightarrow$

$-t = -2 \Rightarrow t = 2 \Rightarrow t = \dfrac{-7}{2}, 2$

33. $0.1x^2 - 0.3x = 1 \Rightarrow x^2 - 3x - 10 = 0 \Rightarrow (x - 5)(x + 2) = 0 \Rightarrow x = -2, 5$

35. $(k - 1)^2 = \dfrac{9}{4} \Rightarrow k^2 - 2k + 1 = \dfrac{9}{4} \Rightarrow k^2 - 2k - \dfrac{5}{4} = 0 \Rightarrow 4k^2 - 8k - 5 = 0 \Rightarrow$

$(2k - 5)(2k + 1) = 0 \Rightarrow 2k - 5 = 0 \Rightarrow 2k = 5 \Rightarrow k = \dfrac{5}{2}$ or $2k + 1 = 0 \Rightarrow 2k = -1 \Rightarrow k = -\dfrac{1}{2} \Rightarrow$

$k = -\dfrac{1}{2}, \dfrac{5}{2}$

37. $x^2 + 2x = 5 \Rightarrow x^2 + 2x + 1 = 5 + 1 \Rightarrow (x + 1)^2 = 6 \Rightarrow x + 1 = \pm\sqrt{6} \Rightarrow x = -1 \pm \sqrt{6}$

39. $2z^2 - 6z - 1 = 0 \Rightarrow z^2 - 3z = \dfrac{1}{2} \Rightarrow z^2 - 3z + \dfrac{9}{4} = \dfrac{1}{2} + \dfrac{9}{4} \Rightarrow \left(z - \dfrac{3}{2}\right)^2 = \dfrac{11}{4} \Rightarrow$

$z - \dfrac{3}{2} = \pm\sqrt{\dfrac{11}{4}} \Rightarrow z = \dfrac{3}{2} \pm \dfrac{\sqrt{11}}{2} \Rightarrow z = \dfrac{3 \pm \sqrt{11}}{2}$

41. $\frac{1}{2}x^2 - 4x + 1 = 0 \Rightarrow x^2 - 8x + 2 = 0 \Rightarrow x^2 - 8x + 16 = -2 + 16 \Rightarrow (x - 4)^2 = 14 \Rightarrow$

    $x - 4 = \pm\sqrt{14} \Rightarrow x = 4 \pm \sqrt{14}$

43. $2x^2 - 3y^2 = 6 \Rightarrow -3y^2 = -2x^2 + 6 \Rightarrow y^2 = \dfrac{2x^2 - 6}{3} \Rightarrow y = \pm\sqrt{\dfrac{2x^2 - 6}{3}}$

    Since there are two solutions for each input, y is not a function of x.

45. (a) $x^2 - 3x + 2 = 0 \Rightarrow (x - 2)(x - 1) = 0 \Rightarrow x = 1, 2$. Using $x = \dfrac{3}{2}$ which is in between 1 and 2 we get:

    $\left(\dfrac{3}{2}\right)^2 - 3\left(\dfrac{3}{2}\right) + 2 = \dfrac{9}{4} - \dfrac{9}{2} + 2 = \dfrac{9}{4} - \dfrac{18}{4} + \dfrac{8}{4} = -\dfrac{1}{4}$ which is less than 0.

    (b) $x^2 - 3x + 2 < 0$ is $(1, 2)$

    (c) $x^2 - 3x + 2 > 0$ is $(-\infty, 1) \cup (2, \infty)$

47. First solve $x^2 - 3x + 2 = 0 \Rightarrow (x - 1)(x - 2) = 0 \Rightarrow x = 1, 2$. Now test $x = \dfrac{3}{2}$ which is inbetween

    1 and 2. $\left(\dfrac{3}{2}\right)^2 - 3\left(\dfrac{3}{2}\right) + 2 = \dfrac{9}{4} - \dfrac{9}{2} + 2 = \dfrac{9}{4} - \dfrac{18}{4} + \dfrac{8}{4} = -\dfrac{1}{4}$ which is less than 0 $\Rightarrow$

    For $x^2 - 3x + 2 \leq 0$ we get $[1, 2]$.

49. First solve $10x^2 + 19x + 5 = 0$. Using the quadratic formula we get:

    $\dfrac{-19 \pm \sqrt{19^2 - 4(10)(5)}}{2(10)} = \dfrac{-19 \pm \sqrt{161}}{20}$. Now test $x = -1$ which is in between $\dfrac{-19 - \sqrt{161}}{20}$ and

    $\dfrac{-19 + \sqrt{161}}{20}$. $10(-1)^2 + 19(-1) + 5 = 10 - 19 + 5 = -4$ which is less than 0 $\Rightarrow$

    or $10x^2 + 19x + 5 < 0$ we get $\left(\dfrac{-19 - \sqrt{161}}{20}, \dfrac{-19 + \sqrt{161}}{20}\right)$

51. First solve $n(n - 2) = 15 \Rightarrow n^2 - 2n - 15 = 0 \Rightarrow (n - 5)(n + 3) = 0 \Rightarrow n = -3, 5$.

    Now test $x = 0$ which is in between -3 and 5. $(0)^2 - 2(0) - 15 = -15$ which is less than 0 $\Rightarrow$

    For $n(n - 2) \geq 15$ we get $(-\infty, -3] \cup [5, \infty)$.

53. (a) The graph of $y = f(x)$ has x-intercepts at $x = -3, 2$. It is above the x-axis between these values, which are

    solutions to $f(x) = 0$, so $f(x) > 0$ when $-3 < x < 2$.

    (b) The graph of $y = f(x)$ is below or intersects the x-axis outside of and including the values where $f(x) = 0$,

    so $f(x) \leq 0$ when $x \leq -3$ or $x \geq 2$.

55. $y = -f(x) = -(2x^2 - 3x + 1) = -2x^2 + 3x - 1$. See Figure 55a.

    $y = f(-x) = 2(-x)^2 - 3(-x) + 1 = 2x^2 + 3x + 1$. See Figure 55b.

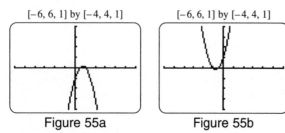

[-6, 6, 1] by [-4, 4, 1]   [-6, 6, 1] by [-4, 4, 1]

Figure 55a   Figure 55b

57. Figure 57 shows the graph of of $y = x^2$ (dashed) and the graph of of $y = x^2 - 4$ (solid).

59. Figure 59 shows the graph of of $y = x^2$ (dashed) and the graph of of $y = (x + 4)^2 - 2$ (solid).

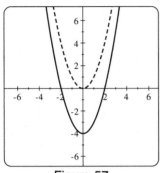

Figure 57

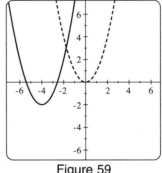

Figure 59

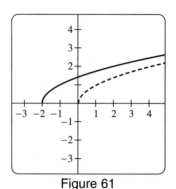

Figure 61

61. Figure 61 shows the graph of of $y = \sqrt{x}$ (dashed) and the graph of of $y = \sqrt{x} + 2$ (solid).

63. Figure 63 shows the graph of of $y = x^2$ (dashed) and the graph of of $y = -2(x - 2)^2 + 3$ (solid).

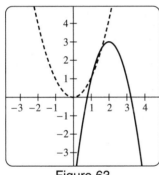

Figure 63

x	1	2	3	4
$f(x)$	−3	3	4	7
$g(x)$	−1	5	6	9

Figure 65

65. To obtain $g(x) = f(x) + 2$, add 2 to each $f(x)$ value. See Figure 65.

67. Since the axis of symmetry is $x = 2$, the point $(2, 3)$ is the vertex of the parabola. Because $(4, -1)$ is a point on

the parabola and $-1 < 3$, the parabola opens downward. $f(x) = a(x - h)^2 + k \Rightarrow f(x) = a(x - 2)^2 + 3$;

$f(4) = -1 \Rightarrow -1 = a(4 - 2)^2 + 3 \Rightarrow a = -1$; thus $f(x) = -(x - 2)^2 + 3$. Therefore a formula for the

function is $f(x) = -(x - 2)^2 + 3$. See Figure 67.

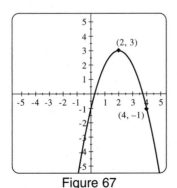

Figure 67

Applications

69. Let $W + L + W = 44 \Rightarrow L = 44 - 2W$. If $A = L \cdot W$ then $A = (44 - 2W)(W) \Rightarrow A = 44W - 2W^2$ or

$A = -2W^2 + 44W$. Now using the vertex formula: $W = -\dfrac{44}{2(-2)} \Rightarrow W = 11$ and $L = 44 - 2(11) \Rightarrow$

$L = 22$. So the dimensions are 11 feet by 22 feet.

71. (a) $h(t) = -16t^2 + 88t + 5 \Rightarrow h(0) = -16(0)^2 + 88(0) + 5 \Rightarrow h(0) = 5$

The stone was held 5 feet above the ground when it was released.

(b) $h(2) = -16(2)^2 + 88(2) + 5 = 117$; the stone was 117 feet high after 2 seconds.

(c) $t = -\dfrac{b}{2a} = -\dfrac{88}{2(-16)} = 2.75$ and $h(2.75) = -16(2.75)^2 + 88(2.75) + 5 = 126$; the maximum height of the

stone was 126 feet.

(d) $h(t) = 117 \Rightarrow 117 = -16t^2 + 88t + 5 \Rightarrow 16t^2 - 88t + 112 = 0 \Rightarrow$

$t = \dfrac{88 \pm \sqrt{(-88)^2 - 4(16)(112)}}{2(16)} \Rightarrow t = 3.5 \text{ or } t = 2$

The stone was 117 feet high after 2 seconds and after 3.5 seconds.

73. (a) $f(1985) = 0.000478(1985)^2 - 1.813(1985) + 1720.1 \approx 4.7$; in 1985 there were about 4.7 billion people.

(b) $f(2000) = 0.000478(2000)^2 - 1.813(2000) + 1720.1 \approx 6.1$; in 2000 there were about 6.1 billion people.

(c) Graph $Y_1 = 0.000478X^2 - 1.813X + 1720.1$ and $Y_2 = 7$; in [1980, 2030, 20] by [5, 9, 1], (not shown).

The intersection point is approximately (2008.7235, 7). The model predicts that the world population may

reach 7 billion in the year 2009.

75. (a) $C(x) = x(103 - 3x)$

(b) $C(6) = 6(103 - 3(6)) = 510$; the cost of renting 6 rooms is $510.

(c) $730 = x(103 - 3x) \Rightarrow 730 = 103x - 3x^2 \Rightarrow 3x^2 - 103x + 730 = 0 \Rightarrow (3x - 73)(x - 10) = 0 \Rightarrow$

$x = \dfrac{73}{3}$ or $x = 10$. Since $x = \dfrac{73}{3}$ has no physical meaning for the number of rooms, $x = 10$ rooms.

(d) $C(x) = 103x - 3x^2 \Rightarrow x = -\dfrac{b}{2a} = -\dfrac{103}{2(-3)} \approx 17$; the maximum cost occurs when $x = 17$ rooms.

The greatest cost is $C(17) = 103(17) - 3(17)^2 = \$884..$

77. (a) The change in debt is not constant for each 4-year period.

(b) Using (1980, 82) as the vertex, let $h = 1980$ and $k = 82$ in $f(x) = a(x - h)^2 + k$. Then the function can be

written $f(x) = a(x - 1980)^2 + 82$. To find the value of a, note that $f(x) = 444$ when $x = 1996$.

By substitution, $444 = a(1996 - 1980)^2 + 82 \Rightarrow 362 = 256a \Rightarrow a = \dfrac{362}{256} \approx 1.4$. When the data is

plotted along with the graph of $f(x) = 1.4(x - 1980)^2 + 82$, it may be reasonable to adjust the value of a.

A good fit appears to be obtained when $f(x) = 1.3(x - 1980)^2 + 82$. To solve $f(x) = 212$, substitute 212

for $f(x)$. $212 = 1.3(x - 1980)^2 + 82 \Rightarrow 130 = 1.3(x - 1980)^2 \Rightarrow 100 = (x - 1980)^2 \Rightarrow$

$\sqrt{100} = x - 1980 \Rightarrow x = 1980 + \sqrt{100} = 1990$. Note that only the positive square root of 100 is

necessary for dates after 1980. Visa and Mastercard debt reached $212 billion in 1990.

Chapter 4: Nonlinear Functions and Equations

4.1: Nonlinear Functions and Their Graphs

Polynomials

1. $f(x) = 2x^3 - x + 5$ is a polynomial; degree: 3; leading coefficient: 2.

3. $f(x) = \sqrt{x}$ is not a polynomial.

5. $f(x) = 1 - 4x - 5x^4$ is a polynomial; degree: 4; leading coefficient: –5.

7. $g(t) = \dfrac{1}{t^2 + 3t - 1}$ is not a polynomial.

9. $g(t) = 22$ is a polynomial; degree: 0; leading coefficient: 22.

Increasing and Decreasing Functions

11. f is increasing on $[2, \infty)$ and decreasing on $(-\infty, 2]$.

13. f is increasing on $(-\infty, -2] \cup [1, \infty)$ and decreasing on $[-2, 1]$.

15. f is increasing on $[-8, 0] \cup [8, \infty)$ and decreasing on $(-\infty, -8] \cup [0, 8]$.

17. When x increases, $f(x)$ remains constant at $(-3) \Rightarrow$ the function is neither increasing or decreasing.

19. The graph of this equation is linear with a slope of $2 \Rightarrow$ it is increasing: $(-\infty, \infty)$ and decreasing: never.

21. The graph of this equation is a parabola with a vertex $(0, -2) \Rightarrow$ it is increasing: $[0, \infty)$ and decreasing: $(-\infty, 0)$.

23. The graph of this equation is a parabola with a vertex $(1, 1)$, because of the negative coefficient before x^2 it opens downward $\Rightarrow$ it is increasing: $(-\infty, 1]$ and decreasing: $[1, \infty)$.

25. The graph of this equation is a parabola with a vertex $(-1, -5) \Rightarrow$ it is increasing: $[-1, \infty)$ and decreasing: $(-\infty, -1]$.

27. The square root equation graph has a starting point $(1, 0)$, all x-values less than 1 are undefined $\Rightarrow$ it is increasing: $[1, \infty)$ and decreasing: never.

29. The absolute value graph has a vertex $(-3, 0) \Rightarrow$ it is increasing: $[-3, \infty)$ and decreasing: $(-\infty, -3]$.

31. The basic x^3 function is always increasing $\Rightarrow$ it is increaing: $(-\infty, \infty)$ and decreasing: never.

33. The graph of this cubic equation has turning points $\left(-2, \dfrac{16}{3}\right)$ and $\left(2, \dfrac{32}{3}\right) \Rightarrow$ it is increasing: $(-\infty, -2]$ or $[2, \infty)$; and decreasing: $[-2, 2]$.

35. The graph of this cubic equation has turning points $(-2, 20)$ and $(1, -7) \Rightarrow$ it is increasing: $(-\infty, -2]$ or $[1, \infty)$; and decreasing: $[-2, 1]$.

37. The graph of this x^4 graph has a negative lead coefficient therefore is reflected through the x-axis has turning points $\left(-1, \dfrac{5}{12}\right)$, $(0, 0)$ and $\left(2, \dfrac{8}{3}\right) \Rightarrow$ it is increasing: $(-\infty, -1]$ or $[0, 2]$; and decreasing: $[-1, 0]$ or $[2, \infty)$.

39. (a) A local maximum of approximately 5.5 occurs when $x \approx -2$. A local minimum of approximately -5.5 occurs when $x \approx 2$. *Answers may vary slightly.*

    (b) There are no absolute extrema.

41. (a) Local maxima of approximately 17 and 27 occur when $x \approx -3$ and 2, respectively. Local minima of approximately -10 and 24 occur when $x \approx -1$ and 3, respectively. *Answers may vary slightly.*

    (b) There are no absolute extrema.

43. (a) A local minimum of approximately -3.2 occurs when $x \approx 0.5$. There are no local maxima. *Answers may vary slightly.*

    (b) The absolute minimum is approximately -3.2. The absolute maximum is 3 and occurs when $x = -2$.

45. (a) Local minima are all -1. Local maxima are 1.

    (b) The absolute minimum is -1 and the absolute maximum is 1.

47. (a) The local minimum is -2. Local maxima are 1 and 2.

    (b) The absolute minimum is -2 and the absolute maximum is 2.

49. (a) The graph of g is linear $\Rightarrow$ no local extrema.

    (b) The graph of g is linear $\Rightarrow$ no absolute extrema.

51. (a) The graph of g is a parabola opening up with a vertex $(0, 1) \Rightarrow$ it has a local minimum: 1; and no local maximum.

    (b) The graph of g is a parabola opening up with a vertex $(0, 1) \Rightarrow$ it has an absolute minimum: 1; and no absolute maximum.

53. (a) The graph of g is a parabola opening down with a vertex $(-3, 4) \Rightarrow$ it has a local maximum: 4; and no local minimum.

    (b) The graph of g is a parabola opening down with a vertex $(-3, 4) \Rightarrow$ it has an absolute maximum: 4; and no absolute minimum.

55. (a) The graph of g is a parabola opening up with a vertex $\left(\frac{3}{4}, -\frac{1}{8}\right) \Rightarrow$ it has a local minimum: $-\frac{1}{8}$; and no local maximum.

    (b) The graph of g is a parabola opening up with a vertex $\left(\frac{3}{4}, -\frac{1}{8}\right) \Rightarrow$ it has an absolute minimum: $-\frac{1}{8}$; and no absolute maximum.

57. (a) The absolute value graph opening up has a vertex $(-3, 0) \Rightarrow$ it has a local minimum: 0; and no local maximum.

    (b) The absolute value graph opening up has a vertex $(-3, 0) \Rightarrow$ it has an absolute minimum: 0; and no absolute maximum.

59. (a) The graph of g is the basic $\sqrt[3]{x}$ graph and has no local extrema.

    (b) The graph of g is the basic $\sqrt[3]{x}$ graph and has no absolute extrema.

61. (a) The cubic function graph of g has turning points $(-1, -2)$ and $(1, 2) \Rightarrow$ it has a local minimum: -2; and a local maximum: 2.

    (b) The cubic function graph of g has turning points $(-1, -2)$ and $(1, 2)$ but then continues on to $-\infty$ and $\infty \Rightarrow$ it has no absolute extrema.

63. (a) A graph of $f(x) = -3x^4 + 8x^3 + 6x^2 - 24x$ is shown in Figure 63. Local maxima of 19 and -8 occur at $x \approx -1$ and $x \approx 2$ respectively. A local minima of -13 occurs when $x \approx 1$.

    (b) The absolute maximum of 19 occurs at $x \approx -1$, and there is no absolute minimum.

65. (a) A graph of $f(x) = 0.5x^4 - 5x^2 + 4.5$ is shown in Figure 65. A local maximum of 4.5 occurs at $x = 0$. A local minimum of -8 occurs when $x \approx \pm 2.236$.

    (b) There is no absolute maximum. An absolute minimum of -8 occurs at $x \approx \pm 2.236$.

[-5, 5, 1] by [-30, 30, 5]   [-10, 10, 1] by [-10, 10, 1]   [-10, 10, 1] by [-10, 10, 1]

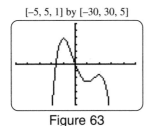
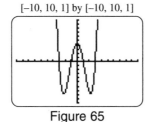
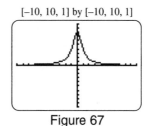

Figure 63   Figure 65   Figure 67

67. (a) A graph of $f(x) = \dfrac{8}{1 + x^2}$ is shown in Figure 67. A local maximum of 8 occurs at $x = 0$. There are no local minima.

    (b) An absolute maximum of 8 occurs at $x = 0$. There is no absolute minimum.

Symmetry

69. From the graph we can see that the function f is odd rather than even. Since $(-8, 6)$ is located on the graph of f, the point $(8, -6)$ must also be located on the graph.

71. The function f is a monomial with only an odd power $\Rightarrow$ it is odd.

73. The function f is a polynomial with an odd power and a constant term, which is an even power $\Rightarrow$ it is neither.

75. The function f is a polynomial with an even power and a constant term, which is also even $\Rightarrow$ it is even.

77. The function f is a polynomial with all even powers and a constant term, which is also even $\Rightarrow$ it is even.

79. The function f is a polynomial with all odd powers $\Rightarrow$ it is odd.

81. The function f is a polynomial with both an odd powered term and an even powered term $\Rightarrow$ it is neither.

83. If $f(x) = \sqrt[3]{x^2}$, then $f(-x) = \sqrt[3]{(-x)^2} = \sqrt[3]{x^2}$, which is equal to $f(x) \Rightarrow$ it is even.

85. If $f(x) = \sqrt{1 - x^2}$, then $f(-x) = \sqrt{1-(-x)^2} = \sqrt{1 - x^2}$, which is equal to $f(x) \Rightarrow$ it is even.

87. If $f(x) = \dfrac{1}{1 + x^2}$, then $f(-x) = \dfrac{1}{1 + (-x)^2} = \dfrac{1}{1 + x^2}$, which is equal to $f(x) \Rightarrow$ it is even.

89. If $f(x) = |x + 2|$, then $f(-x) = |-x + 2|$, and $-f(x) = -|x + 2|$. Since $f(x) \neq f(-x)$ and $f(-x) \neq -f(x)$ It is neither.

91. Notice that $f(-100) = 56$ and $f(100) = -56$. In general, since $f(-x) = -f(x)$ holds for each x in the table, f is odd.

93. Since f is even, the condition $f(-x) = f(x)$ must hold. For example, since $f(-3) = 21$, $f(3)$ must also equal 21. The value assigned to $f(0)$ makes no difference. See Figure 93.

x	-3	-2	-1	0	1	2	3
$f(x)$	21	-12	-25	1	-25	-12	21

Figure 93

95. There are many solutions. Figure 95 shows one possible solution. This is not odd since $f(0) \neq 0$ and it is not even since $f(-2) \neq f(2)$.

x	-2	-1	0	1	2
$f(x)$	5	1	2	3	4

Figure 95

97. Since f is odd, start by plotting the points $(-2, 0)$, $(-1, 1)$, $(0, 0)$, $(1, -1)$ and $(2, 0)$. One can then make a smooth curve as shown in Figure 97.

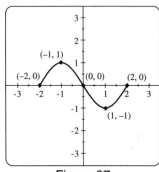
Figure 97

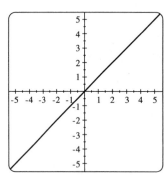
Figure 99

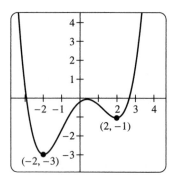
Figure 103

Concepts

99. Sketch any linear function that passes through the origin will work. For example, the graph of $y = x$ is shown in Figure 99.

101. No. If $(2, 5)$ is on the graph of an odd function f, then so is $(-2, -5)$. Since f would pass through $(-3, -4)$ and then $(-2, -5)$, it could not always be increasing.

103. Plot points $(-2, -3)$ and $(2, -1)$ and make them minima. See Figure 103.

105. Plot point $(2, 3)$ and make it a maxima. See Figure 105. Yes, it is a parabola, thus is quadratic.

Translation of Graphs

107. Shift the graph of $y = f(x) = 4x - \dfrac{x^3}{3}$ to the left 1 unit to get the graph of

$$y = f(x + 1) = 4(x + 1) - \frac{(x + 1)^3}{3}.$$ See Figure 107.

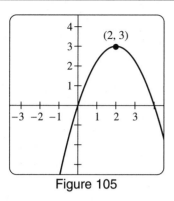
Figure 105

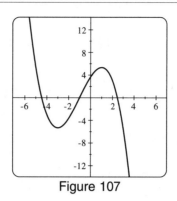
Figure 107

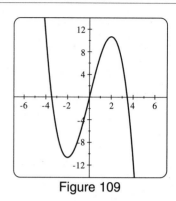
Figure 109

109. Sketch the graph of $y = f(x) = 4x - \dfrac{x^3}{3}$ vertically by multiplying each y-value by 2, that is, each (x, y) point

becomes $(x, 2y)$, to get the graph of $y = 2f(x) = 2\left(4x - \dfrac{x^3}{3}\right) = 8x - \dfrac{2}{3}x^3$. See Figure 109.

111. The graph is transformed 2 units left and 1 unit down $\Rightarrow$ $(-4, 2)$ lies on the graph of $y = -f(x + 2) - 1$.

113. The first equation graph is transformed 1 unit left $\Rightarrow$ $y = f(x + 1) - 2$ increases on $[0, 3]$. The second
equation graph is transformed across the x-axis and transformed 2 units right $\Rightarrow$ $y = -f(x - 2)$ decreases on
$[3, 6]$.

Applications

115. (a) The tides are rising when $[0, 2.2]$; $[8.4, 14.5]$; $[20.7, 26.8]$.

   (b) All local maximum points have $y = 4.2$, meaning that high tide is 4.2 feet; all local minimum have
   $y = 1.8$, meaning that low tide is 1.8 feet.

117. (a) $f(5) = -0.00113(5)^3 + 0.0408(5)^2 - 0.432(5) + 7.66 \approx 8.32$; in 1955 U.S. consumption of energy was
   about 8.32 quadrillion Btu.

   (b) The graph of f is shown in Figure 117b; the energy consumption increased, reached a maximum value,
   and the started to decrease.

   (c) The local maximum of approximately 14.5 when $x \approx 23.5$. In 1973 energy use peaked at 14.5 quadrillion
   Btu. See Figure 117c.

[–5, 5, 1] by [–30, 30, 5]

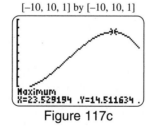
Figure 117b

[–10, 10, 1] by [–10, 10, 1]

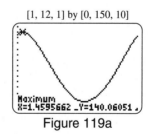
Figure 117c

[1, 12, 1] by [0, 150, 10]

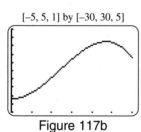
Figure 119a

[1, 12, 1] by [0, 150, 10]

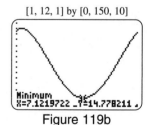
Figure 119b

119. (a) One might expect an absolute maximum to occur in January since it is the coldest month and requires the
   heating. An absolute minimum might occur in July since it is the warmest month.

   (b) The graph f is shown in both Figures 119a & 119b. One can see from the graph that there is one local
   maximum and one local minimum where $1 \leq x \leq 12$. The peak is near the point $(1.46, 140.06)$ and the
   valley is located near $(7.12, 14.78)$. This means that the peak heating expense of \$140 occurs in January
   $(x \approx 1)$. The minimum heating cost is during July $(x \approx 7)$ when it is roughly \$15.   ⚬

121. (a) Since the graph of f is symmetric with respect to the y-axis, it is a graphical representation of an even function.

   (b) If f represents the average temperature in the month x, $f(1) = f(-1)$. Thus, the average monthly temperature in August is also 83°F.

   (c) The average monthly temperatures in March and November are equal, since $f(-4) = f(4)$.

   (d) In Austin the average monthly temperature is symmetric about July. In July the highest average monthly temperature occurs. In January the lowest temperature occurs. The pairs June-August, May-September, April-October, March-November, and February-December have approximately the same average temperatures.

123. (a) Since $h(x) = -16x^2 + h_{max}$ and $h_{max} = 400$. Let $h(x) = -16x^2 + 400$. Then, it follows that $h(-2) = h(2) = -16(\pm 2)^2 + 400 = 336$. Two seconds before $(x = -2)$ and two seconds after $(x = 2)$ the time when the maximum height is reached, the projectile's height is 336 feet. When $x = -2$, the projectile is moving upward, whereas when $x = 2$, the projectile is moving downward.

   (b) $h(x) = -16x^2 + 400 \Rightarrow h(-5) = h(5) = -16(\pm 5)^2 + 400 = 0$. Five seconds before $(x = -5)$ and five seconds after $(x = 5)$ when the maximum height is reached, the projectile's height is 0 feet. When $x = -5$, the object is being shot upward from ground level and when $x = 5$, it is striking the ground.

   (c) The graph of h is shown in Figure 123. h is an even function.

   (d) Since h is an even function, $h(-x) = h(x)$ whenever $-5 \le x \le 5$. This means that the projectile is at exactly the same height either x seconds before or after the time the maximum height is attained.

[–5, 5, 1] by [0, 500, 100]

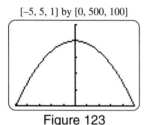

Figure 123

4.2: Polynomial Functions and Models

Graphs of Polynomial Functions

1. (a) Turning points occur near (1.6, 3.6), (3, 1.2), and (4.4, 3.6).

   (b) The turning points represent the times and distances where the runner turned around and jogged the other direction. In this example, after 1.6 minutes the runner is 360 feet from the starting line. The runner turns and jogs toward the starting line. After 3 minutes the runner is 120 feet from the starting line, turns, and jogs away from the starting line. After 4.4 minutes the runner is again 360 feet from the starting line. The runner turns and jogs back to the starting line.

3. (a) It has no turning points and one x-intercept of 0.5.

   (b) The leading coefficient is positive, since the slope of the line is positive.

   (c) Since the graph appears to be a line that is not horizontal, its minimum degree is 1.

5. (a) It has three turning points and three x-intercepts of $-6, -1, 6$.

   (b) Since the graph goes down for large values of x, the leading coefficient is negative.

   (c) Since the graph has three turning points, its minimum degree is 4.

7. (a) It has four turning points and five x-intercepts of $-3, -1, 0, 1$, and 2.

   (b) Since the graph goes up for large values of x, the leading coefficient is positive.

   (c) Since the graph has four turning points, its minimum degree is 5.

9. (a) It has two turning points and one x-intercept of -3.

   (b) Since the graph goes down to the left and up to the right, the leading coefficient is positive.

   (c) The graph has two turning points. The minimum degree of f is 3.

11. (a) It has one turning point and two x-intercepts of -1, and 2.

    (b) Since the graph is a parabola opening up, the leading coefficient is positive.

    (c) The graph has one turning point. The minimum degree of f is 2.

13. (a) Graph d.

    (b) There is one turning point at $(1, 0)$.

    (c) It has one x-intercept: $x = 1$.

    (d) The only local minimum is 0.

    (e) The absolute minimum is 0. There is no absolute maximum.

15. (a) Graph b.

    (b) The turning points are at $(-3, 27)$ and $(1, -5)$.

    (c) The three x-intercepts are $x \approx -4.9$, $x = 0$, and $x \approx 1.9$.

    (d) The local minimum is -5, and the local maximum is 27.

    (e) There are no absolute extrema.

17. (a) Graph a.

    (b) The turning points are at $(-2, 16)$, $(0, 0)$, and $(2, 16)$.

    (c) The three x-intercepts are $x \approx -2.8$, $x = 0$, and $x \approx 2.8$.

    (d) The local minimum is 0, and the local maximum is 16.

    (e) The absolute maximum is 16. There is no absolute minimum.

19. (a) The degree is 1 and the leading coefficient is -2.

    (b) Since the degree of f is 1 and the lead coefficient is negative the graph is a line with a negative slope $\Rightarrow$ it goes up on the left end and down on the right.

21. (a) The degree is 2 and the leading coefficient is 1.

    (b) Since the degree of f is 2 and the lead coefficient is positive the graph is a parabola that opens up on each side.

23. (a) The degree is 3 and the leading coefficient is -2.

   (b) Since the degree of f is odd and the lead coefficient is negative its graph will go up on the left and down on the right.

25. (a) The degree is 3 and the leading coefficient is -1.

   (b) Since the degree of f is odd and the leading coefficient is negative, its graph will go up on the left and down on the right.

27. (a) The degree is 5 and the leading coefficient is 0.1.

   (b) Since the degree of f is odd and the leading coefficient is positive, its graph will go down on the left and up on the right.

29. (a) The degree is 2 and the leading coefficient is $-\dfrac{1}{2}$.

   (b) Since the degree of f is even and the leading coefficient is negative, its graph will go down on both sides.

31. (a) The graph of $f(x) = \dfrac{1}{9}x^3 - 3x$ is shown in Figure 31.

   (b) There are two turning points located at $(-3, 6)$ and $(3, -6)$.

   (c) At $x = -3$ there is a local maximum of 6 and at $x = 3$ there is a local minimum of -6.

\33. (a) The graph of $f(x) = 0.025x^4 - 0.45x^2 - 3$ is shown in Figure 33.

   (b) There are three turning points located at $(-3, -7.025)$, $(0, -5)$, and $(3, -7.025)$.

   (c) At $x = \pm 3$ there is a local minimum of -7.025. At $x = 0$ there is a local maximum of -5.

[–10, 10, 1] by [–10, 10, 1]	[–10, 10, 1] by [–10, 10, 1]	[–10, 10, 1] by [–10, 10, 1]	[–10, 10, 1] by [–10, 10, 1]
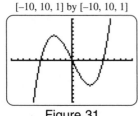	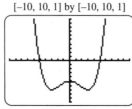	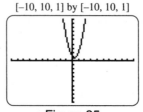	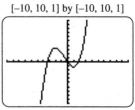
Figure 31	Figure 33	Figure 35	Figure 37

35. (a) The graph of $f(x) = 1 - 2x + 3x^2$ is shown in Figure 35.

   (b) There is one turning point located at $\left(\dfrac{1}{3}, \dfrac{2}{3}\right) \approx (0.333, 0.667)$.

   (c) At $x = \dfrac{1}{3}$ there is a local minimum of $\dfrac{2}{3} \approx 0.667$.

37. (a) The graph of $f(x) = \dfrac{1}{3}x^3 + \dfrac{1}{2}x^2 - 2x$ is shown in Figure 37.

   (b) The turning points are at $\left(-2, \dfrac{10}{3}\right) \approx (-2, 3.333)$ and $\left(1, -\dfrac{7}{6}\right) \approx (1, -1.167)$.

   (c) There is a local minimum of $-\dfrac{7}{6} \approx -1.167$, and there is a local maximum of $\dfrac{10}{3} \approx 3.333$.

39. The graphs of $f(x) = 2x^4$, $g(x) = 2x^4 - 5x^2 + 1$, and $h(x) = 2x^4 + 3x^2 - x - 2$ are shown in Figures 39a-c. As the viewing rectangle increases in size, the graphs begin to look alike. Each formula contains the term $2x^4$. This term determines the end behavior of the graph for large values of $|x|$. The other terms are relatively small in absolute value by comparison when $|x| \geq 100$. End behavior is determined by the leading term.

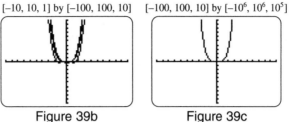

$[-4, 4, 1]$ by $[-4, 4, 1]$ $[-10, 10, 1]$ by $[-100, 100, 10]$ $[-100, 100, 10]$ by $[-10^6, 10^6, 10^5]$

     Figure 39a   Figure 39b   Figure 39c

41. (a) A line graph of the data is shown in Figure 41.

    (b) The data appears linear so f could be degree 1 with one x-intercept.

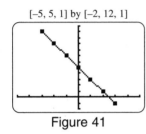 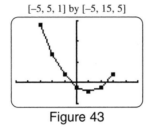

    $[-5, 5, 1]$ by $[-2, 12, 1]$ $[-5, 5, 1]$ by $[-5, 15, 5]$

      Figure 41   Figure 43

43. (a) A line graph of the data is shown in Figure 43.

    (b) The data decreases and then increases. This could not be a linear function. The data appears to be parabolic and the line graph crosses the x-axis twice. Since all zeros are real and between -3 and 3, f must be degree 2 with two real zeros.

Sketching Graphs of Polynomials

45. One possible graph is shown in Figure 45.

47. One possible graph is shown in Figure 47.

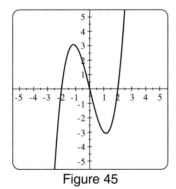 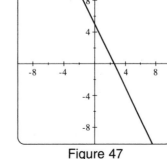

      Figure 45   Figure 47

49. No such graph is possible.

51. No such graph is possible.

53. Start by plotting the points $(-1, 2)$ and $\left(1, \dfrac{2}{3}\right)$. Then sketch a cubic polynomial having these two turning points.

    One possible graph is shown in Figure 53.

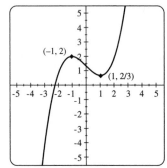
Figure 53

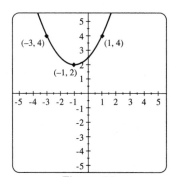
Figure 55

55. Since a quadratic polynomial has only one turning point, $(-1, 2)$ must be the vertex. Plot the points $(-1, 2)$, $(-3, 4)$, and $(1, 4)$. Sketch a parabola passing through these points with a vertex of $(-1, 2)$. The parabola must open up. See Figure 55.

Average Rates of Change

57. For $f(x) = x$ we get $(0, 0)$ and $\left(\dfrac{1}{2}, \dfrac{1}{2}\right)$ $\Rightarrow$ the average rate of change is $\dfrac{\frac{1}{2} - 0}{\frac{1}{2} - 0} = \dfrac{\frac{1}{2}}{\frac{1}{2}} = 1$

    For $g(x) = x^2$ we get $(0, 0)$ and $\left(\dfrac{1}{2}, \dfrac{1}{4}\right)$ $\Rightarrow$ the average rate of change is $\dfrac{\frac{1}{4} - 0}{\frac{1}{2} - 0} = \dfrac{\frac{1}{4}}{\frac{1}{2}} = 0.5$

    For $h(x) = x^3$ we get $(0, 0)$ and $\left(\dfrac{1}{2}, \dfrac{1}{8}\right)$ $\Rightarrow$ the average rate of change is $\dfrac{\frac{1}{8} - 0}{\frac{1}{2} - 0} = \dfrac{\frac{1}{8}}{\frac{1}{2}} = 0.25$

59. (a) For $[1.9, 2.1]$ we get: $(1.9, 6.859)$ and $(2.1, 9.261)$ $\Rightarrow$ the average rate of change is $\dfrac{9.261 - 6.859}{2.1 - 1.9} = 12.01$

    (b) For $[1.99, 2.01]$ we get: $(1.99, 7.880599)$ and $(2.01, 8.120601)$ $\Rightarrow$ the average rate of change is

    $\dfrac{8.120601 - 7.880599}{2.01 - 1.99} = 12.0001$

    (c) For $[1.999, 2.001]$ we get: $(1.999, 7.988005999)$ and $(2.001, 8.012006001)$ $\Rightarrow$ the average rate of change is

    $\dfrac{8.012006001 - 7.988005999}{2.001 - 1.999} = 12.00000101$

    As the interval decreases in length the average rate of change is approaching 12.

61. (a) For $[1.9, 2.1]$ we get: $\frac{1}{4}(1.9)^4 - \frac{1}{3}(1.9)^3 = 0.971691\overline{6}$ or $(1.9, 0.971691\overline{6})$ and $\frac{1}{4}(2.1)^4 - \frac{1}{3}(2.1)^3 =$

   1.775025 or $(2.1, 1.775025)$ $\Rightarrow$ the average rate of change is: $\dfrac{1.775025 - 0.971691\overline{6}}{2.1 - 1.9} = 4.01\overline{6}$

   (b) For $[1.99, 2.01]$ we get: $\frac{1}{4}(1.99)^4 - \frac{1}{3}(1.99)^3 = 1.293731\overline{6}$ or $(1.99, 1.293731\overline{6})$ and

   $\frac{1}{4}(2.01)^4 - \frac{1}{3}(2.01)^3 = 1.373735003$ or $(2.01, 1.373735003)$ $\Rightarrow$ the average rate of change is:

   $\dfrac{1.373735003 - 1.293731\overline{6}}{2.01 - 1.99} = 4.0001\overline{6}$

   (c) For $[1.999, 2.001]$ we get: $\frac{1}{4}(1.999)^4 - \frac{1}{3}(1.999)^3 = 1.329337332$ or $(1.999, 1.329337332)$ and

   $\frac{1}{4}(2.001)^4 - \frac{1}{3}(2.001)^3 = 1.337337335$ or $(2.001, 1.337337335)$ $\Rightarrow$ the average rate of change is:

   $\dfrac{1.337337335 - 1.329337332}{2.001 - 1.999} = 4.000001\overline{6}$

   As the interval decreases in length the average rate of change is approaching 4.

63. Use $\dfrac{f(x + h) - f(x)}{h}$, then $\dfrac{3(x + h)^3 - 3x^3}{h} = \dfrac{3x^3 + 9x^2h + 9xh^2 + 3h^3 - 3x^3}{h} = 9x^2 + 9xh + 3h^2$

65. Use $\dfrac{f(x + h) - f(x)}{h}$, then $\dfrac{1 + (x + h) - (x + h)^3 - (1 + x - x^3)}{h} =$

   $\dfrac{1 + x + h - x^3 - 3x^2h - 3xh^2 - h^3 - 1 - x + x^3}{h} = -3x^2 - 3xh - h^2 + 1$

Piecewise-Defined Functions

67. $f(-2) \approx 5$ and $f(1) \approx 0$

69. $f(-1) \approx -1$, $f(1) \approx 1$, and $f(2) \approx -2$

71. $f(-3) = (-3)^3 - 4(-3)^2 = -27 - 36 = -63$, $f(1) = 3(1)^2 = 3$, and $f(4) = (4)^3 - 54 = 64 - 54 = 10$

73. $f(-2) = (-2)^2 + 2(-2) + 6 = 6$, $f(1) = 1 + 6 = 7$, and $f(2) = 2^3 + 1 = 9$

75. (a) The graph of f is shown in Figure 75.

   (b) The function f is discontinuous at $x = 0$.

   (c) First solve: $4 - x^2 = 0 \Rightarrow -x^2 = -4 \Rightarrow x^2 = 4 \Rightarrow x = -2, 2$; only $x = -2$ is in the domain for the

   equation. Now solve: $x^2 - 4 = 0 \Rightarrow x^2 = 4 \Rightarrow x = -2, 2$; only $x = 2$ is in the domain for the

   equation $\Rightarrow x = -2, 2$.

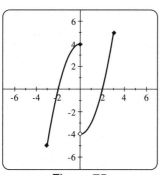

Figure 75

77. (a) The graph of f is shown in Figure 77.

(b) f is continuous.

(c) First solve: $2x = 0 \Rightarrow x = 0$; this is not in the domain for the equation.

$-2 \neq 0 \Rightarrow$ no solutions $-1 \leq x \leq 0$. Finally solve: $x^2 - 2 = 0 \Rightarrow x^2 = 2 \Rightarrow x = \pm\sqrt{2}$; only $\sqrt{2}$ is in the domain for the equation $\Rightarrow x = \sqrt{2}$.

79. (a) The graph of f is shown in Figure 79.

(b) f is continuous on its domain.

(c) First solve: $x^3 + 3 = 0 \Rightarrow x^3 = -3 \Rightarrow x = \sqrt[3]{-3}$ or $-\sqrt[3]{3}$; this is in the domain for the equation.

Now solve: $x + 3 = 0 \Rightarrow x = -3$; this is not in the domain for the equation.

Finally solve: $4 + x - x^2 = 0 \Rightarrow$ using quadratic formula for $-x^2 + x + 4 = 0$ we get:

$$\frac{-1 \pm \sqrt{1 - 4(-1)(4)}}{2(-1)} = \frac{-1 \pm \sqrt{17}}{-2}; \text{ only } \frac{1 + \sqrt{17}}{2} \text{ is in the domain of the equation} \Rightarrow$$

$$x = -\sqrt[3]{3}, \frac{\sqrt{17} + 1}{2}.$$

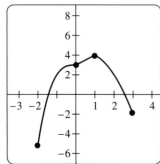

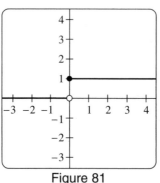

Figure 79   Figure 81

Applications

81. (a) $H(-2) = 0$, because $t < 0$; $H(0) = 1$, because $t \geq 0$; $H(3.5) = 1$, because $t \geq 0$

(b) See Figure 81.

83. (a) There are two turning points that occur near $(1, 13)$ and $(7, 72)$.

(b) The point $(1, 13)$ means that in January the average temperature is 13°F. This is the minimum average monthly temperature in Minneapolis. After this the average temperature begins to increase. The point $(7, 72)$ means that in July the average temperature is 72°F. This is the maximum average monthly temperature in Minneapolis. After this the average temperature begins to decrease.

85. (a) Figure 85a shows the data. It appears to be linear. As the speed increases, so does the minimum passing sight distance.

(b) Since the data appears linear, a linear polynomial would model the data. Let $D(x) = m(x - h) + k$. Start by letting $(h, k) = (20, 810)$ and adjusting the slope until a visually pleasing fit is found. Figure 85b shows the fit with a value of $m \approx 34$. Let $D(x) = 34(x - 20) + 810$ or $D(x) = 34x + 130$. *Answers may vary.*

(c) The minimum passing distance at 43 mph would be $D(43) = 34(43 - 20) + 810 = 1592$ feet. *Answers may vary slightly.*

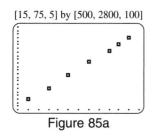

[15, 75, 5] by [500, 2800, 100]

Figure 85a

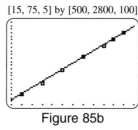

[15, 75, 5] by [500, 2800, 100]

Figure 85b

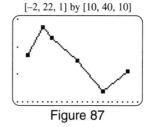

[-2, 22, 1] by [10, 40, 10]

Figure 87

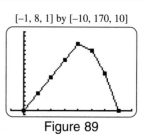

[-1, 8, 1] by [-10, 170, 10]

Figure 89

87. (a) A line graph of the data is shown in Figure 87. Marijuana use among seniors peaked in 1978 ($x = 3$) at
37%. There was a decline until 1990. Marijuana use then increased significantly from 1990 to 1995. The
line graph suggests that a cubic polynomial might be used to model the data. Selecting cubic regression on
the calculator, we get $f(x) \approx 0.02352x^3 - 0.7088x^2 + 4.444x + 27.46$.

(b) In the table, 1997 corresponds to $x = 22$;

$f(22) = 0.02352(22)^3 - 0.7088(22)^2 + 4.444(22) + 27.64 \approx 32.6$, or 32.6%; the estimate is too high
compared to the actual value of 23%.

89. (a) A line graph of the data is shown in Figure 89. The height begins to decrease after 4 seconds. The object
appears to have dropped after 4 seconds.

(b) From 0 to 4 seconds, when the object is being lifted, the height could be modeled with a linear function.
From 4 to 7 seconds, when the object begins to fall, the height can be modeled with a nonlinear function.

(c) The altitude increases by 36 feet each second, during the first four seconds. Therefore, let $m = 36$. When
$x = 4$, the height is 144 feet. Let $b = 144$. Since when $x = 7$, the height was 0,

$a(7 - 4)^2 + 144 = 0 \Rightarrow a = -16$. Thus, $f(x) = \begin{cases} 36x, & \text{if } 0 \le x \le 4 \\ -16(x - 4)^2 + 144, & \text{if } 4 < x \le 7 \end{cases}$

(d) Using $f(x) = 36x$, $f(x) = 100 \Rightarrow 100 = 36x \Rightarrow x \approx 2.8$;

using $f(x) = -16(x - 4)^2 + 144$, $f(x) = 100 \Rightarrow 100 = -16(x - 4)^2 + 144 \Rightarrow -44 = -16(x - 4)^2 \Rightarrow$
$x \approx 5.7$; the object is 100 feet high at about 2.8 seconds and at about 5.7 seconds.

4.3: Real Zeros of Polynomial Functions

Division of Polynomials

1. $\dfrac{5x^4 - 15}{10x} = \dfrac{5x^4}{10x} - \dfrac{15}{10x} = \dfrac{x^3}{2} - \dfrac{3}{2x}$

3. $\dfrac{3x^4 - 2x^2 - 1}{3x^3} = \dfrac{3x^4}{3x^3} - \dfrac{2x^2}{3x^3} - \dfrac{1}{3x^3} = x - \dfrac{2}{3x} - \dfrac{1}{3x^3}$

5. $\dfrac{x^3 - 4}{4x^3} = \dfrac{x^3}{4x^3} - \dfrac{4}{4x^3} = \dfrac{1}{4} - \dfrac{1}{x^3}$

7. $x^3 - 2x^2 - 5x + 6$ divided by $x - 3$ can be performed using synthetic division.

$\begin{array}{r|rrrr} 3 & 1 & -2 & -5 & 6 \\ & & 3 & 3 & -6 \\ \hline & 1 & 1 & -2 & 0 \end{array}$ The quotient is $x^2 + x - 2$ and the remainder is 0.

9. $2x^4 - 7x^3 - 5x^2 - 19x + 17$ divided by $x + 1$ can be performed using synthetic division.

$$
\begin{array}{r|rrrrr}
-1 & 2 & -7 & -5 & -19 & 17 \\
   &   & -2 & 9 & -4 & 23 \\
\hline
   & 2 & -9 & 4 & -23 & 40
\end{array}
$$

The quotient is $2x^3 - 9x^2 + 4x - 23$ and the remainder is 40.

11. $3x^3 - 7x + 10$ divided by $x - 1$ can be performed using synthetic division.

$$
\begin{array}{r|rrrr}
1 & 3 & 0 & -7 & 10 \\
  &   & 3 & 3 & -4 \\
\hline
  & 3 & 3 & -4 & 6
\end{array}
$$

The quotient is $3x^2 + 3x - 4$ and the remainder is 6.

This can also be found using long division as shown.

$$
\begin{array}{r}
3x^2 + 3x - 4 \\
x - 1{\overline{\smash{\big)}\,3x^3 - 0x^2 - 7x + 10}} \\
\underline{3x^3 - 3x^2} \\
3x^2 - 7x \\
\underline{3x^2 - 3x} \\
-4x + 10 \\
\underline{-4x + 4} \\
6
\end{array}
$$

13. The divisor times the quotient will be equal to the dividend, $(x - 2)(x^2 - 6x + 3) = x^3 - 8x^2 + 15x - 6$.

15. We can use synthetic division to divide $x^4 - 3x^3 - x + 3$ by $x - 3$.

$$
\begin{array}{r|rrrrr}
3 & 1 & -3 & 0 & -1 & 3 \\
  &   & 3 & 0 & 0 & -3 \\
\hline
  & 1 & 0 & 0 & -1 & 0
\end{array}
$$

The quotient is $x^3 - 1$ and the remainder is 0.

17. We can use long division to divide $4x^3 - x^2 - 5x + 6$ by $x - 1$.

$$
\begin{array}{r}
4x^2 + 3x - 2 \\
x - 1{\overline{\smash{\big)}\,4x^3 - x^2 - 5x + 6}} \\
\underline{4x^3 - 4x^2} \\
3x^2 - 5x \\
\underline{3x^2 - 3x} \\
-2x + 6 \\
\underline{-2x + 2} \\
4
\end{array}
$$

The quotient is $4x^2 + 3x - 2 + \dfrac{4}{x - 1}$.

19. We can use synthetic division to divide $x^3 + 1$ by $x + 1$.

$$
\begin{array}{r|rrrr}
-1 & 1 & 0 & 0 & 1 \\
   &   & -1 & 1 & -1 \\
\hline
   & 1 & -1 & 1 & 0
\end{array}
$$

The quotient is $x^2 - x + 1$ and the remainder is 0.

21. We can use long division to divide $6x^3 + 5x^2 - 8x + 4$ by $2x - 1$.

$$
\begin{array}{r}
3x^2 + 4x - 2 \\
2x - 1{\overline{\smash{\big)}\,6x^3 + 5x^2 - 8x + 4}} \\
\underline{6x^3 - 3x^2} \\
8x^2 - 8x + 4 \\
\underline{8x^2 - 4x} \\
-4x + 4 \\
\underline{-4x + 2} \\
2
\end{array}
$$

The quotient is $3x^2 + 4x - 2 + \dfrac{2}{2x - 1}$.

23. We can use long division to divide $3x^4 - 7x^3 + 6x - 16$ by $3x - 7$.

$$
\begin{array}{r}
x^3 + 2 \\
3x - 7\overline{)3x^4 - 7x^3 + 0x^2 + 6x - 16} \\
\underline{3x^4 - 7x^3} \\
6x - 16 \\
\underline{6x - 14} \\
-2
\end{array}
$$

The quotient is $x^3 + 2 + \dfrac{-2}{3x - 7}$.

25. We can use long division to divide $5x^4 - 2x^2 + 6$ by $x^2 + 2$.

$$
\begin{array}{r}
5x^2 - 12 \\
x^2 + 2\overline{)5x^4 - 2x^2 + 6} \\
\underline{5x^4 + 10x^2} \\
-12x^2 + 6 \\
\underline{-12x^2 - 24} \\
30
\end{array}
$$

The quotient is $5x^2 - 12 + \dfrac{30}{x^2 + 2}$.

27. We can use long division to divide $8x^3 + 10x^2 - 12x - 15$ by $2x^2 - 3$.

$$
\begin{array}{r}
4x + 5 \\
2x^2 - 3\overline{)8x^3 + 10x^2 - 12x - 15} \\
\underline{8x^3 \quad\quad - 12x} \\
10x^2 \quad\quad - 15 \\
\underline{10x^2 \quad\quad - 15} \\
0
\end{array}
$$

The quotient is $4x + 5$.

29. We can use long division to divide $2x^4 - x^3 + 4x^2 + 8x + 7$ by $2x^2 + 3x + 2$.

$$
\begin{array}{r}
x^2 - 2x + 4 \\
2x^2 + 3x + 2\overline{)2x^4 - x^3 + 4x^2 + 8x + 7} \\
\underline{2x^4 + 3x^3 + 2x^2} \\
-4x^3 + 2x^2 + 8x \\
\underline{-4x^3 - 6x^2 - 4x} \\
8x^2 + 12x + 7 \\
\underline{8x^2 + 12x + 8} \\
-1
\end{array}
$$

The quotient is $x^2 - 2x + 4 + \dfrac{-1}{2x^3 + 3x + 2}$.

Remainder Theorem

31. Using the remainder theorem we find: $f(1)$, $\Rightarrow 5(1)^2 - 3(1) + 1 = 5 - 3 + 1 = 3$

33. Using the remainder theorem we find: $f(-2)$, $\Rightarrow 4(-2)^3 - (-2)^2 + 4(-2) + 2 = -32 - 4 - 8 + 2 = -42$

Factor Theorem

35. $f(2) = (2)^3 - 6(2)^2 + 11(2) - 6 = 0$; since $f(2) = 0$, by the factor theorem,

   $x - 2$ is a factor of $f(x) = x^3 - 6x^2 + 11x - 6$.

37. $f(3) = (3)^4 - 2(3)^3 - 13(3)^2 - 10(3) = -120$; since $f(-3) \neq 0$, by the factor theorem,

   $x - 3$ is not a factor of $x^4 - 2x^3 - 13x^2 - 10x$.

Factoring Polynomials

39. $f(x) = 2x^2 - 25x + 77$ and zeros: $\dfrac{11}{2}$ and $7 \Rightarrow f(x) = 2\left(x - \dfrac{11}{2}\right)(x - 7)$

41. $f(x) = x^3 - 2x^2 - 5x + 6$ and zeros: $-2, 1,$ and $3 \Rightarrow f(x) = (x + 2)(x - 1)(x - 3)$

43. $f(x) = -2x^3 + 3x^2 + 59x - 30$ and zeros: $-5, \dfrac{1}{2},$ and $6 \Rightarrow f(x) = -2(x + 5)\left(x - \dfrac{1}{2}\right)(x - 6)$

45. If $f(-3) = 0$ then the quadratic equation has a factor of $x - (-3)$ or $x + 3,$ likewise if $f(2) = 0$ then the

    quadratic equation has a factor $x - 2.$ If this quadratic equation has a leading coefficient 7, the complete

    factored form of $f(x)$ is $f(x) = 7(x + 3)(x - 2).$

47. From the graph the zeros are $-4, 2,$ and $8.$ $f(x)$ is a cubic with a positive leading coefficient.

    Therefore, $f(x) = (x + 4)(x - 2)(x - 8).$

49. From the graph the zeros are $-8, -4, -2,$ and $4.$ $f(x)$ is a quartic polynomial with a negative leading coefficient.

    Therefore, $f(x) = -1(x + 8)(x + 4)(x + 2)(x - 4).$

51. Since the polynomial has zeros of $-1, 2,$ and $3,$ it has factors $(x + 1)(x - 2)(x - 3).$

    If f passes through $(0, 3)$ then $a(0 + 1)(0 - 2)(0 - 3) = 3 \Rightarrow a(1)(-2)(-3) = 3 \Rightarrow 6a = 3 \Rightarrow a = \dfrac{1}{2}.$

    The complete factored form is: $f(x) = \dfrac{1}{2}(x + 1)(x - 2)(x - 3).$

53. Since f has zeros $-1, 1,$ and $2,$ it has factors $(x + 1)(x - 1)(x - 2).$ If $f(0) = 1$ then $a(1)(-1)(-2) = 1$ or

    $2a = 1 \Rightarrow a = \dfrac{1}{2}.$ The complete factored form is: $f(x) = \dfrac{1}{2}(x + 1)(x - 1)(x - 2).$

55. Since f has zeros $-2, -1, 1,$ and $2,$ it has factors $(x + 2)(x + 1)(x - 1)(x - 2).$ If $f(0) = -8$ then

    $a(2)(1)(-1)(-2) = -8 \Rightarrow 4a = -8 \Rightarrow a = -2.$

    The complete factored form is: $f(x) = -2(x + 2)(x + 1)(x - 1)(x - 2).$

57. A graph of $Y_1 = 10X{\wedge}2 + 17X - 6$ is shown in Figure 57. Its zeros are -2 and $0.3.$ Since the leading

    coefficient is 10, the complete factorization is $f(x) = 10(x + 2)\left(x - \dfrac{3}{10}\right).$

59. A graph of $Y_1 = -3X{\wedge}3 - 3X{\wedge}2 + 18X$ is shown in Figure 59. Its zeros are $-3, 0,$ and $2.$ Since the leading

    coefficient is $-3,$ the complete factorization is $f(x) = -3(x - 0)(x - 2)(x + 3) = -3x(x - 2)(x + 3).$

[-5, 5, 1] by [-20, 20, 5]   [-5, 5, 1] by [-40, 40, 5]   [-5, 5, 1] by [-10, 10, 1]

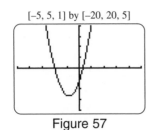   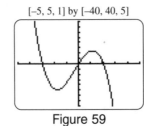   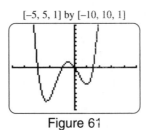

Figure 57   Figure 59   Figure 61

61. A graph of $Y_1 = X{\wedge}4 + (5/2)X{\wedge}3 - 3X{\wedge}2 - (9/2)X$ is shown in Figure 61. Its zeros are $-3, -1, 0,$ and $\dfrac{3}{2}.$

    Since the leading coefficient is 1, the complete factorization is

    $$f(x) = (x + 3)(x + 1)(x - 0)\left(x - \dfrac{3}{2}\right) = x(x + 3)(x + 1)\left(x - \dfrac{3}{2}\right).$$

63. By the factor theorem, since 1 is a zero, $(x - 1)$ is a factor. $x^3 - 9x^2 + 23x - 15$ divided by $x - 1$ can be performed using synthetic division.

$$\begin{array}{r|rrrr} 1 & 1 & -9 & 23 & -15 \\ & & 1 & -8 & 15 \\ \hline & 1 & -8 & 15 & 0 \end{array}$$

The quotient is $x^2 - 8x + 15$ and the remainder is 0.

Thus, $x^3 - 9x^2 + 23x - 15 = (x - 1)(x^2 - 8x + 15) = (x - 1)(x - 3)(x - 5)$.

The complete factored form of $f(x) = x^3 - 9x^2 + 23x - 15$ is $f(x) = (x - 1)(x - 3)(x - 5)$.

65. By the factor theorem, since -4 is a zero, $(x + 4)$ is a factor. $-4x^3 - x^2 + 51x - 36$ divided by $x + 4$ can be performed using synthetic division.

$$\begin{array}{r|rrrr} -4 & -4 & -1 & 51 & -36 \\ & & 16 & -60 & 36 \\ \hline & -4 & 15 & -9 & 0 \end{array}$$

The quotient is $-4x^2 + 15x - 9$ and the remainder is 0.

Thus, $-4x^3 - x^2 + 51x - 36 = (x + 4)(-4x^2 + 15x - 9) = (x + 4)(-4x + 3)(x - 3)$.

The complete factored form of $f(x) = -4x^3 - x^2 + 51x - 36$ is $f(x) = -4(x + 4)\left(x - \dfrac{3}{4}\right)(x - 3)$.

67. By the factor theorem, since -2 is a zero, $(x + 2)$ is a factor. $2x^4 - x^3 - 13x^2 - 6x$ divided by $x + 2$ can be performed using synthetic division.

$$\begin{array}{r|rrrr} -2 & 2 & -1 & -13 & -6 \\ & & -4 & 10 & 6 \\ \hline & 2 & -5 & -3 & 0 \end{array}$$

The quotient is $2x^3 - 5x^2 - 3x$ and the remainder is 0. Thus,

$2x^4 - x^3 - 13x^2 - 6x = (x + 2)(2x^3 - 5x^2 - 3x) = (x + 2)x(2x^2 - 5x - 3) = x(x + 2)(2x + 1)(x - 3)$.

The complete factored form of $f(x) = 2x^4 - x^3 - 13x^2 - 6x$ is $f(x) = 2x(x + 2)\left(x + \dfrac{1}{2}\right)(x - 3)$.

Graphs and Multiple Zeros

69. The zeros of $f(x)$ are 4 and -2. Since the graph does not cross the x-axis at $x = 4$, the zero of 4 has even multiplicity. The graph crosses the x-axis at $x = -2$. The zero of -2 has odd multiplicity. Since the graph levels off, crossing the x-axis at $x = -2$, this zero has at least multiplicity 3, and the zero of 4 has at least multiplicity 2. Thus, the minimum degree of $f(x)$ is $3 + 2$, or 5.

71. The zeros of $f(x)$ are $-6, -1$, and 4. Since the graph does not cross the x-axis at either $x = -6$ or -1, these zeros have even multiplicity. The graph crosses the x-axis at $x = 4$. The zero of 4 has odd multiplicity. Since the graph levels off, crossing the x-axis at $x = 4$, this zero has at least multiplicity 3, and the zeros of -6 and -1 both have at least multiplicity 2. Thus, the minimum degree of $f(x)$ is $3 + 2 + 2$, or 7.

73. Degree: 3; zeros: -1 with multiplicity 2 and 6 with multiplicity 1. $f(x) = (x + 1)^2(x - 6)$

75. Degree: 4; zeros: 2 with multiplicity 3 and 6 with multiplicity 1. $f(x) = (x - 2)^3(x - 6)$

77. The graph shows a cubic polynomial with a positive leading coefficient and zeros of 4 and -2. The zero of 4 has odd multiplicity, whereas the zero of -2 has even multiplicity. Since the graph has a degree three, its complete factored form is $f(x) = (x - 4)(x + 2)^2$.

79. The graph shows a quartic polynomial with a negative leading coefficient and zeros of –3 and 3. Both zeros have even multiplicity. Since the graph has a degree four polynomial, its complete factored form is
$$f(x) = -1(x + 3)^2(x - 3)^2.$$

81. The graph shows a fifth degree polynomial with a positive leading coefficient and zeros of –1 and 1. The –1 has an even multiplicity, whereas the zero of 1 has an odd multiplicity. Since the graph shows a fifth degree polynomial the factors are $(x + 1)^2(x - 1)^3$. To find the leading coefficient we use
$$f(0) = -2 \Rightarrow a(1)^2(-1)^3 = -2 \Rightarrow -a = -2 \Rightarrow a = 2. \text{ Its factored form is } f(x) = 2(x + 1)^2(x - 1)^3.$$

83. (a) From the factors the x-intercepts are: $-2, -1$. To find the y-intercept set $x = 0 \Rightarrow$
$$2(0 + 2)(0 + 1)^2 = 2(2)(1)^2 = 4$$

(b) The zero -2 has multiplicity 1. The zero -1 has multiplicity 2.

(c) See Figure 83.

85. (a) From the factors the x-intercepts are: $-1, 1,$ and 2. To find the y-intercept set $x = 0 \Rightarrow$
$$-(0 + 1)(0 - 1)(0 - 2) = -(1)(-1)(-2) = -2$$

(b) The zero -1 has multiplicity 1. The zero 1 has multiplicity 1. The zero 2 has multiplicity 1.

(c) See Figure 85.

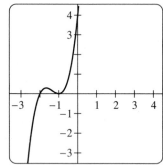
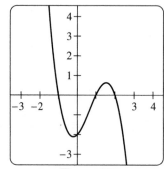
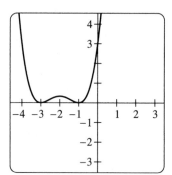

| Figure 83 | Figure 85 | Figure 87 |

87. (a) From the factors the x-intercepts are: $-3, -1$. To find the y-intercept set $x = 0 \Rightarrow$
$$\frac{1}{3}(0 + 3)^2(0 + 1)^2 = \frac{1}{3}(3)^2(1)^2 = 3$$

(b) The zero -3 has multiplicity 2. The zero -1 has multiplicity 2.

(c) See Figure 87.

Rational Zeros

89. $f(x) = 2x^3 + 3x^2 - 8x + 3$

(a) If $\dfrac{p}{q}$ is a rational zero, then p is a factor of 3, which are ± 1 and ± 3 and q is a factor of 2, which are ± 1 or ± 2. Thus, any rational zero must be in the list $\pm\dfrac{1}{2}, \pm 1, \pm\dfrac{3}{2},$ or ± 3. From Figure 89 we see that there are three rational zeros of $\dfrac{1}{2}, 1,$ and -3.

(b) The complete factored form is $f(x) = 2\left(x - \dfrac{1}{2}\right)(x - 1)(x + 3).$

x	$\frac{1}{2}$	$-\frac{1}{2}$	1	-1	$\frac{3}{2}$	$-\frac{3}{2}$	3	-3
$f(x)$	0	$\frac{15}{2}$	0	12	$\frac{9}{2}$	15	60	0

Figure 89

91. $f(x) = 2x^4 + x^3 - 8x^2 - x + 6$

(a) If $\frac{p}{q}$ is a rational zero, then p is a factor of 6, which are $\pm 1, \pm 2, \pm 3$, and ± 6 and q is a factor of 2, which

are ± 1 and ± 2. Thus, any rational zero must be in the list $\pm \frac{1}{2}, \pm 1, \pm \frac{3}{2}, \pm 2, \pm 3,$ or ± 6. By evaluating $f(x)$

at each of these values, we find that the zeros are $-2, -1, 1$, and $\frac{3}{2}$.

(b) The complete factored form is $f(x) = 2(x + 2)(x + 1)(x - 1)\left(x - \frac{3}{2}\right)$.

93. $f(x) = 3x^3 - 16x^2 + 17x - 4$

(a) If $\frac{p}{q}$ is a rational zero, then p is a factor of 4, which are $\pm 1, \pm 2$, and ± 4 and q is a factor of 3, which are

± 1 and ± 3. Thus, any rational zero must be in the list $\pm \frac{1}{3}, \pm 1, \pm \frac{2}{3}, \pm 2, \pm \frac{4}{3},$ or ± 4. By evaluating $f(x)$ at

each of these values, we find that the zeros are $\frac{1}{3}, 1,$ and 4.

(b) The complete factored form is $f(x) = 3\left(x - \frac{1}{3}\right)(x - 1)(x - 4)$.

95. $f(x) = x^3 - x^2 - 7x + 7$

(a) If $\frac{p}{q}$ is a rational zero, then p is a factor of 7, which are ± 1, and ± 7 and q is a factor of 1, which are ± 1.

Thus, any rational zero must be in the list ± 1 or ± 7. By evaluating $f(x)$ at each of these values, we find

that the only rational zero is 1.

(b) In order to find the complete factored form of $f(x)$ we need to divide the factor $(x - 1)$ into

$x^3 - x^2 - 7x + 7$ using synthetic division.

$$
\begin{array}{r|rrrr}
1\!\!\rfloor & 1 & -1 & -7 & 7 \\
  &   & 1 & 0 & -7 \\
\hline
  & 1 & 0 & -7 & 0
\end{array}
\quad \text{Thus } x^3 - x^2 - 7x + 7 = (x - 1)(x^2 - 7).
$$

The complete factored form is $f(x) = (x - 1)(x - \sqrt{7})(x + \sqrt{7})$.

Polynomial Equations

97. (a) $f(x) = x^3 + x^2 - 6x = 0 \Rightarrow x(x^2 + x - 6) = x(x + 3)(x - 2) = 0 \Rightarrow x = 0, -3,$ or 2.

(b) Graph $Y_1 = X\!\wedge\!3 + X\!\wedge\!2 - 6X$ in $[-5, 5, 1]$ by $[-10, 10, 1]$. The x-intercepts are $-3, 0,$ and 2.

(c) Table $Y_1 = X\!\wedge\!3 + X\!\wedge\!2 - 6X$ starting at $x = -4$, incrementing by 1. The zeros or x-intercepts

are $-3, 0,$ and 2.

99. (a) $f(x) = x^4 - 1 = 0 \Rightarrow x^4 = 1 \Rightarrow x = \pm 1$.

(b) Graph $Y_1 = X\!\wedge\!4 - 1$ in $[-2, 2, 1]$ by $[-10, 10, 1]$. The x-intercepts are -1 and 1.

(c) Table $Y_1 = X\!\wedge\!4 - 1$ starting at $x = -2$, incrementing by 1. The zeros or x-intercepts are ± 1.

101. (a) $f(x) = -x^3 + 4x = 0 \Rightarrow -x(x^2 - 4) = x(x + 2)(x - 2) = 0 \Rightarrow x = 0$, or ± 2.

(b) Graph $Y_1 = -X^3 + 4X$ in $[-5, 5, 1]$ by $[-5, 5, 1]$. The x-intercepts are -2, 0, and 2.

(c) Table $Y_1 = -X^3 + 4X$ starting at $x = -3$, incrementing by 1. The zeros or x-intercepts are -2, 0, and 2.

103. $x^3 - 25x = 0 \Rightarrow x(x^2 - 25) = 0 \Rightarrow x(x - 5)(x + 5) = 0 \Rightarrow x = 0$, or ± 5

105. $x^4 - x^2 = 2x^2 + 4 \Rightarrow x^4 - 3x^2 - 4 = 0 \Rightarrow (x^2 - 4)(x^2 + 1) = 0 \Rightarrow (x - 2)(x + 2)(x^2 + 1) = 0 \Rightarrow$

$x = \pm 2$

107. $x^3 - 3x^2 - 18x = 0 \Rightarrow x(x^2 - 3x - 18) = 0 \Rightarrow x(x - 6)(x + 3) = 0 \Rightarrow x = 0, 6$, or -3

109. $2x^3 = 4x^2 - 2x \Rightarrow 2x^3 - 4x^2 + 2x = 0 \Rightarrow 2x(x^2 - 2x + 1) = 0 \Rightarrow 2x(x - 1)(x - 1) = 0 \Rightarrow$

$x = 0$ or 1

111. $12x^3 = 17x^2 + 5x \Rightarrow 12x^3 - 17x^2 - 5x = 0 \Rightarrow x(12x^2 - 17x - 5) = 0 \Rightarrow$

$x(4x + 1)(3x - 5) = 0 \Rightarrow x = 0, -\dfrac{1}{4}$, or $\dfrac{5}{3}$

113. $9x^4 + 4 = 13x^2 \Rightarrow 9x^4 - 13x^2 + 4 = 0 \Rightarrow (9x^2 - 4)(x^2 - 1) = 0 \Rightarrow x = \pm\dfrac{2}{3}, x = \pm 1$

115. $4x^3 + 4x^2 - 3x - 3 = 0 \Rightarrow (4x^3 + 4x^2) - (3x + 3) = 0 \Rightarrow 4x^2(x + 1) - 3(x + 1) = 0 \Rightarrow$

$(4x^2 - 3)(x + 1) = 0$. Set $4x^2 - 3 = 0 \Rightarrow 4x^2 = 3 \Rightarrow x^2 = \dfrac{3}{4} \Rightarrow x = \pm\dfrac{\sqrt{3}}{2}$.

The solutions are; $x = -1 \pm \dfrac{\sqrt{3}}{2}$.

117. $2x^3 + 4 = x(x + 8) \Rightarrow 2x^3 + 4 = x^2 + 8x \Rightarrow 2x^3 - x^2 - 8x + 4 = 0 \Rightarrow$

$(2x^3 - x^2) - (8x - 4) = 0 \Rightarrow x^2(2x - 1) - 4(2x - 1) = 0 \Rightarrow (x^2 - 4)(2x - 1) = 0 \Rightarrow$

$(x + 2)(x - 2)(2x - 1) = 0 \Rightarrow x = -2, 2, \dfrac{1}{2}$

119. $8x^4 - 30x^2 + 27 = 0 \Rightarrow (4x^2 - 9)(2x^2 - 3) = 0 \Rightarrow (2x + 3)(2x - 3)(2x^2 - 3) = 0$.

Set $2x^2 - 3 = 0 \Rightarrow 2x^2 = 3 \Rightarrow x^2 = \dfrac{3}{2} \Rightarrow x = \pm\sqrt{\dfrac{3}{2}} \Rightarrow x = \pm\dfrac{\sqrt{6}}{2}$. The solutions are; $x = \pm\dfrac{3}{2}, \pm\dfrac{\sqrt{6}}{2}$

121. $x^6 - 19x^3 - 216 = 0 \Rightarrow (x^3 + 8)(x^3 - 27) = 0$. set $x^3 + 8 = 0 \Rightarrow x^3 = -8 \Rightarrow x = -2$.

Also set $x^3 - 27 = 0 \Rightarrow x^3 = 27 \Rightarrow x = 3$. The solutions are; $x = -2, 3$.

123. The graph of $f(x) = x^3 - 1.1x^2 - 5.9x + 0.7$ is shown in Figure 123. Its zeros are approximately -2.0095,

0.11639, and 2.9931. The solutions are $x \approx -2.01, 0.12$, or 2.99.

125. The graph of $f(x) = -0.7x^3 - 2x^2 + 4x + 2.5$ is shown in Figure 125. Its zeros are approximately -4.0503,

-0.51594, and 1.7091. The solutions are $x \approx -4.05, -0.52$, or 1.71.

[−10, 10, 1] by [−10, 10, 1]   [−10, 10, 1] by [−10, 10, 1]   [−10, 10, 1] by [−120, 120, 20]

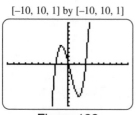
Figure 123

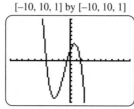
Figure 125

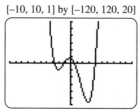
Figure 127

127. The graph of $f(x) = 2x^4 - 1.5x^3 - 24x^2 - 10x + 13$ is shown in Figure 127. Its zeros are approximately $-2.6878, -1.0957, 0.55475,$ and $3.9787.$ The solutions are $x \approx -2.69, -1.10, 0.55$ or $3.98.$

Applications

129. If we divide the Area by the Width we will find the Length:

$$
\begin{array}{r}
4x + 3 \\
3x + 1 \overline{)12x^2 + 13x + 3} \\
\underline{12x^2 + 4x} \\
9x + 3 \\
\underline{9x + 3} \\
0
\end{array}
$$

The length is $4x + 3.$ When $x = 10$, the Length is $4(10) + 3 = 43$ feet.

131. (a) The greater the distance downstream from the plant the less the concentration of copper. This agrees with intuition.

(b) The graph of $y = C(x)$ and the data points are shown in Figure 131b.

(c) We must approximate the distance where the concentration of copper first drops to 10. Graph $Y_1 = C(x)$ and $Y_2 = 10.$ The point of intersection is near $(32.1, 10)$ as shown in Figure 131c. Mussels would not be expected to live between the plant and approximately 32.1 miles downstream, that is when $0 \le x < 32.1$ (approximately).

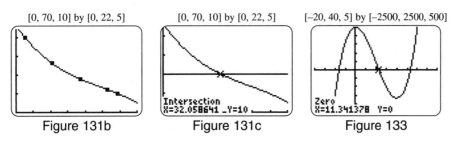

[0, 70, 10] by [0, 22, 5]   [0, 70, 10] by [0, 22, 5]   [−20, 40, 5] by [−2500, 2500, 500]

Figure 131b   Figure 131c   Figure 133

133. The graph of $Y_1 = (\pi/3)X^3 - 10\pi X^2 + ((4000\pi)(0.6))/3$ and the smallest positive zero are shown in Figure 133. The ball with a 20-centimeter diameter will sink approximately 11.34 centimeters into the water.

135. (a) Graph f in $[-10, 15, 1]$ by $[-70, 70, 10]$. It has three zeros of approximately $-6.01, 2.15,$ and $11.7.$ The approximate complete factored form is $-0.184(x + 6.01)(x - 2.15)(x - 11.7).$

(b) The zeros represent the months when the average temperature is 0°F. The zero of -6.01 has no significance since it does not correspond to a month. The zeros of 2.15 and 11.7 mean that in approximately February and December the average temperature in Trout Lake is 0°F.

137. $f(x) = x^3 - 66x^2 + 1052x + 652$ and $f(x) = 2500 \Rightarrow 2500 = x^3 - 66x^2 + 1052x + 652 \Rightarrow$ $x^3 - 66x^2 + 1052x - 1848 = 0$; graph $Y_1 = X^3 - 66X^2 + 1052X - 1848$ in the window $[0, 45, 5]$ by $[-5000, 5000, 1000].$ The zeros are at $x = 2, x = 22,$ and $x = 42.$ Since $x = 1$ corresponds to June 1, there were 2500 birds on approximately June 2, June 22, and July 12.

4.4: The Fundamental Theorem of Algebra

Complex Numbers

1. $\sqrt{-4} = i\sqrt{4} = 2i$

3. $\sqrt{-100} = i\sqrt{100} = 10i$

5. $\sqrt{-23} = i\sqrt{23}$

7. $\sqrt{-12} = i\sqrt{12} = i\sqrt{4}\sqrt{3} = 2i\sqrt{3}$

9. $3i + 5i = (3+5)i = 8i$

11. $(3+i) + (-5-2i) = (3+(-5)) + (1-2)i = -2 - i$

13. $(12-7i) - (-1+9i) = (12-(-1)) + (-7-9)i = 13 - 16i$

15. $(3) - (4-6i) = (3+0i) - (4-6i) = (3-4) + (0+6)i = -1 + 6i$

17. $(2)(2+4i) = 4 + 8i$

19. $(1+i)(2-3i) = (1)(2) + (1)(-3i) + (i)(2) + (i)(-3i) = 2 - i - 3i^2 = 2 - i - 3(-1) = 5 - i$

21. $(-3+2i)(-2+i) = (-3)(-2) + (-3)(i) + (2i)(-2) + (2i)(i) = 6 - 7i + 2i^2 = 6 - 7i + 2(-1) = 4 - 7i$

23. $(-2+3i)^2 = (-2+3i)(-2+3i) = (-2)(-2) + (-2)(3i) + (-2)(3i) + (3i)(3i) =$

    $4 + (-6i) + (-6i) + (9i^2) = 4 + (-12i) + (-9) = -5 - 12i$

25. $2i(1-i)^2 = 2i(1-i)(1-i) = 2i[(1)(1) + (1)(-i) + (1)(-i) + (-i)(-i)] =$

    $2i[1 - i - i + (-1)] = 2i(-2i) = -4i^2 = 4$

27. $\dfrac{1}{1+i} = \dfrac{1}{1+i} \cdot \dfrac{1-i}{1-i} = \dfrac{1-i}{(1+i)(1-i)} = \dfrac{1-i}{1-i^2} = \dfrac{1-i}{2} = \dfrac{1}{2} - \dfrac{1}{2}i$

29. $\dfrac{4+i}{5-i} = \dfrac{4+i}{5-i} \cdot \dfrac{5+i}{5+i} = \dfrac{(4+i)(5+i)}{(5-i)(5+i)} = \dfrac{20+9i+i^2}{25-i^2} = \dfrac{19+9i}{26} = \dfrac{19}{26} + \dfrac{9}{26}i$

31. $\dfrac{2i}{10-5i} = \dfrac{2i}{10-5i} \cdot \dfrac{10+5i}{10+5i} = \dfrac{20i+10i^2}{(10-5i)(10+5i)} = \dfrac{-10+20i}{100-25i^2} = \dfrac{-10+20i}{125} = -\dfrac{2}{25} + \dfrac{4}{25}i$

33. $\dfrac{3}{-i} \cdot \dfrac{-i}{-i} = \dfrac{-3i}{i^2} = \dfrac{-3i}{-1} = 3i$

35. $\dfrac{-2+i}{(1+i)^2} = \dfrac{-2+i}{(1+i)(1+i)} = \dfrac{-2+i}{(1)(1) + (1)(i) + (1)(i) + (i)(i)} = \dfrac{-2+i}{1+i+i+i^2} = \dfrac{-2+i}{1+2i-1} =$

    $\dfrac{-2+i}{2i} \cdot \dfrac{i}{i} = \dfrac{-2i+i^2}{2i^2} = \dfrac{-2i-1}{2(-1)} = \dfrac{-2i-1}{-2} = \dfrac{1}{2} + i$

37. $(23-5.6i) + (-41.5+93i) = -18.5 + 87.4i$

39. $(17.1-6i) - (8.4+0.7i) = 8.7 - 6.7i$

41. $(-12.6-5.7i)(5.1-9.3i) = -117.27 + 88.11i$

43. $\dfrac{17-135i}{18+142i} \approx -0.921 - 0.236i$

Quadratic Equations with Complex Solutions

45. $x^2 + 5 = 0 \Rightarrow x^2 = -5 \Rightarrow x = \pm\sqrt{-5} = \pm i\sqrt{5}$

47. $5x^2 + 1 = 3x^2 \Rightarrow 2x^2 = -1 \Rightarrow x^2 = -\dfrac{1}{2} \Rightarrow x = \pm i\sqrt{\dfrac{1}{2}}$

49. $3x = 5x^2 + 1 \Rightarrow 5x^2 - 3x + 1 = 0$; use the quadratic formula with $a = 5$, $b = -3$, and $c = 1$.

$$x = \frac{-(-3) \pm \sqrt{(-3)^2 - 4(5)(1)}}{2(5)} = \frac{3 \pm \sqrt{9 - 20}}{10} = \frac{3 \pm \sqrt{-11}}{10} = \frac{3 \pm i\sqrt{11}}{10} = \frac{3}{10} \pm \frac{i\sqrt{11}}{10}$$

51. Use the quadratic formula to solve $x^2 - 4x + 5 = 0$ with $a = 1$, $b = -4$, and $c = 5$.

$$x = \frac{-(-4) \pm \sqrt{(-4)^2 - 4(1)(5)}}{2(1)} = \frac{4 \pm \sqrt{16 - 20}}{2} = \frac{4 \pm \sqrt{-4}}{2} = \frac{4 \pm 2i}{2} = 2 \pm i$$

53. Use the quadratic formula to solve $x^2 - 3x + 5 = 0$ with $a = 1$, $b = -3$, and $c = 5$.

$$x = \frac{-(-3) \pm \sqrt{(-3)^2 - 4(1)(5)}}{2(1)} = \frac{3 \pm \sqrt{9 - 20}}{2} = \frac{3 \pm \sqrt{-11}}{2} = \frac{3 \pm i\sqrt{11}}{2} = \frac{3}{2} \pm \frac{i\sqrt{11}}{2}$$

55. Use the quadratic formula to solve $x^2 + 2x + 4 = 0$ with $a = 1$, $b = 2$, and $c = 4$.

$$x = \frac{-2 \pm \sqrt{2^2 - 4(1)(4)}}{2(1)} = \frac{-2 \pm \sqrt{4 - 16}}{2} = \frac{-2 \pm \sqrt{-12}}{2} = \frac{-2 \pm 2i\sqrt{3}}{2} = -1 \pm i\sqrt{3}$$

57. $3x^2 - 4x = x^2 - 3 \Rightarrow 2x^2 - 4x + 3 = 0$ Using the quadratic formula we get:

$$\frac{4 \pm \sqrt{16 - 4(2)(3)}}{2(2)} = \frac{4 \pm \sqrt{-8}}{4} = \frac{4 \pm 2\sqrt{-2}}{4} = 1 + \frac{\sqrt{-2}}{2} = 1 \pm \frac{i\sqrt{2}}{2}$$

59. $2x(x - 2) = x - 4 \Rightarrow 2x^2 - 4x = x - 4 \Rightarrow 2x^2 - 5x + 4 = 0$

Using the quadratic formula we get: $\dfrac{5 \pm \sqrt{25 - 4(2)(4)}}{2(2)} = \dfrac{5 \pm \sqrt{-7}}{4} = \dfrac{5}{4} \pm \dfrac{i\sqrt{7}}{4}$

61. $3x(3 - x) - 8 = x(x - 2) \Rightarrow 9x - 3x^2 - 8 = x^2 - 2x \Rightarrow -4x^2 + 11x - 8 = 0$

Using the quadratic formula we get: $\dfrac{-11 \pm \sqrt{121 - 4(-4)(-8)}}{2(-4)} = \dfrac{-11 \pm \sqrt{-7}}{-8} = \dfrac{11}{8} \pm \dfrac{i\sqrt{7}}{8}$

Zeros of Quadratic Polynomials

63. (a) The graph of $f(x) = 2x^2 - x - 3$ intersects the x-axis twice. Since f is degree 2, both of its zeros are real.

    (b) $x = \dfrac{-b \pm \sqrt{b^2 - 4ac}}{2a} = \dfrac{1 \pm \sqrt{1 - 4(2)(-3)}}{2(2)} = \dfrac{1 \pm 5}{4} = \dfrac{3}{2}, -1$

65. (a) The graph of $f(x) = x^2 + x + 2$ does not intersect the x-axis. Since f is degree 2, both of its zeros are imaginary.

    (b) $x = \dfrac{-b \pm \sqrt{b^2 - 4ac}}{2a} = \dfrac{-1 \pm \sqrt{1^2 - 4(1)(2)}}{2(1)} = \dfrac{-1 \pm \sqrt{-7}}{2} = -\dfrac{1}{2} \pm \dfrac{i\sqrt{7}}{2}$

67. (a) The graph of $f(x) = 5x^2 + 4x + 1$ does not intersect the x-axis. Since f is degree 2, both of its zeros are imaginary.

    (b) $x = \dfrac{-b \pm \sqrt{b^2 - 4ac}}{2a} = \dfrac{-4 \pm \sqrt{4^2 - 4(5)(1)}}{2(5)} = \dfrac{-4 \pm \sqrt{-4}}{10} = -\dfrac{2}{5} \pm \dfrac{1}{10}\sqrt{4}\,i = -\dfrac{2}{5} \pm \dfrac{1}{5}i$

Zeros of Polynomials

69. The graph of $f(x)$ does not intersect the x-axis. Therefore, f has no real zeros. Since f is degree 2, there must be two imaginary zeros.

71. The graph of $f(x)$ intersects the x-axis once. Therefore, f has one real zero. Since f is degree 3, there must be two imaginary zeros.

73. The graph of $f(x)$ intersects the x-axis twice. Therefore, f has two real zeros. Since f is degree 4, there must be two imaginary zeros.

75. The graph of $f(x)$ intersects the x-axis three times. Therefore, f has three real zeros. Since f is degree 5, there must be two imaginary zeros.

77. Degree: 2; leading coefficient: 1; zeros: $6i$ and $-6i$

    (a) $f(x) = (x - 6i)(x + 6i)$

    (b) $(x - 6i)(x + 6i) = x^2 - 36i^2 = x^2 + 36$. Thus, $f(x) = x^2 + 36$.

79. Degree: 3; leading coefficient: -1; zeros: -1, $2i$, and $-2i$

    (a) $f(x) = -1(x + 1)(x - 2i)(x + 2i)$

    (b) $-1(x + 1)(x - 2i)(x + 2i) = -1(x + 1)(x^2 - 4i^2) = -1(x + 1)(x^2 + 4) = -(x^3 + 4x + x^2 + 4) =$
    $-x^3 - x^2 - 4x - 4$. Thus, $f(x) = -x^3 - x^2 - 4x - 4$.

81. Degree: 4; leading coefficient: 10; zeros: 1, -1, $3i$, and $-3i$

    (a) $f(x) = 10(x - 1)(x + 1)(x - 3i)(x + 3i)$

    (b) $10(x - 1)(x + 1)(x - 3i)(x + 3i) = 10(x^2 - 1)(x^2 + 9) = 10(x^4 + 8x^2 - 9) = 10x^4 + 80x^2 - 90$.
    Thus, $f(x) = 10x^4 + 80x^2 - 90$.

83. Degree: 4; leading coefficient: $\dfrac{1}{2}$; zeros: $-i$ and $2i$

    (a) Since $f(x)$ has real coefficients, it must also have a third and fourth zero of i and $-2i$ the conjugate of
    $-i$ and $2i$. Therefore the complete factored form is: $f(x) = \dfrac{1}{2}(x + i)(x - i)(x + 2i)(x - 2i)$

    (b) $\dfrac{1}{2}(x + i)(x - i)(x + 2i)(x - 2i) = \dfrac{1}{2}(x^2 - i^2)(x^2 - 4i^2) = \dfrac{1}{2}(x^2 + 1)(x^2 + 4) =$

    $\dfrac{1}{2}(x^4 + 4x^2 + x^2 + 4) = \dfrac{1}{2}x^4 + \dfrac{5}{2}x^2 + 2$. Thus, $f(x) = \dfrac{1}{2}x^4 + \dfrac{5}{2}x^2 + 2$.

85. Degree: 3; leading coefficient: -2; zeros: $1 - i$ and 3

(a) Since $f(x)$ has real coefficients, it must also have a third zero of $1 + i$ the conjugate of

$1 - i$. Therefore the complete factored form is: $f(x) = -2(x - (1 + i))(x - (1 - i))(x - 3)$

(b) $-2(x - (1 + i))(x - (1 - i))(x - 3) = -2(x^2 - x + xi - x - xi + 1 - i^2)(x - 3) =$

$-2(x^2 - 2x + 2)(x - 3) = -2(x^3 - 2x^2 + 2x - 3x^2 + 6x - 6) = -2(x^3 - 5x^2 + 8x - 6) =$

$-2x^3 + 10x^2 - 16x + 12$. Thus, $f(x) = -2x^3 + 10x^2 - 16x + 12$.

87. First divide:

$$
\begin{array}{r}
3x^2 + 75 \\
x - \tfrac{5}{3}\overline{)3x^3 - 5x^2 + 75x - 125} \\
\underline{3x^3 - 5x^2} \\
75x - 125 \\
\underline{75x - 125} \\
0
\end{array}
$$

Now set $3x^2 + 75 = 0$ and solve. $3x^2 + 75 = 0 \Rightarrow 3x^2 = -75 \Rightarrow x^2 = -25 \Rightarrow x = \pm\sqrt{-25} \Rightarrow x = \pm5i$.

The solutions are $x = \dfrac{5}{3}, \pm5i$.

89. If $-3i$ is a zero then $3i$ is also a zero $\Rightarrow (x + 3i)(x - 3i) = x^2 - 9i^2 = x^2 + 9$. Now divide this into the

equation:

$$
\begin{array}{r}
2x^2 - x + 1 \\
x^2 + 9\overline{)2x^4 - x^3 + 19x^2 - 9x + 9} \\
\underline{2x^4 \qquad\quad + 18x^2} \\
-x^3 + x^2 - 9x + 9 \\
\underline{-x^3 \qquad\quad - 9x} \\
x^2 \qquad\quad + 9 \\
\underline{x^2 \qquad\quad + 9} \\
0
\end{array}
$$

Then use the quadratic formula to solve $2x^2 - x + 1$. $\dfrac{1 \pm \sqrt{1 - 4(2)(1)}}{2(2)} = \dfrac{1 \pm \sqrt{-7}}{4} =$

$\dfrac{1}{4} \pm \dfrac{i\sqrt{7}}{4}$. Therefore the solutions are: $x = \pm3i, \dfrac{1}{4} \pm \dfrac{i\sqrt{7}}{4}$.

91. $x^2 + 25 = 0 \Rightarrow x^2 = -25 \Rightarrow x = \pm5i$. Thus, $f(x) = (x - 5i)(x + 5i)$.

93. $3x^3 + 3x = 0 \Rightarrow 3x(x^2 + 1) = 0 \Rightarrow x = 0$ or $\pm i$. Thus, $f(x) = 3(x - 0)(x - i)(x + i)$.

95. $x^4 + 5x^2 + 4 = 0 \Rightarrow (x^2 + 1)(x^2 + 4) = 0 \Rightarrow x = \pm i$ or $\pm2i$. Thus, $f(x) = (x - i)(x + i)(x - 2i)(x + 2i)$.

97. The graph of $y = x^4 + 2x^3 + x^2 + 8x - 12$ is shown in Figure 97. The x-intercepts appear to be -3 and 1.

We can use synthetic division to help factor f.

$$
\begin{array}{r|rrrrr}
1 & 1 & 2 & 1 & 8 & -12 \\
  &   & 1 & 3 & 4 & 12 \\
\hline
  & 1 & 3 & 4 & 12 & 0
\end{array}
$$

$x^4 + 2x^3 + x^2 + 8x - 12 = (x - 1)(x^3 + 3x^2 + 4x + 12)$

$$
\begin{array}{r|rrrr}
-3 & 1 & 3 & 4 & 12 \\
   &   & -3 & 0 & -12 \\
\hline
   & 1 & 0 & 4 & 0
\end{array}
$$

$x^4 + 2x^3 + x^2 + 8x - 12 = (x - 1)(x^3 + 3x^2 + 4x + 12) = (x - 1)(x + 3)(x^2 + 4) =$

$(x - 1)(x + 3)(x - 2i)(x + 2i)$. Thus, $f(x) = (x - 1)(x + 3)(x - 2i)(x + 2i)$.

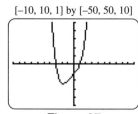

[−10, 10, 1] by [−50, 50, 10]

Figure 97

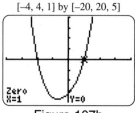

[−4, 4, 1] by [−20, 20, 5]

Figure 107a

[−4, 4, 1] by [−20, 20, 5]

Figure 107b

99. $x^3 + x = 0 \Rightarrow x(x^2 + 1) = 0 \Rightarrow x(x + i)(x - i) = 0 \Rightarrow x = 0, \pm i$

101. Factor or find the zero of 2 graphically.

$x^3 = 2x^2 - 7x + 14 \Rightarrow x^3 - 2x^2 + 7x - 14 = 0 \Rightarrow x^2(x - 2) + 7(x - 2) = 0 \Rightarrow$

$(x - 2)(x^2 + 7) = 0 \Rightarrow (x - 2)(x + i\sqrt{7})(x - i\sqrt{7}) = 0 \Rightarrow x = 2, \pm i\sqrt{7}$

103. $x^4 + 5x^2 = 0 \Rightarrow x^2(x^2 + 5) = x^2(x + i\sqrt{5})(x - i\sqrt{5}) = 0 \Rightarrow x = 0, \pm i\sqrt{5}$

105. $x^4 = x^3 - 4x^2 \Rightarrow x^4 - x^3 + 4x^2 = 0 \Rightarrow x^2(x^2 - x + 4) = 0$

Use the quadratic formula to find the zeros of $x^2 - x + 4$. The solutions are $x = 0, \dfrac{1}{2} \pm \dfrac{i\sqrt{15}}{2}$.

107. Find the zeros of 1 and 2 using the rational zero test or a graph. See Figures 107a & 107b.

$x^4 + x^3 = 16 - 8x - 6x^2 \Rightarrow x^4 + x^3 + 6x^2 + 8x - 16 = 0$

The graph of $x^4 + x^3 + 6x^2 + 8x - 16$ shows that -2 and 1 are zeros, so $(x + 2)$ and $(x - 1)$ are factors; using synthetic division twice gives the missing factor.

$$
\begin{array}{r|rrrrr}
-2 & 1 & 1 & 6 & 8 & -16 \\
   &   & -2 & 2 & -16 & 16 \\
\hline
   & 1 & -1 & 8 & -8 & 0
\end{array}
$$

$$
\begin{array}{r|rrrr}
1 & 1 & -1 & 8 & -8 \\
  &   & 1 & 0 & 8 \\
\hline
  & 1 & 0 & 8 & 0
\end{array}
$$

$x^4 + x^3 + 6x^2 + 8x - 16 = (x + 2)(x - 1)(x^2 + 8) = (x + 2)(x - 1)(x + i\sqrt{8})(x - i\sqrt{8})$

The solutions are $x = -2, 1, \pm i\sqrt{8}$.

109. Find the zero of –2 using the rational zero test or a graph of $y = 3x^3 + 4x^2 - x + 6$. Use synthetic division to factor $3x^3 + 4x^2 - x + 6$.

$$\begin{array}{r|rrrr} -2 & 3 & 4 & -1 & 6 \\ & & -6 & 4 & -6 \\ \hline & 3 & -2 & 3 & 0 \end{array}$$

$3x^3 + 4x^2 - x + 6 = (x + 2)(3x^2 - 2x + 3)$

Use the quadratic formula to find the zeros of $3x^2 - 2x + 3$. The solutions are $x = -2, \dfrac{1}{3} \pm \dfrac{i\sqrt{8}}{3}$.

Applications

111. $Z = \dfrac{V}{I} = \dfrac{50 + 98i}{8 + 5i} = 10 + 6i$

113. $V = IZ = (1 + 2i)(3 - 4i) = 11 + 2i$

115. $I = \dfrac{V}{Z} = \dfrac{27 + 17i}{22 - 5i} = 1 + i$

4.5: Rational Functions and Models

Rational Functions

1. Yes, since the numerator and denominator are both polynomials. Since $4x - 5 \ne 0, D = \left\{ x \;\middle|\; x \ne \dfrac{5}{4} \right\}$.

3. Yes, since f can be written as $f(x) = \dfrac{x^2 - x - 2}{1}$, and $g(x) = 1$ is a polynomial. $D =$ all real numbers.

5. No, since the numerator is not a polynomial. $D = \{x \mid x \ne -1\}$

7. Yes, since the numerator and denominator are both polynomials. Since $x^2 + 1 \ne 0, D =$ all real numbers.

9. No, since the numerator is not a polynomial. Since $x^2 + x \ne 0 \Rightarrow x(x + 1) \ne 0, D = \{x \mid x \ne -1, x \ne 0\}$.

11. Yes, since f can be written as $f(x) = \dfrac{4(x + 1) - 3}{x + 1} \Rightarrow f(x) = \dfrac{4x + 1}{x + 1}$. Since $x + 1 \ne 0, D = \{x \mid x \ne -1\}$.

Asymptotes

13. $f(x) \to \infty$ as $x \to 0^-$; $f(x) \to \infty$ as $x \to 0^+$; $f(x) \to 0$ as $x \to -\infty$; $f(x) \to 0$ as $x \to \infty$.

15. There is a horizontal asymptote of $y = 4$ and a vertical asymptote of $x = 2$; $D = \{x \mid x \ne 2\}$.

17. There is a horizontal asymptote of $y = -4$ and a vertical asymptote of $x = \pm 2$; $D = \{x \mid x \ne \pm 2\}$.

19. There is a horizontal asymptote of $y = 0$ and there is no vertical asymptote; $D =$ all real numbers.

21. Horizontal Asymptotes: The degree of the numerator is equal to the degree of the denominator and the ratio of the leading coefficients is $\dfrac{4}{2} = 2$. Therefore, $y = 2$ is a horizontal asymptote.

Vertical Asymptotes: Find the zeros of the denominator, $2x - 6 = 0 \Rightarrow 2x = 6 \Rightarrow x = 3$. Therefore $x = 3$ is a vertical asymptote. (Note that 3 is not a zero of the numerator.)

23. Horizontal Asymptotes: The degree of the numerator is less than the degree of the denominator, therefore, the x-axis, $y = 0$, is a horizontal asymptote.

    Vertical Asymptotes: Find the zeros of the denominator, $x^2 - 5 = 0 \Rightarrow x^2 = 5 \Rightarrow x = \pm\sqrt{5}$. Therefore $x = \pm\sqrt{5}$ are vertical asymptotes. (Note that $\pm\sqrt{5}$ are not zeros of the numerator.)

25. Horizontal Asymptotes: The degree of the numerator is greater than the degree of the denominator, therefore there is no horizontal asymptote.

    Vertical Asymptotes: Find the zeros of the denominator,

    $x^2 + 3x - 10 = 0 \Rightarrow (x + 5)(x - 2) = 0 \Rightarrow x = -5, 2$. Therefore $x = -5$ and $x = 2$ are the vertical asymptotes. (Note that -5 and 2 are not a zeros of the numerator.)

27. Horizontal Asymptotes: The degree of the numerator is equal to the degree of the denominator and the ratio of the leading coefficients is $\frac{1}{2}$. Therefore, $y = \frac{1}{2}$ is a horizontal asymptote.

    Vertical Asymptotes: Find the zeros of the denominator, $(2x - 5)(x + 1) = 0 \Rightarrow x = \frac{5}{2}$ and $x = -1$.

    Therefore $x = \frac{5}{2}$ is a vertical asymptote. (Note that $\frac{5}{2}$ is not a zero of the numerator and $x = -1$ is a zero of the numerator).

29. Horizontal Asymptotes: The degree of the numerator is equal to the degree of the denominator and the ratio of the leading coefficients is $\frac{3}{1} = 3$. Therefore, $y = 3$ is a horizontal asymptote.

    Vertical Asymptotes: Find the zeros of the denominator, $(x + 2)(x - 1) = 0 \Rightarrow x = -2, 1$. Here, only $x = 1$ is a vertical asymptote. (Note that -2 is a zeros of the numerator.)

31. Horizontal Asymptotes: The degree of the numerator is greater than the degree of the denominator, therefore there is no horizontal asymptote.

    Vertical Asymptotes: Find the zeros of the denominator, $x + 3 = 0 \Rightarrow x = -3$. But $x = -3$ is not a vertical asymptote since -3 is a zero of the numerator and $f(x) = x - 3$ for $x \neq -3$. There are no vertical asymptotes.

33. $f(x) = \dfrac{a}{x - 1}$ has a vertical asymptote of $x = 1$. It has a horizontal asymptote of $y = 0$, since the degree of the numerator is less than the degree of the denominator. The best choice is graph b.

35. $f(x) = \dfrac{x - a}{x + 2}$ has a vertical asymptote of $x = -2$ and a horizontal asymptote of $y = 1$, since the degree of the numerator equals the degree of the denominator and the ratio of the leading coefficients is $\frac{1}{1}$.

    The best choice is graph d.

37. $g(x) = \dfrac{1}{x - 3}$ is $y = \dfrac{1}{x}$ transformed 3 units right $\Rightarrow$ it has a vertical asymptote of $x = 3$. It still has a horizontal asymptote of $y = 0$. See Figure 37. Then $g(x) = f(x - 3)$.

39. $g(x) = \dfrac{1}{x} + 2$ is $y = \dfrac{1}{x}$ transformed 2 units up $\Rightarrow$ it has a horizontal asymptote of $y = 2$. It still has a vertical asymptote of $x = 0$. See Figure 39. Then $g(x) = f(x) + 2$.

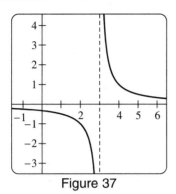

Figure 37

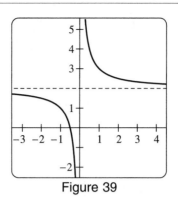

Figure 39

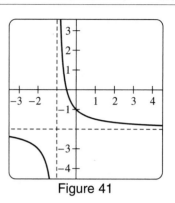

Figure 41

41. $g(x) = \dfrac{1}{x + 1} - 2$ is $y = \dfrac{1}{x}$ transformed 1 unit left and 2 units down. It now has a vertical asymptote of

   $x = -1$, and a horizontal asymptote of $y = -2$. See Figure 41. Then $g(x) = f(x + 1) - 2$.

43. $g(x) = -\dfrac{2}{(x - 1)^2}$ is $y = \dfrac{1}{x^2}$ transformed by reflecting it across the x-axis, then shifting 1 unit right, and

   making it slightly wider. It now has a vertical asymptote of $x = 1$, and it still has a horizontal asymptote of

   $y = 0$. See Figure 43. Then $g(x) = -2h(x - 1)$.

45. $g(x) = \dfrac{1}{(x + 1)^2} - 2$ is $y = \dfrac{1}{x^2}$ transformed 1 unit left and 2 units down. It now has a vertical asymptote of

   $x = -1$, and a horizontal asymptote of $y = -2$. See Figure 45. Then $g(x) = h(x + 1) - 2$.

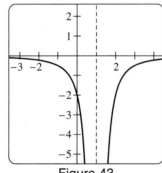

Figure 43

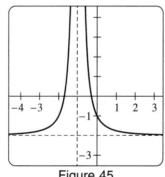

Figure 45

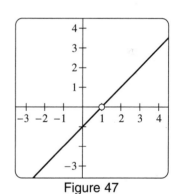

Figure 47

47. First divide:

$$
\begin{array}{r}
x - 1 \\
x - 1 \overline{\smash{)}\, x^2 - 2x + 1} \\
\underline{x^2 - x} \\
-x + 1 \\
\underline{-x + 1} \\
0
\end{array}
$$

Now graph $x - 1$; since the denominator $x - 1 \neq 0$ there is a hole at $x = 1$ in the graph. See Figure 47.

49. First we find the asymptotes. Since the degree of the numinator and denominator is the same, the ratio of the lead coefficients $\frac{1}{1}$ or $y = 1$ is the horizontal asymptote. To find the vertical asymptote we find the zero of the denominator. $x + 1 = 0 \Rightarrow x = -1$. Since $x = -1$ is a vertical asymptote it has no holes. See Figure 49.

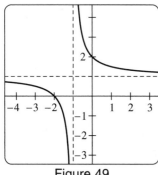
Figure 49

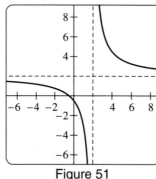
Figure 51

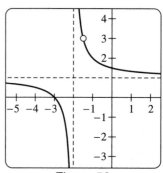
Figure 53

51. First divide: $g(x) = \frac{(2x + 1)(x - 2)}{(x - 2)(x - 2)} = \frac{2x + 1}{x - 2}$. Now graph $\frac{2x + 1}{x - 2}$ by finding it's asymptotes. The ratio of the leading coefficients $\frac{2}{1}$ or $y = 2$ is the horizontal asymptote. To find the vertical asymptote we find the zero of the denominator. $x - 2 = 0 \Rightarrow x = 2$. Since $x = 2$ is an asymptote there are no holes. See Figure 51.

53. Factor the numinator and denominator, then divide: $f(x) = \frac{2x^2 + 9x + 9}{2x^2 + 7x + 6} = \frac{(2x + 3)(x + 3)}{(2x + 3)(x + 2)} = \frac{x + 3}{x + 2}$.

Now graph $\frac{x + 3}{x + 2}$ by finding it's asymptotes. The ratio of the leading coefficients $\frac{1}{1}$ or $y = 1$ is the horizontal asymptote. To find the vertical asymptote we find the zero of the denominator. $x + 2 = 0 \Rightarrow x = -2$.

Since $x = -2$ is an asymptote there is only a hole is at $2x + 3 = 0$ or $x = \frac{-3}{2}$. See Figure 53.

55. Factor the numinator and denominator, then divide: $f(x) = \frac{-2x^2 + 11x - 14}{x^2 - 5x + 6} = \frac{(-2x + 7)(x - 2)}{(x - 3)(x - 2)} = \frac{-2x + 7}{x - 3}$.

Now graph $\frac{-2x + 7}{x - 3}$ by finding it's asymptotes. The ratio of the leading coefficients $\frac{-2}{1}$ or $y = -2$ is the horizontal asymptote. To find the vertical asymptote we find the zero of the denominator.

$x - 3 = 0 \Rightarrow x = 3$. Since $x = 3$ is an asymptote there is only a hole is at $x = 2$. See Figure 55.

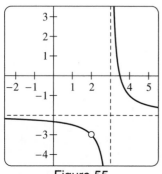
Figure 55

57. (a) $x - 2 = 0 \Rightarrow x = 2; D = \{x \mid x \neq 2\}$

    (b) The graph of f using dot mode is shown in Figure 57b.

    (c) Since the degree of the numerator equals the degree of the denominator and the ratio of the leading

    coefficients is $\dfrac{1}{1} = 1$, the horizontal asymptote is $y = 1$. There is a vertical asymptote at $x = 2$.

    (d) First, sketch the vertical and horizontal asymptotes found in part (c). Then use Figure 57b as a guide to a

    more complete graph of f. The sketch is shown in Figure 57d.

[−8, 8, 1] by [−8, 8, 1]

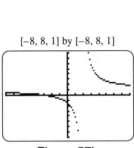

Figure 57b

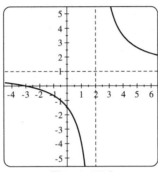

Figure 57d

59. (a) $x^2 - 4 = 0 \Rightarrow x^2 = 4 \Rightarrow x = \pm 2; D = \{x \mid x \neq \pm 2\}$

    (b) The graph of f using dot mode is shown in Figure 59b.

    (c) Since the degree of the numerator is less than the degree of the denominator there is a horizontal asymptote

    at $y = 0$. There are vertical asymptotes at $x = \pm 2$.

    (d) First, sketch the vertical and horizontal asymptotes found in part (c). Then use Figure 59b as a guide to a

    more complete graph of f. The sketch is shown in Figure 59d.

[−8, 8, 1] by [−8, 8, 1]

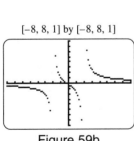

Figure 59b

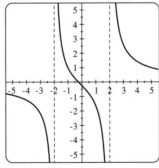

Figure 59d

61. (a) $1 - 0.25x^2 = 0 \Rightarrow 0.25x^2 = 1 \Rightarrow x^2 = 4 \Rightarrow x = \pm 2$; $D = \{x \mid x \neq \pm 2\}$

    (b) The graph of f using dot mode is shown in Figure 61b.

    (c) Since the degree of the numerator is less than the degree of the denominator there is a horizontal asymptote at $y = 0$. There are vertical asymptotes at $x = \pm 2$.

    (d) First, sketch the vertical and horizontal asymptotes found in part (c). Then use Figure 61b as a guide to a more complete graph of f. The sketch is shown in Figure 61d.

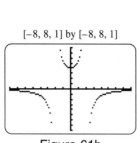

[−8, 8, 1] by [−8, 8, 1]

Figure 61b

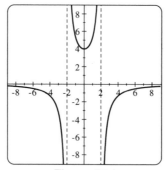

Figure 61d

63. (a) $x - 2 = 0 \Rightarrow x = 2$; $D = \{x \mid x \neq 2\}$

    (b) The graph of f using dot mode is shown in Figure 63b.

    (c) Since the degree of the numerator is greater than the degree of the denominator there is no horizontal asymptote. There are no vertical asymptotes. (The numerator equals 2 when $x = 2$).

    (d) First, sketch the vertical and horizontal asymptotes found in part (c). Then use Figure 63b as a guide to a more complete graph of f. The sketch is shown in Figure 63d.

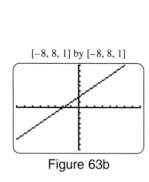

[−8, 8, 1] by [−8, 8, 1]

Figure 63b

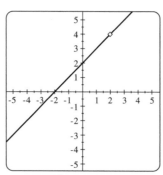

Figure 63d

65. Since the output of Y_1 gets closer to 3 as the input x gets larger, it is reasonable to conjecture that the equation for the horizontal asymptote is $y = 3$.

67. One example of a symbolic representation of a rational function with a vertical asymptote of $x = -3$ and a horizontal asymptote of $y = 1$ is $f(x) = \dfrac{x + 1}{x + 3}$. *Answers may vary.*

69. One example of a symbolic representation of a rational function with vertical asymptotes of $x = \pm 3$ and a horizontal asymptote of $y = 0$ is $f(x) = \dfrac{1}{x^2 - 9}$. *Answers may vary.*

71. There is a vertical asymptote at $x = -1$ since -1 is a zero of the denominator but is not a zero of the numerator.

   To find any slant asymptote we will divide the numerator by the denominator using synthetic division:

$$\begin{array}{r|rrr} -1 & 1 & 0 & 1 \\ & & -1 & 1 \\ \hline & 1 & -1 & 2 \end{array}$$

   Therefore, $f(x) = x - 1 + \dfrac{2}{x + 1}$ and as $|x|$ becomes large, $f(x)$ approaches $y = x - 1$. There is a slant

   asymptote at $y = x - 1$. The graph of f using dot mode is shown in Figure 71a. To sketch this graph, first

   sketch the vertical and slant asymptotes found above. Then use Figure 71a as a guide to a more complete graph

   of f. The sketch is shown in Figure 71b.

[−8, 8, 1] by [−8, 8, 1]

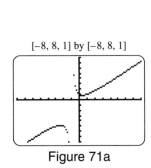

Figure 71a

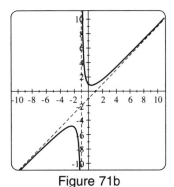

Figure 71b

73. There is a vertical asymptote at $x = -2$ since -2 is a zero of the denominator but is not a zero of the numerator.

   To find any slant asymptote we will divide the numerator by the denominator using synthetic division:

$$\begin{array}{r|rrr} -2 & 0.5 & -2 & 2 \\ & & -1 & 6 \\ \hline & 0.5 & -3 & 8 \end{array}$$

   Therefore, $f(x) = 0.5x - 3 + \dfrac{8}{x + 2}$ and as $|x|$ becomes large, $f(x)$ approaches $y = 0.5x - 3$. There is a slant

   asymptote at $y = 0.5x - 3$. The graph of f using dot mode is shown in Figure 73a. To sketch this graph, first

   sketch the vertical and slant asymptotes found above. Then use Figure 73a as a guide to a more complete graph

   of f. The sketch is shown in Figure 73b.

[−14, 14, 2] by [−14, 14, 2]

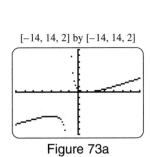

Figure 73a

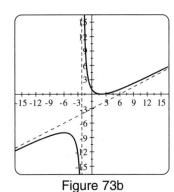

Figure 73b

75. There is a vertical asymptote at $x = 1$ since 1 is a zero of the denominator but is not a zero of the numerator.

To find any slant asymptote we will divide the numerator by the denominator using synthetic division:

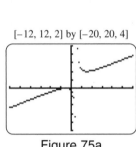

Therefore, $f(x) = x + 3 + \dfrac{4}{x - 1}$ and as $|x|$ becomes large, $f(x)$ approaches $y = x + 3$. There is a slant asymptote at $y = x + 3$. The graph of f using dot mode is shown in Figure 75a. To sketch this graph, first sketch the vertical and slant asymptotes found above. Then use Figure 75a as a guide to a more complete graph of f. The sketch is shown in Figure 75b.

$[-12, 12, 2]$ by $[-20, 20, 4]$

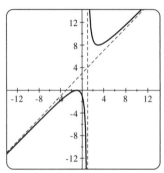

Figure 75a   Figure 75b

77. There is a vertical asymptote at $x = \dfrac{1}{2}$ since $\dfrac{1}{2}$ is a zero of the denominator but is not a zero of the numerator. To find any slant asymptote we will divide the numerator by the denominator.

$$
\begin{array}{r}
2x + 1 + \frac{1}{2x-1} \\
2x - 1 \overline{)\,4x^2 + 0x + 0} \\
\underline{4x^2 - 2x} \\
2x + 0 \\
\underline{2x - 1} \\
1
\end{array}
$$

Therefore, $f(x) = 2x + 1 + \dfrac{1}{2x - 1}$ and as $|x|$ becomes large, $f(x)$ approaches $y = 2x + 1$. There is a slant asymptote at $y = 2x + 1$. The graph of f using dot mode is shown in Figure 77a. To sketch this graph, first sketch the vertical and slant asymptotes found above. Then use Figure 77a as a guide to a more complete graph of f. The sketch is shown in Figure 77b.

$[-3, 7, 1]$ by $[-3, 7, 1]$

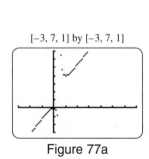
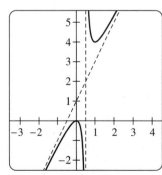

Figure 77a   Figure 77b

Rational Equations

79. (a) $\dfrac{2x}{x + 2} = 6 \Rightarrow 2x = 6(x + 2) \Rightarrow 2x = 6x + 12 \Rightarrow 4x = -12 \Rightarrow x = -3$

   (b) Graph $Y_1 = (2X)/(X + 2)$ and $Y_2 = 6$ in $[-10, 10, 1]$ by $[-10, 10, 1]$ using dot mode. See Figure 79b.

      The intersection point is $(-3, 6)$. The solution is $x = -3$.

   (c) Table $Y_1 = (2X)/(X + 2)$ starting at $x = -5$ and incrementing by 1. See Figure 79c. The solution is $x = -3$.

$[-10, 10, 1]$ by $[-10, 10, 1]$

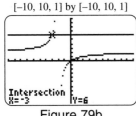
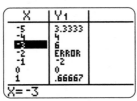

          Figure 79b   Figure 79c

81. (a) $2 - \dfrac{5}{x} + \dfrac{2}{x^2} = 0 \Rightarrow \dfrac{2x^2}{x^2} - \dfrac{5x}{x^2} + \dfrac{2}{x^2} = 0 \Rightarrow \dfrac{2x^2 - 5x + 2}{x^2} = 0 \Rightarrow 2x^2 - 5x + 2 = 0 \Rightarrow$

      $(2x - 1)(x - 2) = 0 \Rightarrow x = \dfrac{1}{2}, 2$

   (b) Graph $Y_1 = 2 - 5/X + 2/X^2$ in $[-10, 10, 1]$ by $[-10, 10, 1]$ using dot mode. The x-intercepts are the

      points is $(0.5, 0)$ and $(2, 0)$. The solutions are $x = \dfrac{1}{2}, 2$.

   (c) Table $Y_1 = 2 - 5/X + 2/X^2$ starting at $x = -0.5$ and incrementing by 0.5. The solutions are $x = \dfrac{1}{2}, 2$.

83. (a) $\dfrac{1}{x + 1} + \dfrac{1}{x - 1} = \dfrac{1}{x^2 - 1} \Rightarrow \dfrac{x - 1}{x^2 - 1} + \dfrac{x + 1}{x^2 - 1} = \dfrac{1}{x^2 - 1} \Rightarrow \dfrac{2x}{x^2 - 1} = \dfrac{1}{x^2 - 1} \Rightarrow 2x = 1 \Rightarrow x = \dfrac{1}{2}$

   (b) Graph $Y_1 = 1/(X + 1) + 1/(X - 1)$ and $Y_2 = 1/(X^2 - 1)$ in $[-2, 2, 1]$ by $[-5, 5, 1]$ using dot mode. The

      intersection point is $\left(\dfrac{1}{2}, \dfrac{4}{3}\right)$. The solution is $x = \dfrac{1}{2}$.

   (c) Table $Y_1 = 1/(X + 1) + 1/(X - 1) - 1/(X^2 - 1)$ starting at $x = -1$ and incrementing by 0.5. Find the

      x-value where $Y_1 = 0$. The solution is $x = \dfrac{1}{2}$.

85. $\dfrac{x + 1}{x - 5} = 0 \Rightarrow x + 1 = 0(x - 5) \Rightarrow x + 1 = 0 \Rightarrow x = -1$; Check: $\dfrac{-1 + 1}{-1 - 5} = \dfrac{0}{-6} = 0$

87. $\dfrac{6(1 - 2x)}{x - 5} = 4 \Rightarrow 6(1 - 2x) = 4(x - 5) \Rightarrow 6 - 12x = 4x - 20 \Rightarrow -16x = -26 \Rightarrow x = \dfrac{13}{8}$

    Check: $\dfrac{6(1 - 2(\frac{13}{8}))}{\frac{13}{8} - 5} = \dfrac{6(-\frac{9}{4})}{-\frac{27}{8}} = \dfrac{-\frac{27}{2}}{-\frac{27}{8}} = -\dfrac{27}{2} \cdot \left(-\dfrac{8}{27}\right) = 4$

89. $\dfrac{1}{x + 2} + \dfrac{1}{x} = 1 \Rightarrow x + (x + 2) = x(x + 2) \Rightarrow 2x + 2 = x^2 + 2x \Rightarrow x^2 = 2 \Rightarrow x = \pm\sqrt{2}$

    Check: $\dfrac{1}{\sqrt{2} + 2} + \dfrac{1}{\sqrt{2}} = \dfrac{\sqrt{2}}{\sqrt{2}(\sqrt{2} + 2)} + \dfrac{\sqrt{2} + 2}{\sqrt{2}(\sqrt{2} + 2)} = \dfrac{2\sqrt{2} + 2}{2 + 2\sqrt{2}} = 1$

    Check: $\dfrac{1}{-\sqrt{2} + 2} + \dfrac{1}{-\sqrt{2}} = \dfrac{-\sqrt{2}}{-\sqrt{2}(-\sqrt{2} + 2)} + \dfrac{-\sqrt{2} + 2}{-\sqrt{2}(-\sqrt{2} + 2)} = \dfrac{2 - 2\sqrt{2}}{2 - 2\sqrt{2}} = 1$

91. $\dfrac{1}{x} - \dfrac{2}{x^2} = 5 \Rightarrow x - 2 = 5x^2 \Rightarrow 5x^2 - x + 2 = 0$; This quadratic equation has no solutions since the

discriminant is negative: $(-1)^2 - 4(5)(2) = 1 - 40 = -39 < 0$. No real solution.

93. $\dfrac{x^3 - 4x}{x^2 + 1} = 0 \Rightarrow x^3 - 4x = 0(x^2 + 1) \Rightarrow x^3 - 4x = 0 \Rightarrow x(x + 2)(x - 2) = 0 \Rightarrow x = 0, -2, 2$

   Check: $\dfrac{(0)^3 - 4(0)}{(0)^2 + 1} = \dfrac{0}{1} = 0$; Check: $\dfrac{(-2)^3 - 4(-2)}{(-2)^2 + 1} = \dfrac{0}{5} = 0$; Check: $\dfrac{(2)^3 - 4(2)}{(2)^2 + 1} = \dfrac{0}{5} = 0$

95. $\dfrac{35}{x^2} = \dfrac{4}{x} + 15 \Rightarrow 35 = 4x + 15x^2 \Rightarrow 15x^2 + 4x - 35 = 0 \Rightarrow (5x - 7)(3x + 5) = 0 \Rightarrow x = \dfrac{-5}{3}, \dfrac{7}{5}$

   Check: $\dfrac{35}{(\frac{-5}{3})^2} = \dfrac{4}{\frac{-5}{3}} + 15 \Rightarrow \dfrac{63}{5} = \dfrac{-12}{5} + \dfrac{75}{5} \Rightarrow \dfrac{63}{5} = \dfrac{63}{5}$;

   Check: $\dfrac{35}{(\frac{7}{5})^2} = \dfrac{4}{\frac{7}{5}} + 15 \Rightarrow \dfrac{125}{7} = \dfrac{20}{7} + \dfrac{105}{7} \Rightarrow \dfrac{125}{7} = \dfrac{125}{7}$

97. $\dfrac{x + 5}{x + 2} = \dfrac{x - 4}{x - 10} \Rightarrow (x + 5)(x - 10) = (x - 4)(x + 2) \Rightarrow x^2 - 5x - 50 = x^2 - 2x - 8 \Rightarrow$

   $3x + 42 = 0 \Rightarrow x = -14$; Check: $\dfrac{(-14) + 5}{(-14) + 2} = \dfrac{(-14) - 4}{(-14) - 10} \Rightarrow \dfrac{-9}{-12} = \dfrac{-18}{-24} \Rightarrow \dfrac{3}{4} = \dfrac{3}{4}$

99. $\dfrac{1}{x - 2} - \dfrac{2}{x - 3} = \dfrac{3}{x^2 - 5x + 6} \Rightarrow \dfrac{1}{x - 2} - \dfrac{2}{x - 3} = \dfrac{3}{(x - 2)(x - 3)} \Rightarrow x - 3 - 2(x - 2) = 3 \Rightarrow$

   $x - 3 - 2x + 4 = 3 \Rightarrow -x - 2 = 0 \Rightarrow x = -2$; Check: $\dfrac{1}{(-2) - 2} - \dfrac{2}{(-2) - 3} = \dfrac{3}{(-2)^2 - 5(-2) + 6} \Rightarrow$

   $-\dfrac{1}{4} - \left(-\dfrac{2}{5}\right) = \dfrac{3}{20} \Rightarrow -\dfrac{5}{20} + \dfrac{8}{20} = \dfrac{3}{20} \Rightarrow \dfrac{3}{20} = \dfrac{3}{20}$

Applications

101. (a) $T(4) = -\dfrac{1}{4 - 8} \Rightarrow T(4) = \dfrac{1}{4} \Rightarrow T(4) = 0.25$; when vehicles leave the ramp at an average rate of 4

   vehicles per minute, the wait is 0.25 minutes or 15 seconds.

   $T(7.5) = \dfrac{1}{7.5 - 8} \Rightarrow T(7.5) = \dfrac{1}{0.5} \Rightarrow T(7.5) = 2.0$; when vehicles leave the ramp at an average rate of

   7.5 vehicles per minute, the wait is 2 minutes.

   (b) The wait increases dramatically.

103. (a) Graph $Y_1 = (10X + 1)/(X + 1)$ and $Y_2 = 10$ in [0, 14, 1] by [0, 14, 1]. Since the degree of the

   numerator and denominator are equal, there is a horizontal asymptote at $y = \dfrac{10}{1} = 10$. See Figure 103.

   (b) The initial population would be $f(0) = 1$ million insects.

   (c) After many months the population starts to level off at 10 million.

   (d) The horizontal asymptote $y = 10$ represents the limiting population after a very long time.

105. (a) Since 15 seconds equals 0.25 minute, graph $Y_1 = (X - 5)(X^2 - 10X)$ and $Y_2 = 0.25$ as shown in Figure 105. The intersection point is near $(12.4, 0.25)$. The gate should admit 12.4 cars per minute on average to keep the wait less than 15 seconds. Note: The reason the answer is greater than 10 cars per minute is because cars are arriving randomly. For some minutes, more than 10 vehicles might arrive at the gate.

(b) $\dfrac{12.4}{5} = 2.48$ or 3 parking attendants must be on duty to keep the average wait less than 15 seconds.

$[0, 14, 1]$ by $[0, 14, 1]$ $[10, 15, 1]$ by $[0, 1, 0.1]$ $[0, 10, 2]$ by $[100, 400, 20]$ $[0, 10, 2]$ by $[100, 400, 20]$

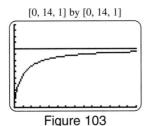   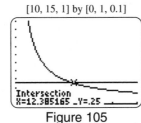   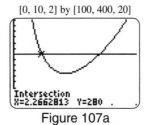   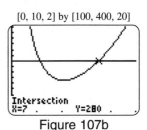

Figure 103   Figure 105   Figure 107a   Figure 107b

107. Let y = height, x = width, and $2x$ = length of the box. To relate the variables, we use the volume formula for a box and the fact that $V = 196$ cubic inches.

$$V = (2x)xy = 2x^2y \Rightarrow y = \frac{V}{2x^2} = \frac{196}{2x^2} = \frac{98}{x^2}$$

The surface area of the box is the sum of the areas of the 6 rectangular sides. We are given that $A = 280$ in^2.

$A = 2(x \cdot y) + 2(2x \cdot y) + 2(x \cdot 2x) \Rightarrow A = 6xy + 4x^2 = 280$; and since $y = \dfrac{98}{x^2}$ we get the equation

$6x\left(\dfrac{98}{x^2}\right) + 4x^2 = 280 \Rightarrow \dfrac{588}{x} + 4x^2 = 280$

Figures 107a and 107b show the intersection points when graphing $Y_1 = 588/X + 4X^2$ and $Y_2 = 280$.

There are two possible solutions (in inches):

width = 7, length = 14, height = $\dfrac{98}{7^2} = 2$ or width ≈ 2.266, length ≈ 4.532, height $\approx \dfrac{98}{2.266^2} \approx 19.086$

109. (a) $V = L \cdot W \cdot H \Rightarrow V = x \cdot x \cdot h \Rightarrow 108 = x^2h \Rightarrow \dfrac{108}{x^2} = h$. The surface area $A(x) = x^2 + 4xh \Rightarrow$

$A(x) = x^2 + 4x\left(\dfrac{108}{x^2}\right) \Rightarrow A(x) = \left(x^2 + \dfrac{432}{x}\right)$. This finds surface area in square inches. To find

square feet we must divide by 144(the number of square inches in a square foot) $\Rightarrow$

$A(x) = \left(x^2 + \dfrac{432}{x}\right) \div 144 \Rightarrow A(x) = \dfrac{x^2}{144} + \dfrac{3}{x}$.

(b) $C = 0.10\left(\dfrac{x^2}{144} + \dfrac{3}{x}\right)$

(c) Graph the equation: $C = 0.10\left(\dfrac{x^2}{144} + \dfrac{3}{x}\right)$. The minimum cost is when $x \approx 6 \Rightarrow h = \dfrac{108}{(6)^2} \Rightarrow h = 3$.

The dimensions are $6 \times 6 \times 3$ inches.

111. (a) $f(x) = \dfrac{2540}{x} \Rightarrow f(400) = \dfrac{2540}{400} = 6.35$ inches. A curve designed for 60 mph with a radius of 400 feet

should have the outer rail elevated 6.35 inches.

(b) Figure 111 shows the graph of $Y_1 = 2540/X$. This means that a sharper curve must be banked more. As

the radius increases, the curve is not as sharp and the elevation of the outer rail decreases.

(c) Since the degree of the numerator is less than the degree of the denominator, the graph of $f(x) = \dfrac{2540}{x}$ has

a horizontal asymptote of $y = 0$. As the radius of the curve increases without bound $(x \rightarrow \infty)$, the tracks

become straight and no elevation or banking $(y \rightarrow 0)$ of the outer rail is necessary.

(d) $f(x) = 12.7 \Rightarrow 12.7 = \dfrac{2540}{x} \Rightarrow x = \dfrac{2540}{12.7} = 200$; a radius of 200 feet requires an elevation of 12.7 inches.

[0, 100, 10] by [0, 10, 1]

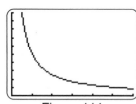
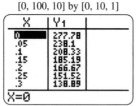

Figure 111   Figure 113

113. (a) $D(0.05) = \dfrac{2500}{30(0.3 + 0.05)} \approx 238$

The braking distance far a car traveling at 50 mph on a 5% uphill grade is about 238 feet.

(b) Table $Y_1 = 2500/(30(0.3 + X))$ starting at $x = 0$ and incrementing by 0.05 as shown in Figure 113.

The table indicates that as the grade x increases, the braking distance decreases. This agrees with intuition.

(c) $220 = \dfrac{2500}{30(0.3 + x)} \Rightarrow 0.3 + x = \dfrac{2500}{30(220)} \Rightarrow x = \dfrac{2500}{30(220)} - 0.3 \approx 0.079$, or 7.9%

115. (a) At $x = 0$, the denominator of $\dfrac{2500}{30x}$ is equal to zero, whereas the numerator is nonzero. There is a vertical

asymptote at $x = 0$. As the coefficient of friction x becomes smaller and approaches 0, the stopping

distance becomes larger and larger without bound. This means that as the road surface becomes very

slippery, the distance required to stop becomes very large. If it were possible for the roadway to be

covered with glare ice having a coefficient of friction of $x = 0$, then the car would continue indefinitely. A

situation similar to this occurs in space where a satellite will travel indefinitely without slowing down.

(b) $340 = \dfrac{2500}{30x} \Rightarrow x = \dfrac{2500}{30(340)} \Rightarrow x = \dfrac{2500}{10,200} = \dfrac{25}{102}$

The coefficient of friction associated with a braking distance of 340 feet is $\dfrac{25}{102} \approx 0.245$.

Variation

117. If $y = \dfrac{k}{x}$ and $y = 2$ when $x = 3$, then $2 = \dfrac{k}{3} \Rightarrow k = 6$.

119. If $y = kx^3$ and $y = 64$ when $x = 2$, then $64 = k(8) \Rightarrow 8k = 64 \Rightarrow k = 8$.

121. If $T = kx^{3/2}$ and $T = 20$ when $x = 4$, then $20 = k(4)^{3/2} \Rightarrow 20 = 8k \Rightarrow k = 2.5$.

    The variation equation becomes $T = 2.5x^{3/2}$. When $x = 16$, $T = 2.5(16)^{3/2} = 2.5(64) = 160$.

123. If $y = \dfrac{k}{x}$ and $y = 5$ when $x = 6$, then $5 = \dfrac{k}{6} \Rightarrow k = 30$.

    The variation equation becomes $y = \dfrac{30}{x}$. When $x = 15$, $y = \dfrac{30}{15} = 2$.

125. If y is inversely proportional to x, then $y = \dfrac{k}{x}$ must hold. So if the value of x is doubled, the right side of this

    equation becomes $\dfrac{k}{2x} = \dfrac{1}{2} \cdot \dfrac{k}{x} = \dfrac{1}{2}y$. Therefore y becomes half its original value.

127. If y is directly proportional to x^3, then $y = kx^3$ must hold. So if the value of x is tripled, the right side of this

    equation becomes $k(3x)^3 = 27 \cdot kx^3 = 27y$. Therefore y becomes 27 times its original value.

129. Since $y = kx^n$, we know that $k = \dfrac{y}{x^n}$ where k is a constant. Using trial and error, let $n = 1$ while calculating

    various values for k using x- and y-values from the table. For example, when $x = 2$ and $y = 2$, the value of k is

    1. But for $x = 4$ and $y = 8$, the value of k is 2. Since the value of k did not remain constant we know that the

    value of n is not 1. Repeat this process for $n = 2$. For each x and y pair in the table, the value of k is 0.5.

    Therefore $k = 0.5$ and $n = 2$.

131. Since $y = \dfrac{k}{x^n}$, we know that $k = yx^n$ where k is a constant. Using trial and error, let $n = 1$ while calculating

    various values for k using x- and y-values from the table. For example, when $x = 2$ and $y = 1.5$, the value of

    k is 3. And for $x = 3$ and $y = 1$, the value of k is also 3. Since the value of k remained constant we know that

    the value of n is 1. For each x and y pair in the table, the value of k is 3. Therefore $k = 3$ and $n = 1$.

133. If y is directly proportional to $x^{1.25}$, then $y = kx^{1.25}$ must hold. Given that $y = 1.9$ when $x = 1.1$, we may

    calculate the value of $k = \dfrac{y}{x^{1.25}} = \dfrac{1.9}{1.1^{1.25}} \approx 1.69$. The variation equation can be written as $y \approx 1.69x^{1.25}$.

    When a fiddler crab has claws weighing 0.75 grams, its body weight will be $y \approx 1.69(0.75)^{1.25} \approx 1.18$ grams.

135. Let $I = $ the intensity of the light, and let $d = $ the distance from a star to the earth. If I is inversely

    proportional to d^2, then $I = \dfrac{k}{d^2}$ must hold. Suppose a ground-based telescope can see a star with intensity I_g.

    Then the Hubble Telescope can see stars of intensity $\dfrac{1}{50} \cdot I_g = \dfrac{1}{50} \cdot \dfrac{k}{d^2} = \dfrac{k}{(d\sqrt{50})^2}$. That is, the Hubble

    Telescope can see $\sqrt{50} \approx 7$ times as far as ground-based telescopes.

137. Since the resistance varies inversely as the square of the diameter of the wire, increasing the diameter of the

    wire by a factor of 1.5 will decrease the resistance by a factor of $\dfrac{1}{1.5^2} = \dfrac{4}{9}$. Hence, a 25 foot wire with a

    diameter of 3 millimeters (1.5 times 2 millimeters) will have a resistance of $\dfrac{4}{9}(0.5\text{ ohm}) = \dfrac{2}{9}$ ohm.

139. In this exercise $F = \dfrac{K\sqrt{T}}{L}$. If both T and L are doubled, $\dfrac{K\sqrt{2T}}{2L} = \dfrac{\sqrt{2}}{2} \cdot \dfrac{K\sqrt{T}}{L} = \dfrac{\sqrt{2}}{2}F$. Therefore F

    decreases by a factor of $\dfrac{\sqrt{2}}{2}$.

4.6: Polynomial and Rational Inequalities

Graphical Solutions

1. (a) The boundary numbers, the x-values for which $f(x) = 0$, are $-4, -2$, and 2.

    (b) $f(x) > 0$ on the interval for which the graph of f is above the x-axis. That is $(-4, -2) \cup (2, \infty)$.

    (c) $f(x) < 0$ on the interval for which the graph of f is below the x-axis. That is $(-\infty, -4) \cup (-2, 2)$.

3. (a) The boundary numbers, the x-values for which $f(x) = 0$, are $-4, -2, 0$, and 2.

    (b) $f(x) > 0$ on the interval for which the graph of f is above the x-axis. That is $(-4, -2) \cup (0, 2)$.

    (c) $f(x) < 0$ on the interval for which the graph of f is below the x-axis. That is $(-\infty, -4) \cup (-2, 0) \cup (2, \infty)$.

5. (a) The boundary numbers, the x-values for which $f(x) = 0$, are $-2, 1$, and 2.

    (b) $f(x) > 0$ on the interval for which the graph of f is above the x-axis. That is $(-\infty, -2) \cup (-2, 1)$.

    (c) $f(x) < 0$ on the interval for which the graph of f is below the x-axis. That is $(1, 2) \cup (2, \infty)$.

7. (a) $f(x)$ is undefined at $x = 0$.

    (b) $f(x) > 0$ on the interval for which the graph of f is above the x-axis. That is $(-\infty, 0) \cup (0, \infty)$.

    (c) $f(x) < 0$ on the interval for which the graph of f is below the x-axis. There are no solutions.

9. (a) $f(x)$ is undefined at $x = 1, f(x) = 0$ at $x = 0$.

    (b) $f(x) > 0$ on the interval for which the graph of f is above the x-axis. That is $(-\infty, 0) \cup (1, \infty)$.

    (c) $f(x) < 0$ on the interval for which the graph of f is below the x-axis. That is $(0, 1)$.

11. (a) $f(x)$ is undefined at $x = \pm 2, f(x) = 0$ at $x = 0$.

    (b) $f(x) > 0$ on the interval for which the graph of f is above the x-axis. That is $(-\infty, -2) \cup (2, \infty)$.

    (c) $f(x) < 0$ on the interval for which the graph of f is below the x-axis. That is $(-2, 0) \cup (0, 2)$.

Polynomial Inequalities

13. (a) First find the boundary numbers by solving $x^3 - x = 0$.

    $$x^3 - x = 0 \Rightarrow x(x^2 - 1) = 0 \Rightarrow x(x + 1)(x - 1) = 0 \Rightarrow x = 0, -1, \text{ or } 1$$

    The boundary values divide the number line into four intervals. Choose a test value from each of these

    intervals and evaluate $x^3 - x$ for these values. See Figure 13a. The solution is $(-1, 0) \cup (1, \infty)$.

    (b) Graph $Y_1 = X^\wedge 3 - X$ as shown in Figure 13b. The x-intercepts are $-1, 0$, and 1. The graph is above the

    x-axis on the interval $(-1, 0) \cup (1, \infty)$.

Interval	Test Value x	$x^3 - x$	Positive or Negative?
$(-\infty, -1)$	-2	-6	Negative
$(-1, 0)$	-0.5	0.375	Positive
$(0, 1)$	0.5	-0.375	Negative
$(1, \infty)$	2	6	Positive

Figure 13a

[-3, 3, 0.5] by [-3, 3, 0.5]

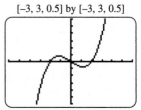

Figure 13b

15. (a) Write the inequality as $x^3 + x^2 - 2x \geq 0$. Then find the boundary numbers by solving $x^3 + x^2 - 2x = 0$.

$x^3 + x^2 - 2x = 0 \Rightarrow x(x^2 + x - 2) = 0 \Rightarrow x(x + 2)(x - 1) = 0 \Rightarrow x = 0, -2,$ or 1

The boundary values divide the number line into four intervals. Choose a test value from each of these intervals and evaluate $x^3 + x^2 - 2x$ for these values. See Figure 15a. The solution is $[-2, 0] \cup [1, \infty)$.

(b) Graph $Y_1 = X^3 + X^2 - 2X$ as shown in Figure 15b. The x-intercepts are $-1, 0,$ and 1. The graph intersects or is above the x-axis on the interval $[-2, 0] \cup [1, \infty)$.

Interval	Test Value x	$x^3 + x^2 - 2x$	Positive or Negative?
$(-\infty, -2)$	-3	-12	Negative
$(-2, 0)$	-1	2	Positive
$(0, 1)$	0.5	-0.625	Negative
$(1, \infty)$	2	8	Positive

Figure 15a

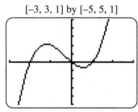

[-3, 3, 1] by [-5, 5, 1]

Figure 15b

17. (a) First find the boundary numbers by solving $x^4 - 13x^2 + 36 = 0$.

$x^4 - 13x^2 + 36 = 0 \Rightarrow (x^2 - 4)(x^2 - 9) = 0 \Rightarrow (x + 2)(x - 2)(x + 3)(x - 3) \Rightarrow x = -3, -2, 2$ or 3

The boundary values divide the number line into five intervals. Choose a test value from each of these intervals and evaluate $x^4 - 13x^2 + 36$ for these values. See Figure 17a. The solution is $(-3, -2) \cup (2, 3)$.

(b) Graph $Y_1 = X^4 - 13X^2 + 36$ as shown in Figure 17b. The x-intercepts are $-3, -2, 2$ and 3. The graph is below the x-axis on the interval $(-3, -2) \cup (2, 3)$.

Interval	Test Value x	$x^4 - 13x^2 + 36$	Positive or Negative?
$(-\infty, -3)$	-4	84	Positive
$(-3, -2)$	-2.5	-6.1875	Negative
$(-2, 2)$	0	36	Positive
$(2, 3)$	2.5	-6.1875	Negative
$(3, \infty)$	4	84	Positive

Figure 17a

[-5, 5, 1] by [-10, 100, 10]

Figure 17b

19. Write the inequality as $7x^4 - 14x^2 \geq 0$. Then find the boundary numbers by solving $7x^4 - 14x^2 = 0$.

$7x^4 - 14x^2 = 0 \Rightarrow 7x^2(x^2 - 2) = 0 \Rightarrow x = 0$ or $\pm\sqrt{2}$

The boundary values divide the number line into four intervals. Choose a test value from each of these intervals and evaluate $7x^4 - 14x^2$ for these values. See Figure 19. The solution is $(-\infty, -\sqrt{2}] \cup [\sqrt{2}, \infty)$.

Interval	Test Value x	$7x^4 - 14x^2$	Positive or Negative?
$(-\infty, -\sqrt{2})$	-2	56	Positive
$(-\sqrt{2}, 0)$	-1	-7	Negative
$(0, \sqrt{2})$	1	-7	Negative
$(\sqrt{2}, \infty)$	2	56	Positive

Figure 19

21. First find the boundary numbers by solving $(x - 1)(x - 2)(x + 2) = 0$.

    $(x - 1)(x - 2)(x + 2) = 0 \Rightarrow x = -2, 1$ or 2

    The boundary values divide the number line into four intervals. Choose a test value from each of these intervals and evaluate $(x - 1)(x - 2)(x + 2)$ for these values. See Figure 21. The solution is $[-2, 1] \cup [2, \infty)$.

Interval	Test Value x	$(x - 1)(x - 2)(x + 2)$	Positive or Negative?
$(-\infty, -2)$	-3	-20	Negative
$(-2, 1)$	0	4	Positive
$(1, 2)$	1.5	-0.875	Negative
$(2, \infty)$	3	10	Positive

Figure 21

23. Write the inequality as $2x^4 + 2x^3 - 12x^2 \le 0$. Find the boundary numbers by solving $2x^4 + 2x^3 - 12x^2 = 0$.

    $2x^4 + 2x^3 - 12x^2 = 0 \Rightarrow 2x^2(x^2 + x - 6) = 0 \Rightarrow 2x^2(x + 3)(x - 2) = 0 \Rightarrow x = -3, 0$ or 2

    The boundary values divide the number line into four intervals. Choose a test value from each of these intervals and evaluate $2x^4 + 2x^3 - 12x^2$ for these values. See Figure 23. The solution is $[-3, 2]$.

Interval	Test Value x	$2x^4 + 2x^3 - 12x^2$	Positive or Negative?
$(-\infty, -3)$	-4	192	Positive
$(-3, 0)$	-1	-12	Negative
$(0, 2)$	1	-8	Negative
$(2, \infty)$	3	108	Positive

Figure 23

25. Write the inequality as $x^3 - 7x^2 + 14x - 8 \le 0$.

    Then Graph $Y_1 = X\wedge 3 - 7X^2 + 14X - 8$ as shown in Figure 25. The x-intercepts are 1, 2, and 4.

    The graph intersects or is below the x-axis on the interval $(-\infty, 1] \cup [2, 4]$.

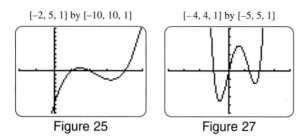

[-2, 5, 1] by [-10, 10, 1]   [-4, 4, 1] by [-5, 5, 1]

Figure 25   Figure 27

27. Graph $Y_1 = 3X\wedge 4 - 7X\wedge 3 - 2X^2 + 8X$ as shown in Figure 27. The x-intercepts are $-1, 0, \dfrac{4}{3}$ and 2.

    The graph is above the x-axis on the interval $(-\infty, -1) \cup \left(0, \dfrac{4}{3}\right) \cup (2, \infty)$.

29. (a) Find the zeros of the numerator and the denominator.

Numerator: No zero; Denominator: $x = 0$

This boundary number divides the number line into two intervals. Choose a test value from each of these intervals and evaluate $\dfrac{1}{x}$ for these values. See Figure 29a. The solution is $(-\infty, 0)$.

(b) Graph $Y_1 = 1/X$ as shown in Figure 29b. The graph has a vertical asymptote at $x = 0$ and no x-intercepts. The graph is below the x-axis on the interval $(-\infty, 0)$.

[-4, 4, 1] by [-2, 2, 0.5]

Interval	Test Value x	$1/x$	Positive or Negative?
$(-\infty, 0)$	-1	-1	Negative
$(0, \infty)$	1	1	Positive

Figure 29a

Figure 29b

31. (a) Find the zeros of the numerator and the denominator.

Numerator: No zero; Denominator: $x + 3 = 0 \Rightarrow x = -3$

This boundary number divides the number line into two intervals. Choose a test value from each of these intervals and evaluate $\dfrac{4}{x + 3}$ for these values. See Figure 31a. The solution is $(-3, \infty)$.

(b) Graph $Y_1 = 4/(X + 3)$ using dot mode as shown in Figure 31b. The graph has a vertical asymptote at $x = -3$ and no x-intercepts. The graph is above the x-axis on the interval $(-3, \infty)$. Note that the value $x = -3$ can not be included in the solution set since $\dfrac{4}{x + 3}$ is undefined for $x = -3$.

[-6, 1, 1] by [-4, 4, 1]

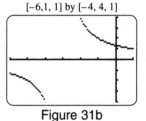

Interval	Test Value x	$4/(x + 3)$	Positive or Negative?
$(-\infty, -3)$	-4	-4	Negative
$(-3, \infty)$	1	1	Positive

Figure 31a

Figure 31b

33. (a) Find the zeros of the numerator and the denominator.

    Numerator: No zero; Denominator: $x^2 - 4 = 0 \Rightarrow x = \pm 2$

    These boundary numbers divide the number line into three intervals. Choose a test value from each of these

    intervals and evaluate $\dfrac{5}{x^2 - 4}$ for these values. See Figure 33a. The solution is $(-2, 2)$.

    (b) Graph $Y_1 = 5/(X^2 - 4)$ using dot mode as shown in Figure 33b. The graph has vertical asymptotes

    at $x = \pm 2$ and no x-intercepts. The graph is below the x-axis on the interval $(-2, 2)$.

Interval	Test Value x	$5/(x^2 - 4)$	Positive or Negative?
$(-\infty, -2)$	-3	1	Positive
$(-2, 2)$	0	-1.25	Negative
$(2, \infty)$	3	1	Positive

Figure 33a

[−4, 4, 1] by [−4, 4, 1]

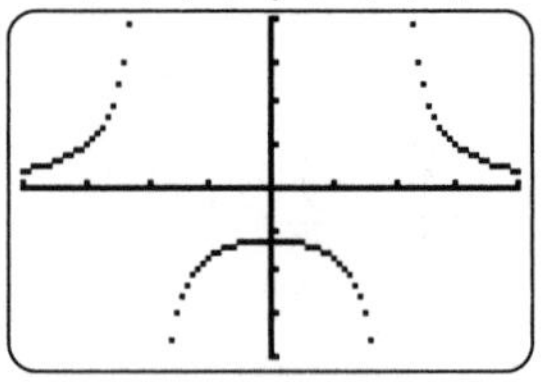

Figure 33b

35. The graph of $Y_1 = (X + 1)^2/(X - 2)$ has a vertical asymptote at $x = 2$ and an x-intercept at $x = -1$.

    The graph of Y_1 intersects or is below the x-axis on the interval $(-\infty, 2)$. Note that 2 can not be included.

37. The graph of $Y_1 = (3 - 2X)/(1 + X)$ has a vertical asymptote at $x = -1$ and an x-intercept at $x = \dfrac{3}{2}$.

    The graph of Y_1 is below the x-axis on the interval $(-\infty, -1) \cup \left(\dfrac{3}{2}, \infty\right)$.

39. The graph of $Y_1 = \dfrac{(x + 1)(x - 2)}{x + 3}$ has a vertical

    asymptote of at $x = -3$ and x-intercepts at $x = -1$ and $x = 2$.

    The graph of Y_1 is below the x-axis on the interval $(-\infty, -3) \cup (-1, 2)$.

41. The graph of $Y_1 = \dfrac{2x - 5}{(x + 1)(x - 1)}$ has vertical asymptotes at $x = -1$ and $x = 1$ and an x-intercept at $x = \dfrac{5}{2}$.

    The graph of Y_1 is on or above the x-axis on the interval $(-1, 1) \cup \left[\dfrac{5}{2}, \infty\right)$.

43. First, rewrite the inequality:

    $$\frac{1}{x - 3} \le \frac{5}{x - 3} \Rightarrow \frac{1}{x - 3} - \frac{5}{x - 3} \le 0 \Rightarrow \frac{-4}{x - 3} \le 0$$

    The graph of $Y_1 = -4/(X - 3)$ has a vertical asymptote at $x = 3$ and no x-intercept.

    The graph of Y_1 intersects or is below the x-axis on the interval $(3, \infty)$. Note that 3 can not be included.

45. First, rewrite the inequality:

    $$2 - \frac{5}{x} + \frac{2}{x^2} \ge 0 \Rightarrow \frac{2x^2 - 5x + 2}{x^2} \ge 0 \Rightarrow \frac{(2x - 1)(x - 2)}{x^2} \ge 0$$

    The graph of $Y_1 = (2X - 1)(X - 2)/X^2$ has a vertical asymptote at $x = 0$ and x-intercepts $x = \dfrac{1}{2}$ and $x = 2$.

    The graph of Y_1 intersects or is above the x-axis on the interval $(-\infty, 0) \cup \left(0, \dfrac{1}{2}\right] \cup [2, \infty)$.

Applications

47. (a) Graph $Y_1 = X^2/(1600 - 40X)$ and $Y_2 = 8$ (not shown). The graphs intersect approximately at (36, 8).

    The graph of Y_1 is equal to or is below the graph of Y_2 when $x \leq 36$ (approximately).

    (b) The average line length is less than or equal to 8 cars when the average arrival rate is 36 cars per hour or less.

49. (a) The graph of $D(x) = \dfrac{120}{x}$ increases as x decreases, which means the braking distance increases as the

    coefficient of friction becomes smaller.

    (b) $\dfrac{120}{x} \geq 400 \Rightarrow \dfrac{120}{400} \geq x \Rightarrow 0.3 \geq x$; the braking distance is 400 feet or more when $0 < x \leq 0.3$

51. The volume of a cube is $V = x^3$, where x is the length of a side.

    $212.8 \leq V \leq 213.2 \Rightarrow 212.8 \leq x^3 \leq 213.2 \Rightarrow \sqrt[3]{212.8} \leq x \leq \sqrt[3]{213.2}$ inches.

    This is approximately $5.97022 \leq x \leq 5.97396$ inches.

4.7: Power Functions and Radical Equations

Properties of Exponents

1. $8^{2/3} = (8^{1/3})^2 = 2^2 = 4$

3. $16^{-3/4} = (16^{1/4})^{-3} = (2)^{-3} = \dfrac{1}{2^3} = \dfrac{1}{8}$

5. $-81^{0.5} = -81^{1/2} = -\sqrt{81} = -9$

7. $64^{1/6} = \sqrt[6]{64} = 2$

9. $(-9^{3/4})^2 = (9^{3/4})^2 = 9^{3/2} = (\sqrt{9})^3 = 3^3 = 27$

11. $\dfrac{8^{5/6}}{8^{1/2}} = 8^{5/6 - 1/2} = 8^{1/3} = 2$

13. $27^{5/6} \cdot 27^{-1/6} = 27^{5/6 + (-1/6)} = 27^{2/3} = \sqrt[3]{27^2} = \sqrt[3]{729} = 9$

15. $(-27)^{-5/3} = \dfrac{1}{(-27)^{5/3}} = \dfrac{1}{\sqrt[3]{(-27)^5}} = \dfrac{1}{\sqrt[3]{-14348907}} = -\dfrac{1}{243}$

17. $(0.5^{-2})^2 = \left(\dfrac{1}{2}\right)^{-4} = 2^4 = 16$

19. $\left(\dfrac{2}{3}\right)^{-2} = \left(\dfrac{3}{2}\right)^2 = \dfrac{9}{4}$

21. $1^{-1} + 2^{-1} + 3^{-1} = 1 + \dfrac{1}{2} + \dfrac{1}{3} = \dfrac{11}{6}$

23. $\sqrt[3]{2x} = (2x)^{1/3}$

25. $\sqrt[3]{z^5} = z^{5/3}$

27. $(\sqrt[4]{y})^{-3} = (y^{1/4})^{-3} = y^{-3/4} = \dfrac{1}{y^{3/4}}$

29. $\sqrt{x} \cdot \sqrt[3]{x} = x^{1/2} \cdot x^{1/3} = x^{1/2 + 1/3} = x^{5/6}$

31. $\sqrt{y} \cdot \sqrt{\sqrt{y}} = (y \cdot y^{1/2})^{1/2} = (y^{3/2})^{1/2} = y^{3/4}$

Power Functions

33. $f(x) = x^{1.62} \Rightarrow f(1.2) = 1.2^{1.62} \approx 1.34$

35. $f(x) = x^{3/2} - x^{1/2} \Rightarrow f(50) = 50^{3/2} - 50^{1/2} \approx 346.48$

37. The graph of $f(x) = x^a$ is increasing, since $a > 0$. However, since $a < b$ its graph increases more slowly than the graph of $f(x) = x^b$. The best choice is graph b.

39. Shift the graph of $f(x) = \sqrt{x}$ up 1 unit to get the graph of $f(x) = \sqrt{x} + 1$. See Figure 39.

41. Shift the graph of $f(x) = x^{2/3}$ down 1 unit to get the graph of $f(x) = x^{2/3} - 1$. See Figure 41.

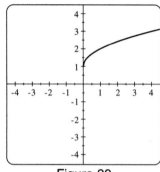
Figure 39

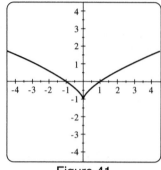
Figure 41

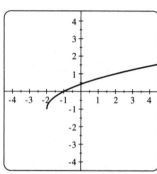
Figure 43

43. Shift the graph of $f(x) = \sqrt{x}$ left 2 units and down 1 unit to get the graph of $f(x) = \sqrt{x + 2} - 1$. See Figure 43.

Equations Involving Rational Exponents

45. $x^3 = 8 \Rightarrow x = \sqrt[3]{8} \Rightarrow x = 2$; Check: $2^3 = 8$

47. $x^{1/4} = 3 \Rightarrow x = 3^4 \Rightarrow x = 81$; Check: $81^{1/4} = \sqrt[4]{81} = 3$

49. $x^{2/5} = 4 \Rightarrow (x^{2/5})^{5/2} = 4^{5/2} \Rightarrow x = (\sqrt{4})^5 \Rightarrow x = (\pm 2)^5 \Rightarrow x = \pm 32$

    Check: $(\pm 32)^{2/5} = (\sqrt[5]{\pm 32})^2 = (\pm 2)^2 = 4$

51. $2(x^{1/5} - 2) = 0 \Rightarrow x^{1/5} - 2 = 0 \Rightarrow x^{1/5} = 2 \Rightarrow (x^{1/5})^5 = 2^5 \Rightarrow x = 32$

    Check: $2(32^{1/5} - 2) = 2(2 - 2) = 2(0) = 0$

53. $2x^{1/3} - 5 = 1 \Rightarrow 2x^{1/3} = 6 \Rightarrow x^{1/3} = 3 \Rightarrow (x^{1/3})^3 = 3^3 \Rightarrow x = 27$

    Check: $2(27)^{1/3} - 5 = 2\sqrt[3]{27} - 5 = 2(3) - 5 = 6 - 5 = 1$

55. $n^{-2} + 3n^{-1} + 2 = 0 \Rightarrow (n^{-1})^2 + 3(n^{-1}) + 2 = 0$, let $u = n^{-1}$, then $u^2 + 3u + 2 = 0 \Rightarrow$

    $(u + 2)(u + 1) = 0 \Rightarrow u = -2$ or $u = -1$. Because $u = n^{-1}$, it follows that $n = u^{-1}$.

    Thus $n = (-2)^{-1} \Rightarrow n = \dfrac{1}{(-2)} \Rightarrow n = -\dfrac{1}{2}$ or $n = (-1)^{-1} \Rightarrow n = \dfrac{1}{(-1)} = -1$. Therefore, $n = -1, -\dfrac{1}{2}$.

57. $5n^{-2} + 13n^{-1} = 28 \Rightarrow 5(n^{-1})^2 + 13(n^{-1}) - 28 = 0$, let $u = n^{-1}$, then $5u^2 + 13u - 28 = 0 \Rightarrow$

    $(5u - 7)(u + 4) = 0 \Rightarrow u = \dfrac{7}{5}$ or $u = -4$. Because $u = n^{-1}$, it follows that $n = u^{-1}$.

    Thus $n = \left(\dfrac{7}{5}\right)^{-1} \Rightarrow n = \dfrac{5}{7}$ or $n = (-4)^{-1} \Rightarrow n = -\dfrac{1}{4}$. Therefore, $n = -\dfrac{1}{4}, \dfrac{5}{7}$.

59. $x^{2/3} - x^{1/3} - 6 = 0 \Rightarrow (x^{1/3})^2 - x^{1/3} - 6 = 0$, let $u = x^{1/3}$, then $u^2 - u - 6 = 0 \Rightarrow$

    $(u - 3)(u + 2) = 0 \Rightarrow u = -2$ or $u = 3$. Because $u = x^{1/3}$, it follows that $x = u^3$.

    Thus $x = (-2)^3 \Rightarrow x = -8$ or $x = 3^3 \Rightarrow x = 27$. Therefore, $x = -8, 27$.

61. $6x^{2/3} - 11x^{1/3} + 4 = 0 \Rightarrow 6(x^{1/3})^2 - 11(x^{1/3}) + 4 = 0$, let $u = x^{1/3}$, then $6u^2 - 11u + 4 = 0$.

    Using the quadratic formula to solve we get: $u = \dfrac{11 \pm \sqrt{121 - 4(6)(4)}}{2(6)} = \dfrac{11 \pm \sqrt{25}}{12} = \dfrac{11 \pm 5}{12} \Rightarrow$

    $u = \dfrac{16}{12} \Rightarrow u = \dfrac{4}{3}$ or $u = \dfrac{6}{12} \Rightarrow u = \dfrac{1}{2}$. Because $u = x^{1/3}$, it follows that $x = u^3$.

    Thus $x = \left(\dfrac{4}{3}\right)^3 \Rightarrow x = \dfrac{64}{27}$ or $x = \left(\dfrac{1}{2}\right)^3 \Rightarrow x = \dfrac{1}{8}$. Therefore, $x = \dfrac{1}{8}, \dfrac{64}{27}$.

63. $x^{3/4} - x^{1/2} - x^{1/4} + 1 = 0 \Rightarrow (x^{1/4})^3 - (x^{1/4})^2 - (x^{1/4}) + 1 = 0$, let $u = x^{1/4}$, then $u^3 - u^2 - u + 1 = 0 \Rightarrow$

    $(u^3 - u^2) - (u - 1) = 0 \Rightarrow u^2(u - 1) - 1(u - 1) = 0 \Rightarrow (u^2 - 1)(u - 1) = 0 \Rightarrow$

    $(u + 1)(u - 1)(u - 1) = 0 \Rightarrow u = -1$ or $u = 1$. Because $u = x^{1/4}$ it follows that $x = u^4$.

    Thus $x = (-1)^4 \Rightarrow x = 1$ or $x = (1)^4 \Rightarrow x = 1$. Therefore, $x = 1$.

Equations Involving Radicals

65. $\sqrt{x + 2} = x - 4 \Rightarrow x + 2 = x^2 - 8x + 16 \Rightarrow x^2 - 9x + 14 = 0 \Rightarrow (x - 2)(x - 7) = 0 \Rightarrow x = 2$ or 7

    Check: $\sqrt{2 + 2} = 2 \neq 2 - 4$ (not a solution); $\sqrt{7 + 2} = 3 = 7 - 4$. The only solution is $x = 7$.

67. $\sqrt{3x + 7} = 3x + 5 \Rightarrow 3x + 7 = 9x^2 + 30x + 25 \Rightarrow 9x^2 + 27x + 18 = 0 \Rightarrow 9(x^2 + 3x + 2) = 0 \Rightarrow$

    $9(x + 2)(x + 1) = 0 \Rightarrow x = -2$ or -1

    Check: $\sqrt{3(-2) + 7} \neq 3(-2) + 5$ (not a solution); $\sqrt{3(-1) + 7} = 2 = 3(-1) + 5$. The only solution is $x = -1$.

69. $\sqrt{5x - 6} = x \Rightarrow 5x - 6 = x^2 \Rightarrow x^2 - 5x + 6 = 0 \Rightarrow (x - 2)(x - 3) = 0 \Rightarrow x = 2$ or 3

    Check: $\sqrt{5(2) - 6} = 2$; $\sqrt{5(3) - 6} = 3$

71. $\sqrt{x + 5} + 1 = x \Rightarrow \sqrt{x + 5} = x - 1 \Rightarrow x + 5 = (x - 1)^2 \Rightarrow x + 5 = x^2 - 2x + 1 \Rightarrow$

    $x^2 - 3x - 4 = 0 \Rightarrow (x - 4)(x + 1) = 0 \Rightarrow x = -1$ or $x = 4$

    Check: $\sqrt{-1 + 5} + 1 = -1 \Rightarrow \sqrt{4} + 1 = -1 \Rightarrow 3 \neq -1$ (not a solution).

    Check: $\sqrt{4 + 5} + 1 = 4 \Rightarrow \sqrt{9} + 1 = 4 \Rightarrow 4 = 4$. The only solution is $x = 4$.

73. $\sqrt{x + 1} + 3 = \sqrt{3x + 4} \Rightarrow (\sqrt{x + 1} + 3)^2 = 3x + 4 \Rightarrow x + 1 + 6\sqrt{x + 1} + 9 = 3x + 4 \Rightarrow$

    $6\sqrt{x + 1} = 2x - 6 \Rightarrow 3\sqrt{x + 1} = x - 3 \Rightarrow (3\sqrt{x + 1})^2 = (x - 3)^2 \Rightarrow 9(x + 1) = x^2 - 6x + 9 \Rightarrow$

    $x^2 - 15x = 0 \Rightarrow x(x - 15) = 0 \Rightarrow x = 0$ or 15

    Check: $\sqrt{(0) + 1} + 3 \neq \sqrt{3(0) + 4}$ (not a solution); $\sqrt{(15) + 1} + 3 = 7 = \sqrt{3(15) + 4}$.

    The only solution is $x = 15$.

75. $\sqrt{2x + 3} - \sqrt{x + 1} = 1 \Rightarrow \sqrt{2x + 3} = \sqrt{x + 1} + 1 \Rightarrow 2x + 3 = (\sqrt{x + 1} + 1)^2 \Rightarrow$

    $2x + 3 = x + 1 + 2\sqrt{x + 1} + 1 \Rightarrow 2x + 3 = x + 2 + 2\sqrt{x + 1} \Rightarrow x + 1 = 2\sqrt{x + 1} \Rightarrow$

    $(x + 1)^2 = 4(x + 1) \Rightarrow x^2 + 2x + 1 = 4x + 4 \Rightarrow x^2 - 2x - 3 = 0 \Rightarrow (x - 3)(x + 1) = 0 \Rightarrow$

    $x = -1$ or $x = 3$

    Check: $\sqrt{2(-1) + 3} - \sqrt{(-1) + 1} = 1 \Rightarrow \sqrt{1} - \sqrt{0} = 1 \Rightarrow 1 = 1$.

    Check: $\sqrt{2(3) + 3} - \sqrt{3 + 1} = 1 \Rightarrow \sqrt{9} - \sqrt{4} = 1 \Rightarrow 3 - 2 = 1 \Rightarrow 1 = 1$. The solution is $x = -1, 3$.

77. $\sqrt[3]{x+1} = -3 \Rightarrow x + 1 = (-3)^3 \Rightarrow x + 1 = -27 \Rightarrow x = -28$

Check: $\sqrt[3]{-28+1} = \sqrt[3]{-27} = -3$

79. $\sqrt[3]{x+1} = \sqrt[3]{2x-1} \Rightarrow x + 1 = 2x - 1 \Rightarrow x = 2$

Check: $\sqrt[3]{2+1} = \sqrt[3]{3} = \sqrt[3]{2(2)-1}$

81. $\sqrt[4]{x-2} + 4 = 20 \Rightarrow \sqrt[4]{x-2} = 16 \Rightarrow x - 2 = (16)^4 \Rightarrow x - 2 = 65{,}536 \Rightarrow x = 65{,}538$

Check: $\sqrt[4]{65{,}538 - 2} + 4 = 20 \Rightarrow \sqrt[4]{65{,}536} + 4 = 20 \Rightarrow 20 = 20$

Applications

83. $S(w) = 3 \Rightarrow 1.27w^{2/3} = 3 \Rightarrow w^{2/3} = \dfrac{3}{1.27} \Rightarrow w = \left(\dfrac{3}{1.27}\right)^{3/2} \Rightarrow w \approx 3.63$

The bird weighs about 3.63 pounds

85. $f(15) = 15^{1.5} \approx 58.1$; The planet would take about 58.1 years to orbit the sun.

87. (a) Following the hint, $f(1) = a(1)^b = 1960 \Rightarrow a = 1960$

(b) Plot the four data points (0.5, 4500), (1, 1960), (2, 850), and (3, 525) together with $f(x) = 1960x^b$ for different values of b. Through trial and error, a value of $b \approx -1.2$ can be found. Figure 87 shows a graph of $Y_1 = 1960X^{\wedge}(-1.2)$ along with the data.

(c) $f(x) = 1960x^{-1.2} \Rightarrow f(4) = 1960(4)^{-1.2} \approx 371$. This means that if the zinc ion concentration in the water reaches 371 milligrams per liter, the rainbow trout will live only 4 minutes on average.

89. (a) $f(2) = 0.445(2)^{1.25} \approx 1.06$ grams

(b) We must solve the equation $0.445x^{1.25} = 0.5$ for x. Graph $Y_1 = 0.445X^{\wedge}1.25$ and $Y_2 = 0.5$. Their graphs intersect near (1.1, 0.5). See Figure 89. When the claw weighs 0.5 grams, the crab weighs about 1.1 grams.

(c) $0.445x^{1.25} = 0.5 \Rightarrow 0.445x^{5/4} = 0.5 \Rightarrow x^{5/4} = \dfrac{0.5}{0.445} \Rightarrow x = \left(\dfrac{0.5}{0.445}\right)^{4/5} \Rightarrow x \approx 1.1$ grams

[0, 5, 1] by [0, 5000, 1000]   [0, 2, 0.2] by [0, 1, 0.2]   [1, 9, 1] by [0, 6, 1]

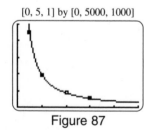
Figure 87

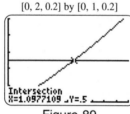
Figure 89

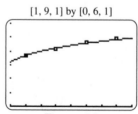
Figure 91

91. Use the power regression feature of your graphing calculator to find the values of a and b in the equation $f(x) = ax^b$. For this exercise, $a \approx 3.20$ and $b \approx 0.20$.

The graph of $Y_1 = 3.20X^{\wedge}0.20$ is shown together with the data in Figure 91.

93. Use the power regression feature of your graphing calculator to find the values of a and b in the equation $f(x) = ax^b$. For this exercise, $a \approx 874.54$ and $b \approx -0.49789$.

Chapter 4 Review Exercises

1. First put in standard order: $f(x) = -7x^3 - 2x^2 + x + 4$, now the degree is: 3, and the leading coefficient is: -7.

3. (a) The graph of $Y_1 = X^2 - X$ is shown in Figure 3.

   (b) The function f is decreasing when $(-\infty, 0.5]$ and increasing when $[0.5, \infty)$.

[–10, 10, 1] by [–10, 10, 1]   [–10, 10, 1] by [–150, 150, 50]

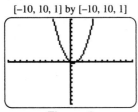 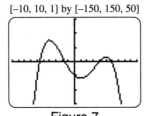

Figure 3   Figure 7

5. (a) There are local minima of -4.5 and -0.5. There is a local maximum of 0.

   (b) The absolute minimum is -4.5. There is no absolute maximum.

7. (a) The graph of $Y_1 = -0.25X^4 + 0.67X^3 + 9.5X^2 - 20X - 50$ is shown in Figure 7. Two local maxima and one local minimum occur on this graph. Local maxima are associated with the points near $(-3.996, 75.12)$ and $(5.007, 15.00)$, while a local minimum corresponds to the point $(0.9995, -60.08)$. Thus, f has two local maxima of approximately 75.12 and 15, and a local minimum of about -60.08.

   (b) The graph of f has an absolute maximum of approximately 75.12 and no absolute minimum.

   (c) Tracing the graph of f from left to right, the y-values increase when $-\infty < x \le -3.996$, decrease when $-3.996 \le x \le 0.9995$, increase for $0.9995 \le x \le 5.007$, and then decrease for $5.007 \le x < \infty$. In interval notation f increases on $(-\infty, -3.996] \cup [0.9995, 5.007]$ and decreases on $[-3.996, 0.9995] \cup [5.007, \infty)$. All value have been rounded to four significant digits.

9. (a) The graph of $f(x) = x^4 + 2x^3 - 9x^2 - 2x + 20$ is shown in Figure 9a.

   There are two local minima, one local maximum and two x-intercepts.

   (b) The graph of $g(x) = -0.4x^4 + 3.6x^2$ is shown in Figure 9b.

   There are two local maxima, one local minimum and three x-intercepts.

   (c) The graph of $h(x) = x^4 + 2x^3 - 21x^2 - 22x + 40$ is shown in Figure 9c.

   There are two local minima, one local maximum and four x-intercepts.

[–10, 10, 1] by [–100, 100, 10]   [–10, 10, 1] by [–10, 10, 1]   [–10, 10, 1] by [–100, 100, 10]

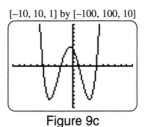

Figure 9a   Figure 9b   Figure 9c

11. $f(x) = 2x^6 - 5x^4 - x^2 \Rightarrow f$ is even since each power of x is even.

13. $f(x) = 7x^5 + 3x^3 - x \Rightarrow f$ is odd since each power of x is odd.

15. f is odd since $f(-x) = -f(x)$ for all x-values in the table.

17. For equation $y = -f(x + 1) - 2$ the function f has been reflected across the x-axis, giving the negative of y, shifted left one unit, and down 2 units $\Rightarrow f(2) = 4$ or point $(2, 4)$ has the same transformations $\Rightarrow (1, -6)$

19. See Figure 19. *Answers may vary.*

21. See Figure 21. *Answers may vary.*

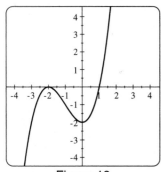

Figure 19

Figure 21

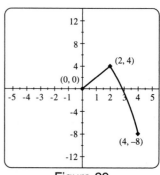

Figure 29

23. (a) 2 turning points; 3 x-intercepts at: $x \approx -2, 0, 1$.

  (b) This is a cubic function going down to the right $\Rightarrow$ negative.

  (c) Since it has 2 turning points, it is a 3rd degree.

25. A negative cubic function $\Rightarrow$ up on the left end and down on the right end;

   $f(x) \rightarrow \infty$ as $x \rightarrow -\infty$; $f(x) \rightarrow -\infty$ as $x \rightarrow \infty$.

27. $f(-2) = (-2)^3 + 1 \Rightarrow f(-2) = -7 \Rightarrow (-2, -7)$ and $f(-1) = (-1)^3 + 1 \Rightarrow f(-1) = 0 \Rightarrow (-1, 0)$.

   To find the average rate of change we use: $\dfrac{y_2 - y_1}{x_2 - x_1} \Rightarrow \dfrac{0 - (-7)}{(-1) - (-2)} \Rightarrow \dfrac{7}{1} \Rightarrow 7$

29. (a) The graph of f has of two different pieces: $y = 2x$ when $0 \le x < 2$ and $y = 8 - x^2$ when $2 \le x \le 4$.

   The graph is shown in Figure 29. The graph of f is continuous.

  (b) To evaluate $f(1)$ we must use the first piece of the function: $f(1) = 2(1) = 2$

   To evaluate $f(3)$ we must use the second piece of the function: $f(3) = 8 - (3)^2 = 8 - 9 = -1$

  (c) We must solve the equation $f(x) = 2$ for each piece of the piecewise-defined function f.

   $2x = 2 \Rightarrow x = 1$ and $8 - x^2 = 2 \Rightarrow x^2 = 6 \Rightarrow x = \pm\sqrt{6}$

   Since $x = -\sqrt{6}$ is not in the domain of f, the solutions are $x = 1$ or $\sqrt{6}$.

31. $\dfrac{14x^3 - 21x^2 - 7x}{7x} = 2x^2 - 3x - 1$

33.
$$
\begin{array}{r}
2x^2 - 3x + 1 \\
2x + 3 \overline{\smash{)}\, 4x^3 - 7x + 4} \\
\underline{4x^3 + 6x^2 } \\
-6x^2 - 7x + 4 \\
\underline{-6x^2 - 9x } \\
2x + 4 \\
\underline{2x + 3} \\
1
\end{array}
$$

   Therefore the quotient is: $2x^2 - 3x + 1 + \dfrac{1}{2x + 3}$

35. Since the leading coefficient of $f(x)$ is $\dfrac{1}{2}$ and the degree of the polynomial is 3, all of the zeros are given, thus:

$$f(x) = \frac{1}{2}(x - 1)(x - 2)(x - 3)$$

37. The graph of $Y_1 = 2X^3 + 3X^2 - 18X + 8$ is shown in Figure 37. From the graph, the x-intercepts of f are

located at $-4, \dfrac{1}{2}$, and 2. Three x-intercepts is the maximum number possible for a cubic polynomial. The

leading coefficient is $a_n = 2$. Thus $f(x) = 2(x + 4)\left(x - \dfrac{1}{2}\right)(x - 2)$.

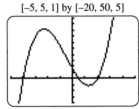

[−5, 5, 1] by [−20, 50, 5]

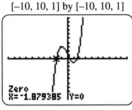

[−10, 10, 1] by [−10, 10, 1]

Figure 37   Figure 55

39. By the factor theorem, since 3 is a zero, $(x - 3)$ is a factor.

$x^3 - 2x^2 - 5x + 6$ divided by $x - 3$ can be performed using synthetic division.

$$\begin{array}{r|rrrr} 3 & 1 & -2 & -5 & 6 \\ & & 3 & 3 & -6 \\ \hline & 1 & 1 & -2 & 0 \end{array}$$

The quotient is $x^2 + x - 2$ and the remainder is 0.

The complete factored form is $f(x) = x^3 - 2x^2 - 5x + 6 = (x - 3)(x^2 + x - 2) = (x - 3)(x + 2)(x - 1)$.

41. The zeros are: $x \approx -3, -1, 1, 2$. Since its ending behavior is down, it has a negative leading coefficient.

Since $f(0) = -6$, then $-6 = a(0 + 3)(0 + 1)(0 - 1)(0 - 2) \Rightarrow -6 = 6a \Rightarrow a = -1$.

The complete factored form is: $f(x) = -(x + 3)(x + 1)(x - 1)(x - 2)$.

43. (a) Once

    (b) Twice

    (c) Three times

45. By the rational zeros test, any rational zero of $2x^3 + x^2 - 13x + 6$ must be one of the following:

$\pm 6, \pm 3, \pm 2, \pm 1, \pm\dfrac{3}{2}, \pm\dfrac{1}{2}$. By evaluating f at each of these values we find that the three rational zeros are

$-3, \dfrac{1}{2}$, and 2..

47. $9x = 3x^3 \Rightarrow 3x^3 - 9x = 0 \Rightarrow 3x(x^2 - 3) = 0 \Rightarrow x = 0, \pm\sqrt{3}$

49. $x^3 - x^2 - 6x = 0 \Rightarrow x(x^2 - x - 6) = 0 \Rightarrow x(x - 3)(x + 2) = 0 \Rightarrow x = -2, 0, 3$

51. $x^4 - 3x^2 + 2 = 0 \Rightarrow (x^2 - 1)(x^2 - 2) = 0 \Rightarrow (x + 1)(x - 1)(x^2 - 2) = 0 \Rightarrow x = \pm 1, \pm\sqrt{2}$

53. $2x^3 + x^2 = 6x + 3 \Rightarrow 2x^3 + x^2 - 6x - 3 = 0 \Rightarrow (2x^3 + x^2) - (6x + 3) = 0 \Rightarrow$

$x^2(2x + 1) - 3(2x + 1) = 0 \Rightarrow (x^2 - 3)(2x + 1) = 0 \Rightarrow x = -\dfrac{1}{2}, \pm\sqrt{3}$

55. The x-intercepts on the graph of $Y_1 = X^3 - 3X + 1$ are approximately $-1.88, 0.35$ and 1.53. See Figure 55.

57. $(2 - 3i) + (-3 + 3i) = (2 + (-3)) + (-3 + 3)i = -1$

59. $(3 + 2i)(-4 - i) = -12 - 3i - 8i - 2i^2 = -12 - 11i - 2(-1) = -10 - 11i$

61. $4x^2 + 9 = 0 \Rightarrow 4x^2 = -9 \Rightarrow x^2 = -\dfrac{9}{4} \Rightarrow x = \pm\sqrt{-\dfrac{9}{4}} = \pm\dfrac{3}{2}i$

63. $x^3 + x = 0 \Rightarrow x(x^2 + 1) = 0 \Rightarrow x = 0$ or $x = \pm i$

65. Graph $Y_1 = X^3 - 3X^2 + 3X - 9$ in $[-5, 5, 1]$ by $[-25, 25, 5]$ (not shown). Its graph crosses the x-axis once. Therefore, f has one real zero. Since f is degree 3, there must be two imaginary zeros..

67. The easiest way to factor a polynomial is to determine its zeros. Then, we can find its complete factorization. If one graphs $f(x) = 2x^2 + 4$, it does not intersect the x-axis. Therefore, there are no real zeros. However, the Fundamental Theorem of Algebra tells us that it must have complex zeros. These can be found by solving for x.

$2x^2 + 4 = 0 \Rightarrow 2x^2 = -4 \Rightarrow x^2 = -2 \Rightarrow x = \pm\sqrt{-2} \Rightarrow x = \pm i\sqrt{2}$

Since $a_n = 2$, the complete factorization is $f(x) = 2(x - i\sqrt{2})(x - (-i\sqrt{2})) = 2(x - i\sqrt{2})(x + i\sqrt{2})$.

69. If $-i$ is a solution then i is also a solution $\Rightarrow (x + i)(x - i)$ or $x^2 + 1$ is a factor. Divide this into the equation:

$$
\begin{array}{r}
x^2 + x + 1 \\
x^2 + 1 \overline{) x^4 + x^3 + 2x^2 + x + 1} \\
\underline{x^4 + x^2} \\
x^3 + x^2 + x + 1 \\
\underline{x^3 + x} \\
x^2 + 1 \\
\underline{x^2 + 1} \\
0
\end{array}
$$

Solving for $x^2 + x + 1$ we use the quadratic formula and get: $\dfrac{-1 \pm \sqrt{1 - 4(1)(1)}}{2(1)} = \dfrac{-1 \pm \sqrt{-3}}{2} = $

$-\dfrac{1}{2} \pm \dfrac{i\sqrt{3}}{2}$. Therefore the solutions are $\pm 1, -\dfrac{1}{2} \pm \dfrac{i\sqrt{3}}{2}$ and the complete factored form is:

$$f(x) = (x + 1)(x - 1)\left(x - \left(-\dfrac{1}{2} + \dfrac{i\sqrt{3}}{2}\right)\right)\left(x - \left(-\dfrac{1}{2} - \dfrac{i\sqrt{3}}{2}\right)\right)$$

71. The $D\left\{x \mid x \neq -\dfrac{4}{5}\right\}$; the horizontal asymptote is the leading coefficient ratio $y = \dfrac{3}{5}$, and the vertical asymptote is the zero of the denominator $x = -\dfrac{4}{5}$.

73. Horizontal asymptote: $y = -1$; Vertical asymptote: $x = -1$ and $x = 1$; Domain: $\{x \mid x \neq -1 \text{ or } x \neq 1\}$

75. $g(x)$ is the function $f(x) = \dfrac{1}{x}$ shifted 1 left and 2 down. See Figure 75.

77. $g(x) = \dfrac{x^2 - 1}{x^2 + 2x + 1} \Rightarrow g(x) = \dfrac{(x + 1)(x - 1)}{(x + 1)(x + 1)} \Rightarrow g(x) = \dfrac{(x - 1)}{(x + 1)} \Rightarrow g(x) = \dfrac{x - 1}{x + 1}$. This is similar to

$f(x) = \dfrac{1}{x}$ with a horizontal asymptote of $y = 1$ and a vertical asymptote of $x = -1$. See Figure 77.

79. One example of a function that has a vertical asymptote at $x = -2$ and a horizontal asymptote at $y = 2$ is shown in Figure 79.

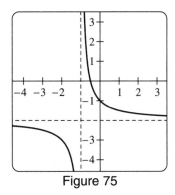
Figure 75

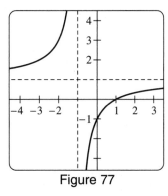
Figure 77

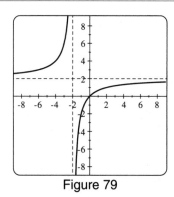
Figure 79

81. $\dfrac{5x + 1}{x + 3} = 3 \Rightarrow 5x + 1 = 3(x + 3) \Rightarrow 5x + 1 = 3x + 9 \Rightarrow 2x = 8 \Rightarrow x = 4$

This answer is supported graphically by graphing $Y_1 = (5X + 1)/(X + 3)$ and $Y_2 = 3$. The intersection point is $(4, 3)$. Making a table of Y_1 starting at 0 and incrementing by 1 will support this result numerically.

83. $\dfrac{1}{x + 2} + \dfrac{1}{x - 2} = \dfrac{1}{x^2 - 4} \Rightarrow (x - 2) + (x + 2) = 1 \Rightarrow 2x = 1 \Rightarrow x = \dfrac{1}{2}$

This answer is supported graphically by graphing $Y_1 = 1/(X + 2) + 1/(X - 2)$ and $Y_2 = 1/(X^2 - 4)$.

The intersection point is near $(0.5, -0.267)$. Making a table of Y_1 and Y_2 starting at 0 and incrementing by 0.5 will support this result numerically.

85. $\dfrac{x + 5}{x - 2} = \dfrac{x - 1}{x + 1} \Rightarrow (x + 5)(x + 1) = (x - 1)(x - 2) \Rightarrow x^2 + 6x + 5 = x^2 - 3x + 2 \Rightarrow 9x = -3 \Rightarrow$

$x = -\dfrac{1}{3}$; Check: $\dfrac{-\frac{1}{3} + 5}{-\frac{1}{3} - 2} = \dfrac{-\frac{1}{3} - 1}{-\frac{1}{3} + 1} \Rightarrow \dfrac{\frac{14}{3}}{-\frac{7}{3}} = \dfrac{-\frac{4}{3}}{\frac{2}{3}} \Rightarrow -2 = -2$

87. $\dfrac{1}{x - 3} - \dfrac{2}{x + 3} = \dfrac{4}{x^2 - 9} \Rightarrow \dfrac{1}{x - 3} - \dfrac{2}{x + 3} = \dfrac{4}{(x - 3)(x + 3)} \Rightarrow (x + 3) - 2(x - 3) = 4 \Rightarrow$

$x + 3 - 2x + 6 = 4 \Rightarrow -x + 9 = 4 \Rightarrow -x = -5 \Rightarrow x = 5$

Check: $\dfrac{1}{5 - 3} - \dfrac{2}{5 + 3} = \dfrac{4}{(5)^2 - 9} \Rightarrow \dfrac{1}{2} - \dfrac{1}{4} = \dfrac{4}{16} \Rightarrow \dfrac{1}{4} = \dfrac{1}{4}$

89. (a) $(-\infty, -4) \cup (-2, 3)$

(b) $(-4, -2) \cup (3, \infty)$

91. Find the boundary numbers by solving $x^3 + x^2 - 6x = 0$.

$x^3 + x^2 - 6x = 0 \Rightarrow x(x^2 + x - 6) = 0 \Rightarrow x(x + 3)(x - 2) = 0 \Rightarrow x = -3, 0$ or 2

The boundary values divide the number line into four intervals. Choose a test value from each of these intervals and evaluate $x^3 + x^2 - 6x$ for these values. See Figure 91. The solution is $(-3, 0) \cup (2, \infty)$.

Interval	Test Value x	$x^3 + x^2 - 6x$	Positive or Negative?
$(-\infty, -3)$	-4	-24	Negative
$(-3, 0)$	-1	6	Positive
$(0, 2)$	1	-4	Negative
$(2, \infty)$	3	18	Positive

Figure 91

$[-4, 4, 1]$ by $[-5, 5, 1]$

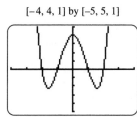

Figure 93

93. Write the inequality as $x^4 - 5x^2 + 4 < 0$.

Then Graph $Y_1 = X^4 - 5X^2 + 4$ as shown in Figure 93. The x-intercepts are $-2, -1, 1$, and 2.

The graph is below the x-axis on the interval $(-2, -1) \cup (1, 2)$.

95. Graph $Y_1 = (2X - 1)/(X + 2)$ using dot mode as shown in Figure 95. The graph has a vertical asymptote at $x = -2$ and an x-intercept at $x = \dfrac{1}{2}$. The graph is above the x-axis on the interval $(-\infty, -2) \cup \left(\dfrac{1}{2}, \infty\right)$.

[–8, 4, 2] by [–5, 5, 1]

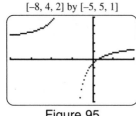

Figure 95

[0, 13, 1] by [0, 100, 10]

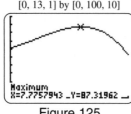

Maximum
X=7.7757943 Y=87.31962

Figure 125

[0, 8, 1] by [0, 500, 100]

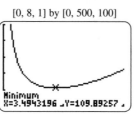

Minimum
X=3.4943196 Y=109.89257

Figure 127

97. $(36^{3/4})^2 = 36^{3/2} = (\sqrt{36})^3 = 6^3 = 216$

99. $(2^{-3/2} \cdot 2^{1/2})^{-3} = (2^{-3/2 + 1/2})^{-3} = (2^{-1})^{-3} = 2^3 = 8$

101. $\sqrt[3]{x^4} = x^{4/3}$

103. $\sqrt[3]{y \cdot \sqrt{y}} = (y \cdot y^{1/2})^{1/3} = (y^{3/2})^{1/3} = y^{1/2}$

105. $D = \{x \mid x \geq 0\}; f(3) = 3^{5/2} \approx 15.59$

107. $x^5 = 1024 \Rightarrow x = \sqrt[5]{1024} = 4$ Check: $4^5 = 1024$

109. $\sqrt{x - 2} = x - 4 \Rightarrow x - 2 = (x - 4)^2 \Rightarrow x - 2 = x^2 - 8x + 16 \Rightarrow x^2 - 9x + 18 = 0 \Rightarrow$

$(x - 6)(x - 3) = 0 \Rightarrow x = 3 \text{ or } 6$

Check: $\sqrt{3 - 2} \neq 3 - 4$ (not a solution); $\sqrt{6 - 2} = 2 = 6 - 4$. The only solution is $x = 6$.

111. $2x^{1/4} + 3 = 6 \Rightarrow 2x^{1/4} = 3 \Rightarrow x^{1/4} = \dfrac{3}{2} \Rightarrow x = \left(\dfrac{3}{2}\right)^4 \Rightarrow x = \dfrac{81}{16}$

Check: $2\left(\dfrac{81}{16}\right)^{1/4} + 3 = 2\left(\dfrac{3}{2}\right) + 3 = 3 + 3 = 6$

113. $\sqrt[3]{2x - 3} + 1 = 4 \Rightarrow \sqrt[3]{2x - 3} = 3 \Rightarrow 2x - 3 = 3^3 \Rightarrow 2x - 3 = 27 \Rightarrow 2x = 30 \Rightarrow x = 15$

Check: $\sqrt[3]{2(15) - 3} + 1 = \sqrt[3]{27} + 1 = 3 + 1 = 4$

115. $2n^{-2} - 5n^{-1} = 3 \Rightarrow 2(n^{-1})^2 - 5(n^{-1}) = 3$, let $u = n^{-1}$, then $2u^2 - 5u - 3 = 0 \Rightarrow$

$(2u + 1)(u - 3) = 0 \Rightarrow u = -\dfrac{1}{2}, 3$. If $u = n^{-1}$, then it follows that $n = u^{-1}$, thus $n = \left(-\dfrac{1}{2}\right)^{-1} \Rightarrow$

$n = -2 \text{ or } n = (3)^{-1} \Rightarrow n = \dfrac{1}{3}$

Check: $2(-2)^{-2} - 5(-2)^{-1} = 3 \Rightarrow 2\left(\dfrac{1}{4}\right) - 5\left(-\dfrac{1}{2}\right) = 3 \Rightarrow \dfrac{1}{2} + \dfrac{5}{2} = 3 \Rightarrow 3 = 3$

Check: $2\left(\dfrac{1}{3}\right)^{-2} - 5\left(\dfrac{1}{3}\right)^{-1} = 3 \Rightarrow 2(9) - 5(3) = 3 \Rightarrow 18 - 15 = 3 \Rightarrow 3 = 3$

The solution are: $n = -2, \dfrac{1}{3}$.

117. $k^{2/3} - 4k^{1/3} - 5 = 0 \Rightarrow (k^{1/3})^2 - 4(k^{1/3}) - 5 = 0$, let $u = k^{1/3}$, then $u^2 - 4u - 5 = 0 \Rightarrow$

$(u - 5)(u + 1) = 0 \Rightarrow u = -1, 5$. If $u = k^{1/3}$, then it follows that $k = u^3$, thus $k = (-1)^3 \Rightarrow$

$k = -1 \text{ or } k = (5)^3 \Rightarrow k = 125$.

Check: $(-1)^{2/3} - 4(-1)^{1/3} - 5 = 0 \Rightarrow 1 - (-4) - 5 = 0 \Rightarrow 0 = 0$

Check: $(125)^{2/3} - 4(125)^{1/3} - 5 = 0 \Rightarrow 25 - 4(5) - 5 = 0 \Rightarrow 0 = 0$. The solutions are: $k = -1, 125$.

119. $\sqrt{x+1}+1 = \sqrt{2x} \Rightarrow (\sqrt{x+1}+1)^2 = 2x \Rightarrow x+1+2\sqrt{x+1}+1 = 2x \Rightarrow 2\sqrt{x+1} = x-2 \Rightarrow$

$4(x+1) = x^2 - 4x + 4 \Rightarrow 4x + 4 = x^2 - 4x + 4 \Rightarrow x^2 - 8x = 0 \Rightarrow x(x-8) = 0 \Rightarrow x = 0, 8$.

Check: $\sqrt{0+1}+1 = \sqrt{2(0)} \Rightarrow 1+1 = 0 \Rightarrow 2 \neq 0 \Rightarrow 0$ is not a solution.

Check: $\sqrt{8+1}+1 = \sqrt{2(8)} \Rightarrow 4 = 4$. The solution is: $x = 8$.

121. (a) Dog: $f(24) = 1607(24)^{-0.75} \approx 148$ bpm; Person: $f(66) = 1607(66)^{-0.75} \approx 69$ bpm

(b) First, we must approximate the length of an animal with a heart rate of 400 bpm. To do this, solve the

equation $f(x) = 400$ or $1607x^{-0.75} = 400$ bpm.

$1607x^{-0.75} = 400 \Rightarrow x^{-0.75} = \dfrac{400}{1607} \Rightarrow x^{0.75} = \dfrac{1607}{400} \Rightarrow x^{3/4} = \dfrac{1607}{400} \Rightarrow x = \left(\dfrac{1607}{400}\right)^{4/3} \approx 6.4$

A six-inch animal such as a large bird or smaller rodent might have a heart rate of 400 bpm.

123. (a) Increasing: $[0, 0.4]$, $[1.6, 2.6]$, and $[3.4, 4]$

(b) Local maxima: Approximately 18.8 and 11.8 at about 12:24 P.M. and 2:36 P.M. the wind reached relative

maximums of 18.8 mph and 11.8 mph; at about 1:36 P.M. and 3:24 P.M. the wind reached relative

minimums of about 1.2 mph and 8.2 mph respectively.

125. (a) May is $T(5) \Rightarrow T(5) = -0.064(5)^3 + 0.56(5)^2 + 2.9(5) + 61 \Rightarrow T(5) = -8 + 14 + 14.5 + 61 \Rightarrow$

$T(5) = 81.5$. The temperature in May is $81.5°$F.

(b) Graph the function. See Figure 125. From the graph the ocean reaches a maximum temperature of about

$87.3°$F in July.

127. Let $x =$ width, then length $= 3x$. If $V = l \cdot w \cdot h$ then $96 = 3x \cdot x \cdot h \Rightarrow 96 = 3x^2h$ and $h = \dfrac{96}{3x^2} \Rightarrow$

$h = \dfrac{32}{x^2}$. The surface area of the box is : $s(x) = l \cdot w + 2lh + 2wh$ or

$s(x) = (3x)(x) + 2(3x)\left(\dfrac{32}{x^2}\right) + 2(x)\left(\dfrac{32}{x^2}\right) \Rightarrow s(x) = 3x^2 + \dfrac{192}{x} + \dfrac{64}{x} \Rightarrow s(x) = 3x^2 + \dfrac{256}{x}$.

Graph the equation, See Figure 127. The minimum value for x is: $x = 3.5$.

Therefore: width $= 3.5$ inches, length $= 3(3.5) = 10.5$ inches, and height $= \dfrac{32}{(3.5)^2} \approx 2.6$ inches.

The box is: $10.5 \times 3.5 \times 2.6$ inches.

129. (a) Using power regression with the given data gives $a \approx 1.8342$ and $b \approx -0.4839$. Thus $f(x) \approx 1.8342x^{-0.4839}$.

(b) $f(3) \approx 1.8342(3)^{-0.4839} \approx 1.08$; the elephant's stepping frequency would be about 1.08 steps per second.

Chapters 1-4 Cumulative Review Exercises

1. Percent change is: $\dfrac{\text{change}}{\text{original quantity}} = \dfrac{54-45}{45} = \dfrac{9}{45} = 0.20$. Therefore a 20% change.

3. $\dfrac{1.2-5.6}{0.4+9^2} = \dfrac{1.2-5.6}{0.4+81} = \dfrac{-4.4}{81.4} = -0.054$

5. Using the distance formula: $D = \sqrt{(x_2-x_1)^2 + (y_2-y_1)^2}$ we get: $D = \sqrt{(3-(-1))^2 + (-9-4)^2} \Rightarrow$

$D = \sqrt{16+169} \Rightarrow D = \sqrt{185}$

7. (a) See Figure 7a.

    (b) See Figure 7b.

    (c) See Figure 7c.

    (d) See Figure 7d.

    (e) See Figure 7e.

    (f) See Figure 7f.

    (g) See Figure 7g.

    (h) See Figure 7h.

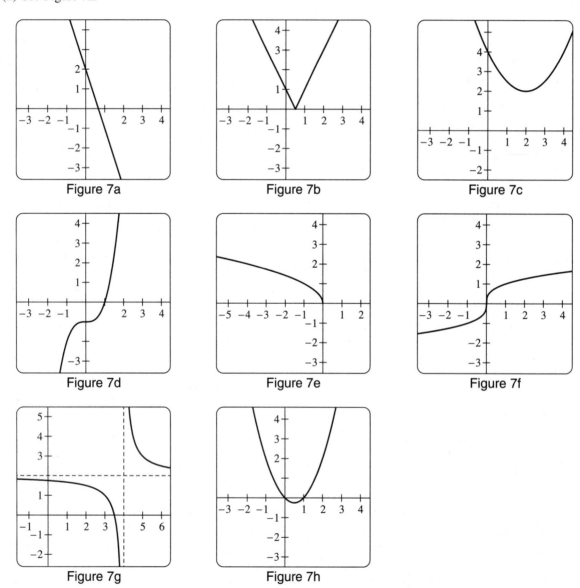

Figure 7a   Figure 7b   Figure 7c

Figure 7d   Figure 7e   Figure 7f

Figure 7g   Figure 7h

9. (a) $D = \{x \mid x \leq -2 \text{ or } x \geq 2\}$, all numbers between -2 and 2 give the unreal solutions, square roots of negative numbers.

    (b) $f(2) = \sqrt{(2)^2 - 4} \Rightarrow f(2) = \sqrt{0} \Rightarrow f(2) = 0$

11. $c(x) = 0.25x + 200$; $c(2000) = 0.25(2000) + 200 \Rightarrow c(2000) = 700$. The monthly cost of driving and maintaining a car driven 2000 miles is $700.

13. $f(-3) = (-3)^3 - (-3) \Rightarrow f(-3) = -24$ or $(-3, -24)$; $f(-2) = (-2)^3 - (-2) \Rightarrow f(-2) = -6$ or $(-2, -6)$

The average rate of change $= \dfrac{-6 - (-24)}{-2 - (-3)} = \dfrac{18}{1} = 18$

15. Slope $= \dfrac{(-4) - 5}{3 - (-2)} = \dfrac{-9}{5}$. Using point slope form $y = m(x - x_1) + y$ and point $(-2, 5)$ we get:

$y = \dfrac{-9}{5}(x + 2) + 5 \Rightarrow y = \dfrac{-9}{5}x - \dfrac{18}{5} + 5 \Rightarrow y = \dfrac{-9}{5}x + \dfrac{7}{5}$

17. All lines parallel to the x-axis have 0 slope and will have graphs of $y =$ (the y-coordinate) $\Rightarrow y = -5$.

19. Each radio costs $15 to manufacture. The fixed cost for this company to produce radios is $2000.

21. $-3(2 - 3x) - (-x - 1) = 1 \Rightarrow -6 + 9x + x + 1 = 1 \Rightarrow 10x - 5 = 1 \Rightarrow 10x = 6 \Rightarrow x = \dfrac{3}{5}$

23. $|3x - 4| + 1 = 5 \Rightarrow |3x - 4| = 4 \Rightarrow 3x - 4 = 4 \Rightarrow 3x = 8 \Rightarrow x = \dfrac{8}{3}$ or $3x - 4 = -4 \Rightarrow 3x = 0 \Rightarrow$

$x = 0$. Therefore $x = 0, \dfrac{8}{3}$.

25. $7x^2 + 9x = 10 \Rightarrow 7x^2 + 9x - 10 = 0 \Rightarrow (7x - 5)(x + 2) = 0 \Rightarrow x = -2, \dfrac{5}{7}$

27. $2x^3 + 4x^2 = 6x \Rightarrow 2x^3 + 4x^2 - 6x = 0 \Rightarrow 2x(x^2 + 2x - 3) = 0 \Rightarrow 2x(x + 3)(x - 1) = 0 \Rightarrow$

$x = -3, 0, 1$

29. $3x^{2/3} + 5x^{1/3} - 2 = 0 \Rightarrow 3(x^{1/3})^2 + 5(x^{1/3}) - 2 = 0$. Let $u = x^{1/3}$, then $3u^2 + 5u - 2 = 0 \Rightarrow$

$(3u - 1)(u + 2) = 0 \Rightarrow u = -2, \dfrac{1}{3}$, If $u = x^{1/3}$ then it follows $x = u^3$, thus $x = (-2)^3 \Rightarrow x = -8$ or

$x = \left(\dfrac{1}{3}\right)^3 \Rightarrow x = \dfrac{1}{27}$. Therefore $x = -8, \dfrac{1}{27}$

31. $\dfrac{2x - 3}{5 - x} = \dfrac{4x - 3}{1 - 2x} \Rightarrow (2x - 3)(1 - 2x) = (4x - 3)(5 - x) \Rightarrow$

$2x - 4x^2 - 3 + 6x = 20x - 4x^2 - 15 + 3x \Rightarrow 15x = 12 \Rightarrow x = \dfrac{4}{5}$

33. $\dfrac{1}{2}x - (4 - x) + 1 = \dfrac{3}{2}x - 5 \Rightarrow \dfrac{1}{2}x - 4 + x + 1 = \dfrac{3}{2}x - 5 \Rightarrow \dfrac{3}{2}x - 3 = \dfrac{3}{2}x - 5 \Rightarrow$

$-3 = -5$, since this is false the equation has no solutions and is a contradiction.

35. $-\dfrac{1}{3}x - (1 + x) > \dfrac{2}{3}x \Rightarrow -\dfrac{1}{3}x - 1 - x > \dfrac{2}{3}x \Rightarrow -\dfrac{4}{3}x - 1 > \dfrac{2}{3}x \Rightarrow -2x > 1 \Rightarrow x < -\dfrac{1}{2} \Rightarrow$

$\left(-\infty, -\dfrac{1}{2}\right)$

37. The solutions to $|5x - 7| \geq 3$ satisfy $x \leq s_1$ or $x \geq s_2$ where s_1 and s_2 are the solutions to $|5x - 7| = 3$.

$|5x - 7| = 3$ is equivalent to $5x - 7 = -3 \Rightarrow x = \dfrac{4}{5}$ and $5x - 7 = 3 \Rightarrow x = 2$.

The interval is $\left(-\infty, \dfrac{4}{5}\right] \cup [2, \infty)$.

39. For $x^3 - 9x \leq 0$, first set $x^3 - 9x = 0 \Rightarrow x(x^2 - 9) = 0 \Rightarrow x(x + 3)(x - 3) = 0 \Rightarrow x = -3, 0, 3$.

These boundary numbers separate the numberline into these intervals: $(-\infty, -3), (-3, 0), (0, 3)$ and $(3, \infty)$.

Checking an x value in each interval we get: for $x = -5, (-5)^3 - 9(-5) \leq 0 \Rightarrow -125 + 45 \leq 0 \Rightarrow -80 \leq 0$,

which is true $\Rightarrow$ a solution; for $x = -1, (-1)^3 - 9(-1) \leq 0 \Rightarrow -1 + 9 \leq 0 \Rightarrow 8 \leq 0$, which is false $\Rightarrow$

not a solution; for $x = 1, (1)^3 - 9(1) \leq 0 \Rightarrow 1 - 9 \leq 0 \Rightarrow -8 \leq 0$, which is true $\Rightarrow$ a solution.

for $x = 5, (5)^3 - 9(5) \leq 0 \Rightarrow 125 - 45 \leq 0 \Rightarrow 80 \leq 0$, which is false $\Rightarrow$ not a solution.

Therefore $(-\infty, -3] \cup [0, 3]$.

41. (a) $x = -3, -1, 1, 2$

(b) $(-\infty, -3) \cup (-1, 1) \cup (2, \infty)$

(c) $[-3, -1] \cup [1, 2]$

43. $f(x) = -\frac{1}{2}x^2 + 3x - 2 \Rightarrow y = -\frac{1}{2}x^2 + 3x - 2 \Rightarrow y + 2 = -\frac{1}{2}x^2 + 3x \Rightarrow y + 2 = -\frac{1}{2}(x^2 - 6x) \Rightarrow$

$y + 2 - \frac{9}{2} = -\frac{1}{2}(x^2 - 6x + 9) \Rightarrow y - \frac{5}{2} = -\frac{1}{2}(x - 3)^2 \Rightarrow y = -\frac{1}{2}(x - 3)^2 + \frac{5}{2} \Rightarrow$

$f(x) = -\frac{1}{2}(x - 3)^2 + \frac{5}{2} \Rightarrow$ the vertex is $\left(3, \frac{5}{2}\right)$.

45. $x^2 - 3x = 1 \Rightarrow x^2 - 3x + \frac{9}{4} = 1 + \frac{9}{4} \Rightarrow \left(x - \frac{3}{2}\right)^2 = \frac{13}{4} \Rightarrow x - \frac{3}{2} = \pm\sqrt{\frac{13}{4}} \Rightarrow$

$x = \frac{3}{2} \pm \frac{\sqrt{13}}{2}$ or $x = \frac{3 \pm \sqrt{13}}{2}$

47. Take the graph of $y = \sqrt{x}$ and shift it one unit left and make it twice as wide. See Figure 47.

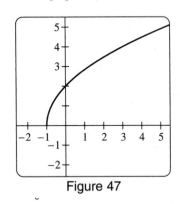

Figure 47

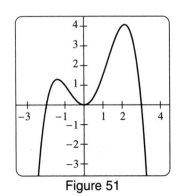

Figure 51

49. (a) It is increasing: $(-\infty, -2]$ and $[1, \infty)$; it is decreasing: $[-2, 1]$.

(b) The zeros are approximately $x = -3.3, 0$ and 1.8.

(c) The turning points are: $(-2, 3)$ and $(1, -1)$.

(d) It has local extrema, maximum: 3, and minimum: -1.

51. Answers may vary. See Figure 51.

53. (a)
$$\begin{array}{r}
a - 2 + \frac{12}{4a^2} \\
4a^2\overline{\smash{)}4a^3 - 8a^2 + 12} \\
\underline{4a^3} \\
-8a^2 + 12 \\
\underline{-8a^2} \\
12
\end{array}$$

Therefore the quotient is: $a - 2 + \dfrac{3}{a^2}$

(b)
$$\begin{array}{r}
2x^2 + 2x - 2 + \frac{-1}{x-1} \\
x - 1\overline{\smash{)}2x^3 \qquad\quad - 4x + 1} \\
\underline{2x^3 - 2x^2} \\
2x^2 - 4x + 1 \\
\underline{2x^2 - 2x} \\
-2x + 1 \\
\underline{-2x + 2} \\
-1
\end{array}$$

Therefore the quotient is: $2x^2 + 2x - 2 + \dfrac{-1}{x-1}$

(c)
$$\begin{array}{r}
x^2 + 2x - 3 + \frac{x+4}{x^2+2} \\
x^2 + 2\overline{\smash{)}x^4 + 2x^3 \quad - x^2 + 5x - 2} \\
\underline{x^4 \qquad\quad + 2x^2} \\
2x^3 - 3x^2 + 5x - 2 \\
\underline{2x^3 \qquad\quad + 4x} \\
-3x^2 + x - 2 \\
\underline{-3x^2 \qquad - 6} \\
x + 4
\end{array}$$

Therefore the quotient is: $x^2 + 2x - 3 + \dfrac{x+4}{x^2+2}$

55. $f(x) = 4(x + 3)(x - 1)^2(x - 4)^3$

57. $2x^3 - x^2 - 6x + 3 = (2x^3 - x^2) - (6x - 3) = x^2(2x - 1) - 3(2x - 1) = (x^2 - 3)(2x - 1) =$

$(x + \sqrt{3})(x - \sqrt{3})(2x - 1) \Rightarrow f(x) = (x + \sqrt{3})(x - \sqrt{3})(2x - 1) \Rightarrow$

the complete factored form is: $f(x) = 2(x + \sqrt{3})(x - \sqrt{3})\left(x - \dfrac{1}{2}\right)$

59. The zeros are: $x = -2, -1$, and $1 \Rightarrow$ we have factors $(x + 2)(x + 1)(x - 1)$. Now to find the leading

coefficient we solve for $f(0) = 4 \Rightarrow a(0 + 2)(0 + 1)(0 - 1) = 4 \Rightarrow a(2)(1)(-1) = 4 \Rightarrow -2a = 4 \Rightarrow$

$a = -2$. Therefore the complete factored form is: $f(x) = -2(x + 2)(x + 1)(x - 1)$.

61. $\dfrac{3 + 4i}{1 - i} \cdot \dfrac{1 + i}{1 + i} = \dfrac{3 + 3i + 4i + 4i^2}{1 - (i)^2} = \dfrac{3 + 7i + 4(-1)}{1 - (-1)} = \dfrac{-1 + 7i}{2}$ or $\dfrac{-1}{2} + \dfrac{7i}{2}$

63. For the domain the denominator cannot equal zero $\Rightarrow$ set $x^2 - 3x - 4 = 0 \Rightarrow (x + 1)(x - 4) = 0$.

    Then $D = \{x \mid x \neq -1, x \neq 4\}$. Since the degree of the numerator is less then the degree of the denominator the horizontal asymptote is $y = 0$. The vertical asymptotes are values when the demoninator equals $0 \Rightarrow$ the vertical asymptotes are: $x = -1, x = 4$.

65. $V = \dfrac{1}{3}\pi r^2 h \Rightarrow 18 = \dfrac{1}{3}\pi\left(\dfrac{5}{4}\right)^2 h \Rightarrow 18 = \dfrac{1}{3} \cdot \pi \cdot \dfrac{25}{16} \cdot h \Rightarrow 54 = \pi \cdot \dfrac{25}{16} \cdot h \Rightarrow 34.56 = \pi \cdot h \Rightarrow 11.00 \approx h$

67. After 0.5 hours the first runner is $8(0.5) - 2$ or 2 miles south of where the second runner starts. After 0.5 hours the second runner is $7(0.5)$ or 3.5 miles west of where they started. Using $(0, 2)$ as the new location of the first runner and $(-3.5, 0)$ as the new location of the second runner and the distance formula we get:

    $d = \sqrt{(-3.5 - 0)^2 + (0 - 2)^2} \Rightarrow d = \sqrt{12.25 + 4} \Rightarrow d = \sqrt{16.25} \Rightarrow d = 4.0$ miles..

69. The first person can paint $\dfrac{1}{10}x$ job per hour, the second $\dfrac{1}{8}x$ job per hour $\Rightarrow$ together they can do one paint job

    in: $\dfrac{1}{10}x + \dfrac{1}{8}x = 1$. $\dfrac{1}{10}x + \dfrac{1}{8}x = 1 \Rightarrow \dfrac{8}{80}x + \dfrac{10}{80}x = 1 \Rightarrow \dfrac{18}{80}x = 1 \Rightarrow x = \dfrac{80}{18} \Rightarrow x = \dfrac{40}{9}$ or

    $x = 4.44$ hours.

71. (a) If $(1990, 250)$ and $(2000, 280)$ then the average rate of change is: $\dfrac{280 - 250}{2000 - 1990} = \dfrac{30}{10} = 3$.

    Then, $P(t) = 3(t - 1990) + 250$

    (b) $300 = 3(t - 1990) + 250 \Rightarrow 50 = 3t - 5970 \Rightarrow 3t = 6020 \Rightarrow t = 2006.7 \Rightarrow 2007$ or late 2006.

73. If the dimension of the rectangular sheet of metal is: x by $x + 6$, then the dimensions of the sides of the box are $x - 4$ by $x + 2$ and the $h = 2 \Rightarrow 2(x - 4)(x + 2) = 270 \Rightarrow 2(x^2 - 2x - 8) = 270 \Rightarrow$

    $2x^2 - 4x - 16 = 270 \Rightarrow 2x^2 - 4x - 286 = 0 \Rightarrow 2(x^2 - 2x - 143) = 0 \Rightarrow 2(x + 11)(x - 13) = 0 \Rightarrow$

    $x = -11, 13$. Since length cannot be negative the dimensions of the sheet of metal are: 13 by 19 inches.

75. Using the lowest values $x = 4$ and $y = 6$ as the vertex of the parabola we get: $f(x) = a(x - 4)^2 + 6$. Using the graph, see Figure 75, and trial and error, $a \approx 2$ produces a reasonable fit for the data $\Rightarrow$

    $f(x) = 2(x - 4)^2 + 6$.

[0, 12, 2] by [0, 90, 10]   [1955, 2005, 5] by [0, 35, 5]

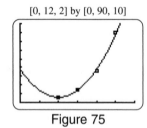   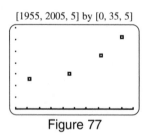

Figure 75   Figure 77

77. (a) Using the lowest values $x = 1961$ and $y = 13$ as the vertex of the parabola we get:

    $f(x) = a(x - 1961)^2 + 13$. Using the graph, see Figure 77, and trial and error, $a \approx 0.01$ produces a reasonable fit for the data $\Rightarrow f(x) = 0.01(x - 1961)^2 + 13$.

    (b) $40 = 0.01(x - 1961)^2 + 13 \Rightarrow 27 = 0.01(x - 1961)^2 \Rightarrow 2700 = (x - 1961)^2 \Rightarrow$

    $\sqrt{2700} = x - 1961 \Rightarrow 51.96 = x - 1961 \Rightarrow 2013 \approx x$, slightly later than the experts prediction of 2009.

Chapter 5: Exponential and Logarithmic Functions

5.1: Combining Functions

Arithmetic Operations on Functions

1. (a) $(f + g)(2) = f(2) + g(2) = 7 + (-2) = 5$

   (b) $(f - g)(4) = f(4) - g(4) = 10 - 5 = 5$

   (c) $(fg)(-2) = f(-2) \cdot g(-2) = (0)(6) = 0$

   (d) $(f/g)(0) = \dfrac{f(0)}{g(0)} = \dfrac{5}{0} \Rightarrow$ undefined

3. For example: $(f - g)(2) = f(2) - g(2) = 7 - (-2) = 9$. See Figure 3. A dash (—) indicates that the value of the function is undefined.

x	-2	0	2	4
$f(x)$	0	5	7	10
$g(x)$	6	0	-2	5
$(f + g)(x)$	6	5	5	15
$(f - g)(x)$	-6	5	9	5
$(fg)(x)$	0	0	-14	50
$(f/g)(x)$	0	—	-3.5	2

Figure 3

5. (a) From the graph $f(2) = 4$ and $g(2) = -2$. Thus, $(f + g)(2) = f(2) + g(2) = 4 + (-2) = 2$.

   (b) From the graph $f(1) = 1$ and $g(1) = -3$. Thus, $(f - g)(1) = f(1) - g(1) = 1 - (-3) = 4$.

   (c) From the graph $f(0) = 0$ and $g(0) = -4$. Thus, $(fg)(0) = f(0) \cdot g(0) = (0)(-4) = 0$.

   (d) From the graph $f(1) = 1$ and $g(1) = -3$. Thus, $(f/g)(1) = \dfrac{f(1)}{g(1)} = \dfrac{1}{-3} = -\dfrac{1}{3}$.

7. (a) From the graph $f(0) = 0$ and $g(0) = 2$. Thus, $(f + g)(0) = f(0) + g(0) = 0 + 2 = 2$.

   (b) From the graph $f(-1) = -2$ and $g(-1) = 1$. Thus, $(f - g)(-1) = f(-1) - g(-1) = -2 - 1 = -3$.

   (c) From the graph $f(1) = 2$ and $g(1) = 1$. Thus, $(fg)(1) = f(1) \cdot g(1) = (2)(1) = 2$.

   (d) From the graph $f(2) = 4$ and $g(2) = -2$. Thus, $(f/g)(2) = \dfrac{f(2)}{g(2)} = \dfrac{4}{-2} = -2$.

9. (a) $f(3) = 2(3) - 3 = 3$ and $g(3) = 1 - 3^2 = -8$. Thus, $(f + g)(3) = f(3) + g(3) = 3 + (-8) = -5$.

   (b) $f(-1) = 2(-1) - 3 = -5$ and $g(-1) = 1 - (-1)^2 = 0$. Thus, $(f - g)(-1) = f(-1) - g(-1) = -5 - 0 = -5$.

   (c) $f(0) = 2(0) - 3 = -3$ and $g(0) = 1 - 0^2 = 1$. Thus, $(fg)(0) = f(0) \cdot g(0) = (-3)(1) = -3$.

   (d) $f(2) = 2(2) - 3 = 1$ and $g(2) = 1 - 2^2 = -3$. Thus, $(f/g)(2) = \dfrac{f(2)}{g(2)} = \dfrac{1}{-3} = -\dfrac{1}{3}$.

11. (a) $f(2) = 2(2) + 1 = 5$ and $g(2) = \dfrac{1}{2}$. Thus, $(f + g)(2) = f(2) + g(2) = 5 + \dfrac{1}{2} = \dfrac{11}{2}$.

(b) $f\left(\dfrac{1}{2}\right) = 2\left(\dfrac{1}{2}\right) + 1 = 2$ and $g\left(\dfrac{1}{2}\right) = \dfrac{1}{\frac{1}{2}} = 2$. Thus, $(f - g)\left(\dfrac{1}{2}\right) = f\left(\dfrac{1}{2}\right) - g\left(\dfrac{1}{2}\right) = 2 - 2 = 0$.

(c) $f(4) = 2(4) + 1 = 9$ and $g(4) = \dfrac{1}{4}$. Thus, $(fg)(4) = f(4) \cdot g(4) = (9)\left(\dfrac{1}{4}\right) = \dfrac{9}{4}$.

(d) $f(0) = 2(0) + 1 = 1$ and $g(0) = \dfrac{1}{0} \Rightarrow$ undefined. Thus, $(f/g)(0) \Rightarrow$ undefined.

13. (a) $(f + g)(x) = f(x) + g(x) = 2x + x^2$; Domain: all real numbers.

(b) $(f - g)(x) = f(x) - g(x) = 2x - x^2$; Domain: all real numbers.

(c) $(fg)(x) = f(x) \cdot g(x) = (2x)(x^2) = 2x^3$; Domain: all real numbers.

(d) $(f/g)(x) = \dfrac{f(x)}{g(x)} = \dfrac{2x}{x^2} = \dfrac{2}{x}$; Domain: $\{x \mid x \neq 0\}$.

15. (a) $(f + g)(x) = f(x) + g(x) = (x^2 - 1) + (x^2 + 1) = 2x^2$; Domain: all real numbers.

(b) $(f - g)(x) = f(x) - g(x) = (x^2 - 1) - (x^2 + 1) = -2$; Domain: all real numbers.

(c) $(fg)(x) = f(x) \cdot g(x) = (x^2 - 1)(x^2 + 1) = x^4 - 1$; Domain: all real numbers.

(d) $(f/g)(x) = \dfrac{f(x)}{g(x)} = \dfrac{x^2 - 1}{x^2 + 1}$; Domain: all real numbers.

17. (a) $(f + g)(x) = f(x) + g(x) = (x - \sqrt{x - 1}) + (x + \sqrt{x - 1}) = 2x$; Domain: $\{x \mid x \geq 1\}$.

(b) $(f - g)(x) = f(x) - g(x) = (x - \sqrt{x - 1}) - (x + \sqrt{x - 1}) = -2\sqrt{x - 1}$; Domain: $\{x \mid x \geq 1\}$.

(c) $(fg)(x) = f(x) \cdot g(x) = (x - \sqrt{x - 1})(x + \sqrt{x - 1}) = x^2 - (x - 1) = x^2 - x + 1$; Domain: $\{x \mid x \geq 1\}$.

(d) $(f/g)(x) = \dfrac{f(x)}{g(x)} = \dfrac{x - \sqrt{x - 1}}{x + \sqrt{x - 1}}$; Domain: $\{x \mid x \geq 1\}$.

19. (a) $(f + g)(x) = f(x) + g(x) = (\sqrt{x} - 1) + (\sqrt{x} + 1) = 2\sqrt{x}$; Domain: $\{x \mid x \geq 0\}$.

(b) $(f - g)(x) = f(x) - g(x) = (\sqrt{x} - 1) - (\sqrt{x} + 1) = -2$; Domain: $\{x \mid x \geq 0\}$.

(c) $(fg)(x) = f(x) \cdot g(x) = (\sqrt{x} - 1)(\sqrt{x} + 1) = x - 1$; Domain: $\{x \mid x \geq 0\}$.

(d) $(f/g)(x) = \dfrac{f(x)}{g(x)} = \dfrac{\sqrt{x} - 1}{\sqrt{x} + 1}$; Domain: $\{x \mid x \geq 0\}$.

21. (a) $(f + g)(x) = f(x) + g(x) = \dfrac{1}{x + 1} + \dfrac{3}{x + 1} = \dfrac{4}{x + 1}$; Domain: $\{x \mid x \neq -1\}$.

(b) $(f - g)(x) = f(x) - g(x) = \dfrac{1}{x + 1} - \dfrac{3}{x + 1} = -\dfrac{2}{x + 1}$; Domain: $\{x \mid x \neq -1\}$.

(c) $(fg)(x) = f(x) \cdot g(x) = \left(\dfrac{1}{x + 1}\right)\left(\dfrac{3}{x + 1}\right) = \dfrac{3}{(x + 1)^2}$; Domain: $\{x \mid x \neq -1\}$.

(d) $(f/g)(x) = \dfrac{f(x)}{g(x)} = \dfrac{\frac{1}{x + 1}}{\frac{3}{x + 1}} = \dfrac{1}{x + 1} \cdot \dfrac{x + 1}{3} = \dfrac{1}{3}$; Domain: $\{x \mid x \neq -1\}$.

23. (a) $(f + g)(x) = f(x) + g(x) = (2x) + (3 - 4x) = 3 - 2x$; Domain: all real numbers.

(b) $(f - g)(x) = f(x) - g(x) = (2x) - (3 - 4x) = 6x - 3$; Domain: all real numbers.

(c) $(fg)(x) = f(x) \cdot g(x) = (2x)(3 - 4x) = 6x - 8x^2$; Domain: all real numbers.

(d) $(f/g)(x) = \dfrac{f(x)}{g(x)} = \dfrac{2x}{3 - 4x}$; Domain: $\left\{x \mid x \neq \dfrac{3}{4}\right\}$.

25. (a) $(f + g)(x) = f(x) + g(x) = \dfrac{1}{2x - 4} + \dfrac{x}{2x - 4} = \dfrac{1 + x}{2x - 4}$; Domain: $\{x \mid x \neq 2\}$.

(b) $(f - g)(x) = f(x) - g(x) = \dfrac{1}{2x - 4} - \dfrac{x}{2x - 4} = \dfrac{1 - x}{2x - 4}$; Domain: $\{x \mid x \neq 2\}$.

(c) $(fg)(x) = f(x) \cdot g(x) = \left(\dfrac{1}{2x - 4}\right)\left(\dfrac{x}{2x - 4}\right) = \dfrac{x}{(2x - 4)^2}$; Domain: $\{x \mid x \neq 2\}$.

(d) $(f/g)(x) = \dfrac{f(x)}{g(x)} = \dfrac{\frac{1}{2x-4}}{\frac{x}{2x-4}} = \dfrac{1}{2x - 4} \cdot \dfrac{2x - 4}{x} = \dfrac{1}{x}$; Domain: $\{x \mid x \neq 0 \text{ and } x \neq 2\}$.

27. (a) $(f + g)(x) = f(x) + g(x) = (x + 2) + (x^2 + 2x) = x^2 + 3x + 2$; Domain: all real numbers.

(b) $(f - g)(x) = f(x) - g(x) = (x + 2) - (x^2 + 2x) = -x^2 - x + 2$; Domain: all real numbers.

(c) $(fg)(x) = f(x) \cdot g(x) = (x + 2)(x^2 + 2x) = x^3 + 4x^2 + 4x$; Domain: all real numbers.

(d) $(f/g)(x) = \dfrac{f(x)}{g(x)} = \dfrac{x + 2}{x^2 + 2x} = \dfrac{x + 2}{x(x + 2)} = \dfrac{1}{x}$; Domain: $\{x \mid x \neq -2 \text{ and } x \neq 0\}$.

29. (a) $(f + g)(x) = f(x) + g(x) = (x^2 - 1) + (|x + 1|) = x^2 - 1 + |x + 1|$; Domain: all real numbers.

(b) $(f - g)(x) = f(x) - g(x) = (x^2 - 1) - (|x + 1|) = x^2 - 1 - |x + 1|$; Domain: all real numbers.

(c) $(fg)(x) = f(x) \cdot g(x) = (x^2 - 1)(|x + 1|) = (x^2 - 1)(|x + 1|)$; Domain: all real numbers.

(d) $(f/g)(x) = \dfrac{f(x)}{g(x)} = \dfrac{x^2 - 1}{|x + 1|}$; Domain: $\{x \mid x \neq -1\}$.

31. (a) $(f + g)(x) = f(x) + g(x) = \dfrac{(x - 1)(x - 2)}{x + 1} + \dfrac{(x + 1)(x - 1)}{x - 2} =$

$\dfrac{(x - 1)(x - 2)(x - 2)}{(x + 1)(x - 2)} + \dfrac{(x - 1)(x + 1)(x + 1)}{(x + 1)(x - 2)} = \dfrac{(x - 1)(x^2 - 4x + 4)}{(x + 1)(x - 2)} + \dfrac{(x - 1)(x^2 + 2x + 1)}{(x + 1)(x - 2)} =$

$\dfrac{(x - 1)(2x^2 - 2x + 5)}{(x + 1)(x - 2)}$; Domain: $\{x \mid x \neq -1, x \neq 2\}$

(b) $(f - g)(x) = f(x) - g(x) = \dfrac{(x - 1)(x - 2)}{x + 1} - \dfrac{(x + 1)(x - 1)}{x - 2} =$

$\dfrac{(x - 1)(x - 2)(x - 2)}{(x + 1)(x - 2)} - \dfrac{(x - 1)(x + 1)(x + 1)}{(x + 1)(x - 2)} = \dfrac{(x - 1)(x^2 - 4x + 4)}{(x + 1)(x - 2)} - \dfrac{(x - 1)(x^2 + 2x + 1)}{(x + 1)(x - 2)} =$

$\dfrac{(x - 1)(-6x + 3)}{(x + 1)(x - 2)} = \dfrac{-3(x - 1)(2x - 1)}{(x + 1)(x - 2)}$; Domain: $\{x \mid x \neq -1, x \neq 2\}$

(c) $(fg)(x) = f(x) \cdot g(x) = \dfrac{(x - 1)(x - 2)}{x + 1} \cdot \dfrac{(x + 1)(x - 1)}{x - 2} = \dfrac{x^2 - 3x + 2}{x + 1} \cdot \dfrac{x^2 - 1}{x - 2} =$

$\dfrac{(x - 2)(x - 1)}{x + 1} \cdot \dfrac{(x + 1)(x - 1)}{x - 2} = (x - 1)(x - 1) = (x - 1)^2$; Domain: $\{x \mid x \neq -1, x \neq 2\}$

(d) $(f/g)(x) = \dfrac{f(x)}{g(x)} = \dfrac{(x - 2)(x - 1)}{x + 1} \div \dfrac{(x + 1)(x - 1)}{x - 2} = \dfrac{(x - 2)(x - 1)}{x + 1} \cdot \dfrac{x - 2}{(x + 1)(x - 1)} = \dfrac{(x - 2)^2}{(x + 1)^2}$;

Domain: $\{x \mid x \neq 1, x \neq -1 \text{ and } x \neq 2\}$

33. (a) $(f + g)(x) = f(x) + g(x) = \dfrac{2}{(x + 1)(x - 1)} + \dfrac{(x + 1)}{(x - 1)(x - 1)} = \dfrac{2(x - 1)}{(x + 1)(x - 1)^2} + \dfrac{(x + 1)^2}{(x + 1)(x - 1)^2} =$

$\dfrac{2x - 2}{(x + 1)(x - 1)^2} + \dfrac{x^2 + 2x + 1}{(x + 1)(x - 1)^2} = \dfrac{x^2 + 4x - 1}{(x + 1)(x - 1)^2};$ Domain: $\{x \mid x \neq -1, x \neq 1\}$

(b) $(f - g)(x) = f(x) - g(x) = \dfrac{2}{(x + 1)(x - 1)} - \dfrac{(x + 1)}{(x - 1)(x - 1)} = \dfrac{2(x - 1)}{(x + 1)(x - 1)^2} - \dfrac{(x + 1)^2}{(x + 1)(x - 1)} =$

$\dfrac{2x - 2}{(x + 1)(x - 1)^2} - \dfrac{x^2 + 2x + 1}{(x + 1)(x - 1)^2} = \dfrac{-x^2 - 3}{(x + 1)(x - 1)^2};$ Domain: $\{x \mid x \neq -1, x \neq 1\}$

(c) $(fg)(x) = f(x) \cdot g(x) = \dfrac{2}{(x + 1)(x - 1)} \cdot \dfrac{(x + 1)}{(x - 1)(x - 1)} = \dfrac{2}{(x - 1)^3};$ Domain: $\{x \mid x \neq -1, x \neq 1\}$

(d) $(f/g)(x) = f(x) \div g(x) = \dfrac{2}{(x + 1)(x - 1)} \div \dfrac{(x + 1)}{(x - 1)(x - 1)} = \dfrac{2}{(x + 1)(x - 1)} \cdot \dfrac{(x - 1)(x - 1)}{(x + 1)} =$

$\dfrac{2(x - 1)}{(x + 1)^2};$ Domain: $\{x \mid x \neq -1, x \neq 1\}$

35. (a) $(f + g)(x) = f(x) + g(x) = (x^{5/2} - x^{3/2}) + (x^{1/2}) = x^{5/2} - x^{3/2} + x^{1/2} = x^{1/2}(x^2 - x + 1);$

Domain: $\{x \mid x \geq 0\}$

(b) $(f - g)(x) = f(x) - g(x) = (x^{5/2} - x^{3/2}) - (x^{1/2}) = x^{5/2} - x^{3/2} - x^{1/2} = x^{1/2}(x^2 - x - 1);$

Domain: $\{x \mid x \geq 0\}$

(c) $(fg)(x) = f(x) \cdot g(x) = (x^{5/2} - x^{3/2})(x^{1/2}) = x^3 - x^2 = x^2(x - 1);$ Domain: $\{x \mid x \geq 0\}$

(d) $(f/g)(x) = f(x) \div g(x) = \left(\dfrac{x^{5/2} - x^{3/2}}{x^{1/2}} \right) = x^2 - x = x(x - 1);$ Domain: $\{x \mid x > 0\}$

37. (a) Let $Y_1 = \sqrt{(X)}$, $Y_2 = X + 1$ and $Y_3 = Y_1 + Y_2$. Their graphs are shown in Figure 37.

(b) To determine the graph of $f + g$ add the y-coordinates of the graphs of f and g.

[0, 9, 1] by [0, 15, 1]

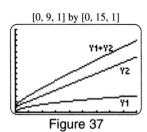

Figure 37

Review of Function Notation

39. (a) $g(x) = 2x + 1 \Rightarrow g(-3) = 2(-3) + 1 \Rightarrow g(-3) = -5$

(b) $g(x) = 2x + 1 \Rightarrow g(b) = 2(b) + 1 \Rightarrow g(b) = 2b + 1$

(c) $g(x) = 2x + 1 \Rightarrow g(x^3) = 2(x^3) + 1 \Rightarrow g(x^3) = 2x^3 + 1$

(d) $g(x) = 2x + 1 \Rightarrow g(2x - 3) = 2(2x - 3) + 1 \Rightarrow g(2x - 3) = 4x - 6 + 1 \Rightarrow g(2x - 3) = 4x - 5$

41. (a) $g(x) = 2(x + 3)^2 - 4 \Rightarrow g(-3) = 2((-3) + 3)^2 - 4 \Rightarrow g(-3) = 2(0) - 4 \Rightarrow g(-3) = -4$

    (b) $g(x) = 2(x + 3)^2 - 4 \Rightarrow g(b) = 2((b) + 3)^2 - 4 \Rightarrow g(b) = 2(b + 3)^2 - 4$

    (c) $g(x) = 2(x + 3)^2 - 4 \Rightarrow g(x^3) = 2((x^3) + 3)^2 - 4 \Rightarrow g(x^3) = 2(x^3 + 3)^2 - 4$

    (d) $g(x) = 2(x + 3)^2 - 4 \Rightarrow g(2x - 3) = 2((2x - 3) + 3)^2 - 4 \Rightarrow g(2x - 3) = 2(2x)^2 - 4 \Rightarrow$

    $g(2x - 3) = 2(4x^2) - 4 \Rightarrow g(2x - 3) = 8x^2 - 4$

43. (a) $g(x) = \dfrac{1}{2}x^2 + 3x - 1 \Rightarrow g(-3) = \dfrac{1}{2}(-3)^2 + 3(-3) - 1 \Rightarrow g(-3) = \dfrac{9}{2} - 9 - 1 \Rightarrow$

    $g(-3) = \dfrac{9}{2} - \dfrac{20}{2} \Rightarrow g(-3) = -\dfrac{11}{2}$

    (b) $g(x) = \dfrac{1}{2}x^2 + 3x - 1 \Rightarrow g(b) = \dfrac{1}{2}(b)^2 + 3(b) - 1 \Rightarrow g(b) = \dfrac{1}{2}b^2 + 3b - 1$

    (c) $g(x) = \dfrac{1}{2}x^2 + 3x - 1 \Rightarrow g(x^3) = \dfrac{1}{2}(x^3)^2 + 3(x^3) - 1 \Rightarrow g(x^3) = \dfrac{1}{2}x^6 + 3x^3 - 1$

    (d) $g(x) = \dfrac{1}{2}x^2 + 3x - 1 \Rightarrow g(2x - 3) = \dfrac{1}{2}(2x - 3)^2 + 3(2x - 3) - 1 \Rightarrow$

    $g(2x - 3) = \dfrac{4x^2 - 12x + 9}{2} + 6x - 9 - 1 \Rightarrow g(2x - 3) = 2x^2 - 6x + \dfrac{9}{2} + 6x - \dfrac{20}{2} \Rightarrow$

    $g(2x - 3) = 2x^2 - \dfrac{11}{2}$

45. (a) $g(x) = \sqrt{x + 4} \Rightarrow g(-3) = \sqrt{(-3) + 4} \Rightarrow g(-3) = \sqrt{1} \Rightarrow g(-3) = 1$

    (b) $g(x) = \sqrt{x + 4} \Rightarrow g(b) = \sqrt{(b) + 4} \Rightarrow g(b) = \sqrt{b + 4}$

    (c) $g(x) = \sqrt{x + 4} \Rightarrow g(x^3) = \sqrt{(x^3) + 4} \Rightarrow g(x^3) = \sqrt{x^3 + 4}$

    (d) $g(x) = \sqrt{x + 4} \Rightarrow g(2x - 3) = \sqrt{(2x - 3) + 4} \Rightarrow g(2x - 3) = \sqrt{2x + 1}$

47. (a) $g(x) = |3x - 1| + 4 \Rightarrow g(-3) = |3(-3) - 1| + 4 \Rightarrow g(-3) = |-10| + 4 \Rightarrow g(-3) = 14$

    (b) $g(x) = |3x - 1| + 4 \Rightarrow g(b) = |3(b) - 1| + 4 \Rightarrow g(b) = |3b - 1| + 4$

    (c) $g(x) = |3x - 1| + 4 \Rightarrow g(x^3) = |3(x^3) - 1| + 4 \Rightarrow g(x^3) = |3x^3 - 1| + 4$

    (d) $g(x) = |3x - 1| + 4 \Rightarrow g(2x - 3) = |3(2x - 3) - 1| + 4 \Rightarrow g(2x - 3) = |6x - 10| + 4$

49. (a) $g(x) = \dfrac{4x}{x + 3} \Rightarrow g(-3) = \dfrac{4(-3)}{(-3) + 3} \Rightarrow g(-3) = \dfrac{-12}{0} \Rightarrow g(-3) = \text{undefined}$

    (b) $g(x) = \dfrac{4x}{x + 3} \Rightarrow g(b) = \dfrac{4(b)}{b + 3} \Rightarrow g(b) = \dfrac{4b}{b + 3}$

    (c) $g(x) = \dfrac{4x}{x + 3} \Rightarrow g(x^3) = \dfrac{4(x^3)}{x^3 + 3} \Rightarrow g(x^3) = \dfrac{4x^3}{x^3 + 3}$

    (d) $g(x) = \dfrac{4x}{x + 3} \Rightarrow g(2x - 3) = \dfrac{4(2x - 3)}{(2x - 3) + 3} \Rightarrow g(2x - 3) = \dfrac{8x - 12}{2x} \Rightarrow g(2x - 3) = \dfrac{4x - 6}{x} \Rightarrow$

    $g(2x - 3) = \dfrac{2(2x - 3)}{x}$

51. (a) $g(x) = 4x^{2/3} \Rightarrow g(-3) = 4(-3)^{2/3} \Rightarrow g(-3) = 4\sqrt[3]{9}$

    (b) $g(x) = 4x^{2/3} \Rightarrow g(b) = 4(b)^{2/3} \Rightarrow g(b) = 4b^{2/3}$

    (c) $g(x) = 4x^{2/3} \Rightarrow g(x^3) = 4(x^3)^{2/3} \Rightarrow g(x^3) = 4x^2$

    (d) $g(x) = 4x^{2/3} \Rightarrow g(2x - 3) = 4(2x - 3)^{2/3} \Rightarrow g(2x - 3) = 4(2x - 3)^{2/3}$

Composition of Functions

53. (a) $(g \circ f)(1) = g(f(1)) = g(4) = 5$

   (b) $(f \circ g)(4) = f(g(4)) = f(5) \Rightarrow$ undefined

   (c) $(f \circ f)(3) = f(f(3)) = f(1) = 4$

55. $g(3) = 4; f(4) = 2$

57. (a) $(f \circ g)(4) = f(g(4)) = f(0) = -4$

   (b) $(g \circ f)(3) = g(f(3)) = g(2) = 2$

   (c) $(f \circ f)(2) = f(f(2)) = f(0) = -4$

59. (a) $(f \circ g)(1) = f(g(1)) = f(2) = -3$

   (b) $(g \circ f)(-2) = g(f(-2)) = g(-3) = -2$

   (c) $(g \circ g)(-2) = g(g(-2)) = g(-1) = 0$

61. (a) $(f \circ g)(2) = f(g(2)) = f(2^2) = f(4) = \sqrt{4+5} = \sqrt{9} = 3$

   (b) $(g \circ f)(-1) = g(f(-1)) = g(\sqrt{-1+5}) = g(2) = 2^2 = 4$

63. (a) $(f \circ g)(-4) = f(g(-4)) = f(|-4|) = f(4) = 5(4) - 2 = 18$

   (b) $(g \circ f)(5) = g(f(5)) = g(5(5) - 2) = g(23) = |23| = 23$

65. (a) $(f \circ g)(x) = f(g(x)) = f(x^2 + 3x - 1) = (x^2 + 3x - 1)^3$; D: all real numbers

   (b) $(g \circ f)(x) = g(f(x)) = g(x^3) = (x^3)^2 + 3(x^3) - 1 = x^6 + 3x^3 - 1$; D: all real numbers

   (c) $(f \circ f)(x) = f(f(x)) = f(x^3) = (x^3)^3 = x^9$; D: all real numbers

67. (a) $(f \circ g)(x) = f(g(x)) = f(\sqrt{1-x}) = (\sqrt{1-x})^2 = 1 - x$; $D = \{x \mid x \le 1\}$

   (b) $(g \circ f)(x) = g(f(x)) = g(x^2) = \sqrt{1-x^2}$; $D = \{x \mid -1 \le x \le 1\}$

   (c) $(f \circ f)(x) = f(f(x)) = f(x^2) = (x^2)^2 = x^4$; D: all real numbers

69. (a) $(f \circ g)(x) = f(g(x)) = f(x^3) = 2 - 3x^3$; D: all real numbers

   (b) $(g \circ f)(x) = g(f(x)) = g(2 - 3x) = (2 - 3x)^3$; D: all real numbers

   (c) $(f \circ f)(x) = f(f(x)) = f(2 - 3x) = 2 - 3(2 - 3x) = 9x - 4$; D: all real numbers

71. (a) $(f \circ g)(x) = f(g(x)) = f(5 - x^2) = (5 - x^2) - 4 = 1 - x^2$; D: all real numbers

   (b) $(g \circ f)(x) = g(f(x)) = g(x - 4) = 5 - (x - 4)^2 = $; D: all real numbers

   (c) $(f \circ f)(x) = f(f(x)) = f(x - 4) = (x - 4) - 4 = x - 8$; D: all real numbers

73. (a) $(f \circ g)(x) = f(g(x)) = f(5x) = \dfrac{1}{5x + 1}$; $D = \left\{x \mid x \ne -\dfrac{1}{5}\right\}$

   (b) $(g \circ f)(x) = g(f(x)) = g\left(\dfrac{1}{x + 1}\right) = 5\left(\dfrac{1}{x + 1}\right) = \dfrac{5}{x + 1}$; $D = \{x \mid x \ne -1\}$

   (c) $(f \circ f)(x) = f(f(x)) = f\left(\dfrac{1}{x + 1}\right) = \dfrac{1}{\frac{1}{x+1} + 1} = \dfrac{1}{\frac{1 + x + 1}{x + 1}} = \dfrac{x + 1}{x + 2}$; $D = \{x \mid x \ne -1 \text{ and } x \ne -2\}$

75. (a) $(f \circ g)(x) = f(g(x)) = f(\sqrt{4 - x^2}) = \sqrt{4 - x^2} + 4$; $D = \{x \mid -2 \le x \le 2\}$

   (b) $(g \circ f)(x) = g(f(x)) = g(x + 4) = \sqrt{4 - (x + 4)^2}$; $D = \{x \mid -6 \le x \le -2\}$

   (c) $(f \circ f)(x) = f(f(x)) = f(x + 4) = (x + 4) + 4 = x + 8$; D: all real numbers

77. (a) $(f \circ g)(x) = f(g(x)) = f(3x) = \sqrt{3x - 1}; D = \left\{ x \mid x \geq \dfrac{1}{3} \right\}$

(b) $(g \circ f)(x) = g(f(x)) = g(\sqrt{x - 1}) = 3\sqrt{x - 1}; D = \{x \mid x \geq 1\}$

(c) $(f \circ f)(x) = f(f(x)) = f(\sqrt{x - 1}) = \sqrt{\sqrt{x - 1} - 1}; D = \{x \mid x \geq 2\}$

79. (a) $(f \circ g)(x) = f(g(x)) = f(2x + 3) = \dfrac{(2x + 3) - 3}{2} = \dfrac{2x + 0}{2} = x; \ D$: all real numbers

(b) $(g \circ f)(x) = g(f(x)) = g\left(\dfrac{x - 3}{2}\right) = 2\left(\dfrac{x - 3}{2}\right) + 3 = x - 3 + 3 = x; \ D$: all real numbers

(c) $(f \circ f)(x) = f(f(x)) = f\left(\dfrac{x - 3}{2}\right) = \dfrac{\frac{x-3}{2} - 3}{2} = \dfrac{\frac{x-3}{2} - \frac{6}{2}}{2} = \dfrac{\frac{x-9}{2}}{2} = \dfrac{x - 9}{2} \cdot \dfrac{1}{2} = \dfrac{x - 9}{4};$

D: all real numbers

81. (a) $(f \circ g)(x) = f(g(x)) = f(x^3 + 1) = \sqrt[3]{(x^3 + 1) - 1} = \sqrt[3]{x^2 + 1 - 1} = x; D$: all real numbers

(b) $(g \circ f)(x) = g(f(x)) = g(\sqrt[3]{x - 1}) = (\sqrt[3]{x - 1})^3 + 1 = x - 1 + 1 = x; D$: all real numbers

(c) $(f \circ f)(x) = f(f(x)) = f(\sqrt[3]{x - 1}) = \sqrt[3]{\sqrt[3]{x - 1} - 1}; D$: all real numbers

83. (a) $(f \circ g)(x) = f(g(x)) = f\left(\dfrac{x}{a} - b\right) = \dfrac{a(\frac{x}{a} - b) + b}{3} = \dfrac{x - ab + b}{3}; \ D$: all real numbers

(b) $(g \circ f)(x) = g(f(x)) = g\left(\dfrac{ax + b}{3}\right) = \dfrac{\frac{ax+b}{3}}{a} - b = \dfrac{ax + b}{3a} - b = \dfrac{ax - 3ab + b}{3a}; \ D$: all real numbers

(c) $(f \circ f)(x) = f(f(x)) = f\left(\dfrac{ax + b}{3}\right) = \dfrac{a(\frac{ax + b}{3}) + b}{3} = \dfrac{a^2 x + ab}{9} + \dfrac{b}{3} = \dfrac{a^2 x + ab + 3b}{9}; \ D$: all real numbers

85. $h(x) = \sqrt{x - 2} \Rightarrow g(x) = \sqrt{x}$ and $f(x) = x - 2$. *Answers may vary.*

87. $h(x) = 5(x + 2)^2 - 4 \Rightarrow g(x) = 5x^2 - 4$ and $f(x) = x + 2$. *Answers may vary.*

89. $h(x) = 4(2x + 1)^3 \Rightarrow g(x) = 4x^3$ and $f(x) = 2x + 1$. *Answers may vary.*

91. $h(x) = (x^3 - 1)^2 \Rightarrow g(x) = x^2$ and $f(x) = x^3 - 1$. *Answers may vary.*

93. $h(x) = -4|x + 2| - 3 \Rightarrow g(x) = -4|x| - 3$ and $f(x) = x + 2$. *Answers may vary.*

95. $h(x) = \dfrac{1}{(x - 1)^2} \Rightarrow g(x) = \dfrac{1}{x^2}$ and $f(x) = x - 1$. *Answers may vary.*

97. $h(x) = x^{3/4} - x^{1/4} \Rightarrow g(x) = x^3 - x$ and $f(x) = x^{1/4}$. *Answers may vary.*

Applications

99. (a) $C(x) = 1.5x + 150{,}000$

(b) $R(x) = 6.5x; \ R(8000) = 6.5(8000) = \$52{,}000$

(c) $P(x) = R(x) - C(x) = 6.5x - (1.5x + 150{,}000) = 5x - 150{,}000$

$P(40{,}000) = 5(40{,}000) - 150{,}000 = \$50{,}000$

(d) $R(x) = C(x) \Rightarrow 6.5x = 1.5x + 150{,}000 \Rightarrow 5x = 150{,}000 \Rightarrow x = 30{,}000$

To break even, 30,000 videos must be sold.

101. (a) $I(x) = 36x$

(b) $C(x) = 2.54x$

(c) $F(x) = (C \circ I)(x)$

(d) $F(x) = 36(2.54x) = 91.44x$

103. (a) $(g \circ f)(1) = g(f(1)) = g(1.5) = 5.25;$

   A 1% decrease in the ozone layer could result in a 5.25% increase in skin cancer.

(b) $(f \circ g)(21) = f(g(21))$, but $g(21)$ is not given by the table. So $(f \circ g)(21)$ is not possible using these tables.

105. (a) $(g \circ f)(1980) = g(f(1980)) = g(3.5) \approx 5;$ In 1980 the average nighttime temperature had risen about

   3.5°C since 1948, which resulted in a 5% increase in peak-demand for electricity.

(b) $(f \circ g)(3)$ is meaningless because the range of g is not the same as the domain of f.

107. (a) When $x = 2$, the amount of water in the pool is approximately 4000 cubic feet. Using 4000 as input into

   function g, we can estimate that $g(4000) \approx 30,000$ gallons.

(b) $(g \circ f)(x)$ computes the gallons of water contained in the pool after x days.

109. (a) In Example 1, the reaction distance for a reaction time of 2.5 seconds was $\dfrac{11}{3}x$.

   The reaction distance for a driver with a reaction time of 1.25 seconds would be half. Thus, $f(x) = \dfrac{11}{6}x$.

(b) $d(x) = f(x) + g(x) = \dfrac{11}{6}x + \dfrac{1}{9}x^2$

(c) $d(60) = \dfrac{11}{6}(60) + \dfrac{1}{9}(60)^2 = 510;$ this car requires 510 feet to stop at 60 mph.

111. The radius of the circular wave at the end of t seconds would be $6t$ inches.

   $C = 2\pi r \Rightarrow C(t) = 2\pi(6t) = 12\pi t$

113. $S = \pi r \sqrt{r^2 + h^2}$ and $h = 2r \Rightarrow S = \pi r \sqrt{r^2 + (2r)^2} \Rightarrow S = \pi r \sqrt{5r^2} \Rightarrow S = \pi r^2 \sqrt{5}$

115. (a) $(f + g)(1970) = f(1970) + g(1970) = 32.4 + 17.6 = 50$

(b) $(f + g)(x)$ computes the total coal and oil SO_2 emissions in the year x.

(c) A tabular representation of $(f + g)(x)$ is given in Figure 115.

x	1860	1900	1940	1970	2000
$(C + O)(x)$	2.4	12.8	26.5	50.0	78.0

Figure 115

117. (a) To find the total emissions we must add the developed and developing countries' emissions for each year.

   The resulting table is shown in Figure 117.

(b) $h(x) = f(x) + g(x)$

x	1990	2000	2010	2020	2030
$f(x)$	27	28	29	30	31
$g(x)$	5	7.5	10	12.5	15
$h(x)$	32	35.5	39	42.5	46

Figure 117

119. $h(x) = f(x) + g(x) = (0.1x - 172) + (0.25x - 492.5) = 0.35x - 664.5$

121. (a) $h(x) = g(x) - f(x)$

(b) $h(1995) = g(1995) - f(1995) = 800 - 600 = 200$; $h(2000) = g(2000) - f(2000) = 950 - 700 = 250$

(c) The function h is linear and passes through the points $(1995, 200)$ and $(2000, 250)$. Therefore, its graph has a slope of $m = \dfrac{250 - 200}{2000 - 1995} = 10$. Thus, $h(x) = 10(x - 1995) + 200 = 10x - 19{,}750$.

123. (a) For $A(s) = \dfrac{\sqrt{3}}{4}s^2$, $A(4s) = \dfrac{\sqrt{3}}{4}(4s)^2 = 16\dfrac{\sqrt{3}}{4}s^2 = 16A(s)$; if the length of a side is quadrupled, the area increases by a factor of 16.

(b) For $A(s) = \dfrac{\sqrt{3}}{4}s^2$, $A(s + 2) = \dfrac{\sqrt{3}}{4}(s + 2)^2 = \dfrac{\sqrt{3}}{4}(s^2 + 4s + 4) = \dfrac{\sqrt{3}}{4}s^2 + \sqrt{3}s + \sqrt{3} = A(s) + \sqrt{3}(s + 1)$; if the length of a side increases by 2, the area increases by $\sqrt{3}(s + 1)$.

125. Let $f(x) = ax + b$ and $g(x) = cx + d$. Then $f(x) + g(x) = (ax + b) + cx + d) = (a + c)x + (b + d)$, which is linear.

127. (a) $f(x) = k$ and $g(x) = ax + b$. Thus, $(f \circ g)(x) = f(g(x)) = f(ax + b) = k$; $f \circ g$ is a constant function.

(b) $f(x) = k$ and $g(x) = ax + b$. Thus, $(g \circ f)(x) = g(f(x)) = g(k) = ak + b$; $g \circ f$ is a constant function.

5.2: Inverse Functions and Their Representations

Inverse Operations

1. Close a window.

3. Closing a book, standing up, and walking out of the classroom.

5. Subtract 2 from x; $x + 2$ and $x - 2$

7. Divide x by 3 and add 2; $3(x - 2)$ and $\dfrac{x}{3} + 2$

9. Subtract 1 from x and cube the result; $\sqrt[3]{x} + 1$ and $(x - 1)^3$

11. Take the reciprocal of x; $\dfrac{1}{x}$ and $\dfrac{1}{x}$

One-to-One Functions

13. Since a horizontal line will intersect the graph at most once, f is one-to-one.

15. Since a horizontal line will intersect the graph two times, f is not one-to-one.

17. Since a horizontal line will intersect the graph two times, f is not one-to-one.

19. Since $f(2) = 3$ and $f(3) = 3$, different inputs result in the same output. Therefore f is not one-to-one. Does not have an inverse.

21. Since different inputs always result in different outputs, f is one-to-one. Does have an inverse.

23. Since the graph of f is a line sloping upward from left to right, a horizontal line can intersect it at most once. Therefore, f is one-to-one.

25. Since the graph of f is a parabola, a horizontal line can intersect it more than once. Therefore, f is not one-to-one.

27. Since the graph of f is a parabola, a horizontal line can intersect it more than once. Therefore, f is not one-to-one.

29. Since the graph of f is a V-shape, a horizontal line can intersect it more than once. Therefore, f is not one-to-one.

31. Since the graph of f is both increasing and decreasing, a horizontal line can intersect it more than once. Therefore, f is not one-to-one.

33. Since the graph of f is both increasing and decreasing, a horizontal line can intersect it more than once. Therefore, f is not one-to-one.

35. Since the graph of f is a line sloping upward from left to right, a horizontal line can intersect it at most once. Therefore f is one-to-one.

37. Since the person goes up and down, there would be several times when the person attained a particular height above the ground. A one-to-one function would not model this situation.

39. Since the population of the United States increased each of these years, a horizontal line would intersect a graph of this population at most once. A one-to-one function would model this situation.

Symbolic Representations of Inverse Functions

41. The inverse operation of taking the cube root of x is cubing x. Therefore, $f^{-1}(x) = x^3$.

43. The inverse operations of multiplying by –2 and adding 10 are subtracting 10 and dividing by 2.

Therefore, $f^{-1}(x) = \dfrac{x - 10}{-2} = -\dfrac{1}{2}x + 5$.

45. The inverse operations of multiplying by 3 and subtracting 1 are adding 1 and dividing by 3.

Therefore, $f^{-1}(x) = \dfrac{x + 1}{3}$.

47. The inverse operations of cubing a number, multiplying by 2 and subtracting 5 are adding 5, dividing by 2 and

taking the cube root of the result. Therefore, $f^{-1}(x) = \sqrt[3]{\dfrac{x + 5}{2}}$.

49. The inverse operations of squaring a number and subtracting 1 are adding 1 and taking the square root.

Therefore, $f^{-1}(x) = \sqrt{x + 1}$.

51. The inverse operations of multiplying by 2 and taking the reciprocal are taking the reciprocal and dividing by 2.

Therefore, $f^{-1}(x) = \dfrac{1}{x} \div 2 = \dfrac{1}{2x}$.

53. The inverse operations of multiplying by -5, adding 4, multiplying by $\dfrac{1}{2}$, and adding 1 are subtracting 1,

dividing by $\dfrac{1}{2}$ (or multipying by 2), subtracting 4, and dividing by -5 $\Rightarrow$

$$f^{-1}(x) = \frac{2(x-1)-4}{-5} = \frac{2x-6}{-5} = -\frac{2(x-3)}{5}$$

55. $y = \dfrac{x}{x+2} \Rightarrow y(x+2) = x \Rightarrow xy + 2y = x \Rightarrow 2y = x - xy \Rightarrow 2y = x(1-y) \Rightarrow x = \dfrac{2y}{1-y} \Rightarrow$

$f^{-1}(x) = \dfrac{2x}{1-x} \Rightarrow f^{-1}(x) = -\dfrac{2x}{x-1}$

57. $y = \dfrac{2x+1}{x-1} \Rightarrow y(x-1) = 2x+1 \Rightarrow xy - y = 2x+1 \Rightarrow xy - 2x = y+1 \Rightarrow x(y-2) = y+1 \Rightarrow$

$x = \dfrac{y+1}{y-2} \Rightarrow f^{-1}(x) = \dfrac{x+1}{x-2}$

59. $y = \dfrac{1}{x} - 3 \Rightarrow y + 3 = \dfrac{1}{x} \Rightarrow x = \dfrac{1}{y+3} \Rightarrow f^{-1}(x) = \dfrac{1}{x+3}$

61. $y = \dfrac{1}{x^3-1} \Rightarrow \dfrac{1}{y} = x^3 - 1 \Rightarrow \dfrac{1}{y} + 1 = x^3 \Rightarrow \dfrac{1}{y} + \dfrac{y}{y} = x^3 \Rightarrow \dfrac{y+1}{y} = x^3 \Rightarrow f^{-1}(x) = \sqrt[3]{\dfrac{x+1}{x}}$

63. $y = 4 - x^2, x \geq 0 \Rightarrow x^2 = 4 - y \Rightarrow x = \sqrt{4-y}$; therefore $f^{-1}(x) = \sqrt{4-x}$

65. $y = (x-2)^2 + 4, x \geq 2 \Rightarrow y - 4 = (x-2)^2 \Rightarrow \sqrt{y-4} = x - 2 \Rightarrow x = \sqrt{y-4} + 2$;

therefore $f^{-1}(x) = \sqrt{x-4} + 2$.

67. $y = x^{2/3} + 1, x \geq 0 \Rightarrow x^{2/3} = y - 1 \Rightarrow x = (y-1)^{3/2}; \Rightarrow f^{-1}(x) = (x-1)^{3/2}$

69. $y = \sqrt{9-2x^2}, x \geq 0 \Rightarrow y^2 = 9 - 2x^2 \Rightarrow y^2 - 9 = -2x^2 \Rightarrow \dfrac{9-y^2}{2} = x^2 \Rightarrow \sqrt{\dfrac{9-y^2}{2}} = x; \Rightarrow$

$f^{-1}(x) = \sqrt{\dfrac{9-x^2}{2}}$

71. $y = 5x - 15 \Rightarrow y + 15 = 5x \Rightarrow \dfrac{y+15}{5} = x \Rightarrow f^{-1}(x) = \dfrac{x+15}{5}$; D and R are all real numbers.

73. $y = \sqrt[3]{x-5} \Rightarrow y^3 = x - 5 \Rightarrow y^3 + 5 = x \Rightarrow f^{-1}(x) = x^3 + 5$; D and R are all real numbers

75. $y = \dfrac{x-5}{4} \Rightarrow 4y = x - 5 \Rightarrow 4y + 5 = x \Rightarrow f^{-1}(x) = 4x + 5$; D and R are all real numbers.

77. $y = \sqrt{x-5}, x \geq 5 \Rightarrow y^2 = x - 5 \Rightarrow y^2 + 5 = x \Rightarrow f^{-1}(x) = x^2 + 5$;

$D = \{x \mid x \geq 0\}$ and $R = \{y \mid y \geq 5\}$

79. $y = \dfrac{1}{x+3} \Rightarrow \dfrac{1}{y} = x + 3 \Rightarrow \dfrac{1}{y} - 3 = x \Rightarrow f^{-1}(x) = \dfrac{1}{x} - 3$; $D = \{x \mid x \neq 0\}$ and $D = \{y \mid y \neq -3\}$

81. $y = 2x^3 \Rightarrow \dfrac{y}{2} = x^3 \Rightarrow \sqrt[3]{\dfrac{y}{2}} = x \Rightarrow f^{-1}(x) = \sqrt[3]{\dfrac{x}{2}}$; D and R are all real numbers.

83. $y = x^2, x \geq 0 \Rightarrow \sqrt{y} = x \Rightarrow f^{-1}(x) = \sqrt{x}$; $D = \{x \mid x \geq 0\}$ and $D = \{y \mid y \geq 0\}$

Numerical Representations of Inverse Functions

85. The domain and range of f are $D = \{1, 2, 3\}$ and $R = \{5, 7, 9\}$. Interchange these to get the domain and range of f^{-1}; $D = \{5, 7, 9\}$ and $R = \{1, 2, 3\}$. See Figure 85.

x	5	7	9
$f^{-1}(x)$	1	2	3

Figure 85

87. The domain and range of f are $D = \{0, 1, 2, 3, 4\}$ and $R = \{0, 1, 4, 9, 16\}$. Interchange these to get the domain and range of f^{-1}; $D = \{0, 1, 4, 9, 16\}$ and $R = \{0, 1, 2, 3, 4\}$. See Figure 87.

x	0	1	4	9	16
$f^{-1}(x)$	0	1	2	3	4

Figure 87

89. Since f adds 5 to x, f^{-1} subtracts 5 from x. See Figure 89.

x	−3	0	3	6
$f^{-1}(x)$	−8	−5	−2	1

Figure 89

91. Since f cubes x, f^{-1} takes the cube root of x. See Figure 91.

x	−8	−1	8	27
$f^{-1}(x)$	−2	−1	2	3

Figure 91

93. $f(1) = 3 \Rightarrow f^{-1}(3) = 1$

95. $g(3) = 4 \Rightarrow g^{-1}(4) = 3$

97. $(f \circ g^{-1})(1) = f(g^{-1}(1)) = f(2) = 5$. Note that $g(2) = 1 \Rightarrow g^{-1}(1) = 2$.

99. $(g \circ f^{-1})(5) = g(f^{-1}(5)) = g(2) = 1$. Note that $f(2) = 5 \Rightarrow f^{-1}(5) = 2$.

Graphs and Inverse Functions

101. (a) $f(1) \approx \$110$

   (b) $f^{-1}(110) \approx 1$ year

   (c) $f^{-1}(160) \approx 5$ years

   The function f^{-1} computes the number of years it takes for this savings account to accumulate x dollars.

103. (a) $f(-1) = 2$

   (b) $f^{-1}(-2) = 3$

   (c) $f^{-1}(0) = 1$

   (d) $(f^{-1} \circ f)(3) = f^{-1}(f(3)) = f^{-1}(-2) = 3$

105. The graph of f passes through the points $(-2, -4)$ and $(1, 2)$. Thus, the graph of f^{-1} passes through the points $(-4, -2)$ and $(2, 1)$. Plot these points and sketch the reflection of f in the line $y = x$ to obtain the graph of f^{-1}. Notice that since the graph of f is a line, its reflection will also be a line. See Figure 105.

107. The graph of f passes through the points $\left(-2, \frac{1}{4}\right)$, $(0, 1)$, $(1, 2)$ and $(2, 4)$. Thus, the graph of f^{-1} passes through the points $\left(\frac{1}{4}, -2\right)$, $(1, 0)$, $(2, 1)$ and $(4, 2)$. Plot these points and sketch the reflection of f in the line $y = x$ to obtain the graph of f^{-1}. See Figure 107.

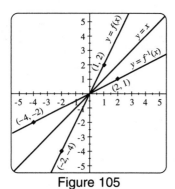

Figure 105

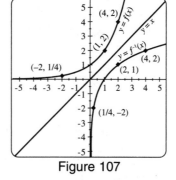

Figure 107

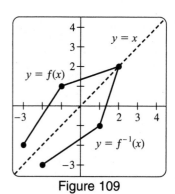

Figure 109

109. The graph of f passes through the points $(-3, -2)$, $(-1, 1)$ and $(2, 2)$. Thus, the graph of f^{-1} passes through the points $(-2, -3)$, $(1, -1)$ and $(2, 2)$. Plot these points and sketch the reflection of f in the line $y = x$ to obtain the graph of f^{-1}. See Figure 109.

111. The graphs of $y = 2x - 1$, $y = \dfrac{x + 1}{2}$, and $y = x$ are shown in Figure 111.

113. The graphs of $y = x^3 - 1$, $y = \sqrt[3]{x + 1}$, and $y = x$ are shown in Figure 113.

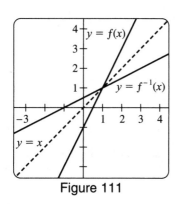

Figure 111

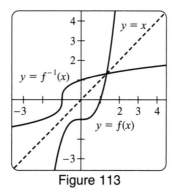

Figure 113

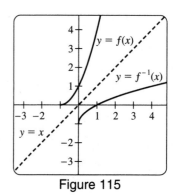

Figure 115

115. The graphs of $y = (x + 1)^2$, $y = \sqrt{x} - 1$, and $y = x$ are shown in Figure 115.

117. The graphs of $Y_1 = 3X - 1$, $Y_2 = (X + 1)/3$ and $Y_3 = X$ are shown in Figure 117.

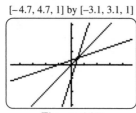

[−4.7, 4.7, 1] by [−3.1, 3.1, 1]   [−4.7, 4.7, 1] by [−3.1, 3.1, 1]

Figure 117   Figure 119

119. The graphs of $Y_1 = X^3/3 - 1$, $Y_2 = \sqrt[3]{(3X + 3)}$ and $Y_3 = X$ are shown in Figure 119.

Applications

121. (a) In 1995 the cost was \$1 million per 30-second commercial. Therefore, $c(1995) = 1$.

(b) The solution to $c(x) = 1$ is the x-value that results in a y-value of 1. This x-value is 1995.

(c) $c^{-1}(1)$ is the solution to the equation $c(x) = 1$. Since $f(1995) = 1$, $c^{-1}(1) = 1995$.

123. (a) Since each volume value is the result of exactly one radius value, V represents a one-to-one function.

(b) The inverse of V computes the radius r of a sphere with volume V.

(c) $V = \dfrac{4}{3}\pi r^3 \Rightarrow r^3 = \dfrac{3V}{4\pi} \Rightarrow r = \sqrt[3]{\dfrac{3V}{4\pi}}$

(d) No; if V and r were interchanged, then r would represent the volume and V would represent the radius.

125. MATH $\rightarrow$ OCVJ; HWPEVKQPU $\rightarrow$ FUNCTIONS

If f is not one-to-one, then f^{-1} does not exist and f^{-1} is needed to decode the message.

127. (a) $W = \dfrac{25}{7}(70) - \dfrac{800}{7} = \dfrac{950}{7} \approx 135.7$ pounds

(b) Yes, since no weight value corresponds to more than one height value.

(c) $W = \dfrac{25}{7}h - \dfrac{800}{7} \Rightarrow \dfrac{25}{7}h = W + \dfrac{800}{7} \Rightarrow h = \dfrac{7}{25}\left(W + \dfrac{800}{7}\right) = \dfrac{7}{25}W + 32 \Rightarrow W^{-1} = \dfrac{7}{25}W + 32$

(d) $W^{-1}(150) = \dfrac{7}{25}(150) + 32 \Rightarrow W^{-1}(150) = 74$; the maximum recommended height for a person weighing 150 pounds is 74 inches.

(e) The inverse computes the maximum recommended height of a person with a given weight.

129. (a) $(F \circ Y)(2) = F(Y(2)) = F(3520) = 10,560$. $(F \circ Y)(2)$ computes the number of feet in 2 miles.

(b) $F^{-1}(26,400) = 8800$. F^{-1} converts feet to yards. There are 8800 yards in 26,400 feet.

(c) $(Y^{-1} \circ F^{-1})(21,120) = Y^{-1}(F^{-1}(21,120)) = Y^{-1}(7040) = 4$. $(Y^{-1} \circ F^{-1})(21,120)$ computes the number of miles in 21,120 feet.

131. (a) $(Q \circ C)(96) = Q(C(96)) = Q(6) = 1.5$. $(Q \circ C)(96)$ computes the number of quarts in 96 tablespoons.

(b) $Q^{-1}(2) = 8$. Q^{-1} converts quarts into cups. There are 8 cups in 2 quarts.

(c) $(C^{-1} \circ Q^{-1})(1.5) = C^{-1}(Q^{-1}(1.5)) = C^{-1}(6) = 96$. $(C^{-1} \circ Q^{-1})(1.5)$ computes the number of tablespoons in 1.5 quarts.

133. (a) $f(x) = 0.06(x - 1930) + 62.5 \Rightarrow f(1930) = 0.06(1930 - 1930) + 62.5 = 62.5$ and

$f(1980) = 0.06(1980 - 1930) + 62.5 = 65.5$. In 1930 there was cloud cover 62.5% of the time, whereas in 1980 it increased to 65.5%.

(b) f^{-1} computes the year when the cloud cover was x percent.

(c) $f(1930) = 62.5 \Rightarrow f^{-1}(62.5) = 1930$. Similarly, $f(1980) = 65.5 \Rightarrow f^{-1}(65.5) = 1980$.

(d) Using the technique shown in Example 6, we can find $f^{-1}(x)$.

$$y = 0.06(x - 1930) + 62.5 \Rightarrow y - 62.5 = 0.06(x - 1930) \Rightarrow \frac{y - 62.5}{0.06} = x - 1930 \Rightarrow$$

$$1930 + \frac{y - 62.5}{0.06} = x \Rightarrow f^{-1}(x) = 1930 + \frac{x - 62.5}{0.06} \text{ or } f^{-1}(x) = \frac{50}{3}(x - 62.5) + 1930$$

5.3: Exponential Functions and Models

Exponents

1. $2^{-3} = \dfrac{1}{2^3} = \dfrac{1}{8}$

3. $3(4)^{1/2} = 3\sqrt{4} = 3(2) = 6$

5. $-2(27)^{2/3} = -2(\sqrt[3]{27})^2 = -2(3)^2 = -2(9) = -18$

7. $\left(\dfrac{1}{8}\right)^{-1} = \left(\dfrac{8}{1}\right)^1 = 8$

9. $4^{1/6}4^{1/3} = 4^{1/6 + 1/3} = 4^{1/2} = \sqrt{4} = 2$

11. $e^x e^x = e^{x+x} = e^{2x}$

13. $3^0 = 1$

15. $(5^{101})^{1/101} = 5^1 = 5$

17. (a) $3^{5/2} \approx 15.59$

(b) $2^\pi \approx 8.82$

(c) $-e^{\sqrt{3}} \approx -5.56$

Linear and Exponential Growth

19. For each unit increase in x, the y-values decrease by 1.25, so the data is linear. Since $y = 2$ when $x = 0$, the function $f(x) = -1.25x + 2$ can model the data.

21. For each unit increase in x, the y-values are multiplied by $\dfrac{1}{2}$, so the data is exponential. Since the initial value is $C = 8$ and $a = \dfrac{1}{2}$, the function $f(x) = 8\left(\dfrac{1}{2}\right)^x$ can model the data.

23. For each 2-unit increase in x, the y-values are multiplied by 4. That is, for each unit increase in x, the y-values are multiplied by 2, so the data is exponential. Since the initial value is $C = 5$ and $a = 2$, the function $f(x) = 5(2^x)$ can model the data.

25. For each 1-year increase, the salary is multiplied by 1.08 to get the new salary. This is an example of exponential growth; $C = \$40,000$ and $a = 1.08 \Rightarrow f(n) = 40,000(1.08)^{n-1}$, where $n = 1, 2, 3, \ldots$.

27. For $x > 4, f(x) > g(x)$; for example $f(10) = 1024$ whereas $g(10) = 100$. That is, $f(x) = 2^x$ becomes larger.

Exponential Functions

29. $f(0) = 5 \Rightarrow C = 5$ and $a = 1.5$

31. $f(0) = 10$ and $f(1) = 20 \Rightarrow C = 10$ and $a = \dfrac{20}{10} = 2$

33. $f(1) = 9$ and $f(2) = 27 \Rightarrow a = \dfrac{27}{9} = 3; C = f(0) = 9 \div 3 = 3$

35. $f(-2) = \dfrac{9}{2}$ and $f(2) = \dfrac{1}{18} \Rightarrow Ca^{-2} = \dfrac{9}{2} \Rightarrow \dfrac{C}{a^2} = \dfrac{9}{2} \Rightarrow C = \dfrac{9a^2}{2}; \ Ca^2 = \dfrac{1}{18} \Rightarrow C = \dfrac{1}{18a^2}$ thus
$\dfrac{1}{18a^2} = \dfrac{9a^2}{2} \Rightarrow 162a^4 = 2 \Rightarrow a^4 = \dfrac{1}{81} \Rightarrow a = \dfrac{1}{3}$; then $C = \dfrac{1}{18(\frac{1}{3})^2} \Rightarrow C = \dfrac{1}{18(\frac{1}{9})} \Rightarrow C = \dfrac{1}{2}$.

37. $C = 5000, a = 2; \ x$ represents time in hours

39. $C = 200,000, a = 0.95; \ x$ represents the number of years after 2000

41. $f(9.5) = 30(0.9)^{9.5} \approx 11$; the tire's pressure is about 11 pounds per square inch after 9.5 minutes.

43. See Figure 43.

45. See Figure 45.

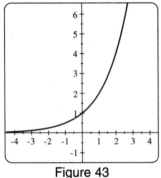
Figure 43

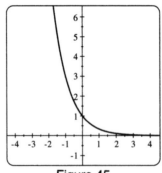
Figure 45

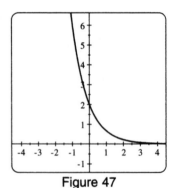
Figure 47

47. See Figure 47.

49. See Figure 49.

51. See Figure 51.

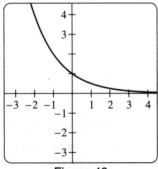
Figure 49

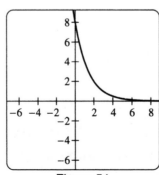
Figure 51

53. Since $y = 1$ when $x = 0$, $C = 1$ and so $y = a^x$.

Since $y = 4$ when $x = -2$, $4 = a^{-2} \Rightarrow \dfrac{1}{a^2} = 4 \Rightarrow a^2 = \dfrac{1}{4} \Rightarrow a = \dfrac{1}{2}$. That is $C = 1$ and $a = \dfrac{1}{2}$.

55. Since $y = \dfrac{1}{2}$ when $x = 0$, $C = \dfrac{1}{2}$ and so $y = \dfrac{1}{2}a^x$.

Since $y = 8$ when $x = 2$, $8 = \dfrac{1}{2}a^2 \Rightarrow a^2 = 16 \Rightarrow a = 4$. That is $C = \dfrac{1}{2}$ and $a = 4$.

57. (a) D: $(-\infty, \infty)$; R: $(0, \infty)$

(b) Decreasing, as x increases $\left(\dfrac{1}{8}\right)^x$ decreases.

(c) $y = 0$ as x increases, $7\left(\dfrac{1}{8}\right)^x$ gets closer and closer to 0.

(d) y-intercept: 7; no x-intercept.

(e) All horizontal lines intersect once $\Rightarrow$ yes; yes.

59. (i) The graph of $y = e^x$ increases faster than the graph of $y = 1.5^x$. The best choice is graph b.

(ii) The graph of $y = 3^{-x}$ decreases faster than the graph of $y = 0.99^x$. The best choice is graph d.

(iii) The graph of $y = 1.5^x$ increases slower than the graph of $y = e^x$. The best choice is graph a.

(iv) The graph of $y = 0.99^x$ is almost a horizontal line since $y = 1^x$ is horizontal. The best choice is graph c.

61. (a) To graph $y = 2^x - 2$, translate the graph of $y = 2^x$ down 2 units. See Figure 61a.

(b) To graph $y = 2^{x-1}$, translate the graph of $y = 2^x$ right 1 unit. See Figure 61b.

(c) To graph $y = 2^{-x}$, reflect the graph of $y = 2^x$ about the y-axis. See Figure 61c.

(d) To graph $y = -2^x$, reflect the graph of $y = 2^x$ about the x-axis. See Figure 61d.

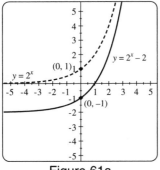

Figure 61a

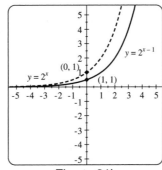

Figure 61b

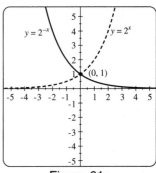

Figure 61c

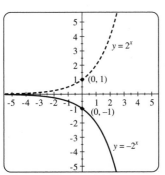

Figure 61d

63. $f(x) = e^x \Rightarrow f(3.1) = e^{3.1} \approx 22.1980$

65. $f(x) = 4e^{-1.2x} \Rightarrow f(-2.4) = 4e^{-1.2(-2.4)} = 4e^{2.88} \approx 71.2571$

67. $f(x) = \dfrac{e^x - e^{-x}}{2} \Rightarrow f(-0.7) = \dfrac{e^{-0.7} - e^{0.7}}{2} \approx -0.7586$

Compound Interest

69. $A_n = A_0(1 + r)^n \Rightarrow A_5 = 600(1 + 0.07)^5 \approx \$841.53$

71. $A_n = A_0\left(1 + \dfrac{r}{m}\right)^{mn} \Rightarrow A_{20} = 950\left(1 + \dfrac{0.03}{365}\right)^{365(20)} \approx \$1730.97$

73. $A_n = A_0 e^{rn} \Rightarrow A_8 = 2000 e^{0.10(8)} \approx \$4451.08$

75. $A_n = A_0\left(1 + \dfrac{r}{m}\right)^{mn} \Rightarrow A_{2.5} = 1600\left(1 + \dfrac{0.104}{12}\right)^{12(2.5)} \approx \$2072.76$

77. $A_{20} = 2000\left(1 + \dfrac{0.10}{12}\right)^{12(20)} \approx \$14,656.15; \quad A_{20} = 2000\left(1 + \dfrac{0.13}{12}\right)^{12(20)} \approx \$26,553.58$

   A 13% interest rate results in considerably more interest than a 10% interest rate.

79. $A_{10} = 8000(1 + 0.06)^{10} \approx \$14,326.78$

81. The account with \$200 will have double the money, since twice the money was deposited.

83. $A_n = A_0(1 + r)^n \Rightarrow A_{30} = 300(1 + 0.0495)^{30} \approx \$1278.2$ billion or \$1.28 trillion.

85. $A_{10} = 50\left[\dfrac{(1 + \frac{0.08}{26})^{260} - 1}{\frac{0.08}{26}}\right] \approx \$19,870.65$

Applications

87. $P = 1\left(\dfrac{1}{2}\right)^{x/1600} \Rightarrow P = 1\left(\dfrac{1}{2}\right)^{3000/1600} \Rightarrow P = \left(\dfrac{1}{2}\right)^{15/8} \Rightarrow P \approx 0.273,$ or 27.3%

89. To find the age of the fossil solve $0.10 = 1\left(\dfrac{1}{2}\right)^{x/5700}$. Graph $Y_1 = 0.10$ and $Y_2 = 0.5^{\wedge}(X/5700)$. The graphs intersect near (18934.99, 0.1), so the fossil is about 18,935 years old.

91. (a) $A(x) = 100e^{-0.02295x} \Rightarrow A(50) = 100e^{-0.02295(50)} \approx 31.7$ milligrams. Since the original sample contained 100 milligrams, the half-life would be the time required for the sample to disintegrate into 50 milligrams. Since 31.7 is less than this, the half-life must be less than 50 years.

   (b) Graph $Y_1 = 100e^{\wedge}(-0.02295X)$ and $Y_2 = 50$ as shown in Figure 91. The graphs intersect near (30.2, 50). Therefore, the half-life of cesium is approximately 30.2 years.

[0, 50, 10] by [0, 100, 10]   [0, 5, 1] by [0, 1, 0.1]   [0, 30, 2] by [0, 10,000, 1000]

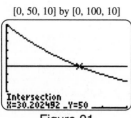

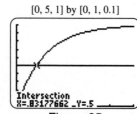

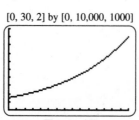

Figure 91   Figure 95   Figure 97

93. (a) $f(0) = 0.72 \Rightarrow C \approx 0.72$, so $f(x) = 0.72a^x$. One possible way to determine the value of a is to let the graph of f pass through the point (20, 1.60). $f(20) = 1.60 \Rightarrow 1.60 = 0.72a^{20} \Rightarrow a^{20} = \dfrac{1.60}{0.72} \Rightarrow$

    $a = \left(\dfrac{1.60}{0.72}\right)^{1/20} \Rightarrow a \approx 1.041$; Thus $f(x) = 0.72(1.041)^x$. *Answers may vary slightly.*

    (b) Let $x = 13$ correspond to 2013, so ; $f(13) = 0.72(1.041)^{13} \approx 1.21$ the CFC-12 concentration in 2013 is about 1.21 ppb. *Answers may vary slightly..*

95. (a) $p(x) = 1 - e^{-5x/6} \Rightarrow p(3) = 1 - e^{-5(3)/6} \Rightarrow p(3) = 1 - e^{-15/6} \approx 0.92$, or 92%

    (b) Find x when $p(x) = 0.5$. That is, solve $0.5 = 1 - e^{-5x/6}$ for x.

    Graph $Y_1 = 0.5$ and $Y_2 = 1 - e^{\wedge}(-5X/6)$ as shown in Figure 95. There is a 50-50 chance of at least one car entering the intersection during an interval of about 0.83 minutes.

97. (a) $P = 1500$, $r = 0.06$ and $t = 30 \Rightarrow A = Pe^{rt} \Rightarrow A = 1500e^{0.06(30)} \approx 9074.47$

    (b) During the last 10 years. See Figure 97.

99. To complete the table for $f(t) = 18.29(1.279)^t$, we must find $f(7.5), f(9)$, and $f(10.5)$. See Figure 99.

t (inches)	6	7.5	9	10.5	12
W (lb × 1000)	80	116	168	242	350

Figure 99

101. (a) The number of *E. coli* was modeled by $N(x) = N_0 e^{0.014x}$, where x is in minutes and $N_0 = 500{,}000$. Since 3 hours is 180 minutes, $N(180) = 500{,}000e^{0.014(180)} \approx 6{,}214{,}000$ bacteria.

    (b) Solve the equation $500{,}000 e^{0.014x} = 10{,}000{,}000$. Graph $Y_1 = 500000e^{\wedge}(0.014X)$ and $Y_2 = 10E6$ as shown in Figure 101. The point of intersection is near (214, 10,000,000). Thus, there will be 10 million *E. coli* after about 214 minutes, or about 3.6 hours.

[0, 300, 100] by [0, 20,000,000, 5,000,000]   [0, 25, 5] by [0, 1, 0.1]

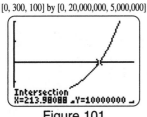
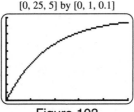

Figure 101   Figure 103

103. (a) $P(2) = 1 - e^{-0.1144(2)} \approx 0.20$ and $P(20) = 1 - e^{-0.1144(20)} \approx 0.90$. There is a 20% chance that at least one tree will be located within a circle having a radius of 2 feet, and there is a 90% chance of at least one tree being located within a circle with a radius of 20 feet.

    (b) The graph of P is shown in Figure 103. The larger the circle in the forest, the more likely it is to contain a tree. It is not very likely that a circle with a radius of 1-foot would contain a tree since trees do not grow that densely. However, there is an excellent chance that a circle with a radius of 100 feet contains at least one tree. Otherwise, it would be more of a clearing than a forest.

    (c) Graph $Y_1 = 1 - e^{\wedge}(-0.1144X)$ and $Y_2 = 0.5$. The graphs intersect near (6.06, 0.5). Thus, a circle with a radius of approximately 6.1 feet will have a 50-50 chance of containing at least one tree.

105. (a) $N(x) = N_0 e^{-0.1x^2/2} \Rightarrow N(0) = N_0 e^0 = N_0(1) = N_0$. Here N_0 represents the initial number of patients that are infected but have no apparent symptoms of AIDS.

   (b) To determine the time lapse for 50% of the patients to develop symptoms of AIDS, solve the equation $100e^{-0.1x^2/2} = 50$, where the initial number of infected individuals is $N_0 = 100$.

   Graph $Y_1 = 100e\,\hat{}\,(-0.1X\hat{}2/2)$ and $Y_2 = 50$ as shown in Figure 105. There is an intersection point near (3.72, 50). Thus, after approximately 3.7 years, 50% of the patients who received tainted blood transfusions developed symptoms of AIDS. (This study indicated that this time delay varied with a person's age.)

[0, 10, 1] by [0, 100, 10]

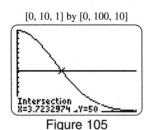

Figure 105

5.4: Logarithmic Functions and Models

Common Logarithms

1. See Figure 1.

x	10^0	10^4	10^8	$10^{1.2}$
$\log x$	0	4	8	1.2

Figure 1

3. (a) $\log(-3)$ is undefined

   (b) $\log \dfrac{1}{100} = \log \dfrac{1}{10^2} = \log 1^{-2} = -2$

   (c) $\log \sqrt{0.1} = \log 10^{-1/2} = -\dfrac{1}{2}$

   (d) $\log 5^0 = 0$

5. (a) $\log 10 = \log 10^1 = 1$

   (b) $\log 10{,}000 = \log 10^4 = 4$

   (c) $20 \log 0.1 = 20 \log 10^{-1} = 20(-1) = -20$

   (d) $\log 10 + \log 0.001 = \log 10^1 + \log 10^{-3} = 1 + (-3) = -2$

7. (a) $2 \log 0.1 + 4 = 2 \log 10^{-1} + 4 = 2(-1) + 4 = -2 + 4 = 2$

   (b) $\log 10^{1/2} = \dfrac{1}{2}$

   (c) $3 \log 100 - \log 1000 = 3 \log 10^2 - \log 10^3 = 3(2) - 3 = 6 - 3 = 3$

   (d) $\log(-10)$ is undefined

9. (a) Since $10^1 \le 79 \le 10^2$, $\log 10^1 \le \log 79 \le \log 10^2 \Rightarrow 1 \le \log 79 \le 2 \Rightarrow n = 1$; $\log 79 \approx 1.898$

(b) Since $10^2 \le 500 \le 10^3$, $\log 10^2 \le \log 500 \le \log 10^3 \Rightarrow 2 \le \log 500 \le 3 \Rightarrow n = 2$; $\log 500 \approx 2.699$

(c) Since $10^0 \le 5 \le 10^1$, $\log 10^0 \le \log 5 \le \log 10^1 \Rightarrow 0 \le \log 5 \le 1 \Rightarrow n = 0$; $\log 5 \approx 0.6990$

(d) Since $10^{-1} \le 0.5 \le 10^0$, $\log 10^{-1} \le \log 0.5 \le \log 10^0 \Rightarrow -1 \le \log 0.5 \le 0 \Rightarrow n = -1$;

$\log 0.5 \approx -0.3010$

11. (a) $x = 3$, since $10^3 = 1000$

(b) $x = 6$, since $10^6 = 1,000,000$

(c) $x = -2$, since $10^{-2} = 0.01$

(d) $x = -4$, since $10^{-4} = 0.0001$

13. (a) $x = 0$, since $10^0 = 1$

(b) $x = \log 50$, since $10^{\log 50} = 50$

(c) $x = -4$, since $10^{-4} = \dfrac{1}{10,000}$

(d) No real solution.

15. (a) $x = \log 300$, since $10^{\log 300} = 300$

(b) $x = \log 5$, since $10^{\log 5} = 5$

(c) $x = \log 0.2$, since $10^{\log 0.2} = 0.2$

17. (a) $\log \sqrt{1000} = \log (10^3)^{1/2} = \log 10^{3/2} = \dfrac{3}{2}$

(b) $\log \sqrt[3]{10} = \log 10^{1/3} = \dfrac{1}{3}$

(c) $\log \sqrt{10} + \log \sqrt[5]{0.1} = \log 10^{1/2} + \log (10^{-1})^{1/5} = \log 10^{1/2} + \log 10^{-1/5} = \dfrac{1}{2} + \left(-\dfrac{1}{5}\right) = \dfrac{3}{10}$

(d) $\log \sqrt{0.01} - \log \sqrt{1} = \log \dfrac{\sqrt{0.01}}{\sqrt{1}} = \log \sqrt{0.01} = \log (10^{-2})^{1/2} = \log 10^{-1} = -1$

Logarithms

19. $\log_2 64 = \log_2 2^6 = 6$

21. $\log_4 2 = \log_4 \sqrt{4} = \log_4 4^{1/2} = \dfrac{1}{2}$

23. $\ln 1 = \ln e^0 = 0$

25. $\ln \sqrt[3]{e} = \ln e^{1/3} = \dfrac{1}{3}$

27. $\log_{1/2}\left(\dfrac{1}{4}\right) = \log_{1/2}\left(\dfrac{1}{2}\right)^2 = 2$

29. $\log_{1/6} 36 = \log_{1/6}\left(\dfrac{1}{6}\right)^{-2} = -2$

31. $\log_a \dfrac{1}{a} = \log_a a^{-1} = -1$

33. $\log_5 5^0 = 0$

35. $\log_2 \dfrac{1}{16} = \log_2 2^{-4} = -4$

37. $2^{\log_2 k} = k$

39. $\log_5 5^\pi = \pi$

41. $3^{\log_3 (x-1)} = x - 1$, for $x > 1$

43. $7^{\log_7 (x^2+2)} = x^2 + 2$

45. See Figure 45.

x	$\frac{1}{16}$	1	8	32
$f(x)$	-4	0	3	5

Figure 45

47. See Figure 47.

x	6	7	21
$f(x)$	0	2	8

Figure 47

Solving Equations

49. $2^x = 72 \Rightarrow \log_2 2^x = \log_2 72 \Rightarrow x = \log_2 72 \approx 6.17$

51. $5^x = 0.25 \Rightarrow \log_5 5^x = \log_5 0.25 \Rightarrow x = \log_5 0.25 \approx -0.86$

53. $e^x = 25 \Rightarrow \ln e^x = \ln 25 \Rightarrow x = \ln 25 \approx 3.219$

55. $e^{-x} = 3 \Rightarrow \ln e^{-x} = \ln 3 \Rightarrow -x = \ln 3 \Rightarrow x = -\ln 3 \approx -1.099$

57. $10^x - 5 = 95 \Rightarrow 10^x = 100 \Rightarrow 10^x = 10^2 \Rightarrow x = 2$

59. $10^{3x} = 100 \Rightarrow 10^{3x} = 10^2 \Rightarrow 3x = 2 \Rightarrow x = \dfrac{2}{3} \approx 0.6667$

61. $e^x + 1 = 24 \Rightarrow e^x = 23 \Rightarrow \ln e^x = \ln 23 \Rightarrow x = \ln 23 \approx 3.1355$

63. $2^x + 1 = 15 \Rightarrow 2^x = 14 \Rightarrow \log_2 2^x = \log_2 14 \Rightarrow x = \log_2 14$ or $x \approx 3.8074$

65. $5e^x + 2 = 20 \Rightarrow 5e^{x=18} \Rightarrow e^x = \dfrac{18}{5} \Rightarrow \ln e^x = \ln \dfrac{18}{5} \Rightarrow x = \ln\left(\dfrac{18}{5}\right) \approx 1.28$

67. $8 - 3(2)^{0.5x} = -40 \Rightarrow -3(2)^{0.5x} = -48 \Rightarrow 2^{0.5x} = 16 \Rightarrow 2^{0.5x} = 2^4 \Rightarrow 0.5x = 4 \Rightarrow x = \dfrac{4}{0.5} \Rightarrow x = 8$

69. $\log x = 2.3 \Rightarrow 10^{\log x} = 10^{2.3} \Rightarrow x = 10^{2.3} \approx 199.5$

71. $\log_2 x = 1.2 \Rightarrow 2^{\log_2 x} = 2^{1.2} \Rightarrow x = 2^{1.2} \approx 2.297$

73. $\ln x = -2 \Rightarrow e^{\ln x} = e^{-2} \Rightarrow x = e^{-2} \approx 0.1353$

75. $2\log x = 6 \Rightarrow \log x = 3 \Rightarrow 10^{\log x} = 10^3 \Rightarrow x = 10^3 = 1000$

77. $2\log 5x = 4 \Rightarrow \log 5x = 2 \Rightarrow 10^{\log 5x} = 10^2 \Rightarrow 5x = 100 \Rightarrow x = 20$

79. $4\ln x = 3 \Rightarrow \ln x = \dfrac{3}{4} \Rightarrow e^{\ln x} = e^{3/4} \Rightarrow x = e^{3/4} \approx 2.1170$

81. $5 \ln x - 1 = 6 \Rightarrow 5 \ln x = 7 \Rightarrow \ln x = \dfrac{7}{5} \Rightarrow e^{\ln x} = e^{7/5} \Rightarrow x = e^{7/5} \approx 4.0552$

83. $4 \log_2 x = 16 \Rightarrow \log_2 x = 4 \Rightarrow 2^{\log_2 x} = 2^4 \Rightarrow x = 2^4 = 16$

85. $5 \ln (2x) + 6 = 12 \Rightarrow 5 \ln (2x) = 6 \Rightarrow \ln (2x) = \dfrac{6}{5} \Rightarrow e^{\ln 2x} = e^{6/5} \Rightarrow 2x = e^{6/5} \Rightarrow x = \dfrac{e^{6/5}}{2} \approx 1.6601$

87. $9 - 3 \log_4 2x = 3 \Rightarrow -3 \log_4 2x = -6 \Rightarrow \log_4 2x = 2 \Rightarrow 4^{\log_4 2x} = 4^2 \Rightarrow 2x = 16 \Rightarrow x = 8$

89. $f(x) = a + b \log x$ and $f(1) = 5 \Rightarrow 5 = a + b \log 1 \Rightarrow 5 = a + b(0) \Rightarrow a = 5$

    $f(x) = 5 + b \log x$ and $f(10) = 7 \Rightarrow 7 = 5 + b \log 10 \Rightarrow 7 = 5 + b(1) \Rightarrow b = 2$

    The function is $f(x) = 5 + 2 \log x$.

Domains of Logarithmic Functions

91. The input to a logarithmic function must be positive. Thus, any element of the domain of f must satisfy

    $x + 3 > 0$, or equivalently, $x > -3$. Thus $D: (-3, \infty)$

93. The input to a logarithmic function must be positive. Thus, any element of the domain of f must satisfy

    $x^2 - 1 > 0 \Rightarrow x^2 > 1 \Rightarrow x < -1$ or $x > 1$ Thus $D: (-\infty, -1) \cup (1, \infty)$

95. The input to a logarithmic function must be positive. Thus, any element of the domain of f must satisfy

    $3^x > 0 \Rightarrow x$ can be any real number. Thus $D: (-\infty, \infty)$

97. The input to a logarithmic function must be positive. Thus, any element of the domain of f must satisfy

    $\sqrt{3 - x} - 1 > 0 \Rightarrow \sqrt{3 - x} > 1 \Rightarrow x < 2$. Thus $D: (-\infty, 2)$

Graphs of Logarithmic Equations

99. Since $f(x) = e^x, f^{-1}(x) = \ln x$. See Figure 99.

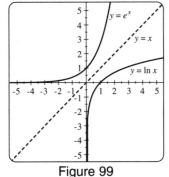

Figure 99

101. Since $\log_2 2 = 1$, the best choice is graph d.b.

103. Since $f(x) = \log (x + 2)$ is a translation of $f(x) = \log x$ left 2 units, the best choice is graph c.

105. The graph is shown in Figure 105. $D = \{x \mid x > -1\}$

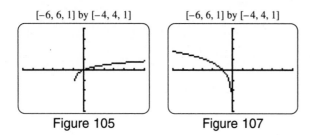

[−6, 6, 1] by [−4, 4, 1]   [−6, 6, 1] by [−4, 4, 1]

Figure 105   Figure 107

107. The graph is shown in Figure 107. $D = \{x \mid x < 0\}$

109. Decreasing. See Figure 109.

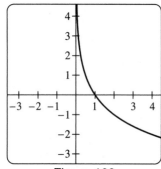

Figure 109

Applications

111. $D(x) = 10 \log (10^{16}x)$, $x = 10^{-11/2} \Rightarrow D(10^{-11/2}) = 10 \log (10^{16}(10^{-11/2})) \Rightarrow 10 \log (10^{21/2}) \Rightarrow$

$10\left(\dfrac{21}{2}\right) \Rightarrow 105$ decibels.

113. (a) $f(x) = a + b \log x$ and $f(1) = 7 \Rightarrow 7 = a + b \log 1 \Rightarrow 7 = a + b(0) \Rightarrow a = 7$

$f(x) = 7 + b \log x$ and $f(10) = 11 \Rightarrow 11 = 7 + b \log 10 \Rightarrow 11 = 7 + b(1) \Rightarrow b = 4$

The function that models the given data is $f(x) = 7 + 4 \log x$.

(b) $f(x) = 16 \Rightarrow 16 = 7 + 4 \log x \Rightarrow 9 = 4 \log x \Rightarrow \log x = \dfrac{9}{4} \Rightarrow 10^{\log x} = 10^{9/4} \Rightarrow x = 10^{9/4} \approx 178$

The island would be about 178 square kilometers.

115. (a) $f(x) = Ca^x$ and $f(0) = 3 \Rightarrow 3 = Ca^0 \Rightarrow 3 = C(1) \Rightarrow C = 3$

$f(x) = 3a^x$ and $f(1) = 6 \Rightarrow 6 = 3a^1 \Rightarrow 2 = a^1 \Rightarrow a = 2$; thus $f(x) = 3(2^x)$ models the data.

(b) $f(x) = 16 \Rightarrow 16 = 3(2^x) \Rightarrow 2^x = \dfrac{16}{3}$. Graph $Y_1 = 2\text{\textasciicircum}X$ and $Y_2 = 16/3$ in [0, 5, 1] by [4, 20, 5].

The graphs intersect near (2.41, 5.33), so there were 16 million bacteria in the colony after about 2.4 days.

117. (a) The graph of L is shown in Figure 117. It is an increasing function. The implication is that heavier planes generally require longer runways.

(b) Since the weight is measured in 1000-pound units, start by evaluating $L(10)$ and $L(100)$.

$L(10) = 3 \log 10 = 3$ and $L(100) = 3 \log 100 = 6$. Thus, a 10,000-pound plane requires approximately 3000 feet of runway, whereas as 100,000-pound plane requires approximately 6000 feet. The distance does not increase by a factor of 10.

(c) If the weight increases tenfold, the runway length increases by 3000 feet.

[0, 50, 10] by [0, 6, 1]   [0, 250, 50] by [25, 30, 1]

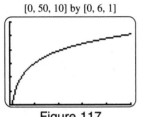 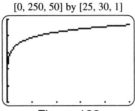

Figure 117   Figure 123

119. (a) Since the pH is computed by $f(x) = -\log x$ where x is the hydrogen ion concentration, we must solve

$-\log x = 4.92 \Rightarrow \log x = -4.92 \Rightarrow 10^{\log x} = 10^{-4.92} \Rightarrow x = 10^{-4.92} \approx 0.000012$

(b) Since the pH is computed by $f(x) = -\log x$ where x is the hydrogen ion concentration, we must solve

$-\log x = 3.9 \Rightarrow \log x = -3.9 \Rightarrow 10^{\log x} = 10^{-3.9} \Rightarrow x = 10^{-3.9} \approx 0.000126$

121. (a) Since $I_0 = 1$, $R(x) = 6.0 \Rightarrow \log x = 6.0 \Rightarrow 10^{\log x} = 10^{6.0} \Rightarrow x = 10^6 = 1{,}000{,}000$

Similarly $R(x) = 8.0 \Rightarrow \log x = 8.0 \Rightarrow 10^{\log x} = 10^{8.0} \Rightarrow x = 10^8 = 100{,}000{,}000$

(b) $\dfrac{10^8}{10^6} = 10^{8-6} = 10^2 = 100$. The Indonesian earthquake was 100 times more intense (powerful) than the Yugoslovian earthquake.

123. (a) $f(x) = 0.48 \ln (x + 1) + 27 \Rightarrow f(0) = 0.48 \ln (0 + 1) + 27 = 0.48 \ln 1 + 27 = 0.48(0) + 27 = 27$ and $f(100) = 0.48 \ln(100 + 1) + 27 \approx 29.2$ inches. At the center, or eye, of the hurricane the pressure is 27 inches of mercury, while 100 miles from the eye the air pressure has risen to 29.2 inches of mercury.

(b) Graph $Y_1 = 0.48 \ln (X + 1) + 27$ as shown in Figure 123. At first, the air pressure rises rapidly as one moves away from the eye. Then, the air pressure starts to level off and does not increase significantly for distances greater than 200 miles.

(c) $f(x) = 28 \Rightarrow 28 = 0.48 \ln (x + 1) + 27 \Rightarrow \ln (x + 1) = \dfrac{1}{0.48} \Rightarrow e^{\ln (x+1)} = e^{1/0.48} \Rightarrow$

$x + 1 = e^{1/0.48} \Rightarrow x = e^{1/0.48} - 1 \approx 7.03$; the air pressure is 28 inches of mercury about 7 miles from the eye of the hurricane.

125. (a) $T(x) = 20 + 80e^{-x} \Rightarrow T(1) = 20 + 80e^{-1} \approx 49.4$; the temperature is about 49.4°C after 1 hour.

(b) $T(x) = 60 \Rightarrow 60 = 20 + 80e^{-x} \Rightarrow 40 = 80e^{-x} \Rightarrow e^{-x} = \dfrac{1}{2} \Rightarrow \ln e^{-x} = \ln\left(\dfrac{1}{2}\right) \Rightarrow$

$-x = \ln\left(\dfrac{1}{2}\right) \Rightarrow x = -\ln\left(\dfrac{1}{2}\right) \approx 0.693$; the water took about 0.69 hours, or 41.4 minutes to cool to 60°C.

127. (a) $f(x) = e^{-x/3} \Rightarrow f(5) = e^{-5/3} \approx 0.189$

The probability that no car enters the intersection during a 5-minute period is about 0.189 or 18.9%.

(b) $f(x) = 0.30 \Rightarrow 0.30 = e^{-x/3} \Rightarrow \ln e^{-x/3} = \ln 0.3 \Rightarrow -\dfrac{x}{3} = \ln 0.3 \Rightarrow x = -3 \ln 0.3 \approx 3.6$

The probability that no car enters the intersection will be 30% during a period of about 3.6 minutes.

5.5: Properties of Logarithms

Properties of Logarithms

1. $\log 4 + \log 7 \approx 1.447$; $\log 28 \approx 1.447$; $\log 4 + \log 7 = \log 28 = \log (4 \cdot 7)$; Property 2

3. $\ln 72 - \ln 8 \approx 2.197$; $\ln 9 \approx 2.197$; $\ln 72 - \ln 8 = \ln 9 = \ln\left(\dfrac{72}{8}\right)$; Property 3

5. $10 \log 2 \approx 3.010$; $\log 1024 \approx 3.010$; $10 \log 2 = \log 1024 = \log 2^{10}$; Property 4

7. $\log_2 ab = \log_2 a + \log_2 b$

9. $\ln 7a^4 = \ln 7 + 4 \ln a$

11. $\log \dfrac{6}{z} = \log 6 - \log z$

13. $\log \dfrac{x^2}{3} = \log x^2 - \log 3 = 2 \log x - \log 3$

15. $\ln \dfrac{2x^7}{3k} = \ln 2x^7 - \ln 3k = \ln 2 + \ln x^7 - (\ln 3 + \ln k) = \ln 2 + 7 \ln x - \ln 3 - \ln k$

17. $\log_2 4k^2x^3 = \log_2 4k^2 + \log_2 x^3 = \log_2 4 + \log_2 k^2 + \log_2 x^3 = 2 + 2 \log_2 k + 3 \log_2 x$

19. $\log_5 \dfrac{25x^3}{y^4} = \log_5 25x^3 - \log_5 y^4 = 2 + 3 \log_5 x - 4 \log_5 y$

21. $\log_4 0.25(x + 2)^3 = \log_4 0.25 + \log_4 (x + 2)^3 = -1 + 3 \log_4 (x + 2)$

23. $\log_5 \dfrac{x^3}{(x - 4)^4} = \log_5 x^3 - \log_5 (x - 4)^4 = 3 \log_5 x - 4 \log_5 (x - 4)$

25. $\log_2 \dfrac{\sqrt{x}}{z^2} = \log_2 \sqrt{x} - \log_2 z^2 = \log_2 x^{1/2} - \log_2 z^2 = \dfrac{1}{2} \log_2 x - 2 \log_2 z$

27. $\ln \sqrt[3]{\dfrac{2x + 6}{(x + 1)^5}} = \ln \left(\dfrac{2x + 6}{(x + 1)^5}\right)^{1/3} = \dfrac{1}{3}[\ln (2x + 6) - 5 \ln (x + 1)] = \dfrac{1}{3} \ln (2x + 6) - \dfrac{5}{3} \ln (x + 1)$

29. $\log_2 \dfrac{\sqrt[3]{x^2 - 1}}{\sqrt{1 + x^2}} = \log_2 \sqrt[3]{x^2 - 1} - \log_2 \sqrt{1 + x^2} = \log_2 (x^2 - 1)^{1/3} - \log_2 (1 + x^2)^{1/2} =$

$\dfrac{1}{3} \log_2 (x^2 - 1) - \dfrac{1}{2} \log_2 (1 + x^2)$

31. $\log 2 + \log 3 = \log (2 \cdot 3) = \log 6$

33. $\ln \sqrt{5} - \ln 25 = \ln \left(\dfrac{5^{1/2}}{5^2}\right) = \ln 5^{-3/2} = -\dfrac{3}{2} \ln 5$

35. $\log 20 + \log \dfrac{1}{10} = \log (20)\left(\dfrac{1}{10}\right) = \log \dfrac{20}{10} = \log 2$

37. $\log 4 + \log 3 - \log 2 = \log (4)\left(\frac{3}{2}\right) = \log \frac{12}{2} = \log 6$

39. $\log_7 5 + \log_7 k^2 = \log_7 (5)(k^2) = \log_7 5k^2$

41. $\ln x^6 - \ln x^3 = \ln \frac{x^6}{x^3} = \ln x^{6-3} = \ln x^3$

43. $\log \sqrt{x} + \log x^2 - \log x = \log\left(\frac{x^{1/2} \cdot x^2}{x}\right) = \log x^{3/2} = \frac{3}{2} \log x$

45. $\ln \frac{1}{e^2} + \ln 2e = \ln\left(\frac{1}{e^2} \cdot 2e\right) = \ln \frac{2}{e}$

47. $2 \ln x - 4 \ln y + \frac{1}{2} \ln z = \ln x^2 - \ln y^4 + \ln z^{1/2} = \ln \frac{x^2}{y^4} + \ln \sqrt{z} = \ln \frac{x^2\sqrt{z}}{y^4}$

49. $\log 4 - \log x + 7 \log \sqrt{x} = \log \frac{4}{x} + \log (\sqrt{x})^7 = \log\left(\frac{4x^{7/2}}{x}\right) = \log 4x^{5/2}$

51. $2 \log (x^2 - 1) + 4 \log (x - 2) - \frac{1}{2} \log y = \log (x^2 - 1)^2 + \log (x - 2)^4 - \log y^{1/2} = \log \frac{(x^2 - 1)^2(x - 2)^4}{\sqrt{y}}$

53. (a) Table $Y_1 = \log (3X) + \log (2X)$ and $Y_2 = \log (6X^2)$ starting at $x = 1$, incrementing by 1. See Figure 53.

From the table we see that $Y_1 = Y_2$, so $f(x) = g(x)$.

(b) By property 2: $\log 3x + \log 2x = \log (3x \cdot 2x) = \log 6x^2$

55. (a) Table $Y_1 = \ln (3X) - \ln (2X)$ and $Y_2 = \ln (X)$ starting at $x = 1$, incrementing by 1. See Figure 55.

From the table we see that $Y_1 \neq Y_2$, so $f(x) \neq g(x)$.

(b) Not possible.

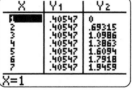

X	Y1	Y2
1	.77815	.77815
2	1.3802	1.3802
3	1.7324	1.7324
4	1.9823	1.9823
5	2.1761	2.1761
6	2.3345	2.3345
7	2.4683	2.4683

X=1

Figure 53

X	Y1	Y2
1	.40547	0
2	.40547	.69315
3	.40547	1.0986
4	.40547	1.3863
5	.40547	1.6094
6	.40547	1.7918
7	.40547	1.9459

X=1

Figure 55

X	Y1	Y2
1	0	0
2	1.2041	1.2041
3	1.9085	1.9085
4	2.4082	2.4082
5	2.7959	2.7959
6	3.1126	3.1126
7	3.3804	3.3804

X=1

Figure 57

X	Y1	Y2
1	0	0
2	1.3863	1.3863
3	2.1972	2.1972
4	2.7726	2.7726
5	3.2189	3.2189
6	3.5835	3.5835
7	3.8918	3.8918

X=1

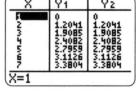

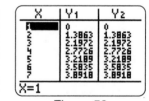

Figure 59

57. (a) Table $Y_1 = \log (X\text{^}4)$ and $Y_2 = 4 \log (X)$ starting at $x = 1$, incrementing by 1. See Figure 57.

From the table we see that $Y_1 = Y_2$, so $f(x) = g(x)$.

(b) By property 4: $\log x^4 = 4 \log x$

59. (a) Table $Y_1 = \ln (X\text{^}4) - \ln (X^2)$ and $Y_2 = 2 \ln (X)$ starting at $x = 1$, incrementing by 1. See Figure 59.

From the table we see that $Y_1 = Y_2$, so $f(x) = g(x)$.

(b) By property 4: $\ln x^4 - \ln x^2 = 4 \ln x - 2 \ln x = 2 \ln x$

(b) Not possible.

61. See Figure 61.

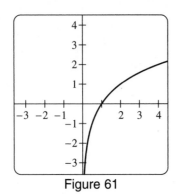

Figure 61

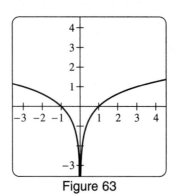

Figure 63

63. See Figure 63.

Change of Base Formula

65. $\log_2 25 = \dfrac{\log 25}{\log 2} \approx 4.644$

67. $\log_5 130 = \dfrac{\log 130}{\log 5} \approx 3.024$

69. $\log_2 5 + \log_2 7 = \dfrac{\log 5}{\log 2} + \dfrac{\log 7}{\log 2} \approx 5.129$

71. $\sqrt{\log_4 46} = \sqrt{\dfrac{\log 46}{\log 4}} \approx 1.662$

73. $\dfrac{\log_2 12}{\log_2 3} = \dfrac{\frac{\log 12}{\log 2}}{\frac{\log 3}{\log 2}} = \dfrac{\log 12}{\log 2} \cdot \dfrac{\log 2}{\log 3} = \dfrac{\log 12}{\log 3} \approx 2.262$

75. Graph $Y_1 = \log (X^\wedge 3 + X^2 + 1)/\log 2$ and $Y_2 = 7$ in [0, 10, 2] by [0, 10, 2]. The graphs intersect near the point (4.714, 7), so the solution to $\log_2 (x^3 + x^2 + 1) = 7$ is $x \approx 4.714$.

77. Graph $Y_1 = \log (X^2 + 1)/\log 2$ and $Y_2 = 5 - \log (X^\wedge 4 + 1)/\log 3$ in [−5, 5, 2] by [−5, 10, 2]. The graphs intersect near the points (−2.035, 2.362) and (2.035, 2.362), so the solutions to the equation $\log_2 (x^2 + 1) = 5 - \log_3 (x^4 + 1)$ are $x \approx \pm 2.035$.

Applications

79. Change base 10 to base e: $L(x) = 3 \log x = \dfrac{3 \cdot \ln x}{\ln 10}$; $L(50) = 3 \log 50 \approx 5.097$ and $L(50) = \dfrac{3 \ln 50}{\ln 10} \approx 5.097$

    Yes, the answers agree.

81. $f(x) = 160 + 10 \log x \Rightarrow f(10x) = 160 + 10 \log (10x) = 160 + 10(\log 10 + \log x) =$

    $160 + 10(1 + \log x) = 160 + 10 + 10 \log x = 160 + 10 \log x + 10$, thus $f(10x) = f(x) + 10$.

    In other words, the decibel level increases by 10 decibels.

83. $\ln I - \ln I_0 = -kx \Rightarrow \ln I = \ln I_0 - kx \Rightarrow e^{\ln I} = e^{\ln I_0 - kx} \Rightarrow e^{\ln I} = e^{\ln I_0} \cdot e^{-kx} \Rightarrow I = I_0 e^{-kx}$

85. $A = Pe^{rt} \Rightarrow \dfrac{A}{P} = e^{rt} \Rightarrow \ln \dfrac{A}{P} = \ln e^{rt} \Rightarrow \ln \dfrac{A}{P} = rt \Rightarrow \dfrac{\ln \frac{A}{P}}{r} = t \Rightarrow t = \dfrac{\ln \frac{A}{P}}{r}$

87. $\log 1 + 2 \log 2 + 3 \log 3 + 4 \log 4 + 5 \log 5 = \log 1 + \log 2^2 + \log 3^3 + \log 4^4 + \log 5^5 =$

   $\log (1 \cdot 2^2 \cdot 3^3 \cdot 4^4 \cdot 5^5) = \log 86{,}400{,}000$

5.6: Exponential and Logarithmic Equations

Solving Exponential Equations

1. (a) The graphs appear to intersect near the point (2, 7.5). Thus, the solution is $x \approx 2$.

   (b) $f(x) = g(x) \Rightarrow e^x = 7.5 \Rightarrow \ln e^x = \ln 7.5 \Rightarrow x = \ln 7.5$ about 2.015

3. (a) The graphs appear to intersect near the point (2, 2.5). Thus, the solution is $x \approx 2$.

   (b) $f(x) = g(x) \Rightarrow 10^{0.2x} = 2.5 \Rightarrow \log 10^{0.2x} = \log 2.5 \Rightarrow 0.2x = \log 2.5 \Rightarrow x = \dfrac{\log 2.5}{0.2}$ about 1.990

5. $4e^x = 5 \Rightarrow e^x = \dfrac{5}{4} \Rightarrow \ln e^x = \ln \dfrac{5}{4} \Rightarrow x = \ln \dfrac{5}{4} \approx 0.2231$

7. $2(10^x) + 5 = 45 \Rightarrow 2(10^x) = 40 \Rightarrow 10^x = 20 \Rightarrow \log 10^x = \log 20 \Rightarrow x = \log 20 \approx 1.301$

9. $2.5e^{-1.2x} = 1 \Rightarrow e^{-1.2x} = \dfrac{1}{2.5} \Rightarrow \ln e^{-1.2x} = \ln \dfrac{1}{2.5} \Rightarrow -1.2x = \ln \dfrac{1}{2.5} \Rightarrow x = \dfrac{\ln \frac{1}{2.5}}{-1.2} \approx 0.7636$

11. $1.2(0.9^x) = 0.6 \Rightarrow 0.9^x = 0.5 \Rightarrow \ln 0.9^x = \ln 0.5 \Rightarrow x \ln 0.9 = \ln 0.5 \Rightarrow x = \dfrac{\ln 0.5}{\ln 0.9} \approx 6.579$

13. $4(1.1^{x-1}) = 16 \Rightarrow 1.1^{x-1} = 4 \Rightarrow \ln 1.1^{x-1} = \ln 4 \Rightarrow (x - 1) \ln 1.1 = \ln 4 \Rightarrow x - 1 = \dfrac{\ln 4}{\ln 1.1} \Rightarrow$

   $x = \dfrac{\ln 4}{\ln 1.1} + 1 \approx 15.55$

15. $5(1.2)^{3x-2} + 94 = 100 \Rightarrow 5(1.2)^{3x-2} = 6 \Rightarrow (1.2)^{3x-2} = 1.2 \Rightarrow \ln 1.2^{3x-2} = \ln 1.2 \Rightarrow$

   $(3x - 2) \ln 1.2 = \ln 1.2 \Rightarrow 3x - 2 = \dfrac{\ln 1.2}{\ln 1.2} \Rightarrow 3x - 2 = 1 \Rightarrow 3x = 3 \Rightarrow x = 1$

17. $5^{3x} = 5^{1-2x} \Rightarrow \log_5 5^{3x} = \log_5 5^{1-2x} \Rightarrow 3x = 1 - 2x \Rightarrow 5x = 1 \Rightarrow x = \dfrac{1}{5}$

19. $10^{x^2} = 10^{3x-2} \Rightarrow \log 10^{x^2} = \log 10^{3x-2} \Rightarrow x^2 = 3x - 2 \Rightarrow x^2 - 3x + 2 = 0 \Rightarrow$

   $(x - 1)(x - 2) = 0 \Rightarrow x = 1$ or $x = 2$

21. No solution since no power of $\dfrac{1}{5}$ will result in a negative value.

23. $4^{x-1} = 3^{2x} \Rightarrow \log 4^{x-1} = \log 3^{2x} \Rightarrow (x - 1) \log 4 = 2x \log 3 \Rightarrow x \log 4 - \log 4 = 2x \log 3 \Rightarrow$

   $x \log 4 - 2x \log 3 = \log 4 \Rightarrow x(\log 4 - 2 \log 3) = \log 4 \Rightarrow x = \dfrac{\log 4}{\log 4 - 2 \log 3} \approx -1.710$

25. $e^{x-3} = 2^{3x} \Rightarrow \ln e^{x-3} = \ln 2^{3x} \Rightarrow x - 3 = 3x \ln 2 \Rightarrow -3 = -x + 3x \ln 2 \Rightarrow$

   $3 = x - 3x \ln 2 \Rightarrow 3 = x(1 - 3 \ln 2) \Rightarrow x = \dfrac{3}{1 - 3 \ln 2} \Rightarrow x = \dfrac{3}{1 - \ln 8} \Rightarrow x \approx -2.779$

27. $3(1.4)^x - 4 = 60 \Rightarrow 3(1.4)^x = 64 \Rightarrow 1.4^x = \dfrac{64}{3} \Rightarrow \log 1.4^x = \log \dfrac{64}{3} \Rightarrow x \log 1.4 = \log \dfrac{64}{3} \Rightarrow$

$x = \dfrac{\log \frac{64}{3}}{\log 1.4} \approx 9.095$

29. $5(1.015)^{x-1980} = 8 \Rightarrow 1.015^{x-1980} = \dfrac{8}{5} \Rightarrow \log 1.015^{x-1980} = \log \dfrac{8}{5} \Rightarrow (x - 1980) \log 1.015 = \log \dfrac{8}{5} \Rightarrow$

$x \log 1.015 - 1980 \log 1.015 = \log \dfrac{8}{5} \Rightarrow x \log 1.015 = \log \dfrac{8}{5} + 1980 \log 1.015 \Rightarrow$

$x = \dfrac{\log \frac{8}{5} + 1980 \log 1.015}{\log 1.015} = \dfrac{\log \frac{8}{5}}{\log 1.015} + 1980 \approx 2012$

31. $4\left(\dfrac{3}{4}\right)^{x+1} = \dfrac{1}{81}\left(\dfrac{3}{2}\right)^{5+x} \Rightarrow 324\left(\dfrac{3}{4}\right)^{x+1} = \left(\dfrac{3}{2}\right)^{5+x} \Rightarrow \ln\left[324\left(\dfrac{3}{4}\right)^{x+1}\right] = \ln\left(\dfrac{3}{2}\right)^{5+x} \Rightarrow$

$\ln 324 + \ln\left(\dfrac{3}{4}\right)^{x+1} = \ln\left(\dfrac{3}{2}\right)^{5+x} \Rightarrow \ln 324 + (x+1)\ln\dfrac{3}{4} = (5+x)\ln\dfrac{3}{2} \Rightarrow$

$\ln 324 + x\ln\dfrac{3}{4} + \ln\dfrac{3}{4} = 5\ln\dfrac{3}{2} + x\ln\dfrac{3}{2} \Rightarrow x\ln\dfrac{3}{4} - x\ln\dfrac{3}{2} = 5\ln\dfrac{3}{2} - \ln 324 - \ln\dfrac{3}{4} \Rightarrow$

$x\left(\ln\dfrac{3}{4} - \ln\dfrac{3}{2}\right) = 5\ln\dfrac{3}{2} - \ln 324 - \ln\dfrac{3}{4} \Rightarrow x = \dfrac{5\ln\frac{3}{2} - \ln 324 - \ln\frac{3}{4}}{\ln\frac{3}{4} - \ln\frac{3}{2}} \Rightarrow x = 5$

Solving Logarithmic Equations

33. $3\log x = 2 \Rightarrow \log x = \dfrac{2}{3} \Rightarrow 10^{\log x} = 10^{2/3} \Rightarrow x = 10^{2/3} = \sqrt[3]{10^2} = \sqrt[3]{100} \approx 4.642$

35. $\ln 2x = 5 \Rightarrow e^{\ln 2x} = e^5 \Rightarrow 2x = e^5 \Rightarrow x = \dfrac{e^5}{2} \approx 74.207$

37. $\log 2x^2 = 2 \Rightarrow 10^{\log 2x^2} = 10^2 \Rightarrow 2x^2 = 100 \Rightarrow x^2 = 50 \Rightarrow x = \pm\sqrt{50} \approx \pm 7.071$

39. $\log_2(3x - 2) = 4 \Rightarrow 2^{\log_2(3x-2)} = 2^4 \Rightarrow 3x - 2 = 16 \Rightarrow 3x = 18 \Rightarrow x = 6$

41. $\log_5(8 - 3x) = 3 \Rightarrow 5^{\log_5(8-3x)} = 5^3 \Rightarrow 8 - 3x = 125 \Rightarrow -3x = 117 \Rightarrow x = -39$

43. $160 + 10\log x = 50 \Rightarrow 10\log x = -110 \Rightarrow \log x = -11 \Rightarrow 10^{\log x} = 10^{-11} \Rightarrow x = 10^{-11}$

45. $\ln x + \ln x^2 = 3 \Rightarrow \ln x + 2\ln x = 3 \Rightarrow 3\ln x = 3 \Rightarrow \ln x = 1 \Rightarrow e^{\ln x} = e^1 \Rightarrow x = e \approx 2.718$

47. $2\log_2 x = 4.2 \Rightarrow \log_2 x = 2.1 \Rightarrow 2^{\log_2 x} = 2^{2.1} \Rightarrow x = 2^{2.1} \approx 4.287$

49. $\log x + \log 2x = 2 \Rightarrow \log(x \cdot 2x) = 2 \Rightarrow \log 2x^2 = 2 \Rightarrow 10^{\log 2x^2} = 10^2 \Rightarrow 2x^2 = 100 \Rightarrow x^2 = 50 \Rightarrow$

$x = \pm\sqrt{50}$. When $x = -\sqrt{50} < 0$, $\log x$ is undefined. Therefore, the only solution is $x = \sqrt{50} \approx 7.071$.

51. $\ln(x - 1) = 1 \Rightarrow e^{\ln(x-1)} = e^1 \Rightarrow x - 1 = e \Rightarrow x = e + 1 \approx 3.718$

53. $2\ln x = \ln(2x + 1) \Rightarrow \ln x^2 = \ln(2x + 1) \Rightarrow e^{\ln x^2} = e^{\ln(2x+1)} \Rightarrow x^2 = 2x + 1 \Rightarrow x^2 - 2x - 1 = 0 \Rightarrow$

$x = \dfrac{2 \pm \sqrt{2^2 - 4(1)(-1)}}{2(1)} = \dfrac{2 \pm \sqrt{8}}{2} = \dfrac{2 \pm 2\sqrt{2}}{2} = 1 \pm \sqrt{2} \Rightarrow x = 1 + \sqrt{2}$ or $x = 1 - \sqrt{2}$

When $x = 1 - \sqrt{2} \approx -0.414$ both $\ln x$ and $\ln(2x + 1)$ are undefined. The only solution is $x = 1 + \sqrt{2}$.

55. $\log(x + 1) + \log(x - 1) = \log 3 \Rightarrow \log[(x + 1)(x - 1)] = \log 3 \Rightarrow \log(x^2 - 1) = \log 3 \Rightarrow$

$10^{\log(x^2-1)} = 10^{\log 3} \Rightarrow x^2 - 1 = 3 \Rightarrow x^2 = 4 \Rightarrow x = -2$ or $x = 2$

When $x = -2$ both $\log(x - 1)$ and $\log(x + 1)$ are undefined. The only solution is $x = 2$.

57. $\log_2 2x = 4 - \log_2(x+2) \Rightarrow \log_2 2x + \log_2(x+2) = 4 \Rightarrow \log_2[2x(x+2)] = 4 \Rightarrow$

    $\log_2(2x^2 + 4x) = 4 \Rightarrow 2^{\log_2(2x^2+4x)} = 4 \Rightarrow 2x^2 + 4x - 16 = 0 \Rightarrow 2(x^2 + 2x - 8) = 0 \Rightarrow$

    $2(x+4)(x-2) = 0 \Rightarrow x = -4, x = 2,$

    since we cannot take the log of a negative number, $x = 2$.

59. $\log_5(x+1) + \log_5(x-1) = \log_5 15 \Rightarrow \log_5(x+1)(x-1) = \log_5 15 \Rightarrow \log_5(x^2-1) = \log_5 15 \Rightarrow$

    $5^{\log_5(x^2-1)} = 5^{\log_5 15} \Rightarrow x^2 - 1 = 15 \Rightarrow x^2 - 16 = 0 \Rightarrow (x+4)(x-4) = 0,$

    since we cannot take the log of a negative number, $x = 4$.

Solving Equations Graphically

61. Graph $Y_1 = 2X + e\wedge(X)$ and $Y_2 = 2$ as shown in Figure 61.

    The graphs intersect near the point $(0.31, 2)$, so the solution to $2x + e^x = 2$ is $x \approx 0.31..$

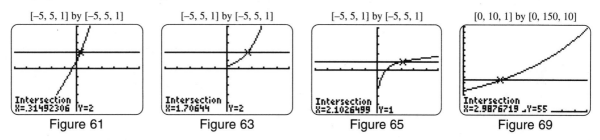

[-5, 5, 1] by [-5, 5, 1]   [-5, 5, 1] by [-5, 5, 1]   [-5, 5, 1] by [-5, 5, 1]   [0, 10, 1] by [0, 150, 10]

Intersection X=.31492306 Y=2   Intersection X=1.70644 Y=2   Intersection X=2.1026499 Y=1   Intersection X=2.9876719 Y=55

Figure 61   Figure 63   Figure 65   Figure 69

63. Graph $Y_1 = X^2 + X\ln(X)$ and $Y_2 = 2$ as shown in Figure 63.

    The graphs intersect near the point $(1.71, 2)$, so the solution to $x^2 - x\ln x = 2$ is $x \approx 1.71$.

65. Graph $Y_1 = Xe\wedge(-X) + \ln(X)$ and $Y_2 = 1$ as shown in Figure 65.

    The graphs intersect near the point $(2.10, 2)$, so the solution to $xe^{-x} + \ln x = 1$ is $x \approx 2.10$.

Applications

67. $4 = 3(1.018)^{x-1960} \Rightarrow \dfrac{4}{3} = 1.018^{x-1960} \Rightarrow \ln 1.33 = \ln 1.018^{x-1960} \Rightarrow \ln 1.33 = (x - 1960)(\ln 1.018) \Rightarrow$

    $\dfrac{\ln 1.33}{\ln 1.018} = x - 1960 \Rightarrow \dfrac{\ln 1.33}{\ln 1.018} + 1960 = x \Rightarrow x = 1976$

69. (a) We must solve the equation $f(x) = 55$.

    $36.2e^{0.14x} = 55 \Rightarrow e^{0.14x} = \dfrac{55}{36.2} \Rightarrow \ln e^{0.14x} = \ln \dfrac{55}{36.2} \Rightarrow 0.14x = \ln \dfrac{55}{36.2} \Rightarrow x = \dfrac{\ln \frac{55}{36.2}}{0.14} \approx 2.99$

    Since $x = 0$ corresponds to 1987, $x \approx 3$ represents $1987 + 3 = 1990$, rounded to the nearest year.

    (b) Graph $Y_1 = 36.2e\wedge(0.14X)$ and $Y_2 = 55$ as shown in Figure 69. The graphs intersect near $(3, 55)$.

    Therefore, Christmas credit card spending was approximately \$55 billion in 1990.

71. We must solve the equation $f(x) = 95$.

    $230(0.881)^x = 95 \Rightarrow 0.881^x = \dfrac{95}{230} \Rightarrow \log 0.881^x = \log \dfrac{95}{230} \Rightarrow x\log 0.881 = \log \dfrac{95}{230} \Rightarrow$

    $x = \dfrac{\log \frac{95}{230}}{\log 0.881} \approx 7$

    Since $x = 0$ corresponds to 1974, $x \approx 7$ represents $1974 + 7 = 1981$, rounded to the nearest year.

73. (a) $P(x) = Ca^{x-2000} \Rightarrow P(2000) = Ca^{2000-2000} \Rightarrow 1 - Ca^0 \Rightarrow C = 1$, so $P(x) = a^{x-2000}$

$P(2025) = a^{2025-2000} \Rightarrow 1.4 = a^{25} \Rightarrow a = 1.4^{1/25} \approx 1.01355$

(b) $P(x) = (1.01355)^{x-2000} \Rightarrow P(2010) = (1.01355)^{10} \approx 1.144$;

in 2010 the population if India will be about1.14 billion.

(c) $P(x) = 1.5 \Rightarrow 1.01355^{x-2000} = 1.5 \Rightarrow \ln 1.01355^{x-2000} = \ln 1.5 \Rightarrow (x - 2000)\ln 1.01355 = \ln 1.5 \Rightarrow$

$x \ln 1.01355 - 2000 \ln 1.01355 = \ln 1.5 \Rightarrow x \ln 1.01355 = \ln 1.5 + 2000 \ln 1.01355 \Rightarrow$

$x = \dfrac{\ln 1.5 + 2000 \ln 1.01355}{\ln 1.01355} = \dfrac{\ln 1.5}{\ln 1.01355} + 2000 \approx 2030.13$;

India's population might reach 1.5 billion in 2030.

75. (a) $P(t) = 750 \Rightarrow 500e^{0.09x} = 750 \Rightarrow e^{0.09x} = 1.5 \Rightarrow \ln e^{0.09x} = \ln 1.5 \Rightarrow 0.09x = \ln 1.5 \Rightarrow$

$t = \dfrac{\ln 1.5}{0.09} \approx 4.505$

(b) If \$500 is invested at 9% compounded continuously, it will grow to approximately \$750 after 4.5 years.

77. $A_0 = \$1000$ and $A_n = \$2000 \Rightarrow 2000 = 1000\left(1 + \dfrac{0.085}{4}\right)^{4n} \Rightarrow 2 = (1.02125)^{4n} \Rightarrow$

$\ln 2 = \ln(1.02125)^{4n} \Rightarrow \ln 2 = 4n \ln 1.02125 \Rightarrow 4n = \dfrac{\ln 2}{\ln 1.02125} \Rightarrow n = \dfrac{\ln 2}{4 \ln 1.02125} \approx 8.25$

The \$1000 doubles in value in about 8.25 years or about 8 years 3 months.

79. $P = 100\left(\dfrac{1}{2}\right)^{t/5700} \Rightarrow 35 = 100\left(\dfrac{1}{2}\right)^{t/5700} \Rightarrow \left(\dfrac{1}{2}\right)^{t/5700} = 0.35 \Rightarrow \ln\left(\dfrac{1}{2}\right)^{t/5700} = \ln 0.35 \Rightarrow$

$\dfrac{t}{5700}\ln\dfrac{1}{2} = \ln 0.35 \Rightarrow \dfrac{t}{5700} = \dfrac{\ln 0.35}{\ln\frac{1}{2}} \Rightarrow t = 5700\left(\dfrac{\ln 0.35}{\ln\frac{1}{2}}\right) \approx 8633$; The fossil is about 8633 years old.

81. (a) $f(x) = 1 - e^{-x} \Rightarrow f(5) = 1 - e^{-5} \approx 0.993$, or 99.3%

(b) $0.40 = 1 - e^{-x} \Rightarrow e^{-x} = 0.6 \Rightarrow \ln e^{-x} = \ln 0.6 \Rightarrow -x = \ln 0.6 \Rightarrow x = -\ln 0.6 \approx 0.51$

During an interval of about 0.5 minute, there is a 40% chance that at least one car enters the intersection.

83. (a) $T_0 = 32, D = 212 - 32 \Rightarrow D = 180$. Solving for a use $70 = 32 + 180a^{1/2} \Rightarrow 38 = 180a^{1/2} \Rightarrow$

$\dfrac{38}{180} = a^{1/2} \Rightarrow a = 0.045$

(b) $T(t) = 32 + 180(0.045)^t \Rightarrow T\left(\dfrac{1}{6}\right) = 32 + 180(0.045)^{1/6} \Rightarrow T\left(\dfrac{1}{6}\right) \approx 139°$

(c) $40 = 32 + 180(0.045)^t \Rightarrow 8 = 180(0.045)^t \Rightarrow \dfrac{8}{180} = 0.045^t \Rightarrow \log\dfrac{8}{180} = \log 0.045^t \Rightarrow$

$\log\dfrac{8}{180} = t \log 0.045 \Rightarrow t = \dfrac{\log\frac{8}{180}}{\log 0.045} \Rightarrow t \approx 1$

Graph $Y_1 = 32 + 180(0.045)^t$ and $Y_2 = 40$. The lines intersect at $t \approx 1$. See Figure 83.

[0, 1.5, 0.5] by [0, 225, 25]

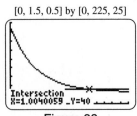

Figure 83

85. (a) $f(1.5) = 20 - 15(0.365)^{1.5} \approx 16.7°C$

(b) $15 = 20 - 15(0.365)^t \Rightarrow 15(0.365)^t = 5 \Rightarrow 0.365^t = \dfrac{1}{3} \Rightarrow \ln 0.365^t = \ln \dfrac{1}{3} \Rightarrow t\ln 0.365 = \ln \dfrac{1}{3} \Rightarrow$

$t = \dfrac{\ln \frac{1}{3}}{\ln 0.365} \approx 1.09$; the soda can warmed to 15°C after about 1.09 minutes or 1 hour 5.4 minutes.

87. $N(t) = 100,000e^{rt} \Rightarrow 200,000 = 100,000e^{r(2)} \Rightarrow 2 = e^{2r} \Rightarrow \ln 2 = \ln e^{2r} \Rightarrow \ln 2 = 2r \Rightarrow \dfrac{\ln 2}{2} = r \Rightarrow$

$r = 0.3466$. Now $350,000 = 100,000e^{0.3466t} \Rightarrow 3.5 = e^{0.3466t} \Rightarrow \ln 3.5 = \ln e^{0.3466t} \Rightarrow$

$\ln 3.5 = 0.3466t \Rightarrow t = \dfrac{\ln 3.5}{0.3466} \Rightarrow t \approx 3.6$ hrs.

89. $A(x) = C\left(\dfrac{1}{2}\right)^{x/k} \Rightarrow 0.04 = 0.05\left(\dfrac{1}{2}\right)^{20/k} \Rightarrow 0.8 = \left(\dfrac{1}{2}\right)^{20/k} \Rightarrow \ln 0.8 = \dfrac{20}{k}\ln \dfrac{1}{2} \Rightarrow \dfrac{\ln 0.8}{\ln \frac{1}{2}} = \dfrac{20}{k} \Rightarrow$

$0.3219 = \dfrac{20}{k} \Rightarrow .03219k = 20 \Rightarrow k = 62.13$. Now $0.025 = 0.05\left(\dfrac{1}{2}\right)^{t/62.13} \Rightarrow 0.5 = \left(\dfrac{1}{2}\right)^{t/62.13} \Rightarrow$

$\ln 0.5 = \dfrac{t}{62.13}\ln \dfrac{1}{2} \Rightarrow \dfrac{\ln 0.5}{\ln \frac{1}{2}} = \dfrac{t}{62.13} \Rightarrow 1 = \dfrac{t}{62.13} \Rightarrow t = 62.13$ or about 62 days.

91. $\dfrac{2 - \log(100 - x)}{0.42} = 2 \Rightarrow 2 - \log(100 - x) = 0.84 \Rightarrow \log(100 - x) = 1.16 \Rightarrow 10^{\log(100-x)} = 10^{1.16} \Rightarrow$

$100 - x = 10^{1.16} \Rightarrow x = 100 - 10^{1.16} \approx 85.546$; after 2 years approximately 85.5% of the robins have died.

93. $280\ln(x + 1) + 1925 = 2300 \Rightarrow 280\ln(x + 1) = 375 \Rightarrow \ln(x + 1) = \dfrac{375}{280} \Rightarrow e^{\ln(x+1)} = e^{375/280} \Rightarrow$

$x + 1 = e^{375/280} \Rightarrow x = e^{375/280} - 1 \approx 2.8$; a person who consumes 2300 calories daily owns about 2.8 acres.

95. When $x = 60$ the equation becomes $\ln(1 - P) = -0.0034 - 0.0053(60) \Rightarrow \ln(1 - P) = -0.3214$.

$\ln(1 - P) = -0.3214 \Rightarrow e^{\ln(1-P)} = e^{-0.3214} \Rightarrow 1 - P = e^{-0.3214} \Rightarrow P = 1 - e^{-0.3214} \approx 0.275$

According to this model, if a $60 tax was placed on each ton of carbon that was burned into the atmosphere, carbon dioxide emissions would be reduced by 27.5%.

97. $P\left(1 + \dfrac{r}{n}\right)^{nt} = A \Rightarrow \left(1 + \dfrac{r}{n}\right)^{nt} = \dfrac{A}{P} \Rightarrow \log\left(1 + \dfrac{r}{n}\right)^{nt} = \log\left(\dfrac{A}{P}\right) \Rightarrow$

$nt\log\left(1 + \dfrac{r}{n}\right) = \log\left(\dfrac{A}{P}\right) \Rightarrow t = \dfrac{\log(A/P)}{n\log(1 + r/n)}$

5.7: Constructing Nonlinear Models

Selecting a Model

1. Growth slows down as x increases; logarithmic..

3. Growth rate increases as x increases; exponential.

5. Exponential; least-squares regression gives $f(x) = 1.2(1.7)^x$.

7. Logarithmic; least-squares regression gives $f(x) = 1.088 + 2.937\ln x$.

Applications

9. (a) See Figure 9.

    (b) $f(x) = 1.568(1.109)^x$ deaths per 100,000.

    (c) $f(80) = 1.568(1.109)^{80} \approx 6164$; about 6164 deaths per 100,000.

[25, 75, 5] by [−100, 2100, 200]   [−2, 32, 5] by [0, 80, 10]

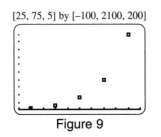   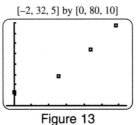

Figure 9   Figure 13

11. (a) $a \approx 1.4734$, $b \approx 0.99986$, or $f(x) = 1.4734(0.99986)^x$.

    (b) $f(7000) = 1.4734(0.99986)^{7000} \approx 0.55$; approximately 0.55 kg/m^3.

13. (a) The data is not linear. See Figure 13.

    (b) $a \approx 12.42$, $b \approx 1.066$, or $f(x) = 12.42(1.066)^x$.

    (c) $f(39) = 12.42(1.066)^{39} \approx 150$; 150 kilograms per hectare. Chemical fertilizer use increased, but at a

    slower rate than predicted by f.

15. (a) $f(x) = \dfrac{4.9955}{1 + 49.7081e^{-0.6998x}}$

    (b) For large x, $f(x) \approx \dfrac{4.9955}{1} \approx 5$; the density after a long time is about 5 thousand per acre.

17. (a) $f(25) = \dfrac{0.9}{1 + 271e^{-0.122 \cdot 25}} \approx 0.065$, $f(65) = \dfrac{0.9}{1 + 271e^{-0.122 \cdot 65}} \approx 0.82$; for people age 25, 6.5% have

    some CHD, whereas for people age 65, 82% have some CHD.

    (b) Solving the equation $\dfrac{0.9}{1 + 271e^{-0.122x}} = 0.5$ graphically, we find $x \approx 48$; at approximately 48 years the

    likelihood of coronary heart disease is 50%.

Chapter 5 Review Exercises

1. (a) $(f + g)(1) = f(1) + g(1) = 7 + 1 = 8$

    (b) $(f - g)(3) = f(3) - g(3) = 9 - 9 = 0$

    (c) $(fg)(-1) = f(-1)g(-1) = 3(-2) = -6$

    (d) $(f/g)(0) = \dfrac{f(0)}{g(0)} = \dfrac{5}{0}$; undefined.

3. (a) $f(x) = x^2 \Rightarrow f(3) = 9$ and $g(x) = 1 - x \Rightarrow g(3) = 1 - 3 = -2.$

Thus, $(f + g)(3) = f(3) + g(3) = 9 + (-2) = 7.$

(b) $f(-2) = 4$ and $g(-2) = 1 - (-2) = 3.$ Thus, $(f - g)(-2) = f(-2) - g(-2) = 4 - 3 = 1.$

(c) $f(1) = 1$ and $g(1) = 1 - 1 = 0.$ Thus, $(fg)(1) = f(1)g(1) = (1)(0) = 0.$

(d) $f(3) = 9$ and $g(3) = 1 - 3 = -2.$ Thus, $(f/g)(3) = \dfrac{f(3)}{g(3)} = \dfrac{9}{-2} = -\dfrac{9}{2}.$

5. (a) $(g \circ f)(-2) = g(f(-2)) = g(1) = 2$

(b) $(f \circ g)(3) = f(g(3)) = f(-2) = 1$

7. (a) $f(x) = \sqrt{x}$ and $g(x) = x^2 + x \Rightarrow (f \circ g)(2) = f(g(2)) = f(6) = \sqrt{6}$

(b) $(g \circ f)(9) = g(f(9)) = g(\sqrt{9}) = g(3) = 3^2 + 3 = 12$

9. $(f \circ g) = f(g(x)) = f\left(\dfrac{1}{x}\right) = \left(\dfrac{1}{x}\right)^3 - \left(\dfrac{1}{x}\right)^2 + 3\left(\dfrac{1}{x}\right) - 2;\ D = \{x\,|\,x \ne 0\}$

11. $(f \circ g) = f(g(x)) = f\left(\dfrac{1}{2}x^3 + \dfrac{1}{2}\right) = \sqrt[3]{2\left(\dfrac{1}{2}x^3 + \dfrac{1}{2}\right) - 1} = \sqrt[3]{x^3 + 1 - 1} = \sqrt[3]{x^3} = x;\ D = $ all real numbers

13. $h(x) = (g \circ f)(x) \Rightarrow h(x) = \sqrt{x^2 + 3} \Rightarrow f(x) = x^2 + 3,\ g(x) = \sqrt{x}.$ *Answers may vary.*

15. Subtract 6 from x and then multiply the results by 10. $\dfrac{x}{10} + 6$ and $10(x - 6).$

17. Since the graph of $f(x) = 3x - 1$ is a line sloping upward from left to right, a horizontal line can intersect it at most once. Therefore, f is one-to-one.

19. Since $f(-2) = 4$ and $f(4) = 4,$ two different inputs result in the same output. Therefore, f is not one-to-one and does not have an inverse.

21. f is not one-to-one. It does not pass the Horizontal line test..

23. (a) $f(1) \approx \$1050$

(b) $f^{-1}(1200) \approx 4$ years; f^{-1} computes the number of years it takes to accumulate x dollars.

25. The inverse operations are add 5 to x and divide by 3. Therefore, $f^{-1}(x) = \dfrac{x + 5}{3}.$

27. $y = \dfrac{3x}{x + 7} \Rightarrow y(x + 7) = 3x \Rightarrow xy + 7y = 3x \Rightarrow 7y = 3x - xy \Rightarrow 7y = x(3 - y) \Rightarrow x = \dfrac{7y}{3 - y};\ \Rightarrow$

$f^{-1}(x) = \dfrac{7x}{3 - x}$

29. $\{x\,|\,x \le 4\},\ y = 2(x - 4)^2 + 3 \Rightarrow \dfrac{y - 3}{2} = (x - 4)^2 \Rightarrow \sqrt{\dfrac{y - 3}{2}} = x - 4 \Rightarrow x = \sqrt{\dfrac{y - 3}{2}} + 4;\ \Rightarrow$

$f^{-1}(x) = \sqrt{\dfrac{x - 3}{2}} + 4;\ x \ge 3$

31. $(f \circ g^{-1})(4) = f(g^{-1}(4)) = f(3) = 1$

33. $e^x e^{-2x} = e^{x - 2x} = e^{-x}$

35. If $f(0) = 3 \Rightarrow Ca^x = 3 \Rightarrow Ca^0 = 3 \Rightarrow a^0 = 1;\ C(1) = 3 \Rightarrow C = 3;$ then $f(3) = 24 \Rightarrow Ca^x = 24 \Rightarrow$

$3a^3 = 24 \Rightarrow a^3 = 8 \Rightarrow a = 2.$ Therefore $C = 3$ and $a = 2.$

37. See Figure 37. D = all real numbers..

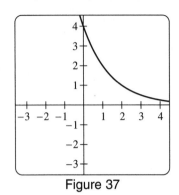

Figure 37

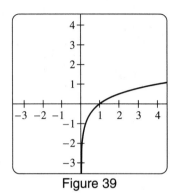

Figure 39

39. See Figure 39. $D = \{x \mid x > 0\}$.

41. $y = Ca^x \Rightarrow y = 2(2)^x \Rightarrow C = 2; \ a = 2$

43. $A = 1200\left(1 + \dfrac{0.09}{2}\right)^{2(3)} \approx \$1562.71$

45. $f(-1.2) = e^{-2(-1.2)} = e^{2.4} \approx 11.023$

47. $e^x = 19 \Rightarrow \ln e^x = \ln 19 \Rightarrow x = \ln 19 \approx 2.9444$; the result can be supported graphically using the
intersection method with the graphs $Y_1 = e{\wedge}X$ and $Y_2 = 19$. The intersection point is near (2.9444, 19). The
result may also be supported numerically with a table of $Y_1 = e{\wedge}X$ starting at 2.5 and incrementing by 0.1.
Here $Y_1 = 19$ when $x \approx 2.9$.

49. $\log 1000 = \log 10^3 = 3$

51. $10 \log 0.01 + \log \dfrac{1}{10} = 10 \log 10^{-2} + \log 10^{-1} = 10(-2) + (-1) = -21$

53. $\log_3 9 = \log_3 3^2 = 2$

55. $\ln e = \ln e^1 = 1$

57. $\log_3 18 = \dfrac{\log 18}{\log 3} \approx 2.631$

59. $10^x = 125 \Rightarrow \log 10^x = \log 125 \Rightarrow x = \log 125 \approx 2.097$

61. $e^{0.1x} = 5.2 \Rightarrow \ln e^{0.1x} = \ln 5.2 \Rightarrow 0.1x = \ln 5.2 \Rightarrow x = 10 \ln 5.2 \approx 16.49$

63. $5^{-x} = 10 \Rightarrow \log 5^{-x} = \log 10 \Rightarrow -x \log 5 = 1 \Rightarrow -x = \dfrac{1}{\log 5} \Rightarrow x = -\dfrac{1}{\log 5} \approx -1.431$

65. $50 - 3(0.78)^{x-10} = 21 \Rightarrow -3(0.78)^{x-10} = -29 \Rightarrow (0.78)^{x-10} = \dfrac{29}{3} \Rightarrow (x-10)\log(0.78) = \log\left(\dfrac{29}{3}\right) \Rightarrow$

$x - 10 = \dfrac{\log\left(\frac{29}{3}\right)}{\log(0.78)} \Rightarrow x = 10 + \dfrac{\log\left(\frac{29}{3}\right)}{\log(0.78)} \approx 0.869$

67. For each unit increase in x, the y-values are multiplied by 2, so the data are exponential. Since $y = 1.5$ when
$x = 0$, the initial value is 1.5, so $f(x) = Ca^x \Rightarrow f(x) = 1.5a^x$. Since $y = 3$ when $x = 1$, $f(1) = 1.5a^1 \Rightarrow$
$3 = 1.5a \Rightarrow a = 2$. Thus the function $f(x) = 1.5(2^x)$ can model the data.

69. $\log x = 1.5 \Rightarrow 10^{\log x} = 10^{1.5} \Rightarrow x = 10^{1.5} \approx 31.62$

71. $\ln x = 3.4 \Rightarrow e^{\ln x} = e^{3.4} \Rightarrow x = e^{3.4} \approx 29.96$

73. $2\log_4(x + 2) + 5 = 12 \Rightarrow 2\log_4(x + 2) = 7 \Rightarrow \log_4(x + 2) = 3.5 \Rightarrow 4^{\log_4(x+2)} = 4^{3.5} \Rightarrow$

$(x + 2) = 4^{3.5} \Rightarrow x = 4^{3.5} - 2 = 126$

75. $\log 6 + \log 5x = \log(6 \cdot 5x) = \log 30x$

77. $\ln\dfrac{4}{x^2} = \ln 4 - \ln x^2 = \ln 4 - 2\ln x$

79. $8\log x = 2 \Rightarrow \log x = \dfrac{1}{4} \Rightarrow 10^{\log x} = 10^{1/4} \Rightarrow x = 10^{1/4} \Rightarrow x = \sqrt[4]{10} \approx 1.778$

81. $2\log 3x + 5 = 15 \Rightarrow 2\log 3x = 10 \Rightarrow \log 3x = 5 \Rightarrow 10^{\log 3x} = 10^5 \Rightarrow 3x = 100{,}000 \Rightarrow$

$x = \dfrac{100{,}000}{3} \approx 33{,}333$

83. $2\log(x + 2) = \log(x + 8) \Rightarrow \log(x + 2)^2 = \log(x + 8) \Rightarrow 10^{\log(x+2)^2} = 10^{\log(x+8)} \Rightarrow$

$(x + 2)^2 = (x + 8) \Rightarrow x^2 + 4x + 4 = x + 8 \Rightarrow x^2 + 3x - 4 = 0 \Rightarrow (x + 4)(x - 1) = 0 \Rightarrow$

$x = -4$ or $x = 1$; since $x = -4$ is undefined in $\log(x + 2)$, the only solution is $x = 1$.

85. If b is the y-intercept, then the point $(0, b)$ is on the graph of f. Therefore, the point $(b, 0)$ is on the graph

of f^{-1}. The point $(b, 0)$ lies on the x-axis and is the x-intercept of the graph of f^{-1}.

Applications

87. (a) $N(t) = N_0 e^{rt} \Rightarrow 6000 = 4000 e^{r(1)} \Rightarrow 1.5 = e^r \Rightarrow \ln 1.5 = \ln e^r \Rightarrow \ln 1.5 = r \Rightarrow r = .04055$

Now $N(2.5) = 4000 e^{.04055(2.5)} \Rightarrow N(2.5) \approx 11{,}022$

(b) $8500 = 4000 e^{0.4055t} \Rightarrow 2.125 = e^{0.4055t} \Rightarrow \ln 2.125 - \ln e^{0.4055t} \Rightarrow \ln 2.125 = 0.4055t \Rightarrow$

$\dfrac{\ln 2.125}{0.4055} = t \Rightarrow t = 1.86$ hours.

89. (a) $T_0 = 20, D = 100 - 20 \Rightarrow D = 80$. Solving for a use $50 = 20 + 80a^{2/3} \Rightarrow 30 = 80a^{2/3} \Rightarrow$

$\dfrac{3}{8} = a^{2/3} \Rightarrow a = \left(\dfrac{3}{8}\right)^{3/2} \Rightarrow a = 0.23$

(b) $T\left(\dfrac{3}{2}\right) = 20 + 80(0.23)^{3/2} \Rightarrow T\left(\dfrac{3}{2}\right) \approx 28.8°\,C$

(c) $30 = 20 + 80(0.23)^t \Rightarrow 10 = 80(0.23)^t \Rightarrow \dfrac{1}{8} = (0.23)^t \Rightarrow \ln\dfrac{1}{8} = t\ln 0.23 \Rightarrow t = \dfrac{\ln\frac{1}{8}}{\ln 0.23} \Rightarrow$

$t = 1.4$ hours.

91. (a) $(g \circ f)(32) = g(f(32)) = g(2) = 1$. $(g \circ f)(32)$ computes the number quarts in 32 fluid ounces, which is

1 quart.

(b) Since $f(16) = 1, f^{-1}(1) = 16$. f^{-1} converts pints into fluid ounces. There are 16 fluid ounces in 1 pint.

(c) $(f^{-1} \circ g^{-1})(1) = f^{-1}(g^{-1}(1)) = f^{-1}(2) = 32$. $(f^{-1} \circ g^{-1})(1)$ computes the number of fluid ounces in 1 quart.

There are 32 ounces in 1 quart.

93. (a) $W(1) = 175.6(1 - 0.66e^{-0.24(1)})^3 \approx 19.5$

(b) Use the intersection method by graphing $Y_1 = 175.6(1 - 0.66e\,^{\wedge}(-0.24X))^{\wedge}3$ and $Y_2 = 50$ in the viewing rectangle [0, 14, 1] by [0, 175, 25]. The intersection is near the point (2.74, 50). See Figure 93. This means that after approximately 3 weeks the fish weighed 50 milligrams.

(c) $175.6(1 - 0.66e^{-0.24x})^3 = 50 \Rightarrow (1 - 0.66e^{-0.24x})^3 = \dfrac{50}{175.6} \Rightarrow 1 - 0.66e^{-0.24x} = \left(\dfrac{50}{175.6}\right)^{1/3} \Rightarrow$

$0.66e^{-0.24x} = 1 - \left(\dfrac{50}{175.6}\right)^{1/3} \Rightarrow e^{-0.24x} = \dfrac{1 - \left(\frac{50}{175.6}\right)^{1/3}}{0.66} \Rightarrow \ln e^{-0.24x} = \ln\left(\dfrac{1 - \left(\frac{50}{175.6}\right)^{1/3}}{0.66}\right) \Rightarrow$

$-0.24x = \ln\left(\dfrac{1 - \left(\frac{50}{175.6}\right)^{1/3}}{0.66}\right) \Rightarrow x = -\dfrac{1}{0.24}\ln\left(\dfrac{1 - \left(\frac{50}{175.6}\right)^{1/3}}{0.66}\right) \approx 2.74$

[0, 14, 1] by [0, 175, 25]

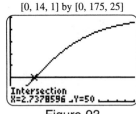

Intersection
X=2.7378596 Y=50

Figure 93

95. $15 = 32e^{-0.2t} \Rightarrow \dfrac{15}{32} = e^{-0.2t} \Rightarrow -0.2t = \ln\left(\dfrac{15}{32}\right) \Rightarrow t = \left(-\dfrac{1}{0.2}\right)\ln\left(\dfrac{15}{32}\right) \approx (-5)(-0.7577) \approx 3.8$; after

about 3.8 minutes.

97. (a)

x	0	15	30	45	60	75	90	125
$g(x)$	7	21	57	111	136	158	164	178

(b) $g(x) = 261 - f(x)$

(c) y_1 models $g(x)$ better. See Figure 97.

(d) $f(x) = 261 - y_1(x) = 261 - \dfrac{171}{1 + 18.6e^{-0.0747x}}$

[−10, 140, 10] by [−20, 200, 10]   0, 5, 1] by [0, 3, 1]

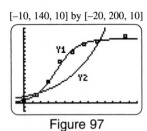

Figure 97   Figure 99

99. See Figure 99. $a \approx 3.50, b \approx 0.74$, or $f(x) = 3.50(0.74)^x$

Chapter 6: Trigonometric Functions

6.1: Angles and Their Measure

Angles

1. (a) See Figure 1a.

   (b) See Figure 1b.

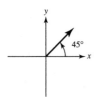

Figure 1a

Figure 1b

Figure 1c

Figure 1d

   (c) See Figure 1c.

   (d) See Figure 1d.

3. Any angle with degree measure between 0° and 90°. See Figure 3. *Answers may vary.*

5. See Figure 5.

Figure 3

Figure 5

Figure 7

Figure 9

7. Any angle with degree measure between 180° and 270°. See Figure 7. *Answers may vary.*

9. Any angle with degree measure of the form $90° - n(360°)$ where $n = 1, 2, 3, \ldots$. See Figure 9. *Answers may vary.*

11. (a) $\dfrac{90°}{360°} = \dfrac{1}{4}$

    (b) $\dfrac{30°}{360°} = \dfrac{1}{12}$

    (c) $\dfrac{\frac{\pi}{3}}{2\pi} = \dfrac{1}{6}$

    (d) $\dfrac{\frac{\pi}{4}}{2\pi} = \dfrac{1}{8}$

13. $150° + 360° = 510°$, $150° - 360° = -210°$ *Answers may vary.*

15. $-72° + 360° = 288°$, $-72° - 360° = -432°$ *Answers may vary.*

17. $\dfrac{\pi}{2} + 2\pi = \dfrac{\pi}{2} + \dfrac{4\pi}{2} = \dfrac{5\pi}{2}$, $\dfrac{\pi}{2} - 2\pi = \dfrac{\pi}{2} - \dfrac{4\pi}{2} = -\dfrac{3\pi}{2}$ *Answers may vary.*

19. $-\dfrac{\pi}{5} + 2\pi = -\dfrac{\pi}{5} + \dfrac{10\pi}{5} = \dfrac{9\pi}{5}$, $-\dfrac{\pi}{5} - 2\pi = -\dfrac{\pi}{5} - \dfrac{10\pi}{5} = -\dfrac{11\pi}{5}$ *Answers may vary.*

21. $\alpha = 90° - 55.9° = 34.1°, \beta = 180° - 55.9° = 124.1°$

23. $\alpha = 90° - 85°23'45'' = 89°59'60'' - 85°23'45'' = 4°36'15''$

$\beta = 180° - 85°23'45'' = 179°59'60'' - 85°23'45'' = 94°36'15''$

25. $\alpha = 90° - 23°40'35'' = 89°59'60'' - 23°40'35'' = 66°19'25''$

$\beta = 180° - 23°40'35'' = 179°59'60'' - 23°40'35'' = 156°19'25''$

27. $125°15' = 125° + \left(\dfrac{15}{60}\right)^° = 125.25°$

29. $108°45'36'' = 108° + \left(\dfrac{45}{60}\right)^° + \left(\dfrac{36}{3600}\right)^° = 108.76°$

31. $125.3° = 125° + \left(\dfrac{3}{10}\right)^° = 125° + \left(\dfrac{18}{60}\right)^° = 125°18'$

33. $51.36° = 51° + \left(\dfrac{36}{100}\right)^° = 51° + \left(\dfrac{35}{100}\right)^° + \left(\dfrac{1}{100}\right)^° = 51° + \left(\dfrac{21}{60}\right)^° + \left(\dfrac{36}{3600}\right)^° = 51°21'36''$

35. $\theta = \dfrac{4}{2} = 2$ radians. The degree measure of θ is $2\left(\dfrac{180°}{\pi}\right) = \dfrac{360°}{\pi} \approx 114.6°.$

37. $\theta = \dfrac{13}{10} = 1.3$ radians. The degree measure of θ is $\dfrac{13}{10}\left(\dfrac{180°}{\pi}\right) = \dfrac{234°}{\pi} \approx 74.5°.$

39. (a) $45°\left(\dfrac{\pi}{180°}\right) = \dfrac{45\pi}{180} = \dfrac{\pi}{4}$

(b) $135°\left(\dfrac{\pi}{180°}\right) = \dfrac{135\pi}{180} = \dfrac{3\pi}{4}$

(c) $-120°\left(\dfrac{\pi}{180°}\right) = -\dfrac{120\pi}{180} = -\dfrac{2\pi}{3}$

(d) $-210°\left(\dfrac{\pi}{180°}\right) = -\dfrac{210\pi}{180} = -\dfrac{7\pi}{6}$

41. (a) $37°\left(\dfrac{\pi}{180°}\right) = \dfrac{37\pi}{180}$

(b) $123.4\left(\dfrac{\pi}{180°}\right) = \dfrac{123.4\pi}{180} \approx 2.15$

(c) $-92°25' = -\left[92 + \left(\dfrac{25}{60}\right)^°\right] \approx -92.42\left(\dfrac{\pi}{180°}\right) = -\dfrac{92.42\pi}{180} \approx -1.61$

(d) $230°17' = 230 + \left(\dfrac{17}{60}\right)^° \approx 230.28\left(\dfrac{\pi}{180°}\right) = \dfrac{230.28\pi}{180} \approx 4.02$

43. (a) $\dfrac{\pi}{6}\left(\dfrac{180°}{\pi}\right) = \dfrac{180°}{6} = 30°$

(b) $\dfrac{\pi}{15}\left(\dfrac{180°}{\pi}\right) = \dfrac{180°}{15} = 12°$

(c) $-\dfrac{5\pi}{3}\left(\dfrac{180°}{\pi}\right) = -\dfrac{900°}{3} = -300°$

(d) $-\dfrac{7\pi}{6}\left(\dfrac{180°}{\pi}\right) = -\dfrac{1260°}{6} = -210°$

45. (a) $\dfrac{\pi}{4}\left(\dfrac{180°}{\pi}\right) = \dfrac{180°}{4} = 45°$

(b) $\dfrac{\pi}{7}\left(\dfrac{180°}{\pi}\right) = \dfrac{180°}{7} \approx 25.71°$

(c) $3.1\left(\dfrac{180°}{\pi}\right) = \dfrac{558°}{\pi} \approx 177.62°$

(d) $-\dfrac{5}{2}\left(\dfrac{180°}{\pi}\right) = -\dfrac{900°}{2\pi} \approx -143.24°$

Arc Length

47. Convert 60° to radians. $60°\left(\dfrac{\pi}{180°}\right) = \dfrac{60\pi}{180} = \dfrac{\pi}{3} \Rightarrow s = 2\left(\dfrac{\pi}{3}\right) = \dfrac{2\pi}{3}$ in.

49. $12 = 5\theta \Rightarrow \theta = \dfrac{12}{5}$ radians

51. $5 = r\pi \Rightarrow r = \dfrac{5}{\pi}$ ft

53. $s = 3\left(\dfrac{\pi}{12}\right) = \dfrac{3\pi}{12} = \dfrac{\pi}{4}$ m

55. Convert 15° to radians. $15°\left(\dfrac{\pi}{180°}\right) = \dfrac{15\pi}{180} = \dfrac{\pi}{12} \Rightarrow s = 12\left(\dfrac{\pi}{12}\right) = \dfrac{12\pi}{12} = \pi$ ft

57. Note that $1°45' = 1° + \left(\dfrac{45}{60}\right)° = 1.75°$.

Convert 1.75° to radians. $1.75°\left(\dfrac{\pi}{180°}\right) = \dfrac{1.75\pi}{180} = \dfrac{7\pi}{720} \Rightarrow s = 2\left(\dfrac{7\pi}{720}\right) = \dfrac{14\pi}{720} = \dfrac{7\pi}{360}$ mi

59. A total of 15 minutes passes between 10:15 A.M. and 10:30 A.M. The minute hand has moved through the

angle of $\dfrac{15}{60} \cdot 2\pi = \dfrac{30\pi}{60} = \dfrac{\pi}{2}$ radians. Therefore, the tip of the minute hand travels

$s = 4\left(\dfrac{\pi}{2}\right) = \dfrac{4\pi}{2} = 2\pi$ inches. The linear speed of the tip of the minute hand is $\dfrac{2\pi}{15}$ in./min.

61. A total of 75 minutes passes between 3:00 P.M. and 4:15 P.M. The minute hand has moved through the

angle of $\dfrac{75}{60} \cdot 2\pi = \dfrac{150\pi}{60} = \dfrac{5\pi}{2}$ radians. Therefore, the tip of the minute hand travels

$s = 4\left(\dfrac{5\pi}{2}\right) = 10\pi$ inches. The linear speed of the tip of the minute hand is $\dfrac{10\pi}{75} = \dfrac{2\pi}{15}$ in./min.

63. The angular velocity is $\omega = 15$ radians/sec. The radius is $r = \dfrac{26}{2} = 13$ inches.

The linear speed is given by $v = r\omega = 13(15) = 195$ in/sec. That is $\dfrac{195 \text{ in}}{\text{sec}} \cdot \dfrac{1 \text{ ft}}{12 \text{ in}} = 16.25$ ft/sec.

And $\dfrac{16.25 \text{ ft}}{\text{sec}} \cdot \dfrac{1 \text{ mi}}{5280 \text{ ft}} \cdot \dfrac{3600 \text{ sec}}{1 \text{ hr}} \approx 11.1$ mi/hr.

Area of a Sector

65. $A = \dfrac{1}{2} \cdot 3^2 \cdot \left(\dfrac{\pi}{3}\right) = \dfrac{3\pi}{2} = 1.5\pi$ in.^{2}

67. Convert 45° to radians. $45°\left(\dfrac{\pi}{180°}\right) = \dfrac{\pi}{4}$. Then, $A = \dfrac{1}{2} \cdot 6^2 \cdot \left(\dfrac{\pi}{4}\right) = \dfrac{9\pi}{2} = 4.5\pi$ in.^{2}

69. $A = \dfrac{1}{2} \cdot 13.1^2 \cdot \left(\dfrac{\pi}{15}\right) = \dfrac{17{,}161\pi}{3000} \approx 5.72\pi$ cm^2

71. Convert 30° to radians. $30°\left(\dfrac{\pi}{180°}\right) = \dfrac{\pi}{6}$. Then, $A = \dfrac{1}{2} \cdot 1.5^2 \cdot \left(\dfrac{\pi}{6}\right) = \dfrac{3\pi}{16} \approx 0.59$ ft^2

73. When rotating from $-45°$ to 90° the arm swings through an angle of 135°. Convert 135° to radians:

$135°\left(\dfrac{\pi}{180°}\right) = \dfrac{3\pi}{4}$. To find the area of the work space, we subtract the area of the inner circular sector from

the area of the outer circular sector. $A = \dfrac{1}{2}\left(\dfrac{3\pi}{4}\right)(26^2 - 6^2) = \dfrac{3\pi}{8}(640) = 240\pi$ in.^{2}

75. When rotating from 15° to 195° the arm swings through an angle of 180°. Convert 180° to radians:

$180°\left(\dfrac{\pi}{180°}\right) = \pi$. To find the area of the work space, we subtract the area of the inner circular sector from

the area of the outer circular sector. $A = \dfrac{1}{2}(\pi)(95^2 - 21^2) = \dfrac{\pi}{2}(8584) = 4292\pi$ cm^2

Applications

77. Since there are 3600 seconds in one degree, it follows that $\pm \dfrac{1}{100{,}000}{}^{\circ} \times 3600 = \pm \dfrac{36}{1000} = \pm 0.036''$

79. (a) The angular velocity is $\omega = 500(2\pi) = 1000\pi \approx 3141.6$ radians/min.

(b) The linear speed is $v = r\omega$; $r = \dfrac{30}{2} = 15$ in. and $\omega = 1000\pi$

therefore $v = 15(1000\pi) \approx 47{,}123.89$ in./min or about 65.5 ft/sec.

81. Convert $40°55' - 29°11' = 11°44'$ to radians: $\left[11° + \left(\dfrac{44}{60}\right)^{\circ}\right]\left(\dfrac{\pi}{180°}\right) \approx 0.2048$.

Then, $s = 3955(0.2048) \approx 809.93 \approx 810$ mi.

83. Convert 75.3° to radians: $75.3°\left(\dfrac{\pi}{180°}\right) \approx 1.314$. Then, $s = 11(1.314) \approx 14.5$ in.

85. (a) In one revolution of the pedals, the chain will travel $s = 3.75(2\pi) = 7.5\pi$ inches. The circumference of the

small gear is $C = 1.5(2\pi) = 3\pi$ inches. When the chain moves 7.5π inches, the smaller gear will rotate

$\dfrac{7.5\pi}{3\pi} = \dfrac{5}{2} = 2.5$ times. Thus, the bicycle tire will rotate 2.5 times for each revolution of the pedals.

(b) When the pedals turn through 2 revolutions per second, the angular velocity of the tire is

$\omega = 2.5(2)(2\pi) = 10\pi$ radians/sec. Thus, the tire of radius 13 inches (diameter = 26 inches) will travel

$v = 13(10\pi) = 130\pi$ in./sec. That is $\dfrac{130\pi \text{ in.}}{\text{sec}} \cdot \dfrac{1 \text{ ft}}{12 \text{ in.}} \approx 34$ ft/sec.

87. (a) $\omega = \dfrac{2\pi}{0.615}$ radians/yr $\Rightarrow \dfrac{\frac{2\pi}{0.615}\,\text{rad}}{1\,\text{yr}} \cdot \dfrac{1\,\text{yr}}{365\,\text{days}} \cdot \dfrac{1\,\text{day}}{24\,\text{hr}} \approx 1.1662741 \times 10^{-3}$ radians/hr

Thus, the orbital velocity of Venus is $v = 67.2 \times 10^{6}(1.1662741 \times 10^{-3}) \approx 78{,}370$ mi/hr.

(b) $\omega = \dfrac{2\pi}{1} = 2\pi$ radians/yr $\Rightarrow \dfrac{2\pi\,\text{rad}}{1\,\text{yr}} \cdot \dfrac{1\,\text{yr}}{365\,\text{days}} \cdot \dfrac{1\,\text{day}}{24\,\text{hr}} \approx 7.172585967 \times 10^{-4}$ radians/hr

Thus, the orbital velocity of Earth is $v = 92.9 \times 10^{6}(7.172585967 \times 10^{-4}) \approx 66{,}630$ mi/hr.

(c) $\omega = \dfrac{2\pi}{11.86}$ radians/yr $\Rightarrow \dfrac{\frac{2\pi}{11.86}\,\text{rad}}{1\,\text{yr}} \cdot \dfrac{1\,\text{yr}}{365\,\text{days}} \cdot \dfrac{1\,\text{day}}{24\,\text{hr}} \approx 6.047711608 \times 10^{-5}$ radians/hr

Thus, the orbital velocity of Jupiter is $v = 483.6 \times 10^{6}(6.047711608 \times 10^{-5}) \approx 29{,}250$ mi/hr.

(d) $\omega = \dfrac{2\pi}{164.8}$ radians/yr $\Rightarrow \dfrac{\frac{2\pi}{164.8}\,\text{rad}}{1\,\text{yr}} \cdot \dfrac{1\,\text{yr}}{365\,\text{days}} \cdot \dfrac{1\,\text{day}}{24\,\text{hr}} \approx 4.35229731 \times 10^{-6}$ radians/hr

Thus, the orbital velocity of Neptune is $v = 2794 \times 10^{6}(4.35229731 \times 10^{-6}) \approx 12{,}160$ mi/hr.

89. Since the bar is 2 meters in length, we may use 2 meters as an estimate of the arc length from P to Q.

Convert $0.835°$ to radians: $0.835°\left(\dfrac{\pi}{180°}\right) = \dfrac{167\pi}{36{,}000}$. Then, $2 = r\left(\dfrac{167\pi}{36{,}000}\right) \Rightarrow r = 2\left(\dfrac{36{,}000}{167\pi}\right) \approx 137.2$ m.

A reasonable approximation for d is 137.2 meters.

91. Find the length of an arc on the surface of the earth intersected by a $0.001°$ angle.

Convert $0.001°$ to radians: $0.001°\left(\dfrac{\pi}{180°}\right) = \dfrac{\pi}{180{,}000}$. Then, $s = 3955\left(\dfrac{\pi}{180{,}000}\right) \approx 0.069$ mi.

Thus, a civilian can locate the north-south position of an object to within ± 0.069 miles, or about 364 feet.

93. To convert degrees to radians we use $\theta°\left(\dfrac{\pi}{180°}\right)$. So the modified formula is $s = r\left[\theta\left(\dfrac{\pi}{180°}\right)\right] = r\theta\left(\dfrac{\pi}{180°}\right)$.

The radian measure formula is simpler.

95. (a) $475{,}000 = r^2\pi \Rightarrow r = \sqrt{\dfrac{475{,}000}{\pi}} \approx 388.8$ m

(b) Convert $70°$ to radians: $70°\left(\dfrac{\pi}{180°}\right) = \dfrac{7\pi}{18}$. So $475{,}000 = \dfrac{1}{2}\left(\dfrac{7\pi}{18}\right)r^2 \Rightarrow r = \sqrt{2(475{,}000)\left(\dfrac{18}{7\pi}\right)} \approx 881.8$ m

97. Convert $7°12'$ to radians: $\left(7 + \dfrac{12}{60}\right)°\left(\dfrac{\pi}{180°}\right) = 7.2°\left(\dfrac{\pi}{180°}\right) = \dfrac{\pi}{25}$. Then,

$496 = r\left(\dfrac{\pi}{25}\right) \Rightarrow r = 496\left(\dfrac{25}{\pi}\right) \approx 3947$ mi. The circumference of the earth is $C \approx 2\pi(3947) \approx 24{,}800$ mi.

6.2: Right Triangle Trigonometry

Basic Concepts

1. See Figure 1.

3. See Figure 3.

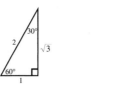

Figure 1

Figure 3

5. $\sin 60° = \dfrac{\text{side opposite}}{\text{hypotenuse}} = \dfrac{\sqrt{3}}{2}$

7. $\cos 30° = \dfrac{\text{side adjacent}}{\text{hypotenuse}} = \dfrac{\sqrt{3}}{2}$

9. $\sec 60° = \dfrac{\text{hypotenuse}}{\text{side adjacent}} = \dfrac{2}{1} = 2$

11. $\tan 45° = \dfrac{\text{side opposite}}{\text{side adjacent}} = \dfrac{1}{1} = 1$

13. $\cot 45° = \dfrac{\text{side adjacent}}{\text{side opposite}} = \dfrac{1}{1} = 1$

15. $\sin 45° = \dfrac{\text{side opposite}}{\text{hypotenuse}} = \dfrac{1}{\sqrt{2}}$

17. Use the Pythagorean theorem to calculate the length of the hypotenuse:

$$h^2 = 3^2 + 4^2 \Rightarrow h^2 = 9 + 16 \Rightarrow h^2 = 25 \Rightarrow h = 5$$

$\sin \theta = \dfrac{4}{5} \qquad \cos \theta = \dfrac{3}{5} \qquad \tan \theta = \dfrac{4}{3} \qquad \csc \theta = \dfrac{5}{4} \qquad \sec \theta = \dfrac{5}{3} \qquad \cot \theta = \dfrac{3}{4}$

19. Use the Pythagorean theorem to calculate the length of the missing leg:

$$13^2 = 12^2 + l^2 \Rightarrow l^2 = 169 - 144 \Rightarrow l^2 = 25 \Rightarrow l = 5$$

$\sin \theta = \dfrac{12}{13} \qquad \cos \theta = \dfrac{5}{13} \qquad \tan \theta = \dfrac{12}{5} \qquad \csc \theta = \dfrac{13}{12} \qquad \sec \theta = \dfrac{13}{5} \qquad \cot \theta = \dfrac{5}{12}$

21. Triangle ABC is a $30° \sim 60°$ right triangle, thus $b = 8$. Find a using the Pythagorean theorem:

$$8^2 + a^2 = 16^2 \Rightarrow a^2 = 256 - 64 \Rightarrow a^2 = 192 \Rightarrow a = 8\sqrt{3} \approx 13.86.$$

23. To find b note that $\tan 50° = \dfrac{6}{b} \Rightarrow b = \dfrac{6}{\tan 50°} \Rightarrow b \approx 5.03.$

To find c note that $\sin 50° = \dfrac{6}{c} \Rightarrow c = \dfrac{6}{\sin 50°} \Rightarrow c \approx 7.83.$

25. To find a note that $\sin 41° = \dfrac{a}{8} \Rightarrow a = 8\sin 41° \Rightarrow a \approx 5.25.$

To find b note that $\cos 41° = \dfrac{b}{8} \Rightarrow b = 8\cos 41° \Rightarrow b \approx 6.04.$

27. To find a note that $\tan 46°15' = \dfrac{a}{16.1} \Rightarrow a = 16.1\tan 46°15' \Rightarrow a \approx 16.82.$

To find c note that $\cos 46°15' = \dfrac{16.1}{c} \Rightarrow c = \dfrac{16.1}{\cos 46°15'} \Rightarrow c \approx 23.28.$

29. To find a note that $\tan 60° = \dfrac{a}{12} \Rightarrow a = 12\tan 60° \Rightarrow a \approx 20.78.$

31. To find c note that $\cos 53°43' = \dfrac{100}{c} \Rightarrow c = \dfrac{100}{\cos 53°43'} \Rightarrow c \approx 168.98.$

33. $\sec \theta = \dfrac{1}{\cos \theta} = \dfrac{1}{\frac{1}{3}} = 3$

35. $\csc \theta = \dfrac{1}{\sin \theta} = \dfrac{1}{\frac{12}{13}} = \dfrac{13}{12}$

37. $\tan \theta = \dfrac{1}{\cot \theta} = \dfrac{1}{\frac{7}{24}} = \dfrac{24}{7}$

39. $\sin 60° \approx 0.866$ $\cos 60° \approx 0.5$ $\tan 60° \approx 1.732$

     $\csc 60° \approx 1.155$ $\sec 60° \approx 2$ $\cot 60° \approx 0.577$

41. $\sin 25° \approx 0.423$ $\cos 25° \approx 0.906$ $\tan 25° \approx 0.466$

     $\csc 25° \approx 2.366$ $\sec 25° \approx 1.103$ $\cot 25° \approx 2.145$

43. $\sin 5°35' \approx 0.097$ $\cos 5°35' \approx 0.995$ $\tan 5°35' \approx 0.098$

     $\csc 5°35' \approx 10.278$ $\sec 5°35' \approx 1.005$ $\cot 5°35' \approx 10.229$

45. $\sin 13°45'30'' \approx 0.238$ $\cos 13°45'30'' \approx 0.971$ $\tan 13°45'30'' \approx 0.245$

     $\csc 13°45'30'' \approx 4.205$ $\sec 13°45'30'' \approx 1.030$ $\cot 13°45'30'' \approx 4.084$

47. $\sin 1.05° \approx 0.018$ $\cos 1.05° \approx 1.000$ $\tan 1.05° \approx 0.018$

     $\csc 1.05° \approx 54.570$ $\sec 1.05° \approx 1.000$ $\cot 1.05° \approx 54.561$

49. To find a note that $\cos 60° = \dfrac{a}{24} \Rightarrow a = 24 \cos 60° \Rightarrow a = 24\left(\dfrac{1}{2}\right) = 12$

     To find b note that $\sin 60° = \dfrac{b}{24} \Rightarrow b = 24 \sin 60° \Rightarrow b = 24\left(\dfrac{\sqrt{3}}{2}\right) = 12\sqrt{3}$

     To find c note that $\cos 45° = \dfrac{24 \sin 60°}{c} \Rightarrow c = \dfrac{24 \sin 60°}{\cos 45°} \Rightarrow c = 24\sqrt{2} \sin 60° \Rightarrow c = 24\sqrt{2}\left(\dfrac{\sqrt{3}}{2}\right) \Rightarrow$

     $c = 12\sqrt{6}$

     To find d note that $\sin 45° = \dfrac{d}{24\sqrt{2} \sin 60°} \Rightarrow d = \sin 45° (24\sqrt{2} \sin 60°) \Rightarrow d = \dfrac{1}{\sqrt{2}}(24\sqrt{2} \sin 60°) \Rightarrow$

     $d = 24 \sin 60° \Rightarrow d = 24\left(\dfrac{\sqrt{3}}{2}\right) \Rightarrow d = 12\sqrt{3}$

51. To find a note that $\sin 60° = \dfrac{7}{a} \Rightarrow a = \dfrac{7}{\sin 60°} \Rightarrow a = \dfrac{7}{\frac{\sqrt{3}}{2}} \Rightarrow a = \dfrac{14\sqrt{3}}{3}$

     To find b note that $\tan 60° = \dfrac{7}{b} \Rightarrow b = \dfrac{7}{\tan 60°} \Rightarrow b = \dfrac{7}{\sqrt{3}} \Rightarrow b = \dfrac{7\sqrt{3}}{3}$

     To find c note that $\tan 45° = \left(\dfrac{7}{\sin 60°} \div c\right) \Rightarrow \tan 45° = \dfrac{7}{c \sin 60°} \Rightarrow (c \sin 60°)(\tan 45°) = 7,$

     $\tan 45° = 1 \Rightarrow (c \sin 60°)(1) = 7 \Rightarrow c = \dfrac{7}{\sin 60°} \Rightarrow c = \dfrac{7}{\frac{\sqrt{3}}{2}} \Rightarrow c = \dfrac{14\sqrt{3}}{3}$

     To find d note that $\sin 45° = \left(\dfrac{7}{\sin 60°} \div d\right) \Rightarrow \sin 45° = \dfrac{7}{d \sin 60°} \Rightarrow (d \sin 60°)(\sin 45°) = 7 \Rightarrow$

     $d = \dfrac{7}{(\sin 60°)(\sin 45°)} \Rightarrow d = \dfrac{7}{(\frac{\sqrt{3}}{2})(\frac{\sqrt{2}}{2})} \Rightarrow d = \dfrac{14\sqrt{6}}{3}$

53. (a) $\sin 70° = \cos (90° - 70°) = \cos 20° \approx 0.9397$

     (b) $\cos 40° = \sin (90° - 40°) = \sin 50° \approx 0.7660$

55. (a) $\csc 49° = \sec (90° - 49°) = \sec 41° \approx 1.3250$

     (b) $\sec 63° = \csc (90° - 63°) = \csc 27° \approx 2.2027$

Applications

57. $\tan 37°30' = \dfrac{h}{1500} \Rightarrow h = 1500 \tan 37°30' \Rightarrow h \approx 1151$ ft

59. $\tan 35° = \dfrac{h}{100} \Rightarrow h = 100 \tan 35° \Rightarrow h \approx 70$ ft

61. $\tan 34° = \dfrac{5 \text{ ft } 3 \text{ in.}}{a}$ or $\tan 34° = \dfrac{63 \text{ in.}}{a} \Rightarrow a = \dfrac{63 \text{ in.}}{\tan 34°} \Rightarrow a \approx 93.4$ in. or about 7.8 ft.

63. From the example we know that the distance d to a star with parallax θ is given by $d = \dfrac{93,000,000}{\sin \theta}$.

    For Barnard's star: $d = \dfrac{93,000,000}{\sin 0.000152} \approx 3.5 \times 10^{13}$ mi or $\dfrac{3.5 \times 10^{13}}{5.9 \times 10^{12}} \approx 5.9$ light-years.

    For Sirus: $d = \dfrac{93,000,000}{\sin 0.000105} \approx 5.1 \times 10^{13}$ mi or $\dfrac{5.1 \times 10^{13}}{5.9 \times 10^{12}} \approx 8.6$ light-years.

    For 61 Cygni: $d = \dfrac{93,000,000}{\sin 0.0000811} \approx 6.6 \times 10^{13}$ mi or $\dfrac{6.6 \times 10^{13}}{5.9 \times 10^{12}} \approx 11.1$ light-years.

    For Procyon: $d = \dfrac{93,000,000}{\sin 0.0000797} \approx 6.7 \times 10^{13}$ mi or $\dfrac{6.7 \times 10^{13}}{5.9 \times 10^{12}} \approx 11.3$ light-years.

65. Using the Earth - Sun distance of 93,000,000 miles, the minimum distance d is given by:

    $\sin 18° = \dfrac{d}{93,000,000} \Rightarrow d = 93,000,000 \sin 18° \Rightarrow d \approx 2.9 \times 10^7$ mi.

    The maximum distance is given by: $\sin 28° = \dfrac{d}{93,000,000} \Rightarrow d = 93,000,000 \sin 28° \Rightarrow d \approx 4.4 \times 10^7$ mi.

67. $d = r\left(\dfrac{1}{\cos \theta} - 1\right) = 3963\left(\dfrac{1}{\cos 76.1°} - 1\right) \approx 12,533.8 \approx 12,534$ mi.

69. $\dfrac{21,000}{2.8} = \dfrac{x}{1.8} \Rightarrow 2.8x = 37,800 \Rightarrow x = 13,500$ ft

71. From the diagram in the text we see that: $\tan 23°45'30'' = \dfrac{PR}{PQ} \Rightarrow PQ = \dfrac{85.62}{\tan 23°45'30''} \Rightarrow PQ = 194.5$ ft.

73. (a) $\tan\left(\dfrac{\theta}{2}\right) = \dfrac{1}{d} \Rightarrow d = \dfrac{1}{\tan\left(\frac{\theta}{2}\right)} = \cot\left(\dfrac{\theta}{2}\right)$

    (b) $\theta = 1°45'15'' \Rightarrow d = \cot\left(\dfrac{1°45'15''}{2}\right) \approx 65.32$ m

75. (a) $r = \dfrac{66^2}{4.5 + 32.2 \tan 3°} \approx 704$ ft

    (b) $r = \dfrac{66^2}{4.5 + 32.2 \tan 5°} \approx 595$ ft

    (c) Larger values of θ correspond to a smaller safe radius. A table of $Y_1 = 66\wedge 2/(4.5 + 32.2 \tan(X))$ starting

    at $x = 0$ and incrementing by 1 is shown in Figure 75.

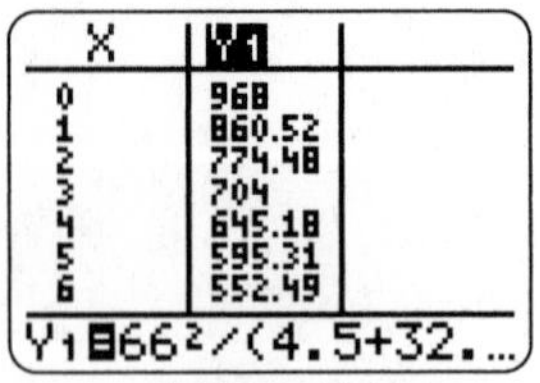

X	Y1
0	968
1	860.52
2	774.48
3	704
4	645.18
5	595.31
6	552.49

$Y_1 = 66^2/(4.5+32...$

Figure 75

77. From the example $d = r\left(\dfrac{1}{\cos \theta} - 1\right)$. Thus, $r = 625$ and $\theta = 54° \Rightarrow d = 625\left(\dfrac{1}{\cos 54°} - 1\right) \approx 438$ ft.

79. The area of a triangle is $\dfrac{b \cdot h}{2}$ and the height is unknown. Height is found by using the following:

$$\cos 30° = \frac{h}{s} \Rightarrow s \cos 30° = h, \cos 30° = \frac{\sqrt{3}}{2} \Rightarrow s \cdot \frac{\sqrt{3}}{2} = \text{ height or } h = \frac{s\sqrt{3}}{2}.$$

Since area equals $\dfrac{b \cdot h}{2}$ and $b = s$, then area $= \left(s \cdot \dfrac{s\sqrt{3}}{2}\right) \div 2$ therefore area $= \dfrac{\sqrt{3}}{4}s^2$

6.3: The Sine and Cosine Functions and Their Graphs

Basic Concepts

1. (a) $r = \sqrt{12^2 + 5^2} = \sqrt{144 + 25} = \sqrt{169} = 13$

 (b) $\sin \theta = \dfrac{y}{r} = \dfrac{5}{13}$ $\cos \theta = \dfrac{x}{r} = \dfrac{12}{13}$

3. (a) $r = \sqrt{(-15)^2 + 8^2} = \sqrt{225 + 64} = \sqrt{289} = 17$

 (b) $\sin \theta = \dfrac{y}{r} = \dfrac{8}{17}$ $\cos \theta = \dfrac{x}{r} = -\dfrac{15}{17}$

5. $r = \sqrt{4^2 + 3^2} = \sqrt{16 + 9} = \sqrt{25} = 5$, $\sin \theta = \dfrac{y}{r} = \dfrac{3}{5}$ $\cos \theta = \dfrac{x}{r} = \dfrac{4}{5}$

7. $r = \sqrt{1^2 + (-2)^2} = \sqrt{1 + 4} = \sqrt{5}$, $\sin \theta = \dfrac{y}{r} = -\dfrac{2}{\sqrt{5}}$ $\cos \theta = \dfrac{x}{r} = \dfrac{1}{\sqrt{5}}$

9. $r = \sqrt{(-15)^2 + (-8)^2} = \sqrt{225 + 64} = \sqrt{289} = 17$, $\sin \theta = \dfrac{y}{r} = -\dfrac{8}{17}$ $\cos \theta = \dfrac{x}{r} = -\dfrac{15}{17}$

11. The length of the legs in a $45° \sim 45°$ right triangle are equal. For convenience we let $r = 1$ as shown in Figure 11a.

Next, find x and y: $x^2 + y^2 = 1^2 \Rightarrow x^2 + x^2 = 1^2 \Rightarrow 2x^2 = 1 \Rightarrow x^2 = \dfrac{1}{2} \Rightarrow x = \dfrac{1}{\sqrt{2}}$. Thus, $x = y = \dfrac{1}{\sqrt{2}}$.

$\sin 45° = \dfrac{1}{\sqrt{2}} \approx 0.7071$ $\cos 45° = \dfrac{1}{\sqrt{2}} \approx 0.7071$

The results are supported in Figure 11b.

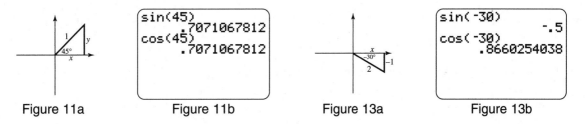

Figure 11a   Figure 11b   Figure 13a   Figure 13b

13. The length of the shorter leg in a $30° \sim 60°$ right triangle is equal to half the length of the hypotenuse. For convenience we let $r = 2$ as shown in Figure 13a. Note that y is negative so $y = -1$.

Next, find x: $x^2 = r^2 - y^2 \Rightarrow x^2 = 2^2 - 1^2 \Rightarrow x^2 = 3 \Rightarrow x = \sqrt{3}$.

$\sin (-30°) = -\dfrac{1}{2}$ $\cos (-30°) = \dfrac{\sqrt{3}}{2} \approx 0.8660$

The results are supported in Figure 13b.

15. Note that $\dfrac{\pi}{3} = 60°$. The length of the shorter leg in a $30° \sim 60°$ right triangle is equal to half the length of the hypotenuse. For convenience we let $r = 2$ as shown in Figure 15a. Note that x is positive so $x = 1$.

Next, find y: $y^2 = r^2 - x^2 \Rightarrow y^2 = 2^2 - 1^2 \Rightarrow y^2 = 3 \Rightarrow y = \sqrt{3}$.

$\sin \dfrac{\pi}{3} = \dfrac{\sqrt{3}}{2} \approx 0.8660 \quad \cos \dfrac{\pi}{3} = \dfrac{1}{2}$ The results are supported in Figure 15b.

Figure 15a

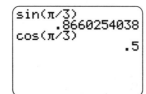

Figure 15b

Figure 17a

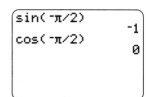

Figure 17b

17. A $-\dfrac{\pi}{2}$ angle is quadrantal. Its terminal side lies on the y-axis. For convenience we let $r = 1$ as shown in Figure 17a. Note that if $r = 1$, then $x = 0$ and $y = -1$.

$\sin\left(-\dfrac{\pi}{2}\right) = -1 \quad \cos\left(-\dfrac{\pi}{2}\right) = 0$ The results are supported in Figure 17b.

19. Note that $\dfrac{7\pi}{6} = 210°$ lies in the third quadrant and it forms a $30° \sim 60°$ right triangle with the x-axis.

For convenience we let $r = 2$ as shown in Figure 19a. Note that y is negative so $y = -1$.

Next find $x^2 = r^2 - y^2 \Rightarrow x^2 = 2^2 - (-1)^2 \Rightarrow x^2 = 3 \Rightarrow x = -\sqrt{3}$.

$\sin \dfrac{7\pi}{6} = -\dfrac{1}{2} \quad \cos \dfrac{7\pi}{6} = -\dfrac{\sqrt{3}}{2} \approx -0.8660$

These results are supported in Figure 19b.

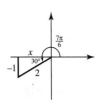

Figure 19a

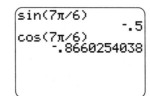

Figure 19b

Figure 21

Figure 23

21. The point $(1, 2)$ lies on the line $y = 2x$ in the first quadrant. This is shown in Figure 21.

So, $r = \sqrt{1^2 + 2^2} = \sqrt{1 + 4} = \sqrt{5}$. It follows that: $\sin \theta = \dfrac{2}{\sqrt{5}}$ and $\cos \theta = \dfrac{1}{\sqrt{5}}$.

23. The point $(1, -3)$ lies on the line $y = -3x$ in the fourth quadrant. This is shown in Figure 23.

So, $r = \sqrt{1^2 + (-3)^2} = \sqrt{1 + 9} = \sqrt{10}$. It follows that: $\sin \theta = -\dfrac{3}{\sqrt{10}}$ and $\cos \theta = \dfrac{1}{\sqrt{10}}$.

25. The point $(-3, 2)$ lies on the line $y = -\dfrac{2}{3}x$ in the second quadrant. This is shown in Figure 25.

So, $r = \sqrt{(-3)^2 + 2^2} = \sqrt{9 + 4} = \sqrt{13}$. It follows that: $\sin \theta = \dfrac{2}{\sqrt{13}}$ and $\cos \theta = -\dfrac{3}{\sqrt{13}}$.

Figure 25

27. $\sin 93.2° \approx 0.9984$ $\cos 93.2° \approx -0.0558$

29. $\sin 123°50' \approx 0.8307$ $\cos 123°50' \approx -0.5568$

31. $\sin(-4) \approx 0.7568$ $\cos(-4) \approx -0.6536$

33. $\sin \dfrac{11\pi}{7} \approx -0.9749$ $\cos \dfrac{11\pi}{7} \approx 0.2225$

35. $\sin \theta = y = \dfrac{3}{5}$ $\cos \theta = x = \dfrac{4}{5}$

37. $\sin \theta = y = -\dfrac{5}{13}$ $\cos \theta = x = \dfrac{12}{13}$

39. An angle θ of $\dfrac{\pi}{2}$ radians in standard position has a terminal side that intersects the unit circle at the point $(0, 1)$.

Therefore $\sin \dfrac{\pi}{2} = 1$ and $\cos \dfrac{\pi}{2} = 0$.

41. An angle θ of $\dfrac{7\pi}{6}$ radians in standard position has a terminal side that intersects the unit circle in the third

quadrant. To find the point of intersection, the hypotenuse of a $30°/60°$ right triangle has length 1 and the

shorter leg has length $y = \dfrac{1}{2}$. Since $x^2 + y^2 = 1$, we can solve for x. $x^2 + \left(\dfrac{1}{2}\right)^2 = 1 \Rightarrow x^2 = \dfrac{3}{4} \Rightarrow x = \dfrac{\sqrt{3}}{2}$.

Since the point (x, y) is located in the third quadrant, the point becomes $\left(-\dfrac{\sqrt{3}}{2}, -\dfrac{1}{2}\right)$. Therefore

$\sin\left(\dfrac{7\pi}{6}\right) = -\dfrac{1}{2}$ and $\cos\left(\dfrac{7\pi}{6}\right) = -\dfrac{\sqrt{3}}{2}$.

43. An angle θ of $-\dfrac{3\pi}{4}$ radians in standard position has a terminal side that intersects the unit circle in the third

quadrant at the point $\left(-\dfrac{1}{\sqrt{2}}, -\dfrac{1}{\sqrt{2}}\right)$. Therefore $\sin\left(-\dfrac{3\pi}{4}\right) = -\dfrac{1}{\sqrt{2}}$ and $\cos\left(-\dfrac{3\pi}{4}\right) = -\dfrac{1}{\sqrt{2}}$.

45. An angle θ of $\dfrac{5\pi}{2}$ radians in standard position has a terminal side that intersects the unit circle at the point $(0, 1)$.

Therefore $\sin \dfrac{5\pi}{2} = 1$ and $\cos \dfrac{5\pi}{2} = 0$.

47. An angle θ of $-\dfrac{\pi}{3}$ radians in standard position has a terminal side that intersects the unit circle in the fourth

quadrant. To find the point of intersection, the hypotenuse of a $30°/60°$ right triangle has length 1 and the

shorter leg has length $x = \dfrac{1}{2}$. Since $x^2 + y^2 = 1$, we can solve for y. $\left(\dfrac{1}{2}\right)^2 + y^2 = 1 \Rightarrow y^2 = \dfrac{3}{4} \Rightarrow y = \dfrac{\sqrt{3}}{2}$.

Since the point (x, y) is located in the fourth quadrant, the point becomes $\left(\dfrac{1}{2}, -\dfrac{\sqrt{3}}{2}\right)$. Therefore

$\sin\left(-\dfrac{\pi}{3}\right) = -\dfrac{\sqrt{3}}{2}$ and $\cos\left(-\dfrac{\pi}{3}\right) = \dfrac{1}{2}$.

Representations of Functions

49. (a) $f(0) = \sin 0 = 0$

    (b) See Figure 49-53.

51. (a) $f\left(\dfrac{\pi}{2}\right) = \sin\dfrac{\pi}{2} = 1$

    (b) See Figure 49-53.

53. (a) $f\left(-\dfrac{\pi}{6}\right) = \sin\left(-\dfrac{\pi}{6}\right) = -\dfrac{1}{2}$

    (b) See Figure 49-53.

$[-2\pi, 2\pi, \pi/2]$ by $[-4, 4, 1]$

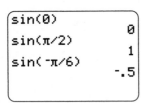

Figure 49-53

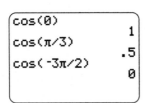

Figure 55-59

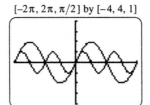

Figure 61a

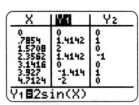

Figure 61b

55. (a) $f(0) = \cos 0 = 1$

    (b) See Figure 55-59.

57. (a) $f\left(\dfrac{\pi}{3}\right) = \cos\dfrac{\pi}{3} = \dfrac{1}{2}$

    (b) See Figure 55-59.

59. (a) $f\left(-\dfrac{3\pi}{2}\right) = \cos\left(-\dfrac{3\pi}{2}\right) = 0$

    (b) See Figure 55-59.

61. Graph $Y_1 = 2\sin(X)$ and $Y_2 = \sin(2X)$ in $[-2\pi, 2\pi, \pi/2]$ by $[-4, 4, 1]$. See Figure 61a.

    Table $Y_1 = 2\sin(X)$ and $Y_2 = \sin(2X)$ starting at $x = 0$, incrementing by $\dfrac{\pi}{4}$. See Figure 61b.

    The two functions are not the same. The function f outputs twice the sine of t, while the function g outputs the sine of twice t.

63. (a) $R = \{y \mid -3 \le y \le 3\}$. See Figure 63a.

    (b) $R = \{y \mid -1 \le y \le 1\}$. See Figure 63b.

$[-2\pi, 2\pi, \pi/2]$ by $[-4, 4, 1]$

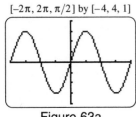

Figure 63a

$[-2\pi, 2\pi, \pi/2]$ by $[-4, 4, 1]$

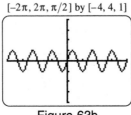

Figure 63b

$[-2\pi, 2\pi, \pi/2]$ by $[-4, 4, 1]$

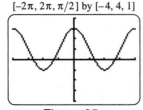

Figure 65a

$[-2\pi, 2\pi, \pi/2]$ by $[-4, 4, 1]$

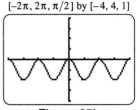

Figure 65b

65. (a) $R = \{y \mid -1 \le y \le 3\}$. See Figure 65a.

    (b) $R = \{y \mid -2 \le y \le 0\}$. See Figure 65b.

Applications

67. (a) The slope of the line is 0.03, so when x increases by 100 feet, y increases by 3 feet. The hill has a 3% grade.

(b) First we must find $\sin \theta$: $r = \sqrt{100^2 + 3^2} = \sqrt{10{,}009}$ and so $\sin \theta = \dfrac{y}{r} = \dfrac{3}{\sqrt{10{,}009}}$.

The grade resistance is $R = W \sin \theta = 25{,}000 \left(\dfrac{3}{\sqrt{10{,}009}} \right) \approx 750$ lb.

69. (a) The graph is sinusoidal and varies between –310 volts and 310 volts. The voltage is changing direction.

(b) $V\left(\dfrac{1}{120}\right) = 310 \sin\left(120\pi \cdot \dfrac{1}{120}\right) = 310 \sin (\pi) = 310(0) = 0$. After $\dfrac{1}{120}$ second the voltage is 0.

(c) The maximum voltage is 310 volts. The root mean square voltage is $\dfrac{310}{\sqrt{2}} \approx 219$ volts. Note that the common electrical rating for electric ranges and ovens is 220 volts.

71. (a) The scatterplot is shown in Figure 71a.

(b) The graph of $Y_1 = 1.9 \sin (0.42X - 1.2) + 5.7$ in $[0, 13, 1]$ by $[3, 8, 1]$ is shown with the data in Figure 71b.

From the close data fit shown, we may conclude that the flying squirrels become active near sunset.

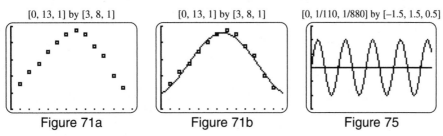

[0, 13, 1] by [3, 8, 1]   [0, 13, 1] by [3, 8, 1]   [0, 1/110, 1/880] by [–1.5, 1.5, 0.5]

      Figure 71a   Figure 71b   Figure 75

73. (a) $D = \dfrac{1.05(88^2 - 44^2)}{27 + 64.4 \sin 3°} \approx 201$ ft

(b) $D = \dfrac{1.05(88^2 - 44^2)}{27 + 64.4 \sin (-3°)} \approx 258$ ft

(c) When the slope is negative (down hill) the car will require a greater stopping distance.

75. (a) Graph $Y_1 = \sin (880\pi X)$ in $[0, 1/110, 1/880]$ by $[-1.5, 1.5, 0.5]$. See Figure 75.

(b) The period can be found by finding the change in the x-coordinate from one peak to the next.

$x_1 \approx 0.00511371$, $x_2 \approx 0.00738592$ $\Rightarrow$

$x_2 - x_1 = 0.00738592 - 0.00511371 \approx 0.00227$ sec $\left(\text{actual: } \dfrac{1}{440} \approx 0.00227 \text{ sec} \right)$

(c) $F = \dfrac{1}{0.00227} \approx 440$ cycles per second (actual: $F = 440$)

6.4: Other Trigonometric Functions and Their Graphs

Basic Concepts

1. $r = \sqrt{5^2 + 12^2} = \sqrt{25 + 144} = \sqrt{169} = 13$

$\sin \theta = \dfrac{12}{13}$ $\cos \theta = \dfrac{5}{13}$ $\tan \theta = \dfrac{12}{5}$

$\csc \theta = \dfrac{13}{12}$ $\sec \theta = \dfrac{13}{5}$ $\cot \theta = \dfrac{5}{12}$

3. $r = \sqrt{7^2 + (-24)^2} = \sqrt{49 + 576} = \sqrt{625} = 25$

$$\sin \theta = -\frac{24}{25} \qquad\qquad \cos \theta = \frac{7}{25} \qquad\qquad \tan \theta = -\frac{24}{7}$$

$$\csc \theta = -\frac{25}{24} \qquad\qquad \sec \theta = \frac{25}{7} \qquad\qquad \cot \theta = -\frac{7}{24}$$

5. $r = \sqrt{5^2 + 0^2} = \sqrt{25} = 5$

$$\sin \theta = \frac{0}{5} = 0 \qquad\qquad \cos \theta = \frac{5}{5} = 1 \qquad\qquad \tan \theta = \frac{0}{5} = 0$$

$$\csc \theta \to \text{undefined} \qquad\qquad \sec \theta = 1 \qquad\qquad \cot \theta \to \text{undefined}$$

7. $\sin \theta = \dfrac{3}{5} \qquad\qquad \cos \theta = \dfrac{4}{5} \qquad\qquad \tan \theta = \dfrac{\sin \theta}{\cos \theta} = \dfrac{\frac{3}{5}}{\frac{4}{5}} = \dfrac{3}{4}$

$$\csc \theta = \frac{1}{\sin \theta} = \frac{1}{\frac{3}{5}} = \frac{5}{3} \qquad \sec \theta = \frac{1}{\cos \theta} = \frac{1}{\frac{4}{5}} = \frac{5}{4} \qquad \cot \theta = \frac{\cos \theta}{\sin \theta} = \frac{\frac{4}{5}}{\frac{3}{5}} = \frac{4}{3}$$

9. $\sin \theta = \dfrac{1}{\csc \theta} = \dfrac{1}{-\frac{17}{15}} = -\dfrac{15}{17} \qquad \cos \theta = \dfrac{1}{\sec \theta} = \dfrac{1}{-\frac{17}{8}} = -\dfrac{8}{17} \qquad \tan \theta = \dfrac{\sin \theta}{\cos \theta} = \dfrac{-\frac{15}{17}}{-\frac{8}{17}} = \dfrac{15}{8}$

$$\csc \theta = -\frac{17}{15} \qquad\qquad \sec \theta = -\frac{17}{8} \qquad\qquad \cot \theta = \frac{\cos \theta}{\sin \theta} = \frac{-\frac{8}{17}}{-\frac{15}{17}} = \frac{8}{15}$$

11. $\sin \theta = \tan \theta \cos \theta = \left(\dfrac{5}{12}\right)\left(\dfrac{12}{13}\right) = \dfrac{5}{13} \qquad \cos \theta = \dfrac{12}{13} \qquad \tan \theta = \dfrac{5}{12}$

$$\csc \theta = \frac{1}{\sin \theta} = \frac{1}{\frac{5}{13}} = \frac{13}{5} \qquad \sec \theta = \frac{1}{\cos \theta} = \frac{1}{\frac{12}{13}} = \frac{13}{12} \qquad \cot \theta = \frac{\cos \theta}{\sin \theta} = \frac{\frac{12}{13}}{\frac{5}{13}} = \frac{12}{5}$$

13. $\sin \theta = -\dfrac{3}{5} \qquad\qquad \cos \theta = \dfrac{4}{5} \qquad\qquad \tan \theta = \dfrac{\sin \theta}{\cos \theta} = \dfrac{-\frac{3}{5}}{\frac{4}{5}} = -\dfrac{3}{4}$

$$\csc \theta = \frac{1}{\sin \theta} = \frac{1}{-\frac{3}{5}} = -\frac{5}{3} \qquad \sec \theta = \frac{1}{\cos \theta} = \frac{1}{\frac{4}{5}} = \frac{5}{4} \qquad \cot \theta = \frac{\cos \theta}{\sin \theta} = \frac{\frac{4}{5}}{-\frac{3}{5}} = -\frac{4}{3}$$

15. $\sin \theta = -\dfrac{3}{5} \qquad\qquad \cos \theta = -\dfrac{4}{5} \qquad\qquad \tan \theta = \dfrac{\sin \theta}{\cos \theta} = \dfrac{-\frac{3}{5}}{-\frac{4}{5}} = \dfrac{3}{4}$

$$\csc \theta = \frac{1}{\sin \theta} = \frac{1}{-\frac{3}{5}} = -\frac{5}{3} \qquad \sec \theta = \frac{1}{\cos \theta} = \frac{1}{-\frac{4}{5}} = -\frac{5}{4} \qquad \cot \theta = \frac{\cos \theta}{\sin \theta} = \frac{-\frac{4}{5}}{-\frac{3}{5}} = \frac{4}{3}$$

17. $\tan \dfrac{3\pi}{4} = \dfrac{\sin \left(\frac{3}{4}\pi\right)}{\cos \left(\frac{3}{4}\pi\right)} = \dfrac{\frac{\sqrt{2}}{2}}{-\frac{\sqrt{2}}{2}} = -1$

19. (a) $(0, 1)$; *Answers may vary*

(b) Distance is 1; *Answers may vary*

(c) $\tan \dfrac{\pi}{2}$ is undefined, $\cot \dfrac{\pi}{2} = \dfrac{\cos \frac{\pi}{2}}{\sin \frac{\pi}{2}} = \dfrac{0}{1} = 0$; $\sec \dfrac{\pi}{2}$ is undefined, $\csc \dfrac{\pi}{2} = \dfrac{1}{\sin \frac{\pi}{2}} = \dfrac{1}{1} = 1$

21. (a) $(-1, -1)$; *Answers may vary*

   (b) Distance $= \sqrt{(-1 - 0)^2 + (-1 - 0)^2} = \sqrt{2}$

   (c) $\tan(-135°) = \dfrac{y}{x} = \dfrac{-1}{-1} = 1$; $\cot(-135°) = \dfrac{x}{y} = \dfrac{-1}{-1} = 1$;

   $\sec(-135°) = \dfrac{r}{x} = \dfrac{\sqrt{2}}{-1} = -\sqrt{2}$; $\csc(-135°) = \dfrac{r}{x} = \dfrac{\sqrt{2}}{-1} = -\sqrt{2}$

23. $\sin\theta = \dfrac{1}{\sqrt{2}}$ $\cos\theta = \dfrac{1}{\sqrt{2}}$ $\tan\theta = 1$ $\csc\theta = \sqrt{2}$ $\sec\theta = \sqrt{2}$ $\cot\theta = 1$

25. $\sin\theta = -\dfrac{12}{13}$ $\cos\theta = \dfrac{5}{13}$ $\tan\theta = -\dfrac{12}{5}$ $\csc\theta = -\dfrac{13}{12}$ $\sec\theta = \dfrac{13}{5}$ $\cot\theta = -\dfrac{5}{12}$

27. The terminal side of a $90°$ angle intersects the unit circle at the point $(0, 1)$.

   $\sin 90° = 1$ $\cos 90° = 0$ $\tan 90° \rightarrow$ undefined

   $\csc 90° = 1$ $\sec 90° \rightarrow$ undefined $\cot 90° = 0$

29. The terminal side of a $-45°$ angle intersects the unit circle at the point $\left(\dfrac{1}{\sqrt{2}}, -\dfrac{1}{\sqrt{2}}\right)$.

   $\sin(-45°) = -\dfrac{1}{\sqrt{2}}$ $\cos(-45°) = \dfrac{1}{\sqrt{2}}$ $\tan(-45°) = -1$

   $\csc(-45°) = -\sqrt{2}$ $\sec(-45°) = \sqrt{2}$ $\cot(-45°) = -1$

31. The terminal side of a π radian angle intersects the unit circle at the point $(-1, 0)$.

   $\sin\pi = 0$ $\cos\pi = -1$ $\tan\pi = 0$

   $\csc\pi \rightarrow$ undefined $\sec\pi = -1$ $\cot\pi \rightarrow$ undefined

33. The terminal side of a $-\dfrac{\pi}{3}$ radian angle intersects the unit circle at the point $\left(\dfrac{1}{2}, -\dfrac{\sqrt{3}}{2}\right)$.

   $\sin\left(-\dfrac{\pi}{3}\right) = -\dfrac{\sqrt{3}}{2}$ $\cos\left(-\dfrac{\pi}{3}\right) = \dfrac{1}{2}$ $\tan\left(-\dfrac{\pi}{3}\right) = -\sqrt{3}$

   $\csc\left(-\dfrac{\pi}{3}\right) = -\dfrac{2}{\sqrt{3}}$ $\sec\left(-\dfrac{\pi}{3}\right) = 2$ $\cot\left(-\dfrac{\pi}{3}\right) = -\dfrac{1}{\sqrt{3}}$

35. The terminal side of a $-\dfrac{\pi}{2}$ radian angle intersects the unit circle at the point $(0, -1)$.

   $\sin\left(-\dfrac{\pi}{2}\right) = -1$ $\cos\left(-\dfrac{\pi}{2}\right) = 0$ $\tan\left(-\dfrac{\pi}{2}\right) \rightarrow$ undefined

   $\csc\left(-\dfrac{\pi}{2}\right) = -1$ $\sec\left(-\dfrac{\pi}{2}\right) \rightarrow$ undefined $\cot\left(-\dfrac{\pi}{2}\right) = 0$

37. The terminal side of a $360°$ angle intersects the unit circle at the point $(1, 0)$.

   $\sin 360° = 0$ $\cos 360° = 1$ $\tan 360° = 0$

   $\csc 360° \rightarrow$ undefined $\sec 360° = 1$ $\cot 360° \rightarrow$ undefined

39. The terminal side of a $\dfrac{\pi}{6}$ radian angle intersects the unit circle at the point $\left(\dfrac{\sqrt{3}}{2}, \dfrac{1}{2}\right)$.

   $\sin\dfrac{\pi}{6} = \dfrac{1}{2}$ $\cos\dfrac{\pi}{6} = \dfrac{\sqrt{3}}{2}$ $\tan\dfrac{\pi}{6} = \dfrac{1}{\sqrt{3}}$

   $\csc\dfrac{\pi}{6} = 2$ $\sec\dfrac{\pi}{6} = \dfrac{2}{\sqrt{3}}$ $\cot\dfrac{\pi}{6} = \sqrt{3}$

41. The point $(-1, 4)$ lies on the line $y = -4x$ in the second quadrant. This is shown in Figure 41.

So $r = \sqrt{(-1)^2 + 4^2} = \sqrt{1 + 16} = \sqrt{17}$.

$$\sin \theta = \frac{4}{\sqrt{17}} \qquad\qquad \cos \theta = -\frac{1}{\sqrt{17}} \qquad\qquad \tan \theta = -4$$

$$\csc \theta = \frac{\sqrt{17}}{4} \qquad\qquad \sec \theta = -\sqrt{17} \qquad\qquad \cot \theta = -\frac{1}{4}$$

The slope of the line is equal to $\tan \theta$.

Figure 41   Figure 43

43. The point $(-1, -6)$ lies on the line $y = 6x$ in the third quadrant. This is shown in Figure 43.

So $r = \sqrt{(-1)^2 + (-6)^2} = \sqrt{1 + 36} = \sqrt{37}$.

$$\sin \theta = -\frac{6}{\sqrt{37}} \qquad\qquad \cos \theta = -\frac{1}{\sqrt{37}} \qquad\qquad \tan \theta = 6$$

$$\csc \theta = -\frac{\sqrt{37}}{6} \qquad\qquad \sec \theta = -\sqrt{37} \qquad\qquad \cot \theta = \frac{1}{6}$$

The slope of the line is equal to $\tan \theta$.

45. (a) $\sin 93.2° \approx 0.9984$

    (b) $\csc 93.2° = \dfrac{1}{\sin 93.2°} \approx 1.0016$

47. (a) $\tan 234°33' \approx 1.4045$

    (b) $\cot 234°33' = \dfrac{1}{\tan 234°33'} \approx 0.7120$

49. (a) $\cot (-4) = \dfrac{1}{\tan (-4)} \approx -0.8637$

    (b) $\tan (-4) \approx -1.1578$

51. (a) $\cos \dfrac{11\pi}{7} \approx 0.2225$

    (b) $\sec \dfrac{11\pi}{7} = \dfrac{1}{\cos \frac{11\pi}{7}} \approx 4.4940$

53. See the sine graph in "Putting it All Together" at the end of section 6.4. $f(t) = \sin t$ has origin symmetry.

55. See the tangent graph in "Putting it All Together" at the end of section 6.4. $f(t) = \tan t$ has origin symmetry.

57. See the secant graph in "Putting it All Together" at the end of section 6.4. $f(t) = \sec t$ has y-axis symmetry.

59. The domain is all real numbers. The range is $\{y \mid -1 \le y \le 1\}$. The period is 2π.

61. The domain is $\left\{ t \;\middle|\; t \ne \pm \dfrac{\pi}{2}, \pm \dfrac{3\pi}{2}, \pm \dfrac{5\pi}{2}, \dots \right\}$. The range is all real numbers. The period is π.

63. The domain is $\left\{ t \;\middle|\; t \ne \pm \dfrac{\pi}{2}, \pm \dfrac{3\pi}{2}, \pm \dfrac{5\pi}{2}, \dots \right\}$. The range is $\{y \mid |y| \ge 1\}$. The period is 2π.

65. Graph $Y_1 = \tan(X)$ in $[-352.5°, 352.5°, 90°]$ by $[-4, 4, 1]$. See Figure 65.

    $f(-90°) \rightarrow$ undefined $f(-45°) = -1$ $f(0°) = 0$ $f(45°) = 1$ $f(90°) \rightarrow$ undefined

67. Graph $Y_1 = 1/\sin(X)$ in $[-352.5°, 352.5°, 90°]$ by $[-4, 4, 1]$. See Figure 67.

    $f(-90°) = -1$ $f(-45°) = -\sqrt{2}$ $f(0°) \rightarrow$ undefined $f(45°) = \sqrt{2}$ $f(90°) = 1$

$[-352.5°, 352.5°, 90°]$ by $[-4, 4, 1]$ $[-352.5°, 352.5°, 90°]$ by $[-4, 4, 1]$ $[-2\pi, 2\pi, \pi/2]$ by $[-4, 4, 1]$

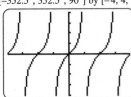
Figure 65

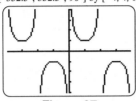
Figure 67

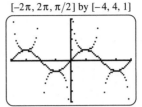
Figure 69

69. Graph $Y_1 = \sin(X)$ and $Y_2 = 1/\sin(X)$ in $[-2\pi, 2\pi, \pi/2]$ by $[-4, 4, 1]$ using dot mode. See Figure 69.

    The zeros of the sine graph correspond to the vertical asymptotes on the cosecant graph.

    If $\sin t = \pm 1$ then $\csc t = \pm 1$.

71. (a) Graph $Y_1 = \sin(X)$ in $[-2\pi, 2\pi, \pi/2]$ by $[-4, 4, 1]$ using dot mode. See Figure 71a. By using the trace

    feature of your calculator we see that $f(t) = 0$ when $\sin t = 0$ at $t = -\pi, 0, \pi, 2\pi$. This can be shown

    numerically by making a table of $Y_1 = \sin(X)$ starting at $x = -2\pi$ incrementing by $\dfrac{\pi}{2}$. See Figure 71b.

    Solutions to parts (b) - (f) are found in a similar manner.

    (b) $f(t) = \cos t = 0$ for $t = \pm\dfrac{\pi}{2}, \pm\dfrac{3\pi}{2}$.

    (c) $f(t) = \tan t = 0$ for $t = -\pi, 0, \pi, 2\pi$.

    (d) $f(t) = \csc t$ is never equal to zero. Zero is not in the range of $f(t) = \csc t$.

    (e) $f(t) = \sec t$ is never equal to zero. Zero is not in the range of $f(t) = \sec t$.

    (f) $f(t) = \cot t = 0$ for $t = \pm\dfrac{\pi}{2}, \pm\dfrac{3\pi}{2}$.

$[-2\pi, 2\pi, \pi/2]$ by $[-4, 4, 1]$ $[0, \pi, \pi/4]$ by $[-4, 4, 1]$

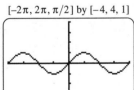
Figure 71a

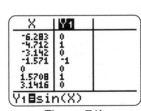
Figure 71b

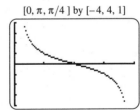
Figure 73

Applications

73. (a) Graph $Y_1 = \cos(X)/\sin(X)$ in $[0, \pi, \pi/4]$ by $[-4, 4, 1]$. See Figure 73.

    (b) At sunrise ($t = 0$) the stick would cast a very long shadow (longer than four meters). As time moves

    toward noon, the length of the shadow decreases until it has length zero at exactly noon $\left(t = \dfrac{\pi}{2}\right)$.

    In the afternoon, the length of the shadow begins to increase again but now the shadow is on the other side

    of the stick (negative values). This continues until sunset ($t = \pi$) when the shadow is again very long.

75. $\theta = 57.35 \tan 17°23'43'' \approx 17.95°$

77. At noon $\theta = 90°$. The sun moves $15°$ per hour. Thus, three hours later $\theta = 135°$. Since $\csc 90° = 1$ and

$$\csc 135° = \frac{1}{\sin 135°} \approx 1.414,$$ we may conclude that sunlight travels through about 41% more atmosphere at

3:00 P.M. than at noon.

79. Graph $Y_1 = -16X^2/(750^2(\cos(30))^2) + X\tan(30)$ in $[0, 16,000, 1000]$ by $[-500, 2500, 500]$.

(a) The maximum height is shown to be about 2197 feet at $(7612, 2197)$. See Figure 79a.

(b) The total distance travelled is shown to be about 15,223 feet. See Figure 79b.

[0, 16,000, 1000] by [−500, 2500, 500]   [0, 16,000, 1000] by [−500, 2500, 500]

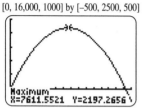

Maximum
X=7611.5521 Y=2197.2656

Zero
X=15223.103 Y=0

Figure 79a   Figure 79b

6.5: Graphing Trigonometric Functions

Graphs of Trigonometric Functions

1. Period is $\dfrac{2\pi}{\frac{1}{2}} = 4\pi$ and Amplitude is $|3| = 3$. See figure 1.

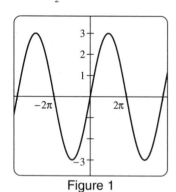

Figure 1

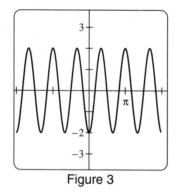

Figure 3

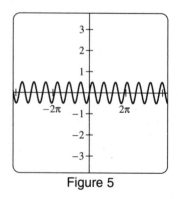

Figure 5

3. Period is $\dfrac{2\pi}{3}$ and Amplitude is $|-2| = 2$. See figure 3.

5. Period is $\dfrac{2\pi}{\pi} = 2$ and Amplitude is $\left|\dfrac{1}{2}\right| = \dfrac{1}{2}$. See figure 5.

7. Shorten the period of the sine graph to π, increase the amplitude to 3, and shift the graph downward 1 unit. See Figure 7.

9. Shift the cosine graph $\dfrac{\pi}{2}$ units to the left, increase the amplitude to 2, and reflect the graph across the x-axis. See Figure 9.

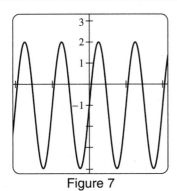

Figure 7

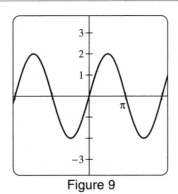

Figure 9

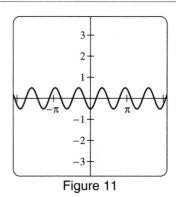

Figure 11

11. Shift the cosine graph 1 unit to the right, shorten the period to 2, and decrease the amplitude to $\frac{1}{2}$.

    See Figure 11.

13. Amplitude is 3; period is $\frac{2\pi}{4} = \frac{\pi}{2}$; phase shift is $\frac{\pi}{4}$; vertical shift is -4.

15. Amplitude is 4; period is $\frac{2\pi}{\frac{\pi}{2}} = 4$; phase shift is 1; vertical shift is 6.

17. Rewrite the equation as $y = \frac{2}{3}\sin\left(6\left(x + \frac{\pi}{2}\right)\right) - \frac{5}{2}$.

    Amplitude is $\frac{2}{3}$; period is $\frac{2\pi}{6} = \frac{\pi}{3}$; phase shift is $-\frac{\pi}{2}$; vertical shift is $-\frac{5}{2}$.

19. Half of the difference between the maximum and minimum values is $0.5(3 - (-3)) = 3$.

    Thus the amplitude is 3. The graph repeats every π units thus $\frac{2\pi}{b} = \pi \Rightarrow b = 2$. So $a = 3$ and $b = 2$.

21. Half of the difference between the maximum and minimum values is $0.5(2 - (-2)) = 2$. Since the sine graph

    is reflected in the x-axis $a = -2$. The graph repeats every 4π units thus $\frac{2\pi}{b} = 4\pi \Rightarrow b = \frac{1}{2}$.

    So $a = -2$ and $b = \frac{1}{2}$.

23. Half of the difference between the maximum and minimum values is $0.5(3 - (-3)) = 3$. Thus the amplitude

    is 3. The graph repeats every 4π units, thus the period is 4π. There is no phase shift for this graph.

25. Half of the difference between the maximum and minimum values is $0.5(4 - (-4)) = 4$. Thus the amplitude

    is 4. The graph repeats every 2 units, thus the period is 2. The phase shift for this graph is 1.

27. Graph c. amplitude $= 2$, period $= \frac{2\pi}{0.5} = 4\pi$, no phase shift

29. Graph d. amplitude $= 3$, period $= \frac{2\pi}{\pi} = 2$, no phase shift

31. Graph a. amplitude $= 1$, period $= \frac{2\pi}{1} = 2\pi$, phase shift $= -\frac{\pi}{2}$

33. $a = 3$, $\frac{2\pi}{b} = 4\pi \Rightarrow b = \frac{1}{2}$, and $c = 0$. Thus $y = 3\sin\left(\frac{1}{2}(x - 0)\right) \Rightarrow y = 3\sin\left(\frac{1}{2}x\right)$.

35. $a = 4$, $\frac{2\pi}{b} = 2 \Rightarrow b = \pi$, and $c = 1$. Thus $y = 4\sin(\pi(x - 1))$.

37. Amplitude = 2, period = 2π, phase shift = 0. See Figure 37.

39. Amplitude is 1; period is $\dfrac{2\pi}{\frac{1}{2}} = 4\pi$; phase shift is 0. See figure 39.

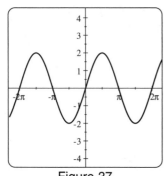

Figure 37

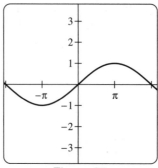

Figure 39

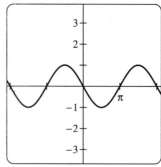

Figure 41

41. Amplitude is 1; period is 2π; phase shift is 0. See figure 41.

43. Amplitude = 1, period = $\dfrac{2\pi}{\pi} = 2$, phase shift = 0, vertical shift up 2 units. See Figure 43.

45. Amplitude = 1, period = $\dfrac{2\pi}{2} = \pi$, phase shift = $-\pi$. See Figure 45.

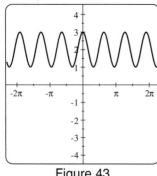

Figure 43

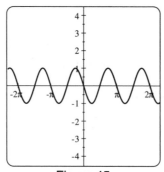

Figure 45

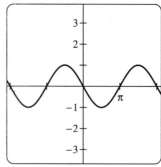

Figure 47

47. Amplitude is 1; period is 2π; phase shift is $\dfrac{\pi}{2}$. See figure 47.

49. Amplitude is $\left| -\dfrac{1}{2} \right| = \dfrac{1}{2}$; period is $\dfrac{2\pi}{2} = \pi$; phase shift is 0. See figure 49.

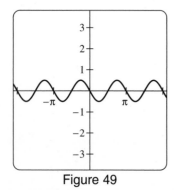

Figure 49

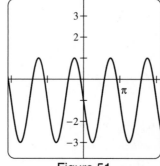

Figure 51

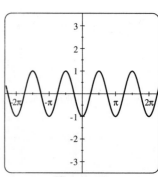

Figure 53

51. Rewrite the equation as $2\cos\left(2\left(x+\dfrac{\pi}{4}\right)\right)-1$. Amplitude is 2; period is $\dfrac{2\pi}{2}=\pi$; phase shift is $-\dfrac{\pi}{4}$.

    See Figure 51.

53. Amplitude $=1$, period $=\dfrac{2\pi}{2}=\pi$, phase shift $=\dfrac{\pi}{2}$. See Figure 53.

55. Graph $Y_1=2\sin(2X)$ in $[-2\pi,2\pi,\pi/2]$ by $[-4,4,1]$. See Figure 55.

    amplitude $=2$, period $=\pi$, phase shift $=0$

57. Graph $Y_1=0.5\cos(3(X+\pi/3))$ in $[-2\pi,2\pi,\pi/2]$ by $[-4,4,1]$. See Figure 57.

    amplitude $=\dfrac{1}{2}$, period $=\dfrac{2\pi}{3}$, phase shift $=-\dfrac{\pi}{3}$

$[-2\pi,2\pi,\pi/2]$ by $[-4,4,1]$ $[-2\pi,2\pi,\pi/2]$ by $[-4,4,1]$ $[-2\pi,2\pi,\pi/2]$ by $[-4,4,1]$ $[-2\pi,2\pi,\pi/2]$ by $[-4,4,1]$

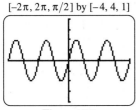
Figure 55

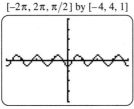
Figure 57

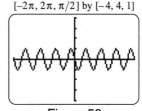
Figure 59

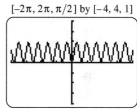
Figure 61

59. Graph $Y_1=-\sin(4X)$ in $[-2\pi,2\pi,\pi/2]$ by $[-4,4,1]$. See Figure 59.

    amplitude $=1$, period $=\dfrac{2\pi}{4}=\dfrac{\pi}{2}$, phase shift $=0$

61. Graph $Y_1=-\cos(2\pi X)+1$ in $[-2\pi,2\pi,\pi/2]$ by $[-4,4,1]$. See Figure 61.

    amplitude $=1$, period $=\dfrac{2\pi}{2\pi}=1$, phase shift $=0$

63. Rewrite the equation as $f(x)=-2\cos\left(2\pi\left(x+\dfrac{1}{8}\right)\right)+1$.

    Graph $Y_1=-2\cos(2\pi(X+1/8))+1$ in $[-2\pi,2\pi,\pi/2]$ by $[-4,4,1]$. See Figure 63.

    Amplitude is $|-2|=2$; period is $\dfrac{2\pi}{2\pi}=1$; phase shift is $-\dfrac{1}{8}$.

$[-2\pi,2\pi,\pi/2]$ by $[-4,4,1]$

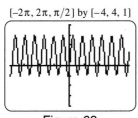
Figure 63

65. If $f(t) = \tan 2t$, then $b = 2$ and $c = 0$. Also, period $= \dfrac{\pi}{b} = \dfrac{\pi}{2}$ and phase shift $= 0$. See Figure 65.

The vertical asymptotes occur at $x = \pm\dfrac{\pi}{4}, \pm\dfrac{3\pi}{4}, \pm\dfrac{5\pi}{4}, \pm\dfrac{7\pi}{4}$.

67. If $f(t) = \tan\left(t - \dfrac{\pi}{2}\right)$, then $b = 1$ and $c = \dfrac{\pi}{2}$. Also, period $= \dfrac{\pi}{b} = \dfrac{\pi}{1} = \pi$ and phase shift $= \dfrac{\pi}{2}$. See Figure 67.

The vertical asymptotes occur at $x = 0, \pm\pi, \pm 2\pi$.

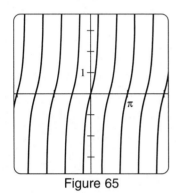

Figure 65

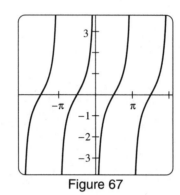

Figure 67

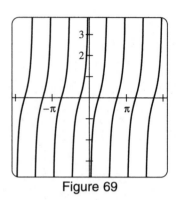

Figure 69

69. If $f(t) = -\cot 2t$, then $b = 2$ and $c = 0$. Also, period $= \dfrac{\pi}{b} = \dfrac{\pi}{2}$ and phase shift $= 0$. See Figure 69.

The vertical asymptotes occur at $x = 0, \pm\dfrac{\pi}{2}, \pm\pi, \pm\dfrac{3\pi}{2}, \pm 2\pi$.

71. If $f(t) = \cot\left(2\left(x - \dfrac{\pi}{4}\right)\right) - 1$, then $b = 2$ and $c = \dfrac{\pi}{4}$. Also, period $= \dfrac{\pi}{2}$ and phase shift $= \dfrac{\pi}{4}$.

The graph is shown in Figure 71. The vertical asymptotes occur at $x = \pm\dfrac{\pi}{4}, \pm\dfrac{3\pi}{4}, \pm\dfrac{5\pi}{4}, \pm\dfrac{7\pi}{4}$.

73. If $f(t) = \sec\left(\dfrac{1}{2}t\right)$, then $b = \dfrac{1}{2}$ and $c = 0$. Also, period $= \dfrac{2\pi}{\frac{1}{2}} = 4\pi$ and phase shift $= 0$. See Figure 73.

The vertical asymptotes occur at $x = \pm\pi$.

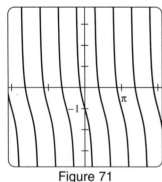

Figure 71

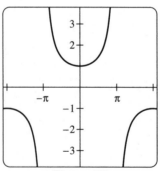

Figure 73

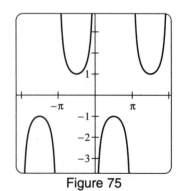

Figure 75

75. If $f(t) = \csc(t - \pi)$, then $b = 1$ and $c = \pi$. Also, period $= \dfrac{2\pi}{1} = 2\pi$ and phase shift $= \pi$. See Figure 75.

The vertical asymptotes occur at $x = 0, \pm\pi, \pm 2\pi$.

77. If $f(t) = \sec\left(\dfrac{1}{3}\left(t - \dfrac{\pi}{6}\right)\right)$, then $b = \dfrac{1}{3}$ and $c = \dfrac{\pi}{6}$. Also, period $= \dfrac{2\pi}{\frac{1}{3}} = 6\pi$ and phase shift $= \dfrac{\pi}{6}$.

See Figure 77. The vertical asymptotes occur at $x = -\dfrac{4\pi}{3}, \dfrac{5\pi}{3}$.

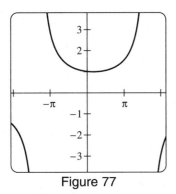

Figure 77

Applications

79. Graph *c* represents The situation best. A person would enter the Ferris wheel near the ground at $t = 0$. Then he or she would climb to a maximum height before returning to the initial ground position. This process would then repeat until the ride ended.

81. Graph *b* represents the situation best. At sunrise the shadow on the wall would have no length. As the time approaches noon, the shadow would then increase in length until, at exactly noon, it would cover the entire length of the wall.

83. (a) The maximum monthly average temperature is found at the peak of the graph: $y = 40°$ F.

    The minimum monthly average temperature is found at the "valley" of the graph: $y = -40°$ F.

    (b) The amplitude is $0.5(40 - (-40)) = 40$. This represents a total temperature swing from $-40°$ F to $40°$ F. Since the average temperatures are cyclic, repeating every 12 months, the period is 12.

    (c) The *x*-intercepts represent the months in which the average temperature is $0°$ F (April and October).

85. (a) Graph $Y_1 = 34\sin((\pi/6)(X - 4.3))$ in $[0, 25, 2]$ by $[-50, 50, 10]$. See Figure 85a.

    $$\text{amplitude} = 34, \text{period} = \frac{2\pi}{\frac{\pi}{6}} = 12, \text{phase shift} = 4.3$$

    (b) May corresponds to $x = 5$, $y \approx 12.2°$ F. See Figure 85b.

    December corresponds to $x = 12$, $y \approx -26.4°$ F. See Figure 85c.

    (c) Since half of the months have average temperatures above zero and half have average temperatures below zero, we would conjecture that the average yearly temperature is $0°$ F.

$[0, 25, 2]$ by $[-50, 50, 10]$ $[0, 25, 2]$ by $[-50, 50, 10]$ $[0, 25, 2]$ by $[-50, 50, 10]$

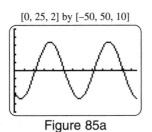
Figure 85a

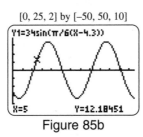
Figure 85b

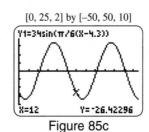
Figure 85c

87. (a) Plot the data is [0, 25, 2] by [0, 80, 10]. See Figure 87a.

(b) The maximum monthly average temperature is 64° F and the minimum is 36° F. The midpoint of these values is $0.5(64 + 36) = 50$, thus $d = 50$. Half the difference between these temperatures is $0.5(64 - 36) = 14$, thus $a = 14$. Since the temperatures cycle every 12 months, $b = \frac{2\pi}{12} = \frac{\pi}{6}$. The maximum of the $y = \sin x$ graph occurs when $x = \frac{\pi}{2}$ while the maximum in the table occurs when $x = 7$.

Thus $\frac{\pi}{6}(7 - c) = \frac{\pi}{2} \Rightarrow 7 - c = 3 \Rightarrow c = 4$. The function is $f(x) = 14 \sin\left(\frac{\pi}{6}(x - 4)\right) + 50$.

(c) See Figure 87c.

[0, 25, 2] by [0, 80, 10]

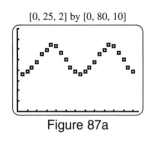

Figure 87a

[0, 25, 2] by [0, 80, 10]

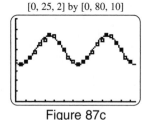

Figure 87c

[0, 25, 2] by [40, 100, 10]

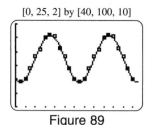

Figure 89

89. Refer to the solution for exercise 87.

(a) $a = 0.5(92 - 58) = 17, b = \frac{\pi}{6}, c = 7, d = 0.5(92 + 58) = 75$. Thus $f(x) = 17 \cos\left(\frac{\pi}{6}(x - 7)\right) + 75$.

The graph of f and the actual data are shown together in [0, 25, 2] by [40, 100, 10]. See Figure 89.

(b) Yes, different values of c are possible of the form $c = 7 + 12n$ where n is an integer.

91. $L = \dfrac{2700}{h + 3\tan\alpha} = \dfrac{2700}{2 + 3\tan 1.5°} \approx 1300$ ft

93. (a) The maximum number of daylight hours is about 18.5 hr. This occurs June 21st - summer solstice.

(b) The minimum number of daylight hours is about 6 hr. This occurs December 21st - winter solstice.

(c) The amplitude represents half the difference in daylight between the longest and shortest days. The period represents one year. *Answers may vary.*

95. (a) The maximum monthly average precipitation is about 8 inches.

The minimum monthly average precipitation is about 0.5 inches.

(b) The amplitude is $0.5(8 - 0.5) = 3.75$. The amplitude represents half of the difference between the maximum and minimum average monthly precipitation.

(c) A graph that models precipitation could touch the *x*-axis but it could not go below the *x*-axis. This is because average precipitation could not be a negative value.

(d) An average month has $0.5(8 + 0.5) = 4.25$ inches of precipitation. Over a one year period this would amount to a yearly average precipitation of $4.25(12) = 51$ inches.

97. Refer to the solution in exercise 87.

(a) $a = 0.5(18 - (-18)) = 18, b = \dfrac{2\pi}{12.4} = \dfrac{\pi}{6.2}, \dfrac{\pi}{6.2}(9.8 - c) = 0 \Rightarrow c = 9.8, d = 0.5(18 + (-18)) = 0.$

Thus $f(x) = 18 \cos\left(\dfrac{\pi}{6.2}(x - 9.8)\right).$

(b) Graph $Y_1 = 18 \cos((\pi/6.2)(X - 9.8))$ in $[0, 24, 4]$ by $[-20, 20, 5]$. See Figure 97.

The canal contains the most water at 3.7 and 16.1 hours after midnight (high tides). At these points water is rushing out of the canal at 18 bk. The canal contains the least water at 9.8 and 22.2 hours after midnight (low tides). At these points water is rushing into the canal at 18 bk.

99. (a) The maximum temperature is approximately 87° F in July.

The minimum temperature is approximately 62° F in January.

(b) The amplitude would increase. See Figure 99.

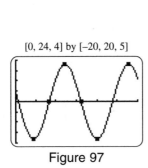

[0, 24, 4] by [−20, 20, 5]

Figure 97

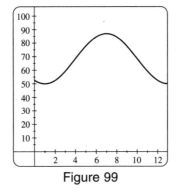

Figure 99

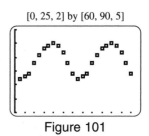

[0, 25, 2] by [60, 90, 5]

Figure 101

101. Refer to the solution in exercise 87.

(a) Plot the data is $[0, 25, 2]$ by $[60, 90, 5]$. See Figure 101.

(b) $a = 0.5(85 - 72) = 6.5, b = \dfrac{\pi}{6}, c = 4, d = 0.5(85 + 72) = 78.5.$

Thus $y = 6.5 \sin\left(\dfrac{\pi}{6}(x - 4)\right) + 78.5.$

103. (a) Graph $Y_1 = 0.022X^2 + 0.55X + 316 + 3.5 \sin(2\pi X)$ in $[20, 35, 5]$ by $[320, 370, 10]$. See Figure 103.

The general trend in carbon dioxide levels has been increasing. During the year there are seasonal variations.

(b) The quadratic function models the trend in carbon dioxide levels, while the sine function models the seasonal fluctuations in carbon dioxide levels.

[20, 35, 5] by [320, 370, 10]

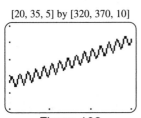

Figure 103

Simple Harmonic Motion

105. (a) Since $s(0) = 2$ inches, $a = 2$. Since the period is 0.5 seconds, the frequency is $F = \dfrac{1}{0.5} = 2$.

      Thus $s(t) = a\cos(2\pi F t) = 2\cos(4\pi t)$.

  (b) $s(1) = 2\cos(4\pi) = 2$. The spring is compressed 2 inches one second after the weight is released. The weight is moving neither upward nor downward.

107. (a) Since $s(0) = -3$ inches, $a = -3$. Since the period is 0.8 seconds, the frequency is $F = \dfrac{1}{0.8} = \dfrac{5}{4}$.

      Thus $s(t) = a\cos(2\pi F t) = -3\cos(2.5\pi t)$.

  (b) $s(1) = -3\cos(2.5\pi) = -3(0) = 0$. The spring is at its natural length one second after the weight is released. The weight is moving upward.

109. First note that $b = 2\pi F = 2\pi(27.5) = 55\pi$. Thus $s(t) = a\cos(55\pi t)$.

    Hence $s(0) = a\cos(0) = 0.21 \Rightarrow a = 0.21$. The equation is $s(t) = 0.21\cos(55\pi t)$. See Figure 109.

111. First note that $b = 2\pi F = 2\pi(55) = 110\pi$. Thus $s(t) = a\cos(110\pi t)$.

    Hence $s(0) = a\cos(0) = 0.14 \Rightarrow a = 0.14$. The equation is $s(t) = 0.14\cos(110\pi t)$. See Figure 111.

[0, 0.05, 0.01] by [−0.3, 0.3, 0.1]   [0, 0.05, 0.01] by [−0.3, 0.3, 0.1]

   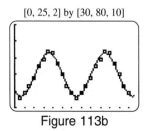

    Figure 109   Figure 111

113. $y \approx 13.2\sin(0.524x - 2.18) + 49.7$. See Figure 113a.

    Graph $Y_1 = 13.2\sin(0.524X - 2.18) + 49.7$ together with the data in $[0, 25, 2]$ by $[30, 80, 10]$.

    See Figure 113b.

[0, 25, 2] by [30, 80, 10]

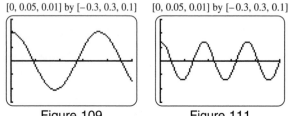

    Figure 113a   Figure 113b

115. $y \approx 16.9\sin(0.522x - 2.09) + 75.4$. See Figure 115a.

    Graph $Y_1 = 16.9\sin(0.522X - 2.09) + 75.4$ together with the data in $[0, 25, 2]$ by $[40, 100, 10]$.

    See Figure 115b.

[0, 25, 2] by [40, 100, 10]

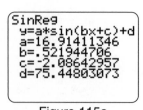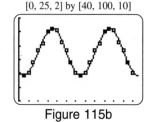

    Figure 115a   Figure 115b

6.6: Inverse Trigonometric Functions

Review of Inverses

1. one-to-one

3. π

5. 0

7. Function f multiplies the input x by 3. The inverse operation in reverse order is:

    divide the input x by 3. $f^{-1}(x) = \dfrac{x}{3}$

9. Function f takes the cube root of input x. The inverse operation in reverse order is:

    cube the input x. $f^{-1}(x) = x^3$

11. Function f subtracts 1 from the input x and then squares the result. The inverse operations in reverse order are:

    take the square root of the input x and then add 1 to the result. Note that the range of f is $f(x) \geq 0$ over the

    given domain. This means that the domain of the inverse function must be $x \geq 0$. $f^{-1}(x) = \sqrt{x} + 1$, for $x \geq 0$

13. Verbal: Function f adds 6 to the input x. The inverse operation in reverse order is: subtract 6 from the input x.

    Numerical: Table $Y_1 = X - 6$ starting at $x = 0$ incrementing by 1. See Figure 13a.

    Graphical: Graph $Y_1 = X - 6$ in $[-10, 10, 1]$ by $[-10, 10, 1]$. See Figure 13b.

    Symbolic: $f^{-1}(x) = x - 6$

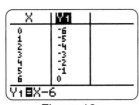

$[-10, 10, 1]$ by $[-10, 10, 1]$

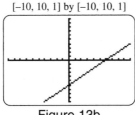

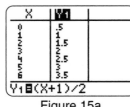

$[-10, 10, 1]$ by $[-10, 10, 1]$

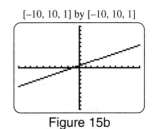

Figure 13a   Figure 13b   Figure 15a   Figure 15b

15. Verbal: Function f divides the input x by 2 and subtracts 1 from the result. The inverse operations in reverse

    order are: add 1 to the input x and then divide the result by 2.

    Numerical: Table $Y_1 = (X + 1)/2$ starting at $x = 0$, incrementing by 1. See Figure 15a.

    Graphical: Graph $Y_1 = (X + 1)/2$ in $[-10, 10, 1]$ by $[-10, 10, 1]$. See Figure 15b.

    Symbolic: $f^{-1}(x) = \dfrac{x + 1}{2}$

17. f is one-to-one on the interval $[-2, 2]$.

Inverse Trigonometric Functions

19. $\dfrac{\pi}{2}$

21. $-\dfrac{\pi}{4}$

23. $\dfrac{2\pi}{3}$

25. (a) The angle in $\left[-\dfrac{\pi}{2}, \dfrac{\pi}{2}\right]$ whose sine is 1 is $\dfrac{\pi}{2}$. Thus $\sin^{-1} 1 = \dfrac{\pi}{2}$ or $90°$.

   (b) The angle in $\left[-\dfrac{\pi}{2}, \dfrac{\pi}{2}\right]$ whose sine is 0 is 0. Thus $\arcsin 0 = 0$ or $0°$.

   (c) The angle in $\left[-\dfrac{\pi}{2}, \dfrac{\pi}{2}\right]$ whose sine is $-\dfrac{\sqrt{3}}{2}$ is $-\dfrac{\pi}{3}$. Thus $\arcsin\left(-\dfrac{\sqrt{3}}{2}\right) = -\dfrac{\pi}{3}$ or $-60°$.

27. (a) The angle in $[0, \pi]$ whose cosine is 0 is $\dfrac{\pi}{2}$. Thus $\cos^{-1} 0 = \dfrac{\pi}{2}$ or $90°$.

   (b) The angle in $[0, \pi]$ whose cosine is -1 is π. Thus $\arccos(-1) = \pi$ or $180°$.

   (c) The angle in $[0, \pi]$ whose cosine is $\dfrac{1}{2}$ is $\dfrac{\pi}{3}$. Thus $\cos^{-1}\left(\dfrac{1}{2}\right) = \dfrac{\pi}{3}$ or $60°$.

29. (a) The angle in $\left(-\dfrac{\pi}{2}, \dfrac{\pi}{2}\right)$ whose tangent is 1 is $\dfrac{\pi}{4}$. Thus $\tan^{-1} 1 = \dfrac{\pi}{4}$ or $45°$.

   (b) The angle in $\left(-\dfrac{\pi}{2}, \dfrac{\pi}{2}\right)$ whose tangent is -1 is $-\dfrac{\pi}{4}$. Thus $\arctan(-1) = -\dfrac{\pi}{4}$ or $-45°$.

   (c) The angle in $\left(-\dfrac{\pi}{2}, \dfrac{\pi}{2}\right)$ whose tangent is $\sqrt{3}$ is $\dfrac{\pi}{3}$. Thus $\tan^{-1} \sqrt{3} = \dfrac{\pi}{3}$ or $60°$.

31. (a) $\sin^{-1} 1.5$ is undefined (Error). See Figure 31a.

   (b) $\tan^{-1} 10 \approx 1.47$ radians or $84.3°$. See Figure 31b.

   (c) $\arccos(-0.75) \approx 2.42$ radians or $138.6°$. See Figure 31c.

$[-\pi/2, \pi/2, \pi/4]$ by $[-2, 2, 1]$ $[-\pi/2, \pi/2, \pi/4]$ by $[-12, 12, 1]$ $[0, \pi, \pi/4]$ by $[-2, 2, 1]$

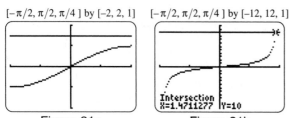

Figure 31a   Figure 31b   Figure 31c

33. (a) Since $\tan \beta = \dfrac{4}{3}$, $\tan^{-1} \dfrac{4}{3} = \beta$.

   (b) Since $\sin \alpha = \dfrac{3}{5}$, $\sin^{-1} \dfrac{3}{5} = \alpha$.

   (c) Since $\cos \beta = \dfrac{3}{5}$, $\arccos \dfrac{3}{5} = \beta$.

35. $\sin(\sin^{-1} 1) = \sin\left(\dfrac{\pi}{2}\right) = 1$

37. $\cos^{-1}\left(\cos \dfrac{5\pi}{4}\right) = \cos^{-1}\left(-\dfrac{1}{\sqrt{2}}\right) = \dfrac{3\pi}{4}$

39. $\tan(\tan^{-1}(-3)) = -3$ since $-3 \in (-\infty, \infty)$

41. (a) $\sin^{-1}\dfrac{3}{5} + \cos^{-1}\dfrac{3}{5} = 90°$

(b) $\sin^{-1}\dfrac{1}{3} + \cos^{-1}\dfrac{1}{3} = 90°$

(c) $\sin^{-1}\dfrac{2}{7} + \cos^{-1}\dfrac{2}{7} = 90°$

Conjecture: $\sin^{-1} x + \cos^{-1} x = 90°$ whenever $-1 \le x \le 1$.

43. Since $\tan \alpha = \dfrac{7}{24}$, $\alpha = \tan^{-1}\dfrac{7}{24} \approx 16.3°$. Since $\tan \beta = \dfrac{24}{7}$, $\beta = \tan^{-1}\dfrac{24}{7} \approx 73.7°$.

By the Pythagorean theorem: $c^2 = 7^2 + 24^2 \Rightarrow c^2 = 625 \Rightarrow c = 25$.

45. Since $\sin \alpha = \dfrac{6}{10}$, $\alpha = \sin^{-1}\dfrac{6}{10} \approx 36.9°$. Since $\cos \beta = \dfrac{6}{10}$, $\beta = \cos^{-1}\dfrac{6}{10} \approx 53.1°$.

By the Pythagorean theorem: $b^2 = 10^2 - 6^2 \Rightarrow b^2 = 64 \Rightarrow b = 8$.

47. β and $55°$ are complimentary, so $\beta = 90° - 55° = 35°$

Since $\tan 55° = \dfrac{5}{b}$, $b = \dfrac{5}{\tan 55°} \approx 3.5$. Since $\sin 55° = \dfrac{5}{c}$, $c = \dfrac{5}{\sin 55°} \approx 6.1$.

49. Verbal: Determine the angle (or real number) θ such that $\sin \theta = 2x$ and $-\dfrac{\pi}{2} \le \theta \le \dfrac{\pi}{2}$.

Numerical: See Figure 49a.

Graphical: See Figure 49b.

[-1, 1, 1] by [-2, 2, 1]

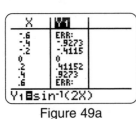

Figure 49a   Figure 49b   Figure 51

51. See Figure 51. By the Pythagorean theorem: $b^2 = 1^2 - x^2 \Rightarrow b = \sqrt{1 - x^2}$. Thus $\tan \theta = \dfrac{x}{\sqrt{1 - x^2}}$.

53. See Figure 53. By the Pythagorean theorem:

$b^2 = (\sqrt{1 + x^2})^2 - x^2 \Rightarrow b^2 = 1 + x^2 - x^2 \Rightarrow b^2 = 1 \Rightarrow b = 1$. Thus $\cos \theta = \dfrac{1}{\sqrt{1 + x^2}}$.

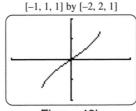

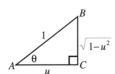

Figure 53   Figure 55

55. Let θ be the angle whose cosine is u. That is $\theta = \cos^{-1} u$. See Figure 55. We find the unknown side x by using

the Pythagorean Theorem: $u^2 + x^2 = 1^2 \Rightarrow x^2 = 1 - u^2 \Rightarrow x = \sqrt{1 - u^2}$.

Therefore $\sin(\cos^{-1} u) = \sin \theta = \dfrac{\sqrt{1 - u^2}}{1} = \sqrt{1 - u^2}$.

57. Let θ be the angle whose cosine is u. That is $\theta = \cos^{-1} u$. See Figure 57. We find the unknown side x by using the Pythagorean Theorem: $u^2 + x^2 = 1^2 \Rightarrow x^2 = 1 - u^2 \Rightarrow x = \sqrt{1 - u^2}$.

Therefore $\tan (\cos^{-1} u) = \tan \theta = \dfrac{\sqrt{1 - u^2}}{u}$.

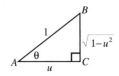

Figure 57   Figure 59

59. Let θ be the angle whose tangent is $\dfrac{1}{u}$. That is $\theta = \tan^{-1} \dfrac{1}{u}$. See Figure 59. We find the unknown side x by using the Pythagorean Theorem: $u^2 + 1^2 = x^2 \Rightarrow \sqrt{u^2 + 1^2} = x \Rightarrow x = \sqrt{u^2 + 1}$.

Therefore $\cot \left(\tan^{-1} \dfrac{1}{u} \right) = \cot \theta = \dfrac{1}{\tan \theta} = \dfrac{1}{\frac{1}{u}} = u$.

Solving Trigonometric Equations

61. $\sin \theta = 1 \Rightarrow \theta = \sin^{-1} 1 = 90°$

63. $\tan \theta = 1 \Rightarrow \theta = \tan^{-1} 1 = 45°$

65. $\cos \theta = 0 \Rightarrow \theta = \tan^{-1} 0 = 90°$

67. $2 \cos \theta = \dfrac{1}{4} \Rightarrow \cos \theta = \dfrac{1}{8} \Rightarrow \theta = \cos^{-1} \dfrac{1}{8} \approx 82.8°$

69. $\tan \theta - 1 = 5 \Rightarrow \tan \theta = 6 \Rightarrow \theta = \tan^{-1} 6 \approx 80.5°$

71. $\sin^2 \theta = 0.87 \Rightarrow \sin \theta \approx 0.9327 \Rightarrow \theta \approx \sin^{-1} 0.9327 \approx 68.9°$

73. $\tan t = -\dfrac{1}{5} \Rightarrow t = \tan^{-1} \left(-\dfrac{1}{5} \right) \approx -0.197$. See Figure 73.

75. $\cos t = 0.452 \Rightarrow t = \cos^{-1} 0.452 \approx 1.102$. See Figure 75.

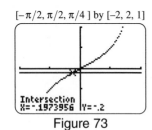 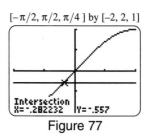 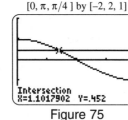 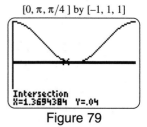

[$-\pi/2, \pi/2, \pi/4$] by [$-2, 2, 1$]   [$0, \pi, \pi/4$] by [$-2, 2, 1$]   [$-\pi/2, \pi/2, \pi/4$] by [$-2, 2, 1$]   [$0, \pi, \pi/4$] by [$-1, 1, 1$]

Figure 73   Figure 75   Figure 77   Figure 79

77. $2 \sin t = -0.557 \Rightarrow \sin t = -0.2785 \Rightarrow t = \sin^{-1} (-0.2785) \approx -0.282$. See Figure 77.

79. $\cos^2 t = \dfrac{1}{25} \Rightarrow \cos t = \pm \dfrac{1}{5} \Rightarrow t = \cos^{-1} \left(\pm \dfrac{1}{5} \right) \approx 1.369, 1.772$. See Figure 79.

Applications

81. $\tan \theta = \dfrac{50}{85}, \theta = \tan^{-1} \dfrac{50}{85} \approx 30.5°$

83. (a) Since $\tan \theta = \dfrac{11}{5}$, $\theta = \tan^{-1} \dfrac{11}{5} \approx 65.6°$

 (b) Since $\tan \theta = -\dfrac{3}{1}$, $\theta = \tan^{-1}\left(-\dfrac{3}{1}\right) \approx -71.6°$

85. Since $\tan \theta = \dfrac{4}{10}$, $\theta = \tan^{-1} \dfrac{4}{10} \approx 21.8°$

87. (a) $F = \dfrac{1}{2}(1 - \cos \theta) \Rightarrow \dfrac{1}{4} = \dfrac{1}{2}(1 - \cos \theta) \Rightarrow \dfrac{1}{2} = 1 - \cos \theta \Rightarrow \cos \theta = \dfrac{1}{2} \Rightarrow \theta = \cos^{-1} \dfrac{1}{2} = 60°$

 (b) $F = \dfrac{1}{2}(1 - \cos \theta) \Rightarrow \dfrac{3}{4} = \dfrac{1}{2}(1 - \cos \theta) \Rightarrow \dfrac{3}{2} = 1 - \cos \theta \Rightarrow \cos \theta = -\dfrac{1}{2} \Rightarrow \theta = \cos^{-1}\left(-\dfrac{1}{2}\right) = 120°$

89. $\theta = \sin^{-1}\left(\sqrt{\dfrac{43^2}{2(43^2) + 64.4(7)}}\right) \approx 41.9°$

91. If $\cos (0.1309H) = -0.4336 \tan L$, then $0.1309H = \cos^{-1}(-0.4336 \tan L) \Rightarrow H = \dfrac{\cos^{-1}(-0.4336 \tan L)}{0.1309}$.

 (a) Convert $40°55'$ to radians: $\left(40 + \dfrac{55}{60}\right) \cdot \dfrac{\pi}{180} \approx 0.714$. Then $H \approx \dfrac{\cos^{-1}(-0.4336 \tan 0.714)}{0.1309} \approx 14.9$ hr.

 (b) Convert $27°46'$ to radians: $\left(27 + \dfrac{46}{60}\right) \cdot \dfrac{\pi}{180} \approx 0.485$ Then $H \approx \dfrac{\cos^{-1}(-0.4336 \tan 0.485)}{0.1309} \approx 13.8$ hr.

 (c) Convert $37°30'$ to radians: $\left(37 + \dfrac{30}{60}\right) \cdot \dfrac{\pi}{180} \approx 0.654$. Then $H \approx \dfrac{\cos^{-1}(-0.4336 \tan 0.654)}{0.1309} \approx 14.6$ hr.

93. $n_1 \sin \theta_1 = n_2 \sin \theta_2 \Rightarrow 1 \sin 40° = 1.33 \sin \theta_2 \Rightarrow \sin \theta_2 = \dfrac{\sin 40°}{1.33} \Rightarrow \theta_2 = \sin^{-1}\left(\dfrac{\sin 40°}{1.33}\right) \approx 28.9°$

95. Let α and β represent the angles of elevation from the shrub to the shorter and taller buildings respectively.

The distance from the shrub to the shorter building is $100 - x$, thus $\alpha = \arctan\left(\dfrac{75}{100 - x}\right)$. Similarly

$\beta = \arctan\left(\dfrac{150}{x}\right)$. Because the angles α, θ, and β form a straight angle, $\theta = \pi - \alpha - \beta$. That is

$\theta = \pi - \arctan\left(\dfrac{75}{100 - x}\right) - \arctan\left(\dfrac{150}{x}\right)$.

Chapter 6 Review Exercises

1. (a) See Figure 1a.

 (b) See Figure 1b.

 (c) See Figure 1c.

 (d) See Figure 1d.

Figure 1a

Figure 1b

Figure 1c

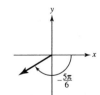

Figure 1d

3. (a) $\dfrac{\pi}{3} \cdot \dfrac{180°}{\pi} = \dfrac{180°}{3} = 60°$

   (b) $\dfrac{\pi}{36} \cdot \dfrac{180°}{\pi} = \dfrac{180°}{36} = 5°$

   (c) $-\dfrac{5\pi}{6} \cdot \dfrac{180°}{\pi} = \dfrac{-5(180°)}{6} = -150°$

   (d) $-\dfrac{7\pi}{4} \cdot \dfrac{180°}{\pi} = \dfrac{-7(180°)}{4} = -315°$

5. Note that $60° = \dfrac{\pi}{3}$ radians, thus $s = \theta r = \dfrac{\pi}{3}(6) = 2\pi$ ft.

7. $\sin 30° = \dfrac{\text{opposite}}{\text{hypotenuse}} = \dfrac{1}{2}$

9. $\cot 60° = \dfrac{\text{adjacent}}{\text{opposite}} = \dfrac{1}{\sqrt{3}}$

11. $\sec 45° = \dfrac{\text{hypotenuse}}{\text{adjacent}} = \dfrac{\sqrt{2}}{1} = \sqrt{2}$

13. $c = \sqrt{8^2 + 9^2} = \sqrt{145}$

   $\sin\theta = \dfrac{8}{\sqrt{145}}$ $\cos\theta = \dfrac{9}{\sqrt{45}}$ $\tan\theta = \dfrac{8}{9}$ $\csc\theta = \dfrac{\sqrt{145}}{8}$ $\sec\theta = \dfrac{\sqrt{145}}{9}$ $\cot\theta = \dfrac{9}{8}$

15. $\csc\theta = \dfrac{1}{\sin\theta} = \dfrac{1}{\frac{1}{3}} = 3$

17. $\sin 25° \approx 0.423$ $\cos 25° \approx 0.906$ $\tan 25° \approx 0.466$

    $\csc 25° \approx 2.366$ $\sec 25° \approx 1.103$ $\cot 25° \approx 2.145$

19. $r = \sqrt{1^2 + (-2)^2} = \sqrt{5}$

   $\sin\theta = -\dfrac{2}{\sqrt{5}}$ $\cos\theta = \dfrac{1}{\sqrt{5}}$ $\tan\theta = -\dfrac{2}{1} = -2$ $\csc\theta = -\dfrac{\sqrt{5}}{2}$ $\sec\theta = \dfrac{\sqrt{5}}{1} = \sqrt{5}$ $\cot\theta = -\dfrac{1}{2}$

21. $\sin(-45°) = -\dfrac{1}{\sqrt{2}}$ $\cos(-45°) = \dfrac{1}{\sqrt{2}}$ $\tan(-45°) = -1$

    $\csc(-45°) = -\sqrt{2}$ $\sec(-45°) = \sqrt{2}$ $\cot(-45°) = -1$

23. $\sin\theta = \dfrac{\sqrt{3}}{2}$ $\cos\theta = -\dfrac{1}{2}$ $\tan\theta = -\sqrt{3}$ $\csc\theta = \dfrac{2}{\sqrt{3}}$ $\sec\theta = -2$ $\cot\theta = -\dfrac{1}{\sqrt{3}}$

25. An angle $\theta = -\dfrac{\pi}{2}$ radians in standard position has terminal side that intersects the unit

   circle at the point $(0, -1)$. Therefore $\sin\left(-\dfrac{\pi}{2}\right) = -1$.

27. An angle $\theta = -3\pi$ radians in standard position has terminal side that intersects the unit

   circle at the point $(-1, 0)$. Therefore $\tan(-3\pi) = 0$.

29. $\sin\theta = -\dfrac{4}{5}$ $\cos\theta = \dfrac{3}{5}$ $\tan\theta = \dfrac{-\frac{4}{5}}{\frac{3}{5}} = -\dfrac{4}{3}$ $\csc\theta = -\dfrac{5}{4}$ $\sec\theta = \dfrac{5}{3}$ $\cot\theta = -\dfrac{3}{4}$

31. $y = \sin x$

33. See Figure 33. Amplitude $= 3$; Period $= \dfrac{2\pi}{2} = \pi$; Phase shift $= 0$

35. See Figure 35. Amplitude $= 1$; Period $= \dfrac{2\pi}{3}$; Phase shift $= \dfrac{\pi}{3}$

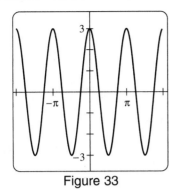

Figure 33

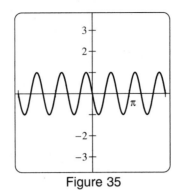

Figure 35

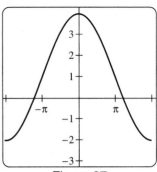

Figure 37

37. See Figure 37. Amplitude = $|-3| = 3$; Period = $\dfrac{2\pi}{\frac{1}{2}} = 4\pi$; Phase shift = π

39. Amplitude = $2 \Rightarrow a = 2$ Period = $\dfrac{2\pi}{3} \Rightarrow \dfrac{2\pi}{3} = \dfrac{2\pi}{b} \Rightarrow b = 3$

41. See Figure 41. Period = $\dfrac{\pi}{b} = \dfrac{\pi}{2}$; Phase shift = 0

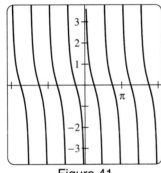

Figure 41

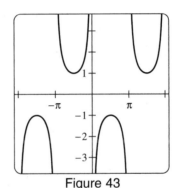

Figure 43

43. See Figure 43. Period = $\dfrac{2\pi}{b} = 2\pi$; Phase shift = $-\pi$

45. (a) $\sin^{-1}(-1) = -\dfrac{\pi}{2}$ or $-90°$

    (b) $\arccos \dfrac{1}{2} = \dfrac{\pi}{3}$ or $60°$

    (c) $\tan^{-1} 1 = \dfrac{\pi}{4}$ or $45°$

47. (a) $\sin^{-1}(-0.6) \approx -0.64$ radians or $-36.9°$

    (b) $\tan^{-1} 5 \approx 1.37$ radians or $78.7°$

    (c) $\arccos(0.12) \approx 1.45$ radians or $83.1°$

49. Since $\tan \alpha = \dfrac{5}{3}$, $\alpha = \tan^{-1} \dfrac{5}{3} \approx 59.0°$. And since $\tan \beta = \dfrac{3}{5}$, $\beta = \tan^{-1} \dfrac{3}{5} \approx 31.0°$.

    By the Pythagorean theorem: $c^2 = 5^2 + 3^2 \Rightarrow c^2 = 34 \Rightarrow c = \sqrt{34}$.

51. $\sin \theta = \dfrac{1}{2} \Rightarrow \theta = \sin^{-1} \dfrac{1}{2} = 30°$

53. $\tan \theta = \dfrac{1}{\sqrt{3}} \Rightarrow \theta = \tan^{-1} \dfrac{1}{\sqrt{3}} \approx 30°$

55. $\cos \theta = \dfrac{1}{5} \Rightarrow \theta = \cos^{-1} \dfrac{1}{5} \approx 78.5°$

57. $2\tan\theta - 1 = 5 \Rightarrow \tan\theta = 3 \Rightarrow \theta = \tan^{-1} 3 \approx 71.6°$

59. $\tan t = -\dfrac{3}{4} \Rightarrow t = \tan^{-1}\left(-\dfrac{3}{4}\right) \approx -0.6435$

Graph $Y_1 = \tan(X)$ together with $Y_2 = -0.75$ in $[-\pi/2, \pi/2, \pi/4]$ by $[-4, 4, 1]$. See Figure 59.

$[-\pi/2, \pi/2, \pi/4]$ by $[-4, 4, 1]$	$[0, \pi, \pi/4]$ by $[-4, 4, 1]$	$[0, 25, 2]$ by $[0, 70, 10]$	$[0, 25, 2]$ by $[0, 70, 10]$

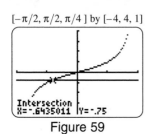
Figure 59

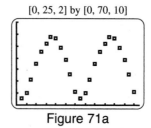
Figure 61

Figure 71a

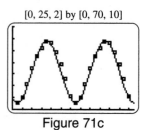
Figure 71c

61. $2\cos t = 1.8 \Rightarrow \cos t = 0.9 \Rightarrow t = \cos^{-1}(0.9) \approx 0.4510$

Graph $Y_1 = 2\cos(X)$ together with $Y_2 = 1.8$ in $[0, \pi, \pi/4]$ by $[-4, 4, 1]$. See Figure 61.

Applications

63. (a) Angular velocity $= \dfrac{\text{angle of rotation}}{\text{time}} = \dfrac{2\pi}{50 \text{ sec}} = \dfrac{\pi}{25}$ radians/sec.

(b) Linear speed $=$ (radius)(angular velocity) $= (25)\left(\dfrac{\pi}{25}\right) = \pi$ ft/sec.

65. Let h represent the height of the tree. Then $\tan 48° = \dfrac{h}{80} \Rightarrow h = 80\tan 48° \approx 89$ ft.

67. (a) $350 = 6000\sin\theta \Rightarrow \sin\theta = \dfrac{7}{120} \Rightarrow \theta = \sin^{-1}\dfrac{7}{120} \approx 3.3°$

(b) $160 = 4500\sin\theta \Rightarrow \sin\theta = \dfrac{8}{225} \Rightarrow \theta = \sin^{-1}\dfrac{8}{225} \approx 2.0°$

69. The two cities are $41°09' - 38°49' = 2°20'$ apart. Convert $2°20'$ to radians: $\left(2 + \dfrac{20}{60}\right)° \cdot \dfrac{\pi}{180°} \approx 0.0407$.

Using the arc length formula: $s \approx 0.0407(3955) \approx 161$ mi.

71. (a) Plot the data in $[0, 25, 2]$ by $[0, 70, 10]$. See Figure 71a.

(b) The maximum monthly average temperature is $58°$F and the minimum is $6°$F. The midpoint of these values is $0.5(58 + 6) = 32$, thus $d = 32$. Half the difference between these temperatures is $0.5(58 - 6) = 26$, thus $a = 26$. Since the temperature cycles every 12 months, $b = \dfrac{2\pi}{12} = \dfrac{\pi}{6}$. The maximum of the $y = \cos x$ graph occurs when $x = 0$, while the maximum in the table occurs when $x = 7$. Thus $c = 7$.

The function is $f(x) = 26\cos\left(\dfrac{\pi}{6}(x - 7)\right) + 32$.

(c) Graph $Y_1 = 26\cos((\pi/6)(X - 7)) + 32$ together with the data in $[0, 25, 2]$ by $[0, 70, 10]$. See Figure 71c.

73. $a = 6$ and $c = 220°$, thus $y = 6\sin(x - 220°)$

Chapters 1-6 Cumulative Review Exercises

1. $\dfrac{400 - 350}{350} = \dfrac{50}{350} = \dfrac{1}{7} = $ about 14.3%

3. $\dfrac{2 - \sqrt{2}}{2 + \sqrt{2}} \cdot \dfrac{2 - \sqrt{2}}{2 - \sqrt{2}} = \dfrac{4 - 2\sqrt{2} - 2\sqrt{2} + 2}{4 - 2} = \dfrac{6 - 4\sqrt{2}}{2} = 0.17$

5. $D: [-3, 2]; \ R = [-2, 2]; \ f(-0.5) = -2$

7. (a) Because $4 - x \not< 0: \ 4 - x \geq 0 \Rightarrow -x \geq -4 \Rightarrow x \leq 4$ or $D: \{x \,|\, x \leq 4\}$

   (b) $f(-1) = \sqrt{4 - (-1)} = \sqrt{5}; \ f(2a) = \sqrt{4 - (2a)}$

9. (a) Using $(0, -1)$ and $(2, 0)$, $m = \dfrac{0 - (-1)}{2 - 0} = \dfrac{1}{2}; \ m = \dfrac{1}{2}; \ y$-intercept $(0, -1): \ -1;$ and x-intercept $(2, 0): \ 2.$

   (b) Using slope-intercept form: $f(x) = \dfrac{1}{2}x - 1.$

   (c) ... Graphical: Since the point $(-2, -2)$ is on the graph of $y = f(x)$ it follows that $f(-2) = -2.$

   (d) $f(x) = 0$ at $x = 2.$

11. $\dfrac{f(x + h) - f(x)}{h} = \dfrac{3(x + h)^2 - 3x^2}{h} = \dfrac{3(x^2 + 2xh + h^2) - 3x^2}{h} = \dfrac{3x^2 + 6xh + 3h^2 - 3x^2}{h} = $

   $\dfrac{6xh + 3h^2}{h} = 6x + 3h$

13. x-intercept: $y = 0: \ -2x + 5(0) = 20 \Rightarrow -2x = 20 \Rightarrow x = -10.$ The x-intercept $-10.$

   y-intercept: $x = 0: \ -2(0) + 5y = 20 \Rightarrow 5y = 20 \Rightarrow y = 4.$ The y-intercept: $4.$

15. (a) $2(1 - 2x) = 5 - (4 - x) \Rightarrow 2 - 4x = 1 + x \Rightarrow 1 = 5x \Rightarrow x = \dfrac{1}{5}$

   (b) $2e^x - 1 = 27 \Rightarrow 2e^x = 28 \Rightarrow e^x = 14 \Rightarrow \ln e^x = \ln 14 \Rightarrow x = \ln 14 \Rightarrow x \approx 2.64$

   (c) $|4 - 5x| = 8 \Rightarrow 4 - 5x = -8$ or $4 - 5x = 8; \ 4 - 5x = -8 \Rightarrow -5x = -12 \Rightarrow x = \dfrac{12}{5}$ or

   $4 - 5x = 8 \Rightarrow -5x = 4 \Rightarrow x = -\dfrac{4}{5}.$ Therefore $x = -\dfrac{4}{5}, \dfrac{12}{5}.$

   (d) $2x^2 + x = 1 \Rightarrow 2x^2 + x - 1 = 0 \Rightarrow (2x - 1)(x + 1) = 0 \Rightarrow x = -1, \dfrac{1}{2}$

   (e) $x^3 - 3x^2 + 2x = 0 \Rightarrow x(x^2 - 3x + 2) = 0 \Rightarrow x(x - 2)(x - 1) = 0 \Rightarrow x = 0, 1, 2$

   (f) $x^4 + 8 = 6x^2 \Rightarrow x^4 - 6x^2 + 8 = 0 \Rightarrow (x^2 - 4)(x^2 - 2) = 0 \Rightarrow (x + 2)(x - 2)(x^2 - 2) = 0 \Rightarrow$

   $x = -2, 2, \pm\sqrt{2}$

   (g) $x^2 - x - 2 = 0 \Rightarrow (x - 2)(x + 1) = 0 \Rightarrow x = -1, 2$

   (h) $\sqrt{2x - 1} = x - 2 \Rightarrow 2x - 1 = x^2 - 4x + 4 \Rightarrow x^2 - 6x + 5 = 0 \Rightarrow (x - 5)(x - 1) = 0 \Rightarrow$

   $x = 1, 5;$ Checking: $x = 1$ we get $\sqrt{2(1) - 1} = 1 - 2 \Rightarrow \sqrt{1} = -1$ which is false therefore $x = 5.$

   (i) $\dfrac{x}{x - 2} = \dfrac{2x - 1}{x + 1} \Rightarrow x(x + 1) = (2x - 1)(x - 2) \Rightarrow x^2 + x = 2x^2 - 5x + 2 \Rightarrow x^2 - 6x + 2 = 0$

   Using the quadratic formula: $\dfrac{6 \pm \sqrt{36 - 4(1)(2)}}{2(1)} = \dfrac{6 \pm \sqrt{28}}{2} = 3 \pm \sqrt{7}$

   (j) $\log_2(x + 1) = 16 \Rightarrow x + 1 = 2^{16} \Rightarrow x + 1 = 65,536 \Rightarrow x = 65,535$

17. (a) $-3(2 - x) < 4 - (2x + 1) \Rightarrow -6 + 3x < 3 - 2x \Rightarrow 5x < 9 \Rightarrow x < \dfrac{9}{5} \Rightarrow \left(-\infty, \dfrac{9}{5}\right)$

(b) $-3 \le 4 - 3x < 6 \Rightarrow -7 \le -3x < 2 \Rightarrow \dfrac{7}{3} \ge x > -\dfrac{2}{3} \Rightarrow \left(-\dfrac{2}{3}, \dfrac{7}{3}\right]$

(c) The solutions to $|4x - 3| \ge 9$ satisfy $x \le s_1$ or $x \ge s_2$ where s_1 and s_2 are the solutions to $|4x - 3| = 9$.

$|4x - 3| = 9$ is equivalent to $4x - 3 = -9 \Rightarrow x = -\dfrac{3}{2}$ and $4x - 3 = 9 \Rightarrow x = 3$.

The interval is $\left(-\infty, -\dfrac{3}{2}\right] \cup [3, \infty)$.

(d) For $x^2 - 5x + 4 \le 0$ first we solve $x^2 - 5x + 4 = 0 \Rightarrow (x - 4)(x - 1) = 0 \Rightarrow x = 1, 4$. These are the

boundary numbers and divide the number line into three sections: $(-\infty, 1], [1, 4]$, and $[4, \infty)$.

Testing a value in each section we get: $x = 0 \Rightarrow 0^2 - 5(0) + 4 \le 0 \Rightarrow 4 \le 0$, which is false;

$x = 2 \Rightarrow 2^2 - 5(2) + 4 \le 0 \Rightarrow 4 - 10 + 4 \le 0 \Rightarrow -2 \le 0$, which is true;

$x = 5 \Rightarrow 5^2 - 5(5) + 4 \le 0 \Rightarrow 25 - 25 + 4 \le 0 \Rightarrow 4 \le 0$, which is false.

Therefore the solution is; $[1, 4]$.

(e) For $t^3 - t > 0$ first solve $t^3 - t = 0 \Rightarrow t(t^2 - 1) = 0 \Rightarrow t(t + 1)(t - 1) = 0 \Rightarrow t = -1, 0, 1$. These

are the boundary numbers and divide the number line into four sections:

$(-\infty, -1), (-1, 0), (0, 1)$, and $(1, \infty)$.

Testing a value in each section we get: $x = -2 \Rightarrow -2^3 - (-2) > 0 \Rightarrow -6 > 0$, which is false;

$x = -\dfrac{1}{2} \Rightarrow \left(-\dfrac{1}{2}\right)^3 - \left(-\dfrac{1}{2}\right) > 0 \Rightarrow -\dfrac{1}{8} + \dfrac{1}{2} > 0 \Rightarrow \dfrac{3}{8} > 0$, which is true;

$x = \dfrac{1}{2} \Rightarrow \left(\dfrac{1}{2}\right)^3 - \dfrac{1}{2} > 0 \Rightarrow \dfrac{1}{8} - \dfrac{1}{2} > 0 \Rightarrow -\dfrac{3}{8} > 0$, which is false;

$x = 2 \Rightarrow 2^3 - 2 > 0 \Rightarrow 8 - 2 > 0 \Rightarrow 6 > 0$, which is true.

Therefore the solution is: $(-1, 0) \cup (1, \infty)$.

(f) One boundary number is: $t + 2 = 0 \Rightarrow t = -2$. For the other we solve: $\dfrac{1}{t + 2} - 3 \ge 0 \Rightarrow$

$1 - 3(t + 2) \ge 0 \Rightarrow 1 - 3t - 6 \ge 0 \Rightarrow -3t \ge 5 \Rightarrow t \le -\dfrac{5}{3}$. The two boundary numbers are:

-2 and $-\dfrac{5}{3} \Rightarrow$ the possible intervals are: $(-\infty, -2], \left[-2, -\dfrac{5}{3}\right]$, and $[-2, \infty]$.

Testing a value in each gives us: $x = -3 \Rightarrow \dfrac{1}{-3 + 2} - 3 \ge 0 \Rightarrow -4 \ge 0$, which is false;

$x = -\dfrac{11}{6} \Rightarrow \dfrac{1}{-\frac{11}{6} + 2} - 3 \ge 0 \Rightarrow 3 \ge 0$, which is true;

$x = 0 \Rightarrow \dfrac{1}{0 + 2} - 3 \ge 0 \Rightarrow -\dfrac{5}{2} \ge 0$, which is false. Therefore the solution is $\left(-2, -\dfrac{5}{3}\right]$.

19. $-2x^2 + 6x - 1 = 0 \Rightarrow -2x^2 + 6x = 1 \Rightarrow -2(x^2 - 3x) = 1 \Rightarrow -2\left(x^2 - 3x + \dfrac{9}{4}\right) = 1 + \left(-2\left(\dfrac{9}{4}\right)\right) \Rightarrow$

$-2\left(x^2 - \dfrac{3}{2}\right)^2 + \dfrac{7}{2} = 0 \Rightarrow f(x) = -2\left(x^2 - \dfrac{3}{2}\right)^2 + \dfrac{7}{2}$

21. (a) $a < 0$ because the parabola opens down which means a negative leading coefficient.

(b) The graph shows zeros at : $x = -2, 1$.

(c) Since there are two real solutions the discriminant $b^2 - 4ac > 0$ or positive.

(d) $f(x) = a(x + 2)(x - 1) \Rightarrow 4 = a(0 + 2)(0 - 1) \Rightarrow 4 = -2a \Rightarrow a = -2$

The complete factored form is: $f(x) = -2(x + 2)(x - 1)$.

23. (a) Shift f one unit right and 2 units up. See Figure 23a.

(b) Make graph shorter by taking $\dfrac{1}{2}$ of all y-coordinates of f. See Figure 23b.

(c) Reflect f across both the x-axis and y-axis. Since the graph is symmetric to the y-axis only the reflection across the x-axis will show a result. See Figure 23c.

(d) Make the graph narrower by taking $\dfrac{1}{2}$ of all x-coordinates of f. See Figure 23d.

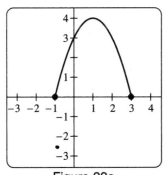

Figure 23a

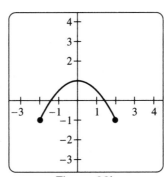

Figure 23b

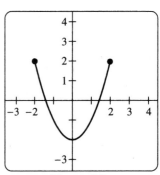

Figure 23c

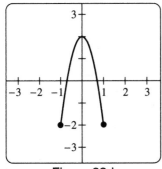

Figure 23d

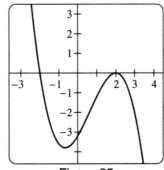

Figure 25

25. See Figure 25. *Answers may vary*

27. (a) $\dfrac{5a^4 - 2a^2 + 4}{2a^2} = \dfrac{5a^4}{2a^2} - \dfrac{2a^2}{2a^2} + \dfrac{4}{2a^2} = \dfrac{5a^2}{2} - 1 + \dfrac{2}{2a^2}$

(b)
$$
\begin{array}{r}
x^3 + 3x^2 + 3x + 1 + \frac{1}{x-1} \\
x - 1\overline{\smash{)}x^4 + 2x^3 \qquad\quad - 2x} \\
\underline{x^4 - x^3} \\
3x^3 \qquad - 2x \\
\underline{3x^3 - 3x^2} \\
3x^2 - 2x \\
\underline{3x^2 - 3x} \\
x \\
\underline{x - 1} \\
1
\end{array}
$$

The quotient is: $x^3 + 3x^2 + 3x + 1 + \dfrac{1}{x-1}$.

(c)
$$
\begin{array}{r}
x - 3 + \frac{4}{x^2+1} \\
x^2 + 1\overline{\smash{)}x^3 - 3x^2 + x + 1} \\
\underline{x^3 \qquad\quad + x} \\
-3x^2 \qquad + 1 \\
\underline{-3x^2 \qquad - 3} \\
4
\end{array}
$$

The quotient is: $x - 3 + \dfrac{4}{x^2+1}$.

29. $f(x) = -2(x+2)(x+1)^2(x-3)^2$

31. Since the graph crosses the x-axis at $x = -2$ and 2, $(x+2)(x-2)$ are two of the four factors. Since the graph touches the x-axis at $x = 0$, x is a factor of even multiplicity. Therefore, $f(x) = x^2(x+2)(x-2)$ and since its end behavior is going down, the complete factored form is: $f(x) = -x^2(x+2)(x-2)$.

33. $3x - 7 \neq 0 \Rightarrow 3x \neq 7 \Rightarrow x \neq \dfrac{7}{3}$. Then $D = \left\{x \mid x \neq \dfrac{7}{3}\right\}$; the vertical asymptote is the denominator $=$

$0 \Rightarrow x = \dfrac{7}{3}$; and the horizontal asymptote is the ratio of the lead coefficients $\Rightarrow y = \dfrac{2}{3}$.

35. (a) $(f+g)(2) = f(2) + g(2) = 2 + 3 = 5$

(b) $(g/f)(4) = g(4)/f(4) = 1/0 \Rightarrow$ undefined

(c) $(f \circ g)(3) = f(g(3)) = f(2) = 2$

(d) $(f^{-1} \circ g)(1) = f^{-1}(g(1)) = f^{-1}(4) = 0$

37. (a) $f(2) = 2^2 + 3(2) - 2 = 4 + 6 - 2 = 8$; $g(2) = 2 - 2 = 0$; $(f+g)(2) = 8 + 0 = 8$

(b) $(g \circ f)(1) = g(f(1)) \Rightarrow f(1) = 1^2 + 3(1) - 2 = 1 + 3 - 2 = 2$. Then $g(2) = 2 - 2 = 0 \Rightarrow$

$(g \circ f)(1) = 0$

(c) $f(x) = x^2 + 3x - 2$ and $g(x) = x - 2 \Rightarrow (f - g)(x) = x^2 + 3x - 2 - (x - 2) \Rightarrow (f - g)(x) = x^2 + 2x$

(d) $(f \circ g)(x) = f(g(x)) \Rightarrow (f \circ g)(x) = (x - 2)^2 + 3(x - 2) - 2 \Rightarrow$

$(f \circ g)(x) = x^2 - 4x + 4 + 3x - 6 - 2 \Rightarrow (f \circ g)(x) = x^2 - x - 4$

39. For $(f + g)(x) = \dfrac{1}{x} + \sqrt{2 - x} \Rightarrow x \neq 0$ and $x \not> 2$ or $D = \{x \mid x \leq 2, x \neq 0\}$.

    For $(f \circ g)(x) = f(g(x)) \Rightarrow (f \circ g)(x) = \dfrac{1}{\sqrt{2 - x}} \Rightarrow x \not\geq 2$ or $D = \{x \mid x < 2\}$.

41. For each unit increase in x, the y-value is multiplied by $\dfrac{2}{3}$. This is an exponential function with $c = 9$ and

    $a = \dfrac{2}{3}$, so $f(x) = 9\left(\dfrac{2}{3}\right)^x$.

43. From the graph when $x = 0$, $y = \dfrac{1}{2}$, $\Rightarrow c = \dfrac{1}{2}$ and when $x = 1$, $y = 1 \Rightarrow$ using $y = Ca^x \Rightarrow 1 = \dfrac{1}{2}a^1 \Rightarrow$

    $a = 2$. So $C = \dfrac{1}{2}$ and $a = 2$.

45. (a) $\log 100 \Rightarrow 10^x = 100 \Rightarrow x = 2 \Rightarrow \log 100 = 2$

    (b) $\log_2 16 \Rightarrow 2^x = 16 \Rightarrow x = 4 \Rightarrow \log_2 16 = 4$

    (c) $\ln \dfrac{1}{e^2} \Rightarrow e^x = \dfrac{1}{e^2} \Rightarrow x = -2 \Rightarrow \ln \dfrac{1}{e^2} = -2$

    (d) $\log_4 \sqrt[3]{4} \Rightarrow 4^x = \sqrt[3]{4}$ or $4^x = 4^{1/3} \Rightarrow x = \dfrac{1}{3} \Rightarrow \log_4 \sqrt[3]{4} = \dfrac{1}{3}$

    (e) $\log 4 + \log 25 = \log 4(25) = \log 100 \Rightarrow 10^x = 100 \Rightarrow x = 2 \Rightarrow \log 4 + \log 25 = 2$

    (f) $\log_6 24 - \log_6 4 = \log_6 \dfrac{24}{4} = \log_6 6 \Rightarrow 6^x = 6 \Rightarrow x = 1 \Rightarrow \log_6 24 - \log_6 4 = 1$

47. $\log_2 \dfrac{\sqrt[3]{x^2 - 4}}{\sqrt{x^2 + 4}} = \log_2 \dfrac{(x^2 - 4)^{1/3}}{(x^2 + 4)^{1/2}} = \log_2 (x^2 - 4)^{1/3} - \log_2 (x^2 + 4)^{1/2} =$

    $\dfrac{1}{3} \log_2 (x^2 - 4) - \dfrac{1}{2} \log_2 (x^2 + 4) = \dfrac{1}{3} \log_2 [(x + 2)(x - 2)] - \dfrac{1}{2} \log_2 (x^2 + 4) =$

    $\dfrac{1}{3}[\log_2 (x + 2) + \log_2 (x - 2)] - \dfrac{1}{2} \log_2 (x^2 + 4) = \dfrac{1}{3} \log_2 (x + 2) + \dfrac{1}{3} \log_2 (x - 2) - \dfrac{1}{2} \log_2 (x^2 + 4)$

49. $\log_3 125 = 4.395$

51. Complimentary angle $\alpha = 90° - 10° \, 34' = 89° \, 60' - 10° \, 34' = 79° \, 26'$

    Supplementary angle $\beta = 180° - 10° \, 34' = 179° \, 60' - 10° \, 34' = 169° \, 26'$

53. $150°\left(\dfrac{\pi}{180°}\right) = \dfrac{5\pi}{6}$

55. Convert $15°$ to radians $\Rightarrow 15°\left(\dfrac{\pi}{180°}\right) = \dfrac{15\pi}{180} = \dfrac{\pi}{12}$; $s = 3\left(\dfrac{\pi}{12}\right) = \dfrac{3\pi}{12} = \dfrac{\pi}{4} \approx 0.79$ ft.

57. $\tan 30° = \dfrac{9}{a} \Rightarrow a \tan 30° = 9 \Rightarrow a = \dfrac{9}{\tan 30°} \Rightarrow a \approx 15.6$

59. $r = \sqrt{(-3)^2 + (-3)^2} = \sqrt{18} = 3\sqrt{2}$

    $\sin \theta = \dfrac{-3}{3\sqrt{2}} = -\dfrac{1}{\sqrt{2}} \qquad \cos \theta = \dfrac{-3}{3\sqrt{2}} = -\dfrac{1}{\sqrt{2}} \qquad \tan \theta = \dfrac{-3}{-3} = 1$

    $\csc \theta = \dfrac{3\sqrt{2}}{-3} = -\sqrt{2} \qquad \sec \theta = \dfrac{3\sqrt{2}}{-3} = -\sqrt{2} \qquad \cot \theta = \dfrac{-3}{-3} = 1$

61. Refer to Figure 61 and use the Pythagorean theorem to find the opposite side.

$5^2 + x^2 = 13^2, 25 + x^2 = 169, x^2 = 144, x = 12.$

$$\sin \theta = -\frac{12}{13} \qquad\qquad \cos \theta = \frac{5}{13} \qquad\qquad \tan \theta = -\frac{12}{5}$$

$$\csc \theta = -\frac{13}{12} \qquad\qquad \sec \theta = \frac{13}{5} \qquad\qquad \cot \theta = -\frac{5}{12}$$

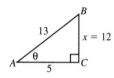

Figure 61

63. (a) See Figure 63a.

(b) See Figure 63b.

(c) See Figure 63c.

(d) See Figure 63d.

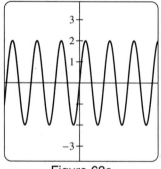

Figure 63a

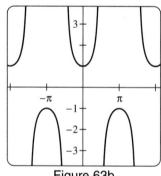

Figure 63b

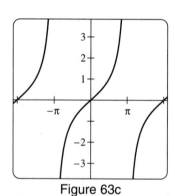

Figure 63c

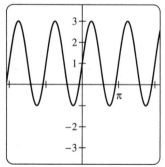

Figure 63d

65. $\sin \alpha = \frac{11}{20} \Rightarrow \alpha = \sin^{-1}\left(\frac{11}{20}\right) \Rightarrow \alpha \approx 33.4°, \cos \beta = \frac{11}{20} \Rightarrow \beta = \cos^{-1}\left(\frac{11}{20}\right) \Rightarrow \beta = 56.6.$

Using the Pythagorean theorem: $11^2 + b^2 = 20^2 \Rightarrow 121 + b^2 = 400 \Rightarrow b^2 = 279 \Rightarrow b \approx 16.7.$

Applications

67. If $V = \pi r^2 h$, then $12 = \pi(1)^2 h \Rightarrow 12 = \pi h \Rightarrow h = \dfrac{12}{\pi} \Rightarrow h = 3.82$ in.

69. (a) $t = 0 \Rightarrow D(t) = 16(0)^2 \Rightarrow (0, 0)$; $t = 1 \Rightarrow D(t) = 16(1)^2 \Rightarrow (1, 16)$

    The average rate of change from 0 to 1 is: $\dfrac{16 - 0}{1 - 0} = 16$ ft./sec.

    $t = 1 \Rightarrow D(t) = 16(1)^2 \Rightarrow (1, 16)$; $t = 2 \Rightarrow D(t) = 16(2)^2 \Rightarrow (2, 64)$

    The average rate of change from 1 to 2 is: $\dfrac{64 - 16}{2 - 1} = 48$ ft./sec.

    (b) During the first second, the average speed is 16 ft./sec. During the next second, the average speed is 48 ft./sec. The object is speeding up.

    (c) The difference quotient is: $\dfrac{16(t + h)^2 - 16t^2}{h} = \dfrac{16(t^2 + 2th + h^2) - 16t^2}{h} =$

    $\dfrac{16t^2 + 32th + 16h^2 - 16t^2}{h} = 32t + 16h$

71. $F = \dfrac{k}{x} \Rightarrow 150 = \dfrac{k}{(4000)^2} \Rightarrow k = 2{,}400{,}000{,}000.$ Now, $F = \dfrac{2{,}400{,}000{,}000}{(10{,}000)^2} \Rightarrow F = \dfrac{2{,}400{,}000{,}000}{100{,}000{,}000} \Rightarrow$

    $F = 24$ lbs.

73. (a) Use (1980, 800) and (2000, 3500), now find $m = \dfrac{3500 - 800}{2000 - 1980} \Rightarrow m = \dfrac{2700}{20} \Rightarrow m = 135.$

    Using $C(t) = m(t - t_1) + y_1$ we get: $C(t) = 135(t - 1980) + 800$

    (b) $3000 = 135(t - 1980) + 800 \Rightarrow 2200 = 135(t - 1980) \Rightarrow 16.3 = t - 1980 \Rightarrow t = 1996.3$ or the year 1996 which does agree with the actual, the formula gives the correct year.

75. (a) As t increases, r also increases.

    (b) $V = \dfrac{4}{3}\pi r^3 \Rightarrow V(t) = \dfrac{4}{3}\pi(\sqrt{t})^3$

    (c) $V(4) = \dfrac{4}{3}\pi(\sqrt{4})^3 \Rightarrow V(4) = \dfrac{4}{3}\pi(2)^3 \Rightarrow V(4) = \dfrac{4}{3}\pi(8) \Rightarrow V(4) = \dfrac{32}{3}\pi \approx 33.5$; after 4 seconds the volume of the balloon is about 33.5 in.

77. (a) $300{,}000 = 200{,}000 e^{k(3)} \Rightarrow \dfrac{3}{2} = e^{3k} \Rightarrow \ln\dfrac{3}{2} = \ln e^{3k} \Rightarrow \ln\dfrac{3}{2} = 3k \Rightarrow \dfrac{\ln\frac{3}{2}}{3} = k \Rightarrow k \approx 0.135$

    The formula that models this is: $N(t) = 200{,}000 e^{0.135t}$

    (b) $N(5) = 200{,}000 e^{0.135(5)} \Rightarrow N(5) = 200{,}000 e^{0.675} \Rightarrow N(5) \approx 392{,}807$; after 5 hours there are about 392,807 bacteria per ml. in the sample.

    (c) $500{,}000 = 200{,}000 e^{0.135t} \Rightarrow \dfrac{5}{2} = e^{0.135t} \Rightarrow \ln\dfrac{5}{2} = \ln e^{0.135t} \Rightarrow \ln\dfrac{5}{2} = 0.135t \Rightarrow \dfrac{\ln\frac{5}{2}}{0.135} = t \Rightarrow$

    $t \approx 6.8$ hrs.

79. $A = P\left(1 + \dfrac{r}{n}\right)^{nt} \Rightarrow \dfrac{A}{P} = \left(1 + \dfrac{r}{n}\right)^{nt} \Rightarrow \dfrac{A}{P} = \left[\left(1 + \dfrac{r}{n}\right)^n\right]^t \Rightarrow \ln\left(\dfrac{A}{P}\right) = \ln\left[\left(1 + \dfrac{r}{n}\right)^n\right]^t \Rightarrow$

    $\ln\left(\dfrac{A}{P}\right) = t \ln\left(1 + \dfrac{r}{n}\right)^n \Rightarrow \dfrac{\ln\left(\frac{A}{P}\right)}{\ln\left(a + \frac{r}{n}\right)^n} = t \Rightarrow t = \dfrac{\ln\left(\frac{A}{P}\right)}{n \ln\left(1 + \frac{r}{n}\right)}$

81. $\tan 43° = \dfrac{6}{x}$, where x = shadow length $\Rightarrow x = \dfrac{6}{\tan 43°} \Rightarrow x \approx 6.4$ feet.

83. The maximum monthly average temperature is 90° F and the minimum is 61° F. The midpoint of these values is $0.5(90 + 61) = 75.5$, thus $d = 75.5$. Half the difference between these temperatures is $0.5(90 - 61) = 14.5$, thus $a = 14.5$. Since the temperatures cycle every 12 months, $b = \dfrac{2\pi}{12} = \dfrac{\pi}{6}$. The maximum of the $y = \sin x$ graph occurs when $x = \dfrac{\pi}{2}$, while the maximum in the table occurs when $x = 7$. Thus $\dfrac{\pi}{6}(7 - c) = \dfrac{\pi}{2} \Rightarrow$ $7 - c = 3 \Rightarrow c = 4$. The function $f(x) = 14.5 \sin\left(\dfrac{\pi}{6}(x - 4)\right) + 75.5$.

Chapter 7: Trigonometric Identities and Equations

7.1: Fundamental Identities

Fundamental Identities

1. $\cot \theta = \dfrac{1}{\tan \theta} = \dfrac{1}{\frac{1}{2}} = 2$

3. $\sec \theta = \dfrac{1}{\cos \theta} = \dfrac{1}{\frac{2}{7}} = \dfrac{7}{2}$

5. $\cos \theta = \dfrac{1}{\sec \theta} = \dfrac{1}{-4} = -\dfrac{1}{4}$

7. $\tan \theta = \dfrac{\sin \theta}{\cos \theta} = \dfrac{\frac{3}{5}}{-\frac{4}{5}} = -\dfrac{3}{4}$ $\qquad \csc \theta = \dfrac{1}{\sin \theta} = \dfrac{1}{\frac{3}{5}} = \dfrac{5}{3}$

$\quad\ \ \sec \theta = \dfrac{1}{\cos \theta} = \dfrac{1}{-\frac{4}{5}} = -\dfrac{5}{4}$ $\qquad \cot \theta = \dfrac{1}{\tan \theta} = \dfrac{1}{-\frac{3}{4}} = -\dfrac{4}{3}$

9. $\cos \theta = \sin \theta \cot \theta = -\dfrac{24}{25}\left(\dfrac{7}{24}\right) = -\dfrac{7}{25}$ $\qquad \tan \theta = \dfrac{1}{\cot \theta} = \dfrac{1}{\frac{7}{24}} = \dfrac{24}{7}$

$\quad\ \ \csc \theta = \dfrac{1}{\sin \theta} = \dfrac{1}{-\frac{24}{25}} = -\dfrac{25}{24}$ $\qquad \sec \theta = \dfrac{1}{\cos \theta} = \dfrac{1}{-\frac{7}{25}} = -\dfrac{25}{7}$

11. $\tan \theta = \dfrac{\sin \theta}{\cos \theta} = \dfrac{-\frac{60}{61}}{-\frac{11}{61}} = \dfrac{60}{11}$ $\qquad \cot \theta = \dfrac{\cos \theta}{\sin \theta} = \dfrac{-\frac{11}{61}}{-\frac{60}{61}} = \dfrac{11}{60}$

$\quad\ \ \csc \theta = \dfrac{1}{\sin \theta} = \dfrac{1}{-\frac{60}{61}} = -\dfrac{61}{60}$ $\qquad \sec \theta = \dfrac{1}{\cos \theta} = \dfrac{1}{-\frac{11}{61}} = -\dfrac{61}{11}$

13. $\sin \theta = \dfrac{1}{\csc \theta} = \dfrac{1}{\sqrt{2}}$ $\qquad\qquad \cos \theta = \dfrac{1}{\sec \theta} = -\dfrac{1}{\sqrt{2}}$

$\quad\ \ \tan \theta = \dfrac{\sin \theta}{\cos \theta} = \dfrac{\frac{1}{\sqrt{2}}}{-\frac{1}{\sqrt{2}}} = -\dfrac{1}{1} = -1$ $\quad \cot \theta = \dfrac{\cos \theta}{\sin \theta} = \dfrac{-\frac{1}{\sqrt{2}}}{\frac{1}{\sqrt{2}}} = -\dfrac{1}{1} = -1$

15. $\sec \theta \cos \theta = \dfrac{1}{\cos \theta} \cdot \cos \theta = 1$

17. $\sin \theta \csc \theta = \sin \theta \cdot \dfrac{1}{\sin \theta} = 1$

19. $(\sin^2 \theta + \cos^2 \theta)^3 = 1^3 = 1$

21. $1 - \sin^2 \theta = 1 - (1 - \cos^2 \theta) = \cos^2 \theta$

23. $\sec^2 \theta - 1 = (1 + \tan^2 \theta) - 1 = \tan^2 \theta$

25. $\dfrac{\sin (-\theta)}{\cos (-\theta)} = \dfrac{-\sin \theta}{\cos \theta} = -\tan \theta$

27. $\dfrac{\sin^2 \theta + \cos^2 \theta}{\cos \theta} = \dfrac{1}{\cos \theta} = \sec \theta$

29. $\dfrac{\sec^2 (-\theta)}{\csc^2 \theta} = \dfrac{\sec^2 \theta}{\csc^2 \theta} = \left(\dfrac{\sec \theta}{\csc \theta}\right)^2 = \left(\dfrac{\frac{1}{\cos \theta}}{\frac{1}{\sin \theta}}\right)^2 = \left(\dfrac{1}{\cos \theta} \cdot \dfrac{\sin \theta}{1}\right)^2 = \left(\dfrac{\sin \theta}{\cos \theta}\right)^2 = \tan^2 \theta$

31. $\dfrac{\cot x}{\csc x} = \dfrac{\frac{\cos x}{\sin x}}{\frac{1}{\sin x}} = \dfrac{\cos x}{\sin x} \cdot \dfrac{\sin x}{1} = \cos x$

33. $(\sin^2 x)(1 + \cot^2 x) = (\sin^2 x)(\csc^2 x) = (\sin^2 x)\left(\dfrac{1}{\sin^2 x}\right) = 1$

35. If $\sin \theta < 0$ and $\cos \theta > 0$ then any point (x, y) on the terminal side of θ must satisfy $y < 0$ and $x > 0$. Thus,

   θ is contained in Quadrant IV. To support this result numerically, table the following in degree mode:

   $Y_1 = \sin (X)$ and $Y_2 = \cos (X)$ starting at $x = 270$, incrementing by 15. See Figure 35.

37. If $\sec \theta < 0$ and $\sin \theta < 0$ then any point (x, y) on the terminal side of θ must satisfy the following conditions:

   $y < 0$ and $\dfrac{r}{x} < 0 \Rightarrow x < 0$. Thus, θ is contained in quadrant III. For numerical support, table the following

   in degree mode: $Y_1 = 1/\cos (X)$ and $Y_2 = \sin (X)$ starting at $x = 180$, incrementing by 15. See Figure 37.

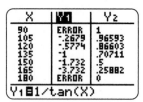

X	Y1	Y2
270	-1	0
285	-.9659	.25882
300	-.866	.5
315	-.7071	.70711
330	-.5	.86603
345	-.2588	.96593
360	0	1

Y1=sin(X)

Figure 35

X	Y1	Y2
180	-1	0
195	-1.035	-.2588
210	-1.155	-.5
225	-1.414	-.7071
240	-2	-.866
255	-3.864	-.9659
270	ERROR	-1

Y1=1/cos(X)

Figure 37

X	Y1	Y2
90	ERROR	1
105	-.2679	.96593
120	-.5774	.86603
135	-1	.70711
150	-1.732	.5
165	-3.732	.25882
180	ERROR	0

Y1=1/tan(X)

Figure 39

39. If $\cot \theta < 0$ and $\sin \theta > 0$ then any point (x, y) on the terminal side of θ must satisfy the following conditions:

   $y > 0$ and $\dfrac{x}{y} < 0 \Rightarrow x < 0$. Thus, θ is contained in quadrant II. For numerical support, table the following

   in degree mode: $Y_1 = 1/\tan (X)$ and $Y_2 = \sin (X)$ starting at $x = 90$, incrementing by 15. See Figure 39.

41. $\sin^2 \theta + \cos^2 \theta = 1 \Rightarrow \sin \theta = \pm\sqrt{1 - \cos^2 \theta}$ and since $\sin \theta < 0$ we know $\sin \theta = -\sqrt{1 - \cos^2 \theta}$.

   $\sin \theta = -\sqrt{1 - \cos^2 \theta} = -\sqrt{1 - \left(\dfrac{1}{2}\right)^2} = -\sqrt{1 - \dfrac{1}{4}} = -\sqrt{\dfrac{3}{4}} = -\dfrac{\sqrt{3}}{2}$ $\tan \theta = \dfrac{\sin \theta}{\cos \theta} = \dfrac{-\frac{\sqrt{3}}{2}}{\frac{1}{2}} = -\sqrt{3}$

   $\csc \theta = \dfrac{1}{\sin \theta} = \dfrac{1}{-\frac{\sqrt{3}}{2}} = -\dfrac{2}{\sqrt{3}}$ $\sec \theta = \dfrac{1}{\cos \theta} = \dfrac{1}{\frac{1}{2}} = 2$ $\cot \theta = \dfrac{1}{\tan \theta} = \dfrac{1}{-\sqrt{3}} = -\dfrac{1}{\sqrt{3}}$

43. $1 + \cot^2 \theta = \csc^2 \theta \Rightarrow \csc \theta = \pm\sqrt{1 + \cot^2 \theta}$ and since $\csc \theta < 0$, $\csc \theta = -\sqrt{1 + \cot^2 \theta}$

   $\cot \theta = \dfrac{1}{\tan \theta} = \dfrac{1}{-\frac{11}{60}} = -\dfrac{60}{11}$

   $\csc \theta = -\sqrt{1 + \cot^2 \theta} = -\sqrt{1 + \left(-\dfrac{60}{11}\right)^2} = -\sqrt{1 + \dfrac{3600}{121}} = -\sqrt{\dfrac{3721}{121}} = -\dfrac{61}{11}$

   $\sin \theta = \dfrac{1}{\csc \theta} = \dfrac{1}{-\frac{61}{11}} = -\dfrac{11}{61}$ $\cos \theta = \dfrac{\sin \theta}{\tan \theta} = \dfrac{-\frac{11}{61}}{-\frac{11}{60}} = \dfrac{60}{61}$ $\sec \theta = \dfrac{1}{\cos \theta} = \dfrac{1}{\frac{60}{61}} = \dfrac{61}{60}$

45. $\sin^2 \theta + \cos^2 \theta = 1 \Rightarrow \cos \theta = \pm\sqrt{1 - \sin^2 \theta}$ and since $\cos \theta > 0$, $\cos \theta = \sqrt{1 - \sin^2 \theta}$

   $\cos \theta = \sqrt{1 - \sin^2 \theta} = \sqrt{1 - \left(\dfrac{7}{25}\right)^2} = \sqrt{1 - \dfrac{49}{625}} = \sqrt{\dfrac{576}{625}} = \dfrac{24}{25}$ $\tan \theta = \dfrac{\sin \theta}{\cos \theta} = \dfrac{\frac{7}{25}}{\frac{24}{25}} = \dfrac{7}{24}$

   $\csc \theta = \dfrac{1}{\sin \theta} = \dfrac{1}{\frac{7}{25}} = \dfrac{25}{7}$ $\sec \theta = \dfrac{1}{\cos \theta} = \dfrac{1}{\frac{24}{25}} = \dfrac{25}{24}$ $\cot \theta = \dfrac{1}{\tan \theta} = \dfrac{1}{\frac{7}{24}} = \dfrac{24}{7}$

47. If $\sin\theta < 0$ and $\sec\theta < 0$, θ lies in quadrant III, $\sin^2\theta + \cos^2\theta = 1 \Rightarrow \left(-\dfrac{1}{3}\right)^2 + \cos^2\theta = 1 \Rightarrow$

$\cos^2\theta = 1 - \dfrac{1}{9} = \dfrac{8}{9} \Rightarrow \cos\theta = \dfrac{\sqrt{8}}{3} \Rightarrow \cos\theta = \dfrac{2\sqrt{2}}{3} \Rightarrow \cos\theta = -\dfrac{2\sqrt{2}}{3}$ in quadrant III.

$\tan\theta = \dfrac{\sin\theta}{\cos\theta} = \dfrac{-\frac{1}{3}}{-\frac{2\sqrt{2}}{3}} = \dfrac{1}{2\sqrt{2}}$ \qquad $\cot\theta = \dfrac{\cos\theta}{\sin\theta} = \dfrac{-\frac{2\sqrt{2}}{3}}{-\frac{1}{3}} = 2\sqrt{2} = \sqrt{8}$

$\csc\theta = \dfrac{1}{\sin\theta} = \dfrac{1}{-\frac{1}{3}} = -3$ \qquad $\csc\theta = \dfrac{1}{\cos\theta} = \dfrac{1}{-\frac{2\sqrt{2}}{3}} = -\dfrac{3}{2\sqrt{2}} = -\dfrac{3}{\sqrt{8}}$

49. $\sin^2\theta + \cos^2\theta = 1 \Rightarrow \sin\theta = \pm\sqrt{1-\cos^2\theta}$ and since θ is in quadrant IV, $\sin\theta = -\sqrt{1-\cos^2\theta}$.

$\cos\theta = \dfrac{1}{\sec\theta} = \dfrac{1}{3}$ \qquad $\sin\theta = -\sqrt{1-\cos^2\theta} = -\sqrt{1-\left(\dfrac{1}{3}\right)^2} = -\sqrt{1-\dfrac{1}{9}} = -\sqrt{\dfrac{8}{9}} = -\dfrac{\sqrt{8}}{3}$

$\tan\theta = \dfrac{\sin\theta}{\cos\theta} = \dfrac{-\frac{\sqrt{8}}{3}}{\frac{1}{3}} = -\sqrt{8}$ \qquad $\csc\theta = \dfrac{1}{\sin\theta} = \dfrac{1}{-\frac{\sqrt{8}}{3}} = -\dfrac{3}{\sqrt{8}}$ \qquad $\cot\theta = \dfrac{1}{\tan\theta} = \dfrac{1}{-\sqrt{8}} = -\dfrac{1}{\sqrt{8}}$

51. $\sin^2\theta + \cos^2\theta = 1 \Rightarrow \cos\theta = \pm\sqrt{1-\sin^2\theta}$ and since θ is in quadrant II, $\cos\theta = -\sqrt{1-\sin^2\theta}$.

$\sin\theta = \dfrac{1}{\csc\theta} = \dfrac{1}{\frac{7}{3}} = \dfrac{3}{7}$ \qquad $\cos\theta = -\sqrt{1-\sin^2\theta} = -\sqrt{1-\left(\dfrac{3}{7}\right)^2} = -\sqrt{1-\dfrac{9}{49}} = -\sqrt{\dfrac{40}{49}} = -\dfrac{\sqrt{40}}{7}$

$\tan\theta = \dfrac{\sin\theta}{\cos\theta} = \dfrac{\frac{3}{7}}{-\frac{\sqrt{40}}{7}} = -\dfrac{3}{\sqrt{40}}$ \qquad $\sec\theta = \dfrac{1}{\cos\theta} = \dfrac{1}{-\frac{\sqrt{40}}{7}} = -\dfrac{7}{\sqrt{40}}$ \qquad $\cot\theta = \dfrac{1}{\tan\theta} = \dfrac{1}{-\frac{3}{\sqrt{40}}} = -\dfrac{\sqrt{40}}{3}$

53. If $\csc x > 0$, $\sin x > 0$. Then $\sin^2 x + \cos^2 x = 1 \Rightarrow \sin^2 x = 1 - \cos^2 x \Rightarrow \sin x = \sqrt{1-\cos^2 x}$, thus

$\tan x = \dfrac{\sin x}{\cos x} \Rightarrow \tan x = \dfrac{\sqrt{1-\cos^2 x}}{\cos x}$.

55. If $\cos x < 0$, $\sec x < 0$. Then $\sec^2 x = 1 + \tan^2 x \Rightarrow \sec x = -\sqrt{1+\tan^2 x}$ and

$\sin x = \dfrac{\tan x}{\sec x} \Rightarrow \sin x = -\dfrac{\tan x}{\sqrt{1+\tan^2 x}}$.

57. If $\cot x < 0$, $\cot^2 x = \csc^2 x - 1 \Rightarrow \cot x = -\sqrt{\csc^2 x - 1}$ and $\cos x = \dfrac{\cot x}{\csc x} \Rightarrow \cos x = -\dfrac{\sqrt{\csc^2 x - 1}}{\csc x}$.

59. $\cos^2\theta + \sin^2\theta = 1 \Rightarrow \cos\theta = \pm\sqrt{1-\sin^2\theta}$ and since θ is acute, $\cos\theta = \sqrt{1-\sin^2\theta}$,

thus, $\cos\theta = \sqrt{1-x^2} = \sqrt{1-0.5126^2} \approx 0.8586$.

61. $1 + \cot^2\theta = \csc^2\theta \Rightarrow \dfrac{1}{\sin^2\theta} = 1 + \cot^2\theta \Rightarrow \sin^2\theta = \dfrac{1}{1+\cot^2\theta} \Rightarrow \sin\theta = \pm\sqrt{\dfrac{1}{1+\cot^2\theta}}$

Since θ is in quadrant III, $\sin\theta = -\sqrt{\dfrac{1}{1+\cot^2\theta}}$, $\sin\theta = -\sqrt{\dfrac{1}{1+x^2}} = -\sqrt{\dfrac{1}{1+0.5126^2}} \approx -0.8899$.

63. If θ is acute, θ lies in quadrant I, $\sin^2\theta + \cos^2\theta = 1 \Rightarrow \cos^2\theta = 1 - \sin\theta \Rightarrow \cos\theta = \sqrt{1-\sin^2\theta}$,

$\tan\theta = \dfrac{\sin\theta}{\sqrt{1-\sin^2\theta}}$. Since $\sin\theta = x \Rightarrow \tan\theta = \dfrac{x}{\sqrt{1-x^2}}$.

65. Since sine is an odd function, $\sin(-13°) = -\sin 13°$.

67. Since tangent is an odd function, $\tan\left(-\dfrac{\pi}{11}\right) = -\tan\dfrac{\pi}{11}$.

69. Since secant is an even function, $\sec\left(-\dfrac{2\pi}{5}\right) = \sec\dfrac{2\pi}{5}$.

71. Yes, since $\sin^2\theta + \cos^2\theta = 1$ and $1 + \tan^2\theta = \sec^2\theta$. For numerical support, table the following in degree mode: $Y_1 = (\sin(X))^2 + (\cos(X))^2 + (\tan(X))^2$ and $Y_2 = (1/\cos(X))^2$ starting at $x = 0$, incrementing by 50. See Figure 71.

73. No, since $(\sin\theta + \cos\theta)^2 = \sin^2\theta + 2\sin\theta\cos\theta + \cos^2\theta = 1 + 2\sin\theta\cos\theta \neq 1$.

    For numerical support, table the following in degree mode:

    $Y_1 = (\sin(X) + (\cos(X))^2$ and $Y_2 = 1$ starting at $x = 0$, incrementing by 50. See Figure 73.

[0, 2, 0.5] by [-1, 40, 8]

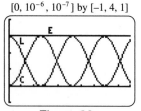

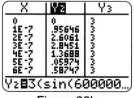

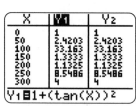

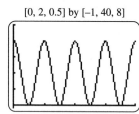

Figure 71   Figure 73   Figure 75   Figure 81

75. Yes, since $1 + \tan^2\theta = \sec^2\theta = \dfrac{1}{\cos^2\theta}$. For numerical support, table the following in degree mode:

    $Y_1 = 1 + (\tan(X))^2$ and $Y_2 = 1/(\cos(X))^2$ starting at $x = 0$, incrementing by 50. See Figure 75.77.

    This is not correct. The variable should be included. For example $\cos^2 0° + \sin^2 90° = 1^2 + 1^2 = 2 \neq 1$.

Applications

79. $d = 93{,}000{,}000 \cdot \dfrac{1}{\sin\theta} = 93{,}000{,}000\,\csc\theta$

81. (a) $P = ky^2 = k(4\cos(2\pi t))^2 = 16k\cos^2(2\pi t)$

    (b) Graph $Y_1 = 32(\cos(2\pi X))^2$ in $[0, 2, 0.5]$ by $[-1, 40, 8]$. See Figure 81. P has a maximum value of 32 when $t = 0, 0.5, 1, 1.5, 2$ and P has a minimum value of 0 when $t = 0.25, 0.75, 1.25, 1.75$. The spring is either stretched or compressed the most when P is maximum.

    (c) $P = 16k\cos^2(2\pi t) = 16k(1 - \sin^2(2\pi t))$

83. (a) Graph $Y_1 = 3(\cos(6{,}000{,}000X))^2$, $Y_2 = 3(\sin(6{,}000{,}000X))^2$ and $Y_3 = Y_1 + Y_2$ in $[0, 10^{-6}, 10^{-7}]$ by $[-1, 4, 1]$. See Figure 83a. The total energy E (the sum of L and C) is always 3.

    (b) Table Y_1, Y_2 and Y_3 starting at $x = 0$, incrementing by 10^{-7}. See Figure 83b.

    (c) $E(t) = 3\cos^2(6{,}000{,}000t) + 3\sin^2(6{,}000{,}000t) = 3(\cos^2(6{,}000{,}000t) + \sin^2(6{,}000{,}000t)) = 3(1) = 3$

$[0, 10^{-6}, 10^{-7}]$ by $[-1, 4, 1]$ $[-6, 6, 1]$ by $[0, 100, 10]$

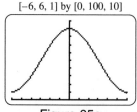

Figure 83a   Figure 83b   Figure 85a   Figure 85b

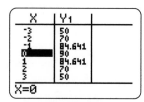

85. (a) In radian mode, the graph has y-axis symmetry. The monthly high temperatures x months before and x months after July are equal. See Figure 85a.

   (b) A table of Y_1 is shown in Figure 85b. f is an even function since the sign of the input does not affect the output.

   (c) Symbolically this symmetry can be expressed as $f(-x) = f(x)$.

7.2: Verifying Identities

Simplifying Expressions

1. (a) $(1 + x)(1 - x) = 1 - x^2$

   (b) $(1 + \sin \theta)(1 - \sin \theta) = 1 - \sin^2 \theta = \cos^2 \theta$

3. (a) $x(x - 1) = x^2 - x$

   (b) $\sec \theta (\sec \theta - 1) = \sec^2 \theta - \sec \theta$

5. (a) $x^2 + 2x + 1 = (x + 1)(x + 1)$

   (b) $\cos^2 \theta + 2 \cos \theta + 1 = (\cos \theta + 1)(\cos \theta + 1)$

7. (a) $x^2 - 2x = x(x - 2)$

   (b) $\sec^2 t - 2 \sec t = \sec t (\sec t - 2)$

9. (a) $\dfrac{1}{1 - x} + \dfrac{1}{1 + x} = \dfrac{1 + x}{(1 - x)(1 + x)} + \dfrac{1 - x}{(1 - x)(1 + x)} = \dfrac{1 + x + 1 - x}{(1 - x)(1 + x)} = \dfrac{2}{1 - x^2}$

   (b) $\dfrac{1}{1 - \cos \theta} + \dfrac{1}{1 + \cos \theta} = \dfrac{1 + \cos \theta}{(1 - \cos \theta)(1 + \cos \theta)} + \dfrac{1 - \cos \theta}{(1 - \cos \theta)(1 + \cos \theta)} =$

   $\dfrac{1 + \cos \theta + 1 - \cos \theta}{(1 - \cos \theta)(1 + \cos \theta)} = \dfrac{2}{1 - \cos^2 \theta} = \dfrac{2}{\sin^2 \theta} = 2 \csc^2 \theta$

11. (a) $\dfrac{x}{y} + \dfrac{y}{x} = \dfrac{x^2}{xy} + \dfrac{y^2}{xy} = \dfrac{x^2 + y^2}{xy}$

   (b) $\dfrac{\cos t}{\sin t} + \dfrac{\sin t}{\cos t} = \dfrac{\cos^2 t}{\cos t \sin t} + \dfrac{\sin^2 t}{\cos t \sin t} = \dfrac{\cos^2 t + \sin^2 t}{\cos t \sin t} = \dfrac{1}{\cos t \sin t} = \sec t \csc t$

13. (a) $\dfrac{1}{\frac{1}{y^2}} + \dfrac{1}{\frac{1}{x^2}} = y^2 + x^2$

   (b) $\dfrac{1}{\csc^2 t} + \dfrac{1}{\sec^2 t} = \sin^2 t + \cos^2 t = 1$

15. (a) $\dfrac{\frac{x}{y}}{\frac{1}{y}} = \dfrac{x}{y} \cdot \dfrac{y}{1} = x$

   (b) $\dfrac{\cot \theta}{\csc \theta} = \dfrac{\frac{\cos \theta}{\sin \theta}}{\frac{1}{\sin \theta}} = \dfrac{\cos \theta}{\sin \theta} \cdot \dfrac{\sin \theta}{1} = \cos \theta$

17. $\cos \theta \tan \theta = \cos \theta \cdot \dfrac{\sin \theta}{\cos \theta} = \sin \theta$

19. $\tan \theta (\cos \theta - \csc \theta) = \dfrac{\sin \theta}{\cos \theta} \left(\cos \theta - \dfrac{1}{\sin \theta} \right) = \sin \theta - \dfrac{1}{\cos \theta} = \sin \theta - \sec \theta$

21. $(1 + \tan t)^2 = 1 + 2 \tan t + \tan^2 t = 2 \tan t + (1 + \tan^2 t) = 2 \tan t + \sec^2 t$

23. $\dfrac{\csc^2 \theta - 1}{\csc^2 \theta} = \dfrac{\csc^2 \theta}{\csc^2 \theta} - \dfrac{1}{\csc^2 \theta} = 1 - \sin^2 \theta = \cos^2 \theta$

25. $1 - \tan^2 \theta = (1 - \tan \theta)(1 + \tan \theta)$

27. $\sec^2 t - \sec t - 6 = (\sec t - 3)(\sec t + 2)$

29. $\tan^4 \theta + 3 \tan^2 \theta + 2 = (\tan^2 \theta + 1)(\tan^2 \theta + 2) = \sec^2 \theta (\tan^2 \theta + 2)$

Verifying Identities

31. $\csc^2 \theta - \cot^2 \theta = (1 + \cot^2 \theta) - \cot^2 \theta = 1$

33. $(1 - \sin t)^2 = (1 - \sin t)(1 - \sin t) = 1 - \sin t - \sin t + \sin^2 t = 1 - 2 \sin t + \sin^2 t$

35. $\dfrac{\sin t + \cos t}{\sin t} = \dfrac{\sin t}{\sin t} + \dfrac{\cos t}{\sin t} = 1 + \cot t$

37. $\sec^2 \theta - 1 = (1 + \tan^2 \theta) - 1 = \tan^2 \theta$

39. $\dfrac{\tan^2 t}{\sec t} = \dfrac{\sec^2 t - 1}{\sec t} = \dfrac{\sec^2 t}{\sec t} - \dfrac{1}{\sec t} = \sec t - \cos t$

41. $\dfrac{\sec t}{1 + \sec t} = \dfrac{\sec t}{1 + \sec t} \cdot \dfrac{\cos t}{\cos t} = \dfrac{1}{\cos t + 1}$

43. $(\sec t - 1)(\sec t + 1) = \sec^2 t - 1 = \tan^2 t$

45. $\dfrac{1 - \sin^2 \theta}{\cos \theta} = \dfrac{\cos^2 \theta}{\cos \theta} = \cos \theta$

47. $\dfrac{\sec t}{\tan t} - \dfrac{\tan t}{\sec t} = \dfrac{\sec^2 t - \tan^2 t}{\sec t \tan t} = \dfrac{(1 + \tan^2 t) - \tan^2 t}{\sec t \tan t} = \dfrac{1}{\sec t \tan t} = \cos t \cot t$

49. $\dfrac{\cot^2 t}{\csc t + 1} = \dfrac{\csc^2 t - 1}{\csc t + 1} = \dfrac{(\csc t - 1)(\csc t + 1)}{\csc t + 1} = \csc t - 1$

51. $\dfrac{\cot t}{\cot t + 1} = \dfrac{\cot t}{\cot t + 1} \cdot \dfrac{\tan t}{\tan t} = \dfrac{\cot t \tan t}{\cot t \tan t + \tan t} = \dfrac{1}{1 + \tan t}$

53. $\dfrac{1}{1 - \sin t} + \dfrac{1}{1 + \sin t} = \dfrac{(1 + \sin t) + (1 - \sin t)}{(1 - \sin t)(1 + \sin t)} = \dfrac{2}{1 - \sin^2 t} = \dfrac{2}{\cos^2 t} = 2 \sec^2 t$

55. $\dfrac{\csc t + \cot t}{\csc t - \cot t} = \dfrac{\csc t + \cot t}{\csc t - \cot t} \cdot \dfrac{\csc t + \cot t}{\csc + \cot t} = \dfrac{(\csc t + \cot t)^2}{\csc^2 t - \cot^2 t} = \dfrac{(\csc t + \cot t)^2}{\csc^2 t - (\csc^2 t - 1)} = (\csc t + \cot t)^2$

57. $\dfrac{\cos^2 t}{1 - \sin t} = \dfrac{1 - \sin^2 t}{1 - \sin t} = \dfrac{(1 + \sin t)(1 - \sin t)}{1 - \sin t} = 1 + \sin t$

59. $\dfrac{1}{1 + \sin \theta} = \dfrac{1}{1 + \sin \theta} \cdot \dfrac{1 - \sin \theta}{1 - \sin \theta} = \dfrac{1 - \sin \theta}{1 - \sin^2 \theta} = \dfrac{1 - \sin \theta}{\cos^2 \theta}$

61. $\sqrt{1 - \sin^2 \theta} = \sqrt{\cos^2 \theta} = \pm \cos \theta = \cos \theta$, since θ is acute.

63. $\dfrac{1 + 2 \sin x + \sin^2 x}{\cos^2 x} = \dfrac{(1 + \sin x)^2}{1 - \sin^2 x} = \dfrac{(1 + \sin x)(1 + \sin x)}{(1 - \sin x)(1 + \sin x)} = \dfrac{1 + \sin x}{1 - \sin x}$

65. $(1 - \cos^2 x)(1 + \cos^2 x) = \sin^2 x(1 + (1 - \sin^2 x)) = \sin^2 x(2 - \sin^2 x) = 2 \sin^2 x - \sin^4 x$

67. $\cot \theta \sin \theta = \dfrac{\cos \theta}{\sin \theta} \cdot \sin \theta = \cos \theta$

   Graph $Y_1 = (\cos (X)/\sin (X)) \sin (X)$ and $Y_2 = \cos (X)$ in $[-2\pi, 2\pi, \pi/2]$ by $[-4, 4, 1]$.

   Table Y_1 and Y_2 together in degree mode starting at $x = 0$, incrementing by 50.

   Graph Y_1 is shown in Figure 67a. Graph Y_2 is shown in Figure 67b. The table is shown in Figure 67c.

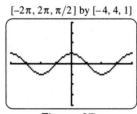

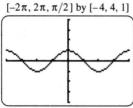
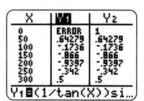

Figure 67a	Figure 67b	Figure 67c

69. $(1 - \cos^2 \theta)(1 + \tan^2 \theta) = \sin^2 \theta \sec^2 \theta = \sin^2 \theta \cdot \dfrac{1}{\cos^2 \theta} = \dfrac{\sin^2 \theta}{\cos^2 \theta} = \tan^2 \theta$

   Graph $Y_1 = (1 - (\cos (X))^2)(1 + (\tan (X))^2)$ and $Y_2 = (\tan (X))^2$ in $[-2\pi, 2\pi, \pi/2]$ by $[-4, 4, 1]$.

   Table Y_1 and Y_2 together in degree mode starting at $x = 0$, incrementing by 50.

   Graph Y_1 is shown in Figure 69a. Graph Y_2 is shown in Figure 69b. The table is shown in Figure 69c.

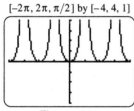
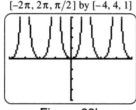

Figure 69a	Figure 69b	Figure 69c

71. $\cos t (\tan t - \sec t) = \cos t \left(\dfrac{\sin t}{\cos t} - \dfrac{1}{\cos t} \right) = \sin t - 1$

   Graph $Y_1 = \cos (X)(\tan (X) - 1/\cos (X))$ and $Y_2 = \sin (X) - 1$ in $[-2\pi, 2\pi, \pi/2]$ by $[-4, 4, 1]$.

   Table Y_1 and Y_2 together in degree mode starting at $x = 0$, incrementing by 50.

   Graph Y_1 is shown in Figure 71a. Graph Y_2 is shown in Figure 71b. The table is shown in Figure 71c.

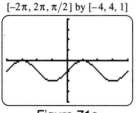
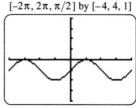

Figure 71a	Figure 71b	Figure 71c

73. $\dfrac{\tan(-\theta)}{\sin(-\theta)} = \dfrac{-\tan\theta}{-\sin\theta} = \dfrac{\frac{\sin\theta}{\cos\theta}}{\frac{\sin\theta}{1}} = \dfrac{\sin\theta}{\cos\theta}\cdot\dfrac{1}{\sin\theta} = \dfrac{1}{\cos\theta} = \sec\theta$

Graph $Y_1 = \tan(-X)/\sin(-X)$ and $Y_2 = 1/\cos(X)$ in $[-2\pi, 2\pi, \pi/2]$ by $[-4, 4, 1]$.

Table Y_1 and Y_2 together in degree mode starting at $x = 0$, incrementing by 50.

Graph Y_1 is shown in Figure 73a. Graph Y_2 is shown in Figure 73b. The table is shown in Figure 73c.

$[-2\pi, 2\pi, \pi/2]$ by $[-4, 4, 1]$	$[-2\pi, 2\pi, \pi/2]$ by $[-4, 4, 1]$		$[0, 1/15, 1/60]$ by $[-30, 30, 10]$
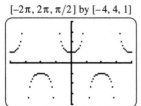	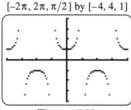	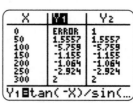	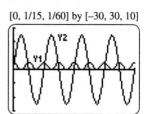
Figure 73a	Figure 73b	Figure 73c	Figure 75

Application

75. (a) $W(t) = 5\cos^2(120\pi t) = 5(1 - \sin^2(120\pi t))$

$V = 25\sin(120\pi t)$ is maximum when $\sin(120\pi t) = 1$ and is minimum when $\sin(120\pi t) = -1$.

Thus, $W = 5(1 - (\pm 1)^2) = 0$.

(b) Graph $Y_1 = 5(\cos(120\pi X))^{\wedge}2$ and $Y_2 = 25\sin(120\pi X)$ in $[0, 1/15, 1/60]$ by $[-30, 30, 10]$.

See Figure 75. The wattage $Y_1 = 0$ whenever the voltage $Y_2 = \pm 160$.

7.3: Trigonometric Equations

Reference Angles

1.　$\theta = 120°$ is in quadrant II, $\theta_R = 180° - 120° = 60°$.

3.　$\theta = 85°$ is in quadrant I, $\theta_R = 85°$.

5.　$\theta = -65°$ is coterminal with $295°$ in quadrant IV, $\theta_R = 360° - 295° = 65°$.

7.　$\theta = \dfrac{5\pi}{6}$ is in quadrant II, $\theta_R = \pi - \dfrac{5\pi}{6} = \dfrac{\pi}{6}$.

9.　$\theta = -\dfrac{2\pi}{3}$ is coterminal with $\dfrac{4\pi}{3}$ in quadrant III, $\theta_R = \dfrac{4\pi}{3} - \pi = \dfrac{\pi}{3}$.

Solving Trigonometric Equations

11.　At the intersection points $t = \dfrac{\pi}{4}$ and $t = \dfrac{5\pi}{4}$ therefore, $\sin t = \cos t$ for $t = \dfrac{\pi}{4}, \dfrac{5\pi}{4}$.

13.　At the intersection points $t = \dfrac{\pi}{3}$ and $t = \dfrac{5\pi}{3}$ therefore, $3\cot t = 2\sin t$ for $t = \dfrac{\pi}{3}, \dfrac{5\pi}{3}$.

15.　(a) Start by solving $\sin\theta_R = 1$: $\theta_R = \sin^{-1} 1 = 90°$. This angle is quadrantal, thus $\theta = 90°$ or $\theta = \dfrac{\pi}{2}$.

(b) Start by solving $\sin\theta_R = 1$: $\theta_R = \sin^{-1} 1 = 90°$. This angle is quadrantal, thus $\theta = 270°$ or $\theta = \dfrac{3\pi}{2}$.

17. (a) Start by solving $\tan \theta_R = \sqrt{3}$: $\theta_R = \tan^{-1} \sqrt{3} = 60°$. The tangent function is positive in quadrants I and III.

Angles in these quadrants with a $60°$ reference angle are $60°$ and $240°$ or $\dfrac{\pi}{3}$ and $\dfrac{4\pi}{3}$.

(b) Start by solving $\tan \theta_R = \sqrt{3}$: $\theta_R = \tan^{-1} \sqrt{3} = 60°$. The tangent function is negative in quadrants II and IV.

Angles in these quadrants with a $60°$ reference angle are $120°$ and $300°$ or $\dfrac{2\pi}{3}$ and $\dfrac{5\pi}{3}$.

19. First note that $\sec \theta = 2$ when $\cos \theta = \dfrac{1}{2}$ and $\sec \theta = -2$ when $\cos \theta = -\dfrac{1}{2}$.

(a) Start by solving $\cos \theta_R = \dfrac{1}{2}$: $\theta_R = \cos^{-1} \dfrac{1}{2} = 60°$. The secant function is positive in quadrants I and IV.

Angles in these quadrants with a $60°$ reference angle are $60°$ and $300°$ or $\dfrac{\pi}{3}$ and $\dfrac{5\pi}{3}$.

(b) Start by solving $\cos \theta_R = \dfrac{1}{2}$: $\theta_R = \cos^{-1} \dfrac{1}{2} = 60°$. The secant function is negative in quadrants II and III.

Angles in these quadrants with a $60°$ reference angle are $120°$ and $240°$ or $\dfrac{2\pi}{3}$ and $\dfrac{4\pi}{3}$.

21. (a) No solution since $\sin \theta \neq 3$ for any θ.

(b) No solution since $\sin \theta \neq -3$ for any θ.

23. (a) $2x - 1 = 0 \Rightarrow 2x = 1 \Rightarrow x = \dfrac{1}{2}$

(b) $2 \sin \theta - 1 = 0 \Rightarrow 2 \sin \theta = 1 \Rightarrow \sin \theta = \dfrac{1}{2}$

Since $\theta_R = \sin^{-1} \dfrac{1}{2} = 30°$ and sine is positive in quadrants I and II, $\theta = 30°, 150°$.

25. (a) $x^2 = x \Rightarrow x^2 - x = 0 \Rightarrow x(x - 1) = 0 \Rightarrow x = 0, 1$

(b) $\sin^2 \theta = \sin \theta \Rightarrow \sin^2 \theta - \sin \theta = 0 \Rightarrow \sin \theta (\sin \theta - 1) = 0 \Rightarrow \sin \theta = 0, 1$

Since $\theta_R = \sin^{-1} 0 = 0°$ or $\theta_R = \sin^{-1} 1 = 90°$ and sine is positive in quadrants I and II, $\theta = 0°, 90°, 180°$.

27. (a) $x^2 + 1 = 2 \Rightarrow x^2 = 1 \Rightarrow x = \pm 1$

(b) $\tan^2 \theta + 1 = 2 \Rightarrow \tan^2 \theta = 1 \Rightarrow \tan \theta = \pm 1$

Since $\theta_R = \tan^{-1} 1 = 45°$ and tangent is positive or negative in all quadrants, $\theta = 45°, 135°, 225°, 315°$.

29. (a) $x^2 + x = 2 \Rightarrow x^2 + x - 2 = 0 \Rightarrow (x + 2)(x - 1) = 0 \Rightarrow x = -2, 1$

(b) $\cos^2 \theta + \cos \theta = 2 \Rightarrow \cos^2 \theta + \cos \theta - 2 = 0 \Rightarrow (\cos \theta + 2)(\cos \theta - 1) = 0 \Rightarrow \cos \theta = -2, 1$

Since $\theta_R = \cos^{-1} 1 = 0°$ ($\cos \theta = -2$ is not possible) and cosine is positive in quadrants I and IV, $\theta = 0°$.

31. $\tan^2 t - 3 = 0 \Rightarrow \tan^2 t = 3 \Rightarrow \tan t = \pm \sqrt{3}$

Since $t_R = \tan^{-1} \sqrt{3} = \dfrac{\pi}{3}$ and tangent is positive or negative in all quadrants, $t = \dfrac{\pi}{3}, \dfrac{2\pi}{3}, \dfrac{4\pi}{3}, \dfrac{5\pi}{3}$.

33. $3 \cos t + 4 = 0 \Rightarrow 3 \cos t = -4 \Rightarrow \cos t = -\dfrac{4}{3}$

$\cos t = -\dfrac{4}{3}$ is not possible. No solution.

35. $\sin t \cos t = \cos t \Rightarrow \sin t \cos t - \cos t = 0 \Rightarrow \cos t (\sin t - 1) = 0 \Rightarrow \cos t = 0$ or $\sin t = 1$

Since $t_R = \cos^{-1} 0 = \dfrac{\pi}{2}$ or $t_R = \sin^{-1} 1 = \dfrac{\pi}{2}$ and the angle is quadrantal, $t = \dfrac{\pi}{2}, \dfrac{3\pi}{2}$.

37. $\csc^2 t = 2\cot t \Rightarrow 1 + \cot^2 t = 2\cot t \Rightarrow \cot^2 t - 2\cot t + 1 = 0 \Rightarrow (\cot t - 1)^2 = 0 \Rightarrow \cot t = 1$

    Since $t_R = \cot^{-1} 1 = \tan^{-1} 1 = \dfrac{\pi}{4}$ and tangent is positive in quadrants I and III, $t = \ldots, -\dfrac{3\pi}{4}, \dfrac{\pi}{4}, \dfrac{5\pi}{4}, \dfrac{9\pi}{4}, \ldots$.

    $t = \dfrac{\pi}{4} + \pi n$ for $n = 0, \pm 1, \pm 2, \pm 3, \ldots$.

39. $\sin^2 t = \dfrac{1}{4} \Rightarrow \sin t = \pm\dfrac{1}{2}$

    Since $t_R = \sin^{-1}\dfrac{1}{2} = \dfrac{\pi}{6}$ and sine is positive or negative in all quadrants, $t = \dfrac{\pi}{6}, \dfrac{5\pi}{6}, \dfrac{7\pi}{6}, \dfrac{11\pi}{6}$.

41. $\sin t \cos t = 0 \Rightarrow \sin t = 0$ or $\cos t = 0$. Since $t_R = \sin^{-1} 0 = 0$ or $t_R = \cos^{-1} 0 = \dfrac{\pi}{2}$, $t = 0, \dfrac{\pi}{2}, \pi, \dfrac{3\pi}{2}$.

43. $\tan^2 t - 1 = 0 \Rightarrow \tan^2 t = 1 \Rightarrow \tan t = \pm 1$

    Since $t_R = \tan^{-1} 1 = \dfrac{\pi}{4}$ and tangent is positive or negative in all quadrants, $t = \pm\dfrac{\pi}{4}, \pm\dfrac{3\pi}{4}, \pm\dfrac{5\pi}{4}, \ldots$.

    $t = \dfrac{\pi}{4} + \dfrac{\pi}{2}n$ for $n = 0, \pm 1, \pm 2, \pm 3, \ldots$.

45. $\sin^2 t + \sin t - 20 = 0 \Rightarrow (\sin t + 5)(\sin t - 4) = 0 \Rightarrow \sin t = -5, 4$

    $\sin t = -5$ and $\sin t = 4$ are both not possible. No solution.

47. $\cos t \sin t = \sin t \Rightarrow \cos t \sin t - \sin t = 0 \Rightarrow \sin t(\cos t - 1) = 0 \Rightarrow \sin t = 0$ or $\cos t = 1$

    Since $t_R = \sin^{-1} 0 = 0$ or $t_R = \cos^{-1} 1 = 0$ and the angle is quadrantal, $t = 0, \pm\pi, \pm 2\pi, \ldots$.

    $t = \pi n$ for $n = 0, \pm 1, \pm 2, \pm 3, \ldots$.

49. $\sec^2 t = 2\tan t \Rightarrow 1 + \tan^2 t = 2\tan t \Rightarrow \tan^2 t - 2\tan t + 1 = 0 \Rightarrow (\tan t - 1)^2 = 0 \Rightarrow \tan t = 1$

    Since $t_R = \tan^{-1} 1$ and tangent is positive in quadrants I and III, $t = \dfrac{\pi}{4}, \dfrac{5\pi}{4}$.

51. $\sin^2 t \cos^2 t = 0 \Rightarrow \sin^2 t = 0$ or $\cos^2 t = 0 \Rightarrow \sin t = 0$ or $\cos t = 0$

    Since $t_R = \sin^{-1} 0 = 0$ or $t_R = \cos^{-1} 0 = \dfrac{\pi}{2}$ and the angle is quadrantal, $t = 0, \pm\dfrac{\pi}{2}, \pm\pi, \pm\dfrac{3\pi}{2}, \ldots$.

    $t = \dfrac{\pi}{2}n$ for $n = 0, \pm 1, \pm 2, \pm 3, \ldots$.

53. $\sin t + \cos t = 1 \Rightarrow \sin^2 t + 2\sin t \cos t + \cos^2 t = 1 \Rightarrow 2\sin t \cos t = 0 \Rightarrow \sin t = 0$ or $\cos t = 0$

    Since $t_R = \sin^{-1} 0 = 0$ or $t_R = \cos^{-1} 0 = \dfrac{\pi}{2}$ and the angle is quadrantal, $t = \ldots, -\dfrac{3\pi}{2}, 0, \dfrac{\pi}{2}, 2\pi, \ldots$.

    $t = 2\pi n, \dfrac{\pi}{2} + 2\pi n$ for $n = 0, \pm 1, \pm 2, \pm 3, \ldots$.

55. $\sin 3t = \dfrac{1}{2}$, let $\theta = 3t$ so then the equation becomes $\sin \theta = \dfrac{1}{2}$ and in radian measure these solutions are

    $\theta = \dfrac{\pi}{6}$ and $\dfrac{5\pi}{6}$. For all real number solutions $\theta = \dfrac{\pi}{6} + 2\pi n$ or $\theta = \dfrac{5\pi}{6} + 2\pi n$. Because $\theta = 3t$, we can

    determine t by substituting $3t$ for $\theta \Rightarrow 3t = \dfrac{\pi}{6} + 2\pi n$ or $3t = \dfrac{5\pi}{6} + 2\pi n$. Therefore $t = \dfrac{\pi}{18} + \dfrac{2\pi}{3}n$ or

    $t = \dfrac{5\pi}{18} + \dfrac{2\pi}{3}n$.

57. $\cos 4t = -\dfrac{\sqrt{3}}{2}$, let $\theta = 4t$ so then the equation becomes $\cos \theta = -\dfrac{\sqrt{3}}{2}$ and in radian measure these solutions

are $\theta = \dfrac{5\pi}{6}$ and $\dfrac{7\pi}{6}$. For all real number solutions $\theta = \dfrac{5\pi}{6} + 2\pi n$ or $\theta = \dfrac{7\pi}{6} + 2\pi n$. Because $\theta = 4t$, we can

determine t by substituting $4t$ for $\theta \Rightarrow 4t = \dfrac{5\pi}{6} + 2\pi n$ or $4t = \dfrac{7\pi}{6} + 2\pi n$. Therefore $t = \dfrac{5\pi}{24} + \dfrac{\pi}{2}n$ or

$t = \dfrac{7\pi}{24} + \dfrac{\pi}{2}n$.

59. $\tan 5t = 1$, let $\theta = 5t$ so then the equation becomes $\tan \theta = 1$ and in radian measure these solutions are

$\theta = \dfrac{\pi}{4}$ and $\dfrac{5\pi}{4}$. For all real number solutions $\theta = \dfrac{\pi}{4} + \pi n$ or $\theta = \dfrac{5\pi}{4} + \pi n$. Since these representations are

equivalent, we use $\theta = \dfrac{\pi}{4} + \pi n$. Because $\theta = 5t$, we can determine t by substituting $5t$ for $\theta \Rightarrow 5t = \dfrac{\pi}{4} + \pi n$.

Therefore $t = \dfrac{\pi}{20} + \dfrac{\pi}{5}n$.

61. $2 \sin 4t = -1 \Rightarrow \sin 4t = -\dfrac{1}{2}$, let $\theta = 4t$ so then the equation becomes $\sin \theta = -\dfrac{1}{2}$ and in radian measure these

solutions are $\theta = \dfrac{7\pi}{6}$ and $\dfrac{11\pi}{6}$. For all real number solutions $\theta = \dfrac{7\pi}{6} + 2\pi n$ or $\theta = \dfrac{11\pi}{6} + 2\pi n$. Because

$\theta = 4t$, we can determine t by substituting $4t$ for $\theta \Rightarrow 4t = \dfrac{7\pi}{6} + 2\pi n$ or $4t = \dfrac{11\pi}{6} + 2\pi n$. Therefore

$t = \dfrac{7\pi}{24} + \dfrac{\pi}{2}n$ or $t = \dfrac{11\pi}{24} + \dfrac{\pi}{2}n$.

63. $-\sec 4t = \sqrt{2} \Rightarrow \sec 4t = -\sqrt{2}$, let $\theta = 4t$ so then the equation becomes $\sec \theta = -\sqrt{2}$ or $\cos \theta = \dfrac{1}{-\sqrt{2}}$ and

in radian measure these solutions are $\theta = \dfrac{3\pi}{4}$ and $\dfrac{5\pi}{4}$. For all real number solutions $\theta = \dfrac{3\pi}{4} + 2\pi n$ or $\theta =$

$\dfrac{5\pi}{4} + 2\pi n$. Because $\theta = 4t$, we can determine t by substituting $4t$ for $\theta \Rightarrow 4t = \dfrac{3\pi}{4} + 2\pi n$ or

$4t = \dfrac{5\pi}{4} + 2\pi n$. Therefore $t = \dfrac{3\pi}{16} + \dfrac{\pi}{2}n$ or $t = \dfrac{5\pi}{16} + \dfrac{\pi}{2}n$.

65. $2 \sin 8t - 3 = -1 \Rightarrow 2 \sin 8t = 2 \Rightarrow \sin 8t = 1$, let $\theta = 8t$ so then the equation becomes $\sin \theta = 1$ and in

radian measure this solution is $\theta = \dfrac{\pi}{2}$. For all real number solutions $\theta = \dfrac{\pi}{2} + 2\pi n$. Because $\theta = 8t$, we can

determine t by substituting $8t$ for $\theta \Rightarrow 8t = \dfrac{\pi}{2} + 2\pi n$. Therefore $t = \dfrac{\pi}{16} + \dfrac{\pi}{4}n$.

67. $\cot 4t + 5 = 6 \Rightarrow \cot 4t = 1$, let $\theta = 4t$ so then the equation becomes $\cot \theta = 1$ or $\tan \theta = 1$ and in radian

measure this solution is $\theta = \dfrac{\pi}{4}$. For all real number solutions $\theta = \dfrac{\pi}{4} + \pi n$. Because $\theta = 4t$, we can determine

t by substituting $4t$ for $\theta \Rightarrow 4t = \dfrac{\pi}{4} + \pi n$. Therefore $t = \dfrac{\pi}{16} + \dfrac{\pi}{4}n$.

69. $5 \cos 3t = 1 \Rightarrow \cos 3t = \dfrac{1}{5}$, let $\theta = 3t$ so then the equation becomes $\cos \theta = \dfrac{1}{5}$ and in radian measure these

solutions are $\theta \approx 1.369$ and 4.914. For all real number solutions $\theta \approx 1.369 + 2\pi n$ or $\theta \approx 4.914 + 2\pi n$.

Because $\theta = 3t$ we can determine t by substituting $3t$ for $\theta \Rightarrow 3t \approx 1.369 + 2\pi n$ or $3t \approx 4.914 + 2\pi n$.

Therefore $t \approx 0.456 + \dfrac{2}{3}\pi n$ or $t \approx 1.638 + \dfrac{2}{3}\pi n$.

71. $\sin 2t = \dfrac{1}{3}$, let $\theta = 2t$ so then the equation becomes $\sin \theta = \dfrac{1}{3}$ and in radian measure these solutions are

$\theta \approx 0.340$ and 2.802. For all real number solutions $\theta \approx 0.340 + 2\pi n$ or $\theta \approx 2.802 + 2\pi n$. Because

$\theta = 2t$ we can determine t by substituting $2t$ for $\theta \Rightarrow 2t \approx 0.340 + 2\pi n$ or $2t \approx 2.802 + 2\pi n$. Therefore

$t \approx 0.170 + \pi n$ or $t \approx 1.401 + \pi n$.

73. Graph $Y_1 = \tan (X)$ and $Y_2 = X$ in dot mode in $[-\pi/6, 2\pi, \pi/2]$ by $[-2, 6, 1]$.

The intersections of the two graphs are shown in Figure 73a and Figure 73b.

Table $Y_1 = \tan (X)$ and $Y_2 = X$ starting at $x = 0$, incrementing by 0.9. See Figure 73c.

The solutions are $x = 0$, $x \approx 4.49$.

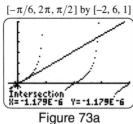
$[-\pi/6, 2\pi, \pi/2]$ by $[-2, 6, 1]$

Figure 73a

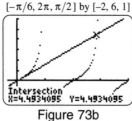
$[-\pi/6, 2\pi, \pi/2]$ by $[-2, 6, 1]$

Figure 73b

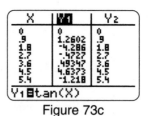

Figure 73c

75. Graph $Y_1 = \sin (X)$ and $Y_2 = (X - 1)^\wedge 2$ in $[-\pi/6, 2\pi, \pi/2]$ by $[-3, 7, 1]$.

The intersections of the two graphs are shown in Figure 75a and Figure 75b.

Table $Y_1 = \sin (X)$ and $Y_2 = (X - 1)^\wedge 2$ starting at $x = 0$, incrementing by 0.4. See Figure 75c.

The solutions are $x \approx 0.39$, $x \approx 1.96$.

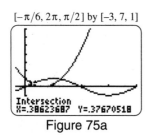
$[-\pi/6, 2\pi, \pi/2]$ by $[-3, 7, 1]$

Figure 75a

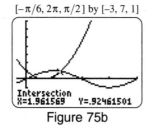
$[-\pi/6, 2\pi, \pi/2]$ by $[-3, 7, 1]$

Figure 75b

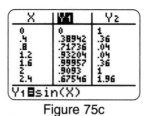

Figure 75c

77. Graph $Y_1 = 2X \cos (X + 1)$ and $Y_2 = \sin (\cos (X))$ in $[-\pi/6, 2\pi, \pi/2]$ by $[-5, 5, 1]$.

The intersection of the two graphs is shown in Figure 77a.

Table $Y_1 = 2X \cos (X + 1)$ and $Y_2 = \sin (\cos (X))$ starting at $x = 3.0$, incrementing by 0.1. See Figure 77b.

The solutions is $x \approx 3.60$.

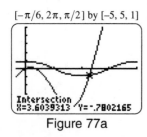
$[-\pi/6, 2\pi, \pi/2]$ by $[-5, 5, 1]$

Figure 77a

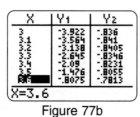

Figure 77b

79. $Y_1 = 0$ for $x = 60°, 120°, 240°, 300°$

We can write all solutions to this equation as $60° + 180°n$, $120° + 180°n$, where $n = 0, \pm 1, \pm 2, \ldots$.

81. $Y_1 = 0$ for $x = 30°, 210°$

We can write all solutions to this equation as $30° + 180°n$, where $n = 0, \pm 1, \pm 2, \ldots$.

Solving Inverse Trigonometric Equations

83. $\sin^{-1} x = \dfrac{\pi}{2} \Rightarrow \sin(\sin^{-1} x) = \sin\dfrac{\pi}{2} \Rightarrow x = 1$

85. $2\cos^{-1} x = \dfrac{5\pi}{3} \Rightarrow \cos^{-1} x = \dfrac{5\pi}{6} \Rightarrow \cos(\cos^{-1} x) = \cos\dfrac{5\pi}{6} \Rightarrow x = -\dfrac{\sqrt{3}}{2} \approx -0.866$

87. $\pi + \tan^{-1} x = \dfrac{3\pi}{4} \Rightarrow \tan^{-1} x = -\dfrac{\pi}{4} \Rightarrow \tan(\tan^{-1} x) = \tan\left(-\dfrac{\pi}{4}\right) \Rightarrow x = -1$

89. $\tan^{-1}(3x + 1) = \dfrac{\pi}{4} \Rightarrow \tan(\tan^{-1}(3x + 1)) = \tan\left(\dfrac{\pi}{4}\right) \Rightarrow 3x + 1 = 1 \Rightarrow 3x = 0 \Rightarrow x = 0$

91. $\cos^{-1} x + 3\cos^{-1} x = \pi \Rightarrow 4\cos^{-1} x = \pi \Rightarrow \cos^{-1} x = \dfrac{\pi}{4} \Rightarrow \cos(\cos^{-1} x) = \cos\dfrac{\pi}{4} \Rightarrow x = \dfrac{1}{\sqrt{2}} \approx 0.707$

Applications

93. For $F = 1$: $1 = \dfrac{1}{2}(1 - \cos\theta) \Rightarrow 2 = 1 - \cos\theta \Rightarrow \cos\theta = -1$. Thus $\theta = 180°$. All solutions can be written in the form $180° + 360° \cdot n$.

For $F = 0.25$: $0.25 = \dfrac{1}{2}(1 - \cos\theta) \Rightarrow 0.5 = 1 - \cos\theta \Rightarrow \cos\theta = 0.5$. Thus $\theta = 60°$ or $300°$.

All solutions can be written in the form $60° + 360° \cdot n$ or $300° + 360° \cdot n$.

95. Graph $Y_1 = 6.5 \sin(\pi/6(X - 3.65)) + 12.4$ and $Y_2 = 9$ in $[0, 13, 1]$ by $[0, 24, 2]$.

The intersections of the two graphs are shown in Figure 95a and Figure 95b.

Table $Y_1 = 6.5 \sin(\pi/6(X - 3.65)) + 12.4$ starting at $x = 2.6$, incrementing by 1.35. See Figure 95c.

The solutions are $x \approx 2.6$ and $x \approx 10.7$; near February 17 and October 22.

[0, 13, 1] by [0, 24, 2] [0, 13, 1] by [0, 24, 2]

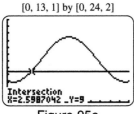
Intersection
X=2.5987042 _Y=9

Figure 95a

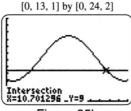
Intersection
X=10.701296 _Y=9

Figure 95b

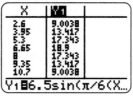

X	Y1
2.6	9.0038
3.95	13.417
5.3	17.343
6.65	18.9
8	17.343
9.35	13.417
10.7	9.0038

Y1■6.5sin(π/6(X...

Figure 95c

97. $6.5 \sin\left(\dfrac{\pi}{6}(x - 3.65)\right) + 12.4 = 9 \Rightarrow 6.5 \sin\left(\dfrac{\pi}{6}(x - 3.65)\right) = -3.4 \Rightarrow \sin\left(\dfrac{\pi}{6}(x - 3.65)\right) = -0.5231$

$\Rightarrow \dfrac{\pi}{6}(x - 3.65) = \sin^{-1}(-0.5231)$

$\Rightarrow \dfrac{\pi}{6}(x - 3.65) \approx -0.5505 \quad$ or $\quad \dfrac{\pi}{6}(x - 3.65) \approx 3.6920$

$\Rightarrow x - 3.65 \approx -1.0513 \quad$ or $\quad x - 3.65 \approx 7.0513$

$\Rightarrow x \approx 2.6 \quad$ or $\quad x \approx 10.7$

99. (a) Graph $Y_1 = 122.3 \sin(0.524X - 1.7) + 367$ along with the data in $[0, 13, 1]$ by $[200, 600, 50]$.

See Figure 99a.

(b) Graph $Y_1 = 122.3 \sin(0.524X - 1.7) + 367$ and $Y_2 = 350$ together in $[0, 13, 1]$ by $[200, 600, 50]$.

The intersections of the two graphs are shown in Figure 99b and Figure 99c.

We see that $f(x) \geq 350$ for $k_1 \leq x \leq k_2$ where $k_1 \approx 2.98$ and $k_2 \approx 9.51$. At $50°$ N latitude the maximum

monthly hours of sunshine is greater than or equal to 350 hours from roughly March through September.

[0, 13, 1] by [200, 600, 50]   [0, 13, 1] by [200, 600, 50]   [0, 13, 1] by [200, 600, 50]

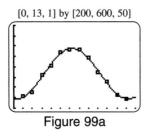   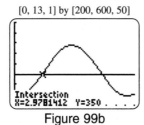   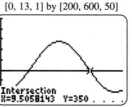

Figure 99a   Figure 99b   Figure 99c

101. Use the intercept method to solve the equation.

Graph $Y_1 = X - 0.26 - 0.017 \sin(X)$ in $[0, 0.5, 0.1]$ by $[-0.2, 0.2, 0.1]$. See Figure 101a.

Table $Y_1 = X - 0.26 - 0.017 \sin(X)$ starting at $x = 0.20$, incrementing by 0.01. See Figure 101b.

The solution is approximately 0.26.

[0, 0.5, 0.1] by [−0.2, 0.2, 0.1]

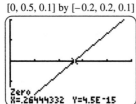   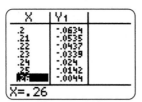

Figure 101a   Figure 101b

103. (a) The peaks occur at the values 0.005, 0.025, 0.045.

(b) At these times, the pressure on an eardrum is maximum.

105. (a) Let $Y_1 = 0.003 \sin(880\pi X - 0.7)$, $Y_2 = 0.002 \sin(880\pi X + 0.6)$.

Graph Y_1, Y_2 and $Y_3 = Y_1 + Y_2$ seperately in $[0, 0.01, 0.005]$ by $[-0.005, 0.005, 0.001]$.

Graph Y_1 is shown in Figure 105a. Graph Y_2 is shown in Figure 105b. Graph Y_3 is shown in Figure 105c.

(b) The maximum pressure is $P \approx 0.004$. This can be found at any peak on the graph in Figure 105c.

(c) No. The maximum of P_1 is 0.003 and the maximum of P_2 is 0.002.

[0, 0.01, 0.005] by [−0.005, 0.005, 0.001]   [0, 0.01, 0.005] by [−0.005, 0.005, 0.001]   [0, 0.01, 0.005] by [−0.005, 0.005, 0.001]

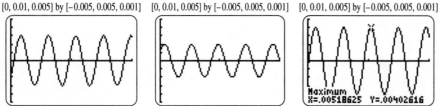

Figure 105a   Figure 105b   Figure 105c

7.4: Sum and Difference Identities

Sum and Difference Identities

1. $\sin 15° = \sin (45° - 30°) = \sin 45° \cos 30° - \cos 45° \sin 30° = \left(\dfrac{\sqrt{2}}{2}\right)\left(\dfrac{\sqrt{3}}{2}\right) - \left(\dfrac{\sqrt{2}}{2}\right)\left(\dfrac{1}{2}\right) = \dfrac{\sqrt{6} - \sqrt{2}}{4}$

3. $\sin 15° = \tan (60° - 45°) = \dfrac{\tan 60° - \tan 45°}{1 + \tan 60° \tan 45°} = \dfrac{\sqrt{3} - 1}{1 + \sqrt{3}} = \dfrac{\sqrt{3} - 1}{\sqrt{3} + 1} \cdot \dfrac{\sqrt{3} - 1}{\sqrt{3} - 1} =$

    $\dfrac{3 - 2\sqrt{3} + 1}{3 - 1} = 2 - \sqrt{3}$

5. $\cos 75° = \cos (45° + 30°) = \cos 45° \cos 30° - \sin 45° \sin 30° = \left(\dfrac{\sqrt{2}}{2}\right)\left(\dfrac{\sqrt{3}}{2}\right) - \left(\dfrac{\sqrt{2}}{2}\right)\left(\dfrac{1}{2}\right) = \dfrac{\sqrt{6} - \sqrt{2}}{4}$

7. $\sin \dfrac{\pi}{12} = \sin \left(\dfrac{\pi}{3} - \dfrac{\pi}{4}\right) = \sin \dfrac{\pi}{3} \cos \dfrac{\pi}{4} - \cos \dfrac{\pi}{3} \sin \dfrac{\pi}{4} = \left(\dfrac{\sqrt{3}}{2}\right)\left(\dfrac{\sqrt{2}}{2}\right) - \left(\dfrac{1}{2}\right)\left(\dfrac{\sqrt{2}}{2}\right) = \dfrac{\sqrt{6} - \sqrt{2}}{4}$

9. $\sin \dfrac{5\pi}{12} = \sin \left(\dfrac{2\pi}{3} - \dfrac{\pi}{4}\right) = \sin \dfrac{2\pi}{3} \cos \dfrac{\pi}{4} - \cos \dfrac{2\pi}{3} \sin \dfrac{\pi}{4} = \left(\dfrac{\sqrt{3}}{2}\right)\left(\dfrac{\sqrt{2}}{2}\right) - \left(-\dfrac{1}{2}\right)\left(\dfrac{\sqrt{2}}{2}\right) = \dfrac{\sqrt{6} + \sqrt{2}}{4}$

11. (a) Graphical: Graph $Y_1 = \sin (X + \pi/2)$ and $Y_2 = \cos (X)$ seperately in $[-2\pi, 2\pi, \pi/2]$ by $[-4, 4, 1]$.

    Graph Y_1 is shown in Figure 11a. Graph Y_2 is shown in Figure 11b. The graphs are the same.

    Verbal: If the sine graph is translated $\dfrac{\pi}{2}$ units left it coincides with the cosine graph.

    (b) $\sin \left(t + \dfrac{\pi}{2}\right) = \sin t \cos \dfrac{\pi}{2} + \cos t \sin \dfrac{\pi}{2} = \sin t (0) + \cos t (1) = \cos t$

$[-2\pi, 2\pi, \pi/2]$ by $[-4, 4, 1]$ $[-2\pi, 2\pi, \pi/2]$ by $[-4, 4, 1]$

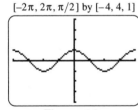
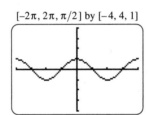

Figure 11a   Figure 11b

13. (a) Graphical: Graph $Y_1 = \cos (X + \pi)$ and $Y_2 = -\cos (X)$ seperately in $[-2\pi, 2\pi, \pi/2]$ by $[-4, 4, 1]$.

    Graph Y_1 is shown in Figure 13a. Graph Y_2 is shown in Figure 13b. The graphs are the same.

    Verbal: If the cosine graph is translated π units left it coincides with the cosine graph reflected across the *x*-axis.

    (b) $\cos (t + \pi) = \cos t \cos \pi - \sin t \sin \pi = \cos t (-1) - \sin t (0) = -\cos t$

$[-2\pi, 2\pi, \pi/2]$ by $[-4, 4, 1]$ $[-2\pi, 2\pi, \pi/2]$ by $[-4, 4, 1]$

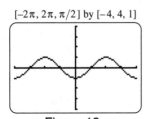
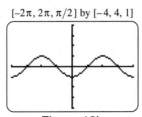

Figure 13a   Figure 13b

15. (a) Graphical: Graph $Y_1 = 1/\cos(X - \pi/2)$ and $Y_2 = 1/\sin(X)$ seperately in $[-2\pi, 2\pi, \pi/2]$ by $[-4, 4, 1]$.

Graph Y_1 is shown in Figure 15a. Graph Y_2 is shown in Figure 15b. The graphs are the same.

Verbal: If the secant graph is translated $\dfrac{\pi}{2}$ units left it coincides with the cosecant graph.

(b) $\sec\left(t - \dfrac{\pi}{2}\right) = \dfrac{1}{\cos\left(t - \frac{\pi}{2}\right)} = \dfrac{1}{\cos t \cos\frac{\pi}{2} + \sin t \sin\frac{\pi}{2}} = \dfrac{1}{\cos t(0) + \sin t(1)} = \dfrac{1}{\sin t} = \csc t$

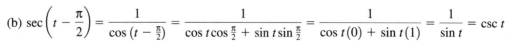

$[-2\pi, 2\pi, \pi/2]$ by $[-4, 4, 1]$ $[-2\pi, 2\pi, \pi/2]$ by $[-4, 4, 1]$

Figure 15a   Figure 15b

17. $\sin\left(\dfrac{\pi}{2} - t\right) = \sin\dfrac{\pi}{2}\cos t - \cos\dfrac{\pi}{2}\sin t = (1)\cos t - (0)\sin t = \cos t$

19. $\sec\left(\dfrac{\pi}{2} - t\right) = \dfrac{1}{\cos\left(\frac{\pi}{2} - t\right)} = \dfrac{1}{\cos\frac{\pi}{2}\cos t + \sin\frac{\pi}{2}\sin t} = \dfrac{1}{(0)\cos t + (1)\sin t} = \dfrac{1}{\sin t} = \csc t$

21. First sketch possible angles for α and β. See Figures 21a & 21b. Note: $\cos\alpha = \dfrac{4}{5}$ and $\cos\beta = \dfrac{12}{13}$.

(a) $\sin(\alpha + \beta) = \sin\alpha\cos\beta + \cos\alpha\sin\beta = \left(\dfrac{3}{5}\right)\left(\dfrac{12}{13}\right) + \left(\dfrac{4}{5}\right)\left(\dfrac{5}{13}\right) = \dfrac{36}{65} + \dfrac{20}{65} = \dfrac{56}{65}$

(b) $\cos(\alpha + \beta) = \cos\alpha\cos\beta - \sin\alpha\sin\beta = \left(\dfrac{4}{5}\right)\left(\dfrac{12}{13}\right) - \left(\dfrac{3}{5}\right)\left(\dfrac{5}{13}\right) = \dfrac{48}{65} - \dfrac{15}{65} = \dfrac{33}{65}$

(c) $\tan(\alpha + \beta) = \dfrac{\sin(\alpha + \beta)}{\cos(\alpha + \beta)} = \dfrac{\frac{56}{65}}{\frac{33}{65}} = \dfrac{56}{65} \cdot \dfrac{65}{33} = \dfrac{56}{33}$

(d) Since both $\sin(\alpha + \beta)$ and $\cos(\alpha + \beta)$ are positive, $\alpha + \beta$ is in quadrant I.

            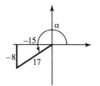

Figure 21a   Figure 21b   Figure 23a   Figure 23b

23. First sketch possible angles for α and β. See Figures 23a & 23b. Note: $\cos\alpha = -\dfrac{15}{17}$ and $\sin\beta = \dfrac{60}{61}$.

(a) $\sin(\alpha + \beta) = \sin\alpha\cos\beta + \cos\alpha\sin\beta = \left(-\dfrac{8}{17}\right)\left(\dfrac{11}{61}\right) + \left(-\dfrac{15}{17}\right)\left(\dfrac{60}{61}\right) = -\dfrac{88}{1037} - \dfrac{900}{1037} = -\dfrac{988}{1037}$

(b) $\cos(\alpha + \beta) = \cos\alpha\cos\beta - \sin\alpha\sin\beta = \left(-\dfrac{15}{17}\right)\left(\dfrac{11}{61}\right) - \left(-\dfrac{8}{17}\right)\left(\dfrac{60}{61}\right) = -\dfrac{165}{1037} + \dfrac{480}{1037} = \dfrac{315}{1037}$

(c) $\tan(\alpha + \beta) = \dfrac{\sin(\alpha + \beta)}{\cos(\alpha + \beta)} = \dfrac{-\frac{988}{1037}}{\frac{315}{1037}} = -\dfrac{988}{1037} \cdot \dfrac{1037}{315} = -\dfrac{988}{315}$

(d) Since $\sin(\alpha + \beta)$ is negative and $\cos(\alpha + \beta)$ is positive, $\alpha + \beta$ is in quadrant IV.

25. First sketch possible angles for α and β. See Figures 25a & 25b. Note: $\sin\alpha = -\dfrac{4}{5}$ and $\sin\beta = -\dfrac{5}{13}$.

(a) $\sin(\alpha + \beta) = \sin\alpha\cos\beta + \cos\alpha\sin\beta = \left(-\dfrac{4}{5}\right)\left(\dfrac{12}{13}\right) + \left(-\dfrac{3}{5}\right)\left(-\dfrac{5}{13}\right) = -\dfrac{48}{65} + \dfrac{15}{65} = -\dfrac{33}{65}$

(b) $\cos(\alpha + \beta) = \cos\alpha\cos\beta - \sin\alpha\sin\beta = \left(-\dfrac{3}{5}\right)\left(\dfrac{12}{13}\right) - \left(-\dfrac{4}{5}\right)\left(-\dfrac{5}{13}\right) = -\dfrac{36}{65} - \dfrac{20}{65} = -\dfrac{56}{65}$

(c) $\tan(\alpha + \beta) = \dfrac{\sin(\alpha + \beta)}{\cos(\alpha + \beta)} = \dfrac{-\frac{33}{65}}{-\frac{56}{65}} = -\dfrac{33}{65}\cdot\left(-\dfrac{65}{56}\right) = \dfrac{33}{56}$

(d) Since both $\sin(\alpha + \beta)$ and $\cos(\alpha + \beta)$ are negative, $\alpha + \beta$ is in quadrant III.

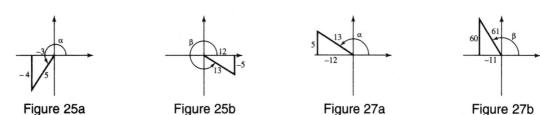

| Figure 25a | Figure 25b | Figure 27a | Figure 27b |

27. First sketch possible angles for α and β. See Figures 27a and 27b.

Note: $\sin\alpha = \dfrac{5}{13}$, $\cos\alpha = -\dfrac{12}{13}$, $\sin\beta = \dfrac{60}{61}$, $\cos\beta = -\dfrac{11}{60}$

(a) $\sin(\alpha + \beta) = \sin\alpha\cos\beta + \cos\alpha\sin\beta = \left(\dfrac{5}{13}\right)\left(-\dfrac{11}{61}\right) + \left(-\dfrac{12}{13}\right)\left(\dfrac{60}{61}\right) = \left(-\dfrac{55}{793}\right) + \left(-\dfrac{720}{793}\right) = -\dfrac{775}{793}$

(b) $\cos(\alpha + \beta) = \cos\alpha\cos\beta - \sin\alpha\sin\beta = \left(-\dfrac{12}{13}\right)\left(-\dfrac{11}{61}\right) - \left(\dfrac{5}{13}\right)\left(\dfrac{60}{61}\right) = \dfrac{132}{793} - \dfrac{300}{793} = -\dfrac{168}{793}$

(c) $\tan(\alpha + \beta) = \dfrac{\sin(\alpha + \beta)}{\cos(\alpha + \beta)} = \dfrac{-\frac{775}{793}}{-\frac{168}{793}} = -\dfrac{775}{793}\cdot\left(-\dfrac{793}{168}\right) = \dfrac{775}{168}$

(d) Since both $\sin(\alpha + \beta)$ and $\cos(\alpha + \beta)$ are negative, $\alpha + \beta$ is in quadrant III.

29. $\cos\left(t - \dfrac{\pi}{4}\right) = \cos t\cos\dfrac{\pi}{4} + \sin t\sin\dfrac{\pi}{4} = \cos t\left(\dfrac{\sqrt{2}}{2}\right) + \sin t\left(\dfrac{\sqrt{2}}{2}\right) = \dfrac{\sqrt{2}}{2}(\cos t + \sin t)$

31. $\tan\left(t + \dfrac{\pi}{4}\right) = \dfrac{\tan t + \tan\frac{\pi}{4}}{1 - \tan t\tan\frac{\pi}{4}} = \dfrac{\tan t + 1}{1 - \tan t(1)} = \dfrac{1 + \tan t}{1 - \tan t}$

33. $\dfrac{\cos(x - y)}{\cos(x + y)} = \dfrac{\cos x\cos y + \sin x\sin y}{\cos x\cos y - \sin x\sin y} = \dfrac{\cos x\cos y + \sin x\sin y}{\cos x\cos y - \sin x\sin y}\cdot\dfrac{\sec x\sec y}{\sec x\sec y} = \dfrac{1 + \tan x\tan y}{1 - \tan x\tan y}$

35. $\dfrac{\cos(\alpha - \beta)}{\cos\alpha\sin\beta} = \dfrac{\cos\alpha\cos\beta + \sin\alpha\sin\beta}{\cos\alpha\sin\beta} = \dfrac{\cos\alpha\cos\beta}{\cos\alpha\sin\beta} + \dfrac{\sin\alpha\sin\beta}{\cos\alpha\sin\beta} = \cot\beta + \tan\alpha = \tan\alpha + \cot\beta$

37. $\sin 2t = \sin(t + t) = \sin t\cos t + \cos t\sin t = 2\sin t\cos t$

39. $\sin(\alpha + \beta) + \sin(\alpha - \beta) = (\sin\alpha\cos\beta + \cos\alpha\sin\beta) + (\sin\alpha\cos\beta - \cos\alpha\sin\beta) = 2\sin\alpha\cos\beta$

41. $\tan(\pi - \theta) = \dfrac{\tan\pi - \tan\theta}{1 + \tan\pi\tan\theta} = \dfrac{0 - \tan\theta}{1 + (0)\tan\theta} = -\tan\theta$

43. $\dfrac{\sin (x - y)}{\sin y} + \dfrac{\cos (x - y)}{\cos y} = \dfrac{\sin x \cos y - \cos x \sin y}{\sin y} + \dfrac{\cos x \cos y + \sin x \sin y}{\cos y} =$

$$\sin x \left(\dfrac{\cos y}{\sin y}\right) - \cos x \left(\dfrac{\sin y}{\sin y}\right) + \cos x \left(\dfrac{\cos y}{\cos y}\right) + \sin x \left(\dfrac{\sin y}{\cos y}\right) =$$

$$\sin x \left(\dfrac{\cos y}{\sin y}\right) - \cos x + \cos x + \sin x \left(\dfrac{\sin y}{\cos y}\right) = \sin x \left(\dfrac{\cos y}{\sin y}\right) + \sin x \left(\dfrac{\sin y}{\cos y}\right) = \sin x \left(\dfrac{\cos y}{\sin y} + \dfrac{\sin y}{\cos y}\right) =$$

$$\sin x \left(\dfrac{\cos^2 y + \sin^2 y}{\sin y \cos y}\right) = \dfrac{\sin x}{\sin y \cos y}$$

45. $\dfrac{\sin (x + y)}{\cos x \cos y} = \dfrac{\sin x \cos y + \cos x \sin y}{\cos x \cos y} = \dfrac{\sin x \cos y}{\cos x \cos y} + \dfrac{\cos x \sin y}{\cos x \cos y} = \dfrac{\sin x}{\cos x} + \dfrac{\sin y}{\cos y} = \tan x + \tan y$

47. $\tan \gamma = \tan (\alpha - \beta) = \dfrac{\tan \alpha - \tan \beta}{1 + \tan \alpha \tan \beta} = \dfrac{\left(\frac{6}{7}\right) - \left(\frac{5}{7}\right)}{1 + \left(\frac{6}{7}\right)\left(\frac{5}{7}\right)} = \dfrac{\frac{1}{7}}{1 + \frac{30}{49}} = \dfrac{\frac{1}{7}}{\frac{79}{49}} = \dfrac{1}{7} \cdot \dfrac{49}{79} = \dfrac{7}{79}$

Lines and Slopes

49. $\tan \theta = \tan (\beta - \alpha) = \dfrac{\tan \beta - \tan \alpha}{1 + \tan \beta \tan \alpha} = \dfrac{m_2 - m_1}{1 + m_2 m_1} = \dfrac{m_2 - m_1}{1 + m_1 m_2}$

51. For $y = 2x - 3$, $m_2 = 2$ and for $y = \dfrac{3}{5}x + 1$, $m_1 = \dfrac{3}{5}$.

$$\tan \theta = \dfrac{2 - \frac{3}{5}}{1 + 2\left(\frac{3}{5}\right)} = \dfrac{\frac{7}{5}}{\frac{11}{5}} = \dfrac{7}{5} \cdot \dfrac{5}{11} = \dfrac{7}{11} \Rightarrow \theta = \tan^{-1} \dfrac{7}{11} \approx 32.5°$$

Applications

53. (a) $F = 2.89W \cos \theta \Rightarrow F = 2.89(200) \cos \dfrac{\pi}{4} \approx 409 \text{ lb}$

(b) $2.89(200)\cos \theta = 400 \Rightarrow \cos \theta = \dfrac{400}{2.89(200)} \Rightarrow \theta = \cos^{-1}\left(\dfrac{2}{2.89}\right) \approx 0.81 \text{ or about } 46.2°.$

55. (a) Graph $Y_1 = 4 \cos (220\pi X) + 3 \sin (220\pi X)$ in $[0, 0.02, 0.001]$ by $[-6, 6, 1]$. See Figure 55.

(b) A maximum occurs at approximately $(0.00093114, 5)$. Thus $a = 5$. Since $\sin \theta$ is maximum when $\theta = \dfrac{\pi}{2}$,

we let $220\pi(0.00093114) + k = \dfrac{\pi}{2}$. Thus $k = \dfrac{\pi}{2} - 220\pi(0.00093114) \approx 0.9272$.

(c) $a \sin (220\pi t + k) \approx 5 \sin (220\pi t + 0.9272) = 5 [\sin (220\pi t) \cos 0.9272 + \cos (220\pi t) \sin 0.9272] \approx$

$5 [\sin (220\pi t)(0.6) + \cos (220\pi t)(0.8)] = 3 \sin (220\pi t) + 4 \cos (220\pi t)$

[0, 0.02, 0.001] by [−6, 6, 1]

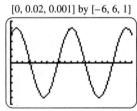

Figure 55

7.5: Multiple-Angle Identities

Double-Angle Identities

1. (a) $\sin 30° + \sin 30° = \dfrac{1}{2} + \dfrac{1}{2} = 1$

    (b) $\sin 60° = \dfrac{\sqrt{3}}{2}$

    The results are not the same.

3. (a) $\cos 60° + \cos 60° = \dfrac{1}{2} + \dfrac{1}{2} = 1$

    (b) $\cos 120° = -\dfrac{1}{2}$

    The results are not the same.

5. (a) $\tan 45° + \tan 45° = 1 + 1 = 2$

    (b) $\tan 90°$ is undefined

    The results are not the same.

7. Graphical: Graph $Y_1 = \tan(2X)$ and $Y_2 = 2\tan(X)$ in dot mode in $[-2\pi, 2\pi, \pi/2]$ by $[-4, 4, 1]$.

    Graph Y_1 is shown in Figure 7a. Graph Y_2 is shown in Figure 7b. Note that the graphs are different.

    Symbolic: $\tan 2\theta = \dfrac{2\tan\theta}{1 - \tan^2\theta} \neq 2\tan\theta$, unless $\tan\theta = 0$

[−2π, 2π, π/2] by [−4, 4, 1]   [−2π, 2π, π/2] by [−4, 4, 1]

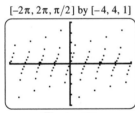

Figure 7a

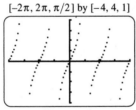

Figure 7b

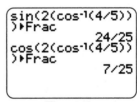

Figure 9a

Figure 9b

9. (a) $\sin 2\theta = 2\sin\theta\cos\theta = 2\left(\dfrac{3}{5}\right)\left(\dfrac{4}{5}\right) = \dfrac{24}{25}$

    $\cos 2\theta = \cos^2\theta - \sin^2\theta = \left(\dfrac{4}{5}\right)^2 - \left(\dfrac{3}{5}\right)^2 = \dfrac{16 - 9}{25} = \dfrac{7}{25}$

    $\tan 2\theta = \dfrac{\sin 2\theta}{\cos 2\theta} = \dfrac{\frac{24}{25}}{\frac{7}{25}} = \dfrac{24}{25} \cdot \dfrac{25}{7} = \dfrac{24}{7}$

    (b) Since θ is in quadrant I, $\theta = \cos^{-1}\dfrac{4}{5}$. Numerical support is shown in Figures 9a and 9b.

11. Since $\sin\theta$ is negative and $\cos\theta$ is positive, θ is in quadrant IV. One possibility is shown in Figure 11a.

We see that $\cos\theta = \dfrac{7}{25}$.

(a) $\sin 2\theta = 2\sin\theta\cos\theta = 2\left(-\dfrac{24}{25}\right)\left(\dfrac{7}{25}\right) = -\dfrac{336}{625}$

$\cos 2\theta = \cos^2\theta - \sin^2\theta = \left(\dfrac{7}{25}\right)^2 - \left(-\dfrac{24}{25}\right)^2 = \dfrac{49 - 576}{625} = -\dfrac{527}{625}$

$\tan 2\theta = \dfrac{\sin 2\theta}{\cos 2\theta} = \dfrac{-\frac{336}{625}}{-\frac{527}{625}} = -\dfrac{336}{625}\cdot\left(-\dfrac{625}{527}\right) = \dfrac{336}{527}$

(b) Since θ is in quadrant IV, $\theta = \sin^{-1}\left(-\dfrac{24}{25}\right)$. Numerical support is shown in Figures 11b and 11c.

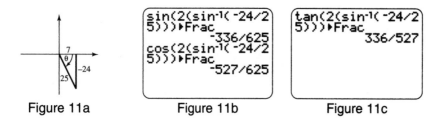

Figure 11a   Figure 11b   Figure 11c

13. Since $\sin\theta$ is negative and $\sec\theta$ is positive, θ is in quadrant IV. One possibility is shown in Figure 13a.

We see that $\cos\theta = \dfrac{60}{61}$.

(a) $\sin 2\theta = 2\sin\theta\cos\theta = 2\left(-\dfrac{11}{61}\right)\left(\dfrac{60}{61}\right) = -\dfrac{1320}{3721}$

$\cos 2\theta = \cos^2\theta - \sin^2\theta = \left(\dfrac{60}{61}\right)^2 - \left(-\dfrac{11}{61}\right)^2 = \dfrac{3600 - 121}{3721} = \dfrac{3479}{3721}$

$\tan 2\theta = \dfrac{\sin 2\theta}{\cos 2\theta} = \dfrac{-\frac{1320}{3721}}{\frac{3479}{3721}} = -\dfrac{1320}{3721}\cdot\dfrac{3721}{3479} = -\dfrac{1320}{3479}$

(b) Since θ is in quadrant IV, $\theta = \sin^{-1}\left(-\dfrac{11}{61}\right)$. Numerical support is shown in Figures 13b and 13c.

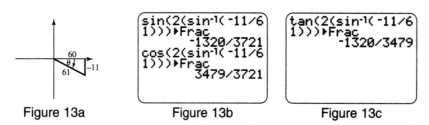

Figure 13a   Figure 13b   Figure 13c

15. Since $\tan \theta$ is positive and $\cos \theta$ is negative, θ is in quadrant III. One possibility is shown in Figure 15a.

We see that $\sin \theta = -\dfrac{7}{25}$ and $\cos \theta = -\dfrac{24}{25}$.

(a) $\sin 2\theta = 2 \sin \theta \cos \theta = 2\left(-\dfrac{7}{25}\right)\left(-\dfrac{24}{25}\right) = \dfrac{336}{625}$

$\cos 2\theta = \cos^2 \theta - \sin^2 \theta = \left(-\dfrac{24}{25}\right)^2 - \left(-\dfrac{7}{25}\right)^2 = \dfrac{576 - 49}{625} = \dfrac{527}{625}$

$\tan 2\theta = \dfrac{\sin 2\theta}{\cos 2\theta} = \dfrac{\frac{336}{625}}{\frac{527}{625}} = \dfrac{336}{625} \cdot \dfrac{625}{527} = \dfrac{336}{527}$

(b) Since θ is in quadrant III, $\theta = 360° - \cos^{-1}\left(-\dfrac{24}{25}\right)$. Numerical support is shown in Figures 15b and 15c.

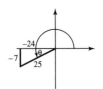
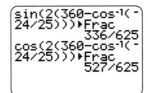
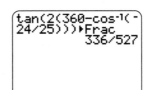

Figure 15a   Figure 15b   Figure 15c

17. Let $\theta = \cos^{-1} 1$. It follows that $\cos \theta = 1$ and $\sin \theta = 0$ (since $\theta = 0°$).

$\sin (2 \cos^{-1} 1) = \sin (2\theta) = 2 \sin \theta \cos \theta = 2(0)(1) = 0$

19. Let $\theta = \sin^{-1} \dfrac{7}{25}$. It follows that $\sin \theta = \dfrac{7}{25}$.

$\cos\left(2 \sin^{-1} \dfrac{7}{25}\right) = \cos (2\theta) = 1 - 2\sin^2 \theta = 1 - 2\left(\dfrac{7}{25}\right)^2 = 1 - \dfrac{98}{625} = \dfrac{527}{625}$

21. Let $\theta = \sin^{-1} \dfrac{5}{13}$. It follows that $\sin \theta = \dfrac{5}{13}$ and $\cos \theta = \dfrac{12}{13}$. See Figure 21.

$\cos\left(3 \sin^{-1} \dfrac{5}{13}\right) = \cos (3\theta) = 4\cos^3 \theta - 3\cos \theta = 4\left(\dfrac{12}{13}\right)^3 - 3\left(\dfrac{12}{13}\right) = \dfrac{828}{2197}$

Figure 21   Figure 25a   Figure 25b

23. Let $\theta = \sin^{-1} x$, where $x > 0$. It follows that $\sin \theta = x$.

$\cos (2 \sin^{-1} x) = \cos (2\theta) = 1 - 2\sin^2 \theta = 1 - 2(x)^2 = 1 - 2x^2$

25. Let $\alpha = \sin^{-1} \dfrac{3}{5}$. It follows that $\sin \alpha = \dfrac{3}{5}$ and $\cos \alpha = \dfrac{4}{5}$. See Figure 25a.

Let $\beta = \sin^{-1} \dfrac{4}{5}$. It follows that $\sin \beta = \dfrac{4}{5}$ and $\cos \beta = \dfrac{3}{5}$. See Figure 25b.

$\cos\left(\sin^{-1} \dfrac{3}{5} - \sin^{-1} \dfrac{4}{5}\right) = \cos (\alpha - \beta) = \cos \alpha \cos \beta + \sin \alpha \sin \beta = \left(\dfrac{4}{5}\right)\left(\dfrac{3}{5}\right) + \left(\dfrac{3}{5}\right)\left(\dfrac{4}{5}\right) = \dfrac{24}{25}$

27. $2 \cos \theta \sin \theta = 2 \sin \theta \cos \theta = \sin 2\theta$

29. $\sin \theta \cos \theta = \dfrac{1}{2} (2 \sin \theta \cos \theta) = \dfrac{1}{2} \sin 2\theta$

31. $2 \cos^2 2\theta - 1 = \cos 4\theta$ (let $t = 2\theta$, then $2 \cos^2 t - 1 = \cos 2t$)

33. $\sin^2 3\theta + \cos^2 3\theta = 1$ (let $t = 3\theta$, then $\sin^2 t + \cos^2 t = 1$)

35. $\csc^2 5x - 1 = \cot^2 5x$ (let $t = 5x$, then $\csc^2 t - 1 = \cot^2 t$)

Power-Reducing Identities and Half-Angle Formulas

37. $\cos^2 (22.5°) = \dfrac{1 + \cos (2 \cdot 22.5°)}{2} = \dfrac{1 + \cos 45°}{2} = \dfrac{1 + \frac{\sqrt{2}}{2}}{2} = \dfrac{\frac{2 + \sqrt{2}}{2}}{2} = \dfrac{2 + \sqrt{2}}{4}$

Numerical support is shown in Figure 37.

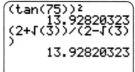

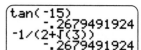

      Figure 37   Figure 39   Figure 41a   Figure 41b

39. $\tan^2 (75°) = \dfrac{1 - \cos (2 \cdot 75°)}{1 + \cos (2 \cdot 75°)} = \dfrac{1 - \cos 150°}{1 + \cos 150°} = \dfrac{1 - (-\frac{\sqrt{3}}{2})}{1 + (-\frac{\sqrt{3}}{2})} = \dfrac{\frac{2 + \sqrt{3}}{2}}{\frac{2 - \sqrt{3}}{2}} = \dfrac{2 + \sqrt{3}}{2 - \sqrt{3}}$

Numerical support is shown in Figure 39.

41. (a) Since the angle is in quadrant I, we use the positive value of the half-angle formula.

$$\cos 15° = \sqrt{\dfrac{1 + \cos 30°}{2}} = \sqrt{\dfrac{1 + \frac{\sqrt{3}}{2}}{2}} = \sqrt{\dfrac{\frac{2 + \sqrt{3}}{2}}{2}} = \sqrt{\dfrac{2 + \sqrt{3}}{4}} = \dfrac{\sqrt{2 + \sqrt{3}}}{2}$$

(b) Since the angle is in quadrant IV, we use the negative value of the half-angle formula.

$$\tan (-15°) = -\sqrt{\dfrac{1 - \cos (-30°)}{1 + \cos (-30°)}} = -\sqrt{\dfrac{1 - \frac{\sqrt{3}}{2}}{1 + \frac{\sqrt{3}}{2}}} = -\sqrt{\dfrac{\frac{2 - \sqrt{3}}{2}}{\frac{2 + \sqrt{3}}{2}}} = -\sqrt{\dfrac{2 - \sqrt{3}}{2 + \sqrt{3}}} \text{ or } -\dfrac{1}{2 + \sqrt{3}} \text{ or } \sqrt{3} - 2$$

Numerical support is shown in Figure 41a and 41b.

43. (a) Since the angle is in quadrant I, we use the positive value of the half-angle formula.

$$\tan \left(\dfrac{\pi}{8}\right) = \sqrt{\dfrac{1 - \cos \frac{\pi}{4}}{1 + \cos \frac{\pi}{4}}} = \sqrt{\dfrac{1 - \frac{\sqrt{2}}{2}}{1 + \frac{\sqrt{2}}{2}}} = \sqrt{\dfrac{\frac{2 - \sqrt{2}}{2}}{\frac{2 + \sqrt{2}}{2}}} = \sqrt{\dfrac{2 - \sqrt{2}}{2 + \sqrt{2}}} \text{ or } \dfrac{\sqrt{2}}{2 + \sqrt{2}} \text{ or } \sqrt{2} - 1$$

(b) Since the angle is in quadrant IV, we use the negative value of the half-angle formula.

$$\sin \left(-\dfrac{\pi}{8}\right) = -\sqrt{\dfrac{1 - \cos (-\frac{\pi}{4})}{2}} = -\sqrt{\dfrac{1 - (\frac{\sqrt{2}}{2})}{2}} = -\sqrt{\dfrac{\frac{2 - \sqrt{2}}{2}}{2}} = -\sqrt{\dfrac{2 - \sqrt{2}}{4}} = -\dfrac{\sqrt{2 - \sqrt{2}}}{2}$$

Numerical support is shown in Figure 43a and 43b.

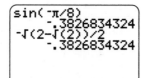
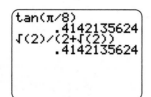

      Figure 43a   Figure 43b

45. $\sqrt{\dfrac{1-\cos 60°}{2}} = \sin\left(\dfrac{60°}{2}\right) = \sin 30°$. Numerical support is shown in Figure 45.

47. $\sqrt{\dfrac{1+\cos 50°}{2}} = \cos\left(\dfrac{50°}{2}\right) = \cos 25°$. Numerical support is shown in Figure 47.

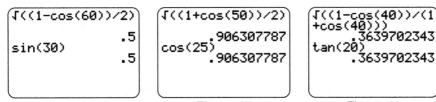

Figure 45   Figure 47   Figure 49

49. $\sqrt{\dfrac{1-\cos 40°}{1+\cos 40°}} = \tan\left(\dfrac{40°}{2}\right) = \tan 20°$. Numerical support is shown in Figure 49.

51. Since $0° < \dfrac{\theta}{2} < 45°$, we know that $\dfrac{\theta}{2}$ is in quadrant I.

$$\sin\frac{\theta}{2} = \sqrt{\frac{1-\cos\theta}{2}} = \sqrt{\frac{1-\frac{4}{5}}{2}} = \sqrt{\frac{\frac{5-4}{5}}{2}} = \sqrt{\frac{1}{10}} = \frac{1}{\sqrt{10}}$$

$$\cos\frac{\theta}{2} = \sqrt{\frac{1+\cos\theta}{2}} = \sqrt{\frac{1+\frac{4}{5}}{2}} = \sqrt{\frac{\frac{5+4}{5}}{2}} = \sqrt{\frac{9}{10}} = \frac{3}{\sqrt{10}}$$

$$\tan\frac{\theta}{2} = \sqrt{\frac{1-\cos\theta}{1+\cos\theta}} = \sqrt{\frac{1-\frac{4}{5}}{1+\frac{4}{5}}} = \sqrt{\frac{\frac{5-4}{5}}{\frac{5+4}{5}}} = \sqrt{\frac{\frac{1}{5}}{\frac{9}{5}}} = \sqrt{\frac{1}{9}} = \frac{1}{3}$$

53. Since $-45° < \dfrac{\theta}{2} < 0°$, we know that $\dfrac{\theta}{2}$ is in quadrant IV.

By sketching an appropriate triangle in quadrant IV, we see that $\cos\theta = \dfrac{12}{13}$.

$$\sin\frac{\theta}{2} = -\sqrt{\frac{1-\cos\theta}{2}} = -\sqrt{\frac{1-\frac{12}{13}}{2}} = -\sqrt{\frac{\frac{13-12}{13}}{2}} = -\sqrt{\frac{1}{26}} = -\frac{1}{\sqrt{26}}$$

$$\cos\frac{\theta}{2} = \sqrt{\frac{1+\cos\theta}{2}} = \sqrt{\frac{1+\frac{12}{13}}{2}} = \sqrt{\frac{\frac{13+12}{13}}{2}} = \sqrt{\frac{25}{26}} = \frac{5}{\sqrt{26}}$$

$$\tan\frac{\theta}{2} = -\sqrt{\frac{1-\cos\theta}{1+\cos\theta}} = -\sqrt{\frac{1-\frac{12}{13}}{1+\frac{12}{13}}} = -\sqrt{\frac{\frac{13-12}{13}}{\frac{13+12}{13}}} = -\sqrt{\frac{\frac{1}{13}}{\frac{25}{13}}} = -\sqrt{\frac{1}{25}} = -\frac{1}{5}$$

55. Since $45° < \dfrac{\theta}{2} < 90°$, we know that $\dfrac{\theta}{2}$ is in quadrant I.

By sketching an appropriate triangle in quadrant II, we see that $\cos\theta = -\dfrac{7}{25}$.

$$\sin\frac{\theta}{2} = \sqrt{\frac{1-\cos\theta}{2}} = \sqrt{\frac{1-\left(-\frac{7}{25}\right)}{2}} = \sqrt{\frac{\frac{25+7}{25}}{2}} = \sqrt{\frac{32}{50}} = \sqrt{\frac{16}{25}} = \frac{4}{5}$$

$$\cos\frac{\theta}{2} = \sqrt{\frac{1+\cos\theta}{2}} = \sqrt{\frac{1+\left(-\frac{7}{25}\right)}{2}} = \sqrt{\frac{\frac{25-7}{25}}{2}} = \sqrt{\frac{18}{50}} = \sqrt{\frac{9}{25}} = \frac{3}{5}$$

$$\tan\frac{\theta}{2} = \sqrt{\frac{1-\cos\theta}{1+\cos\theta}} = \sqrt{\frac{1-\left(-\frac{7}{25}\right)}{1+\left(-\frac{7}{25}\right)}} = \sqrt{\frac{\frac{25+7}{25}}{\frac{25-7}{25}}} = \sqrt{\frac{\frac{32}{25}}{\frac{18}{25}}} = \sqrt{\frac{32}{18}} = \sqrt{\frac{16}{9}} = \frac{4}{3}$$

Verifying Identities

57. $4 \sin 2x = 4(2 \sin x \cos x) = 8 \sin x \cos x$

Graph $Y_1 = 4 \sin (2X)$ and $Y_2 = 8 \sin (X) \cos (X)$ in $[-2\pi, 2\pi, \pi/2]$ by $[-6, 6, 1]$.

Graph Y_1 is shown in Figure 57a. Graph Y_2 is shown in Figure 57b. The graphs are the same.

Table $Y_1 = 4 \sin (2X)$ and $Y_2 = 8 \sin (X) \cos (X)$ starting at $x = 0$, incrementing by $\dfrac{\pi}{6}$. See Figure 57c.

The tables are the same.

$[-2\pi, 2\pi, \pi/2]$ by $[-6, 6, 1]$ $[-2\pi, 2\pi, \pi/2]$ by $[-6, 6, 1]$

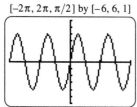   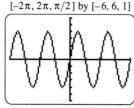   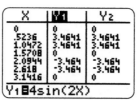

Figure 57a   Figure 57b   Figure 57c

59. $\dfrac{2 - \sec^2 x}{\sec^2 x} = \dfrac{2}{\sec^2 x} - \dfrac{\sec^2 x}{\sec^2 x} = 2 \cos^2 x - 1 = \cos 2x$

Graph $Y_1 = (2 - 1/(\cos (X))^2)/(1/(\cos (X))^2)$ and $Y_2 = \cos (2X)$ in $[-2\pi, 2\pi, \pi/2]$ by $[-2, 2, 1]$.

Graph Y_1 is shown in Figure 59a. Graph Y_2 is shown in Figure 59b. The graphs are the same.

Table $Y_1 = (2 - 1/(\cos (X))^2)/(1/(\cos (X))^2)$ and $Y_2 = \cos (2X)$ starting at $x = 0$, incrementing by $\dfrac{\pi}{6}$.

See Figure 59c. The tables are the same.

$[-2\pi, 2\pi, \pi/2]$ by $[-2, 2, 1]$ $[-2\pi, 2\pi, \pi/2]$ by $[-2, 2, 1]$

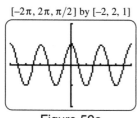   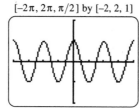

Figure 59a   Figure 59b   Figure 59c

61. $\sec 2x = \dfrac{1}{\cos 2x} = \dfrac{1}{1 - 2 \sin^2 x}$

Graph $Y_1 = 1/\cos (2X)$ and $Y_2 = 1/(1 - 2 (\sin (X))^2)$ in $[-2\pi, 2\pi, \pi/2]$ by $[-4, 4, 1]$.

Graph Y_1 is shown in Figure 61a. Graph Y_2 is shown in Figure 61b. The graphs are the same.

Table $Y_1 = 1/\cos (2X)$ and $Y_2 = 1/(1 - 2 (\sin (X))^2)$ starting at $x = 0$, incrementing by $\dfrac{\pi}{6}$.

See Figure 61c. The tables are the same.

$[-2\pi, 2\pi, \pi/2]$ by $[-4, 4, 1]$ $[-2\pi, 2\pi, \pi/2]$ by $[-4, 4, 1]$

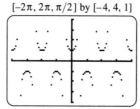

Figure 61a   Figure 61b   Figure 61c

63. $\sin 3\theta = \sin(2\theta + \theta) = \sin 2\theta \cos \theta + \cos 2\theta \sin \theta = (2 \sin \theta \cos \theta) \cos \theta + (1 - 2 \sin^2 \theta) \sin \theta =$

$2 \sin \theta \cos^2 \theta + \sin \theta - 2 \sin^3 \theta = 2 \sin \theta (1 - \sin^2 \theta) + \sin \theta - 2 \sin^3 \theta =$

$2 \sin \theta - 2 \sin^3 \theta + \sin \theta - 2 \sin^3 \theta = 3 \sin \theta - 4 \sin^3 \theta$

Graph $Y_1 = \sin(3X)$ and $Y_2 = 3 \sin(X) - 4(\sin(X))^\wedge 3$ in $[-2\pi, 2\pi, \pi/2]$ by $[-2, 2, 1]$.

Graph Y_1 is shown in Figure 63a. Graph Y_2 is shown in Figure 63b. The graphs are the same.

Table $Y_1 = \sin(3X)$ and $Y_2 = 3 \sin(X) - 4(\sin(X))^\wedge 3$ starting at $x = 0$, incrementing by $\dfrac{\pi}{6}$.

See Figure 63c. The tables are the same.

$[-2\pi, 2\pi, \pi/2]$ by $[-2, 2, 1]$ $[-2\pi, 2\pi, \pi/2]$ by $[-2, 2, 1]$

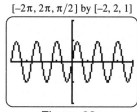   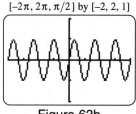   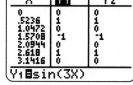

Figure 63a   Figure 63b   Figure 63c

65. $\sin 4\theta = \sin(2 \cdot 2\theta) = 2 \sin 2\theta \cos 2\theta = 2(2 \sin \theta \cos \theta) \cos 2\theta = 4 \sin \theta \cos \theta \cos 2\theta$

Graph $Y_1 = \sin(4X)$ and $Y_2 = 4 \sin(X) \cos(X) \cos(2X)$ in $[-2\pi, 2\pi, \pi/2]$ by $[-2, 2, 1]$.

Graph Y_1 is shown in Figure 65a. Graph Y_2 is shown in Figure 65b. The graphs are the same.

Table $Y_1 = \sin(4X)$ and $Y_2 = 4 \sin(X) \cos(X) \cos(2X)$ starting at $x = 0$, incrementing by $\dfrac{\pi}{6}$.

See Figure 65c. The tables are the same.

$[-2\pi, 2\pi, \pi/2]$ by $[-2, 2, 1]$ $[-2\pi, 2\pi, \pi/2]$ by $[-2, 2, 1]$

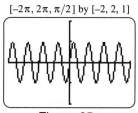   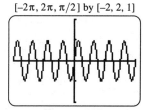

Figure 65a   Figure 65b   Figure 65c

67. $\dfrac{\sin 2\theta}{\sin \theta} = \dfrac{2 \sin \theta \cos \theta}{\sin \theta} = 2 \cos \theta$

69. $2 \cos^2 \dfrac{\theta}{2} = 2\left(\dfrac{1 + \cos \theta}{2}\right) = 1 + \cos \theta$

71. $\cos^4 \theta - \sin^4 \theta = (\cos^2 \theta - \sin^2 \theta)(\cos^2 \theta + \sin^2 \theta) = (\cos 2\theta)(1) = \cos 2\theta$

73. $\csc 2t = \dfrac{1}{\sin 2t} = \dfrac{1}{2 \sin t \cos t} = \dfrac{\csc t}{2 \cos t}$

75. $\tan \dfrac{x}{2} = \dfrac{\sin \frac{x}{2}}{\cos \frac{x}{2}} = \dfrac{2 \sin \frac{x}{2}}{2 \cos \frac{x}{2}} = \dfrac{2 \sin \frac{x}{2} \cos \frac{x}{2}}{2 \cos^2 \frac{x}{2}} = \dfrac{\sin x}{1 + (2 \cos^2 \frac{x}{2} - 1)} = \dfrac{\sin x}{1 + \cos x}$

Product-to-Sum and Sum-to-Product Identities

77. (a) $\cos 50° \sin 20° = \dfrac{1}{2}(\sin(50° + 20°) - \sin(50° - 20°)) = \dfrac{1}{2}(\sin 70° - \sin 30°)$

   (b) $\cos 40° \cos 20° = \dfrac{1}{2}(\cos(40° + 20°) + \cos(40° - 20°)) = \dfrac{1}{2}(\cos 60° + \cos 20°)$

79. (a) $\sin 7\theta \cos 3\theta = \dfrac{1}{2}(\sin(7\theta + 3\theta) + \sin(7\theta - 3\theta)) = \dfrac{1}{2}(\sin 10\theta + \sin 4\theta)$

   (b) $\sin 8x \sin 4x = \dfrac{1}{2}(\cos(8x - 4x) - \cos(8x + 4x)) = \dfrac{1}{2}(\cos 4x - \cos 12x)$

81. (a) $\sin 40° + \sin 30° = 2\sin\left(\dfrac{40° + 30°}{2}\right)\cos\left(\dfrac{40° - 30°}{2}\right) = 2\sin 35° \cos 5°$

   (b) $\cos 45° + \cos 35° = 2\cos\left(\dfrac{45° + 35°}{2}\right)\cos\left(\dfrac{45° - 35°}{2}\right) = 2\cos 40° \cos 5°$

83. (a) $\cos 6\theta + \cos 4\theta = 2\cos\left(\dfrac{6\theta + 4\theta}{2}\right)\cos\left(\dfrac{6\theta - 4\theta}{2}\right) = 2\cos 5\theta \cos \theta$

   (b) $\sin 7x + \sin 4x = 2\sin\left(\dfrac{7x + 4x}{2}\right)\cos\left(\dfrac{7x - 4x}{2}\right) = 2\sin\dfrac{11x}{2}\cos\dfrac{3x}{2}$

Solving Equations

85. (a) Symbolic: $\cos 2\theta = 1 \Rightarrow 2\theta = \cos^{-1}1 \Rightarrow 2\theta = 0°, 360° \Rightarrow \theta = 0°, 180°$

   (b) Graphical: Graph $Y_1 = \cos(2X)$ and $Y_2 = 1$ in degree mode in $[0°, 360°, 30°]$ by $[-2, 2, 1]$.

   In $[0°, 360°)$, the x-coordinates of the intersection points are $x = 0°, 180°$. See Figure 85b.

   (c) Numerical: Table $Y_1 = \cos(2X)$ and $Y_2 = 1$ in degree mode starting at $x = 0$, incrementing by 60.

   In the interval $[0°, 360°)$, the tables match when $x = 0°, 180°$. See Figure 85c.

$[0°, 360°, 30°]$ by $[-2, 2, 1]$ $[0°, 360°, 30°]$ by $[-2, 2, 1]$

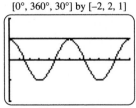
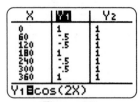
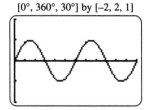
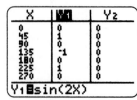

Figure 85b   Figure 85c   Figure 87b   Figure 87c

87. (a) Symbolic: $\sin 2\theta = 0 \Rightarrow 2\theta = \sin^{-1}0 \Rightarrow 2\theta = 0°, 180°, 360°, 540° \Rightarrow \theta = 0°, 90°, 180°, 270°$

   (b) Graphical: Graph $Y_1 = \sin(2X)$ and $Y_2 = 0$ in degree mode in $[0°, 360°, 30°]$ by $[-2, 2, 1]$.

   In $[0°, 360°)$, the x-coordinates of the intersection points are $x = 0°, 90°, 180°, 270°$. See Figure 87b.

   (c) Numerical: Table $Y_1 = \sin(2X)$ and $Y_2 = 0$ in degree mode starting at $x = 0$, incrementing by 45.

   In the interval $[0°, 360°)$, the tables match when $x = 0°, 90°, 180°, 270°$. See Figure 87c.

89. (a) Symbolic: $\sin\dfrac{\theta}{2} = 1 \Rightarrow \dfrac{\theta}{2} = \sin^{-1} 1 \Rightarrow \dfrac{\theta}{2} = 90° \Rightarrow \theta = 180°$

    (b) Graphical: Graph $Y_1 = \sin(X/2)$ and $Y_2 = 1$ in degree mode in $[0°, 360°, 30°]$ by $[-2, 2, 1]$.

    In $[0°, 360°)$, the x-coordinate of the intersection point is $x = 180°$. See Figure 89b.

    (c) Numerical: Table $Y_1 = \sin(X/2)$ and $Y_2 = 1$ in degree mode starting at $x = 0$, incrementing by 60.

    In the interval $[0°, 360°)$, the tables match when $x = 180°$. See Figure 89c.

$[0°, 360°, 30°]$ by $[-2, 2, 1]$

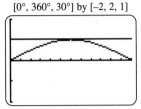

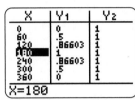

$[0°, 360°, 30°]$ by $[-2, 2, 1]$

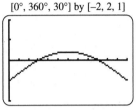

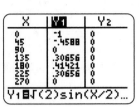

| Figure 89b | Figure 89c | Figure 91b | Figure 91c |

91. (a) Symbolic: $\sqrt{2}\sin\dfrac{\theta}{2} - 1 = 0 \Rightarrow \sqrt{2}\sin\dfrac{\theta}{2} = 1 \Rightarrow \sin\dfrac{\theta}{2} = \dfrac{\sqrt{2}}{2} \Rightarrow \dfrac{\theta}{2} = \sin^{-1}\dfrac{\sqrt{2}}{2} \Rightarrow$

$\dfrac{\theta}{2} = 45°, 135° \Rightarrow \theta = 90°, 270°$

    (b) Graphical: graph $Y_1 = \sqrt{2}\sin(X/2) - 1$ and $Y_2 = 0$ in degree mode in $[0°, 360°, 30°]$ by $[-2, 2, 1]$. In $[0°, 360°)$, the x coordinates of the intersection points are $x = 90°, 270°$. See Figure 91b.

    (c) Numeric: table $Y_1 = \sqrt{2}\sin(X/2) - 1$ and $Y_2 = 0$ in degree mode starting at $x = 0$, incrementing by 45. In the interval $[0, 360)$, the tables match when $x = 90°, 270°$. See Figure 91c.

93. $\sin 2t + \sin t = 0 \Rightarrow 2\sin t\cos t + \sin t = 0 \Rightarrow \sin t(2\cos t + 1) = 0$

$\Rightarrow \sin t = 0$ or $2\cos t + 1 = 0$

$\Rightarrow \sin t = 0$ or $\cos t = -\dfrac{1}{2}$

$\Rightarrow t = \sin^{-1} 0$ or $t = \cos^{-1}\left(-\dfrac{1}{2}\right)$

$\Rightarrow t = \pi n, \dfrac{2\pi}{3} + 2\pi n, \dfrac{4\pi}{3} + 2\pi n$

95. $2\sin\dfrac{t}{2} - 1 = 0 \Rightarrow \sin\dfrac{t}{2} = \dfrac{1}{2} \Rightarrow \dfrac{t}{2} = \sin^{-1}\dfrac{1}{2}$

$\Rightarrow \dfrac{t}{2} = \dfrac{\pi}{6} + 2\pi n, \dfrac{5\pi}{6} + 2\pi n$

$\Rightarrow t = \dfrac{\pi}{3} + 4\pi n, \dfrac{5\pi}{3} + 4\pi n$

97. $\cos 2t = \sin t \Rightarrow 1 - 2\sin^2 t = \sin t \Rightarrow 2\sin^2 t + \sin t - 1 = 0 \Rightarrow (2\sin t - 1)(\sin t + 1) = 0$

$\Rightarrow 2\sin t - 1 = 0$ or $\sin t + 1 = 0$

$\Rightarrow \sin t = \dfrac{1}{2}$ or $\sin t = -1$

$\Rightarrow t = \sin^{-1}\dfrac{1}{2}$ or $t = \sin^{-1}(-1)$

$\Rightarrow t = \dfrac{\pi}{6} + 2\pi n, \dfrac{5\pi}{6} + 2\pi n, \dfrac{3\pi}{2} + 2\pi n$

99. $\tan 2t = 1 \Rightarrow 2t = \tan^{-1} 1 \Rightarrow 2t = \dfrac{\pi}{4} + 2\pi n$ or $2t = \dfrac{5\pi}{4} + 2\pi n \Rightarrow t = \dfrac{\pi}{8} + \pi n$ or $t = \dfrac{5\pi}{8} + \pi n$

   or equivalently, $t = \dfrac{\pi}{8} + \dfrac{\pi n}{2}$

101. $2\cos\dfrac{t}{2} = 1 \Rightarrow \cos\dfrac{t}{2} = \dfrac{1}{2} \Rightarrow \dfrac{t}{2} = \cos^{-1}\dfrac{1}{2} \Rightarrow \dfrac{t}{2} = \dfrac{\pi}{3} + 2\pi n$ or $\dfrac{t}{2} = \dfrac{5\pi}{3} + 2\pi n \Rightarrow$

   $t = \dfrac{2\pi}{3} + 4\pi n$ or $t = \dfrac{10\pi}{3} + 4\pi n$

103. $\cos 2t = 2\sin t \cos t$, let $2\sin t \cos t = \sin 2t$ and let $\theta = 2t$. So then the equation becomes $\cos\theta = \sin\theta$,

   divide both sides by $\cos\theta \Rightarrow \dfrac{\cos\theta}{\cos\theta} = \dfrac{\sin\theta}{\cos\theta} \Rightarrow 1 = \tan\theta \Rightarrow \theta = \dfrac{\pi}{4}$ and $\dfrac{5\pi}{4}$. For all real number solutions

   $\theta = \dfrac{\pi}{4} + 2\pi n$ or $\theta = \dfrac{5\pi}{4} + 2\pi n$. Because $\theta = 2t$, we can determine t by substituting $2t$ for $\theta \Rightarrow$

   $2t = \dfrac{\pi}{4} + 2\pi n$ or $2t = \dfrac{5\pi}{4} + 2\pi n$. Therefore $t = \dfrac{\pi}{8} + \pi n$ or $t = \dfrac{5\pi}{8} + \pi n$ or equivalently, $\dfrac{\pi}{8} + \dfrac{\pi n}{2}$.

105. $2\sin^2 2t + \sin 2t - 1 = 0 \Rightarrow (2\sin 2t - 1)(\sin 2t + 1) = 0$

   $\Rightarrow 2\sin 2t - 1 = 0 \quad$ or $\quad \sin 2t + 1 = 0$

   $\Rightarrow \sin 2t = \dfrac{1}{2} \quad\quad$ or $\quad \sin 2t = -1$

   Let $\theta = 2t$, then

   $\Rightarrow \sin\theta = \dfrac{1}{2} \quad\quad$ or $\quad \sin\theta = -1$

   $\Rightarrow \theta = \sin^{-1}\left(\dfrac{1}{2}\right) \quad$ or $\quad \theta = \sin^{-1}(-1)$

   $\Rightarrow \theta = \dfrac{\pi}{6}$ or $\dfrac{5\pi}{6} \quad$ or $\quad \theta = \dfrac{3\pi}{2}$

   For all real number solutions $\theta = \dfrac{\pi}{6} + 2\pi n$ or $\theta = \dfrac{5\pi}{6} + 2\pi n$ or $\theta = \dfrac{3\pi}{2} + 2\pi n$. Because $\theta = 2t$, we can

   determine t by substituting $2t$ for $\theta \Rightarrow 2t = \dfrac{\pi}{6} + 2\pi n$ or $2t = \dfrac{5\pi}{6} + 2\pi n$ or $2t = \dfrac{3\pi}{2} + 2\pi n$.

   Therefore $t = \dfrac{\pi}{12} + \pi n$ or $t = \dfrac{5\pi}{12} + \pi n$ or $t = \dfrac{3\pi}{4} + \pi n$.

107. $\sin t + \sin 2t = \cos t$. Graph $Y_1 = \sin(X) + \sin(2X)$ and $Y_2 = \cos(X)$ in radian mode in $[0, 2\pi, \pi/2]$ by

   $[-3, 3, 1]$. See Figure 107. The x-coordinates of the intersection points are approximately 0.333 and 4.379.

   $[0, 2\pi, \pi/2]$ by $[-3, 3, 1]$

   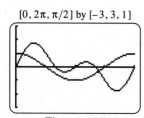

   Figure 107

Applications

109. (a) Graph $Y_1 = (310\sin(120\pi X))(7\sin(120\pi X))$ in $[0, 0.04, 0.01]$ by $[-500, 2500, 500]$. See Figure 109.

(b) $W = (310\sin(120\pi x))(7\sin(120\pi x)) = 2170\sin^2(120\pi x) = 2170 \cdot \dfrac{1 - \cos(240\pi x)}{2} =$

$1085(1 - \cos(240\pi x)) = 1085 - 1085\cos(240\pi x) = -1085\cos(240\pi x) + 1085$

$\Rightarrow a = -1085, k = 240, d = 1085$

[0, 0.04, 0.01] by [−500, 2500, 500]

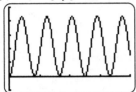

Figure 109

111. We must solve the equation $V(t) = 160$ for t: $320\sin(120\pi t) = 160 \Rightarrow \sin(120\pi t) = \dfrac{1}{2}$.

Let $\theta = 120\pi t$ then $\sin\theta = \dfrac{1}{2} \Rightarrow \theta = \dfrac{\pi}{6} + 2\pi n$ or $\theta = \dfrac{5\pi}{6} + 2\pi n$. Substituting $120\pi t$ for θ yields

$\Rightarrow 120\pi t = \dfrac{\pi}{6} + 2\pi n$ or $120\pi t = \dfrac{5\pi}{6} + 2\pi n$

$\Rightarrow t = \dfrac{1}{720} + \dfrac{n}{60}$ or $t = \dfrac{5}{720} + \dfrac{n}{60}$ sec

113. (a) $d = 600\left(1 - \cos\dfrac{80°}{2}\right) = 600(1 - \cos 40°) \approx 140.4$ ft

(b) By bisecting β, we create a diagram as shown in Figure 113.

Here $r = d + a$ and from right triangle trigonometry, $\cos\dfrac{\beta}{2} = \dfrac{a}{r} \Rightarrow a = r\cos\dfrac{\beta}{2}$.

Thus, $r = d + r\cos\dfrac{\beta}{2} \Rightarrow d = r - r\cos\dfrac{\beta}{2} \Rightarrow d = r\left(1 - \cos\dfrac{\beta}{2}\right)$

(c) No, $\cos\dfrac{\beta}{2} = \pm\sqrt{\dfrac{1 + \cos\beta}{2}} \neq \dfrac{1}{2}\cos\beta$

[0, 0.005, 0.001] by [−2, 2, 1]

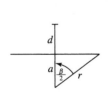

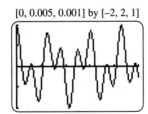

Figure 113   Figure 115

115. (a) $f(t) = \cos(2\pi(697)t) + \cos(2\pi(1633)t) = \cos(1394\pi t) + \cos(3266\pi t)$

(b) $f(t) = 2\cos\left(\dfrac{1394\pi t + 3266\pi t}{2}\right)\cos\left(\dfrac{1394\pi t - 3266\pi t}{2}\right) = 2\cos(2330\pi t)\cos(-936\pi t) =$

$2\cos(2330\pi t)\cos(936\pi t)$

(c) Graph $Y_1 = 2\cos(2330\pi X)\cos(936\pi X)$ in $[0, 0.005, 0.001]$ by $[-2, 2, 1]$. See Figure 115.

117. Let $f(t)$ model the tone for number 3 and let $g(t)$ model the tone for number 4.

(a) $f(t) = \cos(2(697)\pi t) + \cos(2(1477)\pi t) = \cos(1394\pi t) + \cos(2954\pi t)$

$g(t) = \cos(2(770)\pi t) + \cos(2(1209)\pi t) = \cos(1540\pi t) + \cos(2418\pi t)$

(b) $f(t) = 2\cos\left(\dfrac{1394\pi t + 2954\pi t}{2}\right)\cos\left(\dfrac{1394\pi t - 2954\pi t}{2}\right) = 2\cos(2174\pi t)\cos(-780\pi t) =$

$2\cos(2174\pi t)\cos(780\pi t)$

$g(t) = 2\cos\left(\dfrac{1540\pi t + 2418\pi t}{2}\right)\cos\left(\dfrac{1540\pi t - 2418\pi t}{2}\right) = 2\cos(1979\pi t)\cos(-439\pi t) =$

$2\cos(1979\pi t)\cos(439\pi t)$

(c) Graph $Y_1 = 2\cos(2174\pi X)\cos(780\pi X)$ in $[0, 0.005, 0.001]$ by $[-2, 2, 1]$. See Figure 117a.

Graph $Y_2 = 2\cos(1979\pi X)\cos(439\pi X)$ in $[0, 0.005, 0.001]$ by $[-2, 2, 1]$. See Figure 117b.

The graphs are different. These two numbers sound different so their graphs should be different.

$[0, 0.005, 0.001]$ by $[-2, 2, 1]$ $[0, 0.005, 0.001]$ by $[-2, 2, 1]$

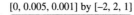

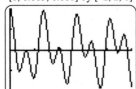   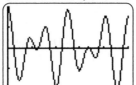

Figure 117a   Figure 117b

Chapter 7 Review Exercises

1. If $\sec\theta < 0$ and $\sin\theta > 0$ then any point (x, y) on the terminal side of θ must satisfy

$y > 0$ and $\dfrac{r}{x} < 0 \Rightarrow x < 0$. Thus, θ is contained in quadrant II.

3. $\tan\theta = \dfrac{\sin\theta}{\cos\theta} = \dfrac{\frac{3}{5}}{-\frac{4}{5}} = -\dfrac{3}{4}$ 　　　　$\csc\theta = \dfrac{1}{\sin\theta} = \dfrac{1}{\frac{3}{5}} = \dfrac{5}{3}$

$\sec\theta = \dfrac{1}{\cos\theta} = \dfrac{1}{-\frac{4}{5}} = -\dfrac{5}{4}$ 　　　　$\cot\theta = \dfrac{1}{\tan\theta} = \dfrac{1}{-\frac{3}{4}} = -\dfrac{4}{3}$

5. $\sin\theta = \cos\theta\tan\theta = \dfrac{24}{25}\left(-\dfrac{7}{24}\right) = -\dfrac{7}{25}$ 　　$\csc\theta = \dfrac{1}{\sin\theta} = \dfrac{1}{-\frac{7}{25}} = -\dfrac{25}{7}$

$\sec\theta = \dfrac{1}{\cos\theta} = \dfrac{1}{\frac{24}{25}} = \dfrac{25}{24}$ 　　　　$\cot\theta = \dfrac{1}{\tan\theta} = \dfrac{1}{-\frac{7}{24}} = -\dfrac{24}{7}$

7. Since the sine function is an odd function $\sin(-13°) = -\sin 13°$.

9. Since the secant function is an even function $\sec\left(-\dfrac{3\pi}{7}\right) = \sec\dfrac{3\pi}{7}$.

11. An identity is an equation which is valid for all meaningful values of the variable. Example: $\sin^2\theta + \cos^2\theta = 1$.

A conditional equation is valid only for certain values of the variable. Example: $\tan\theta = 1$.

13. No. Whenever a trigonometric function is positive, the corresponding reciprocal function is also positive.

15. $\sec\theta\cot\theta\sin\theta = \dfrac{1}{\cos\theta}\cdot\dfrac{\cos\theta}{\sin\theta}\cdot\dfrac{\sin\theta}{1} = 1$

17. $(\sec^2 t - 1)(\csc^2 t - 1) = (\tan^2 t)(\cot^2 t) = 1$

19. $\dfrac{\csc \theta \sin \theta}{\sec \theta} = \dfrac{1}{\sec \theta} = \cos \theta$

21. $\tan \theta = 1.2367 \Rightarrow \theta = \tan^{-1} 1.2367$ $\qquad$ $\sin \theta = \sin(\tan^{-1} 1.2367) \approx 0.7776$

$\cos \theta = \cos(\tan^{-1} 1.2367) \approx 0.6288$ $\qquad$ $\csc \theta = \dfrac{1}{\sin(\tan^{-1} 1.2367)} \approx 1.2860$

$\sec \theta = \dfrac{1}{\cos(\tan^{-1} 1.2367)} \approx 1.5904$ $\qquad$ $\cot \theta = \dfrac{1}{\tan \theta} \approx 0.8086$

23. $\cos \theta = -0.4544 \Rightarrow \theta = \cos^{-1}(-0.4544)$

$\sin \theta = \sin(\cos^{-1}(-0.4544)) \approx 0.8908$ $\qquad$ $\tan \theta = \tan(\cos^{-1}(-0.4544)) \approx -1.9604$

$\csc \theta = \dfrac{1}{\sin \theta} \approx 1.1226$ $\qquad$ $\sec \theta = \dfrac{1}{\cos \theta} \approx -2.2007$ $\qquad$ $\cot \theta = \dfrac{1}{\tan \theta} \approx -0.5101$

25. $\sin^2 \theta + 2\sin \theta + 1 = (\sin \theta + 1)(\sin \theta + 1)$

27. $\tan^2 \theta - 9 = (\tan \theta + 3)(\tan \theta - 3)$

29. $(\sec \theta - 1)(\sec \theta + 1) = \sec^2 \theta - 1 = \tan^2 \theta$

31. $(1 + \tan t)^2 = 1 + 2\tan t + \tan^2 t = \sec^2 t + 2\tan t$

33. $\sin(x - \pi) = \sin x \cos \pi - \cos x \sin \pi = \sin x(-1) - \cos x(0) = -\sin x$

35. $\sin 8x = \sin(2 \cdot 4x) = 2\sin 4x \cos 4x$

37. $\sec 2x = \dfrac{1}{\cos 2x} = \dfrac{1}{2\cos^2 x - 1}$

39. $\cos^4 x \sin^3 x = \cos^4 x(\sin^2 x)\sin x = \cos^4 x(1 - \cos^2 x)\sin x = (\cos^4 x - \cos^6 x)\sin x$

41. $\sec^4 \theta - \tan^4 \theta = (\sec^2 \theta - \tan^2 \theta)(\sec^2 \theta + \tan^2 \theta) = (1)((1 + \tan^2 \theta) + \tan^2 \theta) = 1 + 2\tan^2 \theta$

43. $\theta = 240°$ is in quadrant III, $\theta_R = 240° - 180° = 60°$

45. $\theta = \dfrac{9\pi}{7}$ is in quadrant III, $\theta_R = \dfrac{9\pi}{7} - \pi = \dfrac{2\pi}{7}$

47. Estimate: $t = \dfrac{\pi}{6}, \dfrac{\pi}{2}, \dfrac{5\pi}{6}, \dfrac{3\pi}{2}$

Symbolic: $2\sin t \cos t - \cos t = 0 \Rightarrow \cos t(2\sin t - 1) = 0$

$\Rightarrow \cos t = 0$ $\qquad$ or $\qquad$ $2\sin t - 1 = 0$

$\Rightarrow \cos t = 0$ $\qquad$ or $\qquad$ $\sin t = \dfrac{1}{2}$

$\Rightarrow t = \cos^{-1} 0$ $\qquad$ or $\qquad$ $t = \sin^{-1}\dfrac{1}{2} \Rightarrow t = \dfrac{\pi}{6}, \dfrac{\pi}{2}, \dfrac{5\pi}{6}, \dfrac{3\pi}{2}$

49. (a) $\tan \theta = \sqrt{3} \Rightarrow \theta = \tan^{-1} \sqrt{3} = 60°, 240°$

(b) $\cot \theta = -\sqrt{3} \Rightarrow \tan \theta = -\dfrac{1}{\sqrt{3}} \Rightarrow \theta = \tan^{-1}\left(-\dfrac{1}{\sqrt{3}}\right) = 150°, 330°$

51. $2\cos \theta - 1 = 0 \Rightarrow \cos \theta = \dfrac{1}{2} \Rightarrow \theta = \cos^{-1}\dfrac{1}{2} = 60°, 300°$

53. $2\sin^2 \theta + \sin \theta - 3 = 0 \Rightarrow (\sin \theta - 1)(2\sin \theta + 3)$

$\Rightarrow \sin \theta - 1 = 0$ $\qquad$ or $\qquad$ $2\sin \theta + 3 = 0$

$\Rightarrow \sin \theta = 1$ $\qquad$ or $\qquad$ $\sin \theta = -\dfrac{3}{2}$ (not possible)

$\Rightarrow \theta = \sin^{-1} 1 = 90°$

55. $\tan^2 t - 2\tan t + 1 = 0 \Rightarrow (\tan t - 1)(\tan t - 1) = 0 \Rightarrow \tan t = 1 \Rightarrow t = \tan^{-1} 1 \Rightarrow t = \dfrac{\pi}{4}, \dfrac{5\pi}{4}$

57. $3\tan^2 t - 1 = 0 \Rightarrow \tan^2 t = \dfrac{1}{3} \Rightarrow \tan t = \pm\dfrac{1}{\sqrt{3}} \Rightarrow t = \tan^{-1}\left(\pm\dfrac{1}{\sqrt{3}}\right)$

$t = \dfrac{\pi}{6} + \pi n, \; -\dfrac{\pi}{6} + \pi n$ or $t = 30° + 180°n, \; -30° + 180°n$

59. $\sin 2t + 3\cos t = 0 \Rightarrow 2\sin t\cos t + 3\cos t = 0 \Rightarrow \cos t(2\sin t + 3) = 0$

$\Rightarrow \cos t = 0$ or $2\sin t + 3 = 0$

$\Rightarrow \cos t = 0$ or $\sin t = -\dfrac{3}{2}$ (not possible)

$\Rightarrow t = \cos^{-1} 0$

$t = \dfrac{\pi}{2} + \pi n$ or $t = 90° + 180°n$

61. $\cos 105° = \cos\dfrac{210°}{2} = -\sqrt{\dfrac{1 + \cos 210°}{2}} = -\sqrt{\dfrac{1 + \left(-\frac{\sqrt{3}}{2}\right)}{2}} = -\sqrt{\dfrac{\frac{2 - \sqrt{3}}{2}}{2}} = -\sqrt{\dfrac{2 - \sqrt{3}}{4}} = -\dfrac{\sqrt{2 - \sqrt{3}}}{2}.$

See Figure 61.

63. Graph $Y_1 = \tan(X)$ and $Y_2 = X + 1$ using dot mode in $[-2\pi, 2\pi, \pi/2]$ by $[-8, 8, 1]$. See Figure 63.

The x-coordinates of the intersection points are approximately $-4.43, 1.13$ and 4.53.

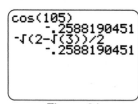

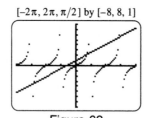
$[-2\pi, 2\pi, \pi/2]$ by $[-8, 8, 1]$

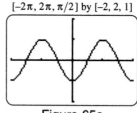
$[-2\pi, 2\pi, \pi/2]$ by $[-2, 2, 1]$

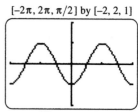
$[-2\pi, 2\pi, \pi/2]$ by $[-2, 2, 1]$

Figure 61   Figure 63   Figure 65a   Figure 65b

65. (a) If the sine graph is translated $\dfrac{\pi}{2}$ units right, it corresponds to the cosine graph reflected across the x-axis.

(b) $\sin\left(t - \dfrac{\pi}{2}\right) = \sin t\cos\dfrac{\pi}{2} - \cos t\sin\dfrac{\pi}{2} = \sin t(0) - \cos t(1) = -\cos t$

(c) Graph $Y_1 = \sin(X - \pi/2)$ and $Y_2 = -\cos(X)$ in $[-2\pi, 2\pi, \pi/2]$ by $[-2, 2, 1]$.

     Graph Y_1 is shown in Figure 65a. Graph Y_2 is shown in Figure 65b. The graphs are the same.

67. If $\cos\alpha = \dfrac{3}{5}$ and α is in quadrant I, then $\sin\alpha = \dfrac{4}{5}$. If $\cos\beta = \dfrac{12}{13}$ and β is in quadrant I, then $\sin\beta = \dfrac{5}{13}$.

(a) $\sin(\alpha + \beta) = \sin\alpha\cos\beta + \cos\alpha\sin\beta = \left(\dfrac{4}{5}\right)\left(\dfrac{12}{13}\right) + \left(\dfrac{3}{5}\right)\left(\dfrac{5}{13}\right) = \dfrac{48 + 15}{65} = \dfrac{63}{65}$

(b) $\cos(\alpha + \beta) = \cos\alpha\cos\beta - \sin\alpha\sin\beta = \left(\dfrac{3}{5}\right)\left(\dfrac{12}{13}\right) - \left(\dfrac{4}{5}\right)\left(\dfrac{5}{13}\right) = \dfrac{36 - 20}{65} = \dfrac{16}{65}$

(c) $\tan(\alpha + \beta) = \dfrac{\sin(\alpha + \beta)}{\cos(\alpha + \beta)} = \dfrac{\frac{63}{65}}{\frac{16}{65}} = \dfrac{63}{65} \cdot \dfrac{65}{16} = \dfrac{63}{16}$

(d) Since $\sin(\alpha + \beta)$ and $\cos(\alpha + \beta)$ are positive, $\alpha + \beta$ is in quadrant I.

69. Since $\tan\theta = \dfrac{\sin\theta}{\cos\theta}$, $\cos\theta = \dfrac{\sin\theta}{\tan\theta} = \dfrac{\frac{4}{5}}{-\frac{4}{3}} = \dfrac{4}{5}\cdot\left(-\dfrac{3}{4}\right) = -\dfrac{3}{5}$

(a) $\sin 2\theta = 2\sin\theta\cos\theta = 2\left(\dfrac{4}{5}\right)\left(-\dfrac{3}{5}\right) = -\dfrac{24}{25}$ $\qquad\cos 2\theta = 1 - 2\sin^2\theta = 1 - 2\left(\dfrac{4}{5}\right)^2 = 1 - \dfrac{32}{25} = -\dfrac{7}{25}$

$\tan 2\theta = \dfrac{\sin 2\theta}{\cos 2\theta} = \dfrac{-\frac{24}{25}}{-\frac{7}{25}} = -\dfrac{24}{25}\cdot\left(-\dfrac{25}{7}\right) = \dfrac{24}{7}$

(b) Since θ is in quadrant II, $\theta = \cos^{-1}\left(-\dfrac{3}{5}\right)$. Numerical support is shown in Figures 69a and 69b.

```
sin(2cos-1(-3/5))
▶Frac
              -24/25
cos(2cos-1(-3/5))
▶Frac
               -7/25
```

```
tan(2cos-1(-3/5))
▶Frac
               24/7
```

Figure 69a   Figure 69b

71. Since $\cos\theta = \dfrac{1}{4}$ and θ is in quadrant I, $\sin\theta = \sqrt{1 - \cos^2\theta} = \sqrt{1 - \left(\dfrac{1}{4}\right)^2} = \sqrt{\dfrac{15}{16}} = \dfrac{\sqrt{15}}{4}$.

(a) $\sin\dfrac{\theta}{2} = \sqrt{\dfrac{1 - \cos\theta}{2}} = \sqrt{\dfrac{1 - \frac{1}{4}}{2}} = \sqrt{\dfrac{\frac{4-1}{4}}{2}} = \sqrt{\dfrac{3}{8}}$

$\cos\dfrac{\theta}{2} = \sqrt{\dfrac{1 + \cos\theta}{2}} = \sqrt{\dfrac{1 + \frac{1}{4}}{2}} = \sqrt{\dfrac{\frac{4+1}{4}}{2}} = \sqrt{\dfrac{5}{8}}$

$\tan\dfrac{\theta}{2} = \dfrac{\sin\frac{\theta}{2}}{\cos\frac{\theta}{2}} = \dfrac{\sqrt{\frac{3}{8}}}{\sqrt{\frac{5}{8}}} = \sqrt{\dfrac{3}{8}}\cdot\sqrt{\dfrac{8}{5}} = \sqrt{\dfrac{3}{5}}$

(b) Since θ is in quadrant I, $\theta = \cos^{-1}\dfrac{1}{4}$. Numerical support is shown in Figures 71a, 71b and 71c.

```
sin(cos-1(1/4)/2)
▶Frac
          .6123724357
√(3/8)
          .6123724357
```

```
cos(cos-1(1/4)/2)
▶Frac
          .790569415
√(5/8)
          .790569415
```

```
tan(cos-1(1/4)/2)
▶Frac
          .7745966692
√(3/5)
          .7745966692
```

Figure 71a   Figure 71b   Figure 71c

73. Let $\theta = \tan^{-1}\dfrac{11}{60}$. It follows that $\sin\theta = \dfrac{11}{61}$. See Figure 73.

$\cos\left(2\tan^{-1}\dfrac{11}{60}\right) = \cos(2\theta) = 1 - 2\sin^2\theta = 1 - 2\left(\dfrac{11}{61}\right)^2 = 1 - \dfrac{242}{3721} = \dfrac{3479}{3721}$

Figure 73

75. Graph $Y_1 = 1.2\cos(\pi/6(X - 0.7)) + 12.1$ and $Y_2 = 11.5$ in $[0, 13, 1]$ by $[10, 15, 1]$. See Figure 75a and 75b.

The x-coordinates of the intersection points are approximately 4.7 and 8.7. The number of daylight hours equals 11.5 on about April 21 and again on about August 22.

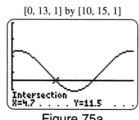

[0, 13, 1] by [10, 15, 1]

Figure 75a

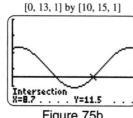

[0, 13, 1] by [10, 15, 1]

Figure 75b

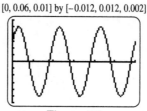

[0, 0.06, 0.01] by [−0.012, 0.012, 0.002]

Figure 77

77. (a) Graph $Y_1 = 0.006\cos(100\pi X) + 0.008\sin(100\pi X)$ in $[0, 0.06, 0.01]$ by $[-0.012, 0.012, 0.002]$. See Figure 77.

(b) The peaks of the graph are located at $y = 0.01$, thus $a = 0.01$. The x-value at the first peak is

$$x \approx 0.00295089. \quad 100\pi(0.00295089) + k = \frac{\pi}{2} \Rightarrow k \approx 0.6435.$$

(c) $P = 0.01\sin(100\pi t + 0.6435) = 0.01[\sin(100\pi t)\cos(0.6435) + \cos(100\pi t)\sin(0.6435)] \approx$
$0.01[\sin(100\pi t)(0.8) + \cos(100\pi t)(0.6)] \approx 0.008\sin(100\pi t) + 0.006\cos(100\pi t)$

79. $W(t) = 7\cos^2(240\pi t) \Rightarrow W(t) = 7(1 - \sin^2(240\pi t)) \Rightarrow W(t) = 7 - 7\sin^2(240\pi t)$

When V is maximum or minimum, $W = 0$.

81. Let $F = 250 \Rightarrow 250 = 289\cos\theta \Rightarrow \cos\theta = \dfrac{250}{289}$ or $\theta = \cos^{-1}\left(\dfrac{250}{289}\right) \Rightarrow \theta \approx 30.11°$ or 0.53 radians.

Chapter 8: Further Topics in Trigonometry

8.1: Law of Sines

Recognizing the Ambiguous Case

1. No. This situation is AAS.

3. No. This situation is SSS.

5. Yes. This situation is SSA.

7. Yes. This situation is SSA.

Solving Triangles

9. This triangle is of the form AAS. There is only one solution.

$$\beta = 180° - 45° - 75° = 60°$$

$$\frac{a}{\sin \alpha} = \frac{c}{\sin \gamma} \Rightarrow a = \frac{c \sin \alpha}{\sin \gamma} = \frac{4 \sin 75°}{\sin 45°} \approx 5.46 \qquad \frac{b}{\sin \beta} = \frac{c}{\sin \gamma} \Rightarrow b = \frac{c \sin \beta}{\sin \gamma} = \frac{4 \sin 60°}{\sin 45°} \approx 4.90$$

$$\beta = 60°, \ a \approx 5.5, \ b \approx 4.9$$

11. This triangle is of the form ASA. There is only one solution.

$$\beta = 180° - 120° - 25° = 35°$$

$$\frac{a}{\sin \alpha} = \frac{b}{\sin \beta} \Rightarrow a = \frac{b \sin \alpha}{\sin \beta} = \frac{10 \sin 120°}{\sin 35°} \approx 15.1 \qquad \frac{c}{\sin \gamma} = \frac{b}{\sin \beta} \Rightarrow c = \frac{b \sin \gamma}{\sin \beta} = \frac{10 \sin 25°}{\sin 35°} \approx 7.37$$

$$\beta = 35°, \ a \approx 15.1, \ c \approx 7.4$$

13. This triangle is of the form SSA. It is ambiguous.

$$\frac{\sin \beta}{b} = \frac{\sin \alpha}{a} \Rightarrow \sin \beta = \frac{b \sin \alpha}{a} = \frac{7 \sin 46°}{6} \Rightarrow \beta_R = \sin^{-1}\left(\frac{7 \sin 46°}{6}\right) \approx 57.1°$$

Thus, $\beta \approx 57.1°$ or $\beta \approx 180° - 57.1° \approx 122.9°$.

Solution 1: Let $\beta \approx 57.1°$. Then $\gamma \approx 180° - 46° - 57.1° \approx 76.9°$.

$$\frac{c}{\sin \gamma} = \frac{a}{\sin \alpha} \Rightarrow c = \frac{a \sin \gamma}{\sin \alpha} = \frac{6 \sin 76.9°}{\sin 46°} \approx 8.12 \qquad \beta \approx 57.1°, \ \gamma \approx 76.9°, \ c \approx 8.1$$

Solution 2: Let $\beta \approx 122.9°$. Then $\gamma \approx 180° - 46° - 122.9° \approx 11.1°$.

$$\frac{c}{\sin \gamma} = \frac{a}{\sin \alpha} \Rightarrow c = \frac{a \sin \gamma}{\sin \alpha} = \frac{6 \sin 11.1°}{\sin 46°} \approx 1.61 \qquad \beta \approx 122.9°, \ \gamma \approx 11.1°, \ c \approx 1.6$$

15. This triangle is of the form SSA. It is ambiguous.

$$\frac{\sin \alpha}{a} = \frac{\sin \beta}{b} \Rightarrow \sin \alpha = \frac{a \sin \beta}{b} = \frac{5 \sin 52°}{3} \approx 1.31 \qquad \text{This is not possible. No solution.}$$

17. This triangle is of the form SSA. It is ambiguous.

$$\alpha = 50° \ 12' = 50° + \left(\frac{12}{60}\right)° = 50.2°$$

$$\frac{\sin \beta}{b} = \frac{\sin \alpha}{a} \Rightarrow \sin \beta = \frac{b \sin \alpha}{a} = \frac{12 \sin 50.2°}{10} \approx 0.9219 \Rightarrow \beta_R \approx \sin^{-1}(0.9219) \approx 67.2° \text{ or } \beta_1 \approx 67° \ 13'$$

Then $\gamma_1 = 180° - 50.2° - 67.2° \approx 62.6°$ or $\gamma_1 \approx 62° \ 35'$

$$\frac{c_1}{\sin \gamma} = \frac{a}{\sin \alpha} \Rightarrow c_1 = \frac{a \sin \gamma}{\sin \alpha} = \frac{10 \sin 62.6°}{\sin 50.2°} \approx 11.6 \text{ or } c_1 \approx 11.6 \qquad \beta_1 \approx 67°13', \ \gamma_1 \approx 62°35', \ c_1 \approx 11.6$$

However if $\beta_2 = 180° - 67.2° = 112.8°$ or $112° \ 48'$ then $\gamma_2 = 180° - 50.2° - 112.8° = 17°$ or $\gamma_2 \approx 17° \ 0'$

$$\frac{c_2}{\sin \gamma} = \frac{a}{\sin \alpha} \Rightarrow c_2 = \frac{a \sin \gamma_2}{\sin \alpha} = \frac{10 \sin 17°}{\sin 50.2°} \approx 3.8 \text{ or } c_2 \approx 3.8 \qquad \beta_2 \approx 112°48', \ \gamma_2 \approx 17°0', \ c_2 \approx 3.8$$

19. This triangle is of the form AAS. There is only one solution.

$$\gamma = 180° - 32° - 55° = 93°$$

$$\frac{a}{\sin \alpha} = \frac{b}{\sin \beta} \Rightarrow a = \frac{b \sin \alpha}{\sin \beta} = \frac{12 \sin 32°}{\sin 55°} \approx 7.76 \qquad \frac{c}{\sin \gamma} = \frac{b}{\sin \beta} \Rightarrow c = \frac{b \sin \gamma}{\sin \beta} = \frac{12 \sin 93°}{\sin 55°} \approx 14.6$$

$\gamma = 93°, \ a \approx 7.8, \ c \approx 14.6$

21. This triangle is of the form SSA. It is ambiguous.

$$\frac{\sin \beta}{b} = \frac{\sin \alpha}{a} \Rightarrow \sin \beta = \frac{b \sin \alpha}{a} = \frac{9 \sin 20°}{7} \Rightarrow \beta_R = \sin^{-1}\left(\frac{9 \sin 20°}{7}\right) \approx 26.1°$$

Thus, $\beta \approx 26.1°$ or $\beta \approx 180° - 26.1° \approx 153.9°$.

Solution 1: Let $\beta \approx 26.1°$. Then $\gamma \approx 180° - 20° - 26.1° \approx 133.9°$.

$$\frac{c}{\sin \gamma} = \frac{a}{\sin \alpha} \Rightarrow c = \frac{a \sin \gamma}{\sin \alpha} = \frac{7 \sin 133.9°}{\sin 20°} \approx 14.7 \qquad \beta \approx 26.1°, \ \gamma \approx 133.9°, \ c \approx 14.7$$

Solution 2: Let $\beta \approx 153.9°$. Then $\gamma \approx 180° - 20° - 153.9° \approx 6.1°$.

$$\frac{c}{\sin \gamma} = \frac{a}{\sin \alpha} \Rightarrow c = \frac{a \sin \gamma}{\sin \alpha} = \frac{7 \sin 6.1°}{\sin 20°} \approx 2.17 \qquad \beta \approx 153.9°, \ \gamma \approx 6.1°, \ c \approx 2.2$$

23. This triangle is of the form SSA. It is ambiguous.

$$\frac{\sin \gamma}{c} = \frac{\sin \beta}{b} \Rightarrow \sin \gamma = \frac{c \sin \beta}{b} = \frac{20 \sin 30°}{10} = 1 \Rightarrow \gamma_R = \sin^{-1} 1 = 90°$$

Thus, $\gamma = 90°$. Then $\alpha = 180° - 30° - 90° = 60°$.

$$\frac{a}{\sin \alpha} = \frac{b}{\sin \beta} \Rightarrow a = \frac{b \sin \alpha}{\sin \beta} = \frac{10 \sin 60°}{\sin 30°} = 10\sqrt{3} \approx 17.3 \qquad \gamma = 90°, \ \alpha = 60°, \ a = 10\sqrt{3} \approx 17.3$$

25. This triangle is of the form SSA. It is ambiguous.

$$\frac{\sin \alpha}{a} = \frac{\sin \gamma}{c} \Rightarrow \sin \alpha = \frac{a \sin \gamma}{c} = \frac{42.1 \sin 102°}{51.6} \Rightarrow \alpha_R = \sin^{-1}\left(\frac{42.1 \sin 102°}{51.6}\right) \approx 52.9°$$

Thus, $\alpha \approx 52.9°$ or $\alpha \approx 180° - 52.9° \approx 127.1°$ (which is not possible).

So $\alpha \approx 52.9°$. Then $\beta \approx 180° - 102° - 52.9° \approx 25.1°$.

$$\frac{b}{\sin \beta} = \frac{c}{\sin \gamma} \Rightarrow b = \frac{c \sin \beta}{\sin \gamma} = \frac{51.6 \sin 25.1°}{\sin 102°} \approx 22.4 \qquad \alpha \approx 52.9°, \ \beta \approx 25.1°, \ b \approx 22.4$$

27. This triangle is of the form ASA. There is only one solution.

$\beta = 180° - 55.2° - 114.8° = 10°$

$$\frac{a}{\sin \alpha} = \frac{b}{\sin \beta} \Rightarrow a = \frac{b \sin \alpha}{\sin \beta} = \frac{19.5 \sin 55.2°}{\sin 10°} \approx 92.2$$

$$\frac{c}{\sin \gamma} = \frac{b}{\sin \beta} \Rightarrow c = \frac{b \sin \gamma}{\sin \beta} = \frac{19.5 \sin 114.8°}{\sin 10°} \approx 101.9$$

$\beta = 10°,\ a \approx 92.2,\ c \approx 101.9$

29. This triangle is of the form SSA. It is ambiguous.

$$\frac{\sin \gamma}{c} = \frac{\sin \beta}{b} \Rightarrow \sin \gamma = \frac{c \sin \beta}{b} = \frac{7.4 \sin 73°}{6.2} \approx 1.14 \quad \text{This is not possible. No solution.}$$

31. This triangle is of the form SSA. It is ambiguous.

$$\frac{\sin \beta}{b} = \frac{\sin \alpha}{a} \Rightarrow \sin \beta = \frac{b \sin \alpha}{a} = \frac{12 \sin 35° \ 15'}{5} \approx 1.39 \quad \text{This is not possible. No solution.}$$

33. This triangle is of the form SSA. It is ambiguous.

$$\frac{\sin \alpha}{a} = \frac{\sin \beta}{b} \Rightarrow \sin \alpha = \frac{a \sin \beta}{b} = \frac{6 \sin 46° \ 45'}{5} \approx 0.8740 \Rightarrow \alpha_R \approx \sin^{-1}(0.8740) \approx 60.93°$$

or $\alpha_1 \approx 60° \ 56'$ then $\gamma_1 \approx 180° - 46.75° - 60.93° \approx 72.32°$ or $\gamma_1 \approx 72° \ 19'$

$$\frac{c_1}{\sin \gamma_1} = \frac{b}{\sin \beta} \Rightarrow c_1 = \frac{b \sin \gamma_1}{\sin \beta} = \frac{5 \sin 72.32°}{\sin 46.75°} \approx 6.54 \text{ or } c_1 \approx 6.5$$

However, if $\alpha_2 \approx 180° - 60.93° \approx 119.07°$ or $119° \ 4'$ then $\gamma_2 \approx 180° - 46.75° - 119.07° \approx 14.18°$ or

$\gamma_2 = 14° \ 11'.$ $\quad \dfrac{c_2}{\sin \gamma_2} = \dfrac{b}{\sin \beta} \Rightarrow c_2 = \dfrac{b \sin \gamma_2}{\sin \beta} = \dfrac{5 \sin 14.18°}{\sin 46.75°} \approx 1.68 \text{ or } c_2 \approx 1.7$

35. This triangle is of the form ASA. There is only one solution. Note that $\alpha = 56° \ 30'$ or $56.5°$ and

$\beta = 23° \ 45'$ or $23.75°$.

$\gamma = 180° - 56.5° - 23.75° = 99.75°$ or $99° \ 45'$

$$\frac{a}{\sin \alpha} = \frac{c}{\sin \gamma} \Rightarrow a = \frac{c \sin \alpha}{\sin \gamma} = \frac{100 \sin 56.5°}{\sin 99.75°} \approx 84.6$$

$$\frac{b}{\sin \beta} = \frac{c}{\sin \gamma} \Rightarrow b = \frac{c \sin \beta}{\sin \gamma} = \frac{100 \sin 23.75°}{\sin 99.75°} \approx 40.9$$

$\gamma = 99° \ 45',\ a \approx 84.6,\ b \approx 40.9$

Applications

37. Using standard labels we have the following: $\beta = 52° - 7° = 45°$ and $\alpha = 180° - 70° - 45° = 65°$.

$$\frac{c}{\sin 70°} = \frac{3500}{\sin 65°} \Rightarrow c = \frac{3500 \sin 70°}{\sin 65°} \approx 3629 \text{ ft.} \quad \text{The ground distance is about 3629 feet.}$$

39. Using standard labels we have the following:

$\alpha = 90° - 54.3° = 35.7°,\ \beta = 325.2° - 270° = 55.2°$ and $\gamma = 180° - 35.7° - 55.2° = 89.1°$.

$$\frac{a}{\sin \alpha} = \frac{c}{\sin \gamma} \Rightarrow a = \frac{c \sin \alpha}{\sin \gamma} = \frac{15 \sin 35.7°}{\sin 89.1°} \approx 8.75$$

Then the perpendicular distance d from the ship to the shore can be found as follows:

$$\sin 55.2° = \frac{d}{8.75} \Rightarrow d = 8.75 \sin 55.2° \approx 7.2 \text{ mi.} \quad \text{The ship is about 7.2 miles from the shore.}$$

41. Note that $\beta = 112°\ 10'$ or $112.17°$ and $\gamma = 15°\ 20'$ or $15.33°$.

    Therefore $\alpha = 180° - 112.17° - 15.33° = 52.5°$ or $52°\ 30'$

    $\dfrac{c}{\sin \gamma} = \dfrac{a}{\sin \alpha} \Rightarrow c = \dfrac{a \sin \gamma}{\sin \alpha} = \dfrac{354 \sin 15.33°}{\sin 52.5°} \approx 118$ meters   Therefore $c \approx 118$ meters from point A to B.

43. Using standard labels we have the distance $AB = c$, $\gamma = 55.1°$, $\alpha = 75.7°$ and $b = 97.3$.

    $\beta = 180° - 75.7° - 55.1° = 49.2°$ $\dfrac{c}{\sin \gamma} = \dfrac{b}{\sin \beta} \Rightarrow c = \dfrac{b \sin \gamma}{\sin \beta} = \dfrac{97.3 \sin 55.1°}{\sin 49.2°} \approx 105.4$

    The distance AB is approximately 105.4 feet.

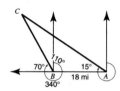

Figure 45

45. From the information given we may form a triangle with standard labels such that the ship's initial position is A. The location of the ship after one hour will be B and the position of the lighthouse is C. See Figure 45. Note that $\alpha = 15°$, $\beta = 110°$ and $\gamma = 180° - 15° - 110° = 55°$. Also, the distance traveled in one hour is $c = 18$.

    We wish to find the distance a: $\dfrac{a}{\sin \alpha} = \dfrac{c}{\sin \gamma} \Rightarrow a = \dfrac{c \sin \alpha}{\sin \gamma} = \dfrac{18 \sin 15°}{\sin 55°} \approx 5.69$

    The distance between the ship and the lighthouse is approximately 5.69 miles.

47. Angle DAB has measure $180° - 118° - 28° = 34°$. $\dfrac{BD}{\sin 34°} = \dfrac{24.2}{\sin 28°} \Rightarrow BD = \dfrac{24.2 \sin 34°}{\sin 28°} \approx 28.8$ ft

49. Since $\alpha = 38°$, angle CAE is $90° - 38° = 52°$ and angle ECA is $180° - 52° - 75° = 53°$.

    Thus, angle BCD is also $53°$.

    Then since $\beta = 15°$, angle CBD is $90° - 15° = 75°$ and angle BDC is $180° - 75° - 53° = 52°$.

    $\dfrac{AC}{\sin 75°} = \dfrac{480}{\sin 53°} \Rightarrow AC = \dfrac{480 \sin 75°}{\sin 53°} \approx 580.5$ and $\dfrac{CB}{\sin 52°} = \dfrac{480}{\sin 53°} \Rightarrow CB = \dfrac{480 \sin 52°}{\sin 53°} \approx 473.6$

    Finally, $AB = AC + CB = 580.5 + 473.6 \approx 1054$ ft.

51. Represent the top of the arch with point C and use standard labels.

    Note that $\alpha = 180° - 64.91° = 115.09°$. Then $\gamma = 180° - 115.09° - 60.81° = 4.1°$

    First we find side $b = AC$: $\dfrac{b}{\sin \beta} = \dfrac{c}{\sin \gamma} \Rightarrow b = \dfrac{c \sin \beta}{\sin \gamma} = \dfrac{57 \sin 60.81°}{\sin 4.1°} \approx 696$ ft.

    Label a point D on the ground directly below the top of the arch and let h represent the height of the arch.

    From triangle ADC: $\sin 64.91° \approx \dfrac{h}{696} \Rightarrow h \approx 696 \sin 64.91° \approx 630$ ft.

8.2: Law of Cosines

Determining a Method to Solve a Triangle

1. (a) This triangle is SAS.

   (b) Law of cosines should be used.

3. (a) This triangle is SSA.

    (b) This case is ambiguous. Law of sines should be used.

5. (a) This triangle is ASA.

    (b) Law of sines should be used.

7. (a) This triangle is ASA.

    (b) Law of sines should be used.

Solving Triangles

9. $a^2 = b^2 + c^2 - 2bc \cos \alpha = 4^2 + 6^2 - 2(4)(6) \cos 61° \Rightarrow a = \sqrt{52 - 48 \cos 61°} \approx 5.35996$

    $c^2 = a^2 + b^2 - 2ab \cos \gamma \Rightarrow \cos \gamma = \dfrac{c^2 - a^2 - b^2}{-2ab} \Rightarrow \gamma = \cos^{-1}\left(\dfrac{6^2 - 5.35996^2 - 4^2}{-2(5.35996)(4)}\right) \approx 78.25399°$

    $\beta = 180° - 61° - 78.25399° \approx 40.74601°$ $a \approx 5.4,\ \beta \approx 40.7°,\ \gamma \approx 78.3°$

11. $a^2 = b^2 + c^2 - 2bc \cos \alpha \Rightarrow \cos \alpha = \dfrac{a^2 - b^2 - c^2}{-2bc} \Rightarrow \alpha = \cos^{-1}\left(\dfrac{4^2 - 10^2 - 8^2}{-2(10)(8)}\right) \approx 22.33165°$

    $b^2 = a^2 + c^2 - 2ac \cos \beta \Rightarrow \cos \beta = \dfrac{b^2 - a^2 - c^2}{-2ac} \Rightarrow \beta = \cos^{-1}\left(\dfrac{10^2 - 4^2 - 8^2}{-2(4)(8)}\right) \approx 108.20996°$

    $\gamma = 180° - 22.33165° - 108.20996° \approx 49.45839°$ $\alpha \approx 22.3°,\ \beta \approx 108.2°,\ \gamma \approx 49.5°$

13. $a^2 = b^2 + c^2 - 2bc \cos \alpha \Rightarrow \cos \alpha = \dfrac{a^2 - b^2 - c^2}{-2bc} \Rightarrow \alpha = \cos^{-1}\left(\dfrac{5^2 - 7^2 - 9^2}{-2(7)(9)}\right) \approx 33.55731°$

    $b^2 = a^2 + c^2 - 2ac \cos \beta \Rightarrow \cos \beta = \dfrac{b^2 - a^2 - c^2}{-2ac} \Rightarrow \beta = \cos^{-1}\left(\dfrac{7^2 - 5^2 - 9^2}{-2(5)(9)}\right) \approx 50.70352°$

    $\gamma = 180° - 33.55731° - 50.70352° \approx 95.73917°$ $\alpha \approx 33.6°,\ \beta \approx 50.7°,\ \gamma \approx 95.7°$

15. $c^2 = a^2 + b^2 - 2ab \cos \gamma = 45^2 + 24^2 - 2(45)(24) \cos 35° \Rightarrow c = \sqrt{2601 - 2160 \cos 35°} \approx 28.83802$

    $a^2 = b^2 + c^2 - 2bc \cos \alpha \Rightarrow \cos \alpha = \dfrac{a^2 - b^2 - c^2}{-2bc} \Rightarrow \alpha = \cos^{-1}\left(\dfrac{45^2 - 24^2 - 28.83802^2}{-2(24)(28.83802)}\right) \approx 116.48753°$

    $\beta = 180° - 35° - 116.48753° \approx 28.51247°$ $c \approx 28.8,\ \alpha \approx 116.5°,\ \beta \approx 28.5°$

17. $a^2 = b^2 + c^2 - 2bc \cos \alpha \Rightarrow \cos \alpha = \dfrac{a^2 - b^2 - c^2}{-2bc} \Rightarrow \alpha = \cos^{-1}\left(\dfrac{2.4^2 - 1.7^2 - 1.4^2}{-2(1.7)(1.4)}\right) \approx 101.02145°$

    $b^2 = a^2 + c^2 - 2ac \cos \beta \Rightarrow \cos \beta = \dfrac{b^2 - a^2 - c^2}{-2ac} \Rightarrow \beta = \cos^{-1}\left(\dfrac{1.7^2 - 2.4^2 - 1.4^2}{-2(2.4)(1.4)}\right) \approx 44.04863°$

    $\gamma = 180° - 101.02145° - 44.04863° \approx 34.92992°$ $\alpha \approx 101.0°,\ \beta \approx 44.0°,\ \gamma \approx 34.9°$

    (The angles do not sum to 180° due to rounding)

19. $a^2 = b^2 + c^2 - 2bc \cos \alpha = 24.1^2 + 15.8^2 - 2(24.1)(15.8) \cos 10°30' \Rightarrow$

    $a = \sqrt{830.45 - 761.56 \cos 10°30'} \approx 9.03562$

    $b^2 = a^2 + c^2 - 2ac \cos \beta \Rightarrow \cos \beta = \dfrac{b^2 - a^2 - c^2}{-2ac} \Rightarrow$

    $\beta = \cos^{-1}\left(\dfrac{24.1^2 - 9.03562^2 - 15.8^2}{-2(9.03562)(15.8)}\right) \approx 150.91785°$

    $\gamma = 180° - 10.5° - 150.91785° \approx 18.58215°$ $a \approx 9.0,\ \beta \approx 150.9°,\ \gamma \approx 18.6°$

21. $a^2 = b^2 + c^2 - 2bc \cos \alpha \Rightarrow \cos \alpha = \dfrac{a^2 - b^2 - c^2}{-2bc} \Rightarrow \alpha = \cos^{-1}\left(\dfrac{10.6^2 - 25.8^2 - 20.6^2}{-2(25.8)(20.6)}\right) \approx 23.11284°$

$b^2 = a^2 + c^2 - 2ac \cos \beta \Rightarrow \cos \beta = \dfrac{b^2 - a^2 - c^2}{-2ac} \Rightarrow \beta = \cos^{-1}\left(\dfrac{25.8^2 - 10.6^2 - 20.6^2}{-2(10.6)(20.6)}\right) \approx 107.16957°$

$\gamma = 180° - 23.11284° - 107.16957° \approx 49.71759°$ $\alpha \approx 23.1°,\ \beta \approx 107.2°,\ \gamma \approx 49.7°$

23. $b^2 = a^2 + c^2 - 2ac \cos \beta = 20^2 + 15^2 - 2(20)(15) \cos 122° 10' \Rightarrow$

$b = \sqrt{625 - 600 \cos 122° 10'} \approx 30.73207$

$a^2 = b^2 + c^2 - 2bc \cos \alpha \Rightarrow \cos \alpha = \dfrac{a^2 - b^2 - c^2}{-2bc} \Rightarrow$

$\alpha = \cos^{-1}\left(\dfrac{20^2 - 30.73207^2 - 15^2}{-2(30.73207)(15)}\right) \approx 33.42685°$ or $33° 26'$

$\gamma \approx 180° - 122° 10' - 33° 26' \approx 24° 24'$ $b \approx 30.7,\ \alpha \approx 33° 26',\ \gamma \approx 24° 24'$

25. $a^2 = b^2 + c^2 - 2bc \cos \alpha \Rightarrow \cos \alpha = \dfrac{a^2 - b^2 - c^2}{-2bc} \Rightarrow \alpha = \cos^{-1}\left(\dfrac{5.3^2 - 6.7^2 - 7.1^2}{-2(6.7)(7.1)}\right) \approx 45.05460°$

$b^2 = a^2 + c^2 - 2ac \cos \beta \Rightarrow \cos \beta = \dfrac{b^2 - a^2 - c^2}{-2ac} \Rightarrow \beta = \cos^{-1}\left(\dfrac{6.7^2 - 5.3^2 - 7.1^2}{-2(5.3)(7.1)}\right) \approx 63.47520°$

$\gamma = 180° - 45.05460° - 63.47520° \approx 71.4702°$ $\alpha \approx 45.1,\ \beta = 63.5°,\ \gamma \approx 71.5°$

(The angles do not sum to 180° due to rounding)

Does This Triangle Exist?

27. No, since $a + b < c$.

29. No, since $89° + 112° > 180°$.

31. Yes, since we are given ASA and $\alpha + \gamma < 180°$.

Area of Triangles

33. Area $= K = \dfrac{1}{2}(18)(15)\sin 40° \approx 86.8$

35. $s = \dfrac{1}{2}(3 + 4 + 6) = 6.5 \Rightarrow$ Area $= K = \sqrt{6.5(6.5 - 3)(6.5 - 4)(6.5 - 6)} \approx 5.3$

37. Area $= K = \dfrac{1}{2}(10)(12)\sin 58° \approx 50.9$

39. Area $= K = \dfrac{1}{2}(5.5)(6.8)\sin 78° \approx 18.3$

41. Note that $\gamma = 180° - 31° - 54° = 95°$. Then $\dfrac{b}{\sin \beta} = \dfrac{a}{\sin \alpha} \Rightarrow b = \dfrac{a \sin \beta}{\sin \alpha} = \dfrac{2.6 \sin 31°}{\sin 54°} \approx 1.65522.$

Area $= K \approx \dfrac{1}{2}(1.65522)(2.6)\sin 95° \approx 2.1$

43. $s = \dfrac{1}{2}(5.5 + 6.7 + 9.2) = 10.7 \Rightarrow$ Area $= K = \sqrt{10.7(10.7 - 5.5)(10.7 - 6.7)(10.7 - 9.2)} \approx 18.3$

45. $s = \dfrac{1}{2}(11 + 13 + 20) = 22 \Rightarrow$ Area $= K = \sqrt{22(22 - 11)(22 - 13)(22 - 20)} = 66$

47. To apply the formula $K = \frac{1}{2}ac \sin \beta$, we need to find β, or to use the formula $K = \frac{1}{2}bc \sin \alpha$, we need to find b.

This triangle is in the form of SSA. It is ambiguous.

$$\frac{\sin \gamma}{c} = \frac{\sin \alpha}{a} \Rightarrow \sin \gamma = \frac{c \sin \alpha}{a} = \frac{16 \sin 42°}{21} \approx 0.50981 \Rightarrow \gamma_R = \sin^{-1}(0.50981) \approx 30.65°$$

$\gamma_1 = 30.65°$ then $\beta = 180° - 42° - 30.65° = 107.35°$

Since $\beta = 107.35$, then Area $= K = \frac{1}{2}(21)(16) \sin 107.35° \approx 160.4°$

Applications

49. Using standard labels, $a = BC$, $b = AC$, $c = AB$ and $\gamma =$ angle ACB.

$$c^2 = a^2 + b^2 - 2ab \cos \gamma \Rightarrow c = \sqrt{123^2 + 143^2 - 2(123)(143) \cos 78°35'} \approx 169 \text{ ft}$$

51. (a) $f(3) = 0.585(3)^2 = 5.265$, the actual height of the hill is $96 + 5.265 \approx 101.3$ feet.

(b) Two sides of the triangle are 3 miles $= 15,840$ feet in length and the third side is 5.265 feet in length.

From the law of cosines:

$$5.265^2 = 15,840^2 + 15,840^2 - 2(15,840)(15,840) \cos \theta \Rightarrow \theta = \cos^{-1}\left(\frac{5.265^2 - 2(15,840^2)}{-2(15,840^2)}\right) \approx 0.01904°$$

The correct angle is about 0.019°.

53. After 1.5 hours, the first ship has gone 30 miles and the second ship has gone 21 miles. See Figure 53.

$$d^2 = 30^2 + 21^2 - 2(30)(21) \cos 69° \Rightarrow d = \sqrt{1341 - 1260 \cos 69°} \approx 29.82$$

After 1.5 hours, the ships are about 29.8 miles apart.

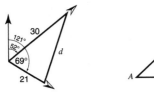

Figure 53

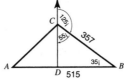

Figure 57

55. The shorter diagonal and the two given sides form a triangle of the form SAS. Using the law of cosines we will be able to find the length of the shorter diagonal. We will use $\alpha = 56°$, $b = 5.2$ and $c = 5.3 \Rightarrow$

$a^2 = b^2 + c^2 - 2bc \cos \alpha = 5.2^2 + 3.5^2 - 2(5.2)(3.5) \cos 56° \Rightarrow$

$a = \sqrt{39.29 - 36.4 \cos 56°} \approx 4.35148$ or $a = 4.4$ feet. The longer diagonal and the two given sides form a triangle of the form SAS. Since adjacent angles of a parallelogram are supplementary, $\alpha = 180° - 56° = 124°$. Using the law of cosines we will be able to find the length of the longer diagonal. We will use

$\alpha = 124°$, $b = 5.2$ and $c = 5.3 \Rightarrow a^2 = b^2 + c^2 - 2bc \cos \alpha = 5.2^2 + 3.5^2 - 2(5.2)(3.5) \cos 124° \Rightarrow$

$a = \sqrt{39.29 - 36.4 \cos 124°} \approx 7.72299$ or $a \approx 7.7$ feet.

57. Refer to Figure 57. Mark point D directly south of C on the segment joining A and B. Then triangle CDB is a right triangle. Since the bearing from C to B is 125°, angle $DCB = 180° - 125° = 55°$.

Thus $\beta = 90° - 55° = 35°$. We will use $a = 357$, $c = 515$ and $\beta = 35°$.

$b^2 = a^2 + c^2 - 2ac \cos \beta = 357^2 + 515^2 - 2(357)(515) \cos 35° \Rightarrow b = \sqrt{392,674 - 367,710 \cos 35°} \approx$

302 miles. Thus, the distance between A and C is approximately 302 miles.

59. The painter's triagular region is in the form of SAS where we will use $c = 25$, $b = 15$ and $\alpha = 128°$.

   If $K = \dfrac{1}{2}bc \sin \alpha$, then $K = \dfrac{1}{2}(15)(25) \sin 128° \approx 147.8$. Thus, the area of the region is 147.8 ft^2.

61. (a) $L = 6$ and $n = 3 \Rightarrow A = \dfrac{3(6)^2}{4} \cot \dfrac{\pi}{3} = 27\left(\dfrac{1}{\sqrt{3}}\right) = 9\sqrt{3} \approx 15.6$ in^2

   (b) $s = \dfrac{1}{2}(6 + 6 + 6) = 9 \Rightarrow$ Area $= K = \sqrt{9(9 - 6)(9 - 6)(9 - 6)} \approx 15.6$ in^2

   The results are equal.

63. $13^2 = 16^2 + 20^2 - 2(16)(20) \cos \theta \Rightarrow \theta = \cos^{-1}\left(\dfrac{13^2 - 16^2 - 20^2}{-2(16)(20)}\right) \approx 40.5°$

65. Area $= K = \dfrac{1}{2}(500)(600) \sin 85° \approx 149{,}429$ ft^2

67. In triangle ADE,

   $s = \dfrac{1}{2}(105 + 190 + 125) = 210 \Rightarrow K_1 = \sqrt{210(210 - 105)(210 - 190)(210 - 125)} \approx 6122.5$ ft^2

   In triangle ABD, $s = \dfrac{1}{2}(190 + 110 + 195) = 247.5 \Rightarrow$

   $K_2 = \sqrt{247.5(247.5 - 190)(247.5 - 110)(247.5 - 195)} \approx 10{,}135.7$ ft^2

   In triangle BCD,

   $s = \dfrac{1}{2}(195 + 75 + 150) = 210 \Rightarrow K_3 = \sqrt{210(210 - 195)(210 - 75)(210 - 150)} \approx 5051.2$ ft^2

   The total area is $K_1 + K_2 + K_3 = 6122.5 + 10{,}135.7 + 5051.2 \approx 21{,}309$ ft^2.

69. Since the satellite makes a complete orbit in 120 minutes, it forms a central angle of $\dfrac{3}{120} \cdot 360 = 9°$ in 3

   minutes. Let d represent the distance between the satellite and the tracking station at 12:03 P.M. The other

   two sides of the triangle shown in the diagram have lengths 6400 km and $6400 + 1600 = 8000$ km.

   Using the law of cosines $d^2 = 6400^2 + 8000^2 - 2(6400)(8000)\cos 9° \Rightarrow$

   $d = \sqrt{6400^2 + 8000^2 - 2(6400)(8000)\cos 9°} \approx 1954.7$. The distance is about 2000 km.

71. $AB = \sqrt{(500 - 100)^2 + (200 - 300)^2} \approx 412.31056$ pixels or $412.31056(0.015) \approx 6.18$ in.

   $AC = \sqrt{(320 - 100)^2 + (600 - 300)^2} \approx 372.02150$ pixels or $372.02150(0.015) \approx 5.58$ in.

   $BC = \sqrt{(500 - 320)^2 + (200 - 600)^2} \approx 438.63424$ pixels or $438.63424(0.015) \approx 6.58$ in.

   Continue by using Heron's formula: $s = \dfrac{1}{2}(6.18 + 5.58 + 6.58) = 9.17 \Rightarrow$

   Area $= K = \sqrt{9.17(9.17 - 6.18)(9.17 - 5.58)(9.17 - 6.58)} \approx 15.96678 \approx 16.0$ in^2

8.3: Vectors

Representing Vectors and Their Magnitudes

1. (a) $a_1 \approx 3$, $a_2 \approx 4$

   (b) $\|\mathbf{v}\| = \sqrt{3^2 + 4^2} = \sqrt{9 + 16} = \sqrt{25} = 5$

3. (a) $a_1 \approx -5$, $a_2 \approx -12$

   (b) $\|\mathbf{v}\| = \sqrt{(-5)^2 + (-12)^2} = \sqrt{25 + 144} = \sqrt{169} = 13$

5. (a) See Figure 5.

   (b) $\mathbf{v} = \langle 0, -20 \rangle$

   (c) $2\mathbf{v} = 2\langle 0, -20 \rangle = \langle 0, -40 \rangle$; this represents a 40 mph north wind.

$$-\frac{1}{2}\mathbf{v} = -\frac{1}{2}\langle 0, -20 \rangle = \langle 0, 10 \rangle; \text{ this represents a 10 mph south wind.}$$

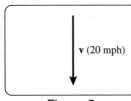

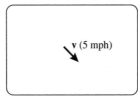

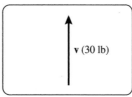

| Figure 5 | Figure 7 | Figure 9 |

7. (a) See Figure 7.

   (b) We may think of this vector as a hypotenuse of an isosceles right triangle. Since the legs l of such a triangle have the same length, we have the following:

$$l^2 + l^2 = 5^2 \Rightarrow 2l^2 = 25 \Rightarrow l^2 = \frac{25}{2} \Rightarrow l = \sqrt{\frac{25}{2}} \Rightarrow l = \frac{5}{\sqrt{2}}.$$

The vector can be represented by $\mathbf{v} = \left\langle \dfrac{5}{\sqrt{2}}, -\dfrac{5}{\sqrt{2}} \right\rangle$ or $\left\langle \dfrac{5}{2}\sqrt{2}, -\dfrac{5}{2}\sqrt{2} \right\rangle$.

   (c) $2\mathbf{v} = 2\left\langle \dfrac{5}{\sqrt{2}}, -\dfrac{5}{\sqrt{2}} \right\rangle = \left\langle \dfrac{10}{\sqrt{2}}, -\dfrac{10}{\sqrt{2}} \right\rangle$ or $\langle 5\sqrt{2}, -5\sqrt{2} \rangle$; this represents a 10 mph northwest wind.

$$-\frac{1}{2}\mathbf{v} = -\frac{1}{2}\left\langle \frac{5}{2}\sqrt{2}, -\frac{5}{2}\sqrt{2} \right\rangle = \left\langle -\frac{5}{4}\sqrt{2}, \frac{5}{4}\sqrt{2} \right\rangle; \text{ this represents a 2.5 mph southeast wind.}$$

9. (a) See Figure 9.

   (b) $\mathbf{v} = \langle 0, 30 \rangle$

   (c) $2\mathbf{v} = 2\langle 0, 30 \rangle = \langle 0, 60 \rangle$; this represents a 60 lb force upward.

$$-\frac{1}{2}\mathbf{v} = -\frac{1}{2}\langle 0, 30 \rangle = \langle 0, -15 \rangle; \text{ this represents a 15 lb force downward.}$$

11. (a) Horizontal $= 1$, vertical $= 1$

   (b) $\|\mathbf{v}\| = \sqrt{1^2 + 1^2} = \sqrt{1 + 1} = \sqrt{2}$; $\mathbf{v}$ is not a unit vector.

   (c) $\|\mathbf{v}\|$ represents the length of $\mathbf{v}$. See Figure 11.

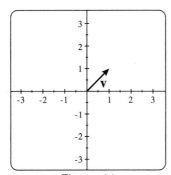

Figure 11

13. (a) Horizontal = 3, vertical = −4

(b) $\|\mathbf{v}\| = \sqrt{3^2 + (-4)^2} = \sqrt{9 + 16} = \sqrt{25} = 5$; **v** is not a unit vector.

(c) $\|\mathbf{v}\|$ represents the length of **v**. See Figure 13.

15. (a) Horizontal = 1, vertical = 0

(b) $\|\mathbf{v}\| = \sqrt{1^2 + 0^2} = \sqrt{1 + 0} = \sqrt{1} = 1$; **v** is a unit vector.

(c) $\|\mathbf{v}\|$ represents the length of **v**. See Figure 15.

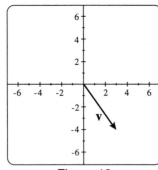

Figure 13

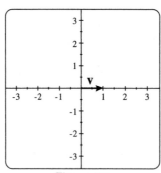

Figure 15

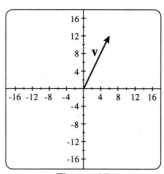

Figure 17

17. (a) Horizontal = 5, vertical = 12

(b) $\|\mathbf{v}\| = \sqrt{5^2 + 12^2} = \sqrt{25 + 144} = \sqrt{169} = 13$; **v** is not a unit vector.

(c) $\|\mathbf{v}\|$ represents the length of **v**. See Figure 17.

19. (a) See Figure 19.

(b) $\mathbf{v} = \langle -1 - 0,\ 2 - 0 \rangle = \langle -1,\ 2 \rangle$

(c) $\|\overrightarrow{PQ}\| = \sqrt{(-1)^2 + (2)^2} = \sqrt{5}$

21. (a) See Figure 21.

(b) $\mathbf{v} = \langle 3 - 1,\ 6 - 2 \rangle = \langle 2,\ 4 \rangle$

(c) $\|\overrightarrow{PQ}\| = \sqrt{(2)^2 + (4)^2} = \sqrt{20}$

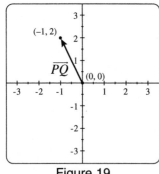

Figure 19

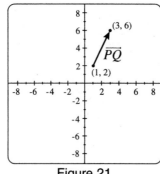

Figure 21

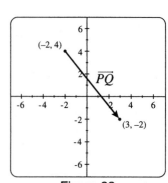

Figure 23

23. (a) See Figure 23.

(b) $\mathbf{v} = \langle 3 - (-2),\ -2 - 4 \rangle = \langle 5,\ -6 \rangle$

(c) $\|\overrightarrow{PQ}\| = \sqrt{(5)^2 + (-6)^2} = \sqrt{61}$

Operations on Vectors

25. (a) $\mathbf{a} + \mathbf{b} = \langle -8,\ 8 \rangle + \langle 4,\ 8 \rangle = \langle -8 + 4,\ 8 + 8 \rangle = \langle -4,\ 16 \rangle$

    (b) $\mathbf{a} - \mathbf{b} = \langle -8,\ 8 \rangle - \langle 4,\ 8 \rangle = \langle -8 - 4,\ 8 - 8 \rangle = \langle -12,\ 0 \rangle$

    (c) $-\mathbf{a} = -\langle -8,\ 8 \rangle = \langle 8,\ -8 \rangle$

27. (a) $\mathbf{a} + \mathbf{b} = \langle 4,\ 8 \rangle + \langle 4,\ -8 \rangle = \langle 4 + 4,\ 8 + (-8) \rangle = \langle 8,\ 0 \rangle$

    (b) $\mathbf{a} - \mathbf{b} = \langle 4,\ 8 \rangle - \langle 4,\ -8 \rangle = \langle 4 - 4,\ 8 - (-8) \rangle = \langle 0,\ 16 \rangle$

    (c) $-\mathbf{a} = -\langle 4,\ 8 \rangle = \langle -4,\ -8 \rangle$

29. (a) $\mathbf{a} + \mathbf{b} = \langle -8,\ 4 \rangle + \langle 8,\ 8 \rangle = \langle -8 + 8,\ 4 + 8 \rangle = \langle 0,\ 12 \rangle$

    (b) $\mathbf{a} - \mathbf{b} = \langle -8,\ 4 \rangle - \langle 8,\ 8 \rangle = \langle -8 - 8,\ 4 - 8 \rangle = \langle -16,\ -4 \rangle$

    (c) $-\mathbf{a} = -\langle -8,\ 4 \rangle = \langle 8,\ -4 \rangle$

31. (a) $\mathbf{a} + \mathbf{b} = \langle 0,\ 2 \rangle + \langle 3,\ 0 \rangle = \langle 0 + 3,\ 2 + 0 \rangle = \langle 3,\ 2 \rangle$

    (b) $\mathbf{a} - \mathbf{b} = \langle 0,\ 2 \rangle - \langle 3,\ 0 \rangle = \langle 0 - 3,\ 2 - 0 \rangle = \langle -3,\ 2 \rangle$

33. (a) $\mathbf{a} + \mathbf{b} = (2\mathbf{i} + \mathbf{j}) + (\mathbf{i} - 2\mathbf{j}) = 3\mathbf{i} - \mathbf{j}$

    (b) $\mathbf{a} - \mathbf{b} = (2\mathbf{i} + \mathbf{j}) - (\mathbf{i} - 2\mathbf{j}) = \mathbf{i} + 3\mathbf{j}$

35. (a) $\mathbf{a} + \mathbf{b} = \left\langle -\sqrt{2},\ \dfrac{1}{2} \right\rangle + \left\langle \sqrt{2},\ -\dfrac{3}{4} \right\rangle = \left\langle -\sqrt{2} + \sqrt{2},\ \dfrac{1}{2} + \left(-\dfrac{3}{4}\right) \right\rangle = \left\langle 0,\ -\dfrac{1}{4} \right\rangle$

    (b) $\mathbf{a} - \mathbf{b} = \left\langle -\sqrt{2},\ \dfrac{1}{2} \right\rangle - \left\langle \sqrt{2},\ -\dfrac{3}{4} \right\rangle = \left\langle -\sqrt{2} - \sqrt{2},\ \dfrac{1}{2} - \left(-\dfrac{3}{4}\right) \right\rangle = \left\langle -2\sqrt{2},\ \dfrac{5}{4} \right\rangle$

37. (a) $\mathbf{a} + \mathbf{b} = \left[\left(\cos\dfrac{\pi}{4}\right)\mathbf{i} + \left(\sin\dfrac{\pi}{4}\right)\mathbf{j} \right] + \left[\left(\cos\dfrac{\pi}{2}\right)\mathbf{i} + \left(\sin\dfrac{\pi}{2}\right)\mathbf{j} \right] =$

$\left[\left(\cos\dfrac{\pi}{4}\right)\mathbf{i} + \left(\cos\dfrac{\pi}{2}\right)\mathbf{i},\ \left(\sin\dfrac{\pi}{4}\right)\mathbf{j} + \left(\sin\dfrac{\pi}{2}\right)\mathbf{j} \right] = \left(\dfrac{\sqrt{2}}{2}\mathbf{i} + 0\mathbf{i}\right) + \left(\dfrac{\sqrt{2}}{2}\mathbf{j} + 1\mathbf{j}\right) = \dfrac{\sqrt{2}}{2}\mathbf{i} + \dfrac{\sqrt{2}+2}{2}\mathbf{j}$

    (b) $\mathbf{a} - \mathbf{b} = \left[\left(\cos\dfrac{\pi}{4}\right)\mathbf{i} + \left(\sin\dfrac{\pi}{4}\right)\mathbf{j} \right] - \left[\left(\cos\dfrac{\pi}{2}\right)\mathbf{i} + \left(\sin\dfrac{\pi}{2}\right)\mathbf{j} \right] =$

$\left[\left(\cos\dfrac{\pi}{4}\right)\mathbf{i} - \left(\cos\dfrac{\pi}{2}\right)\mathbf{i},\ \left(\sin\dfrac{\pi}{4}\right)\mathbf{j} - \left(\sin\dfrac{\pi}{2}\right)\mathbf{j} \right] = \left(\dfrac{\sqrt{2}}{2}\mathbf{i} - 0\mathbf{i}\right) + \left(\dfrac{\sqrt{2}}{2}\mathbf{j} - 1\mathbf{j}\right) = \dfrac{\sqrt{2}}{2}\mathbf{i} + \dfrac{\sqrt{2}-2}{2}\mathbf{j}$

39. (a) Graphical: See Figure 39a. The length appears to be about 2.

    Symbolic: $\|\mathbf{a}\| = \|2\mathbf{i}\| = \sqrt{2^2} = \sqrt{4} = 2$

    (b) Graphical: $2\mathbf{a} = 4\mathbf{i}$. See Figure 39b.

    Symbolic: $2\mathbf{a} = 2(2\mathbf{i}) = 4\mathbf{i}$

    (c) Graphical: $2\mathbf{a} + 3\mathbf{b} = 7\mathbf{i} + 3\mathbf{j}$. See Figure 39c.

    Symbolic: $2\mathbf{a} + 3\mathbf{b} = 2(2\mathbf{i}) + 3(\mathbf{i} + \mathbf{j}) = 4\mathbf{i} + 3\mathbf{i} + 3\mathbf{j} = 7\mathbf{i} + 3\mathbf{j}$

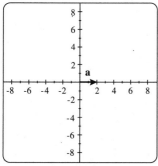

Figure 39a

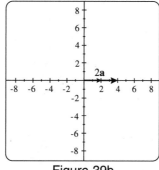

Figure 39b

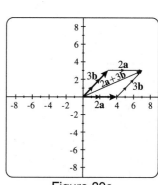

Figure 39c

41. (a) Graphical: Graphical: See Figure 41a. The length appears to be about 2.25.

    Symbolic: $\|\mathbf{a}\| = \sqrt{(-1)^2 + 2^2} = \sqrt{5}$

    (b) Graphical: $2\mathbf{a} = \langle -2, 4 \rangle$. See Figure 41b.

    Symbolic: $2\mathbf{a} = 2\langle -1, 2 \rangle = \langle -2, 4 \rangle$

    (c) Graphical: $2\mathbf{a} + 3\mathbf{b} = \langle 7, 4 \rangle$. See Figure 41c.

    Symbolic: $2\mathbf{a} + 3\mathbf{b} = 2\langle -1, 2 \rangle + 3\langle 3, 0 \rangle = \langle -2, 4 \rangle + \langle 9, 0 \rangle = \langle -2 + 9, 4 + 0 \rangle = \langle 7, 4 \rangle$

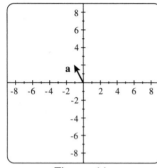

Figure 41a

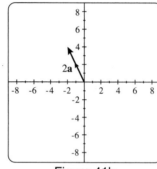

Figure 41b

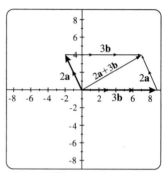

Figure 41c

43. horizontal: $\cos 40° = \dfrac{v_1}{4} \Rightarrow v_1 = 4\cos 40° \approx 3.06$

    vertical: $\sin 40° = \dfrac{v_2}{4} \Rightarrow v_2 = 4\sin 40° \approx 2.57$

45. horizontal: $\cos(-35°) = \dfrac{v_1}{5} \Rightarrow v_1 = 5\cos(-35°) \approx 4.10$

    vertical: $\sin(-35°) = \dfrac{v_2}{5} \Rightarrow v_2 = 5\sin(-35°) \approx -2.87$

47. Complete the parallelogram and find the unknown angle: $180° - 40° = 140°$.

    Find the magnitude of the resultant force by using the law of cosines to find the length of the diagonal d.

    $d^2 = 40^2 + 60^2 - 2(40)(60)\cos 140° \Rightarrow d = \sqrt{5200 - 4800\cos 140°} \approx 94.2 \text{ lb}$

49. Complete the parallelogram and find the unknown angle: $180° - 110° = 70°$.

    Find the magnitude of the resultant force by using the law of cosines to find the length of the diagonal d.

    $d^2 = 15^2 + 25^2 - 2(15)(25)\cos 70° \Rightarrow d = \sqrt{850 - 750\cos 70°} \approx 24.4 \text{ lb}$

Dot Product and Work

51. (a) $\mathbf{a} \cdot \mathbf{b} = (1)(3) + (-2)(1) = 3 - 2 = 1$

    (b) $\|\mathbf{a}\| = \sqrt{1^2 + (-2)^2} = \sqrt{1 + 4} = \sqrt{5}$, $\|\mathbf{b}\| = \sqrt{3^2 + 1^2} = \sqrt{9 + 1} = \sqrt{10}$

    $\theta = \cos^{-1}\left(\dfrac{\mathbf{a} \cdot \mathbf{b}}{\|\mathbf{a}\|\|\mathbf{b}\|} \right) = \cos^{-1}\left(\dfrac{1}{\sqrt{5}\sqrt{10}} \right) = \cos^{-1}\left(\dfrac{1}{\sqrt{50}} \right) \approx 81.9°$

    (c) The vectors are neither parallel nor perpendicular.

53. (a) $\mathbf{a} \cdot \mathbf{b} = (6)(-4) + (8)(3) = -24 + 24 = 0$

   (b) $\|\mathbf{a}\| = \sqrt{6^2 + 8^2} = \sqrt{36 + 64} = \sqrt{100} = 10$, $\|\mathbf{b}\| = \sqrt{(-4)^2 + 3^2} = \sqrt{16 + 9} = \sqrt{25} = 5$

   $\theta = \cos^{-1}\left(\dfrac{\mathbf{a} \cdot \mathbf{b}}{\|\mathbf{a}\|\|\mathbf{b}\|}\right) = \cos^{-1}\left(\dfrac{0}{10 \cdot 5}\right) = \cos^{-1} 0 = 90°$

   (c) The vectors are perpendicular.

55. (a) $\mathbf{a} \cdot \mathbf{b} = (5)(10) + (6)(12) = 50 + 72 = 122$

   (b) $\|\mathbf{a}\| = \sqrt{5^2 + 6^2} = \sqrt{25 + 36} = \sqrt{61}$, $\|\mathbf{b}\| = \sqrt{10^2 + 12^2} = \sqrt{100 + 144} = \sqrt{244}$

   $\theta = \cos^{-1}\left(\dfrac{\mathbf{a} \cdot \mathbf{b}}{\|\mathbf{a}\|\|\mathbf{b}\|}\right) = \cos^{-1}\left(\dfrac{122}{\sqrt{61}\,\sqrt{244}}\right) = \cos^{-1}\left(\dfrac{122}{\sqrt{14{,}884}}\right) = \cos^{-1}\left(\dfrac{122}{122}\right) = \cos^{-1} 1 = 0°$

   (c) The vectors are parallel and they point in the same direction.

57. (a) $\mathbf{a} \cdot \mathbf{b} = (1)(0.5) + (3)(-1.5) = 0.5 - 4.5 = -4$

   (b) $\|\mathbf{a}\| = \sqrt{1^2 + 3^2} = \sqrt{1 + 9} = \sqrt{10}$, $\|\mathbf{b}\| = \sqrt{0.5^2 + (-1.5)^2} = \sqrt{0.25 + 2.25} = \sqrt{2.5}$

   $\theta = \cos^{-1}\left(\dfrac{\mathbf{a} \cdot \mathbf{b}}{\|\mathbf{a}\|\|\mathbf{b}\|}\right) = \cos^{-1}\left(\dfrac{-4}{\sqrt{10}\,\sqrt{2.5}}\right) = \cos^{-1}\left(\dfrac{-4}{\sqrt{25}}\right) = \cos^{-1}\left(\dfrac{-4}{5}\right) \approx 143.1°$

   (c) The vectors are neither parallel nor perpendicular.

59. $W = 30 \cdot 5 = 150$ ft-lb

61. $W = 100 \cdot 1000 = 100{,}000$ ft-lb

63. $W = \mathbf{F} \cdot \mathbf{D} = \langle 10, 20 \rangle \cdot \langle 15, 22 \rangle = (10)(15) + (20)(22) = 590$ ft-lb

   $\|\mathbf{F}\| = \sqrt{10^2 + 20^2} = \sqrt{100 + 400} = \sqrt{500} = 10\sqrt{5} \approx 22.4$ lb

65. $W = \mathbf{F} \cdot \mathbf{D} = (5\mathbf{i} - 3\mathbf{j}) \cdot (3\mathbf{i} - 4\mathbf{j}) = (5)(3) + (-3)(-4) = 27$ ft-lb

   $\|\mathbf{F}\| = \sqrt{5^2 + (-3)^2} = \sqrt{25 + 9} = \sqrt{34} \approx 5.8$ lb

67. $\overrightarrow{PQ} = \langle 1 - (-2), 6 - 3 \rangle = \langle 3, 3 \rangle = 3\mathbf{i} + 3\mathbf{j}$

   $W = \mathbf{F} \cdot \mathbf{D} = (5\mathbf{i} + 3\mathbf{j}) \cdot (3\mathbf{i} + 3\mathbf{j}) = (5)(3) + (3)(3) = 24$

69. $\overrightarrow{PQ} = \langle 4 - 2, -5 - (-3) \rangle = \langle 2, -2 \rangle = 2\mathbf{i} - 2\mathbf{j}$

   $W = \mathbf{F} \cdot \mathbf{D} = (5\mathbf{i} + 3\mathbf{j}) \cdot (2\mathbf{i} - 2\mathbf{j}) = (5)(2) + (3)(-2) = 4$

Applications

71. The swimmer's velocity is modeled by $\mathbf{a} = \langle 0, 3 \rangle$. The velocity of the current is modeled by $\mathbf{b} = \langle 2, 0 \rangle$.

   The resultant velocity of the swimmer is modeled by $\mathbf{a} + \mathbf{b} = \langle 0, 3 \rangle + \langle 2, 0 \rangle = \langle 0 + 2, 3 + 0 \rangle = \langle 2, 3 \rangle$.

   The speed of the swimmer is $\|\mathbf{a} + \mathbf{b}\| = \sqrt{2^2 + 3^2} = \sqrt{4 + 9} = \sqrt{13} \approx 3.6$ mph.

73. The velocity of the airplane is modeled by $\mathbf{a} = \langle -400, 0 \rangle$. Refer to the referenced example when finding

   the appropriate vector to represent the wind. The velocity of the wind can be modeled by $\mathbf{b} = \left\langle \dfrac{50}{\sqrt{2}}, -\dfrac{50}{\sqrt{2}} \right\rangle$.

   The resultant velocity of the airplane can be found by vector addition and is modeled by

   $\mathbf{a} + \mathbf{b} = \langle -400, 0 \rangle + \left\langle \dfrac{50}{\sqrt{2}}, -\dfrac{50}{\sqrt{2}} \right\rangle = \left\langle -400 + \dfrac{50}{\sqrt{2}}, 0 + \left(-\dfrac{50}{\sqrt{2}}\right) \right\rangle \approx \langle -364.6, -35.4 \rangle$.

   The ground speed of the airplane is $\|\mathbf{a} + \mathbf{b}\| \approx \sqrt{(-364.6)^2 + (-35.4)^2} \approx 366.3$ mph.

   The bearing of the airplane is $270° - \tan^{-1}\left(\dfrac{-35.4}{-364.6}\right) \approx 264.5°$.

75. Since the bearing is 160°, the plane is headed for a location which makes a −70° angle with the horizontal.

    Thus the vector location is modeled by $\langle 450\cos(-70°), 450\sin(-70°)\rangle \approx \langle 153.9, -422.9\rangle$. The wind can be

    modeled by the vector $\langle 0, 20\rangle$. The vector **a** that the plane should take can be found as follows:

    $\mathbf{a} + \langle 0, 20\rangle \approx \langle 153.9, -422.9\rangle \Rightarrow \mathbf{a} \approx \langle 153.9, -402.9\rangle$. The airplane's speed is the magnitude of this vector:

    $\|\mathbf{a}\| = \sqrt{153.9^2 + (-402.9)^2} \approx 431.3$ mph = groundspeed. The angle from the horizontal of vector **a** is

    $\tan^{-1}\left(\dfrac{-402.9}{153.9}\right) \approx -69.1°$. Thus the final bearing should be about $90° + 69.1° = 159.1°$.

77. Since the bearing is 75°, the plane is headed for a location which makes a 15° angle with the horizontal. The

    airspeed is found by the use of tangent $\Rightarrow \tan 15° = \dfrac{40}{x} \Rightarrow 0.2679x = 40 \Rightarrow x = 149.3$ mph. The

    groundspeed is found by the use of sine $\Rightarrow \sin 15° = \dfrac{40}{x} \Rightarrow 0.2588x = 40 \Rightarrow x = 154.6$ mph.

79. (a) $\|\mathbf{R}\| = \sqrt{1^2 + (-2)^2} = \sqrt{1+4} = \sqrt{5} \approx 2.2$, $\|\mathbf{A}\| = \sqrt{0.5^2 + 1^2} = \sqrt{0.25 + 1} = \sqrt{1.25} \approx 1.1$

    About 2.2 inches of rain fell. The area of the opening of the rain gauge was about 1.1 in².

    (b) $V = |\mathbf{R} \cdot \mathbf{A}| = |(1)(0.5) + (-2)(1)| = |-1.5| = 1.5$. The volume of the rain collected was 1.5 in³.

    (c) **R** and **A** should be parallel and point in opposite directions.

81. (a) $\mathbf{c} = \mathbf{a} + \mathbf{b} = \langle 3, 2\rangle + \langle -2, 2\rangle = \langle 1, 4\rangle$

    (b) $\|\mathbf{c}\| = \sqrt{1^2 + 4^2} = \sqrt{1 + 16} = \sqrt{17} \approx 4.1$ ft

    (c) $3\mathbf{a} + 0.5\mathbf{b} = 3\langle 3, 2\rangle + 0.5\langle -2, 2\rangle = \langle 9, 6\rangle + \langle -1, 1\rangle = \langle 8, 7\rangle$

83. (a) Represent the point $(-2, 4)$ by the vector $\mathbf{a} = \langle -2, 4\rangle$.

    The new location is given by $\mathbf{b} = \mathbf{a} + \mathbf{v} = \langle -2, 4\rangle + \langle 4, -2\rangle = \langle -2 + 4, 4 + (-2)\rangle = \langle 2, 2\rangle \Rightarrow (2, 2)$

    (b) See Figure 83.

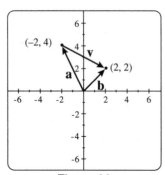

Figure 83

85. Note that 1.5 miles = 7920 feet. The x-component of the displacement vector is $d_1 = 7920\cos 15°$.

    The y-component of the displacement vector is $d_2 = 7920\sin 15°$.

    $W = \mathbf{F} \cdot \mathbf{D} = \langle 0, 145\rangle \cdot \langle 7920\cos 15°, 7920\sin 15°\rangle = (0)(7920\cos 15°) + (145)(7920\sin 15°) \approx 297{,}228$ ft-lb

87. Since the bearing is 135°, the city is located on a vector which makes a $-45°$ angle with the horizontal. Thus the vector to the city is modeled by $\langle 200\cos(-45°),\ 200\sin(-45°)\rangle \approx \langle 141.4,\ -141.4\rangle$. The wind can be modeled by the vector $\langle 0,\ -30\rangle$. The vector **a** that the pilot should take can be found as follows:

$\mathbf{a} + \langle 0,\ -30\rangle \approx \langle 141.4,\ -141.4\rangle \Rightarrow \mathbf{a} \approx \langle 141.4,\ -111.4\rangle$. The airplane's speed is the magnitude of this vector:

$\|\mathbf{a}\| = \sqrt{141.4^2 + (-111.4)^2} \approx 180$ mph. The angle from horizontal of vector **a** is $\tan^{-1}\left(\dfrac{-111.4}{141.4}\right) \approx -38.2°$.

Thus, the bearing should be about $90° + 38.2° = 128.2°$. The pilot should fly approximately 180 mph on a bearing of about 128.2°.

8.4: Parametric Equations

Graphs of Parametric Equations

1. (a)

t	0	1	2	3
x	−1	0	1	2
y	0	2	4	6

(b) See Figure 1.

(c) This curve is a line segment.

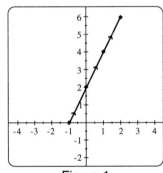
Figure 1

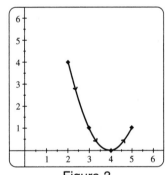
Figure 3

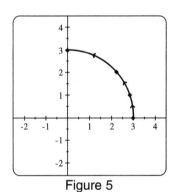
Figure 5

3. (a)

t	0	1	2	3
x	2	3	4	5
y	4	1	0	1

(b) See Figure 3.

(c) This curve is the lower portion of a parabola.

5. (a)

t	0	1	2	3
x	3	$\sqrt{8}$	$\sqrt{5}$	0
y	0	1	2	3

(b) See Figure 5.

(c) This curve is a portion of a circle with radius 3.

7. (a)

t	0	1	2	3
x	0	1	2	3
y	3	$\sqrt{8}$	$\sqrt{5}$	0

(b) See Figure 7.

(c) This curve is a portion of a circle with radius 3.

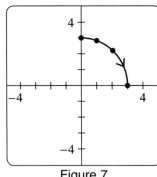

Figure 7

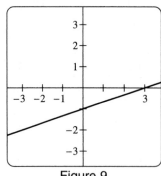

Figure 9

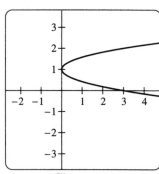

Figure 11

9. $x = 3t \Rightarrow t = \dfrac{1}{3}x$, substituting for t in $y = t - 1$ results in $y = \dfrac{1}{3}x - 1$. This curve is a line segment.

See Figure 9.

11 $y = t + 1 \Rightarrow t = y - 1$, substituting for t in $x = 3t^2$ results in $x = 3(y - 1)^2$. This curve is a parabola.

See Figure 11.

13. $x = \sqrt{1 - t^2} \Rightarrow x^2 = 1 - t^2 \Rightarrow x^2 - 1 = -t^2 \Rightarrow t^2 = 1 - x^2 \Rightarrow t = \sqrt{1 - x^2}$, substituting for t in $y = t$ results in $y = \sqrt{1 - x^2}$. This curve is a portion of a circle with radius 1. See Figure 13.

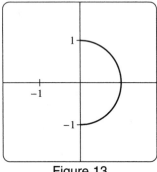

Figure 13

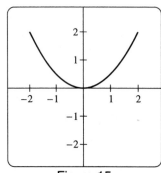

Figure 15

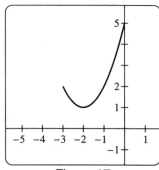

Figure 17

15. $x = t \Rightarrow y = \dfrac{1}{2}x^2$

This curve is a portion of a parabola. See Figure 15.

17. $x = t - 2 \Rightarrow x + 2 = t \Rightarrow y = (x + 2)^2 + 1 \Rightarrow y = x^2 + 4x + 5$

This curve is a portion of a parabola. See Figure 17.

19. $x^2 + y^2 = (3 \sin t)^2 + (3 \cos t)^2 = 9 \sin^2 t + 9 \cos^2 t = 9(\sin^2 t + \cos^2 t) = 9 \Rightarrow x^2 + y^2 = 9$

This curve is a circle with radius 3. See Figure 19.

21. See Figure 21.

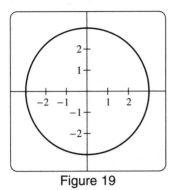

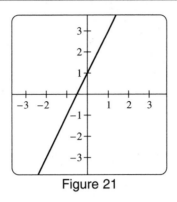

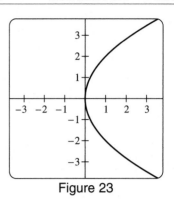

Figure 19   Figure 21   Figure 23

23. See Figure 23.

25. See Figure 25.

27. See Figure 27.

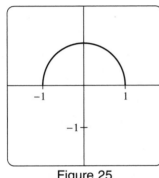

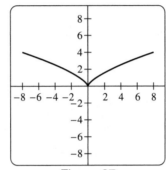

Figure 25   Figure 27

29. See Figure 29.

31. See Figure 31.

[0, 6, 1] by [−2, 2, 1]   [−1, 10, 1] by [−1, 3, 1]   [−4.7, 4.7, 1] by [−3.1, 3.1, 1]   [−1.5, 1.5, 0.5] by [−1, 1, 0.5]

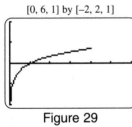

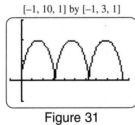

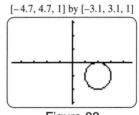

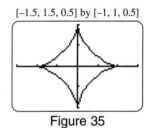

Figure 29   Figure 31   Figure 33   Figure 35

33. See Figure 33

35. See Figure 35.

37. See Figure 37.

[−4.7, 4.7, 1] by [−3.1, 3.1, 1]

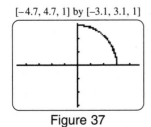

Figure 37

39. Let $x = t$. Then $2x + y = 4 \Rightarrow 2t + y = 4 \Rightarrow y = 4 - 2t$.

    One representation in parametric form is $x = t$, $y = 4 - 2t$. *Answers may vary.*

41. Let $x = t$. Then $y = 4 - x^2 \Rightarrow y = 4 - t^2$.

    One representation in parametric form is $x = t$, $y = 4 - t^2$. *Answers may vary.*

43. Let $y = t$. Then $x = y^2 + y - 3 \Rightarrow x = t^2 + t - 3$.

    One representation in parametric form is $x = t^2 + t - 3$, $y = t$. *Answers may vary.*

45. Since $\cos^2 t + \sin^2 t = 1$ for all t, $4\cos^2 t + 4\sin^2 t = 4 \Rightarrow (2\cos t)^2 + (2\sin t)^2 = 4$.

    One representation in parametric form is $x = 2\cos t$, $y = 2\sin t$, $0 \le t \le 2\pi$. *Answers may vary.*

47. Let $x = t$. Then $\ln y = 0.1x^2 \Rightarrow \ln y = 0.1t^2 \Rightarrow y = e^{0.1t^2}$.

    One representation in parametric form is $x = t$, $y = e^{0.1t^2}$. *Answers may vary.*

49. Let $y = t$. Then $x = y^2 - 2y + 1 \Rightarrow x = t^2 - 2t + 1$. One representation in parametric form is
    $x = t^2 - 2t + 1$, $y = t$. *Answers may vary.*

51. (a) The curve traces a circle of radius 3 once counterclockwise, starting at $(3, 0)$. See Figure 51a.

    (b) The curve traces a circle of radius 3 twice counterclockwise, starting at $(3, 0)$. See Figure 51b.

[-4.7, 4.7, 1] by [-3.1, 3.1, 1]   [-4.7, 4.7, 1] by [-3.1, 3.1, 1]   [-4.7, 4.7, 1] by [-3.1, 3.1, 1]   [-4.7, 4.7, 1] by [-3.1, 3.1, 1]

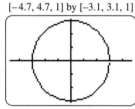 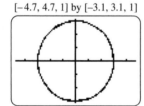 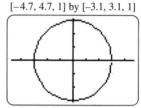 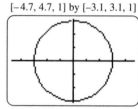

Figure 51a   Figure 51b   Figure 53a   Figure 53b

53. (a) The curve traces a circle of radius 3 once counterclockwise, starting at $(3, 0)$. See Figure 53a.

    (b) The curve traces a circle of radius 3 once clockwise, starting at $(0, 3)$. See Figure 53b.

55. (a) The curve traces a circle of radius 1 centered at $(-1, 2)$. See Figure 55a.

    (b) The curve traces a circle of radius 1 centered at $(1, 2)$. See Figure 55b.

[-4.7, 4.7, 1] by [-3.1, 3.1, 1]   [-4.7, 4.7, 1] by [-3.1, 3.1, 1]   [0, 6, 1] by [0, 4, 1]   [0, 6, 1] by [0, 4, 1]

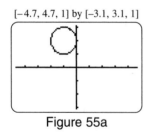 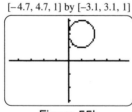 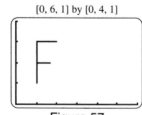 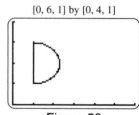

Figure 55a   Figure 55b   Figure 57   Figure 59

Designing Shapes and Figures

57. This is the letter F. See Figure 57.

59. This is the letter D. See Figure 59.

61. One possible solution is $x_1 = 0$, $y_1 = 2t$; $x_2 = t$, $y_2 = 0$ for $0 \le t \le 1$. *Answers may vary.*

63. One possible solution is $x_1 = \sin t$, $y_1 = \cos t$; $x_2 = 0$, $y_2 = t - 2$ for $0 \le t \le \pi$. *Answers may vary.*

65. One possible solution is $x_1 = 2\cos t$, $y_1 = 2\sin t$; $x_2 = 0.75 + 0.2\cos t$, $y_2 = 0.75 + 0.4\sin t$;

    $x_3 = -0.75 + 0.2\cos t$, $y_3 = 0.75 + 0.4\sin t$; $x_4 = 0.75\cos 0.5t$, $y_4 = -0.7 - 0.3\sin 0.5t$ for $0 \le t \le 2\pi$.

    Add pupils by plotting the points $(0.75, 0.6)$ and $(-0.75, 0.6)$. See Figure 65. *Answers may vary.*

[-3, 3, 1] by [-2, 2, 1]

Figure 65

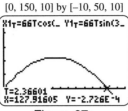

[0, 150, 10] by [-10, 50, 10]

Figure 67a

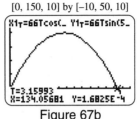

[0, 150, 10] by [-10, 50, 10]

Figure 67b

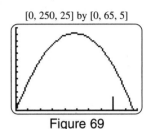

[0, 250, 25] by [0, 65, 5]

Figure 69

Applications

67. The parametric equations which model the flight of the golf ball hit at a 35° angle are given by

    $x_1 = 66\cos 35°(t)$, $y_1 = 66\sin 35°(t) - 16t^2$ for $0 \le t \le 4$. This ball travels about 128 feet. See Figure 67a.

    The parametric equations which model the flight of the golf ball hit at a 50° angle are given by

    $x_2 = 66\cos 50°(t)$, $y_2 = 66\sin 50°(t) - 16t^2$ for $0 \le t \le 4$. This ball travels about 134 feet. See Figure 67b.

69. The parametric equations $x_1 = 88\cos 45°(t)$, $y_1 = 88\sin 45°(t) - 16t^2$ for $0 \le t \le 5$ model the flight of this

    ball. The fence can be modeled by the parametric equations $x_2 = 200$, $y_2 = 2t$ for $0 \le t \le 5$. The ball will

    go over the fence. See Figure 69.

71. Let $v = 88$, $\theta = 45°$, and $h = 50$. The parametric equations are initially

    $x = (88\cos 45°)t$ and $y = (88\sin 45°)t - 16t^2 + 50$.

    Since $\cos 45° = \sin 45° = \dfrac{1}{\sqrt{2}}$, these equations become $x = \left(\dfrac{88}{\sqrt{2}}\right)t$ and $y = \left(\dfrac{88}{\sqrt{2}}\right)t - 16t^2 + 50$.

    Since the ball hits the ground when $y = 0$, we can determine how long the ball was in flight by solving the

    following equation: $-16t^2 + \left(\dfrac{88}{\sqrt{2}}\right)t + 50 = 0$. By the quadratic formula, the solution is $t \approx 4.573$ seconds.

    After 4.573 seconds the ball has traveled horizontally $x = \left(\dfrac{88}{\sqrt{2}}\right)(4.573) \approx 285$ feet.

73. See Figure 73.

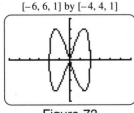

[-6, 6, 1] by [-4, 4, 1]

Figure 73

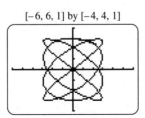

[-6, 6, 1] by [-4, 4, 1]

Figure 75

75. See Figure 75.

8.5: Polar Equations

Polar Coordinates

1. See Figure 1.

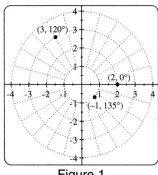

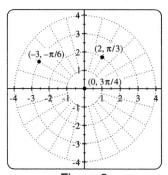

Figure 1   Figure 3

3. See Figure 3.

5. Yes. The angles $180°$ and $-180°$ are coterminal.

7. No. The angles $45°$ and $-45°$ are not coterminal.

9. Yes. The points are both located at the pole.

11. $x = r\cos\theta = 3\cos 45° = \dfrac{3}{\sqrt{2}}$, $y = r\sin\theta = 3\sin 45° = \dfrac{3}{\sqrt{2}}$; $\left(\dfrac{3}{\sqrt{2}}, \dfrac{3}{\sqrt{2}}\right)$ or $\left(\dfrac{3\sqrt{2}}{2}, \dfrac{3\sqrt{2}}{2}\right)$

13. $x = r\cos\theta = 10\cos 90° = 0$, $y = r\sin\theta = 10\sin 90° = 10$; $(0, 10)$

15. $x = r\cos\theta = 5\cos 2\pi = 5$, $y = r\sin\theta = 5\sin 2\pi = 0$; $(5, 0)$

17. $x = r\cos\theta = -3\cos 60° = -3\left(\dfrac{1}{2}\right) = -\dfrac{3}{2}$, $y = r\sin\theta = -3\sin 60° = -3\left(\dfrac{\sqrt{3}}{2}\right) = -\dfrac{3\sqrt{3}}{2}$; $\left(-\dfrac{3}{2}, -\dfrac{3\sqrt{3}}{2}\right)$

19. $x = r\cos\theta = -2\cos\left(-\dfrac{3\pi}{2}\right) = -2(0) = 0$, $y = r\sin\theta = -2\sin\left(-\dfrac{3\pi}{2}\right) = -2(1) = -2$; $(0, -2)$

21. (a) $r^2 = x^2 + y^2 \Rightarrow r^2 = (0)^2 + (3)^2 \Rightarrow r^2 = 9 \Rightarrow r = \pm 3 \Rightarrow r = 3$ (since $r > 0$)

   $\tan\theta = \dfrac{y}{x} \Rightarrow \tan\theta = \dfrac{3}{0} \Rightarrow \tan\theta$ is undefined $\Rightarrow \theta = 90°$ or $270°$ (since $0° \le \theta < 360°$)

   Since $(0, 3)$ is located on the positive y-axis, $\theta = 90°$. The polar coordinates are $(3, 90°)$.

   (b) $r^2 = x^2 + y^2 \Rightarrow r^2 = (0)^2 + (3)^2 \Rightarrow r^2 = 9 \Rightarrow r = \pm 3 \Rightarrow r = -3$ (since $r < 0$)

   $\tan\theta = \dfrac{y}{x} \Rightarrow \tan\theta = \dfrac{3}{0} \Rightarrow \tan\theta$ is undefined $\Rightarrow \theta = -90°$ or $-270°$ (since $-180° < \theta \le 180°$)

   Since $(0, 3)$ is located on the positive y-axis, $\theta = -90°$. The polar coordinates are $(-3, -90°)$.

23. (a) $r^2 = x^2 + y^2 \Rightarrow r^2 = (-1)^2 + (-\sqrt{3})^2 \Rightarrow r^2 = 4 \Rightarrow r = \pm 2 \Rightarrow r = 2$ (since $r > 0$)

$\tan\theta = \dfrac{y}{x} \Rightarrow \tan\theta = \dfrac{-\sqrt{3}}{-1} \Rightarrow \theta = \tan^{-1}\sqrt{3} \Rightarrow \theta = 60°$ or $240°$ (since $0° \le \theta < 360°$)

Since $(-1, -\sqrt{3})$ is located in quadrant III, $\theta = 240°$. The polar coordinates are $(2, 240°)$.

(b) $r^2 = x^2 + y^2 \Rightarrow r^2 = (-1)^2 + (-\sqrt{3})^2 \Rightarrow r^2 = 4 \Rightarrow r = \pm 2 \Rightarrow r = -2$ (since $r < 0$)

$\tan\theta = \dfrac{y}{x} \Rightarrow \tan\theta = \dfrac{-\sqrt{3}}{-1} \Rightarrow \theta = \tan^{-1}\sqrt{3} \Rightarrow \theta = -120°$ or $60°$ (since $-180° < \theta \le 180°$)

Since $(-1, -\sqrt{3})$ is located in quadrant III, $\theta = 60°$. The polar coordinates are $(-2, 60°)$.

25. (a) $r^2 = x^2 + y^2 \Rightarrow r^2 = (3)^2 + (-3)^2 \Rightarrow r^2 = 18 \Rightarrow r = \pm\sqrt{18} \Rightarrow r = \sqrt{18}$ (since $r > 0$)

$\tan\theta = \dfrac{y}{x} \Rightarrow \tan\theta = \dfrac{-3}{3} \Rightarrow \theta = \tan^{-1}(-1) \Rightarrow \theta = 135°$ or $315°$ (since $0° \le \theta < 360°$)

Since $(3, -3)$ is located in quadrant IV, $\theta = 315°$. The polar coordinates are $(\sqrt{18}, 315°)$.

(b) $r^2 = x^2 + y^2 \Rightarrow r^2 = (3)^2 + (-3)^2 \Rightarrow r^2 = 18 \Rightarrow r = \pm\sqrt{18} \Rightarrow r = -\sqrt{18}$ (since $r < 0$)

$\tan\theta = \dfrac{y}{x} \Rightarrow \tan\theta = \dfrac{-3}{3} \Rightarrow \theta = \tan^{-1}(-1) \Rightarrow \theta = -45°$ or $135°$ (since $-180° < \theta \le 180°$)

Since $(3, -3)$ is located in quadrant IV, $\theta = 135°$. The polar coordinates are $(-\sqrt{18}, 135°)$.

27. $r^2 = x^2 + y^2 \Rightarrow r^2 = (7)^2 + (24)^2 \Rightarrow r^2 = 625 \Rightarrow r = \pm 25 \Rightarrow r = 25$ (since $r > 0$)

$\tan\theta = \dfrac{y}{x} \Rightarrow \tan\theta = \dfrac{24}{7} \Rightarrow \theta = \tan^{-1}\dfrac{24}{7} \Rightarrow \theta \approx 1.29$ or $\pi + 1.29 \approx 4.43$ (since $0 \le \theta < 2\pi$)

Since $(7, 24)$ is located in quadrant I, $\theta = 1.29$. The polar coordinates are $(25, 1.29)$.

29. $r^2 = x^2 + y^2 \Rightarrow r^2 = (-5)^2 + (12)^2 \Rightarrow r^2 = 169 \Rightarrow r = \pm 13 \Rightarrow r = 13$ (since $r > 0$)

$\tan\theta = \dfrac{y}{x} \Rightarrow \tan\theta = \dfrac{12}{-5} \Rightarrow \theta = \tan^{-1}\left(-\dfrac{12}{5}\right) \Rightarrow \theta \approx 1.97$ or $\pi + 1.97 \approx 5.11$ (since $0 \le \theta < 2\pi$)

Since $(-5, 12)$ is located in quadrant II, $\theta = 1.97$. The polar coordinates are $(13, 1.97)$.

Graphs of Polar Equations

31. See Figure 31.

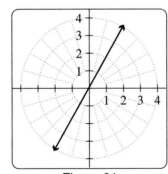

Figure 31

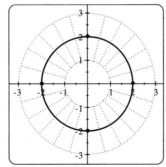

Figure 33

33. (a)

θ	0°	90°	180°	270°
r	2	2	2	2

(b) See Figure 33.

35. (a)

θ	0°	90°	180°	270°
r	0	3	0	−3

(b) See Figure 35.

37. (a)

θ	0°	90°	180°	270°
r	2	4	2	0

(b) See Figure 37.

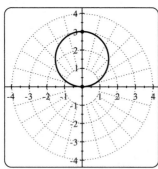

Figure 35

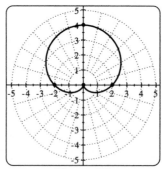

Figure 37

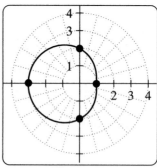

Figure 39

39. (a)

θ	0°	90°	180°	270°
r	1	2	3	2

(b) See Figure 39.

41. See Figure 41.

43. See Figure 43.

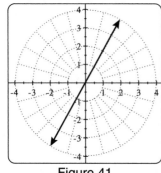

Figure 41

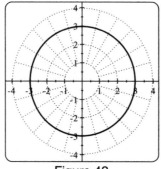

Figure 43

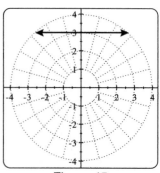

Figure 45

45. See Figure 45.

47. See Figure 47.

49. See Figure 49.

51. See Figure 51.

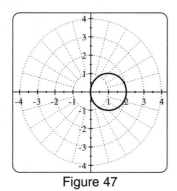

Figure 47

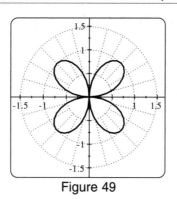

Figure 49

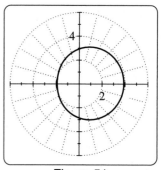

Figure 51

53. See Figure 53.

55. See Figure 55.

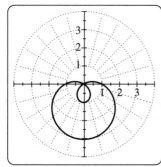

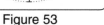

Figure 53

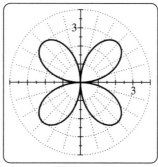

Figure 55

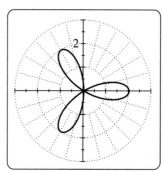

Figure 57

57. See Figure 57.

59. $y = 3 \Rightarrow r\sin\theta = 3$ or $r = 3\csc\theta$

61. $y = x \Rightarrow r\sin\theta = r\cos\theta \Rightarrow \dfrac{r\sin\theta}{r\cos\theta} = 1 \Rightarrow \tan\theta = 1$ or $\theta = \dfrac{\pi}{4}$

63. $x^2 + y^2 = 9 \Rightarrow (r\cos\theta)^2 + (r\sin\theta)^2 = 9 \Rightarrow r^2(\cos^2\theta + \sin^2\theta) = 9 \Rightarrow r^2 = 9$ or $r = 3$

65. $x^2 + y^2 = 9 \Rightarrow (r\cos\theta)^2 + (r\sin\theta)^2 = 2(r\cos\theta) \Rightarrow r^2(\cos^2\theta + \sin^2\theta) = 2(r\cos\theta) \Rightarrow$

    $r^2 = 2(r\cos\theta) \Rightarrow r = 2\cos\theta$

67. $r = 3 \Rightarrow r^2 = 9 \Rightarrow x^2 + y^2 = 9$

69. $r = 2\sec\theta \Rightarrow r = \dfrac{2}{\cos\theta} \Rightarrow r\cos\theta = 2 \Rightarrow x = 2$

71. $r = \dfrac{3}{2\cos\theta + 4\sin\theta} \Rightarrow 2r\cos\theta + 4r\sin\theta = 3 \Rightarrow 2x + 4y = 3$

73. $r = \cos\theta \Rightarrow r^2 = r\cos\theta \Rightarrow x^2 + y^2 = x$

75. Graph $r_1 = 3 + 3\cos(\theta)$ for $0° \le \theta \le 360°$ in $[-9.4, 9.4, 1]$ by $[-6.2, 6.2, 1]$. See Figure 75.

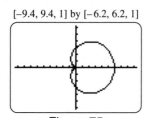

$[-9.4, 9.4, 1]$ by $[-6.2, 6.2, 1]$

Figure 75

77. Graph $r_1 = 3 - 2\sin(\theta)$ for $0° \le \theta \le 360°$ in $[-9.4, 9.4, 1]$ by $[-6.2, 6.2, 1]$. See Figure 77.

79. Graph $r_1 = 2 - 4\cos(\theta)$ for $0° \le \theta \le 360°$ in $[-9.4, 9.4, 1]$ by $[-6.2, 6.2, 1]$. See Figure 79.

$[-9.4, 9.4, 1]$ by $[-6.2, 6.2, 1]$ $[-9.4, 9.4, 1]$ by $[-6.2, 6.2, 1]$ $[-9.4, 9.4, 1]$ by $[-6.2, 6.2, 1]$ $[-4.7, 4.7, 1]$ by $[-3.1, 3.1, 1]$

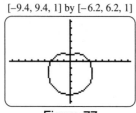 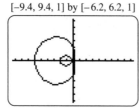 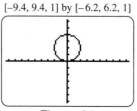 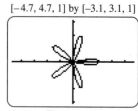

Figure 77   Figure 79   Figure 81   Figure 83

81. Graph $r_1 = 4\sin(\theta)$ for $0° \le \theta \le 180°$ in $[-9.4, 9.4, 1]$ by $[-6.2, 6.2, 1]$. See Figure 81.

83. Graph $r_1 = 2\cos(5\theta)$ for $0° \le \theta \le 180°$ in $[-4.7, 4.7, 1]$ by $[-3.1, 3.1, 1]$. See Figure 83.

85. Graph $r_1 = \theta/2$ for $0 \le \theta \le \dfrac{9\pi}{2}$ in $[-9.4, 9.4, 1]$ by $[-6.2, 6.2, 1]$. See Figure 85.

$[-9.4, 9.4, 1]$ by $[-6.2, 6.2, 1]$ $[-3, 3, 1]$ by $[-2, 2, 1]$

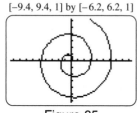 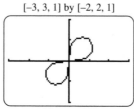

Figure 85   Figure 87

87. Let $r_1 = \sqrt{2}\sin 2\theta$ and $r_2 = -\sqrt{2}\sin 2\theta$

Graph r_1 and r_2 for $0° \le \theta \le 180°$ in $[-3, 3, 1]$ by $[-2, 2, 1]$. See Figure 87.

Solving Equations in Polar Coordinates

89. $r_1 = r_2 \Rightarrow 3 = 2 + 2\sin\theta \Rightarrow 1 = 2\sin\theta \Rightarrow \sin\theta = \dfrac{1}{2} \Rightarrow \theta_R = \sin^{-1}\dfrac{1}{2} = 30° \Rightarrow \theta = 30°, 150°$

91. (a) $r_1 = r_2 \Rightarrow 3 = 2 - 2\sin\theta \Rightarrow 2\sin\theta = -1 \Rightarrow \sin\theta = -\dfrac{1}{2} \Rightarrow \theta_R = \sin^{-1}\dfrac{1}{2} = 30° \Rightarrow \theta = 210°, 330°$

(b) Graph $r_1 = 3$ and $r_2 = 2 - 2\sin(\theta)$ for $0° \le \theta \le 360°$ in $[-9.4, 9.4, 1]$ by $[-6.2, 6.2, 1]$.

Use the calculator's trace feature to find the solutions: $\theta = 210°, 330°$. See Figures 91a and 91b.

(c) Table $r_1 = 3$ and $r_2 = 2 - 2\sin(\theta)$ starting at $\theta = 180$, incrementing by 30.

The solutions are $\theta = 210°, 330°$. See Figure 91c.

$[-9.4, 9.4, 1]$ by $[-6.2, 6.2, 1]$ $[-9.4, 9.4, 1]$ by $[-6.2, 6.2, 1]$

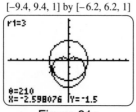 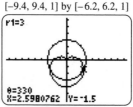

Figure 91a   Figure 91b   Figure 91c

93. (a) $r_1 = r_2 \Rightarrow 1 = 2\sin\theta \Rightarrow \sin\theta = \dfrac{1}{2} \Rightarrow \theta_R = \sin^{-1}\dfrac{1}{2} = 30° \Rightarrow \theta = 30°, 150°$

(b) Graph $r_1 = 1$ and $r_2 = 2\sin(\theta)$ for $0° \le \theta \le 360°$ in $[-4.7, 4.7, 1]$ by $[-3.1, 3.1, 1]$.

   Use the calculator's trace feature to find the solutions: $\theta = 30°, 150°$. See Figures 93a and 93b.

(c) Table $r_1 = 1$ and $r_2 = 2\sin(\theta)$ starting at $\theta = 0$, incrementing by 30.

   The solutions are $\theta = 30°, 150°$. See Figure 93c.

$[-4.7, 4.7, 1]$ by $[-3.1, 3.1, 1]$ $[-4.7, 4.7, 1]$ by $[-3.1, 3.1, 1]$

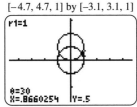   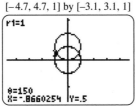   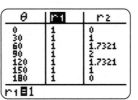

Figure 93a   Figure 93b   Figure 93c

Applications

95. For Saturn $r_1 = \dfrac{9.54(1 - 0.056^2)}{1 + 0.056\cos\theta} = (9.54(1 - 0.056^2))/(1 + 0.056\cos(\theta))$

   For Uranus $r_2 = \dfrac{19.2(1 - 0.047^2)}{1 + 0.047\cos\theta} = (19.2(1 - 0.047^2))/(1 + 0.047\cos(\theta))$

   Graph r_1 and r_2 for $0° \le \theta \le 360°$ in $[-30, 30, 10]$ by $[-20, 20, 10]$. See Figure 95.

$[-30, 30, 10]$ by $[-20, 20, 10]$ $[-3, 3, 1]$ by $[-2, 2, 1]$ $[-300, 300, 100]$ by $[-200, 200, 100]$

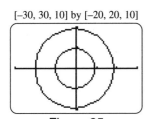   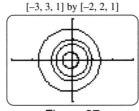   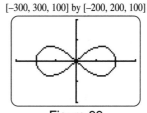

Figure 95   Figure 97   Figure 99

97. For Mercury $r_1 = \dfrac{0.39(1 - 0.206^2)}{1 + 0.206\cos\theta} = (0.39(1 - 0.206^2))/(1 + 0.206\cos(\theta))$

   For Venus $r_2 = \dfrac{0.78(1 - 0.007^2)}{1 + 0.007\cos\theta} = (0.78(1 - 0.007^2))/(1 + 0.007\cos(\theta))$

   For Earth $r_3 = \dfrac{1.00(1 - 0.017^2)}{1 + 0.017\cos\theta} = (1.00(1 - 0.017^2))/(1 + 0.017\cos(\theta))$

   For Mars $r_4 = \dfrac{1.52(1 - 0.093^2)}{1 + 0.093\cos\theta} = (1.52(1 - 0.093^2))/(1 + 0.093\cos(\theta))$

   Graph r_1, r_2, r_3 and r_4 for $0° \le \theta \le 360°$ in $[-3, 3, 1]$ by $[-2, 2, 1]$. See Figure 97.

99. Let $r_1 = \sqrt{40{,}000\cos 2\theta}$ and $r_2 = -\sqrt{40{,}000\cos 2\theta}$

   Graph r_1 and r_2 for $0° \le \theta \le 180°$ in $[-300, 300, 100]$ by $[-200, 200, 100]$. See Figure 99.

   The radio signal is received inside the "figure eight". This region is generally in an east-west direction from the two towers with a maximum distance of 200 miles.

8.6: Trigonometric Form and Roots of Complex Numbers

The Complex Plane

1. See Figure 1.

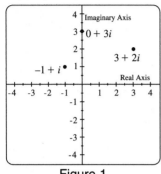

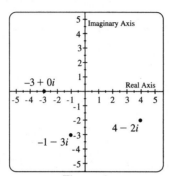

Figure 1   Figure 3

3. See Figure 3.

Trigonometric Form

5. $5(\cos 180° + i\sin 180°) = 5(-1 + i(0)) = -5$

7. $2(\cos 45° + i\sin 45°) = 2\left(\dfrac{1}{\sqrt{2}} + i\left(\dfrac{1}{\sqrt{2}}\right)\right) = \sqrt{2} + i\sqrt{2}$

9. $4\left(\cos \dfrac{3\pi}{2} + i\sin \dfrac{3\pi}{2}\right) = 4(0 + i(-1)) = -4i$

11. $3(\cos 2\pi + i\sin 2\pi) = 3(1 + i(0)) = 3$

13. $|1 + i| = \sqrt{1^2 + 1^2} = \sqrt{1 + 1} = \sqrt{2}$

15. $|12 - 5i| = \sqrt{12^2 + (-5)^2} = \sqrt{144 + 25} = \sqrt{169} = 13$

17. $|-6 + 0i| = \sqrt{(-6)^2 + 0^2} = \sqrt{36} = 6$

19. $|2 - 3i| = \sqrt{2^2 + (-3)^2} = \sqrt{4 + 9} = \sqrt{13}$

21. $r = \sqrt{1^2 + 1^2} = \sqrt{1 + 1} = \sqrt{2}$, $\tan\theta = \dfrac{1}{1} \Rightarrow \theta_R = \tan^{-1} 1 = 45°$

   Since $1 + i$ is located in quadrant I of the complex plane, $\theta = 45°$; $\sqrt{2}(\cos 45° + i\sin 45°)$

23. $r = \sqrt{5^2 + 0^2} = \sqrt{25} = 5$, $\tan\theta = \dfrac{0}{5} \Rightarrow \theta_R = \tan^{-1} 0 = 0°$

   Since 5 is positive and quadrantal in the complex plane, $\theta = 0°$; $5(\cos 0° + i\sin 0°)$

25. $r = \sqrt{0^2 + 4^2} = \sqrt{16} = 4$, $\tan\theta = \dfrac{4}{0}$ is undefined $\Rightarrow \theta_R = 90°$

   Since $4i$ is positive and quadrantal in the complex plane, $\theta = 90°$; $4(\cos 90° + i\sin 90°)$

27. $r = \sqrt{(-1)^2 + (\sqrt{3})^2} = \sqrt{1 + 3} = \sqrt{4} = 2$, $\tan\theta = \dfrac{\sqrt{3}}{-1} \Rightarrow \theta_R = \tan^{-1} \sqrt{3} = 60°$

   Since $-1 + i\sqrt{3}$ is located in quadrant II of the complex plane, $\theta = 120°$; $2(\cos 120° + i\sin 120°)$

29. $r = \sqrt{(\sqrt{3})^2 + 1^2} = \sqrt{3 + 1} = \sqrt{4} = 2$, $\tan \theta = \dfrac{1}{\sqrt{3}} \Rightarrow \theta_R = \tan^{-1} \dfrac{1}{\sqrt{3}} = 30°$

Since $\sqrt{3} + i$ is located in quadrant I of the complex plane, $\theta = 30°$; $2(\cos 30° + i \sin 30°)$

31. $r = \sqrt{(-2)^2 + 0^2} = \sqrt{4} = 2$, $\tan \theta = \dfrac{0}{-2} \Rightarrow \theta_R = \tan^{-1} 0 = 0$

Since -2 is negative and quadrantal in the complex plane, $\theta = \pi$; $2(\cos \pi + i \sin \pi)$

33. $r = \sqrt{(-2)^2 + 2^2} = \sqrt{4 + 4} = \sqrt{8}$, $\tan \theta = \dfrac{2}{-2} \Rightarrow \theta_R = \tan^{-1} 1 = \dfrac{\pi}{4}$

Since $-2 + 2i$ is located in quadrant II of the complex plane, $\theta = \dfrac{3\pi}{4}$; $\sqrt{8}\left(\cos \dfrac{3\pi}{4} + i \sin \dfrac{3\pi}{4}\right)$

35. $z_1 z_2 = (9 \cdot 3)(\cos (45° + 15°) + i \sin (45° + 15°)) = 27(\cos 60° + i \sin 60°) =$

$27\left(\dfrac{1}{2} + i\left(\dfrac{\sqrt{3}}{2}\right)\right) = \dfrac{27}{2} + \dfrac{27\sqrt{3}}{2} i$

$\dfrac{z_1}{z_2} = \left(\dfrac{9}{3}\right)(\cos (45° - 15°) + i \sin (45° - 15°)) = 3(\cos 30° + i \sin 30°) = 3\left(\dfrac{\sqrt{3}}{2} + i\left(\dfrac{1}{2}\right)\right) = \dfrac{3\sqrt{3}}{2} + \dfrac{3}{2} i$

37. $z_1 z_2 = (6 \cdot 1)\left(\cos \left(\dfrac{3\pi}{4} + \dfrac{\pi}{4}\right) + i \sin \left(\dfrac{3\pi}{4} + \dfrac{\pi}{4}\right)\right) = 6(\cos \pi + i \sin \pi) = 6(-1 + i(0)) = -6$

$\dfrac{z_1}{z_2} = \left(\dfrac{6}{1}\right)\left(\cos \left(\dfrac{3\pi}{4} - \dfrac{\pi}{4}\right) + i \sin \left(\dfrac{3\pi}{4} - \dfrac{\pi}{4}\right)\right) = 6\left(\cos \dfrac{\pi}{2} + i \sin \dfrac{\pi}{2}\right) = 6(0 + i(1)) = 6i$

39. $z_1 z_2 = (1 \cdot 1)(\cos (15° + (-45°)) + i \sin (15° + (-45°))) = 1(\cos (-30°) + i \sin (-30°)) = \dfrac{\sqrt{3}}{2} - \dfrac{1}{2} i$

$\dfrac{z_1}{z_2} = \left(\dfrac{1}{1}\right)(\cos (15° - (-45°)) + i \sin (15° - (-45°))) = 1(\cos 60° + i \sin 60°) = \dfrac{1}{2} + \dfrac{\sqrt{3}}{2} i$

Powers of Complex Numbers

41. $(2(\cos 30° + i \sin 30°))^3 = 2^3(\cos (3 \cdot 30°) + i \sin (3 \cdot 30°)) = 8(\cos 90° + i \sin 90°) = 8(0 + i(1)) = 8i$

43. $(\cos 10° + i \sin 10°)^{36} = 1^{36}(\cos (36 \cdot 10°) + i \sin (36 \cdot 10°)) = 1(\cos 360° + i \sin 360°) = 1 + i(0) = 1$

45. $(5(\cos 60° + i \sin 60°))^2 = 5^2(\cos (2 \cdot 60°) + i \sin (2 \cdot 60°)) = 25(\cos 120° + i \sin 120°) = -\dfrac{25}{2} + \dfrac{25\sqrt{3}}{2} i$

47. $(1 + i)^3 = (\sqrt{2}(\cos 45° + i \sin 45°))^3 = (\sqrt{2})^3(\cos (3 \cdot 45°) + i \sin (3 \cdot 45°)) =$

$2\sqrt{2}(\cos 135° + i \sin 135°) = -2 + 2i$ This result is verified with a calculator in Figure 47.

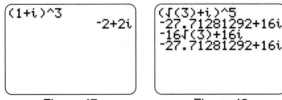

Figure 47   Figure 49

49. $(\sqrt{3} + i)^5 = 2(\cos 30° + i \sin 30°))^5 = 2^5(\cos (5 \cdot 30°) + i \sin (5 \cdot 30°)) =$

$32(\cos 150° + i \sin 150°) = -16\sqrt{3} + 16i$ This result is verified with a calculator in Figure 49.

Roots of Complex Numbers

51. $w_0 = \sqrt{4}\left(\cos\dfrac{120° + 360° \cdot 0}{2} + i\sin\dfrac{120° + 360° \cdot 0}{2}\right) = 2(\cos 60° + i\sin 60°) = 1 + i\sqrt{3}$

$w_1 = \sqrt{4}\left(\cos\dfrac{120° + 360° \cdot 1}{2} + i\sin\dfrac{120° + 360° \cdot 1}{2}\right) = 2(\cos 240° + i\sin 240°) = -1 - i\sqrt{3}$

This result is verified with a calculator in Figure 51.

53. $w_0 = \sqrt[3]{1}\left(\cos\dfrac{180° + 360° \cdot 0}{3} + i\sin\dfrac{180° + 360° \cdot 0}{3}\right) = 1(\cos 60° + i\sin 60°) = \dfrac{1}{2} + \dfrac{\sqrt{3}}{2}i$

$w_1 = \sqrt[3]{1}\left(\cos\dfrac{180° + 360° \cdot 1}{3} + i\sin\dfrac{180° + 360° \cdot 1}{3}\right) = 1(\cos 180° + i\sin 180°) = -1$

$w_2 = \sqrt[3]{1}\left(\cos\dfrac{180° + 360° \cdot 2}{3} + i\sin\dfrac{180° + 360° \cdot 2}{3}\right) = 1(\cos 300° + i\sin 300°) = \dfrac{1}{2} - \dfrac{\sqrt{3}}{2}i$

These results are verified with a calculator in Figure 53a and 53b.

```
4(cos(120)+isin(
120))
      -2+3.464101615i
(1+i√(3))²
      -2+3.464101615i
(-1-i√(3))²
      -2+3.464101615i
```

Figure 51

```
cos(180)+isin(18
0)
              -1
(1/2+√(3)/2i)^3
              -1
```

Figure 53a

```
(-1)^3
              -1
(1/2-√(3)/2i)^3
              -1
```

Figure 53b

```
(√(2)/2+√(2)/2i)
²
              i
(-√(2)/2-√(2)/2i
)²
              i
```

Figure 55

55. $i = \cos 90° + i\sin 90°$

$w_0 = \sqrt{1}\left(\cos\dfrac{90° + 360° \cdot 0}{2} + i\sin\dfrac{90° + 360° \cdot 0}{2}\right) = 1(\cos 45° + i\sin 45°) = \dfrac{\sqrt{2}}{2} + \dfrac{\sqrt{2}}{2}i$

$w_1 = \sqrt{1}\left(\cos\dfrac{90° + 360° \cdot 1}{2} + i\sin\dfrac{90° + 360° \cdot 1}{2}\right) = 1(\cos 225° + i\sin 225°) = -\dfrac{\sqrt{2}}{2} - \dfrac{\sqrt{2}}{2}i$

These results are verified with a calculator in Figure 55.

57. $-8 = 8(\cos 180° + i\sin 180°)$

$w_0 = \sqrt[3]{8}\left(\cos\dfrac{180° + 360° \cdot 0}{3} + i\sin\dfrac{180° + 360° \cdot 0}{3}\right) = 2(\cos 60° + i\sin 60°) = 1 + i\sqrt{3}$

$w_1 = \sqrt[3]{8}\left(\cos\dfrac{180° + 360° \cdot 1}{3} + i\sin\dfrac{180° + 360° \cdot 1}{3}\right) = 2(\cos 180° + i\sin 180°) = -2$

$w_2 = \sqrt[3]{8}\left(\cos\dfrac{180° + 360° \cdot 2}{3} + i\sin\dfrac{180° + 360° \cdot 2}{3}\right) = 2(\cos 300° + i\sin 300°) = 1 - i\sqrt{3}$

These results are verified with a calculator in Figure 57.

```
(1+i√(3))^3
              -8
(-2)^3
              -8
(1-i√(3))^3
              -8
```

Figure 57

59. $64i = 64(\cos 90° + i\sin 90°)$

$$w_0 = \sqrt[3]{64}\left(\cos\frac{90° + 360° \cdot 0}{3} + i\sin\frac{90° + 360° \cdot 0}{3}\right) = 4(\cos 30° + i\sin 30°) = 2\sqrt{3} + 2i$$

$$w_1 = \sqrt[3]{64}\left(\cos\frac{90° + 360° \cdot 1}{3} + i\sin\frac{90° + 360° \cdot 1}{3}\right) = 4(\cos 150° + i\sin 150°) = -2\sqrt{3} + 2i$$

$$w_2 = \sqrt[3]{64}\left(\cos\frac{90° + 360° \cdot 2}{3} + i\sin\frac{90° + 360° \cdot 2}{3}\right) = 4(\cos 270° + i\sin 270°) = -4i$$

These results are verified with a calculator in Figure 59.

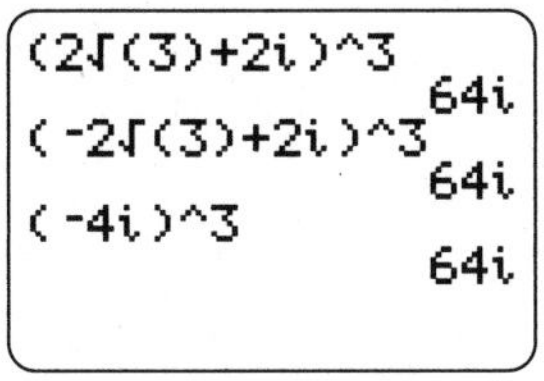

```
(2√(3)+2i)^3
            64i
(-2√(3)+2i)^3
            64i
(-4i)^3
            64i
```

```
3^4
            81
(3i)^4
            81
```

```
(-3)^4
            81
(-3i)^4
            81
```

Figure 59 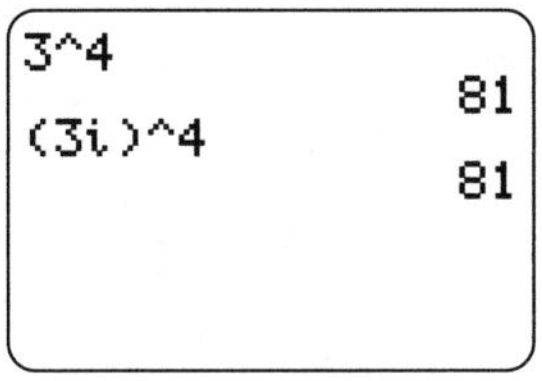   Figure 61a 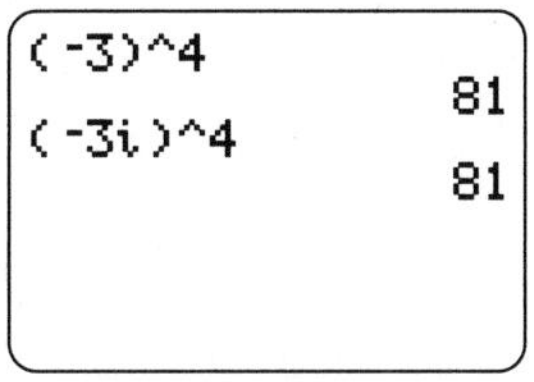   Figure 61b

61. $81 = 81(\cos 0° + i\sin 0°)$

$$w_0 = \sqrt[4]{81}\left(\cos\frac{0° + 360° \cdot 0}{4} + i\sin\frac{0° + 360° \cdot 0}{4}\right) = 3(\cos 0° + i\sin 0°) = 3$$

$$w_1 = \sqrt[4]{81}\left(\cos\frac{0° + 360° \cdot 1}{4} + i\sin\frac{0° + 360° \cdot 1}{4}\right) = 3(\cos 90° + i\sin 90°) = 3i$$

$$w_2 = \sqrt[4]{81}\left(\cos\frac{0° + 360° \cdot 2}{4} + i\sin\frac{0° + 360° \cdot 2}{4}\right) = 3(\cos 180° + i\sin 180°) = -3$$

$$w_3 = \sqrt[4]{81}\left(\cos\frac{0° + 360° \cdot 3}{4} + i\sin\frac{0° + 360° \cdot 3}{4}\right) = 3(\cos 270° + i\sin 270°) = -3i$$

These results are verified with a calculator in Figure 61a and 61b.

Fractals

63. $z_0 = -0.4i$

$$z_1 = (-0.4i)^2 + (-0.4i) = -0.16 - 0.4i \Rightarrow |z_1| \approx 0.4308$$

$$z_2 = (-0.16 - 0.4i)^2 + (-0.4i) = -0.1344 - 0.272i \Rightarrow |z_2| \approx 0.3034$$

$$z_3 = (-0.1344 - 0.272i)^2 + (-0.4i) \approx -0.0559 - 0.3269i \Rightarrow |z_3| \approx 0.3316$$

The modulus of each consecutive z_k never exceeds 2. Thus, $-0.4i$ is in the Mandelbrot set.

65. $z_0 = 1 + i$

$$z_1 = (1 + i)^2 + (1 + i) = 1 + 3i \Rightarrow |z_1| = \sqrt{10} \approx 3.162 > 2$$

The modulus of z_1 exceeds 2. Thus, $1 + i$ is not in the Mandelbrot set.

Applications

67. (a) $Z = 50 + 60 + 15i + 17i = 110 + 32i$

(b) $|110 + 32i| = \sqrt{110^2 + 32^2} = \sqrt{12,100 + 1024} = \sqrt{13,124} \approx 114.6$ ohms

Chapter 8 Review Exercises

1. This triangle is of the form ASA. There is only one solution.

$$\gamma = 180° - 70° - 40° = 70°$$

$$\frac{a}{\sin \alpha} = \frac{c}{\sin \gamma} \Rightarrow a = \frac{c \sin \alpha}{\sin \gamma} = \frac{10.1 \sin 70°}{\sin 70°} = 10.1 \qquad \frac{b}{\sin \beta} = \frac{c}{\sin \gamma} \Rightarrow b = \frac{c \sin \beta}{\sin \gamma} = \frac{10.1 \sin 40°}{\sin 70°} \approx 6.91$$

$$\gamma = 70°, \; a = 10.1, \; b \approx 6.9$$

3. This triangle is of the form SAS. There is only one solution.

$$a^2 = b^2 + c^2 - 2bc \cos \alpha = 7^2 + 8^2 - 2(7)(8) \cos 42° \Rightarrow a = \sqrt{113 - 112 \cos 42°} \approx 5.45599$$

$$b^2 = a^2 + c^2 - 2ac \cos \beta \Rightarrow \cos \beta = \frac{b^2 - a^2 - c^2}{-2ac} \Rightarrow \beta = \cos^{-1}\left(\frac{7^2 - 5.45599^2 - 8^2}{-2(5.45599)(8)}\right) \approx 59.14756°$$

$$\gamma = 180° - 42° - 59.14756° \approx 78.85244° \qquad a \approx 5.5, \; \beta \approx 59.1°, \; \gamma \approx 78.9°$$

5. This triangle is of the form AAS. There is only one solution.

$$\gamma = 180° - 19° - 46° = 115°$$

$$\frac{a}{\sin \alpha} = \frac{b}{\sin \beta} \Rightarrow a = \frac{b \sin \alpha}{\sin \beta} = \frac{13 \sin 19°}{\sin 46°} \approx 5.9 \qquad \frac{c}{\sin \gamma} = \frac{b}{\sin \beta} \Rightarrow c = \frac{b \sin \gamma}{\sin \beta} = \frac{13 \sin 115°}{\sin 46°} \approx 16.4$$

$$\gamma = 115°, \; a \approx 5.9, \; c \approx 16.4$$

7. This triangle is of the form SSA. It is ambiguous.

$$\frac{\sin \beta}{b} = \frac{\sin \gamma}{c} \Rightarrow \sin \beta = \frac{b \sin \gamma}{c} = \frac{8 \sin 20°}{11} \Rightarrow \beta_R = \sin^{-1}\left(\frac{8 \sin 20°}{11}\right) \approx 14.4°$$

Thus, $\beta \approx 14.4°$ or $\beta \approx 180° - 14.4° \approx 165.6°$ (which is not possible).

So $\beta \approx 14.4°$. Then $\alpha \approx 180° - 20° - 14.4° \approx 145.6°$.

$$\frac{a}{\sin \alpha} = \frac{c}{\sin \gamma} \Rightarrow a = \frac{c \sin \alpha}{\sin \gamma} = \frac{11 \sin 145.6°}{\sin 20°} \approx 18.2 \qquad \alpha \approx 145.6°, \; \beta \approx 14.4°, \; a \approx 18.2$$

9. This triangle is of the form SAS. There is only one solution.

$$c^2 = a^2 + b^2 - 2ab \cos \gamma = 18^2 + 23^2 - 2(18)(23) \cos 35° \Rightarrow c = \sqrt{853 - 828 \cos 35°} \approx 13.21901$$

$$a^2 = b^2 + c^2 - 2bc \cos \alpha \Rightarrow \cos \alpha = \frac{a^2 - b^2 - c^2}{-2bc} \Rightarrow \alpha = \cos^{-1}\left(\frac{18^2 - 23^2 - 13.21901^2}{-2(23)(13.21901)}\right) \approx 51.35453°$$

$$\beta = 180° - 35° - 51.35453° \approx 93.64547° \qquad c \approx 13.2, \; \alpha \approx 51.4°, \; \beta \approx 93.6°$$

11. Area $= K = \dfrac{1}{2}(12.3)(13.7) \sin 39° \approx 53.0$

13. $s = \dfrac{1}{2}(34 + 67 + 53) = 77 \Rightarrow$ Area $= K = \sqrt{77(77 - 34)(77 - 67)(77 - 53)} \approx 891.4$

15. (a) Horizontal $= 3$, vertical $= 4$

    (b) $\|\mathbf{v}\| = \sqrt{3^2 + 4^2} = \sqrt{9 + 16} = \sqrt{25} = 5$

    (c) $\|\mathbf{v}\|$ represents the length of $\mathbf{v}$. See Figure 15.

17. (a) See Figure 17.

    (b) $\mathbf{v} = \langle -2 - 0, \; -4 - 0 \rangle = \langle -2, -4 \rangle = -2\mathbf{i} - 4\mathbf{j}$

    (c) $\|\overrightarrow{PQ}\| = \sqrt{(-2)^2 + (-4)^2} = \sqrt{4 + 16} = \sqrt{20}$

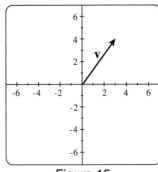

Figure 15

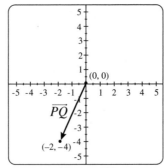

Figure 17

19. (a) $2\mathbf{a} = 2\langle 3, -2\rangle = \langle 6, -4\rangle$

(b) $\mathbf{a} - 3\mathbf{b} = \langle 3, -2\rangle - 3\langle 1, 1\rangle = \langle 3, -2\rangle - \langle 3, 3\rangle = \langle 3 - 3, -2 - 3\rangle = \langle 0, -5\rangle$

(c) $\mathbf{a} \cdot \mathbf{b} = \langle 3, -2\rangle \cdot \langle 1, 1\rangle = (3)(1) + (-2)(1) = 3 - 2 = 1$

(d) $\|\mathbf{a}\| = \sqrt{3^2 + (-2)^2} = \sqrt{9 + 4} = \sqrt{13}$, $\|\mathbf{b}\| = \sqrt{1^2 + 1^2} = \sqrt{1 + 1} = \sqrt{2}$

$\theta = \cos^{-1}\left(\dfrac{\mathbf{a} \cdot \mathbf{b}}{\|\mathbf{a}\|\|\mathbf{b}\|}\right) = \cos^{-1}\left(\dfrac{1}{\sqrt{13}\sqrt{2}}\right) = \cos^{-1}\left(\dfrac{1}{\sqrt{26}}\right) \approx 78.7°$

21. (a) $2\mathbf{a} = 2(2\mathbf{i} + 2\mathbf{j}) = 4\mathbf{i} + 4\mathbf{j}$

(b) $\mathbf{a} - 3\mathbf{b} = (2\mathbf{i} + 2\mathbf{j}) - 3(\mathbf{i} + \mathbf{j}) = 2\mathbf{i} + 2\mathbf{j} - 3\mathbf{i} - 3\mathbf{j} = -\mathbf{i} - \mathbf{j}$

(c) $\mathbf{a} \cdot \mathbf{b} = (2\mathbf{i} + 2\mathbf{j}) \cdot (\mathbf{i} + \mathbf{j}) = (2)(1) + (2)(1) = 2 + 2 = 4$

(d) $\|\mathbf{a}\| = \sqrt{2^2 + 2^2} = \sqrt{4 + 4} = \sqrt{8}$, $\|\mathbf{b}\| = \sqrt{1^2 + 1^2} = \sqrt{1 + 1} = \sqrt{2}$

$\theta = \cos^{-1}\left(\dfrac{\mathbf{a} \cdot \mathbf{b}}{\|\mathbf{a}\|\|\mathbf{b}\|}\right) = \cos^{-1}\left(\dfrac{4}{\sqrt{8}\sqrt{2}}\right) = \cos^{-1}\left(\dfrac{4}{\sqrt{16}}\right) = \cos^{-1} 1 = 0°$

23. Complete the parallelogram and find the unknown angle: $180° - 52° = 128°$.

Find the magnitude of the resultant force by using the law of cosines to find the length of the diagonal d.

$d^2 = 100^2 + 130^2 - 2(100)(130)\cos 128° \Rightarrow d = \sqrt{26{,}900 - 26{,}000\cos 128°} \approx 207.1$ lb

25. See Figure 25.

[−4.7, 4.7, 1] by [−3.1, 3.1, 1]   [−4.7, 4.7, 1] by [−3.1, 3.1, 1]   [−4.7, 4.7, 1] by [−3.1, 3.1, 1]   [−4.7, 4.7, 1] by [−3.1, 3.1, 1]

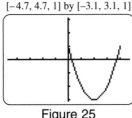

Figure 25

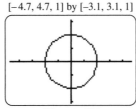

Figure 27

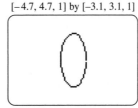

Figure 29a

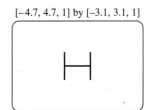

Figure 29b

27. See Figure 27.

29. (a) One possible solution is $x_1 = \cos t$, $y_1 = 2\sin t$ for $0 \le t \le 2\pi$. See Figure 29a. *Answers may vary.*

(b) One possible solution is $x_1 = -1 + \dfrac{t}{\pi}$, $y_1 = 0$; $x_2 = -1$, $y_2 = -1 + \dfrac{t}{\pi}$; $x_3 = 1$, $y_3 = -1 + \dfrac{t}{\pi}$

for $0 \le t \le 2\pi$. See Figure 29b. *Answers may vary.*

31. (a)

θ	0°	90°	180°	270°
r	2	1	0	1

(b) See Figure 31.

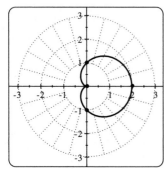

Figure 31

33. Graph $r_1 = 3\sin(3\theta)$ for $0 \le \theta \le \pi$ in $[-4.7, 4.7, 1]$ by $[-3.1, 3.1, 1]$. See Figure 33.

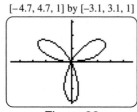

Figure 33

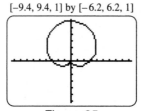

Figure 35

35. Graph $r_1 = 3 + 3\sin(\theta)$ for $0 \le \theta \le 2\pi$ in $[-9.4, 9.4, 1]$ by $[-6.2, 6.2, 1]$. See Figure 35.

37. See Figure 37.

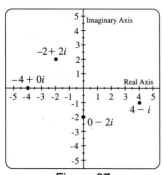

Figure 37

39. $z_1 z_2 = (4 \cdot 2)(\cos(150° + 30°) + i\sin(150° + 30°)) = 8(\cos 180° + i\sin 180°) = 8(-1 + i(0)) = -8$

$\dfrac{z_1}{z_2} = \left(\dfrac{4}{2}\right)(\cos(150° - 30°) + i\sin(150° - 30°)) = 2(\cos 120° + i\sin 120°) =$

$2\left(-\dfrac{1}{2} + i\left(\dfrac{\sqrt{3}}{2}\right)\right) = -1 + i\sqrt{3}$

41. $w_0 = \sqrt{4}\left(\cos\dfrac{60° + 360° \cdot 0}{2} + i\sin\dfrac{60° + 360° \cdot 0}{2}\right) = 2(\cos 30° + i\sin 30°) = \sqrt{3} + i$

$w_1 = \sqrt{4}\left(\cos\dfrac{60° + 360° \cdot 1}{2} + i\sin\dfrac{60° + 360° \cdot 1}{2}\right) = 2(\cos 210° + i\sin 210°) = -\sqrt{3} - i$

This result is verified with a calculator in Figure 41.

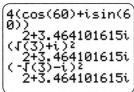

Figure 41   Figure 43

Applications

43. Refer to Figure 43.

$d^2 = 350^2 + 500^2 - 2(350)(500)\cos 110° \Rightarrow d = \sqrt{372{,}500 - 350{,}000\cos 110°} \approx 701.6$

The plane is about 701.6 miles from its takeoff point.

45. Let the airplane be located at position C and use standard labels.

Note that $\beta = 180° - 57° = 123°$. Then $\gamma = 180° - 123° - 52° = 5°$.

First we find side $a = BC$: $\dfrac{a}{\sin\alpha} = \dfrac{c}{\sin\gamma} \Rightarrow a = \dfrac{c\sin\alpha}{\sin\gamma} = \dfrac{950\sin 52°}{\sin 5°} \approx 8589.34$ ft.

Label a point D on the ground directly below the airplane and let h represent the height of the airplane.

From triangle BDC: $\sin 57° = \dfrac{h}{8589.34} \Rightarrow h = 8589.34\sin 57° \approx 7204$ ft.

47. In triangle ABC, $s = \dfrac{1}{2}(150 + 140 + 200) = 245 \Rightarrow$

$K_1 = \sqrt{245(245 - 150)(245 - 140)(245 - 200)} \approx 10{,}486.9$ ft^2

In triangle ACD, $s = \dfrac{1}{2}(100 + 125 + 200) = 212.5 \Rightarrow$

$K_2 = \sqrt{212.5(212.5 - 100)(212.5 - 125)(212.5 - 200)} \approx 5113.5$ ft^2

The total area is $K_1 + K_2 = 10{,}486.9 + 5113.5 \approx 15{,}600$ ft^2.

49. (a) $\|\mathbf{I}\| = \sqrt{40^2 + (-180)^2} = \sqrt{34{,}000} \approx 184.4$, $\|\mathbf{S}\| = \sqrt{2^2 + 5^2} = \sqrt{29} \approx 5.4$

The sun's intensity was about 184.4 watts per square meter. The area of the solar panel was about 5.4 m^2.

(b) $W = |\mathbf{I} \cdot \mathbf{S}| = |(40)(2) + (-180)(5)| = |-820| = 820$. The total number of watts collected was 820 watts.

51. Note that 0.75 miles = 3960 feet. The x-component of the displacement vector is $d_1 = 3960\cos 15°$.

The y-component of the displacement vector is $d_2 = 3960\sin 15°$.

$W = \mathbf{F} \cdot \mathbf{D} = \langle 0, 200\rangle \cdot \langle 3960\cos 15°, 3960\sin 15°\rangle = (0)(3960\cos 15°) + (200)(3960\sin 15°) \approx 204{,}985$ ft-lb

53. Label a point D on the ground directly below the airplane. Then the angle $DAB = 65°$.

From right triangle DAB: $\tan 65° = \dfrac{BD}{5000} \Rightarrow BD = 5000 \tan 65° \approx 10{,}722.5$ ft.

From right triangle DAC: $\tan 5° = \dfrac{CD}{5000} \Rightarrow CD = 5000 \tan 5° \approx 437.4$ ft.

The ground distance is $10{,}722.5 - 437.4 \approx 10{,}285$ ft.

Chapters 1-8 Cumulative Review Exercises

1. Move the decimal piont four places to the left; $91{,}200 \Rightarrow 9.12 \times 10^4$

    Move the decimal piont three places to the left; $6.734 \times 10^{-3} \Rightarrow 0.006734$

3. $D = \sqrt{[7-3]^2 + [(-9)-(-2)]^2} = \sqrt{16 + 49} = \sqrt{65}$

5   The domain of f includes all real numbers less than or equal to $4 \Rightarrow D = \{x \mid x \le 4\}$; $f(-5) = \sqrt{4 - (-5)} = \sqrt{9} = 3$.

7. $f(1) = 3(1)^2 - 2(1) = 3 - 2 = 1 \Rightarrow (1, 1)$; $f(3) = 3(3)^2 - 2(3) = 27 - 6 = 21 \Rightarrow (3, 21)$

    Find slope $m = \dfrac{21 - 1}{3 - 1} = \dfrac{20}{2} = 10$, the average rate of change is 10.

9. The slope-intercept form is $y = mx + b$, where m is the slope and b is the y-intercept. $m = \dfrac{-3 - 4}{1 - (-1)} = -\dfrac{7}{2}$,

    using the point-slope form $y = -\dfrac{7}{2}(x - 1) + (-3) \Rightarrow y = -\dfrac{7}{2}x + \dfrac{7}{2} - 3 \Rightarrow y = -\dfrac{7}{2}x + \dfrac{1}{2}$.

11. The tank initially contains 300 gallons of water and the amount of water is decreasing at a rate of 10 gallons per second.

13. See Figure 13. Because there is a continuous line, then f is considered continuous; $f(-1) = -1 + 2 = 1$

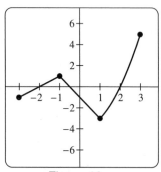

Figure 13

15. (a) The x-intercepts are -2, -1, and 1.

    (b) The graph of f is above the x-axis on the interval $(-\infty, -2) \cup (-1, 1)$.

    (c) The graph of f is below the x-axis on the interval $[-2, -1] \cup [1, \infty)$.

17. $x = \dfrac{-b}{2a} = \dfrac{-(-4)}{2(2)} = \dfrac{4}{4} = 1$; $y = f(1) = 2(1)^2 - 4(1) + 1 = -1$. The vertex is $(1, -1)$.

19. To shift the graph of $y = f(x)$ 2 units to the left and 3 units down, find $g(x) = f(x + 2) - 3$.

$g(x) = (x + 2)^2 + 2(x + 2) - 3 - 3 = x^2 + 4x + 4 + 2x + 4 - 6 = x^2 + 6x + 2$

That is, $g(x) = x^2 + 6x + 2$.

21. (a) It is increasing on $[-2, -1]$. It is decreasing on $(-\infty, -2] \cup [1, \infty)$.

(b) The zeros are approximately $-3.2, -0.4$, and 2.

(c) The turning points are $(-2, -15)$ and $(1, 12)$.

(d) It has local minimum -15 and local maximum 12.

23. Since the highest power of f is even and the leading coefficient is negative, the graph goes down on both ends.

$f(x) \to -\infty$ as $x \to -\infty$, $f(x) \to -\infty$ as $x \to \infty$.

25. $f(x) = 4(x + 2)(x - 3)(x - 5)$

27. $f(x) = 6(x - 1)(x + 1)(x - i)(x + i); f(x) = 6x^4 - 6$

29. $(2 + 3i)(4 - 2i) = 8 - 4i + 12i - 6i^2 = 8 + 8i - 6i^2 = 14 + 8i$

31. The denominator cannot equal zero $\Rightarrow 2 - 3x \neq 0 \Rightarrow x \neq \dfrac{2}{3}$. Then $D = \left\{ x \mid x \neq \dfrac{2}{3} \right\}$. The vertical

asymptotes are values when the function is undefined. The vertical asymptote is $x = \dfrac{2}{3}$. The horizontal

asymptote is $y = \dfrac{3x}{-3x} = -1 \Rightarrow y = -1$.

33. (a) $(f + g)(2) = f(2) + g(2) = 1 + 5 = 6$

(b) $(g \circ f)(2) \Rightarrow g(f(2)) = g(1) = 3$

(c) $(g/f)(x) = \dfrac{g(x)}{f(x)} = \dfrac{2x + 1}{\frac{1}{x^2 - 3}} = (2x + 1)(x^2 - 3) = 2x^3 - 6x + x^2 - 3$

(d) $(g \circ f)(x) = f(g(x)) = f(2x + 1) = \dfrac{1}{(2x + 1)^2 - 3}$

35. For each unit measure in x, the y-value is multiplies by 3. This is an exponential function with $c = 2$ and

$a = 3$, so $f(x) = 2(3)^x$.

37. Using the compound interest formula: $A_n = A_0 \left(1 + \dfrac{r}{m} \right)^{mn} \Rightarrow A_n = 1000 \left(1 + \dfrac{0.07}{4} \right)^{4(8)} \approx \$1742.21$

39. $\ln \sqrt[3]{\dfrac{x^3 y}{z^2}} = \ln \left(\dfrac{x^3 y}{z^2} \right)^{1/3} = \ln \dfrac{xy^{1/3}}{z^{2/3}} = \ln x + \ln y^{1/3} - \ln z^{2/3} = \ln x + \dfrac{1}{3} \ln y - \dfrac{2}{3} \ln z$

41. $\alpha = 90° - 22° 55' = 67° 5'$ and $\beta = 180° - 22° 55' = 157° 5'$

43. $225° \left(\dfrac{\pi}{180} \right) = \dfrac{5\pi}{4}$

45. $c = \sqrt{5^2 + 12^2} = \sqrt{169} = 13$

$\sin \theta = \dfrac{5}{13}$ $\cos \theta = \dfrac{12}{13}$ $\tan \theta = \dfrac{5}{12}$

$\csc \theta = \dfrac{13}{5}$ $\sec \theta = \dfrac{13}{12}$ $\cot \theta = \dfrac{12}{5}$

47. $c = \sqrt{11^2 + 60^2} = \sqrt{3721} = 61$

$$\sin\theta = -\frac{60}{61} \qquad \cos\theta = \frac{11}{61} \qquad \tan\theta = -\frac{60}{11}$$

$$\csc\theta = -\frac{61}{60} \qquad \sec\theta = \frac{61}{11} \qquad \cot\theta = -\frac{11}{60}$$

49. Since $\sec\theta < 0$, $\cos\theta = -\sqrt{1 - \sin^2\theta} = -\sqrt{1 - \left(-\frac{7}{25}\right)^2} = -\sqrt{1 - \frac{49}{625}} = -\sqrt{\frac{576}{625}} = -\frac{24}{25}$

$$\sin\theta = -\frac{7}{25} \qquad \cos\theta = -\frac{24}{25} \qquad \tan\theta = -\frac{7}{24}$$

$$\csc\theta = -\frac{25}{7} \qquad \sec\theta = -\frac{25}{24} \qquad \cot\theta = \frac{24}{7}$$

51. $c = \sqrt{5^2 + 12^2} = \sqrt{169} = 13.$ $\sin\theta = \frac{5}{13} \Rightarrow \sin^{-1}\left(\frac{5}{13}\right) \approx 22.6° \Rightarrow \theta \approx 22.6°$

$180° - 90° - 22.9° = 67.4° \Rightarrow \beta = 67.4°$

53. $\cot^2\theta - 2\cot\theta + 1 = (\cot\theta - 1)(\cot\theta - 1) = (\cot\theta - 1)^2$

55. If $\sin\alpha = \frac{5}{13}$ and α is in QII, $\cos\alpha = -\sqrt{1 - \left(\frac{5}{13}\right)^2} = -\sqrt{1 - \frac{25}{169}} = -\sqrt{\frac{144}{169}} = -\frac{12}{13}$.

If $\cos\beta = -\frac{4}{5}$ and β is in QII, $\sin\beta = \sqrt{1 - \left(-\frac{4}{5}\right)^2} = \sqrt{1 - \frac{16}{25}} = \sqrt{\frac{9}{25}} = \frac{3}{5}$.

Note: $\sin\alpha = \frac{5}{13}$, $\cos\alpha = -\frac{12}{13}$, $\sin\beta = \frac{3}{5}$, $\cos\beta = -\frac{4}{5}$

$$\cos(\alpha - \beta) = \cos\alpha\cos\beta + \sin\alpha\sin\beta = \left(-\frac{12}{13}\right)\left(-\frac{4}{5}\right) + \left(\frac{5}{13}\right)\left(\frac{3}{5}\right) = \frac{48}{65} + \frac{15}{65} = \frac{63}{65}$$

57. (a) This triangle is in the form ASA. There is only one solution. $\beta = 180° - 31° - 53° = 96°$

$$\frac{a}{\sin\alpha} = \frac{b}{\sin\beta} \Rightarrow a = \frac{b\sin\alpha}{\sin\beta} = \frac{15\sin 31°}{\sin 96°} \approx 7.8$$

$$\frac{c}{\sin\gamma} = \frac{b}{\sin\beta} \Rightarrow c = \frac{b\sin\gamma}{\sin\beta} = \frac{15\sin 53°}{\sin 96°} \approx 12.0 \quad \beta = 96°, a \approx 7.8, c \approx 12.0$$

(b) This triangle is in the form SSA. It is ambiguous.

$$\frac{\sin\beta}{b} = \frac{\sin\alpha}{a} \Rightarrow \sin\beta = \frac{b\sin\alpha}{a} = \frac{5\sin 31°}{6} = 0.42920 \Rightarrow \beta_R \approx \sin^{-1}(0.42920) \approx 25.4°$$

Thus $\beta = 25.4°$. Then $\gamma = 180° - 31° - 25.4° \approx 123.6°$

$$\frac{c}{\sin\gamma} = \frac{a}{\sin\alpha} \Rightarrow c = \frac{a\sin\gamma}{\sin\alpha} = \frac{6\sin 123.6°}{\sin 31°} \approx 9.7 \quad \beta \approx 25.4°, \gamma \approx 123.6°, c \approx 9.7$$

(c) $b^2 = a^2 + c^2 - 2ac\cos\beta = 6^2 + 8^2 - 2(6)(8)\cos 56° \Rightarrow b\sqrt{100 - 96\cos 56°} \approx 6.8$

$$a^2 = b^2 + c^2 - 2bc\cos\alpha \Rightarrow \cos\alpha = \frac{a^2 - b^2 - c^2}{-2bc} \Rightarrow \alpha = \cos^{-1}\left(\frac{6^2 - (6.8)^2 - 8^2}{-2(6.8)(8)}\right) \approx 47.0°$$

$\gamma = 180° - 56° - 47.0° \approx 77.0°$ $b \approx 6.8, \alpha \approx 47.0°, \gamma \approx 77.0°$

(d) $a^2 = b^2 + c^2 - 2bc\cos\alpha \Rightarrow \cos\alpha = \frac{a^2 - b^2 - c^2}{-2bc} \Rightarrow \cos^{-1}\left(\frac{6^2 - 7^2 - 8^2}{-2(7)(8)}\right) \approx 46.6°$

$$b^2 = a^2 + c^2 - 2ac\cos\beta \Rightarrow \cos\beta = \frac{b^2 - a^2 - c^2}{-2ac} \Rightarrow \beta = \cos^{-1}\left(\frac{7^2 - 6^2 - 8^2}{-2(6)(8)}\right) \approx 57.9°$$

$\gamma = 180° - 46.6° - 57.9° \approx 75.5°$ $\alpha \approx 46.6°, \beta \approx 57.9°, \gamma \approx 75.5°$

59. (a) $\|\mathbf{b}\| = \sqrt{7^2 + (-24)^2} = \sqrt{49 + 576} = \sqrt{625} = 25$

(b) $2\mathbf{a} - 3\mathbf{b} = 2\langle -5, 12 \rangle - 3\langle 7, -24 \rangle = \langle -10, 24 \rangle - \langle 21, -72 \rangle \Rightarrow \langle -10 - 21, 24 - (-72) \rangle \Rightarrow \langle -31, 96 \rangle$

(c) $\mathbf{a} \cdot \mathbf{b} = (-5)(7) + (12)(-24) = -35 + (-288) = -323$

(d) $\|\mathbf{a}\| = \sqrt{(-5)^2 + (12)^2} = \sqrt{25 + 144} = 13, \|\mathbf{b}\| = \sqrt{7^2 + (-24)^2} = \sqrt{49 + 576} = \sqrt{625} = 25$

$$\theta = \cos^{-1}\left(\frac{\mathbf{a} \cdot \mathbf{b}}{\|\mathbf{a}\|\|\mathbf{b}\|}\right) = \cos^{-1}\left(\frac{-323}{(13)(25)}\right) = \cos^{-1}\left(\frac{-323}{325}\right) \Rightarrow \theta \approx 173.6°$$

61. See Figure 61.

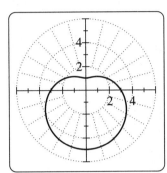

Figure 61

Applications

63. To find slope, we need to find a change in y over a change in x: point 1 = $(0, 0)$, point 2 = $(0.4, 2)$,

point 3 = $(0.6, 2)$, point 4 = $(1, 6)$ and point 5 = $(1.6, 2)$.

From point 1 to point 2 $\Rightarrow m_1 = \left(\dfrac{2 - 0}{0.4 - 0}\right) = 5$: the jogger is moving away from home at 5 mph.

From point 2 to point 3 $\Rightarrow m_2 = \left(\dfrac{2 - 2}{0.6 - 0.4}\right) = \dfrac{0}{0.2} = 0$: the jogger has stopped.

From point 3 to point 4 $\Rightarrow m_3 = \left(\dfrac{6 - 2}{1 - 0.6}\right) = \dfrac{4}{0.4} = 10$: the jogger is moving away from home at 10 mph.

From point 4 to point 5 $\Rightarrow m_4 = \left(\dfrac{2 - 6}{1.6 - 1}\right) = \dfrac{-4}{0.6} = -6.\overline{6}$: the jogger is moving toward home at $6.\overline{6}$ mph.

65. $x = \$3.50$ per pound type of candy, $y = \$4.50$ per pound type of candy $\Rightarrow x + y = 5$ and $3.50x + 4.50y = $

21.25. By the use of substitution, we let $x = 5 - y$ and substitute for x in the second equation

$3.50(5 - y) + 4.50y = 21.25 \Rightarrow 17.50 - 3.50y + 4.50y = 21.25 \Rightarrow 17.50 + y = 21.25 \Rightarrow$

$y = 3.75$, therefore the first equation becomes $x + 3.75 = 5 \Rightarrow x = 1.25$. Therefore 1.25 lb. of $3.50 candy

and 3.75 lb. of $4.50 candy.

67. When the data points are plotted they take on a parabolic shape with a vertex located at $(2, 1)$. Thus $h = 2$ and

$k = 1$. Since $f(0) = -11$, $a(0 - 2)^2 + 1 = -11 \Rightarrow 4a + 1 = -11 \Rightarrow 4a = -12 \Rightarrow a = -3$

$f(x) = -3(x - 2)^2 + 1$

69. Given $V = 20$ and diameter = 3.44, then radius = $3.44 \div 2$ or $1.72 \Rightarrow 20 = \pi(1.72)^2 h \Rightarrow (1.72)^2\pi = h \Rightarrow$

$h = 2.15$ inches.

71. (a) If $(1980, 3600)$ and $(2000, 16{,}200)$ then the average rate of change is: $\dfrac{16{,}200 - 3600}{2000 - 1980} = \dfrac{12{,}600}{20} = 630.$

   Then $c(t) = 630(t - 1980) + 3600.$

   (b) $11{,}000 = 630(t - 1980) + 3600 \Rightarrow 11{,}000 = 630t - 1{,}247{,}400 + 3600 \Rightarrow$

   $11{,}000 = 630t - 1{,}243{,}800 \Rightarrow 1{,}254{,}800 = 630t \Rightarrow t = 1991.7$ or close to the year 1992. This is only

   one year off from the actual year of 1993.

73. The angular velocity is $\omega = 25$ radians/sec. The radius is $r = \dfrac{2}{2} = 1$ foot. The linear speed is given by

   $v = r\omega = (1)(25) = 25$ feet/sec.

75. The maximum monthly average temperature is $75°$ F and the minimum is $25°$ F. The midpoint of these values

   is $0.5(75 + 25) = 50$, thus $d = 50$. Half the difference between these temperatures is $0.5(75 - 25) = 25$, thus

   $a = 25$. Since the temperatures cycle every 12 monthes, $b = \dfrac{2\pi}{12} = \dfrac{\pi}{6}.$

   The maximum of the $y = \sin x$ graph occurs when $x = \dfrac{\pi}{2}$ while the maximum in the table occurs when $x = 7$.

   Thus $\dfrac{\pi}{6}(7 - c) = \dfrac{\pi}{2} \Rightarrow 7 - c = 3 \Rightarrow c = 4$. The function is $f(x) = 25 \sin\left(\dfrac{\pi}{6}(x - 4)\right) + 50.$

77. $76 = 17.5 \sin\left(\dfrac{\pi}{6}(x - 4)\right) + 67.5 \Rightarrow 8.5 = 17.5 \sin\left(\dfrac{\pi}{6}(x - 4)\right) \Rightarrow \dfrac{8.5}{17.5} = \sin\left(\dfrac{\pi}{6}(x - 4)\right) \Rightarrow$

   $\sin^{-1}\left(\dfrac{8.5}{17.5}\right) = \dfrac{\pi}{6}(x - 4)$. Note that $\sin^{-1}\left(\dfrac{8.5}{17.5}\right) \approx 0.507$ or 2.634. Therefore $0.507 = \dfrac{\pi}{6}(x - 4) \Rightarrow$

   $\left(\dfrac{6}{\pi}\right)(0.507) = x - 4 \Rightarrow 0.968 + 4 = x \Rightarrow x \approx 5$, which equates to the month of May.

   $2.634 = \dfrac{\pi}{6}(x - 4) = \dfrac{6}{\pi}(2.634) = x - 4 \Rightarrow 5.031 + 4 = x \Rightarrow x \approx 9$, which equates to the month of

   September.

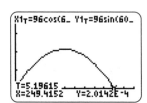

Figure 81

79. (a) $c^2 = a^2 + b^2 - 2ab \cos \gamma = 242^2 + 165^2 - 2(242)(165) \cos 72° \Rightarrow$

   $c = \sqrt{85{,}789 - 79{,}860 \cos 72°} \approx 247°$

   (b) Area $= K = \dfrac{1}{2}(242)(165) \sin 72° \approx 18{,}988$ ft^2

81. Graph the parametic equations $x_1 = 96 \cos 60° (t)$, $y_1 = 96 \sin 60° 9t) - 16t^2$ for $0 \le t \le 9$ in $[0, 300, 25]$ by

   $[-20, 200, 50]$. See Figure 81. About 249.4 feet of horizontal distance was traveled when it strikes the ground.

Chapter 9: Systems of Equations and Inequalities

9.1: Functions and Equations in Two Variables

Functions of More Than One Input

1. $f(5, 8) = \dfrac{1}{2}(5)(8) = 20$. The area of a triangle with a base of 5 and height of 8 is 20 square units.

3. $f(2, -3)$ if $f(x, y) = x^2 + y^2 \Rightarrow f(2, -3) = 2^2 + (-3)^2 = 4 + 9 = 13$

5. $f(-2, 3)$ if $f(x, y) = 3x - 4y \Rightarrow f(-2, 3) = 3(-2) - 4(3) = -6 - 12 = -18$

7. $f\left(\dfrac{1}{2}, -\dfrac{7}{4}\right)$ if $f(x, y) = \dfrac{2x}{y + 3} \Rightarrow f\left(\dfrac{1}{2}, -\dfrac{7}{4}\right) = \dfrac{2\left(\frac{1}{2}\right)}{\left(-\frac{7}{4}\right) + 3} = \dfrac{1}{\frac{5}{4}} = \dfrac{4}{5}$

9. The sum of y and twice x is computed by $f(x, y) = y + 2x$.

11. The product of x and y divided by $1 + x$ is computed by $f(x, y) = \dfrac{xy}{1 + x}$.

13. $3x - 4y = 7 \Rightarrow 3x = 4y + 7 \Rightarrow x = \dfrac{4y + 7}{3}$; $\ 3x - 4y = 7 \Rightarrow -4y = -3x + 7 \Rightarrow y = \dfrac{3x - 7}{4}$

15. $x - y^2 = 5 \Rightarrow x = y^2 + 5$; $\ x - y^2 = 5 \Rightarrow -y^2 = -x + 5 \Rightarrow y^2 = x - 5 \Rightarrow y = \pm\sqrt{x - 5}$

17. $\dfrac{2x - y}{3y} = 1 \Rightarrow 2x - y = 3y \Rightarrow 2x = 4y \Rightarrow x = 2y$; $\ \dfrac{2x - y}{3y} = 1 \Rightarrow 2x - y = 3y \Rightarrow 2x = 4y \Rightarrow y = \dfrac{x}{2}$

19. (a) $A = \dfrac{1}{2}bh \Rightarrow 2A = bh \Rightarrow b = \dfrac{2A}{h}$

    (b) $A = \dfrac{1}{2}bh \Rightarrow 2A = bh \Rightarrow h = \dfrac{2A}{b}$

21. (a) $z = \dfrac{2x^2}{y + 1} \Rightarrow 2x^2 = z(y + 1) \Rightarrow x^2 = \dfrac{z(y + 1)}{2} \Rightarrow x = \pm\sqrt{\dfrac{z(y + 1)}{2}}$

    (b) $z = \dfrac{2x^2}{y + 1} \Rightarrow y + 1 = \dfrac{2x^2}{z} \Rightarrow y = \dfrac{2x^2}{z} - 1$

Systems of Equations

23. The only ordered pair that satisfies both equations is $(2, 1)$.

    $2(2) + 1 = 5\star$ $2(-2) + 1 = -3$ $2(1) + 0 = 2$

    $2 + 1 = 3\star$ $-2 + 1 = -1$ $1 + 0 = 1$

25. The only ordered pair that satisfies both equations is $(4, -3)$.

    $4^2 + (-3)^2 = 25\star$ $0^2 + 5^2 = 25\star$ $4^2 + 3^2 = 25\star$

    $2(4) + 3(-3) = -1\star$ $2(0) + 3(5) = 15$ $2(4) + 3(3) = 17$

27. From the graph the solution is $(2, 2)$. The solution satisfies both $x - y = 0$ and $x + y = 4$.

29. From the graph the solution is $\left(\dfrac{1}{2}, -2\right)$. The solution satisfies both $6x + 4y = -5$ and $2x - 3y = 7$.

31. Solve the first equation for x: $x + 2y = 0 \Rightarrow x = -2y$. Substitute this into the second equation.

    $3x + 7y = 1 \Rightarrow 3(-2y) + 7y = 1 \Rightarrow -6y + 7y = 1 \Rightarrow y = 1$.

    If $y = 1$, then $x = -2(1) = -2$. The solution is $(-2, 1)$.

    Check: $x + 2y = 0 \Rightarrow (-2) + 2(1) = 0 \Rightarrow 0 = 0$; $\ x = -2y \Rightarrow -2 = -2(1) \Rightarrow -2 = -2$

33. Solve the second equation for y: $-4x + y = 28 \Rightarrow y = 28 + 4x$. Substitute this into the first equation.

    $3x - 5y = -38 \Rightarrow 3x - 5(28 + 4x) = -38 \Rightarrow 3x - 140 - 20x = -38 \Rightarrow -17x = 102 \Rightarrow x = -6$.

    If $x = -6$, then $y = 28 + 4x = 28 + 4(-6) = 4$. The solution is $(-6, 4)$.

    Check: $-4x + y = 28 \Rightarrow -4(-6) + 4 = 28 \Rightarrow 24 + 4 = 28 \Rightarrow 28 = 28$;

    $y = 28 + 4x \Rightarrow 4 = 28 + 4(-6) \Rightarrow 4 = 28 + (-24) \Rightarrow 4 = 4$

35. Solve the first equation for x: $2x - 9y = -17 \Rightarrow 2x = 9y - 17 \Rightarrow x = \dfrac{9y - 17}{2}$. Substitute this into the

    second equation. $8x + 5y = 14 \Rightarrow 8\left(\dfrac{9y - 17}{2}\right) + 5y = 14 \Rightarrow 4(9y - 17) + 5y = 14 \Rightarrow$

    $36y - 68 + 5y = 14 \Rightarrow 41y = 82 \Rightarrow y = 2$. If $y = 2$, then $x = \dfrac{9(2) - 17}{2} = \dfrac{1}{2}$. The solution is $(0.5, 2)$.

    Check: $2x - 9y = -17 \Rightarrow 2(0.5) - 9(2) = -17 \Rightarrow 1 - 18 = -17 \Rightarrow -17 = -17$;

    $8x + 5y = 14 \Rightarrow 8(0.5) + 5(2) = 14 \Rightarrow 4 + 10 = 14 \Rightarrow 14 = 14$

37. Solve the second equation for x: $x + \dfrac{1}{2}y = 10 \Rightarrow x = -\dfrac{1}{2}y + 10$. Substitute this into the first equation.

    $\dfrac{1}{2}x - y = -5 \Rightarrow \dfrac{1}{2}\left(-\dfrac{1}{2}y + 10\right) - y = -5 \Rightarrow -\dfrac{1}{4}y + 5 - y = -5 \Rightarrow -\dfrac{5}{4}y = -10 \Rightarrow y = 8$.

    If $y = 8$, then $x = -\dfrac{1}{2}(8) + 10 = 6$. The solution is $(6, 8)$.

    Check: $x + \dfrac{1}{2}y = 10 \Rightarrow 6 + \dfrac{1}{2}(8) = 10 \Rightarrow 6 + 4 = 10 \Rightarrow 10 = 10$

    $\dfrac{1}{2}x - y = -5 \Rightarrow \dfrac{1}{2}(6) - 8 = -5 \Rightarrow 3 - 8 = -5 \Rightarrow -5 = -5$

39. Solve the first equation for y: $3x - 2y = 5 \Rightarrow -2y = -3x + 5 \Rightarrow y = \dfrac{3}{2}x - \dfrac{5}{2}$. Substitute this into the

    second equation. $-6x + 4\left(\dfrac{3}{2}x - \dfrac{5}{2}\right) = -10 \Rightarrow -6x + 6x - 10 = -10 \Rightarrow -10 = -10 \Rightarrow$ there are

    infinitely many solutions.

41. Solve the first equation for x: $2x - 7y = 8 \Rightarrow 2x = 7y + 8 \Rightarrow x = \dfrac{7}{2}y + 4$. Substitute this into the second

    equation. $-3\left(\dfrac{7}{2}y + 4\right) + \dfrac{21}{2}y = 5 \Rightarrow -\dfrac{21}{2}y - 12 + \dfrac{21}{2}y = 5 \Rightarrow -12 = 5 \Rightarrow$ there are no real solutions.

43. $0.2x - 0.1y = 0.5 \Rightarrow 2x - y = 5$ and $0.4x + 0.3y = 2.5 \Rightarrow 4x + 3y = 25$

    Solve the first equation for y: $2x - y = 5 \Rightarrow y = 2x - 5$. Substitute this into the second equation.

    $4x + 3y = 25 \Rightarrow 4x + 3(2x - 5) = 25 \Rightarrow 4x + 6x - 15 = 25 \Rightarrow 10x = 40 \Rightarrow x = 4$.

    If $x = 4$, then $y = 2(4) - 5 \Rightarrow y = 3$. The solution is $(4, 3)$.

45. Solve the second equation for x: $-5x - 20y = -30 \Rightarrow -5x = 20y - 30 \Rightarrow x = -4y + 6$. Substitute this into

    the first equation. $20x - 10y = 30 \Rightarrow 20(-4y + 6) - 10y = 30 \Rightarrow -80y + 120 - 10y = 30 \Rightarrow$

    $-90y = -90 \Rightarrow y = 1$. If $y = 1$, then $x = -4(1) + 6 = 2$. The solution is $(2, 1)$.

47. (a) $2x + y = 1 \Rightarrow y = 1 - 2x$ and $x - 2y = 3 \Rightarrow y = \dfrac{1}{2}(x - 3)$

    Graph $Y_1 = 1 - 2X$ and $Y_2 = 0.5(X - 3)$. Their graphs intersect at the point $(1, -1)$, which is the solution. See Figure 47a.

    (b) Table $Y_1 = 1 - 2X$ and $Y_2 = 0.5(X - 3)$ starting at 0 and incrementing by 0.5. See Figure 47b.

    Here $Y_1 = Y_2 = -1$ when $x = 1$. The solution is $(1, -1)$.

    (c) Substituting $y = 1 - 2x$ into the equation $x - 2y = 3$ gives $x - 2(1 - 2x) = 3 \Rightarrow x - 2 + 4x = 3 \Rightarrow$

    $5x = 5 \Rightarrow x = 1$. If $x = 1$, then $y = 1 - 2(1) = -1$. The solution is $(1, -1)$.

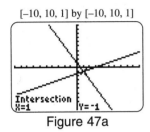

[–10, 10, 1] by [–10, 10, 1]

Figure 47a

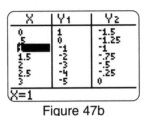

Figure 47b

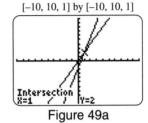

[–10, 10, 1] by [–10, 10, 1]

Figure 49a

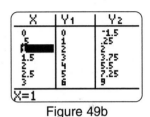

Figure 49b

49. (a) $-2x + y = 0 \Rightarrow y = 2x$ and $7x - 2y = 3 \Rightarrow 7x - 3 = 2y \Rightarrow y = \dfrac{7x - 3}{2}$

    Graph $Y_1 = 2X$ and $Y_2 = (7X - 3)/2$. Their graphs intersect at the point $(1, 2)$, which is the solution. See Figure 49a.

    (b) Table $Y_1 = 2X$ and $Y_2 = (7X - 3)/2$ starting at 0 and incrementing by 0.5. See Figure 49b.

    Here $Y_1 = Y_2 = 2$ when $x = 1$. The solution is $(1, 2)$.

    (c) Substituting $y = 2x$ into the second equation gives $7x - 2(2x) = 3 \Rightarrow 3x = 3 \Rightarrow x = 1$.

    If $x = 1$, then $y = 2(1) = 2$. The solution is $(1, 2)$.

51. Solve the second equation for y: $2x + y = 0 \Rightarrow y = -2x$. Substitute this into the first equation.

    $x^2 - y = 0 \Rightarrow x^2 - (-2x) = 0 \Rightarrow x^2 + 2x = 0 \Rightarrow x(x + 2) = 0 \Rightarrow x = 0$ or $x = -2$.

    When $x = 0$, $y = -2(0) = 0$ and when $x = -2$, $y = -2(-2) = 4$. The solutions are $(0, 0)$ and $(-2, 4)$.

53. Solve the second equation for y: $x + y = 6 \Rightarrow y = 6 - x$. Substitute this into the first equation.

    $xy = 8 \Rightarrow x(6 - x) = 8 \Rightarrow 6x - x^2 = 8 \Rightarrow x^2 - 6x + 8 = 0 \Rightarrow (x - 4)(x - 2) = 0 \Rightarrow x = 4$ or $x = 2$.

    When $x = 4$, $y = 6 - 4 = 2$ and when $x = 2$, $y = 6 - 2 = 4$. The solutions are $(4, 2)$ and $(2, 4)$.

55. Substitute the second equation, $y = 2x$, into the first equation.

    $x^2 + y^2 = 20 \Rightarrow x^2 + (2x)^2 = 20 \Rightarrow x^2 + 4x^2 = 20 \Rightarrow 5x^2 = 20 \Rightarrow x^2 = 4 \Rightarrow x = \pm 2$.

    When $x = -2$, $y = 2(-2) = -4$ and when $x = 2$, $y = 2(2) = 4$. The solutions are $(-2, -4)$ and $(2, 4)$.

57. Solve the second equation for y: $\sqrt{x} - y = 0 \Rightarrow y = \sqrt{x}$. Substitute this into the first equation.

    $x^2 + y^2 = 6 \Rightarrow x^2 + (\sqrt{x})^2 = 6 \Rightarrow x^2 + x - 6 = 0 \Rightarrow (x + 3)(x - 2) = 0 \Rightarrow x = -3$ or $x = 2$.

    When $x = -3$, $y = \sqrt{-3}$ is undefined, and when $x = 2$, $y = \sqrt{2}$. The solution is $(2, \sqrt{2})$.

59. Solve the second equation for x: $x - y = -2 \Rightarrow x = y - 2$. Substitute this into the first equation.

$\sqrt{y - 2} - 2y = 0 \Rightarrow \sqrt{y - 2} = 2y \Rightarrow y - 2 = 4y^2 \Rightarrow 4y^2 - y + 2 = 0$. Using the quadratic formula to

solve we get: $\dfrac{1 \pm \sqrt{1 - 4(4)(2)}}{2(4)} = \dfrac{1 \pm \sqrt{-31}}{8} \Rightarrow$ no real solutions.

61. Solve the first equation for y: $2x^2 - y = 5 \Rightarrow -y = -2x^2 + 5 \Rightarrow y = 2x^2 - 5$. Substitute this into the

second equation. $-4x^2 + 2(2x^2 - 5) = -10 \Rightarrow -4x^2 + 4x^2 - 10 = -10 \Rightarrow -10 = -10 \Rightarrow$ there are

infinitely many solutions.

63. Solve the second equation for y: $x^2 + y = 4 \Rightarrow y = 4 - x^2$. Substitute this into the first equation.

$x^2 - y = 4 \Rightarrow x^2 - (4 - x^2) = 4 \Rightarrow 2x^2 = 8 \Rightarrow x^2 = 4 \Rightarrow x = \pm 2$.

When $x = -2$, $y = 4 - (-2)^2 = 0$ and when $x = 2$, $y = 4 - (2)^2 = 0$. The solutions are $(-2, 0)$ and $(2, 0)$.

65. Solve the second equation for y: $x - y = 0 \Rightarrow y = x$. Substitute this into the first equation.

$x^3 - x = 3y \Rightarrow x^3 - x = 3x \Rightarrow x^3 - 4x = 0 \Rightarrow x(x + 2)(x - 2) = 0 \Rightarrow x = 0, x = -2,$ or $x = 2$.

When $x = 0$, $y = 0$, when $x = -2$, $y = -2$, and when $x = 2$, $y = 2$.

The solutions are $(-2, -2)$, $(0, 0)$, and $(2, 2)$.

67. $x^2 + y^2 = 16 \Rightarrow y = \pm\sqrt{16 - x^2}$ and $x - y = 0 \Rightarrow y = x$

Graph $Y_1 = \sqrt{(16 - X^2)}$, $Y_2 = -\sqrt{(16 - X^2)}$, and $Y_3 = X$. Their graphs intersect near the points

$(-2.828, -2.828)$ and $(2.828, 2.828)$. See Figures 67a & 67b.

Substituting $y = x$ into the first equation gives $x^2 + x^2 = 16 \Rightarrow 2x^2 = 16 \Rightarrow x^2 = 8 \Rightarrow x = \pm\sqrt{8}$

Since $y = x$, the solutions are $(-\sqrt{8}, -\sqrt{8})$ and $(\sqrt{8}, \sqrt{8})$

[−9, 9, 1] by [−6, 6, 1]	[−9, 9, 1] by [−6, 6, 1]	[−10, 10, 2] by [−12, 8, 1]	[−10, 10, 2] by [−12, 8, 1]
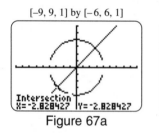	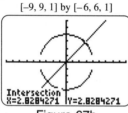	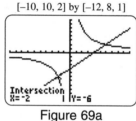	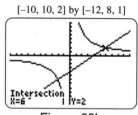
Figure 67a	Figure 67b	Figure 69a	Figure 69b

69. $xy = 12 \Rightarrow y = \dfrac{12}{x}$ and $x - y = 4 \Rightarrow y = x - 4$. Graph $Y_1 = \dfrac{12}{X}$, $Y_2 = X - 4$.

Their graphs intersect near the points $(6, 2)$ and $(-2, -6)$. See Figures 69a & 69b.

Substituting $y = x - 4$ into the first equation gives $x(x - 4) = 12 \Rightarrow x^2 - 4x - 12 = 0 \Rightarrow$

$(x + 2)(x - 6) = 0 \Rightarrow x = -2$ or 6. Since $y = x - 4$, the solutions are $(-2, -6)$ and $(6, 2)$.

71. The given equations result in the following nonlinear system of equations.

$A(l, w) = 35 \Rightarrow lw = 35$ and $P(l, w) = 24 \Rightarrow 2l + 2w = 24$

Begin solving the second equation for l. $2l + 2w = 24 \Rightarrow l + w = 12 \Rightarrow l = 12 - w$. Substitute this into

the first equation. $lw = 35 \Rightarrow (12 - w)w = 35 \Rightarrow 12w - w^2 = 35 \Rightarrow w^2 - 12w + 35 = 0$. This is a

quadratic equation that can be solved by factoring, graphing, or the quadratic formula. The solutions to this

quadratic are found using factoring. $w^2 - 12w + 35 = 0 \Rightarrow (w - 5)(w - 7) = 0 \Rightarrow w = 5$ or 7.

Since $l = 12 - w$, if $w = 5$, then $l = 7$, and if $w = 7$, then $l = 5$. If the length is greater than the width, the

solution is $l = 7$ and $w = 5$. A rectangle with length 7 and width 5 has an area of 35 and a perimeter of 24.

73. $x^3 - 3x + y = 1 \Rightarrow y = 1 + 3x - x^3$ and $x^2 + 2y = 3 \Rightarrow y = \dfrac{3 - x^2}{2}$. Graph $Y_1 = 1 + 3X - X\char`\^3$

and $Y_2 = (3 - X\char`\^2)/2$. See Figure 73. There are three points of intersection. The coordinates of these points

are near $(-1.588, 0.240)$, $(0.164, 1.487)$, and $(1.924, -0.351)$.

75. $2x^3 - x^2 = 5y \Rightarrow y = \dfrac{2x^3 - x^2}{5}$ and $2^{-x} - y = 0 \Rightarrow y = 2^{-x}$. Graph $Y_1 = (2X\char`\^3 - X\char`\^2)/5$

and $Y_2 = 2\char`\^(-X)$. See Figure 75. There is one point of intersection. The coordinates of this point are near

$(1.220, 0.429)$.

[−4, 4, 1] by [−4, 4, 1]	[−5, 5, 1] by [−3, 3, 1]	[−3, 3, 1] by [−5, 5, 1]	[0, 800,000, 100,000] by [0, 800,000, 100,000]
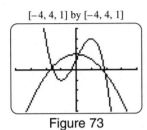	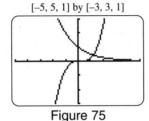	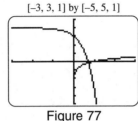	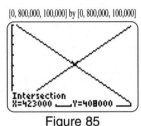
Figure 73	Figure 75	Figure 77	Figure 85

77. $e^{2x} + y = 4 \Rightarrow y = 4 - e^{2x}$ and $\ln x - 2y = 0 \Rightarrow y = \dfrac{\ln x}{2}$. Graph $Y_1 = 4 - e\char`\^(2X)$

and $Y_2 = \ln(X)/2$. See Figure 77. There is one point of intersection. The coordinates of this point are near

$(0.714, -0.169)$.

Applications

79. $S(w, h) = 0.007184(w^{0.425})(h^{0.725}) \Rightarrow S(86, 185) = 0.007184(86^{0.425})(185^{0.725}) \approx 2.1 \text{ m}^2$

81. $w = 132 \text{ lb} \approx \dfrac{132}{2.2} \text{ kg} = 60 \text{ kg}; \quad h = 62 \text{ inches} \approx 62 \cdot 2.54 \text{ cm} = 157.48 \text{ cm}$

   $S(w, h) = 0.007184(w^{0.425})(h^{0.725}) \Rightarrow S(60, 157.48) = 0.007184(60^{0.425})(157.48^{0.725}) \approx 1.6 \text{ m}^2$

83. $f(1960, \text{Male}) = 66.6$. A male born in 1960 had a life expectancy at birth of 66.6 years.

85. (a) Let x represent the number of robberies in 2000 and let y represent the number of robberies in 2001.

   The required system of equations is $x + y = 831,000$ and $x - y = 15,000$.

   (b) Adding the two equations results in $2x = 846,000 \Rightarrow x = 423,000$. From the first equation,

   $423,000 + y = 831,000 \Rightarrow y = 408,000$. There were 423,000 robberies in 2000 and 408,000 in 2001.

   (c) Solve each equation for y and graph $Y_1 = 831,000 - x$ and $Y_2 = x - 15,000$ as shown in Figure 85.

   The solution is the intersection point $(423,000, 408,000)$.

87. Let $x =$ hours spent on the internet in 2001. Then $0.13x = 1.3 \Rightarrow x = 10$. Therefore 10 hours were spent on

   the internet in 2001 and $10 + 1.3$ or 11.3 hours in 2002.

89. We must solve the system of nonlinear equations: $\pi r^2 h = 50$ and $2\pi rh = 65$.

Solving each equation for h results in the following: $\pi r^2 h = 50 \Rightarrow h = \dfrac{50}{\pi r^2}$ and $2\pi rh = 65 \Rightarrow h = \dfrac{65}{2\pi r}$.

Graph $Y_1 = 50/(\pi X^2)$ and $Y_2 = 65/(2\pi X)$. Their graphs intersect near $(1.538, 6.724)$. See Figure 89.

A cylinder with approximate measurements of $r \approx 1.538$ inches and $h \approx 6.724$ inches has a volume of 50 cubic inches and a lateral surface area of 65 square inches.

91. Let $H = 180$ in each equation and solve for y.

$$180 = 0.491x + 0.468y + 11.2 \Rightarrow y = \frac{1}{0.468}(180 - 0.491x - 11.2)$$

$$180 = -0.981x + 1.872y + 26.4 \Rightarrow y = \frac{1}{1.872}(180 + 0.981x - 26.4)$$

Since we are solving this system graphically, it is not necessary to simplify them. The graph of both equations and their point of intersection near $(177.1, 174.9)$ is shown in Figure 91. This means that at exhaustion, if an athlete achieves a maximum heart rate of 180 beats per minute, then 5 seconds after stopping his or her heart rate would be approximately 177.1 and after 10 seconds it would be approximately 174.9.

[0, 4, 1] by [0, 20, 1]   [0, 300, 50] by [0, 300, 50]   [0, 200, 50] by [0, 200, 50]

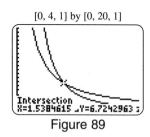
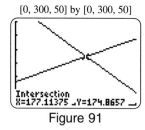
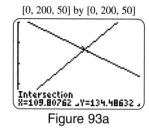
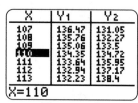

Figure 89   Figure 91   Figure 93a   Figure 93b

93. $W_1 + \sqrt{2}W_2 = 300 \Rightarrow W_2 = \dfrac{300 - W_1}{\sqrt{2}}$ and $\sqrt{3}W_1 - \sqrt{2}W_2 = 0 \Rightarrow W_2 = \dfrac{\sqrt{3}W_1}{\sqrt{2}}$

Graph $Y_1 = (300 - X)/\sqrt{(2)}$ and $Y_2 = \sqrt{(3)}X/\sqrt{(2)}$. Their graphs intersect near the point $(109.81, 134.49)$ as shown in Figure 93a. The forces on the rafters are approximately 110 and 134 pounds. To find the solution numerically, table Y_1 and Y_2 starting at 107 and incrementing by 1. We find that $Y_1 = Y_2 \approx 134$ when $x \approx 110$. See Figure 93b.

We can find the solution symbolically by using the substitution method. Since $W_1 + \sqrt{2}W_2 = 300 \Rightarrow$

$W_1 = 300 - \sqrt{2}W_2$, we will substitute into the other equation. $\sqrt{3}(300 - \sqrt{2}W_2) - \sqrt{2}W_2 = 0 \Rightarrow$

$$300\sqrt{3} - \sqrt{6}W_2 - \sqrt{2}W_2 = 0 \Rightarrow 300\sqrt{3} = (\sqrt{6} + \sqrt{2})W_2 \Rightarrow W_2 = \frac{300\sqrt{3}}{\sqrt{6} + \sqrt{2}} \text{ and}$$

$$W_1 = 300 - \sqrt{2}\left[\frac{300\sqrt{3}}{\sqrt{6} + \sqrt{2}}\right] \Rightarrow W_1 = 300 - \frac{300\sqrt{3}}{1 + \sqrt{3}} \Rightarrow W_1 = \frac{300}{1 + \sqrt{3}}$$

$$W_1 = \frac{300}{1 + \sqrt{3}} \approx 109.8 \text{ lbs}, \quad W_2 = \frac{300\sqrt{3}}{\sqrt{6} + \sqrt{2}} \approx 134.5 \text{ lbs}$$

95. Let x = the length of each side of the base and let y = the height of the box. Since the volume = 576 in^3,

$x^2y = 576$, and so $y = \dfrac{576}{x^2}$. Since the surface are is 336 in^2, $x^2 + 4xy = 336$. Substituting for y in this

equation yields $x^2 + 4x \cdot \dfrac{576}{x^2} = 336$. Simplifying we get: $x^3 - 336x + 2304 = 0$ By graphing the left side

of the equation, the x-intercepts are approximately 9.1 and 12. See Figure 95. When $x \approx 9.1$, the dimensions

are: 9.1 by 9.1 by $\dfrac{576}{(9.1)^2} \approx 6.96$ inches. When $x = 12$, the dimensions are: 12 by 12 by $\dfrac{576}{(12)^2} = 4$ inches.

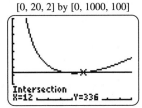

[0, 20, 2] by [0, 1000, 100]

Figure 95

97. Let x = amount invested at 5% and y = amount invested at 7%, then $x + y = 5000$ and $0.05x + 0.07y = 325$.

Multiply $x + y = 5000$ by 0.05 and subtract from $0.05x + 0.07y = 325$.

$$0.05x + 0.07y = 325$$
$$\underline{0.05x + 0.05y = 250}$$
$$0.02y = 75 \Rightarrow y = 3750$$

Substitute $y = 3750$ into $x + y = 5000 \Rightarrow x + 3750 = 5000 \Rightarrow x = 1250$. Therefore \$1250 was invested

at 5% and \$3750 was invested at 7%.

Joint Variation

99. Since $z = kx^2y^3$ and $z = 31.9$ when $x = 2$ and $y = 2.5$.

$$31.9 = k(2)^2(2.5)^3 \Rightarrow 31.9 = 62.5k \Rightarrow k = \dfrac{31.9}{62.5} \approx 0.51$$

101. $z = k\sqrt{x} \cdot \sqrt[3]{y} \Rightarrow 10.8 = k\sqrt{4} \cdot \sqrt[3]{8} \Rightarrow 10.8 = k \cdot 2 \cdot 2 \Rightarrow 10.8 = 4k \Rightarrow k = 2.7$

    Therefore: $z = 2.7\sqrt{x} \cdot \sqrt[3]{y}$ and now $z = 2.7\sqrt{16} \cdot \sqrt[3]{27} \Rightarrow z = 2.7(4)(3) \Rightarrow z = 32.4$.

103. Let d represent the diameter of the blades, v represent the wind velocity, and w represent the watts of power

generated by the windmill. Then $w = kd^2v^3$ and $w = 2405$ when $d = 8$ and $v = 10$.

$2405 = k(8)^2(10)^3 \Rightarrow 2405 = 64,000k \Rightarrow k = \dfrac{481}{12,800}$. The variation equation becomes

$w = \dfrac{481}{12,800}d^2v^3$. Thus, when $d = 6$ and $v = 20$; $w = \dfrac{481}{12,800}(6)^2(20)^3 = 10,822.5$.

With six-foot blades and a 20 mile-per-hour wind, the windmill will produce about 10,823 watts.

105. From the example $V = 0.00132h^{1.12}d^{1.98}$. When $h = 105$ and $d = 38$ we have

$V = 0.00132(105)^{1.12}(38)^{1.98} \approx 325.295$. A tree which is 105 feet tall with a diameter of 38 inches contains

approximately 325.295 cubic feet of wood. To find the number of cords, divide this result by 128:

$\dfrac{325.295}{128} \approx 2.54$ cords.

107. From example 2, $S(w, h) = 0.007184(w^{0.425})(h^{0.725})$ where w is weight in kilograms and h is height in centimeters. Let $S = 1.77$, $w = 154$, and $h = 65$. Then, $1.77 = k(154^{0.425})(65^{0.725})$, which results in $k \approx 0.0101$. Thus, $S(w, h) = 0.0101(w^{0.425})(h^{0.725})$, where w is in pounds, h is in inches, and S is in square meters.

9.2: Systems of Equations and Inequalities in Two Variables

Consistent and Inconsistent Linear Systems

1. Since the lines intersect, the system is consistent with a unique solution at $(2, 2)$.

    $x + y = 4 \Rightarrow y = 4 - x$; We substitute $4 - x$ for y in the other equation.

    $2x - (4 - x) = 2 \Rightarrow 3x = 6 \Rightarrow x = 2$, then $2 + y = 4 \Rightarrow y = 2$. The solution is $(2, 2)$.

3. Since the lines are parallel, the system is inconsistent with no solutions.

The Elimination Method

5. Add the two equations together to eliminate the y-variable.

$$x + y = 20$$
$$\underline{x - y = 8}$$
$$2x = 28 \Rightarrow x = 14$$

    Since $x + y = 20$, it follows that $y = 6$. The unique solution is $(14, 6)$. The system is consistent and independent.

    Graphical and numerical support are shown in Figure 5a & 5b, where $Y_1 = 20 - X$ and $Y_2 = X - 8$.

[0, 24, 4] by [0, 16, 4]

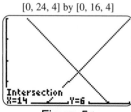
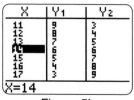

Figure 5a   Figure 5b

7. Subtract the two equations to eliminate the x-variable.

$$x + 3y = 10$$
$$\underline{x - 2y = -5}$$
$$5y = 15 \Rightarrow y = 3$$

    Since $x + 3y = 10$, it follows that $x = 1$. The solution is $(1, 3)$. The system is consistent and independent.

9. Multiply the second equation by 20 and subtract to eliminate both variables.

$$x + y = 500$$
$$\underline{x + y = 500}$$
$$ 0 = 0 \Rightarrow \text{infinite number of solutions}$$

    The solution is $\{(x, y) \mid x + y = 500\}$. The system is consistent but dependent.

11. Multiply the first equation by 3 and add to eliminate the y-variable.

$$15x + 3y = -15$$
$$\underline{7x - 3y = -29}$$
$$22x = -44 \Rightarrow x = -2$$

Since $5x + y = -5$, it follows that $y = 5$. The solution is $(-2, 5)$. The system is consistent and independent.

13. Multiply the second equation by 2 and add. This eliminates both variables.

$$2x + 4y = 7$$
$$\underline{-2x - 4y = 10}$$
$$0 = 17 \Rightarrow \text{no solution}$$

Since $0 \neq 17$, the system is inconsistent.

15. Multiply the second equation by 2 and subtract to eliminate the x-variable.

$$2x + 3y = 2$$
$$\underline{2x - 4y = -10}$$
$$7y = 12 \Rightarrow y = \frac{12}{7}$$

Since $2x + 3y = 2$, it follows that $x = \dfrac{2 - 3y}{2} \Rightarrow x = -\dfrac{11}{7}$. The solution is $\left(-\dfrac{11}{7}, \dfrac{12}{7}\right)$. The system is consistent and independent.

17. Multiply the second equation by $-\dfrac{1}{2}$ and add to eliminate the x-variable.

$$\frac{1}{2}x - y = 5$$
$$\underline{-\frac{1}{2}x + \frac{1}{4}y = -2}$$
$$-\frac{3}{4}y = 3 \Rightarrow y = -4$$

Then, $x - \dfrac{1}{2}y = 4 \Rightarrow x = 4 + \dfrac{1}{2}y \Rightarrow x = 2$. The solution is $(2, -4)$.

19. Multiply the first equation by -3 and add to eliminate the x-variable.

$$-6x + 9y = -5$$
$$\underline{6x - 2y = \frac{8}{3}}$$
$$7y = -\frac{7}{3} \Rightarrow y = -\frac{1}{3}$$

Then, $2x - 3y = \dfrac{5}{3} \Rightarrow x = \dfrac{\frac{5}{3} + 3y}{2} \Rightarrow x = \dfrac{\frac{5}{3} + 3(-\frac{1}{3})}{2} \Rightarrow x = \dfrac{\frac{2}{3}}{2} = \dfrac{1}{3}$. The solution is $\left(\dfrac{1}{3}, -\dfrac{1}{3}\right)$.

21. Multiply the first equation by 3 and add to eliminate both variables.

$$21x - 9y = -51$$
$$\underline{-21x + 9y = 51}$$
$$0 = 0 \Rightarrow \text{infinite number of solutions}$$

There are infinitely many solutions of the form $\{(x, y) \mid 7x - 3y = -17\}$.

23. Multiply the first equation by 3 and add to eliminate both variables.

$$2x + 4y = 1$$
$$-2x - 4y = 5$$
$$\overline{}$$
$$0 = 5 \Rightarrow \text{ no solutions}$$

25. Multiply the second equation by 5 and add to eliminate the x-variable.

$$-5x + 2y = 15$$
$$5x - 30y = -85$$
$$\overline{}$$
$$-28y = -70 \Rightarrow y = \frac{5}{2}$$

Then, $x - 6y = -17 \Rightarrow x = -17 + 6y \Rightarrow x = -17 + 6\left(\dfrac{5}{2}\right) \Rightarrow x = -2$. The solution is $\left(-2, \dfrac{5}{2}\right)$.

27. Clear decimals: $0.2x + 0.3y = 8 \Rightarrow 2x + 3y = 80$ and $-0.4x + 0.2y = 0 \Rightarrow -4x + 2y = 0$.

Multiply the first equation by 2 and add to eliminate the x-variable.

$$4x + 6y = 160$$
$$-4x + 2y = 0$$
$$\overline{}$$
$$8y = 160 \Rightarrow y = 20$$

Then, $-4x + 2y = 0 \Rightarrow 4x = 2y \Rightarrow x = \dfrac{1}{2}y \Rightarrow x = \dfrac{1}{2}(20) = 10$. The solution is $(10, 20)$.

29. (a) Let x be the width and y the height. The following linear system describes this situation.

$x - y = 2 \Rightarrow y = x - 2$ and $2x + 2y = 38 \Rightarrow x + y = 19 \Rightarrow y = 19 - x$

Graph $Y_1 = X - 2$ and $Y_2 = 19 - X$. Their graphs intersect at $(10.5, 8.5)$ as shown in Figure 29.

The screen is 10.5 inches wide and 8.5 inches high.

(b) $x - y = 2$ and $2x + 2y = 38$. Multiply the second equation by $\dfrac{1}{2}$ and add to eliminate the y-variable.

$$x - y = 2$$
$$x + y = 19$$
$$\overline{}$$
$$2x = 21 \Rightarrow x = 10.5$$

Since, $x - y = 2, y = 8.5$. The solution is $(10.5, 8.5)$.

[0, 20, 5] by [0, 20, 5]   [0, 100, 10] by [0, 100, 10]

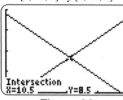
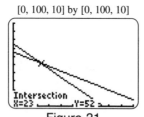

Figure 29   Figure 31

31. (a) Let x be the number of child tickets and y the adult tickets. The following linear system describes this situation.

$x + y = 75 \Rightarrow y = 75 - x$ and $4x + 7y = 456 \Rightarrow y = \dfrac{456 - 4x}{7}$

Graph $Y_1 = 75 - X$ and $Y_2 = (456 - 4X)/7$. Their graphs intersect at $(23, 52)$ as shown in Figure 31.

There were 23 child tickets and 52 adult tickets sold.

(b) $x + y = 75$ and $4x + 7y = 456$. Multiply the first equation by 7 and subtract to eliminate the y-variable.

$$7x + 7y = 525$$
$$4x + 7y = 456$$
$$\overline{}$$
$$3x = 69 \Rightarrow x = 23$$

Since, $x + y = 75, y = 52$. The solution is $(23, 52)$.

33. Add the two equations:

$$x^2 + y = 12$$
$$\underline{x^2 - y = 6}$$
$$2x^2 \quad\ = 18 \Rightarrow x^2 = 9 \Rightarrow x = \pm 3.$$

If $x = 3$, then $3^2 + y = 12 \Rightarrow y = 3$, and if $x = -3$ then $(-3)^2 + y = 12 \Rightarrow y = 3$. Therefore the solutions are: $(3, 3)$ and $(-3, 3)$.

35. Subtract the two equations:

$$x^2 + y^2 = 25$$
$$\underline{x^2 + 7y = 37}$$
$$y^2 - 7y = -12 \Rightarrow y^2 - 7y + 12 = 0 \Rightarrow (y - 3)(y - 4) = 0 \Rightarrow y = 3, 4..$$

If $y = 3$, then $x^2 + 7(3) = 37 \Rightarrow x^2 = 16 \Rightarrow x = \pm 4$, and if $y = 4$ then $x^2 + 7(4) = 37 \Rightarrow x^2 = 9 \Rightarrow$ $x = \pm 3$. Therefore the solutions are: $(4, 3)$, $(-4, 3)$, $(3, 4)$ and $(-3, 4)$.

37. Subtract the two equations:

$$x^2 + y^2 = 4$$
$$\underline{2x^2 + y^2 = 8}$$
$$-x^2 \quad\ = -4 \Rightarrow x^2 = 4 \Rightarrow x = \pm 2.$$

If $x = -2$, then $(-2)^2 + y^2 = 4 \Rightarrow 4 + y^2 = 4 \Rightarrow y^2 = 0 \Rightarrow y = 0$, and if $x = 2$ then $(2)^2 + y^2 = 4 \Rightarrow$ $y^2 = 0 \Rightarrow y = 0$. Therefore the solutions are: $(-2, 0)$ and $(2, 0)$.

39. Multiply the first equation by 2 and add the two equations:

$$2\sqrt{x} - 2y = 0$$
$$\underline{x + 2y = 3}$$
$$2\sqrt{x} + \ x = 3 \Rightarrow 2\sqrt{x} = 3 - x \Rightarrow 4x = (3 - x)^2 \Rightarrow 4x = 9 - 6x + x^2 \Rightarrow x^2 - 10x + 9 = 0 \Rightarrow$$
$(x - 9)(x - 1) = 0 \Rightarrow x = 1, 9$. If $x = 1$, then $1 + 2y = 3 \Rightarrow 2y = 2 \Rightarrow y = 1$, and if $x = 9$ then $9 + 2y = 3 \Rightarrow 2y = -6 \Rightarrow y = -3$. Therefore the solutions are: $(1, 1)$ and $(9, -3)$.

Checking: $(9, -3)$ we get $2\sqrt{9} - 2(-3) = 0 \Rightarrow 6 - (-6) = 0 \Rightarrow 12 = 0 \Rightarrow (9, -3)$ is not a solution.

Therefore the solution is: $(1, 1)$.

Inequalities

41. Graph the boundary line $y = x$. Choose $(-1, -1)$ as a test point. Since $-1 \le -1$, the region containing $(-1, -1)$ is part of the solution set. The inequality $x \ge y$ includes this region and the line $y = x$. See Figure 41.

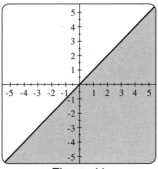

Figure 41

43. Graph the boundary line $x = 1$. Choose $(0, 0)$ as a test point. Since $0 < 1$, the region containing $(0, 0)$ is the solution set. The boundary line $x = 1$ is not part of the solution set. See Figure 43.

45. Graph the boundary line $x + y = 2 \Rightarrow y = 2 - x$. Choose $(0, 0)$ as a test point. Since $0 + 0 \le 2$, the region containing $(0, 0)$ is part of the solution set. The boundary line $y = 2 - x$ is also part of the solution set. See Figure 45.

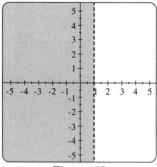

Figure 43

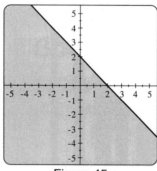

Figure 45

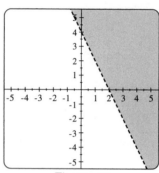

Figure 47

47. Graph the boundary line $2x + y = 4 \Rightarrow y = 4 - 2x$. Choose $(3, 0)$ as a test point. Since $0 > 4 - 2(3)$, the region containing $(3, 0)$ is the solution set. The boundary line $y = 4 - 2x$ is not part of the solution set. See Figure 47.

49. Graph the circle determined by $x^2 + y^2 = 4$. Choose $(0, 0)$ as a test point. Since $0 > 4$, the region inside the circle is not part of the solution $\Rightarrow$ shade the region outside the circle. Note, the circle is not part of the solution. See Figure 49.

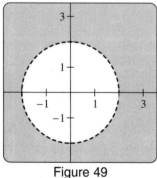

Figure 49

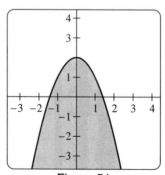

Figure 51

51. Graph the boundary parabola determined by $x^2 + y = 2 \Rightarrow y = -x^2 + 2$. Choose $(0, 0)$ as a test point. Since $0 \le 2$, the region inside the parabola is part of the solution $\Rightarrow$ shade the region inside of the parabola. Note, the parabola is part of the solution. See Figure 51.

53. $x + y \ge 2 \Rightarrow y \ge 2 - x$ is above the line $y = 2 - x$. It includes the line.

$x - y \le 1 \Rightarrow y \ge x - 1$ is above the line $y = x - 1$. It includes the line.

The solution is above two lines, which matches Figure c. One solution is $(2, 3)$. *Answers may vary.*

55. $\frac{1}{2}x^3 - y > 0 \Rightarrow y < \frac{1}{2}x^3$ is below the curve $y = \frac{1}{2}x^3$. It does not include the curve.

    $2x - y \le 1 \Rightarrow y \ge 2x - 1$ is above the line $y = 2x - 1$. It includes the line.

    The solution is below a dotted curve and above a solid line, which matches Figure d. One solution is $(-1, -1)$.

    Answers may vary.

57. The solution region is above the parabola $y = x^2$ and below the line $y = 6 - x$. It includes the boundary. See Figure 57. One solution is $(0, 2)$. *Answers may vary.*

59. The solution region lies between the parallel lines $y = -\frac{1}{2}x - 1$ and $y = -\frac{1}{2}x + \frac{5}{2}$. It does not include the boundary. See Figure 59. One solution is $(0, 0)$. *Answers may vary.*

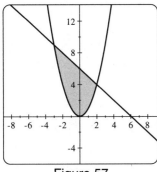

Figure 57

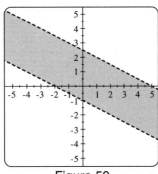

Figure 59

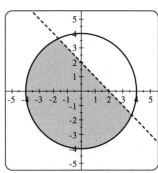

Figure 61

61. The solution region lies inside the circle centered at the origin with radius 4 and below the line $y = -x + 2$. It does not include the boundary determined by the line. See Figure 61. One solution is $(-1, 1)$.

    Answers may vary.

63. $x + 2y \le 4 \Rightarrow y \le -\frac{1}{2}x + 2$ and $2x - y \ge 6 \Rightarrow y \le 2x - 6$. Graph the boundary lines $y = -\frac{1}{2}x + 2$ and

    $y = 2x - 6$. The region satisfying the system is below the line $x + 2y = 4$ and below the line $2x - y = 6$.

    Because equality is included, the boundaries are part of the region. See Figure 63.

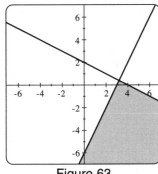

Figure 63

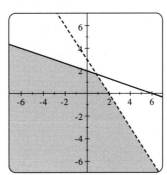

Figure 65

65. $3x + 2y < 6 \Rightarrow y < -\frac{3}{2}x + 3$ and $x + 3y \le 6 \Rightarrow y \le -\frac{1}{3}x + 2$. Graph the boundary lines $y = -\frac{3}{2}x + 3$

    and $y = -\frac{1}{3}x + 2$. The region satisfying the system is below the line $3x + 2y = 6$, not including the boundary

    and below the line $x + 3y = 6$, including the boundary. See Figure 65.

67. $x - 2y \geq 0 \Rightarrow y \leq \dfrac{1}{2}x$ and $x - 3y \leq 3 \Rightarrow y \geq \dfrac{1}{3}x - 1$. Graph the boundary lines $y = \dfrac{1}{2}x$ and $y = \dfrac{1}{3}x - 1$.

The region satisfying the system is below the line $y = \dfrac{1}{2}x$ and above the line $y = \dfrac{1}{3}x - 1$. Because equality is

included, the boundaries are part of the region. See Figure 67.

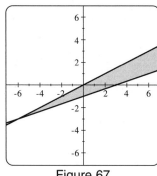

Figure 67

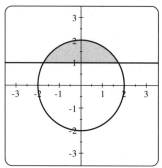

Figure 69

69. $x^2 + y^2 \leq 4$ and $y \geq 1$. Graph the boundary circle $x^2 + y^2 = 4$ and the boundary line $y = 1$. The region

satisfying the system is inside the circle $x^2 + y^2 = 4$ and above the line $y = 1$. Because equality is included,

the boundaries are part of the region. See Figure 69.

Applications

71. (a) To solve this problem start by letting x represent the amount of the 8% loan and y the 10% loan. Since the

total of both loans is $\$3000$, the equation $x + y = 3000$ must be satisfied. The annual interest rate for the

8% loan is given by $0.08x$, while the annual interest rate for the 10% loan is expressed by $0.10y$. the total

interest for both loans is their sum $0.08x + 0.10y = 264$.

(b) Thus, to determine a solution, the following linear system of equations could be solved.

$x + y = 3000$ and $0.08x + 0.10y = 264$. Start by solving each for y. $x + y = 3000 \Rightarrow y = 3000 - x$

and $0.08x + 0.10y = 264 \Rightarrow 0.10y = 264 - 0.08x \Rightarrow y = \dfrac{264 - 0.08x}{0.10} \Rightarrow y = 2640 - 0.8x$. The two

equations can be solved using the intersection-of-graphs method. Let $Y_1 = 3000 - X$ and

$Y_2 = 2640 - 0.8X$. Since the system of equations is linear, each graph is a line. The lines are not parallel

and intersect at the point $(1800, 1200)$, as shown in Figure 71. Thus, the 8% loan is for $\$1800$ and the 10%

loan is for $\$1200$.

[0, 3000, 1000] by [0, 3000, 1000]   [0, 3000, 1000] by [0, 3000, 1000]

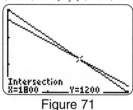

Figure 71   Figure 73

73. With these conditions the system of equations becomes $x + y = 3000$ and $0.10x + 0.10y = 264$.

    Solving each equation for y provides the following results, $y = 3000 - x$ and $y = 2640 - x$.

    Graphs of $Y_1 = 3000 - X$ and $Y_2 = 2640 - X$ are shown in Figure 73. Notice that their graphs are parallel lines with slope –1 that do not intersect. There is no solution. This means that there is no way to have two loans totaling $3000, both with an interest rate of 10%, and only pay $264 in interest each year. the interest must be 10% of $3000 or $300. This system of equations is inconsistent - there is no solution. Graphs of inconsistent systems in two variables consist of parallel lines.

75. (a) Let x represent the number of television sets sold with stereo and y represent the number of television sets sold without stereo. Thus, the system we want to solve is $x + y = 32$ and $\dfrac{19}{10}x = y$.

    (b) Since, $\dfrac{19}{10}x = y \Rightarrow \dfrac{19}{10}x - y = 0,$ we add the two equations to eliminate the y-variable.

    $$x + y = 32$$
    $$\frac{19}{10}x - y = 0$$
    $$\frac{29}{10}x = 32 \Rightarrow x = \frac{320}{29} \approx 11.03$$

    $$x + y = 32 \Rightarrow \frac{320}{29} + y = 32 \Rightarrow y = \frac{628}{29} \approx 20.97$$

    In 1999, about 11 million television sets were sold with stereo sound and about 21 million sets were sold without stereo sound.

77. Let x represent the number of cigarettes sold in 1990 and y the number of cigarettes sold in 2000, then $x + y = 10.9$ and $y = x + 0.1$. Substituting the second equation into the first we get:

    $x + (x + 0.1) = 10.9 \Rightarrow 2x + 0.1 = 10.9 \Rightarrow 2x = 10.8 \Rightarrow x = 5.4,$ then $y = 5.5$. Thus there were 5.4 trillion cigarettes sold in 1990 and 5.5 trillion cigarettes sold in 2000.

79. Let x represent the air speed of the plane and y the wind speed. Traveling with the wind, the average ground speed of the plane is $\dfrac{1680}{3} = 560$ mph, while its ground speed against the wind was $\dfrac{1680}{3.5} = 480$ mph. Thus,

    $$x + y = 560$$
    $$x - y = 480$$
    $$2x = 1040 \Rightarrow x = 520$$

    Thus, $y = 560 - x = 40$. The air speed of the plane is 520 mph and the wind speed is 40 mph.

81. The total number of vehicles entering intersection A is $500 + 150 = 650$ vehicles per hour. The expression $x + y$ represents the number of vehicles leaving intersection A each hour. Therefore, we have $x + y = 650$. The total number of vehicles leaving intersection B is $50 + 400 = 450$. There are 100 vehicles entering intersection B from the south and y vehicles entering intersection B from the west. Thus, $y + 100 = 450$. We must solve the system;

    $$x + y = 650$$
    $$y + 100 = 450$$
    $$x - 100 = 200 \Rightarrow x = 300$$

    Thus, $y = 350$ and $x = 300$. At intersection A, a stoplight should allow for 300 vehicles per hour to travel south and 350 vehicles per hour to continue traveling east.

83. This region corresponds to weights that are less and heights that are greater than recommended. This individual has weight that is less than recommended for his or her height..

85. The upper left boundary is given by $25h - 7w = 800$ or $h = \dfrac{7w + 800}{25}$. The region is below and includes this line, which is described by $h \le \dfrac{7w + 800}{25}$ or $25h - 7w \le 800$. The lower right boundary of this region is given by $5h - w = 170$. The region is above and includes this line, which is described by $5h - w \ge 170$. Thus, the region can be described by the system of inequalities: $25h - 7w \le 800$, $5h - w \ge 170$.

Linear Programming

87. See Figure 87.

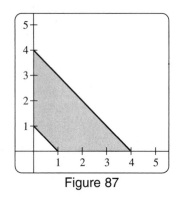

Figure 87   Figure 89

89. See Figure 89.

91. To find the maximum and minimum values of P in the region, we must evaluate P at each of the vertices. These values are shown in the table. See Figure 91. From the table the maximum is 65 and the minimum is 8.

93. To find the maximum and minimum values of C in the region, we must evaluate C at each of the vertices. These values are shown in the table. See Figure 93. From the table the maximum is 66 and the minimum is 3.

Vertex	$P = 3x + 5y$
(1, 1)	$3(1) + 5(1) = 8$
(6, 3)	$3(6) + 5(3) = 33$
(5, 10)	$3(5) + 5(10) = 65$
(2, 7)	$3(2) + 5(7) = 41$

Figure 91

Vertex	$C = 3x + 5y$
(1, 0)	$3(1) + 5(0) = 3$
(7, 6)	$3(7) + 5(6) = 51$
(7, 9)	$3(7) + 5(9) = 66$
(1, 10)	$3(1) + 5(10) = 53$

Figure 93

Vertex	$C = 10y$
(1, 0)	$10(0) = 0$
(7, 6)	$10(6) = 60$
(7, 9)	$10(9) = 90$
(1, 10)	$10(10) = 100$

Figure 95

95. To find the maximum and minimum values of C in the region, we must evaluate C at each of the vertices. These values are shown in the table. See Figure 95. From the table the maximum is 100 and the minimum is 0.

97. The line that goes through the points (0, 4) and (4, 0) has the equation $x + y = 4$. The shaded region is also bounded by the line $x = 0$ and the line $y = 0$. Thus, the shaded region is described by the system: $x + y \le 4$, $x \ge 0$, and $y \ge 0$.

99. The region of feasible solutions is shown in Figure 99a. The vertices of this region are (3, 0), (6, 0), (0, 4), and

(0, 3). To find the minimum value of C in the region, we must evaluate C at each of the vertices. These values

are shown in Figure 99b. From the table the minimum is 6 at the point (0, 3).

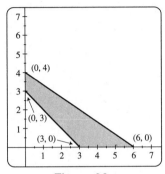

Figure 99a

Vertex	$C = 4x + 2y$
(3, 0)	$4(3) + 2(0) = 12$
(6, 0)	$4(6) + 2(0) = 24$
(0, 4)	$4(0) + 2(4) = 8$
(0, 3)	$4(0) + 2(3) = 6$

Figure 99b

101. The region of feasible solutions is shown in Figure 101a. The vertices of this region are (0, 4), (0, 8), (4, 0),

and (8, 0). To find the maximum and minimum values of $z = 7x + 6y$ in the region, we must evaluate z at

each of the vertices. These values are shown in Figure 101b. From the table the maximum is 56 and the

minimum is 24.

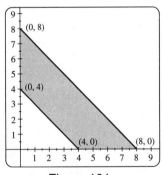

Figure 101a

Vertex	$z = 7x + 6y$
(0, 4)	$7(0) + 6(4) = 24$
(0, 8)	$7(0) + 6(8) = 48$
(4, 0)	$7(4) + 6(0) = 28$
(8, 0)	$7(8) + 6(0) = 56$

Figure 101b

103. The new objective equation is $P = 20x + 15y$. To find the maximum profit we must evaluate P at each of the

vertices. These values are shown in Figure 103. From the table the maximum is 950 at the vertex (25, 30).

The maximum profit will be $950 when 25 radios and 30 CD players are manufactured.

Vertex	$P = 20x + 15y$
(5, 5)	$20(5) + 15(5) = 175$
(25, 25)	$20(25) + 15(25) = 875$
(25, 30)	$20(25) + 15(30) = 950$
(5, 30)	$20(5) + 15(30) = 550$

Figure 103

105. Make a table to list the information given. See Figure 105a. Using the table, we can write the linear programming problem as follows;

Cost: $C = 80x + 50y$, Protein: $15x + 20y \geq 60$, Fat: $10x + 5y \geq 30$, $x \geq 0$, and $y \geq 0$.

Brand	Units	Protein	Fat	Cost
A	x	15	10	80¢
B	y	20	5	50¢
Minimum		60	30	

Figure 105a

The region of feasible solutions is shown in Figure 105b. The vertices of this region are (0, 6), (2.4, 1.2), and (4, 0). To find the minimum value of C in the region, we must evaluate C at each of the vertices. These values are shown in Figure 105c. The minimum cost occurs when 2.4 units of Brand A and 1.2 units of Brand B are mixed, to give a cost of $2.52 per serving.

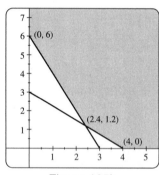

Figure 105b

Vertex	$C = 80x + 50y$
(0, 6)	$80(0) + 50(6) = 300$
(2.4, 1.2)	$80(2.4) + 50(1.2) = 252$
(4, 0)	$80(4) + 50(0) = 320$

Figure 105c

107. Let x and y represent the number of hamsters and mice respectively. Since the total number of animals cannot exceed 50, $x + y \leq 50$. Because no more than 20 hamsters can be raised, $x \leq 20$. Here the revenue function is $R = 15x + 10y$. From the graph of the region of feasible solutions (not shown), the vertices are (0, 50), (20, 30) and (20, 0). To find (20, 30) solve the equations $x + y = 50$ and $x = 20$. The maximum value of R occurs at one of the vertices. For (0, 50), $R = 15(0) + 10(50) = 500$.

For (20, 30), $R = 15(20) + 10(30) = 600$. For (20, 0), $R = 15(20) + 10(0) = 300$.

The maximum revenue is $600.

109. The number of hours on machine A needed to manufacture x units of part X is $4x$ while the number of hours on machine A needed to manufacture y units of part Y is $1y$. Since machine A is only available for 40 hours each week we have the constraint $4x + y \leq 40$. Similarly, the number of hours on machine B needed to manufacture x units of part X is $2x$ while the number of hours on machine B to make y units of part Y is $3y$. Since machine B is only available for 30 hours each week we have the constraint $2x + 3y \leq 30$. It should be noted that the number of parts of each type cannot be negative. This gives the constraint $x \geq 0$ and $y \geq 0$. The profit earned on x units of part X is $500x$ while the profit earned or y units of part Y is $600y$. Thus, the total weekly profit is $P = 500x + 600y$. This is our objective function. Graph the constraints and shade the region. The region of feasible solutions is shown in Figure 109a. The vertices of this region are $(0, 0)$, $(10, 0)$, $(9, 4)$, and $(0, 10)$. To find the maximum value of P in the region, we must evaluate P at each of the vertices. These values are shown in Figure 109b. From the table the maximum is 6900 at the point $(9, 4)$. The maximum profit is \$6900 when there are 9 parts of type X and 4 parts of type Y manufactured.

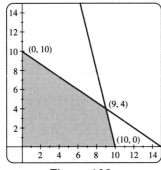

Figure 109a

Vertex	$P = 500x + 600y$
$(0, 0)$	$500(0) + 600(0) = 0$
$(10, 0)$	$500(10) + 600(0) = 5000$
$(9, 4)$	$500(9) + 600(4) = 6900$
$(0, 10)$	$500(0) + 600(10) = 6000$

Figure 109b

9.3 Systems of Linear Equations in Three Variables

1. No, systems of linear equations can have zero, one, or infinitely many solutions.

3. 2, the same as the number of variables.

5. Testing $(0, 2, -2)$: $0 + 2 - (-2) = 4$, true; $-0 + 2 + (-2) = 2$, false; $0 + 2 + (-2) = 0$, true; since the second equation is false, $(0, 2, -2)$ is not a solution for the system.

   Testing $(-1, 3, -2)$: $-1 + 3 - (-2) = 4$, true; $-(-1) + 3 + (-2) = 2$, true; $-1 + 3 + (-2) = 0$, true; since all equation are true, $(-1, 3, -2)$ is a solution for the system.

7. Testing $\left(-\dfrac{5}{11}, \dfrac{20}{11}, -2\right)$: $-\dfrac{5}{11} + 3\left(\dfrac{20}{11}\right) - 2(-2) = 9 \Rightarrow -\dfrac{5}{11} + \dfrac{60}{11} + \dfrac{44}{11} = 9 \Rightarrow \dfrac{99}{11} = 9$, true;

$-3\left(-\dfrac{5}{11}\right) + 2\left(\dfrac{20}{11}\right) + 4(-2) = -3 \Rightarrow \dfrac{15}{11} + \dfrac{40}{11} - \dfrac{88}{11} = -3 \Rightarrow -\dfrac{33}{11} = -3$, true;

$-2\left(-\dfrac{5}{11}\right) + 5\left(\dfrac{20}{11}\right) + 2(-2) = 6 \Rightarrow \dfrac{10}{11} + \dfrac{100}{11} - \dfrac{44}{11} = 6 \Rightarrow \dfrac{66}{11} = 6$, true; since all of the equations are

true, $\left(-\dfrac{5}{11}, \dfrac{20}{11}, -2\right)$ is a solution for the system.

Testing $(1, 2, -1)$: $1 + 3(2) - 2(-1) = 9 \Rightarrow 1 + 6 + 2 = 9 \Rightarrow 9 = 9$, true;

$-3(1) + 2(2) + 4(-1) = -3 \Rightarrow -3 + 4 + (-4) = -3 \Rightarrow -3 = -3$, true;

$-2(1) + 5(2) + 2(-1) = 6 \Rightarrow (-2) + 10 + (-2) = 6 \Rightarrow 6 = 6$, true; since all equations are true, $(1, 2, -1)$

is a solution for the system.   Therefore they are both solutions.

9. Add The first two equations:

$$
\begin{array}{r}
x + y + z = 6 \\
-x + 2y + z = 6 \\
\hline
3y + 2z = 12.
\end{array}
$$

Subtract 2 times the third equation from this equation:

$$
\begin{array}{r}
3y + 2z = 12 \\
2y + 2z = 10 \\
\hline
y \quad\quad = 2.
\end{array}
$$

Substitute $y = 2$ into the equation: $3y + 2z = 12 \Rightarrow 3(2) + 2z = 12 \Rightarrow 6 + 2z = 12 \Rightarrow 2z = 6 \Rightarrow z = 3$.

Finally, substitute $y = 2$ and $z = 3$ into one of the original equations: $x + 2 + 3 = 6 \Rightarrow x = 1$.

The solution is $(1, 2, 3)$.

11. Multiply the first equation by 2 and subtract the second equation:

$$
\begin{array}{r}
2x + 4y + 6z = 8 \\
2x + y + 3z = 5 \\
\hline
3y + 3z = 3.
\end{array}
$$

Subtract the third equation from the first:

$$
\begin{array}{r}
x + 2y + 3z = 4 \\
x - y + z = 2 \\
\hline
3y + 2z = 2.
\end{array}
$$

Now subtract these two equations:

$$
\begin{array}{r}
3y + 3z = 3 \\
3y + 2z = 2 \\
\hline
z = 1.
\end{array}
$$

Substitute $z = 1$ into $3y + 2z = 2$: $3y + 2(1) = 2 \Rightarrow 3y = 0 \Rightarrow y = 0$.

Finally substitute $y = 0$ and $z = 1$ into the original equation $x - y + z = 2$: $x - 0 + 1 = 2 \Rightarrow x = 1$.

The solution is: $(1, 0, 1)$.

13. Add the second and third equations:

$$4x + 2y + z = 1$$
$$2x - 2y - z = 2$$
$$\overline{6x \qquad\quad\; = 3} \Rightarrow x = \frac{1}{2}.$$

Subtract the first two equations:

$$3x + y + z = 0$$
$$4x + 2y + z = 1$$
$$\overline{-x - y \qquad\; = -1.}$$

Substitute $x = \frac{1}{2}$ into $-x - y = -1$: $-\frac{1}{2} - y = -1 \Rightarrow -y = -\frac{1}{2} \Rightarrow y = \frac{1}{2}$.

Substitute $x = \frac{1}{2}$ and $y = \frac{1}{2}$ into $3x + y + z = 0$: $3\left(\frac{1}{2}\right) + \frac{1}{2} + z = 0 \Rightarrow \frac{3}{2} + \frac{1}{2} + z = 0 \Rightarrow 2 + z = 0 \Rightarrow$

$z = -2$.

The solution is: $\left(\frac{1}{2}, \frac{1}{2}, -2\right)$.

15. Subtract the third equation from the first:

$$x + 3y + z = 6$$
$$x - y - z = 0$$
$$\overline{\qquad 4y + 2z = 6.}$$

Multiply the third equation by 3 and subtract it from the second equation:

$$3x + y - z = 6$$
$$3x - 3y - 3z = 0$$
$$\overline{\qquad 4y + 2z = 6.}$$

Subtracting these two equations we get:

$$4y + 2z = 6$$
$$4y + 2z = 6$$
$$\overline{\qquad 0 = 0.}$$

Therefore infinitely many solutions and $4y + 2z = 6 \Rightarrow 4y = -2z + 6 \Rightarrow y = \dfrac{-z + 3}{2}$.

Adding the last two original equations we get:

$$3x + y - z = 6$$
$$x - y - z = 0$$
$$\overline{4x \qquad - 2z = 6} \Rightarrow 4x = 2z + 6 \Rightarrow x = \dfrac{z + 3}{2}.$$

We have infinitely many solutions: $\left(\dfrac{z + 3}{2}, \dfrac{-z + 3}{2}, z\right)$.

17. Add the first two equations:

$$
\begin{array}{r}
x - 4y + 2z = -2 \\
\underline{x + 2y - 2z = -3} \\
2x - 2y = -5.
\end{array}
$$

Multiply the third equation by 2 and subtract from $2x - 2y = -5$:

$$
\begin{array}{r}
2x - 2y = -5 \\
\underline{2x - 2y = 8} \\
0 = -13.
\end{array}
$$

Therefore we have no solutions.

19. Add the last two equations:

$$
\begin{array}{r}
2a + b - c = -11 \\
\underline{2a - 2b + c = 3} \\
4a - b = -8.
\end{array}
$$

Multiply the second equation by 2 and add to the first equation:

$$
\begin{array}{r}
4a - b + 2c = 0 \\
\underline{4a + 2b - 2c = -22} \\
8a + b = -22.
\end{array}
$$

Now add the equation $4a - b = -8$ to $8a + b = -8$.:

$$
\begin{array}{r}
4a - b = -8 \\
\underline{8a + b = -22} \\
12a = -30 \Rightarrow a = \dfrac{-30}{12} \Rightarrow a = \dfrac{-5}{2}.
\end{array}
$$

Substitute $a = \dfrac{-5}{2}$: $8\left(\dfrac{-5}{2}\right) + b = -22 \Rightarrow -20 + b = -22 \Rightarrow b = -2.$

Substitute $a = \dfrac{-5}{2}$ and $b = -2$ into $4\left(\dfrac{-5}{2}\right) - (-2) + 2c = 0 \Rightarrow -10 + 2 + 2c = 0 \Rightarrow 2c = 8 \Rightarrow c = 4.$

The solution is: $\left(\dfrac{-5}{2}, -2, 4\right)$.

21. Subtract the first and third equations:

$$
\begin{array}{r}
a + b + c = 0 \\
\underline{a + 3b + 3c = 5} \\
-2b - 2c = -5.
\end{array}
$$

Subtract the second and third equations:

$$
\begin{array}{r}
a - b - c = 3 \\
\underline{a + 3b + 3c = 5} \\
-4b - 4c = -2.
\end{array}
$$

Multiply $-2b - 2c = -5$ by 2 and subtract $-4b - 4c = -2$:

$$
\begin{array}{r}
-4b - 4c = -10 \\
\underline{-4b - 4c = -2} \\
0 = -8.
\end{array}
$$

Therefore, we have no solution.

23. Add the first two equations:

$$3x + 2y + z = -1$$
$$\underline{3x + 4y - z = \;\; 1}$$
$$6x + 6y \quad\;\;\; = \;\; 0.$$

Add the last two equations:

$$3x + 4y - z = 1$$
$$\underline{\;x + 2y + z = 0}$$
$$4x + 6y \quad\;\;\; = 1.$$

Now, subtract $6x + 6y = 0$ and $4x + 6y = 1$:

$$6x + 6y = \;\; 0$$
$$\underline{4x + 6y = \;\; 1}$$
$$2x \quad\quad\;\; = -1 \Rightarrow x = \frac{-1}{2}.$$

Substitute $x = \dfrac{-1}{2}$ into $6x + 6y = 0$: $6\left(\dfrac{-1}{2}\right) + 6y = 0 \Rightarrow 6y = 3 \Rightarrow y = \dfrac{1}{2}.$

Substitute $x = \dfrac{-1}{2}$ and $y = \dfrac{1}{2}$ into $x + 2y + x = 0 \Rightarrow \dfrac{-1}{2} + 2\left(\dfrac{1}{2}\right) + z = 0 \Rightarrow z = \dfrac{-1}{2}.$

The solution is: $\left(\dfrac{-1}{2}, \dfrac{1}{2}, \dfrac{-1}{2}\right).$

25. Multiply the first equation by 2 and add the second equation:

$$-2x + 6y + 2z = \;\; 6$$
$$\underline{\;\;2x + 7y + 4z = 13}$$
$$13y + 6z = 19.$$

Multiply the second equation by 2 and subtract the third equation:

$$4x + 14y + 8z = 26$$
$$\underline{4x + \;\;\; y + 2z = \;\; 7}$$
$$13y + 6z = 19.$$

Subtracting the two new equations we get:

$$13y + 6z = 19$$
$$\underline{13y + 6z = 19}$$
$$0 = 0.$$

Therefore, we have infinitely many solutions.

$$13y + 6z = 19 \Rightarrow 13y = -6z + 19 \Rightarrow y = \frac{-6z + 19}{13}.$$

Multiply the third equation by 3 and subtract from the first equation:

$$-x + 3y + \;\; z = \;\;\; 3$$
$$\underline{12x + 3y + 6z = \;\; 21}$$
$$-13x \quad\quad - 5z = -18 \Rightarrow x = \frac{-5z + 18}{13}.$$

We have infinitely many solutions: $\left(\dfrac{-5z + 18}{13}, \dfrac{-6z + 19}{13}, z\right).$

27. Subtract the second and third equations:

$$y + 4z = -13$$
$$\underline{3x + y \quad\quad = \quad 13}$$
$$-3x \quad\quad + 4z = -26.$$

Multiply the first equation by 3 and subtract $-3x + 4z = -26$ from it:

$$-3x \quad + 6z = -27$$
$$\underline{-3x \quad + 4z = -26}$$
$$2z = \quad -1 \Rightarrow z = \frac{-1}{2}.$$

Substitute $z = \frac{-1}{2}$ into $y + 4z = -13$: $y + 4\left(\frac{-1}{2}\right) = -13 \Rightarrow y = -11$.

Substitute $z = \frac{-1}{2}$ into $-x + 2z = -9 \Rightarrow -x + 2\left(\frac{-1}{2}\right) = -9 \Rightarrow -x = -8 \Rightarrow x = 8$.

The solution is: $\left(8, -11, \frac{-1}{2}\right)$.

29. Multiply the first equation by 2 and subtract the second equation:

$$x - 2y + \quad z = \quad -8$$
$$\underline{x + 2y - 3z = \quad 20}$$
$$-4y + 4z = -28.$$

Add the first and third equations:

$$\tfrac{1}{2}x - \quad y + \tfrac{1}{2}z = -4$$
$$\underline{-\tfrac{1}{2}x + 3y + 2z = 0}$$
$$2y + \tfrac{5}{2}z = -4.$$

Multiply $2y + \dfrac{5}{2}z = -4$ by 2 and add $-4y + 4z = -28$:

$$4y + 5z = \quad -8$$
$$\underline{-4y + 4z = -28}$$
$$9z = -36 \Rightarrow z = -4.$$

Substitute $z = -4$ into $-4y + 4z = -28$: $-4y + 4(-4) = -28 \Rightarrow -4y - 16 = -28 \Rightarrow -4y = -12 \Rightarrow$

$y = 3$.

Substitute $y = 3$ and $z = -4$ into $x + 2y - 3z = 20$: $x + 2(3) - 3(-4) = 20 \Rightarrow x + 18 = 20 \Rightarrow x = 2$

The solution is: $(2, 3, -4)$.

Applications

31. Let $x =$ children tickets sold, $y =$ student tickets sold, and $z =$ adult tickets sold. Then: $x + y + z = 500$,

    $5x + 7y + 10z = 3560$, and $y = z + 180$ or $y - z = 180$.

    Multiply the first equation by 5 and subtract the second equation:

    $$5x + 5y + 5z = 2500$$
    $$\underline{5x + 7y + 10z = 3560}$$
    $$-2y - 5z = -1060.$$

    Now, multiply $y - z = 180$ by 2 and add $-2y - 5z = -1060$:.

    $$2y - 2z = 360$$
    $$\underline{-2y - 5z = -1060}$$
    $$-7z = -700 \Rightarrow z = 100.$$

    Substitute $z = 100$ into $y - z = 180$: $y - 100 = 180 \Rightarrow y = 280$.

    Substitute $y = 280$ and $z = 100$ into $x + y + z = 500$: $x + 280 + 100 = 500 \Rightarrow x = 120$.

    There were 120 children tickets, 280 student tickets, and 100 adult tickets sold.

33. Let $x =$ cost of a hamburger, $y =$ cost of fries, and $z =$ cost of a soda. Then $2x + 2y + z = 9$;

    $x + y + z = 5$; and $x + y = 5$.

    Multiply the equation $x + y + z = 5$ by 2 and subtract from the equation $2x + 2y + z = 9$:

    $$2x + 2y + z = 9$$
    $$\underline{2x + 2y + 2z = 10}$$
    $$-z = -1 \Rightarrow z = 1.$$

    Subtract $x + y + z = 5$ and $x + y = 5$:

    $$x + y + z = 5$$
    $$\underline{x + y = 5}$$
    $$ z = 0.$$

    z cannot equal both 0 and 1 therefore there is no solution, at least one student was charged incorrectly.

35. (a) $x + y + z = 180$; $x = z + 25 \Rightarrow x - z = 25$; and $y + z = x + 30 \Rightarrow -x + y + z = 30$.

    (b) Add $x - z = 25$ to $-x + y + z = 30$:

    $$x - z = 25$$
    $$\underline{-x + y + z = 30}$$
    $$ y = 55.$$

    Add $x + y + z = 180$ to $x - z = 25$:

    $$x + y + z = 180$$
    $$\underline{x - z = 25}$$
    $$2x + y = 205.$$

    Substitute $y = 55$ into $2x + y = 205$: $2x + 55 = 205 \Rightarrow 2x = 150 \Rightarrow x = 75$.

    Substitute $x = 75$ and $y = 55$ into $x + y + z = 180$: $75 + 55 + z = 180 \Rightarrow z = 50$.

    The angles are: 75°, 55°, and 50°.

    Check: $75 + 55 + 50 = 180 \Rightarrow 180 = 180$; $75 - 50 = 25 \Rightarrow 25 = 25$; $-75 + 55 + 50 = 30 \Rightarrow$

    $30 = 30$.

37. Let x, y, and z equal the amounts invested in the three mutual funds.

    Then $x + y + z = 20{,}000$; $0.05x + 0.07y + 0.10z = 1650$; and $4x = z \Rightarrow 4x - z = 0$.

    Multiply $x + y + z = 20{,}000$ by 0.07 and subtract $0.05x + 0.07y + 0.10z = 1650$:

    $$0.07x + 0.07y + 0.07z = 1400$$
    $$\underline{0.05x + 0.07y + 0.10z = 1650}$$
    $$0.02x \qquad\quad - 0.03z = -250.$$

    Now multiply $0.02x - 0.03z = -250$ by 200 and subtract $4x - z = 0$:

    $$4x - 6z = -50{,}000$$
    $$\underline{4x - \ z = \qquad 0}$$
    $$-5z = -50{,}000 \Rightarrow z = 10{,}000.$$

    Substitute $z = 10{,}000$ into $4x - z = 0$: $4x - 10{,}000 = 0 \Rightarrow 4x = 10{,}000 \Rightarrow x = 2500$.

    Substitute $x = 2500$ and $z = 10{,}000$ into $x + y + z = 20{,}000$: $2500 + y + 10{,}000 = 20{,}000 \Rightarrow y = 7500$.

    The fund amounts are: $2500 at 5%, $7500 at 7%, and $10,000 at 10%.

39. (a) $N + P + K = 80$
    $$N + P - K = 8$$
    $$9P - K = 0$$

    (b) Using technology to solve the system, the solution is $(40, 4, 36)$.

    The sample contains 40 pounds of nitrogen, 4 pounds of phosphorus and 36 pounds of potassium.

9.4 Solutions to Linear Systems Using Matrices

Dimensions of Matrices and Augmented Matrices

1. (a) Since there are three rows and one column, its dimension is 3×1.

   (b) Since there are two rows and three columns, its dimension is 2×3.

   (c) Since there are two rows and two columns, its dimension is 2×2.

3. This system can be written using a 2×3 matrix:
$$\left[\begin{array}{cc|c} 5 & -2 & 3 \\ -1 & 3 & -1 \end{array}\right]$$

5. This system can be written using a 3×4 matrix:
$$\left[\begin{array}{ccc|c} -3 & 2 & 1 & -4 \\ 5 & 0 & -1 & 9 \\ 1 & -3 & -6 & -9 \end{array}\right]$$

7. $3x + 2y = 4$ and $y = 5$

9. $3x + y + 4z = 0$, $5y + 8z = -1$, and $-7z = 1$

Row-Echelon Form

11. (a) Yes

    (b) No, since $a_{22} = -1$ and $a_{32} \neq 0$. The diagonal is not all 1's and there are not all 0's below the diagonal.

    (c) Yes

13. The system can be written as $x + 2y = 3$ and $y = -1$. Substituting $y = -1$ into the first equation gives

$x + 2(-1) = 3 \Rightarrow x = 5$. The solution is $(5, -1)$.

15. The system can be written as $x - y = 2$ and $y = 0$. Substituting $y = 0$ into the first equation gives

$x - 0 = 2 \Rightarrow x = 2$. The solution is $(2, 0)$.

17. The system can be written as $x + y - z = 4$, $y - z = 2$, and $z = 1$. Substituting $z = 1$ into the second

equation gives $y - (1) = 2 \Rightarrow y = 3$. Substituting $y = 3$ and $z = 1$ into the first equation gives

$x + (3) - (1) = 4 \Rightarrow x = 2$. The solution is $(2, 3, 1)$.

19. The system can be written as $x + 2y - z = 5$, $y - 2z = 1$, and $0 = 0$. Since $0 = 0$, there are an infinite

number of solutions. The second equation gives $y = 1 + 2z$. Substituting this into the first equation gives

$x + 2(1 + 2z) - z = 5 \Rightarrow x = 3 - 3z$. The solution can be written as

$\{(3 - 3z, 1 + 2z, z) \,|\, z$ is a real number$\}$.

21. The system can be written as $x + 2y + z = -3$, $y - 3z = \dfrac{1}{2}$, and $0 = 4$. Since $0 = 4$ is false, there are no

solutions.

Solving Systems with Gaussian Elimination

23.
$$\begin{array}{c} (1/2)R_1 \to \\ \\ (1/4)R_3 \to \end{array} \begin{bmatrix} 1 & -2 & 3 & | & 5 \\ -3 & 5 & 3 & | & 2 \\ 1 & 2 & 1 & | & -2 \end{bmatrix}$$

25.
$$\begin{array}{c} \\ R_2 + R_1 \to \\ R_3 - R_1 \to \end{array} \begin{bmatrix} 1 & -1 & 1 & | & 2 \\ 0 & 1 & -1 & | & 2 \\ 0 & 8 & -1 & | & 3 \end{bmatrix}$$

27. The system can be written as follows:

$$\begin{bmatrix} 1 & 2 & | & 3 \\ -1 & -1 & | & 7 \end{bmatrix} R_2 + R_1 \to \begin{bmatrix} 1 & 2 & | & 3 \\ 0 & 1 & | & 10 \end{bmatrix}$$

The solution is $y = 10$ and $x + 2y = 3 \Rightarrow x + 2(10) = 3 \Rightarrow x = -17$. The solution is $(-17, 10)$.

29. The system can be written as follows:

$$\begin{bmatrix} 1 & 2 & 1 & | & 3 \\ 1 & 1 & -1 & | & 3 \\ -1 & -2 & 1 & | & -5 \end{bmatrix} \begin{array}{c} \\ R_2 - R_1 \to \\ R_3 + R_1 \to \end{array} \begin{bmatrix} 1 & 2 & 1 & | & 3 \\ 0 & -1 & -2 & | & 0 \\ 0 & 0 & 2 & | & -2 \end{bmatrix} \begin{array}{c} \\ (-1)R_2 \to \\ (1/2)R_3 \to \end{array} \begin{bmatrix} 1 & 2 & 1 & | & 3 \\ 0 & 1 & 2 & | & 0 \\ 0 & 0 & 1 & | & -1 \end{bmatrix}$$

Back substitution produces $z = -1$; $y + 2z = 0 \Rightarrow y = 2$; $x + 2y + z = 3 \Rightarrow x = 0$.

The solution is $(0, 2, -1)$.

31. The system can be written as follows:

$$\begin{bmatrix} 3 & 1 & 3 & | & 14 \\ 1 & 1 & 1 & | & 6 \\ -2 & -2 & 3 & | & -7 \end{bmatrix} \begin{array}{c} R_2 \to \\ R_1 \to \\ \end{array} \begin{bmatrix} 1 & 1 & 1 & | & 6 \\ 3 & 1 & 3 & | & 14 \\ -2 & -2 & 3 & | & -7 \end{bmatrix} \begin{array}{c} \\ R_2 - 3R_1 \to \\ R_3 + 2R_1 \to \end{array} \begin{bmatrix} 1 & 1 & 1 & | & 6 \\ 0 & -2 & 0 & | & -4 \\ 0 & 0 & 5 & | & 5 \end{bmatrix} \begin{array}{c} \\ (-1/2)R_2 \to \\ (1/5)R_3 \to \end{array} \begin{bmatrix} 1 & 1 & 1 & | & 6 \\ 0 & 1 & 0 & | & 2 \\ 0 & 0 & 1 & | & 1 \end{bmatrix}$$

Back substitution produces $z = 1$; $y = 2$; $x + y + z = 6 \Rightarrow x = 3$. The solution is $(3, 2, 1)$.

33. The system can be written as follows:

$$\begin{bmatrix} 1 & 2 & -1 & | & 2 \\ 2 & 5 & 1 & | & 8 \\ 3 & 7 & 0 & | & 5 \end{bmatrix} \begin{matrix} \\ R_2 - 2R_1 \rightarrow \\ R_3 - 3R_1 \rightarrow \end{matrix} \begin{bmatrix} 1 & 2 & -1 & | & 2 \\ 0 & 1 & 3 & | & 4 \\ 0 & 1 & 3 & | & -1 \end{bmatrix} \begin{matrix} \\ \\ R_3 - R_2 \rightarrow \end{matrix} \begin{bmatrix} 1 & 2 & -1 & | & 2 \\ 0 & 1 & 3 & | & 4 \\ 0 & 0 & 0 & | & -5 \end{bmatrix}$$

The last equation indicates that $0 = -5$, which is false. Therefore, there are no solutions.

35. The system can be written as follows:

$$\begin{bmatrix} -1 & 2 & 4 & | & 10 \\ 3 & -2 & -2 & | & -12 \\ 1 & 2 & 6 & | & 8 \end{bmatrix} \begin{matrix} (-1)R_1 \rightarrow \\ R_2 + 3R_1 \rightarrow \\ R_1 + R_3 \rightarrow \end{matrix} \begin{bmatrix} 1 & -2 & -4 & | & -10 \\ 0 & 4 & 10 & | & 18 \\ 0 & 4 & 10 & | & 18 \end{bmatrix} \begin{matrix} R_1 + (1/2)R_2 \rightarrow \\ (1/4)R_2 \rightarrow \\ R_2 - R_3 \rightarrow \end{matrix} \begin{bmatrix} 1 & 0 & 1 & | & -1 \\ 0 & 1 & \frac{5}{2} & | & \frac{9}{2} \\ 0 & 0 & 0 & | & 0 \end{bmatrix}$$

The last equation indicates that $0 = 0$, which is true. Therefore, there is an infinite number of solutions.

The second equation gives: $y + \dfrac{5}{2}z = \dfrac{9}{2} \Rightarrow y = \dfrac{-5z + 9}{2}$. The first equation gives: $x + z = -1 \Rightarrow$

$x = -1 - z$. Therefore there are infinitely many solutions which can be written as $\left(-1 - z, \dfrac{-5z + 9}{2}, z \right)$.

37. $\begin{bmatrix} 1 & -1 & 1 & | & 1 \\ 1 & 2 & -1 & | & 2 \\ 0 & 1 & -1 & | & 0 \end{bmatrix} \begin{matrix} \\ R_2 - R_1 \rightarrow \\ \\ \end{matrix} \begin{bmatrix} 1 & -1 & 1 & | & 1 \\ 0 & 3 & -2 & | & 1 \\ 0 & 1 & -1 & | & 0 \end{bmatrix} \begin{matrix} \\ (1/3)R_2 \rightarrow \\ \\ \end{matrix} \begin{bmatrix} 1 & -1 & 1 & | & 1 \\ 0 & 1 & -\frac{2}{3} & | & \frac{1}{3} \\ 0 & 1 & -1 & | & 0 \end{bmatrix}$

$$\begin{matrix} \\ \\ R_3 - R_2 \rightarrow \end{matrix} \begin{bmatrix} 1 & -1 & 1 & | & 1 \\ 0 & 1 & -\frac{2}{3} & | & \frac{1}{3} \\ 0 & 0 & -\frac{1}{3} & | & -\frac{1}{3} \end{bmatrix} \begin{matrix} \\ \\ -3R_3 \rightarrow \end{matrix} \begin{bmatrix} 1 & -1 & 1 & | & 1 \\ 0 & 1 & -\frac{2}{3} & | & \frac{1}{3} \\ 0 & 0 & 1 & | & 1 \end{bmatrix}$$

The matrix is now in row-echelon form. We see that $z = 1$. Thus, $y - \dfrac{2}{3}z = \dfrac{1}{3} \Rightarrow$

$y - \dfrac{2}{3}(1) = \dfrac{1}{3} \Rightarrow y = 1$ and $x - y + z = 1 \Rightarrow x - 1 + 1 = 1 \Rightarrow x = 1$. The solution is $(1, 1, 1)$.

39. $\begin{bmatrix} 2 & -4 & 2 & | & 11 \\ 1 & 3 & -2 & | & -9 \\ 4 & -2 & 1 & | & 7 \end{bmatrix} \begin{matrix} (1/2)R_1 \rightarrow \\ \\ \\ \end{matrix} \begin{bmatrix} 1 & -2 & 1 & | & \frac{11}{2} \\ 1 & 3 & -2 & | & -9 \\ 4 & -2 & 1 & | & 7 \end{bmatrix} \begin{matrix} \\ R_2 - R_1 \rightarrow \\ R_3 - 4R_1 \rightarrow \end{matrix} \begin{bmatrix} 1 & -2 & 1 & | & \frac{11}{2} \\ 0 & 5 & -3 & | & -\frac{29}{2} \\ 0 & 6 & -3 & | & -15 \end{bmatrix}$

$$(1/5)R_2 \rightarrow \begin{bmatrix} 1 & -2 & 1 & | & \frac{11}{2} \\ 0 & 1 & -\frac{3}{5} & | & -\frac{29}{10} \\ 0 & 6 & -3 & | & -15 \end{bmatrix} \begin{matrix} \\ \\ R_3 - 6R_2 \rightarrow \end{matrix} \begin{bmatrix} 1 & -2 & 1 & | & \frac{11}{2} \\ 0 & 1 & -\frac{3}{5} & | & -\frac{29}{10} \\ 0 & 0 & \frac{3}{5} & | & \frac{24}{10} \end{bmatrix} \begin{matrix} \\ \\ (5/3)R_3 \rightarrow \end{matrix} \begin{bmatrix} 1 & -2 & 1 & | & \frac{11}{2} \\ 0 & 1 & -\frac{3}{5} & | & -\frac{29}{10} \\ 0 & 0 & 1 & | & 4 \end{bmatrix}$$

The matrix is now in row-echelon form. We see that $z = 4$. Thus, $y - \dfrac{3}{5}z = -\dfrac{29}{10} \Rightarrow$

$y - \dfrac{3}{5}(4) = -\dfrac{29}{10} \Rightarrow y = -\dfrac{1}{2}$ and $x - 2y + z = \dfrac{11}{2} \Rightarrow x - 2\left(-\dfrac{1}{2} \right) + 4 = \dfrac{11}{2} \Rightarrow x = \dfrac{1}{2}$.

The solution is $\left(\dfrac{1}{2}, -\dfrac{1}{2}, 4 \right)$.

41. $\begin{bmatrix} 3 & -2 & 2 & | & -18 \\ -1 & 2 & -4 & | & 16 \\ 4 & -3 & -2 & | & -21 \end{bmatrix}$ $(1/3)R_1 \rightarrow \begin{bmatrix} 1 & -\frac{2}{3} & \frac{2}{3} & | & -6 \\ -1 & 2 & -4 & | & 16 \\ 4 & -3 & -2 & | & -21 \end{bmatrix}$ $\begin{matrix} \\ R_2 + R_1 \rightarrow \\ R_3 - 4R_1 \rightarrow \end{matrix} \begin{bmatrix} 1 & -\frac{2}{3} & \frac{2}{3} & | & -6 \\ 0 & \frac{4}{3} & -\frac{10}{3} & | & 10 \\ 0 & -\frac{1}{3} & -\frac{14}{3} & | & 3 \end{bmatrix}$

$\begin{matrix} \\ (3/4)R_2 \rightarrow \\ 4R_3 + R_2 \rightarrow \end{matrix} \begin{bmatrix} 1 & -\frac{2}{3} & \frac{2}{3} & | & -6 \\ 0 & 1 & -\frac{5}{2} & | & \frac{15}{2} \\ 0 & 0 & -22 & | & 22 \end{bmatrix}$ $(-1/22)R_3 \rightarrow \begin{bmatrix} 1 & -\frac{2}{3} & \frac{2}{3} & | & -6 \\ 0 & 1 & -\frac{5}{2} & | & \frac{15}{2} \\ 0 & 0 & 1 & | & -1 \end{bmatrix}$

The matrix is now in row-echelon form. We see that $z = -1$. Thus, $y - \frac{5}{2}z = \frac{15}{2} \Rightarrow$

$y - \frac{5}{2}(-1) = \frac{15}{2} \Rightarrow y = 5$ and $x - \frac{2}{3}y + \frac{2}{3}z = -6 \Rightarrow x - \frac{2}{3}(5) + \frac{2}{3}(-1) = -6 \Rightarrow x = -2.$

The solution is $(-2, 5, -1)$.

43. $\begin{bmatrix} 1 & -4 & 3 & | & 26 \\ -1 & 3 & -2 & | & -19 \\ 0 & -1 & 1 & | & 10 \end{bmatrix}$ $R_2 + R_1 \rightarrow \begin{bmatrix} 1 & -4 & 3 & | & 26 \\ 0 & -1 & 1 & | & 7 \\ 0 & -1 & 1 & | & 10 \end{bmatrix}$ $\begin{matrix} \\ -R_2 \rightarrow \\ R_3 - R_2 \rightarrow \end{matrix} \begin{bmatrix} 1 & -4 & 3 & | & 26 \\ 0 & 1 & -1 & | & -7 \\ 0 & 0 & 0 & | & 3 \end{bmatrix}$

The last equation indicates that $0 = 3$, which is always false. Therefore, the system are no solutions.

45. $\begin{bmatrix} 5 & 0 & 4 & | & 7 \\ 2 & -4 & 0 & | & 6 \\ 0 & 3 & 3 & | & 3 \end{bmatrix}$ $\begin{matrix} (1/5)R_1 \rightarrow \\ 2R_1 - 5R_2 \rightarrow \\ (1/3)R_3 \rightarrow \end{matrix} \begin{bmatrix} 1 & 0 & \frac{4}{5} & | & \frac{7}{5} \\ 0 & 20 & 8 & | & -16 \\ 0 & 1 & 1 & | & 1 \end{bmatrix}$ $\begin{matrix} \\ (1/20)R_2 \rightarrow \\ R_2 - 20R_3 \rightarrow \end{matrix} \begin{bmatrix} 1 & 0 & \frac{4}{5} & | & \frac{7}{5} \\ 0 & 1 & \frac{2}{5} & | & -\frac{4}{5} \\ 0 & 0 & -12 & | & -36 \end{bmatrix}$

$(-1/12)R_3 \rightarrow \begin{bmatrix} 1 & 0 & \frac{4}{5} & | & \frac{7}{5} \\ 0 & 1 & \frac{2}{5} & | & -\frac{4}{5} \\ 0 & 0 & 1 & | & 3 \end{bmatrix}$ $\begin{matrix} (-4/5)R_3 + R_1 \rightarrow \\ (-2/5)R_3 + R_2 \rightarrow \\ \end{matrix} \begin{bmatrix} 1 & 0 & 0 & | & -1 \\ 0 & 1 & 0 & | & -2 \\ 0 & 0 & 1 & | & 3 \end{bmatrix}$ The solution is $(-1, -2, 3)$.

47. $\begin{bmatrix} 5 & -2 & 1 & | & 5 \\ 1 & 1 & -2 & | & -2 \\ 4 & -3 & 3 & | & 7 \end{bmatrix}$ $\begin{matrix} R_2 \rightarrow \\ 5R_2 - R_1 \rightarrow \\ 4R_2 - R_3 \rightarrow \end{matrix} \begin{bmatrix} 1 & 1 & -2 & | & -2 \\ 0 & 7 & -11 & | & -15 \\ 0 & 7 & -11 & | & -15 \end{bmatrix}$ $\begin{matrix} \\ (1/7)R_2 \rightarrow \\ R_2 - R_3 \rightarrow \end{matrix} \begin{bmatrix} 1 & 1 & -2 & | & -2 \\ 0 & 1 & -\frac{11}{7} & | & -\frac{15}{7} \\ 0 & 0 & 0 & | & 0 \end{bmatrix}$

$(-1)R_2 + R_1 \rightarrow \begin{bmatrix} 1 & 0 & -\frac{3}{7} & | & \frac{1}{7} \\ 0 & 1 & -\frac{11}{7} & | & -\frac{15}{7} \\ 0 & 0 & 0 & | & 0 \end{bmatrix}$

The last equation indicates that $0 = 0$, which is true. Therefore, there is an infinite number of solutions.

The second equation gives: $y - \frac{11}{7}z = -\frac{15}{7} \Rightarrow y = \frac{11z - 15}{7}$. The first equation gives: $x - \frac{3}{7}z = \frac{1}{7} \Rightarrow$

$x = \frac{3z + 1}{7}$. Therefore, there are infinitely many solutions which can be written as: $\left(\frac{3z + 1}{7}, \frac{11z - 15}{7}, z \right)$.

49. The equations are: $x = 12$ and $y = 3 \Rightarrow (12, 3)$.

51. The equations are: $x = -2, y = 4$, and $z = \frac{1}{2} \Rightarrow \left(-2, 4, \frac{1}{2} \right)$.

53. The last equation indicates that $0 = 0$, which is true. Therefore, there is an infinite number of solutions.

The first equation gives: $x + 2z = 4 \Rightarrow x = -2z + 4$. The second equation gives: $y - z = -3 \Rightarrow$

$y = z - 3$. The system has infinitely many solutions which can be written as: $(-2z + 4, z - 3, z)$.

55. The last equation indicates that $0 = \dfrac{2}{3}$, which is always false. Therefore, there are no solutions.

57. $\begin{bmatrix} 1 & -1 & | & 1 \\ 1 & 1 & | & 5 \end{bmatrix} \begin{array}{c} \\ R_2 - R_1 \to \end{array} \begin{bmatrix} 1 & -1 & | & 1 \\ 0 & 2 & | & 4 \end{bmatrix} (1/2)R_2 \to \begin{bmatrix} 1 & -1 & | & 1 \\ 0 & 1 & | & 2 \end{bmatrix} \begin{array}{c} R_1 + R_2 \to \\ \\ \end{array} \begin{bmatrix} 1 & 0 & | & 3 \\ 0 & 1 & | & 2 \end{bmatrix}$

The solution is (3, 2)

59. $\begin{bmatrix} 1 & 2 & 1 & | & 3 \\ 0 & 1 & -1 & | & -2 \\ -1 & -2 & 2 & | & 6 \end{bmatrix} \begin{array}{c} \\ \\ R_1 + R_3 \to \end{array} \begin{bmatrix} 1 & 2 & 1 & | & 3 \\ 0 & 1 & -1 & | & -2 \\ 0 & 0 & 3 & | & 9 \end{bmatrix} \begin{array}{c} R_1 - 2R_2 \to \\ \\ (1/3)R_3 \to \end{array} \begin{bmatrix} 1 & 0 & 3 & | & 7 \\ 0 & 1 & -1 & | & -2 \\ 0 & 0 & 1 & | & 3 \end{bmatrix}$

$\begin{array}{c} R_1 - 3R_3 \to \\ R_2 + R_3 \to \\ \\ \end{array} \begin{bmatrix} 1 & 0 & 0 & | & -2 \\ 0 & 1 & 0 & | & 1 \\ 0 & 0 & 1 & | & 3 \end{bmatrix}$ The solution is (–2, 1, 3).

61. $\begin{bmatrix} 1 & -1 & 2 & | & 7 \\ 2 & 1 & -4 & | & -27 \\ -1 & 1 & -1 & | & 0 \end{bmatrix} \begin{array}{c} \\ R_2 - 2R_1 \to \\ R_1 + R_3 \to \end{array} \begin{bmatrix} 1 & -1 & 2 & | & 7 \\ 0 & 3 & -8 & | & -41 \\ 0 & 0 & 1 & | & 7 \end{bmatrix} (1/3)R_2 \to \begin{bmatrix} 1 & -1 & 2 & | & 7 \\ 0 & 1 & -\frac{8}{3} & | & -\frac{41}{3} \\ 0 & 0 & 1 & | & 7 \end{bmatrix}$

$\begin{array}{c} R_2 + R_1 \to \\ \\ \\ \end{array} \begin{bmatrix} 1 & 0 & -\frac{2}{3} & | & -\frac{20}{3} \\ 0 & 1 & -\frac{8}{3} & | & -\frac{41}{3} \\ 0 & 0 & 1 & | & 7 \end{bmatrix} \begin{array}{c} (2/3)R_3 + R_1 \to \\ (8/3)R_3 + R_2 \to \\ \\ \end{array} \begin{bmatrix} 1 & 0 & 0 & | & -2 \\ 0 & 1 & 0 & | & 5 \\ 0 & 0 & 1 & | & 7 \end{bmatrix}$ The solution is (–2, 5, 7).

63. $\begin{bmatrix} 2 & 1 & -1 & | & 2 \\ 1 & -2 & 1 & | & 0 \\ 1 & 3 & -2 & | & 4 \end{bmatrix} \begin{array}{c} R_2 \to \\ 2R_2 - R_1 \to \\ R_2 - R_3 \to \end{array} \begin{bmatrix} 1 & -2 & 1 & | & 0 \\ 0 & -5 & 3 & | & -2 \\ 0 & -5 & 3 & | & -4 \end{bmatrix} \begin{array}{c} \\ (-1/5)R_2 \to \\ R_2 - R_3 \to \end{array} \begin{bmatrix} 1 & -2 & 1 & | & 0 \\ 0 & 1 & -\frac{3}{5} & | & \frac{2}{5} \\ 0 & 0 & 0 & | & 2 \end{bmatrix}$

The last equation indicates that $0 = 2$, which is always false. Therefore, there are no solutions.

65. Enter the coefficients of the linear system into a 3×4 matrix, as shown in Figure 65a. Reduce the matrix to reduced row-echelon form, as shown in Figure 65b. The solution is the ordered triple (–9.266, –9.167, 2.440).

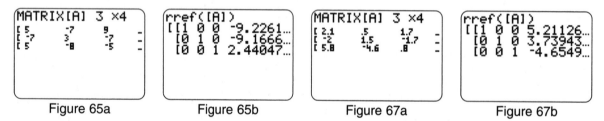

| Figure 65a | Figure 65b | Figure 67a | Figure 67b |

67. Enter the coefficients of the linear system into a 3×4 matrix, as shown in Figure 67a. Reduce the matrix to reduced row-echelon form, as shown in Figure 67b. The solution is the ordered triple (5.211, 3.739, –4.655).

69. Enter the coefficients of the linear system into a 3×4 matrix, as shown in Figure 69a. Reduce the matrix to reduced row-echelon form, as shown in Figure 69b. The solution is the ordered triple (7.993, 1.609, –0.401).

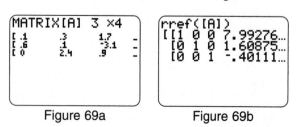

| Figure 69a | Figure 69b |

Applications

71. (a) The constants a, b, and c should satisfy the following three equations:

$1300 = a(1800) + b(5000) + c, 5300 = a(3200) + b(12,000) + c$, and $6500 = a(4500) + b(13,000) + c$

This can be written:

$1800a + 5000b + c = 1300, 3200a + 12,000b + c = 5300$, and $4500a + 13,000b + c = 6500$

The associated augmented matrix is: $\begin{bmatrix} 1800 & 5000 & 1 & 1300 \\ 3200 & 12,000 & 1 & 5300 \\ 4500 & 13,000 & 1 & 6500 \end{bmatrix}$

Using technology, the solution is $a \approx 0.5714$, $b \approx 0.4571$, and $c \approx -2014$. See Figure 71. Thus, the equation modeling the data can be expressed (approximately) as $F = 0.5714N + 0.4571R - 2014$.

(b) To predict the food costs for a shelter that serves 3500 people and receives charitable receipts of $\$12,500$, let $N = 3500$ and $R = 12,500$ and evaluate the equation.

$F = 0.5714(3500) + 0.4571(12,500) - 2014 = 5699.65$. This model predicts monthly food costs of approximately $\$5700$.

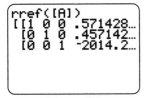

Figure 71

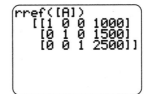

Figure 73   Figure 75

73. Let x represent the fraction of the pool that the first pump can empty each hour, and y the fraction of the pool that the second pump can empty each hour, and z the fraction for the third pump. Since the first pump is twice as fast we have $x = 2y$ or $x - 2y = 0$. Since the first two pumps can empty the pool in 8 hours, it follows that they can empty $\frac{1}{8}$ of the pool in an hour. Thus, $x + y = \frac{1}{8}$. Similarly, all three pumps can empty the pool in 6 hours, so $x + y + x = \frac{1}{6}$. Putting these equations in an augmented matrix results in the following:

$\begin{bmatrix} 1 & -2 & 0 & 0 \\ 1 & 1 & 0 & \frac{1}{8} \\ 1 & 1 & 1 & \frac{1}{6} \end{bmatrix}$ $\begin{bmatrix} 1 & -2 & 0 & 0 \\ 0 & 3 & 0 & \frac{1}{8} \\ 0 & 3 & 1 & \frac{1}{6} \end{bmatrix}$ $\begin{bmatrix} 1 & -2 & 0 & 0 \\ 0 & 1 & 0 & \frac{1}{24} \\ 0 & 0 & 1 & \frac{1}{24} \end{bmatrix}$ $\begin{bmatrix} 1 & 0 & 0 & \frac{1}{12} \\ 0 & 1 & 0 & \frac{1}{24} \\ 0 & 0 & 1 & \frac{1}{24} \end{bmatrix}$

Thus, $x = \frac{1}{12}, y = z = \frac{1}{24}$. The first pump could empty $\frac{1}{12}$ of the pool in one hour or the entire pool in 12 hours, while the second and third pumps individually could empty the pool in 24 hours. Using technology, this solution can be obtained as shown in Figure 73.

75. Let x equal the amount invested at 8%, y equal the amount invested at 11% and z equal the amount invested at 14%.

(a) $x + y + z = 5000, x + y - z = 0$, and $0.08x + 0.11y + 0.14z = 595$

(b) Enter the coefficients of the following 3×4 augmented matrix into a calculator.

$\begin{bmatrix} 1 & 1 & 1 & 5000 \\ 1 & 1 & -1 & 0 \\ 0.08 & 0.11 & 0.14 & 595 \end{bmatrix}$

Using technology, the solution is $x = 1000$, $y = 1500$, and $z = 2500$. See Figure 75. Thus, $\$1000$ needs to be invested at 8%, $\$1500$ at 11%, and $\$2500$ at 14% in order to earn the total amount interest of $\$595$.

77. $I_1 = I_2 + I_3 \Rightarrow I_1 - I_2 - I_3 = 0$, $15 + 4I_3 = 14I_2 \Rightarrow -14I_2 + 4I_3 = -15$, and

$10 + 4I_3 = 5I_1 \Rightarrow -5I_1 + 4I_3 = 10$

Therefore the matrix is:

$$\begin{bmatrix} 1 & -1 & -1 & 0 \\ 0 & -14 & 4 & -15 \\ -5 & 0 & 4 & -10 \end{bmatrix} \begin{matrix} \\ (-1/14)R_2 \to \\ 5R_1 + R_3 \to \end{matrix} \begin{bmatrix} 1 & -1 & -1 & 0 \\ 0 & 1 & -\frac{2}{7} & \frac{15}{14} \\ 0 & -5 & -1 & -10 \end{bmatrix} 5R_2 + R_3 \to \begin{bmatrix} 1 & -1 & -1 & 0 \\ 0 & 1 & -\frac{2}{7} & \frac{15}{14} \\ 0 & 0 & -\frac{17}{7} & -\frac{65}{14} \end{bmatrix}$$

$$\begin{matrix} \\ \\ (-7/17)R_3 \to \end{matrix} \begin{bmatrix} 1 & -1 & -1 & 0 \\ 0 & 1 & -\frac{2}{7} & \frac{15}{14} \\ 0 & 0 & 1 & \frac{65}{34} \end{bmatrix} \begin{matrix} R_1 + R_3 \to \\ (2/7)R_3 + R_2 \to \end{matrix} \begin{bmatrix} 1 & -1 & 0 & \frac{65}{34} \\ 0 & 1 & 0 & \frac{385}{238} \\ 0 & 0 & 1 & \frac{65}{34} \end{bmatrix} R_2 + R_1 \to \begin{bmatrix} 1 & 0 & 0 & \frac{840}{238} \\ 0 & 1 & 0 & \frac{385}{238} \\ 0 & 0 & 1 & \frac{65}{34} \end{bmatrix}$$

The solution is: $(3.53, 1.62, 1.91)$.

79. (a) At intersection A incoming traffic is equal to $x + 5$. The outgoing traffic is given by $y + 7$. Therefore,

$x + 5 = y + 7$, which is the first equation. The incoming traffic at intersection B is $z + 6$ and the

outgoing traffic is $x + 3$, so $z + 6 = x + 3$. Finally at intersection C, the incoming flow is $y + 3$ and the

outgoing flow is $z + 4$, so $y + 3 = z + 4$.

(b) These three equations can be written as: $x - y = 2$, $x - z = 3$, and $y - z = 1$.

The system of linear equations can be represented by the following augmented matrix:

$$\begin{bmatrix} 1 & -1 & 0 & 2 \\ 1 & 0 & -1 & 3 \\ 0 & 1 & -1 & 1 \end{bmatrix}$$

Begin by subtracting the first row from the second, followed by subtracting the second row from the third.

Gaussian elimination results in the following augmented matrix:

$$\begin{bmatrix} 1 & -1 & 0 & 2 \\ 0 & 1 & -1 & 1 \\ 0 & 0 & 0 & 0 \end{bmatrix}$$

The last row of zeros indicates that the linear system is dependent and has an infinite number of solutions.

Back solving produces $y - z = 1 \Rightarrow y = z + 1$. Substituting into the first equation gives

$x - (z + 1) = 2 \Rightarrow x = z + 3$. Thus, the solution can be written

$\{(z + 3, z + 1, z) \mid z \text{ is any nonnegative real number}\}$.

(c) There are an infinite number of solutions to the system. However, solutions such as $z = 1000$, $x = 1003$,

and $y = 1001$ are likely, unless a large number of people are simply driving around the block. In reality there is

an average traffic flow rate for z that could be measured. From this, values for both x and y could be determined.

81. (a) Let $f(x) = ax^2 + bx + c$. The constants a, b, and c must satisfy the following equations:

$f(1990) = a(1990)^2 + b(1990) + c = 11$, $f(2010) = a(2010)^2 + b(2010) + c = 10$, and

$f(2030) = a(2030)^2 + b(2030) + c = 6$. This system of equations can be represented by the following

augmented matrix:

$$\begin{bmatrix} 1990^2 & 1990 & 1 & | & 11 \\ 2010^2 & 2010 & 1 & | & 10 \\ 2030^2 & 2030 & 1 & | & 6 \end{bmatrix}$$

(b) Using technology to solve the system results in the reduced row-echelon form shown in Figure 81b. The

solution is $a = -0.00375$, $b = 14.95$, and $c = -14{,}889.125$. Therefore, the symbolic representation of

f is $f(x) = -0.00375x^2 + 14.95x - 14{,}889.125$.

(c) The data and f are graphed in Figure 81c. Notice that the graph of f passes through each data point.

(d) *Answers may vary.* For example, in 2015 the ratio could be $f(2015) \approx 9.3$.

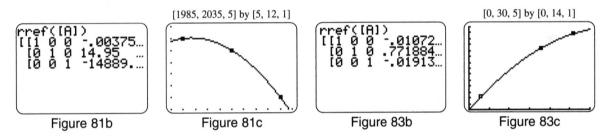

[1985, 2035, 5] by [5, 12, 1]   [0, 30, 5] by [0, 14, 1]

Figure 81b   Figure 81c   Figure 83b   Figure 83c

83. (a) The constants a, b, and c must satisfy the following equations:

$f(3) = a(3)^2 + b(3) + c = 2.2$, $f(18) = a(18)^2 + b(18) + c = 10.4$, and

$f(26) = a(26)^2 + b(26) + c = 12.8$. This system of equations can be represented by the following

augmented matrix:

$$\begin{bmatrix} 3^2 & 3 & 1 & | & 2.2 \\ 18^2 & 18 & 1 & | & 10.4 \\ 26^2 & 26 & 1 & | & 12.8 \end{bmatrix}$$

(b) Using technology to solve the system results in the reduced row-echelon form shown in Figure 83b. An

approximate solution is $a \approx -0.010725$, $b \approx 0.77188$, and $c \approx -0.019130$. Therefore, an approximate

representation of f is $f(x) = -0.010725x^2 + 0.77188x - 0.019130$.

(c) The data and f are graphed in Figure 83c.

(d) *Answers may vary.* For example, in 1990 this percentage was $f(20) \approx 11.1$.

9.5 Properties and Applications of Matrices

Elements of Matrices

1. (a) $a_{12} = 2$ and $a_{21} = 4$. The element a_{32} is undefined since there is no row three.

(b) $a_{11}a_{22} + 3a_{23} = (1)(5) + 3(6) = 23$

3. (a) $a_{12} = -1$, $a_{21} = 3$ and $a_{32} = 0$

   (b) $a_{11}a_{22} + 3a_{23} = (1)(-2) + 3(5) = 13$

5. $x = 1$ and $y = 1$

7. The matrices are not the same size, therefore not possible.

Addition, Subtraction, and Scalar Multiples

9. (a) $\begin{bmatrix} 4 & -1 \\ -1 & 4 \end{bmatrix} + \begin{bmatrix} -1 & 4 \\ 4 & -1 \end{bmatrix} = \begin{bmatrix} 3 & 3 \\ 3 & 3 \end{bmatrix}$

   (b) $\begin{bmatrix} -1 & 4 \\ 4 & -1 \end{bmatrix} + \begin{bmatrix} 4 & -1 \\ -1 & 4 \end{bmatrix} = \begin{bmatrix} 3 & 3 \\ 3 & 3 \end{bmatrix}$

   (c) $\begin{bmatrix} 4 & -1 \\ -1 & 4 \end{bmatrix} - \begin{bmatrix} -1 & 4 \\ 4 & -1 \end{bmatrix} = \begin{bmatrix} 5 & -5 \\ -5 & 5 \end{bmatrix}$

11. (a) $\begin{bmatrix} 3 & 4 & -1 \\ 0 & -3 & 2 \\ -2 & 5 & 10 \end{bmatrix} + \begin{bmatrix} 11 & 5 & -2 \\ 4 & -7 & 12 \\ 6 & 6 & 6 \end{bmatrix} = \begin{bmatrix} 14 & 9 & -3 \\ 4 & -10 & 14 \\ 4 & 11 & 16 \end{bmatrix}$

   (b) $\begin{bmatrix} 11 & 5 & -2 \\ 4 & -7 & 12 \\ 6 & 6 & 6 \end{bmatrix} + \begin{bmatrix} 3 & 4 & -1 \\ 0 & -3 & 2 \\ 2 & 5 & 10 \end{bmatrix} = \begin{bmatrix} 14 & 9 & -3 \\ 4 & -10 & 14 \\ 4 & 11 & 16 \end{bmatrix}$

   (c) $\begin{bmatrix} 3 & 4 & -1 \\ 0 & -3 & 2 \\ -2 & 5 & 10 \end{bmatrix} - \begin{bmatrix} 11 & 5 & -2 \\ 4 & -7 & 12 \\ 6 & 6 & 6 \end{bmatrix} = \begin{bmatrix} -8 & -1 & 1 \\ -4 & 4 & -10 \\ -8 & -1 & 4 \end{bmatrix}$

13. (a) $A + B = \begin{bmatrix} 2 & -6 \\ 3 & 1 \end{bmatrix} + \begin{bmatrix} -1 & 0 \\ -2 & 3 \end{bmatrix} = \begin{bmatrix} 1 & -6 \\ 1 & 4 \end{bmatrix}$

   (b) $3A = 3\begin{bmatrix} 2 & -6 \\ 3 & 1 \end{bmatrix} = \begin{bmatrix} 6 & -18 \\ 9 & 3 \end{bmatrix}$

   (c) $2A - 3B = 2\begin{bmatrix} 2 & -6 \\ 3 & 1 \end{bmatrix} - 3\begin{bmatrix} -1 & 0 \\ -2 & 3 \end{bmatrix} = \begin{bmatrix} 7 & -12 \\ 12 & -7 \end{bmatrix}$

15. (a) $A + B$ is undefined since A is 3×3 and B is 2×3. They do not have the same dimension.

   (b) $3A = 3\begin{bmatrix} 1 & -1 & 0 \\ 1 & 5 & 9 \\ -4 & 8 & -5 \end{bmatrix} = \begin{bmatrix} 3 & -3 & 0 \\ 3 & 15 & 27 \\ -12 & 24 & -15 \end{bmatrix}$

   (c) $2A - 3B$ is undefined since A is 3×3 and B is 2×3. They do not have the same dimension.

17. (a) $A + B = \begin{bmatrix} -2 & -1 \\ -5 & 1 \\ 2 & -3 \end{bmatrix} + \begin{bmatrix} 2 & -1 \\ 3 & 1 \\ 7 & -5 \end{bmatrix} = \begin{bmatrix} 0 & -2 \\ -2 & 2 \\ 9 & -8 \end{bmatrix}$

   (b) $3A = 3\begin{bmatrix} -2 & -1 \\ -5 & 1 \\ 2 & -3 \end{bmatrix} = \begin{bmatrix} -6 & -3 \\ -15 & 3 \\ 6 & -9 \end{bmatrix}$

   (c) $2A - 3B = 2\begin{bmatrix} -2 & -1 \\ -5 & 1 \\ 2 & -3 \end{bmatrix} - 3\begin{bmatrix} 2 & -1 \\ 3 & 1 \\ 7 & -5 \end{bmatrix} = \begin{bmatrix} -10 & 1 \\ -19 & -1 \\ -17 & 9 \end{bmatrix}$

19. $2\begin{bmatrix} 2 & -1 \\ 5 & 1 \\ 0 & 3 \end{bmatrix} + \begin{bmatrix} 5 & 0 \\ 7 & -3 \\ 1 & 1 \end{bmatrix} - \begin{bmatrix} 9 & -4 \\ 4 & 4 \\ 1 & 6 \end{bmatrix} = \begin{bmatrix} 0 & 2 \\ 13 & -5 \\ 0 & 1 \end{bmatrix}$

21. $\begin{bmatrix} 4 & 6 \\ 3 & -7 \end{bmatrix} - 2\begin{bmatrix} 1 & 0 \\ -4 & 1 \end{bmatrix} = \begin{bmatrix} 2 & 6 \\ 11 & -9 \end{bmatrix}$

23. $2\begin{bmatrix} 2 & -1 & -1 \\ -1 & 2 & -1 \\ -1 & -1 & 2 \end{bmatrix} + 3\begin{bmatrix} 1 & 2 & 3 \\ 2 & 1 & 3 \\ 2 & 3 & 1 \end{bmatrix} = \begin{bmatrix} 7 & 4 & 7 \\ 4 & 7 & 7 \\ 4 & 7 & 7 \end{bmatrix}$

Matrices and Digital Photography

25. The "1" is dark gray and the background is light gray.

$A = \begin{bmatrix} 1 & 2 & 1 \\ 1 & 2 & 1 \\ 1 & 2 & 1 \end{bmatrix}$

27. To enhance the contrast, change light gray to white and change dark gray to black. This could be accomplished

by adding the 3×3 matrix B to A.

$B = \begin{bmatrix} -1 & 1 & -1 \\ -1 & 1 & -1 \\ -1 & 1 & -1 \end{bmatrix}; A + B = \begin{bmatrix} 1 & 2 & 1 \\ 1 & 2 & 1 \\ 1 & 2 & 1 \end{bmatrix} + \begin{bmatrix} -1 & 1 & -1 \\ -1 & 1 & -1 \\ -1 & 1 & -1 \end{bmatrix} = \begin{bmatrix} 0 & 3 & 0 \\ 0 & 3 & 0 \\ 0 & 3 & 0 \end{bmatrix}$

29. To make a negative image, subtract the matrix A from a completely black image matrix B.

$B = \begin{bmatrix} 3 & 3 & 3 \\ 3 & 3 & 3 \\ 3 & 3 & 3 \end{bmatrix}; B - A = \begin{bmatrix} 3 & 3 & 3 \\ 3 & 3 & 3 \\ 3 & 3 & 3 \end{bmatrix} - \begin{bmatrix} 0 & 3 & 0 \\ 0 & 3 & 0 \\ 0 & 3 & 0 \end{bmatrix} = \begin{bmatrix} 3 & 0 & 3 \\ 3 & 0 & 3 \\ 3 & 0 & 3 \end{bmatrix}$

31. $A = \begin{bmatrix} 3 & 3 & 3 & 3 \\ 3 & 0 & 0 & 0 \\ 3 & 3 & 3 & 0 \\ 3 & 0 & 0 & 0 \\ 3 & 0 & 0 & 0 \end{bmatrix}$

33. (a) One possible solution for a "Z" is $A = \begin{bmatrix} 3 & 3 & 3 & 3 \\ 0 & 0 & 3 & 0 \\ 0 & 3 & 0 & 0 \\ 3 & 3 & 3 & 3 \end{bmatrix}$

(b) If A is the matrix in part (a) then

$B = \begin{bmatrix} 3 & 3 & 3 & 3 \\ 3 & 3 & 3 & 3 \\ 3 & 3 & 3 & 3 \\ 3 & 3 & 3 & 3 \end{bmatrix}$ and $B - A = \begin{bmatrix} 3 & 3 & 3 & 3 \\ 3 & 3 & 3 & 3 \\ 3 & 3 & 3 & 3 \\ 3 & 3 & 3 & 3 \end{bmatrix} - \begin{bmatrix} 3 & 3 & 3 & 3 \\ 0 & 0 & 3 & 0 \\ 0 & 3 & 0 & 0 \\ 3 & 3 & 3 & 3 \end{bmatrix} = \begin{bmatrix} 0 & 0 & 0 & 0 \\ 3 & 3 & 0 & 3 \\ 3 & 0 & 3 & 3 \\ 0 & 0 & 0 & 0 \end{bmatrix}$

35. (a) One possible solution for a "L" is $A = \begin{bmatrix} 3 & 0 & 0 & 0 \\ 3 & 0 & 0 & 0 \\ 3 & 0 & 0 & 0 \\ 3 & 3 & 3 & 3 \end{bmatrix}$

(b) If A is the matrix in part (a) then

$$B = \begin{bmatrix} 3 & 3 & 3 & 3 \\ 3 & 3 & 3 & 3 \\ 3 & 3 & 3 & 3 \\ 3 & 3 & 3 & 3 \end{bmatrix} \text{ and } B - A = \begin{bmatrix} 3 & 3 & 3 & 3 \\ 3 & 3 & 3 & 3 \\ 3 & 3 & 3 & 3 \\ 3 & 3 & 3 & 3 \end{bmatrix} - \begin{bmatrix} 3 & 0 & 0 & 0 \\ 3 & 0 & 0 & 0 \\ 3 & 0 & 0 & 0 \\ 3 & 3 & 3 & 3 \end{bmatrix} = \begin{bmatrix} 0 & 3 & 3 & 3 \\ 0 & 3 & 3 & 3 \\ 0 & 3 & 3 & 3 \\ 0 & 0 & 0 & 0 \end{bmatrix}$$

Matrix Multiplication

37. (a) These tables can be represented by the matrices A and B where $A = \begin{bmatrix} 12 & 4 \\ 8 & 7 \end{bmatrix}$ and $B = \begin{bmatrix} 55 \\ 70 \end{bmatrix}$.

(b) The product AB of these matrices calculates tuition cost for each student.

$$AB = \begin{bmatrix} 12 & 4 \\ 8 & 7 \end{bmatrix}\begin{bmatrix} 55 \\ 70 \end{bmatrix} = \begin{bmatrix} 12(55) + 4(70) \\ 8(55) + 7(70) \end{bmatrix} = \begin{bmatrix} 940 \\ 930 \end{bmatrix}$$

Student 1 is taking 12 credits at \$55 each and 4 credits at \$70 each. The total tuition for student 1 is $12(\$55) + 4(\$70) = \$940$. Similarly, the tuition for student 2 is \$930.

39. (a) These tables can be represented by the matrices A and B where $A = \begin{bmatrix} 10 & 5 \\ 9 & 8 \\ 11 & 3 \end{bmatrix}$ and $B = \begin{bmatrix} 60 \\ 70 \end{bmatrix}$.

(b) The product AB of these matrices calculates tuition cost for each student.

$$AB = \begin{bmatrix} 10 & 5 \\ 9 & 8 \\ 11 & 3 \end{bmatrix}\begin{bmatrix} 60 \\ 70 \end{bmatrix} = \begin{bmatrix} 10(60) + 5(70) \\ 9(60) + 8(70) \\ 11(60) + 3(70) \end{bmatrix} = \begin{bmatrix} 950 \\ 1100 \\ 870 \end{bmatrix}$$

The total tuition for student 1 is $10(\$60) + 5(\$70) = \$950$. Similarly, the tuition for student 2 is \$1100 and the tuition for student 3 is \$870.

41. A and B are both 2×2 so AB and BA are also both 2×2.

$$AB = \begin{bmatrix} 1 & -1 \\ 2 & 0 \end{bmatrix}\begin{bmatrix} -2 & 3 \\ 1 & 2 \end{bmatrix} = \begin{bmatrix} -3 & 1 \\ -4 & 6 \end{bmatrix}; \; BA = \begin{bmatrix} -2 & 3 \\ 1 & 2 \end{bmatrix}\begin{bmatrix} 1 & -1 \\ 2 & 0 \end{bmatrix} = \begin{bmatrix} 4 & 2 \\ 5 & -1 \end{bmatrix}$$

43. Since both A and B are 3×2, the number of rows in B is not equal to the number of columns in A, so AB is undefined. Also, the number of rows in A is not equal to the number of columns in B so BA is undefined.

45. $AB = \begin{bmatrix} 3 & -1 \\ 1 & 0 \\ -2 & -4 \end{bmatrix}\begin{bmatrix} -2 & 5 & -3 \\ 9 & -7 & 0 \end{bmatrix} = \begin{bmatrix} 3(-2) + (-1)(9) & 3(5) + (-1)(-7) & 3(-3) + (-1)(0) \\ 1(-2) + 0(9) & 1(5) + 0(-7) & 1(-3) + 0(0) \\ -2(-2) + (-4)(9) & -2(5) + (-4)(-7) & -2(-3) + (-4)(0) \end{bmatrix} \Rightarrow$

$$AB = \begin{bmatrix} -15 & 22 & -9 \\ -2 & 5 & -3 \\ -32 & 18 & 6 \end{bmatrix}$$

$$BA = \begin{bmatrix} -2 & 5 & -3 \\ 9 & -7 & 0 \end{bmatrix}\begin{bmatrix} 3 & -1 \\ 1 & 0 \\ -2 & -4 \end{bmatrix} = \begin{bmatrix} -2(3) + 5(1) + (-3)(-2) & -2(-1) + (5)(0) + (-3)(-4) \\ 9(3) + (-7)(1) + 0(-2) & 9(-1) + (-7)(0) + 0(-4) \end{bmatrix} \Rightarrow$$

$$BA = \begin{bmatrix} 5 & 14 \\ 20 & -9 \end{bmatrix}$$

47. *AB* is undefined, we cannot multiply a 3 × 3 by a 2 × 3.

$$BA = \begin{bmatrix} -1 & 3 & -1 \\ 7 & -7 & 1 \end{bmatrix} \begin{bmatrix} 1 & -1 & 0 \\ 2 & -1 & 5 \\ 6 & 1 & -4 \end{bmatrix} =$$

$$\begin{bmatrix} -1(1) + 3(2) + (-1)(6) & -1(-1) + 3(-1) + (-1)(1) & -1(0) + 3(5) + (-1)(-4) \\ 7(1) + (-7)(2) + 1(6) & 7(-1) + (-7)(-1) + 1(1) & 7(0) + (-7)(5) + 1(-4) \end{bmatrix} \Rightarrow$$

$$BA = \begin{bmatrix} -1 & -3 & 19 \\ -1 & 1 & -39 \end{bmatrix}$$

49. Since *A* is 2 × 2 and *B* is 3 × 1, the number of rows in *B* is not equal to the number of columns in *A*, so *AB* is

undefined. Also, the number of rows in *A* is not equal to the number of columns in *B* so *BA* is undefined.

51. *A* and *B* are both 3 × 3 so *AB* and *BA* are also both 3 × 3.

$$AB = \begin{bmatrix} 2 & -1 & 3 \\ 0 & 1 & 0 \\ 2 & -2 & 3 \end{bmatrix} \begin{bmatrix} 1 & 5 & -1 \\ 0 & 1 & 3 \\ -1 & 2 & 1 \end{bmatrix} = \begin{bmatrix} -1 & 15 & -2 \\ 0 & 1 & 3 \\ -1 & 14 & -5 \end{bmatrix}; BA = \begin{bmatrix} 1 & 5 & -1 \\ 0 & 1 & 3 \\ -1 & 2 & 1 \end{bmatrix} \begin{bmatrix} 2 & -1 & 3 \\ 0 & 1 & 0 \\ 2 & -2 & 3 \end{bmatrix} = \begin{bmatrix} 0 & 6 & 0 \\ 6 & -5 & 9 \\ 0 & 1 & 0 \end{bmatrix}$$

53. *A* is 2 × 2 and *B* is 2 × 1 so *AB* is 2 × 1. However *BA* is undefined since the number of rows in *A* is not equal

to the number of columns in *B*.

$$AB = \begin{bmatrix} 2 & -1 \\ 3 & 1 \end{bmatrix} \begin{bmatrix} 1 \\ 3 \end{bmatrix} = \begin{bmatrix} -1 \\ 6 \end{bmatrix}$$

55. *A* is 2 × 2 and *B* is 2 × 3 so *AB* is 2 × 3. However *BA* is undefined since the number of rows in *A* is not equal

to the number of columns in *B*.

$$AB = \begin{bmatrix} -3 & 1 \\ 2 & -4 \end{bmatrix} \begin{bmatrix} 1 & 0 & -2 \\ -4 & 8 & 1 \end{bmatrix} = \begin{bmatrix} -7 & 8 & 7 \\ 18 & -32 & -8 \end{bmatrix}$$

57. *A* is 3 × 3 and *B* is 3 × 1 so *AB* is 3 × 1. However *BA* is undefined since the number of rows in *A* is not equal

to the number of columns in *B*.

$$AB = \begin{bmatrix} 1 & 0 & -2 \\ 3 & -4 & 1 \\ 2 & 0 & 5 \end{bmatrix} \begin{bmatrix} 1 \\ -1 \\ 3 \end{bmatrix} = \begin{bmatrix} -5 \\ 10 \\ 17 \end{bmatrix}$$

59. $AB = \begin{bmatrix} 3 & 4 & 8 \\ 5 & 6 & 2 \end{bmatrix} \begin{bmatrix} 10 \\ 20 \\ 30 \end{bmatrix} = \begin{bmatrix} 350 \\ 230 \end{bmatrix}$; The total cost of Order 1 is \$350, and the total cost of Order 2 is \$250.

61. (a) $B + C = \begin{bmatrix} 6 & 2 & 7 \\ 3 & -4 & -5 \\ 7 & 1 & 0 \end{bmatrix} + \begin{bmatrix} 1 & 4 & -3 \\ 8 & 1 & -1 \\ 4 & 6 & -2 \end{bmatrix} = \begin{bmatrix} 7 & 6 & 4 \\ 11 & -3 & -6 \\ 11 & 7 & -2 \end{bmatrix}$

$A(B + C) = \begin{bmatrix} 2 & -1 & 3 \\ 1 & 3 & -5 \\ 0 & -2 & 1 \end{bmatrix} \begin{bmatrix} 7 & 6 & 4 \\ 11 & -3 & -6 \\ 11 & 7 & -2 \end{bmatrix} = \begin{bmatrix} 36 & 36 & 8 \\ -15 & -38 & -4 \\ -11 & 13 & 10 \end{bmatrix}$

(b) $AB = \begin{bmatrix} 2 & -1 & 3 \\ 1 & 3 & -5 \\ 0 & -2 & 1 \end{bmatrix} \begin{bmatrix} 6 & 2 & 7 \\ 3 & -4 & -5 \\ 7 & 1 & 0 \end{bmatrix} = \begin{bmatrix} 30 & 11 & 19 \\ -20 & -15 & -8 \\ 1 & 9 & 10 \end{bmatrix}$

$AC = \begin{bmatrix} 2 & -1 & 3 \\ 1 & 3 & -5 \\ 0 & -2 & 1 \end{bmatrix} \begin{bmatrix} 1 & 4 & -3 \\ 8 & 1 & -1 \\ 4 & 6 & -2 \end{bmatrix} = \begin{bmatrix} 6 & 25 & -11 \\ 5 & -23 & 4 \\ -12 & 4 & 0 \end{bmatrix}$; $AB + AC = \begin{bmatrix} 36 & 36 & 8 \\ -15 & -38 & -4 \\ -11 & 13 & 10 \end{bmatrix}$

$A(B + C) = AB + AC$, which indicates that the distributive property holds for matrices.

63. (a) $(A - B)^2 = \begin{bmatrix} -4 & -3 & -4 \\ -2 & 7 & 0 \\ -7 & -3 & 1 \end{bmatrix} \begin{bmatrix} -4 & -3 & -4 \\ -2 & 7 & 0 \\ -7 & -3 & 1 \end{bmatrix} = \begin{bmatrix} 50 & 3 & 12 \\ -6 & 55 & 8 \\ 27 & -3 & 29 \end{bmatrix}$

(b) $A^2 - AB - BA + B^2 =$

$\begin{bmatrix} 3 & -11 & 14 \\ 5 & 18 & -17 \\ -2 & -8 & 11 \end{bmatrix} - \begin{bmatrix} 30 & 11 & 19 \\ -20 & -15 & -8 \\ 1 & 9 & 10 \end{bmatrix} - \begin{bmatrix} 14 & -14 & 15 \\ 2 & -5 & 24 \\ 15 & -4 & 16 \end{bmatrix} + \begin{bmatrix} 91 & 11 & 32 \\ -29 & 17 & 41 \\ 45 & 10 & 44 \end{bmatrix} = \begin{bmatrix} 50 & 3 & 12 \\ -6 & 55 & 8 \\ 27 & -3 & 29 \end{bmatrix}$

$(A - B)^2 = A^2 - AB - BA + B^2$, which indicates that matrices seem to conform to common rules of algebra except for the commutative property since $AB \neq BA$, in general.

9.6 Inverses of Matrices

Identifying Inverse Matrices

1. B is the inverse of A.

$AB = \begin{bmatrix} 4 & 3 \\ 5 & 4 \end{bmatrix} \begin{bmatrix} 4 & -3 \\ -5 & 4 \end{bmatrix} = \begin{bmatrix} 1 & 0 \\ 0 & 1 \end{bmatrix}$ and $BA = \begin{bmatrix} 4 & -3 \\ -5 & 4 \end{bmatrix} \begin{bmatrix} 4 & 3 \\ 5 & 4 \end{bmatrix} = \begin{bmatrix} 1 & 0 \\ 0 & 1 \end{bmatrix}$

3. B is the inverse of A.

$AB = \begin{bmatrix} 1 & -1 & 2 \\ 0 & 1 & -1 \\ 1 & 0 & 2 \end{bmatrix} \begin{bmatrix} 2 & 2 & -1 \\ -1 & 0 & 1 \\ -1 & -1 & 1 \end{bmatrix} = \begin{bmatrix} 1 & 0 & 0 \\ 0 & 1 & 0 \\ 0 & 0 & 1 \end{bmatrix}$; $BA = \begin{bmatrix} 2 & 2 & -1 \\ -1 & 0 & 1 \\ -1 & -1 & 1 \end{bmatrix} \begin{bmatrix} 1 & -1 & 2 \\ 0 & 1 & -1 \\ 1 & 0 & 2 \end{bmatrix} = \begin{bmatrix} 1 & 0 & 0 \\ 0 & 1 & 0 \\ 0 & 0 & 1 \end{bmatrix}$

5. $AB = \begin{bmatrix} 1 & 1 \\ 1 & 2 \end{bmatrix} \begin{bmatrix} 2 & -1 \\ -1 & k \end{bmatrix} = \begin{bmatrix} 1 & -1 + k \\ 0 & 2k - 1 \end{bmatrix}$

We must have $-1 + k = 0$ and $2k - 1 = 1$. The solution to both equations is $k = 1$.

7. $AB = \begin{bmatrix} 1 & 3 \\ -1 & -5 \end{bmatrix} \begin{bmatrix} k & 1.5 \\ -0.5 & -0.5 \end{bmatrix} = \begin{bmatrix} k - 1.5 & 0 \\ -k + 2.5 & 1 \end{bmatrix}$

We must have $k - 1.5 = 1$ and $-k + 2.5 = 0$. The solution to both equations is $k = 2.5$.

9. I_2 multiplied by any 2×2 matrix A is equal to A, that is, $I_2A = AI_2 = A$.

11. I_3 multiplied by any 3×3 matrix A is equal to A, that is, $I_3A = AI_3 = A$.

Interpreting Inverses

13. (a) Since $h = 2$ and $k = 3$, the matrix A will translate the point $(0, 1)$ to the right 2 units and up 3 units. Its

     new location will be $(0 + 2, 1 + 3) = (2, 4)$. This is verified by the following computation:

$$AX = \begin{bmatrix} 1 & 0 & 2 \\ 0 & 1 & 3 \\ 0 & 0 & 1 \end{bmatrix} \begin{bmatrix} 0 \\ 1 \\ 1 \end{bmatrix} = \begin{bmatrix} 2 \\ 4 \\ 1 \end{bmatrix}$$

     (b) $A^{-1}Y = X$. That is, A^{-1} will translate $(2, 4)$ back to $(0, 1)$ by moving it left 2 units and down 3 units.

     Thus $h = -2$ and $k = -3$ in A^{-1}.

$$A^{-1}Y = \begin{bmatrix} 1 & 0 & -2 \\ 0 & 1 & -3 \\ 0 & 0 & 1 \end{bmatrix} \begin{bmatrix} 2 \\ 4 \\ 1 \end{bmatrix} = \begin{bmatrix} 0 \\ 1 \\ 1 \end{bmatrix}$$

     (c) The product $AA^{-1} = I_3 = A^{-1}A$, since they are 3×3 inverse matrices.

15. Three units left implies that $h = -3$ and five units down implies that $k = -5$.

$$A = \begin{bmatrix} 1 & 0 & -3 \\ 0 & 1 & -5 \\ 0 & 0 & 1 \end{bmatrix} \text{ and } A^{-1} = \begin{bmatrix} 1 & 0 & 3 \\ 0 & 1 & 5 \\ 0 & 0 & 1 \end{bmatrix}$$

A^{-1} will translate a point 3 units to the right and 5 units up.

17. (a) $BX = \begin{bmatrix} \dfrac{1}{\sqrt{2}} & \dfrac{1}{\sqrt{2}} & 0 \\ -\dfrac{1}{\sqrt{2}} & \dfrac{1}{\sqrt{2}} & 0 \\ 0 & 0 & 1 \end{bmatrix} \begin{bmatrix} -\sqrt{2} \\ -\sqrt{2} \\ 1 \end{bmatrix} = \begin{bmatrix} -2 \\ 0 \\ 1 \end{bmatrix} = Y$

     (b) $B^{-1}Y = \begin{bmatrix} \dfrac{1}{\sqrt{2}} & -\dfrac{1}{\sqrt{2}} & 0 \\ \dfrac{1}{\sqrt{2}} & \dfrac{1}{\sqrt{2}} & 0 \\ 0 & 0 & 1 \end{bmatrix} \begin{bmatrix} -2 \\ 0 \\ 1 \end{bmatrix} = \begin{bmatrix} -\sqrt{2} \\ -\sqrt{2} \\ 1 \end{bmatrix} = X$

     B^{-1} rotates the point represented by Y counterclockwise $45°$ about the origin.

19. (a) In the computation ABX, B translates $(1, 1)$ left 3 units and up 3 units to $(-2, 4)$. Then, A translates $(-2, 4)$ right 4 units and down 2 units to $(2, 2)$.

$$ABX = \begin{bmatrix} 1 & 0 & 4 \\ 0 & 1 & -2 \\ 0 & 0 & 1 \end{bmatrix} \begin{bmatrix} 1 & 0 & -3 \\ 0 & 1 & 3 \\ 0 & 0 & 1 \end{bmatrix} \begin{bmatrix} 1 \\ 1 \\ 1 \end{bmatrix} = \begin{bmatrix} 2 \\ 2 \\ 1 \end{bmatrix} = Y; \text{ This represents the point } (2, 2) \text{ as expected.}$$

(b) The net result of A and B is to translate a point 1 unit right and 1 unit up. Therefore, it is reasonable to expect that $h = 1$ and $k = 1$ in the matrix of the form AB.

$$AB = \begin{bmatrix} 1 & 0 & 4 \\ 0 & 1 & -2 \\ 0 & 0 & 1 \end{bmatrix} \begin{bmatrix} 1 & 0 & -3 \\ 0 & 1 & 3 \\ 0 & 0 & 1 \end{bmatrix} = \begin{bmatrix} 1 & 0 & 1 \\ 0 & 1 & 1 \\ 0 & 0 & 1 \end{bmatrix}$$

(c) Yes. If a point is translated left 3 units and up 3 units followed by right 4 units and down 2 units, the final result will be the same as the translation obtained when the point is first translated right 4 units and down 2 units followed by left 3 units and up 3. Therefore, we might expect that $AB = BA$.

$$BA = \begin{bmatrix} 1 & 0 & -3 \\ 0 & 1 & 3 \\ 0 & 0 & 1 \end{bmatrix} \begin{bmatrix} 1 & 0 & 4 \\ 0 & 1 & -2 \\ 0 & 0 & 1 \end{bmatrix} = \begin{bmatrix} 1 & 0 & 1 \\ 0 & 1 & 1 \\ 0 & 0 & 1 \end{bmatrix} = AB$$

(d) Since AB translates a point 1 unit right and 1 unit up, the inverse of AB would translate a point 1 unit left and 1 unit down. So $h = -1$ and $k = -1$ in $(AB)^{-1}$.

$$(AB)^{-1} = \begin{bmatrix} 1 & 0 & -1 \\ 0 & 1 & -1 \\ 0 & 0 & 1 \end{bmatrix}; \text{ Notice that } (AB)(AB)^{-1} = I_3 \text{ as expected.}$$

Calculating Inverses

21. $A \mid I_2 = \begin{bmatrix} 1 & 2 & | & 1 & 0 \\ 1 & 3 & | & 0 & 1 \end{bmatrix} \begin{matrix} \\ R_2 - R_1 \rightarrow \end{matrix} \begin{bmatrix} 1 & 2 & | & 1 & 0 \\ 0 & 1 & | & -1 & 1 \end{bmatrix} \begin{matrix} R_1 - 2R_2 \rightarrow \\ \end{matrix} \begin{bmatrix} 1 & 0 & | & 3 & -2 \\ 0 & 1 & | & -1 & 1 \end{bmatrix}; A^{-1} = \begin{bmatrix} 3 & -2 \\ -1 & 1 \end{bmatrix}$

23. $A \mid I_2 = \begin{bmatrix} -1 & 2 & | & 1 & 0 \\ 3 & -5 & | & 0 & 1 \end{bmatrix} \begin{matrix} -1R_1 \rightarrow \\ \end{matrix} \begin{bmatrix} 1 & -2 & | & -1 & 0 \\ 3 & -5 & | & 0 & 1 \end{bmatrix} \begin{matrix} \\ R_2 - 3R_1 \rightarrow \end{matrix} \begin{bmatrix} 1 & -2 & | & -1 & 0 \\ 0 & 1 & | & 3 & 1 \end{bmatrix}$

$R_1 + 2R_2 \rightarrow \begin{bmatrix} 1 & 0 & | & 5 & 2 \\ 0 & 1 & | & 3 & 1 \end{bmatrix}; A^{-1} = \begin{bmatrix} 5 & 2 \\ 3 & 1 \end{bmatrix}$

25. $A \mid I_2 = \begin{bmatrix} 8 & 5 & | & 1 & 0 \\ 2 & 1 & | & 0 & 1 \end{bmatrix} \begin{matrix} (1/8)R_1 \rightarrow \\ R_1 - 4R_2 \rightarrow \end{matrix} \begin{bmatrix} 1 & \frac{5}{8} & | & \frac{1}{8} & 0 \\ 0 & 1 & | & 1 & -4 \end{bmatrix} \begin{matrix} (-5/8)R_2 + R_1 \rightarrow \\ \end{matrix} \begin{bmatrix} 1 & 0 & | & -\frac{1}{2} & \frac{5}{2} \\ 0 & 1 & | & 1 & -4 \end{bmatrix};$

$A^{-1} = \begin{bmatrix} -\frac{1}{2} & \frac{5}{2} \\ 1 & -4 \end{bmatrix}$

27. $A \mid I_3 = \begin{bmatrix} 0 & 0 & 1 & | & 1 & 0 & 0 \\ 1 & 0 & 0 & | & 0 & 1 & 0 \\ 0 & 1 & 0 & | & 0 & 0 & 1 \end{bmatrix} \begin{matrix} R_2 \rightarrow \\ R_3 \rightarrow \\ R_1 \rightarrow \end{matrix} \begin{bmatrix} 1 & 0 & 0 & | & 0 & 1 & 0 \\ 0 & 1 & 0 & | & 0 & 0 & 1 \\ 0 & 0 & 1 & | & 1 & 0 & 0 \end{bmatrix}; A^{-1} = \begin{bmatrix} 0 & 1 & 0 \\ 0 & 0 & 1 \\ 1 & 0 & 0 \end{bmatrix}$

29. $A \mid I_3 = \begin{bmatrix} 1 & 0 & 1 & | & 1 & 0 & 0 \\ 2 & 1 & 3 & | & 0 & 1 & 0 \\ -1 & 1 & 1 & | & 0 & 0 & 1 \end{bmatrix} \begin{matrix} \\ R_2 - 2R_1 \rightarrow \\ R_3 + R_1 \rightarrow \end{matrix} \begin{bmatrix} 1 & 0 & 1 & | & 1 & 0 & 0 \\ 0 & 1 & 1 & | & -2 & 1 & 0 \\ 0 & 1 & 2 & | & 1 & 0 & 1 \end{bmatrix} R_3 - R_2 \rightarrow \begin{bmatrix} 1 & 0 & 1 & | & 1 & 0 & 0 \\ 0 & 1 & 1 & | & -2 & 1 & 0 \\ 0 & 0 & 1 & | & 3 & -1 & 1 \end{bmatrix}$

$\begin{matrix} R_1 - R_3 \rightarrow \\ R_2 - R_3 \rightarrow \end{matrix} \begin{bmatrix} 1 & 0 & 0 & | & -2 & 1 & -1 \\ 0 & 1 & 0 & | & -5 & 2 & -1 \\ 0 & 0 & 1 & | & 3 & -1 & 1 \end{bmatrix}$; $A^{-1} = \begin{bmatrix} -2 & 1 & -1 \\ -5 & 2 & -1 \\ 3 & -1 & 1 \end{bmatrix}$

31. $\begin{bmatrix} 1 & 2 & -1 & | & 1 & 0 & 0 \\ 2 & 5 & 0 & | & 0 & 1 & 0 \\ -1 & -1 & 2 & | & 0 & 0 & 1 \end{bmatrix} \begin{matrix} 2R_1 - R_2 \rightarrow \\ R_3 + R_1 \rightarrow \end{matrix} \begin{bmatrix} 1 & 2 & -1 & | & 1 & 0 & 0 \\ 0 & -1 & -2 & | & 2 & -1 & 0 \\ 0 & 1 & 1 & | & 1 & 0 & 1 \end{bmatrix} \begin{matrix} (-1)R_2 \rightarrow \\ R_2 + R_3 \rightarrow \end{matrix} \begin{bmatrix} 1 & 2 & -1 & | & 1 & 0 & 0 \\ 0 & 1 & 2 & | & -2 & 1 & 0 \\ 0 & 0 & -1 & | & 3 & -1 & 1 \end{bmatrix}$

$\begin{matrix} R_1 - R_3 \rightarrow \\ 2R_3 + R_2 \rightarrow \\ (-1)R_3 \rightarrow \end{matrix} \begin{bmatrix} 1 & 2 & 0 & | & -2 & 1 & -1 \\ 0 & 1 & 0 & | & 4 & -1 & 2 \\ 0 & 0 & 1 & | & -3 & 1 & -1 \end{bmatrix} R_1 - 2R_2 \rightarrow \begin{bmatrix} 1 & 0 & 0 & | & -10 & 3 & -5 \\ 0 & 1 & 0 & | & 4 & -1 & 2 \\ 0 & 0 & 1 & | & -3 & 1 & -1 \end{bmatrix}$; $A^{-1} = \begin{bmatrix} -10 & 3 & -5 \\ 4 & -1 & 2 \\ -3 & 1 & -1 \end{bmatrix}$

33. $\begin{bmatrix} -2 & 1 & -3 & | & 1 & 0 & 0 \\ 0 & 1 & 2 & | & 0 & 1 & 0 \\ 1 & -2 & 1 & | & 0 & 0 & 1 \end{bmatrix} \begin{matrix} (-1/2)R_1 \rightarrow \\ \\ 2R_3 + R_1 \rightarrow \end{matrix} \begin{bmatrix} 1 & -\frac{1}{2} & \frac{3}{2} & | & -\frac{1}{2} & 0 & 0 \\ 0 & 1 & 2 & | & 0 & 1 & 0 \\ 0 & -3 & -1 & | & 1 & 0 & 2 \end{bmatrix} 3R_2 + R_3 \rightarrow \begin{bmatrix} 1 & -\frac{1}{2} & \frac{3}{2} & | & -\frac{1}{2} & 0 & 0 \\ 0 & 1 & 2 & | & 0 & 1 & 0 \\ 0 & 0 & 5 & | & 1 & 3 & 2 \end{bmatrix}$

$(1/5)R_3 \rightarrow \begin{bmatrix} 1 & -\frac{1}{2} & \frac{3}{2} & | & -\frac{1}{2} & 0 & 0 \\ 0 & 1 & 2 & | & 0 & 1 & 0 \\ 0 & 0 & 1 & | & \frac{1}{5} & \frac{3}{5} & \frac{2}{5} \end{bmatrix} \begin{matrix} R_1 - (3/2)R_3 \rightarrow \\ (-2)R_3 + R_2 \rightarrow \end{matrix} \begin{bmatrix} 1 & -\frac{1}{2} & 0 & | & -\frac{4}{5} & -\frac{9}{10} & -\frac{3}{5} \\ 0 & 1 & 0 & | & -\frac{2}{5} & -\frac{1}{5} & -\frac{4}{5} \\ 0 & 0 & 1 & | & \frac{1}{5} & \frac{3}{5} & \frac{2}{5} \end{bmatrix}$

$R_1 + (1/2)R_2 \rightarrow \begin{bmatrix} 1 & 0 & 0 & | & -1 & -1 & -1 \\ 0 & 1 & 0 & | & -\frac{2}{5} & -\frac{1}{5} & -\frac{4}{5} \\ 0 & 0 & 1 & | & \frac{1}{5} & \frac{3}{5} & \frac{2}{5} \end{bmatrix}$; $A^{-1} = \begin{bmatrix} -1 & -1 & -1 \\ -\frac{2}{5} & -\frac{1}{5} & -\frac{4}{5} \\ \frac{1}{5} & \frac{3}{5} & \frac{2}{5} \end{bmatrix}$

35. $A = \begin{bmatrix} 0.5 & -1.5 \\ 0.2 & -0.5 \end{bmatrix} \Rightarrow A^{-1} = \begin{bmatrix} -10 & 30 \\ -4 & 10 \end{bmatrix}$ as shown in Figure 35.

37. $A = \begin{bmatrix} 1 & 2 & 0 \\ -1 & 4 & -1 \\ 2 & -1 & 0 \end{bmatrix} \Rightarrow A^{-1} = \begin{bmatrix} 0.2 & 0 & 0.4 \\ 0.4 & 0 & -0.2 \\ 1.4 & -1 & -1.2 \end{bmatrix}$ as shown in Figure 37.

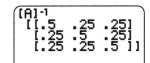

Figure 35	**Figure 37**	**Figure 39**	**Figure 41**

39. $A = \begin{bmatrix} 2 & -2 & 1 \\ 0 & 5 & 8 \\ 0 & 0 & -1 \end{bmatrix} \Rightarrow A^{-1} = \begin{bmatrix} 0.5 & 0.2 & 2.1 \\ 0 & 0.2 & 1.6 \\ 0 & 0 & -1 \end{bmatrix}$ as shown in Figure 39.

41. $A = \begin{bmatrix} 3 & -1 & -1 \\ -1 & 3 & -1 \\ -1 & -1 & 3 \end{bmatrix} \Rightarrow A^{-1} = \begin{bmatrix} 0.5 & 0.25 & 0.25 \\ 0.25 & 0.5 & 0.25 \\ 0.25 & 0.25 & 0.5 \end{bmatrix}$ as shown in Figure 41.

43. $A = \begin{bmatrix} 1 & -1 & 0 & 0 \\ -1 & 5 & -1 & 0 \\ 0 & -1 & 5 & -1 \\ 0 & 0 & -1 & 1 \end{bmatrix} \Rightarrow A^{-1} = \begin{bmatrix} 1.2\overline{6} & 0.2\overline{6} & 0.0\overline{6} & 0.0\overline{6} \\ 0.2\overline{6} & 0.2\overline{6} & 0.0\overline{6} & 0.0\overline{6} \\ 0.0\overline{6} & 0.0\overline{6} & 0.2\overline{6} & 0.2\overline{6} \\ 0.0\overline{6} & 0.0\overline{6} & 0.2\overline{6} & 1.2\overline{6} \end{bmatrix}$ as shown in Figure 43.

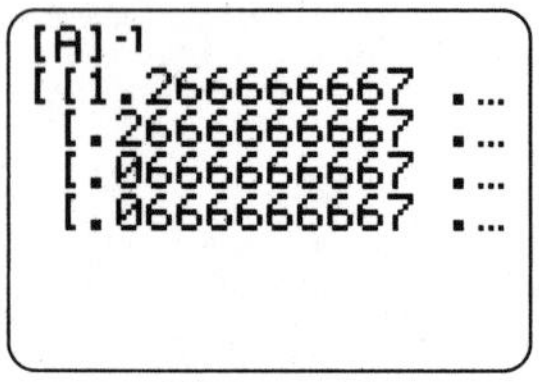

```
[A]⁻¹
[[1.266666667 ....
 [.2666666667 ....
 [.0666666667 ....
 [.0666666667 ....
```

Figure 43

45. $\begin{aligned} 2x - 3y &= 7 \\ -3x - 4y &= 9 \end{aligned} \Rightarrow AX = \begin{bmatrix} 2 & -3 \\ -3 & -4 \end{bmatrix} \begin{bmatrix} x \\ y \end{bmatrix} = \begin{bmatrix} 7 \\ 9 \end{bmatrix} = B$

47. $\begin{aligned} \tfrac{1}{2}x - \tfrac{3}{2}y &= \tfrac{1}{4} \\ -x + 2y &= 5 \end{aligned} \Rightarrow AX = \begin{bmatrix} \tfrac{1}{2} & -\tfrac{3}{2} \\ -1 & 2 \end{bmatrix} \begin{bmatrix} x \\ y \end{bmatrix} = \begin{bmatrix} \tfrac{1}{4} \\ 5 \end{bmatrix} = B$

49. $\begin{aligned} x - 2y + z &= 5 \\ 3y - z &= 6 \\ 5x - 4y - 7z &= 0 \end{aligned} \Rightarrow AX = \begin{bmatrix} 1 & -2 & 1 \\ 0 & 3 & -1 \\ 5 & -4 & -7 \end{bmatrix} \begin{bmatrix} x \\ y \\ z \end{bmatrix} = \begin{bmatrix} 5 \\ 6 \\ 0 \end{bmatrix} = B$

51. $\begin{aligned} 4x - y + 3z &= -2 \\ x + 2y + 5z &= 11 \\ 2x - 3y &= -1 \end{aligned} \Rightarrow AX = \begin{bmatrix} 4 & -1 & 3 \\ 1 & 2 & 5 \\ 2 & -3 & 0 \end{bmatrix} \begin{bmatrix} x \\ y \\ z \end{bmatrix} = \begin{bmatrix} -2 \\ 11 \\ -1 \end{bmatrix} = B$

53. (a) $\begin{aligned} x + 2y &= 3 \\ x + 3y &= 6 \end{aligned} \Rightarrow AX = \begin{bmatrix} 1 & 2 \\ 1 & 3 \end{bmatrix} \begin{bmatrix} x \\ y \end{bmatrix} = \begin{bmatrix} 3 \\ 6 \end{bmatrix} = B$

(b) If $AX = B \Rightarrow \begin{bmatrix} 1 & 2 \\ 1 & 3 \end{bmatrix} \begin{bmatrix} x \\ y \end{bmatrix} = \begin{bmatrix} 3 \\ 6 \end{bmatrix} \Rightarrow X = A^{-1}B \Rightarrow \begin{bmatrix} x \\ y \end{bmatrix} = \begin{bmatrix} 3 & -2 \\ -1 & 1 \end{bmatrix} \begin{bmatrix} 3 \\ 6 \end{bmatrix} = \begin{bmatrix} -3 \\ 3 \end{bmatrix}$

The solution to the system is $(-3, 3)$.

55. (a) $\begin{aligned} -x + 2y &= 5 \\ 3x - 5y &= -2 \end{aligned} \Rightarrow AX = \begin{bmatrix} -1 & 2 \\ 3 & -5 \end{bmatrix} \begin{bmatrix} x \\ y \end{bmatrix} = \begin{bmatrix} 5 \\ -2 \end{bmatrix} = B$

(b) If $AX = B \Rightarrow \begin{bmatrix} -1 & 2 \\ 3 & -5 \end{bmatrix} \begin{bmatrix} x \\ y \end{bmatrix} = \begin{bmatrix} 5 \\ -2 \end{bmatrix} \Rightarrow X = A^{-1}B \Rightarrow \begin{bmatrix} x \\ y \end{bmatrix} = \begin{bmatrix} 5 & 2 \\ 3 & 1 \end{bmatrix} \begin{bmatrix} 5 \\ -2 \end{bmatrix} = \begin{bmatrix} 21 \\ 13 \end{bmatrix}$

The solution to the system is $(21, 13)$.

57. (a) $\begin{aligned} x + z &= -7 \\ 2x + y + 3z &= -13 \\ -x + y + z &= -4 \end{aligned} \Rightarrow AX = \begin{bmatrix} 1 & 0 & 1 \\ 2 & 1 & 3 \\ -1 & 1 & 1 \end{bmatrix} \begin{bmatrix} x \\ y \\ z \end{bmatrix} = \begin{bmatrix} -7 \\ -13 \\ -4 \end{bmatrix} = B$

(b) If $AX = B \Rightarrow \begin{bmatrix} 1 & 0 & 1 \\ 2 & 1 & 3 \\ -1 & 1 & 1 \end{bmatrix} \begin{bmatrix} x \\ y \\ z \end{bmatrix} = \begin{bmatrix} -7 \\ -13 \\ -4 \end{bmatrix} \Rightarrow X = A^{-1}B \Rightarrow \begin{bmatrix} x \\ y \\ z \end{bmatrix} = \begin{bmatrix} -2 & 1 & -1 \\ -5 & 2 & -1 \\ 3 & -1 & 1 \end{bmatrix} \begin{bmatrix} -7 \\ -13 \\ -4 \end{bmatrix} = \begin{bmatrix} 5 \\ 13 \\ -12 \end{bmatrix}$

The solution to the system is $(5, 13, -12)$.

$$x + 2y - z = 2$$

59. (a) $2x + 5y \quad\quad = -1 \Rightarrow AX = \begin{bmatrix} 1 & 2 & -1 \\ 2 & 5 & 0 \\ -1 & -1 & 2 \end{bmatrix} \begin{bmatrix} x \\ y \\ z \end{bmatrix} = \begin{bmatrix} 2 \\ -1 \\ 0 \end{bmatrix} = B$

$$-x - y + 2z = 0$$

(b) If $AX = B \Rightarrow \begin{bmatrix} 1 & 2 & -1 \\ 2 & 5 & 0 \\ -1 & -1 & 2 \end{bmatrix} \begin{bmatrix} x \\ y \\ z \end{bmatrix} = \begin{bmatrix} 2 \\ -1 \\ 0 \end{bmatrix} \Rightarrow X = A^{-1}B = \begin{bmatrix} x \\ y \\ z \end{bmatrix} = \begin{bmatrix} -10 & 3 & -5 \\ 4 & -1 & 2 \\ -3 & 1 & -1 \end{bmatrix} \begin{bmatrix} 2 \\ -1 \\ 0 \end{bmatrix} =$

$\begin{bmatrix} -23 \\ 9 \\ -7 \end{bmatrix}$ The solution to the system is $(-23, 9, -7)$.

Solving Linear Systems

61. (a) $AX = B \Rightarrow \begin{bmatrix} 1.5 & 3.7 \\ -0.4 & -2.1 \end{bmatrix} \begin{bmatrix} x \\ y \end{bmatrix} = \begin{bmatrix} 0.32 \\ 0.36 \end{bmatrix}$

(b) See Figure 61. $X = A^{-1}B \Rightarrow X = \begin{bmatrix} 1.2 \\ -0.4 \end{bmatrix}$

63. (a) $AX = B \Rightarrow \begin{bmatrix} 0.08 & -0.7 \\ 1.1 & -0.05 \end{bmatrix} \begin{bmatrix} x \\ y \end{bmatrix} = \begin{bmatrix} -0.504 \\ 0.73 \end{bmatrix}$

(b) See Figure 63. $X = A^{-1}B \Rightarrow X = \begin{bmatrix} 0.7 \\ 0.8 \end{bmatrix}$

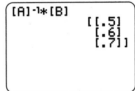

Figure 61

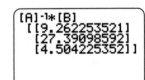

Figure 63   Figure 65   Figure 67

65. (a) $AX = B \Rightarrow \begin{bmatrix} 3.1 & 1.9 & -1 \\ 6.3 & 0 & -9.9 \\ -1 & 1.5 & 7 \end{bmatrix} \begin{bmatrix} x \\ y \\ z \end{bmatrix} = \begin{bmatrix} 1.99 \\ -3.78 \\ 5.3 \end{bmatrix}$

(b) See Figure 65. $X = A^{-1}B \Rightarrow X = \begin{bmatrix} 0.5 \\ 0.6 \\ 0.7 \end{bmatrix}$

67. (a) $AX = B \Rightarrow \begin{bmatrix} 3 & -1 & 1 \\ 5.8 & -2.1 & 0 \\ -1 & 0 & 2.9 \end{bmatrix} \begin{bmatrix} x \\ y \\ z \end{bmatrix} = \begin{bmatrix} 4.9 \\ -3.8 \\ 3.8 \end{bmatrix}$

(b) See Figure 67. $X = A^{-1}B \Rightarrow X \approx \begin{bmatrix} 9.26 \\ 27.39 \\ 4.50 \end{bmatrix}$

Applications

69. Let x be the number of CDs of type A purchased, let y be the CDs of type B and let z be the CDs of type C. The first row in the table implies that $2x + 3y + 4z = 120.91$. The other rows can be interpreted similarly. The system in matrix form is shown below.

$$AX = B \Rightarrow \begin{bmatrix} 2 & 3 & 4 \\ 1 & 4 & 0 \\ 2 & 1 & 3 \end{bmatrix} \begin{bmatrix} x \\ y \\ z \end{bmatrix} = \begin{bmatrix} 120.91 \\ 62.95 \\ 79.94 \end{bmatrix}$$

Use a graphing calculator to find the solution as shown in Figure 69. Type A CDs cost 10.99, type B cost 12.99, and type C cost 14.99.

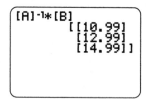

Figure 69

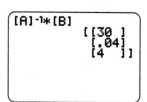

Figure 71   Figure 73

71. (a) At intersection A, incoming traffic is $x_1 + 5$ and outgoing traffic is $4 + 6$. Thus, $x_1 + 5 = 4 + 6$.

At intersection B, incoming traffic is $x_2 + 6$ and outgoing traffic is $x_1 + 3$. Thus, $x_2 + 6 = x_1 + 3$.

At intersection C, incoming traffic is $x_3 + 4$ and outgoing traffic is $x_2 + 7$. Thus, $x_3 + 4 = x_2 + 7$.

At intersection D, incoming traffic is $6 + 5$ and outgoing traffic is $x_3 + x_4$. Thus, $6 + 5 = x_3 + x_4$.

(b) $AX = B \Rightarrow \begin{bmatrix} 1 & 0 & 0 & 0 \\ -1 & 1 & 0 & 0 \\ 0 & -1 & 1 & 0 \\ 0 & 0 & 1 & 1 \end{bmatrix} \begin{bmatrix} x_1 \\ x_2 \\ x_3 \\ x_4 \end{bmatrix} = \begin{bmatrix} 5 \\ -3 \\ 3 \\ 11 \end{bmatrix}$

The solution is given by $X = A^{-1}B$ and can be computed using a graphing calculator. See Figure 71.

Since $X = A^{-1}B = \begin{bmatrix} 5 \\ 2 \\ 5 \\ 6 \end{bmatrix}$, the solution is $x_1 = 5, x_2 = 2, x_3 = 5,$ and $x_4 = 6$.

(c) The traffic traveling west from intersection B to intersection A has a rate of $x_1 = 5$ cars per minute. The values for $x_2, x_3,$ and x_4 can be interpreted in a similar manner.

73. (a) The following three equations must be solved, using the equation $P = a + bS + cC$.

$a + b(1500) + c(\ 8) = 122$
$a + b(2000) + c(\ 5) = 130$
$a + b(2200) + c(10) = 158$

These equations can be written in matrix form as follows:

$$AX = B \Rightarrow \begin{bmatrix} 1 & 1500 & 8 \\ 1 & 2000 & 5 \\ 1 & 2200 & 10 \end{bmatrix} \begin{bmatrix} a \\ b \\ c \end{bmatrix} = \begin{bmatrix} 122 \\ 130 \\ 158 \end{bmatrix}$$

The solution is given by $X = A^{-1}B$ as shown in Figure 73. It is $a = 30, b = 0.04,$ and $c = 4$.

That is, P is given by $P = 30 + 0.04S + 4C$

(b) $P = 30 + 0.04(1800) + 4(7) = 130$, or $130,000$

9.7 Determinants

Calculating Determinants

1. $\det A = \det\begin{bmatrix} 4 & 3 \\ 5 & 4 \end{bmatrix} = (4)(4) - (5)(3) = 1 \neq 0;$ A is invertible.

3. $\det A = \det\begin{bmatrix} -4 & 6 \\ -8 & 12 \end{bmatrix} = (-4)(12) - (-8)(6) = 0;$ A is not invertible.

5. Deleting the first row and second column gives $M_{12} = \det A = \det\begin{bmatrix} 2 & -2 \\ 0 & 5 \end{bmatrix} = (2)(5) - (0)(-2) = 10.$

   The cofactor is $A_{12} = (-1)^{1+2}M_{12} = -1(10) = -10.$

7. Deleting the second row and second column gives $M_{22} = \det A = \det\begin{bmatrix} 7 & 1 \\ 1 & -2 \end{bmatrix} = (7)(-2) - (1)(1) = -15.$

   The cofactor is $A_{22} = (-1)^{2+2}M_{22} = 1(-15) = -15.$

9. $\det A = a_{11}A_{11} + a_{21}A_{21} + a_{32}A_{31} = a_{11}M_{11} - a_{21}M_{21} + a_{31}M_{31} \Rightarrow$

   $\det A = (1)\det\begin{bmatrix} 2 & -3 \\ -1 & 3 \end{bmatrix} - (0)\det\begin{bmatrix} 4 & -7 \\ -1 & 3 \end{bmatrix} + (0)\det\begin{bmatrix} 4 & -7 \\ 2 & -3 \end{bmatrix} = (1)(6-3) = 3$

   Since $\det A = 3 \neq 0$, A^{-1} exists.

11. $\det A = a_{11}A_{11} + a_{21}A_{21} + a_{32}A_{31} = a_{11}M_{11} - a_{21}M_{21} + a_{31}M_{31} \Rightarrow$

    $\det A = (5)\det\begin{bmatrix} -2 & 0 \\ 4 & 0 \end{bmatrix} - (0)\det\begin{bmatrix} 1 & 6 \\ 4 & 0 \end{bmatrix} + (0)\det\begin{bmatrix} 1 & 6 \\ -2 & 0 \end{bmatrix} = (5)(0-0) = 0$

    Since $\det A = 0$, A^{-1} does not exists.

13. Expanding about the first column results in $\det A = 2\det\begin{bmatrix} 3 & 0 \\ 0 & 5 \end{bmatrix} - 0 + 0 = (2)(15-0) = 30.$

15. Expanding about the first row results in $\det A = 0.$

17. Expanding about the first column results in

    $\det A = 3\det\begin{bmatrix} 5 & 7 \\ 0 & -1 \end{bmatrix} - 0 + 1\det\begin{bmatrix} -1 & 2 \\ 5 & 7 \end{bmatrix} = (3)(-5-0) + (1)(-7-10) = -32.$

19. Expanding about the first column results in

    $\det A = 1\det\begin{bmatrix} 1 & 3 \\ 4 & -2 \end{bmatrix} - (-7)\det\begin{bmatrix} -5 & 2 \\ 4 & -2 \end{bmatrix} + 0 = (1)(-2-12) - (-7)(10-8) = 0.$

21. $\det A = \det\begin{bmatrix} 11 & -32 \\ 1.2 & 55 \end{bmatrix} = 643.4$

23. $\det A = \det\begin{bmatrix} 2.3 & 5.1 & 2.8 \\ 1.2 & 4.5 & 8.8 \\ -0.4 & -0.8 & -1.2 \end{bmatrix} = -4.484$

Calculating Area

25. Enter the vertices as columns in a counterclockwise direction.

    $D = \dfrac{1}{2}\det\begin{bmatrix} 0 & 4 & 1 \\ 0 & 2 & 4 \\ 1 & 1 & 1 \end{bmatrix} = 7;$ The area of the triangle is 7 square units.

27. A line segment between (1, 3) and (3, 2) divides the quadrangle into two triangles whose areas can be found using determinants. Enter the vertices of each triangle as columns in a counterclockwise direction.

$$D = \frac{1}{2}\det\begin{bmatrix} 0 & 3 & 1 \\ 0 & 2 & 3 \\ 1 & 1 & 1 \end{bmatrix} + \frac{1}{2}\det\begin{bmatrix} 1 & 3 & 5 \\ 3 & 2 & 4 \\ 1 & 1 & 1 \end{bmatrix} = 6.5; \text{ The area of the quadrangle is 6.5 square units.}$$

Cramer's Rule

29. By Cramer's rule, the solution can be found as follows:

$$E = \det\begin{bmatrix} 5 & 2 \\ 1 & 3 \end{bmatrix} = 13; \quad F = \det\begin{bmatrix} -1 & 5 \\ 3 & 1 \end{bmatrix} = -16; \quad D = \det\begin{bmatrix} -1 & 2 \\ 3 & 3 \end{bmatrix} = -9$$

Thus $x = \dfrac{E}{D} = \dfrac{13}{-9}$ and $y = \dfrac{F}{D} = \dfrac{-16}{-9} = \dfrac{16}{9}$. The solution is $\left(-\dfrac{13}{9}, \dfrac{16}{9}\right)$.

31. By Cramer's rule, the solution can be found as follows:

$$E = \det\begin{bmatrix} 8 & 3 \\ 3 & -5 \end{bmatrix} = -49; \quad F = \det\begin{bmatrix} -2 & 8 \\ 4 & 3 \end{bmatrix} = -38; \quad D = \det\begin{bmatrix} -2 & 3 \\ 4 & -5 \end{bmatrix} = -2$$

Thus $x = \dfrac{E}{D} = \dfrac{-49}{-2} = \dfrac{49}{2}$ and $y = \dfrac{F}{D} = \dfrac{-38}{-2} = 19$. The solution is $\left(\dfrac{49}{2}, 19\right)$.

33. By Cramer's rule, the solution can be found as follows:

$$E = \det\begin{bmatrix} 23 & 4 \\ 70 & -5 \end{bmatrix} = -395; \quad F = \det\begin{bmatrix} 7 & 23 \\ 11 & 70 \end{bmatrix} = 237; \quad D = \det\begin{bmatrix} 7 & 4 \\ 11 & -5 \end{bmatrix} = -79$$

Thus $x = \dfrac{E}{D} = \dfrac{-395}{-79} = 5$ and $y = \dfrac{F}{D} = \dfrac{237}{-79} = -3$. The solution is (5, –3).

35. By Cramer's rule, the solution can be found as follows:

$$E = \det\begin{bmatrix} -0.91 & -2.5 \\ 0.423 & 0.9 \end{bmatrix} = 0.2385; \quad F = \det\begin{bmatrix} 1.7 & -0.91 \\ -0.4 & 0.423 \end{bmatrix} = 0.3551; \quad D = \det\begin{bmatrix} 1.7 & -2.5 \\ -0.4 & 0.9 \end{bmatrix} = 0.53$$

Thus $x = \dfrac{E}{D} = \dfrac{0.2385}{0.53} = 0.45$ and $y = \dfrac{F}{D} = \dfrac{0.3551}{0.53} = 0.67$. The solution is (0.45, 0.67).

Applying a Concept

37. If the three points form a triangle with no area ($D = 0$ using determinants), then the points must be collinear.

$$D = \frac{1}{2}\det\begin{bmatrix} 1 & -3 & 2 \\ 3 & 11 & 1 \\ 1 & 1 & 1 \end{bmatrix} = 0; \text{ The points are collinear.}$$

39. If the three points form a triangle with no area ($D = 0$ using determinants), then the points must be collinear.

$$D = \frac{1}{2}\det\begin{bmatrix} -2 & 4 & 2 \\ -5 & 4 & 3 \\ 1 & 1 & 1 \end{bmatrix} = 6 \neq 0; \text{ The points are not collinear.}$$

Chapter 9 Review Exercises

1. $f(b, h) = \dfrac{1}{2}bh \Rightarrow f(3, 6) = \dfrac{1}{2}(3)(6) = 9$

3. $3x - 4y = -2 \Rightarrow -4y = -3x - 2 \Rightarrow y = \dfrac{3x + 2}{4}$

5. (a) $3x + y = 1 \Rightarrow y = 1 - 3x$ and $2x - 3y = 8 \Rightarrow y = \dfrac{2x - 8}{3}$

   Graph $Y_1 = 1 - 3X$ and $Y_2 = (2X - 8)/3$. The graphs intersect at the point $(1, -2)$. See Figure 5.

   (b) Substituting $y = 1 - 3x$ into the second equation gives $2x - 3(1 - 3x) = 8 \Rightarrow 2x - 3 + 9x = 8 \Rightarrow$

   $11x = 11 \Rightarrow x = 1$. If $x = 1$, then $y = 1 - 3(1) = -2$. The solution is $(1, -2)$.

[−10, 10, 1] by [−10, 10, 1]   [−9.4, 9.4, 1] by [−6.2, 6.2, 1]

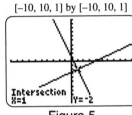
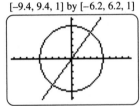

Figure 5   Figure 7

7. (a) $x^2 + y^2 = 25 \Rightarrow y^2 = -x^2 + 25 \Rightarrow y = \sqrt{-x^2 + 25}$ and $4x - 3y = 0 \Rightarrow -3y = -4x \Rightarrow y = \dfrac{4x}{3}$

   Graph $Y_1 = \sqrt{(-X^2 + 25)}$, $Y_2 = -\sqrt{(-X^2 + 5)}$ and $Y_3 = \dfrac{4X}{3}$. The graphs intersect at the points

   $(3, 4)$ and $(-3, -4)$. See Figure 7.

   (b) Substituting $y = \dfrac{4x}{3}$ into the first equation gives $x^2 + \left(\dfrac{4x}{3}\right)^2 = 25 \Rightarrow x^2 + \dfrac{16x^2}{9} = 25 \Rightarrow$

   $9x^2 + 16x^2 = 225 \Rightarrow 25x^2 = 225 \Rightarrow x^2 = 9 \Rightarrow x = \pm 3$. If $x = 3$ then $y = \dfrac{4(3)}{3} \Rightarrow y = 4$ and if

   $x = -3$ then $y = \dfrac{4(-3)}{3} \Rightarrow y = -4$. The solutions are$(3, 4)$ and $(-3, -4)$.

9. Multiply the first equation by 2 and add to eliminate the y-variable.

   $\begin{array}{r} 4x + 2y = 14 \\ x - 2y = -4 \\ \hline 5x \quad\quad = 10 \end{array} \Rightarrow x = 2$

   Since $2x + y = 7$, it follows that $y = 3$. The solution is $(2, 3)$. The system is consistent.

11. Multiply the first equation by 2, the second equation by 3, and add to eliminate the y-variable.

   $\begin{array}{r} 12x - 30y = 24 \\ -12x + 30y = -24 \\ \hline 0 = 0 \end{array} \Rightarrow$ infinitely many solutions

   Since $0 = 0$, the system is consistent and there are infinitely many solutions.

13. Subtract the equations to eliminate the x-variable.

$$x^2 - 3y = 3$$
$$\underline{x^2 + 2y^2 = 5}$$

$$-3y - 2y^2 = -2 \Rightarrow 0 = 2y^2 + 3y - 2 \Rightarrow (2y - 1)(y + 2) = 0 \Rightarrow y = -2, \frac{1}{2}.$$

If $y = -2$ then $x^2 - 3(-2) = 3 \Rightarrow x^2 + 6 = 3 \Rightarrow x^2 = -3$. There is no real solution to this $\Rightarrow y \neq -2$.

If $y = \frac{1}{2}$ then $x^2 - 3\left(\frac{1}{2}\right) = 3 \Rightarrow x^2 = \frac{9}{2} \Rightarrow x = \pm\frac{3}{\sqrt{2}} \Rightarrow x = \pm\frac{3\sqrt{2}}{2}$.

The solutions are $\left(\frac{3\sqrt{2}}{2}, \frac{1}{2}\right)$ and $\left(\frac{-3\sqrt{2}}{2}, \frac{1}{2}\right)$.

15. See Figure 15.

17. $2x - y < 4 \Rightarrow -y < -2x + 4 \Rightarrow y > 2x - 4$. See Figure 17.

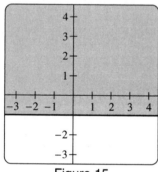

Figure 15

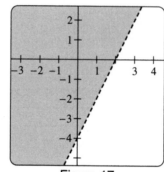

Figure 17

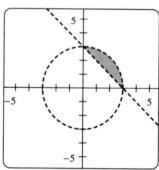

Figure 19

19. The solution region is inside the circle or radius 3 and above the line $y = 3 - x$. Their graphs intersect at the points $(0, 3)$ and $(3, 0)$. It does not include the boundary. See Figure 19. One solution to the system is $(2, 2)$.

21. The solution region lies below the line $x + y = 4$ and above the line $x + 3y = 3$. Their graphs intersect at $\left(\frac{9}{2}, \frac{-1}{2}\right)$. The region includes the boundary. See Figure 21. One solution to the system is $(0, 2)$.

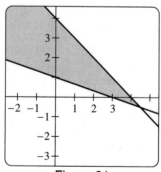

Figure 21

23. First, subtract the first two equations to eliminate the x-variable.

$$\begin{aligned} x - \ y + \ \ z &= -2 \\ x + 2y - \ \ z &= 2 \\ \hline -3y + 2z &= -4 \end{aligned}$$

Multiply this equation by 2 and multiply the last equation by 3 to eliminate the y-variable.

$$\begin{aligned} -6y + 4z &= -8 \\ 6y + 9z &= 21 \\ \hline 13z = 13 &\Rightarrow z = 1 \end{aligned}$$

Substituting $z = 1$ into $2y + 3z = 7$ gives $2y + 3 = 7 \Rightarrow 2y = 4 \Rightarrow y = 2$. Substituting $y = 2$ and $z = 1$

into the original first equation gives $x - 2 + 1 = -2 \Rightarrow x - 1 = -2 \Rightarrow x = -1$.

The solution to the system is $(-1, 2, 1)$.

25. Add the first two equations to eliminate the x-variable.

$$\begin{aligned} -x + 2y + 2z &= \ 9 \\ x + \ y - 3z &= \ 6 \\ \hline 3y - \ z &= 15 \end{aligned}$$

Subtract this new equation from the third equation.

$$\begin{aligned} 3y - z &= \ 8 \\ 3y - z &= 15 \\ \hline 0 &= -7 \end{aligned}$$

Since $0 \neq -7$, there is no solution.

27. The system is $x + 5y = 6$ and $y = 3$.

Substituting $y = 3$ into the first equation gives $x + 5(3) = 6 \Rightarrow x = -9$. The solution is $(-9, 3)$.

29. The augmented matrix is in Row-Echelon Form $\Rightarrow x = -2, y = 3, z = 0 \Rightarrow (-2, 3, 0)$.

31. The system can be written as follows:

$$\begin{bmatrix} 1 & 3 & | & 8 \\ -1 & 1 & | & 4 \end{bmatrix} R_2 + R_1 \rightarrow \begin{bmatrix} 1 & 3 & | & 8 \\ 0 & 4 & | & 12 \end{bmatrix} (1/4)R_2 \rightarrow \begin{bmatrix} 1 & 3 & | & 8 \\ 0 & 1 & | & 3 \end{bmatrix}$$

Thus $y = 3$ and $x + 3(3) = 8 \Rightarrow x = -1$. The solution is $(-1, 3)$.

33. $\begin{bmatrix} 2 & -1 & 2 & | & 10 \\ 1 & -2 & 1 & | & 8 \\ 3 & -1 & 2 & | & 11 \end{bmatrix} \begin{array}{l} (1/2)R_1 \rightarrow \\ R_1 - 2R_2 \rightarrow \\ 3R_2 - R_3 \rightarrow \end{array} \begin{bmatrix} 1 & -\frac{1}{2} & 1 & | & 5 \\ 0 & 3 & 0 & | & -6 \\ 0 & -5 & 1 & | & 13 \end{bmatrix} (1/3)R_2 \rightarrow \begin{bmatrix} 1 & -\frac{1}{2} & 1 & | & 5 \\ 0 & 1 & 0 & | & -2 \\ 0 & -5 & 1 & | & 13 \end{bmatrix}$

$5R_2 + R_3 \rightarrow \begin{bmatrix} 1 & -\frac{1}{2} & 1 & | & 5 \\ 0 & 1 & 0 & | & -2 \\ 0 & 0 & 1 & | & 3 \end{bmatrix}$

Backward substitution produces $z = 3; y = -2; x - \dfrac{1}{2}(-2) + 3 = 5 \Rightarrow x + 1 + 3 = 5 \Rightarrow x = 1$.

The solution is $(1, -2, 3)$.

35. (a) $a_{12} = 3$ and $a_{22} = 2 \Rightarrow a_{12} + a_{22} = 3 + 2 \Rightarrow a_{12} + a_{22} = 5$

(b) $a_{11} = -2$ and $a_{23} = 4 \Rightarrow a_{11} - 2a_{23} = -2 - 2(4) \Rightarrow a_{11} - 2a_{23} = -10$

37. (a) $A + 2B = \begin{bmatrix} 1 & -3 \\ 2 & -1 \end{bmatrix} + 2\begin{bmatrix} 3 & 2 \\ -5 & 1 \end{bmatrix} = \begin{bmatrix} 7 & 1 \\ -8 & 1 \end{bmatrix}$

(b) $A - B = \begin{bmatrix} 1 & -3 \\ 2 & -1 \end{bmatrix} - \begin{bmatrix} 3 & 2 \\ -5 & 1 \end{bmatrix} = \begin{bmatrix} -2 & -5 \\ 7 & -2 \end{bmatrix}$

(c) $-4A = -4\begin{bmatrix} 1 & -3 \\ 2 & -1 \end{bmatrix} = \begin{bmatrix} -4 & 12 \\ -8 & 4 \end{bmatrix}$

39. A and B are both 2×2 so AB and BA are also both 2×2.

$AB = \begin{bmatrix} 2 & 0 \\ -5 & 3 \end{bmatrix}\begin{bmatrix} -1 & -2 \\ 4 & 7 \end{bmatrix} = \begin{bmatrix} -2 & -4 \\ 17 & 31 \end{bmatrix}$; $BA = \begin{bmatrix} -1 & -2 \\ 4 & 7 \end{bmatrix}\begin{bmatrix} 2 & 0 \\ -5 & 3 \end{bmatrix} = \begin{bmatrix} 8 & -6 \\ -27 & 21 \end{bmatrix}$

41. A is 2×3 and B is 3×2 so AB is 2×2 and BA is 3×3.

$AB = \begin{bmatrix} 2 & -1 & 3 \\ 2 & 4 & 0 \end{bmatrix}\begin{bmatrix} 1 & 0 \\ -1 & 2 \\ 0 & 3 \end{bmatrix} = \begin{bmatrix} 3 & 7 \\ -2 & 8 \end{bmatrix}$; $BA = \begin{bmatrix} 1 & 0 \\ -1 & 2 \\ 0 & 3 \end{bmatrix}\begin{bmatrix} 2 & -1 & 3 \\ 2 & 4 & 0 \end{bmatrix} = \begin{bmatrix} 2 & -1 & 3 \\ 2 & 9 & -3 \\ 6 & 12 & 0 \end{bmatrix}$

43. B is the inverse of A.

$AB = \begin{bmatrix} 8 & 5 \\ 6 & 4 \end{bmatrix}\begin{bmatrix} 2 & -2.5 \\ -3 & 4 \end{bmatrix} = \begin{bmatrix} 1 & 0 \\ 0 & 1 \end{bmatrix}$ and $BA = \begin{bmatrix} 2 & -2.5 \\ -3 & 4 \end{bmatrix}\begin{bmatrix} 8 & 5 \\ 6 & 4 \end{bmatrix} = \begin{bmatrix} 1 & 0 \\ 0 & 1 \end{bmatrix}$

45. (a) $\begin{array}{l} x - 3y = 4 \\ 2x - y = 3 \end{array} \Rightarrow AX = \begin{bmatrix} 1 & -3 \\ 2 & -1 \end{bmatrix}\begin{bmatrix} x \\ y \end{bmatrix} = \begin{bmatrix} 4 \\ 3 \end{bmatrix} = B$

(b) $X = A^{-1}B \Rightarrow \begin{bmatrix} x \\ y \end{bmatrix} = \begin{bmatrix} -\frac{1}{5} & \frac{3}{5} \\ -\frac{2}{5} & \frac{1}{5} \end{bmatrix}\begin{bmatrix} 4 \\ 3 \end{bmatrix} = \begin{bmatrix} 1 \\ -1 \end{bmatrix}$

47. (a) $AX = B \Rightarrow \begin{bmatrix} 11 & 31 \\ 37 & -19 \end{bmatrix}\begin{bmatrix} x \\ y \end{bmatrix} = \begin{bmatrix} -27.6 \\ 240 \end{bmatrix}$

(b) See Figure 47. $X = A^{-1}B \Rightarrow X = \begin{bmatrix} 5.1 \\ -2.7 \end{bmatrix}$

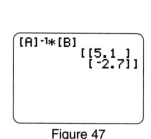

Figure 47

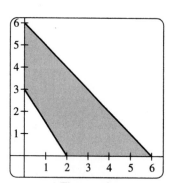

Figure 49

49. See Figure 49.

51. $A \mid I_2 = \begin{bmatrix} 1 & -2 & | & 1 & 0 \\ -1 & 1 & | & 0 & 1 \end{bmatrix} R_2 + R_1 \rightarrow \begin{bmatrix} 1 & -2 & | & 1 & 0 \\ 0 & -1 & | & 1 & 1 \end{bmatrix} (-1)R_2 \rightarrow \begin{bmatrix} 1 & -2 & | & 1 & 0 \\ 0 & 1 & | & -1 & -1 \end{bmatrix}$

$R_1 + 2R_2 \rightarrow \begin{bmatrix} 1 & 0 & | & -1 & -2 \\ 0 & 1 & | & -1 & -1 \end{bmatrix}$; $A^{-1} = \begin{bmatrix} -1 & -2 \\ -1 & -1 \end{bmatrix}$

53. Expanding about the first column results in

$\det A = 2\det\begin{bmatrix} 3 & 4 \\ 0 & 5 \end{bmatrix} - 0 + 1\det\begin{bmatrix} 1 & 3 \\ 3 & 4 \end{bmatrix} = (2)(15 - 0) + (1)(4 - 9) = 25.$

55. $\det A = \det \begin{bmatrix} 13 & 22 \\ 55 & -57 \end{bmatrix} = (13)(-57) - (55)(22) = -1951 \neq 0;$ A is invertible.

Applications

57. (a) $\dfrac{249}{99} \approx 2.5$; the sun is approximately 2.5 times stronger in terms of UV-B light.

(b) They are both approximately equal to 143.

59. The given equations result in the following nonlinear system of equations: $A(l, w) = 77 \Rightarrow lw = 77$ and

$P(l, w) = 36 \Rightarrow 2l + 2w = 36$. Begin by solving the second equation for l.

$2l + 2w = 36 \Rightarrow 2l = 36 - 2w \Rightarrow l = 18 - w$. Substitute this into the first equation.

$lw = 77 \Rightarrow (18 - w)w = 77 \Rightarrow 18w - w^2 = 77 \Rightarrow w^2 - 18w + 77 = 0$. This is a quadratic equation that

can be solved by factoring. $w^2 - 18w + 77 = 0 \Rightarrow (w - 7)(w - 11) = 0 \Rightarrow w = 7$ or 11.

Since $l = 18 - w$, if $w = 7$, then $l = 11$ and if $w = 11$ then $l = 7$. For this rectangle $l = 11$ and $w = 7$.

61. (a) Let x represent the amount of the 7% loan and let y represent the amount of the 9% loan. Then the system

of equations is $x + y = 2000$ and $0.07x + 0.09y = 156$. Multiply the first equation by 0.09 and subtract.

$$0.09x + 0.09y = 180$$
$$\underline{0.07x + 0.09y = 156}$$
$$0.02x \qquad\quad = 24 \ \Rightarrow x = 1200 \text{ and } y = 2000 - 1200 = 800$$

The loan amounts are $1200 at 7% and $800 at 9%.

(b) Graph $Y_1 = 2000 - X$ and $Y_2 = (156 - 0.07X)/0.09$. The graph intersect at (1200, 800). See Figure 61.

The loan amounts are $1200 at 7% and $800 at 9%.

[0, 2000, 100] by [0, 2000, 100]   [0, 7, 1] by [0, 7, 1]   [0, 7, 1] by [0, 7, 1]

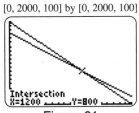
Figure 61

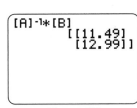
Figure 63

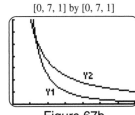
Figure 67b

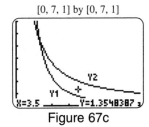
Figure 67c

63. Let x be the number of CDs of type A purchased and let y be the number of CDs of type B. Then from the table

we see that $1x + 2y = 37.47$ and $2x + 3y = 61.95$.

$$AX = B \Rightarrow \begin{bmatrix} 1 & 2 \\ 2 & 3 \end{bmatrix} \begin{bmatrix} x \\ y \end{bmatrix} = \begin{bmatrix} 37.47 \\ 61.95 \end{bmatrix}$$

Use a graphing calculator to solve the system as shown in Figure 63. Type A CDs cost $11.49, type B cost $12.99.

65. Enter the vertices as columns in a counterclockwise direction.

$$D = \frac{1}{2} \det \begin{bmatrix} 0 & 5 & 2 \\ 0 & 2 & 5 \\ 1 & 1 & 1 \end{bmatrix} = 10.5;$$ The area of the triangle is 10.5 square units.

67. (a) The volume satisfies $V = \pi r^2 h \geq 30$ and the surface area satisfies $S = 2\pi r h \leq 45$.

(b) $\pi r^2 h \geq 30 \Rightarrow h \geq \dfrac{30}{\pi r^2}$ and $2\pi r h \leq 45 \Rightarrow h \leq \dfrac{45}{2\pi r}$. Graph $Y_1 = 30/(\pi X^2)$ and $Y_2 = 45/(2\pi X)$.

The solution is between the two graphs (above Y_1 and below Y_2) and includes the graphs. See Figure 67b.

(c) From Figure 67c we can see that the free-moving curser is located in the region of the solutions. One

solution occurs at (3.5, 1.35), that is, a solution to the system of inequalities is $r = 3.5$ and $h = 1.35$.

69. Since P varies jointly as the square of x and the cube of y, the variation equation $P = kx^2y^3$ must hold.

    If $P = 432$ when $x = 2$ and $y = 3$, then $432 = k(2)^2(3)^3 \Rightarrow 432 = 108k \Rightarrow k = 4$.

    Our variation equation becomes $P = 4x^2y^3$. Thus, when $x = 3$ and $y = 5$, $P = 4(3)^2(5)^3 = 4500$.

Chapter 10: Conic Sections

10.1: Parabolas

Parabolas with Vertex (0, 0)

1. See Figure 1.

3. See Figure 3.

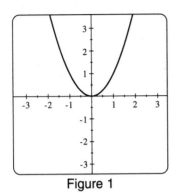
Figure 1

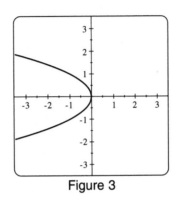
Figure 3

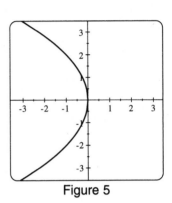
Figure 5

5. See Figure 5.

7. See Figure 7.

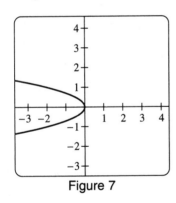
Figure 7

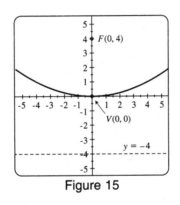
Figure 15

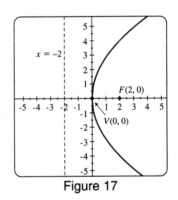
Figure 17

9. Opens upward; e

11. Opens to the left; a

13. Opens to the right, passes through (2, 2); d

15. The equation $16y = x^2$ is in the form $x^2 = 4py$, and the vertex is $V(0, 0)$. Thus, $16 = 4p$ or $p = 4$. The focus is $F(0, 4)$, the equation of the directrix is $y = -4$, and the parabola opens upward. The graph of $y = \dfrac{1}{16}x^2$ is shown in Figure 15.

17. The equation $x = \dfrac{1}{8}y^2$ can be written as $y^2 = 8x$, which is in the form $y^2 = 4px$. The vertex is $V(0, 0)$. Thus, $8 = 4p$ or $p = 2$. The focus is $F(2, 0)$, the equation of the directrix is $x = -2$, and the parabola opens to the right. The graph of $x = \dfrac{1}{8}y^2$ is shown in Figure 17.

19. The equation $-4x = y^2$ can be written as $y^2 = -4x$, which is in the form $y^2 = 4px$. The vertex is $V(0, 0)$. Thus, $-4 = 4p$ or $p = -1$. The focus is $F(-1, 0)$, the equation of the directrix is $x = 1$, and the parabola opens to the left. The graph of $x = -\dfrac{1}{4}y^2$ is shown in Figure 19.

21. The equation $x^2 = -8y$ is in the form $x^2 = 4py$, and the vertex is $V(0, 0)$. Thus, $-8 = 4p$ or $p = -2$. The focus is $F(0, -2)$, the equation of the directrix is $y = 2$, and the parabola opens downward. The graph of $y = -\dfrac{1}{8}x^2$ is shown in Figure 21.

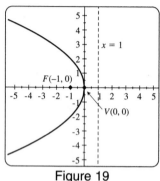

Figure 19

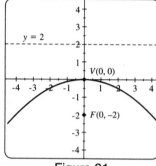

Figure 21

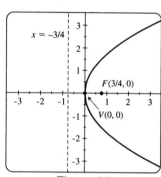

Figure 23

23. The equation $2y^2 = 6x$ can be written as $y^2 = 3x$, which is in the form $y^2 = 4px$. The vertex is $V(0, 0)$. Thus, $3 = 4p$ or $p = \dfrac{3}{4}$. The focus is $F\left(\dfrac{3}{4}, 0\right)$, the equation of the directrix is $x = -\dfrac{3}{4}$, and the parabola opens to the right. The graph of $x = \dfrac{1}{3}y^2$ is shown in Figure 23.

25. The equation $2y^2 = -8x$ can be written as $y^2 = -4x$, which is in the form $y^2 = 4px$. The vertex is $V(0, 0)$. Thus, $-4 = 4p$ or $p = -1$. The focus is $F(-1, 0)$, the equation of the directrix is $x = 1$, and the parabola opens to the left. The graph of $x = -\dfrac{1}{4}y^2$ is shown in Figure 25.

27. The focus is $F(0, 1)$ and the vertex is $V(0, 0)$. The distance between these points is 1. Since the focus is above the directrix, the parabola opens upward, so $p = 1$. Since the line passing through F and V is vertical, the parabola has a vertical axis. Its equation is given by $x^2 = 4py$ or $x^2 = 4y$. The graph of $y = \dfrac{1}{4}x^2$ is shown in Figure 27.

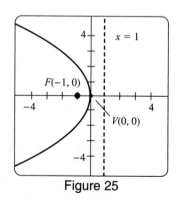

Figure 25

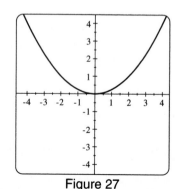

Figure 27

29. The focus is $F(-3, 0)$ and the vertex is $V(0, 0)$. The distance between these points is 3. Since the focus is left of the directrix, the parabola opens to the left, so $p = -3$. Since the line passing through F and V is horizontal, the parabola has a horizontal axis. Its equation is given by $y^2 = 4px$ or $y^2 = -12x$. The graph of $x = -\dfrac{1}{12}y^2$ is shown in Figure 29.

31. If the vertex is $V(0, 0)$ and the focus is $F\left(0, \dfrac{3}{4}\right)$, then the parabola opens upward and $p = \dfrac{3}{4}$. Thus,

$x^2 = 4py \Rightarrow x^2 = 3y$. Its graph is shown in Figure 31.

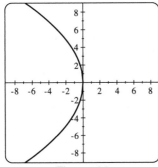
Figure 29

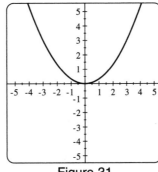
Figure 31

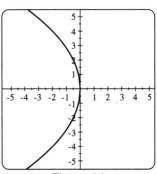
Figure 33

33. If the vertex is $V(0, 0)$ and the directrix is $x = 2$, then the parabola opens to the left and $p = -2$. Thus, $y^2 = 4px \Rightarrow y^2 = -8x$. Its graph is shown in Figure 33.

35. If the vertex is $V(0, 0)$ and the focus is $F(1, 0)$, then the parabola opens to the right and $p = 1$. Thus, $y^2 = 4px \Rightarrow y^2 = 4x$. Its graph is shown in Figure 35.

37. If the vertex is $V(0, 0)$ and the directrix is $x = \dfrac{1}{4}$, then the parabola opens to the left and $p = -\dfrac{1}{4}$. Thus, $y^2 = 4px \Rightarrow y^2 = -x$. Its graph is shown in Figure 37.

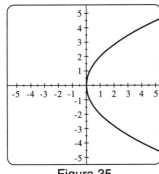
Figure 35

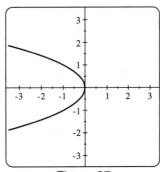
Figure 37

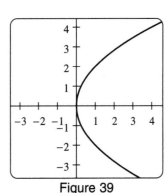
Figure 39

39. If the vertex is $V(0, 0)$ and the parabola has a horizontal axis, the equation is in the form $y^2 = 4px$. Find the value of p by using the fact that the parabola passes through $(1, -2)$. Thus, $(-2)^2 = 4p(1) \Rightarrow p = 1$. The equation is $y^2 = 4x$. Its graph is shown in Figure 39.

41. If the focus is $F(0, -3)$ and the equation of the directrix is $y = 3$, the vertex is $V(0, 0)$, the parabola opens downward, and $p = -3$. Thus, $x^2 = 4py \Rightarrow x^2 = -12y$.

43. If the focus is $F(-1, 0)$ and the equation of the directrix is $x = 1$, the vertex is $V(0, 0)$, the parabola opens to the left, and $p = -1$. Thus, $y^2 = 4px \Rightarrow y^2 = -4x$.

Parabolas with Vertex (h, k)

45. See Figure 45.

47. See Figure 47.

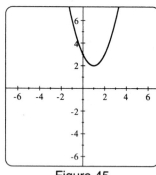

Figure 45

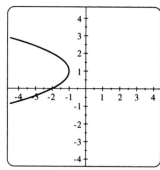

Figure 47

49. Vertex at $(1, 1)$; c

51. Vertex at $(0, 2)$; a

53. The equation $(x - 2)^2 = 8(y + 2)$ is in the form $(x - h)^2 = 4p(y - k)$, with $(h, k) = (2, -2)$ and $p = 2$. The parabola opens upward, with vertex at $(2, -2)$, focus at $(2, 0)$, and equation of directrix $y = -4$. The graph is shown in Figure 53.

55. The equation $x = -\dfrac{1}{4}(y + 3)^2 + 2$ can be written as $(y + 3)^2 = -4(x - 2)$, which is in the form $(y - k)^2 = 4p(x - h)$, with $(h, k) = (2, -3)$ and $p = -1$. The parabola opens to the left, with vertex at $(2, -3)$, focus at $(1, -3)$, and equation of directrix $x = 3$. The graph is shown in Figure 55.

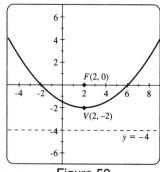

Figure 53

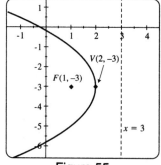

Figure 55

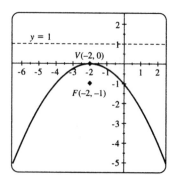

Figure 57

57. The equation $y = -\dfrac{1}{4}(x + 2)^2$ can be written as $(x + 2)^2 = -4y$, which is in the form $(x - h)^2 = 4p(y - k)$, with $(h, k) = (-2, 0)$ and $p = -1$. The parabola opens downward, with vertex at $(-2, 0)$, focus at $(-2, -1)$, and equation of directrix $y = 1$. The graph is shown in Figure 57.

59. If the focus is at $(0, 2)$ and the vertex at $(0, 1)$, the parabola opens upward and $p = 1$. Substituting in $(x - h)^2 = 4p(y - k)$, we get $(x - 0)^2 = 4(1)(y - 1)$ or $x^2 = 4(y - 1)$.

61. If the focus is at $(0, 0)$ and the directrix has equation $x = -2$, the vertex is at $(-1, 0)$, $p = 1$, and the parabola opens to the right. Substituting in $(y - k)^2 = 4p(x - h)$, we get $(y - 0)^2 = 4(1)(x - (-1))$ or $y^2 = 4(x + 1)$.

63. If the focus is at $(-1, 3)$ and the directrix has equation $y = 7$, the vertex is at $(-1, 5)$, $p = -2$, and the parabola opens downward. Substituting in $(x - h)^2 = 4p(y - k)$, we get $(x + 1)^2 = 4(-2)(y - 5)$ or $(x + 1)^2 = -8(y - 5)$.

65. Since the parabola has a horizontal axis, the equation is in the form $(y - k)^2 = a(x - h)$. Find the value of a by using the fact that the parabola passes through $(-4, 0)$ and the vertex is $V(-2, 3)$.

   Substituting $x = -4$, $y = 0$, $h = -2$ and $k = 3$ yields $(0 - 3)^2 = a(-4 - (-2)) \Rightarrow a = -\dfrac{9}{2}$.
   The equation is $(y - 3)^2 = -\dfrac{9}{2}(x + 2)$.

67. $-2x = y^2 + 6x + 10 \Rightarrow y^2 = -8x - 10 \Rightarrow (y - 0)^2 = -8\left(x + \dfrac{5}{4}\right)$

69. $x = 2y^2 + 4y - 1 \Rightarrow 2y^2 + 4y = x + 1 \Rightarrow y^2 + 2y = \dfrac{1}{2}(x + 1) \Rightarrow y^2 + 2y + 1 = \dfrac{1}{2}(x + 1) + 1 \Rightarrow$
   $(y + 1)^2 = \dfrac{1}{2}(x + 1 + 2) \Rightarrow (y + 1)^2 = \dfrac{1}{2}(x + 3)$

71. $x^2 - 3x + 4 = 2y \Rightarrow x^2 - 3x = 2y - 4 \Rightarrow x^2 - 3x + \dfrac{9}{4} = 2y - 4 + \dfrac{9}{4} \Rightarrow \left(x - \dfrac{3}{2}\right)^2 = 2y - \dfrac{7}{4} \Rightarrow$
   $\left(x - \dfrac{3}{2}\right)^2 = 2\left(y - \dfrac{7}{8}\right)$

73. $4y^2 + 4y - 5 = 5x \Rightarrow 4y^2 + 4y = 5x + 5 \Rightarrow y^2 + y = \dfrac{5}{4}(x + 1) \Rightarrow y^2 + y + \dfrac{1}{4} = \dfrac{5}{4}(x + 1) + \dfrac{1}{4} \Rightarrow$
   $\left(y + \dfrac{1}{2}\right)^2 = \dfrac{5}{4}\left(x + 1 + \dfrac{1}{5}\right) \Rightarrow \left(y + \dfrac{1}{2}\right)^2 = \dfrac{5}{4}\left(x + \dfrac{6}{5}\right)$

Graphing Parabolas with Technology

75. $y = \pm\sqrt{2x}$; See Figure 75.

77. $y = -0.75 \pm \sqrt{-3x}$; See Figure 77.

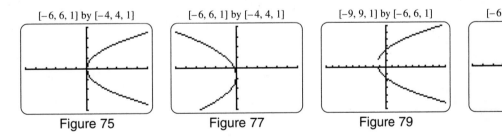

[−6, 6, 1] by [−4, 4, 1]	[−6, 6, 1] by [−4, 4, 1]	[−9, 9, 1] by [−6, 6, 1]	[−6, 6, 1] by [−4, 4, 1]
Figure 75	Figure 77	Figure 79	Figure 81

79. $y = 0.5 \pm \sqrt{3.1(x + 1.3)}$; See Figure 79. Note: If a break in the graph appears near the vertex, it should not be there. It is a result of the low resolution of the graphing calculator screen.

81. $y = -1 \pm \sqrt{\dfrac{x}{2.3}}$; See Figure 81.

Applications

83. Substitute the point $(3, 0.75)$ into $x^2 = 4py$ and solve for p; $9 = 4p(0.75) \Rightarrow 9 = 3p \Rightarrow p = 3$.

    The receiver should be 3 feet from the vertex.

85. (a) Substitute the point $(105, 32)$ into $y = ax^2$ and solve for a; $32 = a(105)^2 \Rightarrow a = \dfrac{32}{11{,}025}$.

    The equation is $y = \dfrac{32}{11{,}025}x^2$.

    (b) Rewriting the answer in (a) we have $x^2 = \dfrac{11{,}025}{32}y$, so $4p = \dfrac{11{,}025}{32}$ and $p = \dfrac{11{,}025}{128} \approx 86.1$.

    The receiver should be located about 86.1 feet from the vertex.

87. Substitute the point $\left(4, \dfrac{5}{2}\right)$ into $y^2 = 4px$ and solve for p; $\left(\dfrac{5}{2}\right)^2 = 4p(4) \Rightarrow p = \dfrac{25}{64}$.

    The bulb should be $\dfrac{25}{64}$ inches from the vertex.

10.2: Ellipses

Ellipses with Center (0, 0)

1. $\dfrac{x^2}{4} + \dfrac{y^2}{9} = 1 \Rightarrow a = 3$ and $b = 2$. $a^2 - b^2 = 3^2 - 2^2 = 5 = c^2 \Rightarrow c = \sqrt{5}$. The foci are $(0, \pm\sqrt{5})$, the

    endpoints of the major axis (vertices) are $(0, \pm 3)$, while the endpoints of the minor axis are $(\pm 2, 0)$. The ellipse

    is graphed in Figure 1.

3. $\dfrac{x^2}{36} + \dfrac{y^2}{16} = 1 \Rightarrow a = 6$ and $b = 4$. $a^2 - b^2 = 6^2 - 4^2 = 20 = c^2 \Rightarrow c = \sqrt{20}$. The foci are $(\pm\sqrt{20}, 0)$,

    the endpoints of the major axis (vertices) are $(\pm 6, 0)$, while the endpoints of the minor axis are $(0, \pm 4)$. The

    ellipse is graphed in Figure 3.

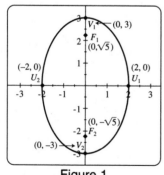
Figure 1

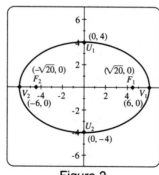
Figure 3

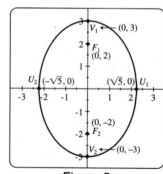
Figure 5

5. $9x^2 + 5y^2 = 45 \Rightarrow \dfrac{x^2}{5} + \dfrac{y^2}{9} = 1 \Rightarrow a = 3$ and $b = \sqrt{5}$. $a^2 - b^2 = 9 - 5 = 4 = c^2 \Rightarrow c = 2$. The foci

    are $(0, \pm 2)$, the endpoints of the major axis (vertices) are $(0, \pm 3)$, while the endpoints of the minor axis are

    $(\pm\sqrt{5}, 0)$. The ellipse is graphed in Figure 5.

7. $25x^2 + 9y^2 = 225 \Rightarrow \dfrac{x^2}{9} + \dfrac{y^2}{25} = 1 \Rightarrow a = 5$ and $b = 3$. $a^2 - b^2 = 25 - 9 = 16 = c^2 \Rightarrow c = 4$. The

foci are $(0, \pm 4)$, the endpoints of the major axis (vertices) are $(0, \pm 5)$, while the endpoints of the minor axis are

$(\pm 3, 0)$. The ellipse is graphed in Figure 7.

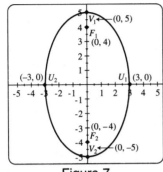

Figure 7

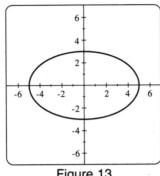

Figure 13

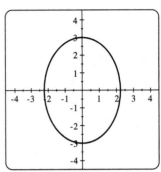

Figure 15

9. Vertices $(0, \pm 6)$; b

11. Vertices $(\pm 4, 0)$; c

13. To sketch a graph of an ellipse centered at the origin, it is helpful to plot the vertices and the endpoints of the

minor axis. The vertices are $V(\pm 5, 0)$ so $a = 5$, the foci are $F(\pm 4, 0)$ so $c = 4$, and the endpoints of the minor

axis are $U(0, \pm 3)$ so $b = 3$. A graph of the ellipse is shown in Figure 13. Its equation is $\dfrac{x^2}{25} + \dfrac{y^2}{9} = 1$.

15. The vertices are $V(0, \pm 3)$ so $a = 3$, the foci are $F(0, \pm 2)$ *so* $c = 2$, and the endpoints of the minor axis are

$U(\pm \sqrt{5}, 0)$ so $b = \sqrt{5}$. A graph of the ellipse is shown in Figure 15. Its equation is $\dfrac{x^2}{5} + \dfrac{y^2}{9} = 1$.

17. Foci of $F(0, \pm 2) \Rightarrow c = 2$ and $V(0, \pm 4) \Rightarrow a = 4$. The major axis lies on the y-axis. The value of b is as

follows: $a^2 - b^2 = c^2 \Rightarrow a^2 - c^2 = b^2 \Rightarrow 4^2 - 2^2 = b^2 \Rightarrow b^2 = 12 \Rightarrow b = \sqrt{12}$. The equation of the

ellipse is $\dfrac{x^2}{12} + \dfrac{y^2}{16} = 1$. Its graph is shown in Figure 17.

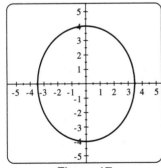

Figure 17

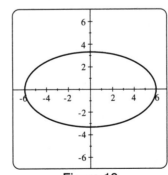

Figure 19

19. Foci of $F(\pm 5, 0) \Rightarrow c = 5$ and $V(\pm 6, 0) \Rightarrow a = 6$. The major axis lies on the x-axis. The value of b is as

follows: $a^2 - b^2 = c^2 \Rightarrow a^2 - c^2 = b^2 \Rightarrow 6^2 - 5^2 = b^2 \Rightarrow b^2 = 11 \Rightarrow b = \sqrt{11}$. The equation of the

ellipse is $\dfrac{x^2}{36} + \dfrac{y^2}{11} = 1$. Its graph is shown in Figure 19.

21. Horizontal major axis of length 8 $\Rightarrow a = 4$. Minor axis of length 6 $\Rightarrow b = 3$. The major axis lies on the
    x-axis. The equation of the ellipse is $\dfrac{x^2}{16} + \dfrac{y^2}{9} = 1$. Its graph is shown in Figure 21.

23. $e = \dfrac{c}{a} = \dfrac{2}{3} \Rightarrow 3c = 2a \Rightarrow c = \dfrac{2}{3}a$. Since the major axis is length 6 $\Rightarrow a = 3$. Thus, $c = \dfrac{2}{3}(3) = 2$.
    Then the value of b is given by the following: $a^2 - b^2 = c^2 \Rightarrow a^2 - c^2 = b^2 \Rightarrow 3^2 - 2^2 = b^2 \Rightarrow$
    $b^2 = 5 \Rightarrow b = \sqrt{5}$. The equation of the ellipse is $\dfrac{x^2}{9} + \dfrac{y^2}{5} = 1$. Its graph is shown in Figure 23.

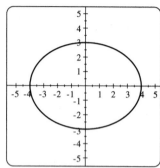

Figure 21

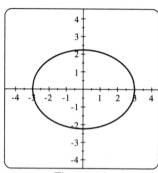

Figure 23

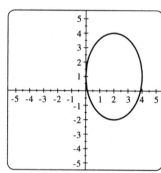

Figure 25

Ellipses with Center (h, k)

25. The ellipse is centered at $(2, 1)$. The major axis has length $2a = 6$ and the length of the minor axis is $2b = 4$.
    The major axis is parallel to the y-axis. The graph is shown in Figure 25.

27. The ellipse is centered at $(-1, -2)$. The major axis has length $2a = 10$ and the length of the minor axis is
    $2b = 8$. The graph is shown in Figure 27.

29. The ellipse is centered at $(-2, 0)$. The horizontal major axis has length $2a = 4$ and the vertical minor axis has
    length $2b = 2$. The graph is shown in Figure 29.

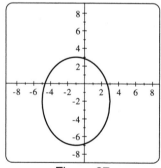

Figure 27

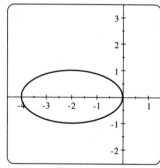

Figure 29

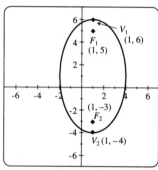

Figure 35

31. Center at $(2, -4)$; d

33. Center at $(-1, 1)$; c

35. Center at $(1, 1)$, $a = 5$, $b = 3$, major axis vertical. $c^2 = a^2 - b^2 = 25 - 9 = 16 \Rightarrow c = 4$.
    Foci: $(1, 1 \pm 4)$; veritices: $(1, 1 \pm 5)$; the graph is shown in Figure 35.

37. Center at $(-4, 2)$, $a = 4$, $b = 3$, major axis horizontal. $c^2 = a^2 - b^2 = 16 - 9 = 7 \Rightarrow c = \sqrt{7}$.

Foci: $(-4 \pm \sqrt{7}, 2)$; veritices: $(-4 \pm 4, 2)$; the graph is shown in Figure 37.

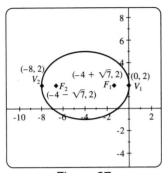

Figure 37

39. Since the center is $(2, 1)$ and a focus is $(2, 3)$, $c = 2$. Since the vertex is $(2, 4)$, $a = 3$;

$b^2 = a^2 - c^2 = 9 - 4 = 5$; the major axis is vertical. The equation is $\dfrac{(x - 2)^2}{5} + \dfrac{(y - 1)^2}{9} = 1$.

41. The center is halfway between the vertices at $(0, 2)$; $a = 3$ and $c = 2$; $b^2 = a^2 - c^2 = 9 - 4 = 5$; the major

axis is horizontal. The equation is $\dfrac{x^2}{9} + \dfrac{(y - 2)^2}{5} = 1$.

43. Center at $(2, 4)$, $a = 4$ and $b = 2$, major axis parallel to the x-axis; the equation is $\dfrac{(x - 2)^2}{16} + \dfrac{(y - 4)^2}{4} = 1$.

45. $9x^2 + 18x + 4y^2 - 8y - 23 = 0 \Rightarrow 9(x^2 + 2x) + 4(y^2 - 2y) = 23 \Rightarrow$

$9(x^2 + 2x + 1) + 4(y^2 - 2y + 1) = 23 + 9 + 4 \Rightarrow 9(x + 1)^2 + 4(y - 1)^2 = 36 \Rightarrow$

$\dfrac{(x + 1)^2}{4} + \dfrac{(y - 1)^2}{9} = 1$; The center is $(-1, 1)$. The vertices are $(-1, 1 - 3)$, $(-1, 1 + 3)$ or $(-1, -2)$, $(-1, 4)$.

47. $4x^2 + 8x + y^2 + 2y + 1 = 0 \Rightarrow 4(x^2 + 2x) + (y^2 + 2y) = -1 \Rightarrow$

$4(x^2 + 2x + 1) + (y^2 + 2y + 1) = -1 + 4 + 1 \Rightarrow 4(x + 1)^2 + (y + 1)^2 = 4 \Rightarrow$

$\dfrac{(x + 1)^2}{1} + \dfrac{(y + 1)^2}{4} = 1$; The center is $(-1, -1)$. The vertices are $(-1, -1 - 2)$, $(-1, -1 + 2)$ or $(-1, -3)$, $(-1, 1)$.

49. $4x^2 + 16x + 5y^2 - 10y + 1 = 0 \Rightarrow 4(x^2 + 4x) + 5(y^2 - 2y) = -1 \Rightarrow$

$4(x^2 + 4x + 4) + 5(y^2 - 2y + 1) = -1 + 16 + 5 \Rightarrow 4(x + 2)^2 + 5(y - 1)^2 = 20 \Rightarrow$

$\dfrac{(x + 2)^2}{5} + \dfrac{(y - 1)^2}{4} = 1$; The center is $(-2, 1)$. The vertices are $(-2 - \sqrt{5}, 1)$, $(-2 + \sqrt{5}, 1)$.

51. $16x^2 - 16x + 4y^2 + 12y = 51 \Rightarrow 16(x^2 - x) + 4(y^2 + 3y) = 51 \Rightarrow$

$16\left(x^2 - x + \dfrac{1}{4}\right) + 4\left(y^2 + 3y + \dfrac{9}{4}\right) = 51 + 4 + 9 \Rightarrow 16\left(x - \dfrac{1}{2}\right)^2 + 4\left(y + \dfrac{3}{2}\right)^2 = 64 \Rightarrow$

$\dfrac{(x - \frac{1}{2})^2}{4} + \dfrac{(y + \frac{3}{2})^2}{16} = 1$

The center is $\left(\dfrac{1}{2}, -\dfrac{3}{2}\right)$. The vertices are $\left(\dfrac{1}{2}, -\dfrac{3}{2} - 4\right)$, $\left(\dfrac{1}{2}, -\dfrac{3}{2} + 4\right)$ or $\left(\dfrac{1}{2}, -\dfrac{11}{2}\right)$, $\left(\dfrac{1}{2}, \dfrac{5}{2}\right)$.

Circles

53. The equation is $x^2 + y^2 = 16$. See Figure 53.

55. The equation is $(x - 3)^2 + (y + 4)^2 = 1$. See Figure 55.

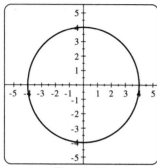

Figure 53

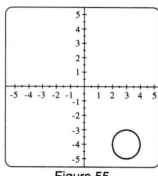

Figure 55

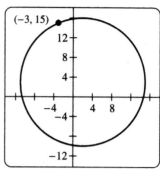

Figure 57

57. The center is (2, 3) and the circle passes through (−3, 15), the radius is equal to the distance between these two points: $r = \sqrt{(-3 - 2)^2 + (15 - 3)^2} = \sqrt{25 + 144} = 13$. The equation of the circle is $(x - 2)^2 + (y - 3)^2 = 169$. See Figure 57.

59. $x^2 - 4x + y^2 - 2y = 11 \Rightarrow (x^2 - 4x + 4) + (y^2 - 2y + 1) = 11 + 4 + 1 \Rightarrow (x - 2)^2 + (y - 1)^2 = 16$. The center is (2, 1) and the radius is $r = 4$.

61. $x^2 + y^2 + 10y = 0 \Rightarrow x^2 + (y^2 + 10y + 25) = 0 + 25 \Rightarrow (x - 0)^2 + (y + 5)^2 = 25$. The center is (0, −5) and the radius is $r = 5$.

Graphing Ellipses with Technology

63. $y = \pm\sqrt{10\left(1 - \dfrac{x^2}{15}\right)}$; See Figure 63.

[−6, 6, 1] by [−4, 4, 1]   [−4.7, 4.7, 1] by [−3.1, 3.1, 1]

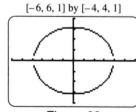

Figure 63   Figure 65

65. $y = \pm\sqrt{\dfrac{25 - 4.1x^2}{6.3}}$; See Figure 65.

Solving Equations and Inequalities

67. $\dfrac{x^2}{4} + \dfrac{y^2}{9} = 1 \Rightarrow 9x^2 + 4y^2 = 36 \Rightarrow 9x^2 + 4(3 - x)^2 = 36 \Rightarrow 9x^2 + 4(9 - 6x + x^2) = 36 \Rightarrow$

$13x^2 - 24x = 0$. Then $x(13x - 24) = 0 \Rightarrow x = 0$ or $x = \dfrac{24}{13}$. Since $y = 3 - x$, the corresponding y values

are $3 - 0 = 3$ and $3 - \dfrac{24}{13} = \dfrac{15}{13}$. The solutions are (0, 3) and $\left(\dfrac{24}{13}, \dfrac{15}{13}\right)$. The system is graphed in Figure 67.

69. $x^2 + y^2 = 9 \Rightarrow x^2 = 9 - y^2$; then $4(9 - y^2) + 16y^2 = 64 \Rightarrow 36 - 4y^2 + 16y^2 = 64 \Rightarrow$

$y^2 = \frac{7}{3} \Rightarrow y = \pm\sqrt{\frac{7}{3}}$. Substituting in $x^2 + y^2 = 9$ we find $x^2 + \frac{7}{3} = 9, x^2 = \frac{27}{3} - \frac{7}{3} = \frac{20}{3}$, so $x = \pm\sqrt{\frac{20}{3}}$.

There are four solutions: $\left(\pm\sqrt{\frac{20}{3}}, \pm\sqrt{\frac{7}{3}}\right)$. The system is graphed in Figure 69.

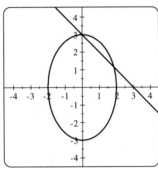

Figure 67

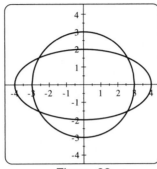

Figure 69

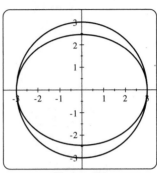

Figure 71

71. $x^2 + y^2 = 9 \Rightarrow y^2 = 9 - x^2$; then $2x^2 + 3(9 - x^2) = 18 \Rightarrow 2x^2 + 27 - 3x^2 = 18 \Rightarrow x^2 = 9 \Rightarrow x = \pm3$;

Substituting in $x^2 + y^2 = 9$ we find $9 + y^2 = 9$, so $y = 0$. There are two solutions: $(\pm3, 0)$. The system is

graphed in Figure 71.

73. $\frac{x^2}{2} + \frac{y^2}{4} = 1 \Rightarrow 2x^2 + y^2 = 4 \Rightarrow 2(2y - 4) + y^2 = 4 \Rightarrow 4y - 8 + y^2 = 4 \Rightarrow y^2 + 4y - 12 = 0$

Then $(y + 6)(y - 2) = 0 \Rightarrow y = -6$ or $y = 2$. Since $x = \pm\sqrt{2y - 4}$, the corresponding x values are

$\pm\sqrt{2(-6) - 4}$, which is undefined, and $\pm\sqrt{2(2) - 4} = 0$. The solution is $(0, 2)$.

75. From the first equation $\frac{x^2}{2} + \frac{y^2}{4} = 1 \Rightarrow 2x^2 + y^2 = 4 \Rightarrow y^2 = 4 - 2x^2$.

From the second equation $\frac{x^2}{4} + \frac{y^2}{2} = 1 \Rightarrow x^2 + 2y^2 = 4 \Rightarrow y^2 = 2 - \frac{1}{2}x^2$. That is $4 - 2x^2 = 2 - \frac{1}{2}x^2$.

$4 - 2x^2 = 2 - \frac{1}{2}x^2 \Rightarrow \frac{3}{2}x^2 = 2 \Rightarrow x^2 = \frac{4}{3} \Rightarrow x = \pm\frac{2}{\sqrt{3}} = \pm\frac{2\sqrt{3}}{3}$

Since $y = \pm\sqrt{4 - 2x^2}$, the y values are $\pm\sqrt{4 - 2\left(\frac{2}{\sqrt{3}}\right)^2} = \pm\sqrt{4 - \frac{8}{3}} = \pm\sqrt{\frac{4}{3}} = \pm\frac{2}{\sqrt{3}} = \pm\frac{2\sqrt{3}}{3}$.

There are four solutions: $\left(\pm\frac{2\sqrt{3}}{3}, \pm\frac{2\sqrt{3}}{3}\right)$.

77. Subtracting the second equation from the first equation yields $(x - 2)^2 - x^2 = 0$.

$(x - 2)^2 - x^2 = 0 \Rightarrow x^2 - 4x + 4 - x^2 = 0 \Rightarrow -4x + 4 = 0 \Rightarrow -4x = -4 \Rightarrow x = 1$

Since $y = \pm\sqrt{9 - x^2}$, the y values are $\pm\sqrt{9 - (1)^2} = \pm\sqrt{8} = \pm2\sqrt{2}$.

The solutions are $(1, -2\sqrt{2}), (1, 2\sqrt{2})$.

79. The system is $(x - 1)^2 + (y + 1)^2 < 4$ and $(x + 1)^2 + y^2 > 1$. See Figure 79.

81. The system is $\dfrac{x^2}{4} + \dfrac{y^2}{9} \leq 1$ and $x + y \geq 2$. See Figure 81.

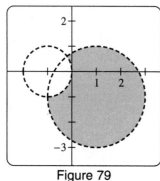
Figure 79

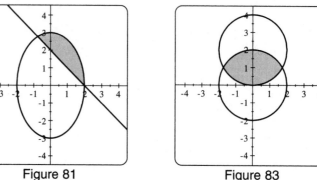
Figure 81

Figure 83

83. The system is $x^2 + y^2 \leq 4$ and $x^2 + (y - 2)^2 \leq 4$. See Figure 83.

85. The system is $x^2 + y^2 \leq 4$ and $(x + 1)^2 - y \leq 0$. See Figure 85.

87. The inequality can be written $\dfrac{x^2}{9} + \dfrac{y^2}{4} \leq 1$. The shaded region is shown in Figure 87.

Here $a = 3$ and $b = 2$. The area is $A = \pi ab = \pi(3)(2) = 6\pi \approx 18.85$ ft^2.

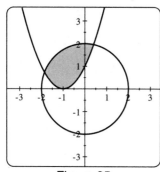
Figure 85

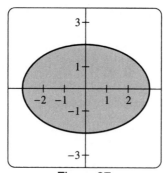
Figure 87

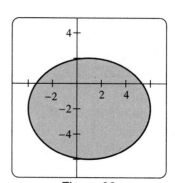
Figure 89

89. The shaded region is shown in Figure 89.

Here $a = 5$ and $b = 4$. The area is $A = \pi ab = \pi(5)(4) = 20\pi \approx 62.83$ ft^2.

Applications

91. $e = \dfrac{c}{a} = 0.206 \Rightarrow c = 0.206a$. Since $a = 0.387$, it follows that $c = 0.206(0.387) \approx 0.0797$. Then, the value

of b is given by the following: $a^2 - b^2 = c^2 \Rightarrow a^2 - c^2 = b^2 \Rightarrow 0.387^2 - 0.0797^2 = b^2 \Rightarrow$

$b^2 = 0.1434 \Rightarrow b = 0.379$. The major axis could be located on either the x- or y-axis. We will choose the

x-axis. Thus, the equation of the orbit is $\dfrac{x^2}{0.387^2} + \dfrac{y^2}{0.379^2} = 1$. The sun can be located on either of the foci.

We will locate the sun at $(0.0797, 0)$. To graph the orbit with a graphing calculator, solve the equation for the

ellipse for the variable y: $\dfrac{x^2}{0.387^2} + \dfrac{y^2}{0.379^2} = 1 \Rightarrow y = \pm 0.379\sqrt{1 - \dfrac{x^2}{0.387^2}}$. Graph each of the equations

and plot the sun at $(0.0797, 0)$. $Y_1 = 0.379\sqrt{(1 - X\text{\textasciicircum}2/0.387\text{\textasciicircum}2}$, $Y_2 = -Y_1$. See Figure 91.

[−0.6, 0.6, 0.1] by [−0.4, 0.4, 0.1]

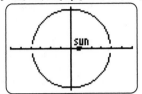

Figure 91

93. The source and the stone are at the two foci of the ellipse, so the distance between them is $2c$.

$c^2 = a^2 - b^2 = 4^2 - (2.5)^2 = 16 - 6.25 = 9.75 \Rightarrow c = \sqrt{9.75}$. Thus, $2c = 2\sqrt{9.75} \approx 6.245$. The stone should be 6.245 inches from the source.

95. (a) $\dfrac{x^2}{17.95^2} + \dfrac{y^2}{4.44^2} = 1$

(b) $c = \sqrt{17.95^2 - 4.44^2} \approx 17.39$, so the sun is at $(17.39, 0)$.

(c) The minimum distance is $a - c = 17.95 - 17.39 = 0.56$, or 0.56 units, which is about 52 million miles.

The maximum distance is $a + c = 17.95 + 17.39 = 35.34$, or 35.34 units, which is about 3.3 billion miles.

97. The equation of the ellipse is $\dfrac{x^2}{30^2} + \dfrac{y^2}{25^2} = 1$. Solving for y we get $y = 25\sqrt{1 - \dfrac{x^2}{900}}$. When $x = 15$,

$y = 25\sqrt{1 - \dfrac{225}{900}} \approx 21.65$ feet.

99. The minimum height is $4464 - (3960 + 164) = 340$ miles; the maximum height is

$4464 - (3960 - 164) = 668$ miles.

Section 10.3: Hyperbolas

Hyperbolas with Center (0, 0)

1. The transverse axis is horizontal with $a = 3$ and $b = 7$. The asymptotes are $y = \pm\dfrac{7}{3}x$. See Figure 1.

Since $c^2 = a^2 + b^2 \Rightarrow c = \pm\sqrt{9 + 49} \Rightarrow \sqrt{58}$, the foci are $(\pm\sqrt{58}, 0)$.

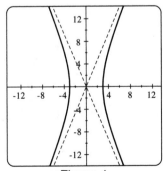

Figure 1

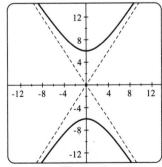

Figure 3

3. The transverse axis is vertical with $a = 6$ and $b = 4$. The asymptotes are $y = \pm\dfrac{3}{2}x$. See Figure 3.

Since $c^2 = a^2 + b^2 \Rightarrow c = \pm\sqrt{36 + 16} \Rightarrow \sqrt{52}$, the foci are $(0, \pm\sqrt{52})$.

5. $x^2 - y^2 = 9 \Rightarrow \dfrac{x^2}{9} - \dfrac{y^2}{9} = 1$. The transverse axis is horizontal with $a = 3$ and $b = 3$. The asymptotes are

$y = \pm x$. See Figure 5. Since $c^2 = a^2 + b^2 \Rightarrow c = \pm\sqrt{9 + 9} \Rightarrow \sqrt{18}$, the foci are $(\pm\sqrt{18}, 0)$.

7. $9y^2 - 16x^2 = 144 \Rightarrow \dfrac{y^2}{16} - \dfrac{x^2}{9} = 1$. The transverse axis is vertical with $a = 4$ and $b = 3$. The asymptotes

are $y = \pm\dfrac{4}{3}x$. See Figure 7. Since $c^2 = a^2 + b^2 \Rightarrow c = \pm\sqrt{16 + 9} \Rightarrow \pm 5$, the foci are $(0, \pm 5)$.

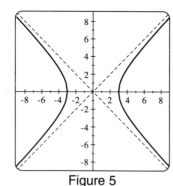

Figure 5

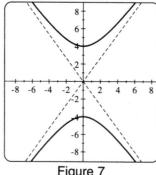

Figure 7

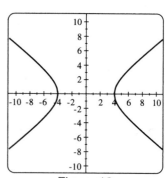

Figure 13

9. Horizontal transverse axis, vertices $(\pm 2, 0)$; d

11. Vertical transverse axis, vertices $(0, \pm 3)$; a

13. Since the foci and the vertices lie on the x-axis, the hyperbola has a horizontal transverse axis with an equation

of the form $\dfrac{x^2}{a^2} - \dfrac{y^2}{b^2} = 1$. $F(\pm 5, 0) \Rightarrow c = 5$ and $V(\pm 4, 0) \Rightarrow a = 4$. $c^2 = a^2 + b^2 \Rightarrow$

$b^2 = c^2 - a^2 = 25 - 16 = 9 \Rightarrow b = 3$. The equation of the hyperbola is $\dfrac{x^2}{16} - \dfrac{y^2}{9} = 1$, and the asymptotes

have the equation $y = \pm\dfrac{b}{a}x$ or $y = \pm\dfrac{3}{4}x$. The hyperbola is graphed in Figure 13.

15. Since the foci and the vertices lie on the y-axis, the hyperbola has a vertical transverse axis with an equation

of the form $\dfrac{y^2}{a^2} - \dfrac{x^2}{b^2} = 1$. $F(0, \pm 10) \Rightarrow c = 10$ and $V(0, \pm 6) \Rightarrow a = 6$. $c^2 = a^2 + b^2 \Rightarrow$

$b^2 = c^2 - a^2 = 100 - 36 = 64 \Rightarrow b = 8$. The equation of the hyperbola is $\dfrac{y^2}{36} - \dfrac{x^2}{64} = 1$, and the

asymptotes have the equation $y = \pm\dfrac{a}{b}x$ or $y = \pm\dfrac{3}{4}x$. The hyperbola is graphed in Figure 15.

17. Since the foci and the vertices lie on the y-axis, the hyperbola has a vertical transverse axis with an equation

of the form $\dfrac{y^2}{a^2} - \dfrac{x^2}{b^2} = 1$. $F(0, \pm 13) \Rightarrow c = 13$ and $V(0, \pm 12) \Rightarrow a = 12$. $c^2 = a^2 + b^2 \Rightarrow$

$b^2 = c^2 - a^2 = 169 - 144 = 25 \Rightarrow b = 5$. The equation of the hyperbola is $\dfrac{y^2}{144} - \dfrac{x^2}{25} = 1$, and the

asymptotes have the equation $y = \pm\dfrac{a}{b}x$ or $y = \pm\dfrac{12}{5}x$. The hyperbola is graphed in Figure 17.

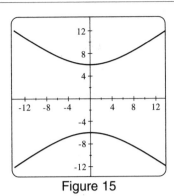

Figure 15

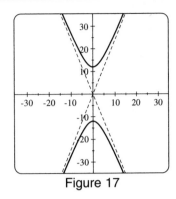

Figure 17

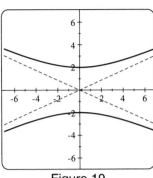

Figure 19

19. Since the foci are $(0, \pm 5)$, $c = 5$. Since the transverse axis is vertical of length 4, $a = 2$. The equation has the

form $\dfrac{y^2}{a^2} - \dfrac{x^2}{b^2} = 1$. $c^2 = a^2 + b^2 \Rightarrow b^2 = c^2 - a^2 = 25 - 4 = 21 \Rightarrow b = \sqrt{21} \approx 4.58$. The equation of

the hyperbola is $\dfrac{y^2}{4} - \dfrac{x^2}{21} = 1$, and the asymptotes have the equation $y = \pm \dfrac{a}{b}x$ or $y = \pm \dfrac{2}{\sqrt{21}}x$. The

hyperbola is graphed in Figure 19.

21. Since the vertices lie on the x-axis and are $(\pm 3, 0)$, $a = 3$. The equation has the form $\dfrac{x^2}{a^2} - \dfrac{y^2}{b^2} = 1$. Since

$y = \pm \dfrac{b}{a}x = \pm \dfrac{2}{3}x$ and $a = 3$, it follows that $b = 2$. The equation of the hyperbola is $\dfrac{x^2}{9} - \dfrac{y^2}{4} = 1$. The

hyperbola is graphed in Figure 21.

23. Since the endpoints of the conjugate axis are $(0, \pm 3)$, $b = 3$. The vertices $(\pm 4, 0)$ lie on the x-axis so $a = 4$

and the equation of the hyperbola is $\dfrac{x^2}{a^2} - \dfrac{y^2}{b^2} = 1$ or $\dfrac{x^2}{16} - \dfrac{y^2}{9} = 1$. The asymptotes have the equation

$y = \pm \dfrac{b}{a}x$ or $y = \pm \dfrac{3}{4}x$. The hyperbola is graphed in Figure 23.

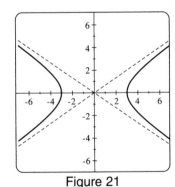

Figure 21

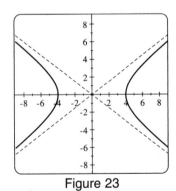

Figure 23

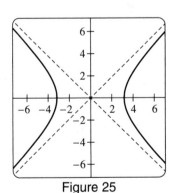

Figure 25

25. Since the vertices lie on the x-axis and are $(\pm \sqrt{10}, 0)$, $a^2 = 10$. The equation has the form $\dfrac{x^2}{a^2} - \dfrac{y^2}{b^2} = 1$.

The value of b^2 can be found by substituting $a^2 = 10$, $x = 10$, and $y = 9$ in this equation.

$\dfrac{(10)^2}{10} - \dfrac{(9)^2}{b^2} = 1 \Rightarrow \dfrac{100}{10} - \dfrac{81}{b^2} = 1 \Rightarrow 10 - \dfrac{81}{b^2} = 1 \Rightarrow 9 = \dfrac{81}{b^2} \Rightarrow b^2 = \dfrac{81}{9} \Rightarrow b^2 = 9$

The equation of the hyperbola is $\dfrac{x^2}{10} - \dfrac{y^2}{9} = 1$. The asymptotes have the equation $y = \pm \dfrac{b}{a}x$ or $y = \pm \dfrac{3}{\sqrt{10}}x$.

The hyperbola is graphed in Figure 25.

Hyperbolas with Center (h, k)

27. The equation is $\dfrac{(x-1)^2}{16} - \dfrac{(y-2)^2}{4} = 1$. See Figure 27.

29. The equation is $\dfrac{(y-2)^2}{36} - \dfrac{(x+2)^2}{4} = 1$. See Figure 29.

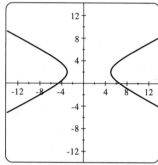

Figure 27

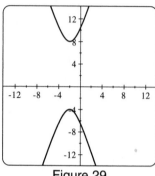

Figure 29

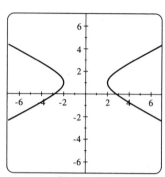

Figure 31

31. The equation is $\dfrac{x^2}{4} - (y-1)^2 = 1$. See Figure 31.

33. Center at $(2, -4)$; b

35. Center at $(2, -1)$; c

37. The hyperbola is centered at $(4, -4)$. The vertical transverse axis has length 8 so $a = 4$. The conjugate axis has length 4, so $b = 2$. Its equation is $\dfrac{(y+4)^2}{16} - \dfrac{(x-4)^2}{4} = 1$.

39. The hyperbola has a horizontal transverse axis, and its center is $(1, 1)$. Since $a^2 = 4, b^2 = 4$, and $c^2 = a^2 + b^2 = 8, a = 2, b = 2$, and $c = \sqrt{8}$. Thus, the vertices are $(1 \pm 2, 1)$ and the foci are $(1 \pm \sqrt{8}, 1)$. The asymptotes are the lines $y = \pm(x - 1) + 1$. See Figure 39.

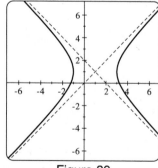

Figure 39

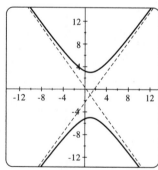

Figure 41

41. The hyperbola has a vertical transverse axis, and its center is $(1, -1)$. Since $a^2 = 16, b^2 = 9$, and $c^2 = a^2 + b^2 = 25, a = 4, b = 3$, and $c = 5$. Thus, the vertices are $(1, -1 \pm 4)$ and the foci are $(1, -1 \pm 5)$. The asymptotes are the lines $y = \pm\dfrac{4}{3}(x - 1) - 1$. See Figure 41.

43. Center $(2, -2) \Rightarrow h = 2$ and $k = -2$. Rewrite the coordinates for the given vertex: $(3, -2) \Rightarrow (1 + 2, -2)$. Rewrite the coordinates for the given focus: $(4, -2) \Rightarrow (2 + 2, -2)$. Thus, the transverse axis is horizontal with $a = 1$ and $c = 2$. $b^2 = c^2 - a^2 = 3$. Using the standard equation form, we get the equation:
$$\frac{(x - h)^2}{a^2} - \frac{(y - k)^2}{b^2} = 1 \Rightarrow (x - 2)^2 - \frac{(y + 2)^2}{3} = 1.$$

45. Since the vertices $(-1, \pm 1)$ and the foci $(-1, \pm 3)$ have the same x-coordinate, $h = -1$ and the transverse axis is vertical. Rewrite the vertices: $(-1, \pm 1) \Rightarrow (-1, 0 \pm 1)$, so $k = 0$ and $a = 1 \Rightarrow a^2 = 1$. Rewrite the foci: $(-1, \pm 3) \Rightarrow (-1, 0 \pm 3)$, so $k = 0$ and $c = 3$. So, $b^2 = c^2 - a^2 = 8$. Using the standard equation form, we get the equation: $\frac{(y - k)^2}{a^2} - \frac{(x - h)^2}{b^2} = 1 \Rightarrow y^2 - \frac{(x + 1)^2}{8} = 1.$

47. $x^2 - 2x - y^2 + 2y = 4 \Rightarrow (x^2 - 2x + 1) - (y^2 - 2y + 1) = 4 + 1 - 1 \Rightarrow (x - 1)^2 - (y - 1)^2 = 4 \Rightarrow$
$\frac{(x - 1)^2}{4} - \frac{(y - 1)^2}{4} = 1.$ The center is $(1, 1)$. The vertices are $(1 - 2, 1), (1 + 2, 1)$ or $(-1, 1), (3, 1)$.

49. $3y^2 + 24y - 2x^2 + 12x + 24 = 0 \Rightarrow 3(y^2 + 8y) - 2(x^2 - 6x) = -24 \Rightarrow$
$3(y^2 + 8y + 16) - 2(x^2 - 6x + 9) = -24 + 48 - 18 \Rightarrow 3(y + 4)^2 - 2(x - 3)^2 = 6 \Rightarrow$
$\frac{(y + 4)^2}{2} - \frac{(x - 3)^2}{3} = 1.$ The center is $(3, -4)$. The vertices are $(3, -4 - \sqrt{2}), (3, -4 + \sqrt{2})$.

51. $x^2 - 6x - 2y^2 + 7 = 0 \Rightarrow (x^2 - 6x + 9) - 2y^2 = -7 + 9 \Rightarrow (x - 3)^2 - 2(y - 0)^2 = 2 \Rightarrow$
$\frac{(x - 3)^2}{2} - \frac{(y - 0)^2}{1} = 1.$ The center is $(3, 0)$. The vertices are $(3 - \sqrt{2}, 0), (3 + \sqrt{2}, 0)$.

53. $4y^2 + 32y - 5x^2 - 10x + 39 = 0 \Rightarrow 4(y^2 + 8y) - 5(x^2 + 2x) = -39 \Rightarrow$
$4(y^2 + 8y + 16) - 5(x^2 + 2x + 1) = -39 + 64 - 5 \Rightarrow 4(y + 4)^2 - 5(x + 1)^2 = 20 \Rightarrow$
$\frac{(y + 4)^2}{5} - \frac{(x + 1)^2}{4} = 1.$ The center is $(-1, -4)$. The vertices are $(-1, -4 - \sqrt{5}), (-1, -4 + \sqrt{5})$.

Graphing Hyperbolas with Technology

55. Solve for y: $\frac{(y - 1)^2}{11} - \frac{x^2}{5.9} = 1 \Rightarrow \frac{(y - 1)^2}{11} = 1 + \frac{x^2}{5.9} \Rightarrow (y - 1)^2 = 11\left(1 + \frac{x^2}{5.9}\right) \Rightarrow$
$y - 1 = \pm\sqrt{11\left(1 + \frac{x^2}{5.9}\right)} \Rightarrow y = 1 \pm \sqrt{11\left(1 + \frac{x^2}{5.9}\right)}.$ Graph $Y_1 = 1 + \sqrt{(11 \cdot (1 + (X^2/5.9)))}$ and
$Y_2 = 1 - \sqrt{(11 \cdot (1 + (X^2/5.9)))}.$ See Figure 55.

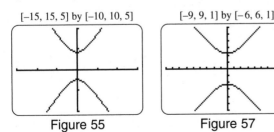

[-15, 15, 5] by [-10, 10, 5]   [-9, 9, 1] by [-6, 6, 1]

Figure 55   Figure 57

57. Solve for y: $3y^2 - 4x^2 = 15 \Rightarrow 3y^2 = 4x^2 + 15 \Rightarrow y^2 = \frac{4x^2 + 15}{3} \Rightarrow y = \pm\sqrt{\frac{4x^2 + 15}{3}}.$
Graph $Y_1 = \sqrt{((4X^2 + 15)/3)}$ and $Y_2 = -\sqrt{((4X^2 + 15)/3)}.$ See Figure 57.

Solving Equations

59. Add both equations together to eliminate the y^2-term:

$$x^2 - y^2 = 4$$
$$\underline{x^2 + y^2 = 9}$$
$$2x^2 \quad\;\; = 13 \Rightarrow x^2 = \frac{13}{2} \Rightarrow x = \pm\sqrt{\frac{13}{2}}$$

Substitute $x^2 = \frac{13}{2}$ into the first equation and solve for y: $x^2 - y^2 = 4 \Rightarrow \frac{13}{2} - y^2 = 4 \Rightarrow y^2 = \frac{5}{2} \Rightarrow$

$y = \pm\sqrt{\frac{5}{2}}$ There are four solutions to the system:

$\left(\sqrt{\frac{13}{2}}, \sqrt{\frac{5}{2}}\right), \left(-\sqrt{\frac{13}{2}}, \sqrt{\frac{5}{2}}\right), \left(\sqrt{\frac{13}{2}}, -\sqrt{\frac{5}{2}}\right),$ and $\left(-\sqrt{\frac{13}{2}}, -\sqrt{\frac{5}{2}}\right).$ See Figure 59.

61. Solve the second equation for y: $x + y = 2 \Rightarrow y = 2 - x$. Substitute this result into the first equation and

solve for x: $\frac{x^2}{4} - \frac{y^2}{9} = 1 \Rightarrow \frac{x^2}{4} - \frac{(2 - x)^2}{9} = 1 \Rightarrow 9x^2 - 4(4 - 4x + x^2) = 36 \Rightarrow$

$9x^2 - 16 + 16x - 4x^2 = 36 \Rightarrow 5x^2 + 16x - 52 = 0 \Rightarrow (5x + 26)(x - 2) = 0 \Rightarrow$

$x = -\frac{26}{5} = -5.2$ or $x = 2$. Substitute for x to find y: $y = 2 - x \Rightarrow y = 2 - (-5.2) = 7.2$ and

$y = 2 - x \Rightarrow y = 2 - 2 = 0$. There are two solutions to the system: $(2, 0)$ and $(-5.2, 7.2)$. See Figure 61.

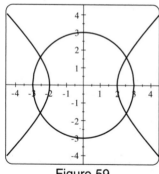

Figure 59

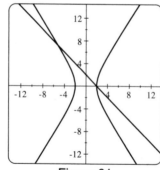

Figure 61

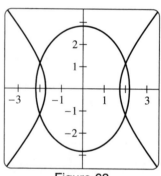

Figure 63

63. Multiply the second equation by 2 and add the equations to eliminate the y^2-term:

$$8x^2 - 6y^2 = 24$$
$$\underline{10x^2 + 6y^2 = 48}$$
$$18x^2 \quad\;\; = 72 \Rightarrow x^2 = 4 \Rightarrow x = \pm 2$$

Substitute $x^2 = 4$ into the first equation and solve for y: $8x^2 - 6y^2 = 24 \Rightarrow 8(4) - 6y^2 = 24 \Rightarrow 6y^2 = 8 \Rightarrow$

$y^2 = \frac{4}{3} \Rightarrow y = \pm\sqrt{\frac{4}{3}} \Rightarrow y = \pm\frac{2}{\sqrt{3}} = \pm\frac{2\sqrt{3}}{3}$

There are four solutions to the system: $\left(-2, -\frac{2\sqrt{3}}{3}\right), \left(-2, \frac{2\sqrt{3}}{3}\right), \left(2, -\frac{2\sqrt{3}}{3}\right), \left(2, \frac{2\sqrt{3}}{3}\right).$ See Figure 63.

65. Solve the second equation for y: $3x - y = 0 \Rightarrow y = 3x$. Substitute this result into the first equation and

solve for x: $\dfrac{y^2}{3} - \dfrac{x^2}{4} = 1 \Rightarrow \dfrac{(3x)^2}{3} - \dfrac{x^2}{4} = 1 \Rightarrow \dfrac{9x^2}{3} - \dfrac{x^2}{4} = 1 \Rightarrow 3x^2 - \dfrac{x^2}{4} = 1 \Rightarrow 12x^2 - x^2 = 4 \Rightarrow$

$11x^2 = 4 \Rightarrow x^2 = \dfrac{4}{11} \Rightarrow x = \pm\dfrac{2}{\sqrt{11}} = \pm\dfrac{2\sqrt{11}}{11}$

Substitute for x to find y: When $x = \dfrac{2\sqrt{11}}{11}, y = 3x \Rightarrow y = 3\left(\dfrac{2\sqrt{11}}{11}\right) \Rightarrow y = \dfrac{6\sqrt{11}}{11}$.

When $x = -\dfrac{2\sqrt{11}}{11}, y = 3x \Rightarrow y = 3\left(-\dfrac{2\sqrt{11}}{11}\right) \Rightarrow y = -\dfrac{6\sqrt{11}}{11}$.

The two solutions are: $\left(\dfrac{2\sqrt{11}}{11}, \dfrac{6\sqrt{11}}{11}\right), \left(-\dfrac{2\sqrt{11}}{11}, -\dfrac{6\sqrt{11}}{11}\right)$. See Figure 65.

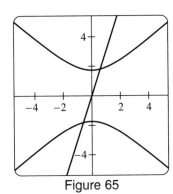

Figure 65

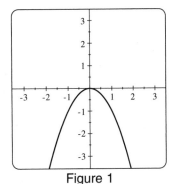

Figure 1

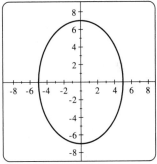

Figure 3

Applications

67. (a) $k = 2.82 \times 10^7$ and $D = 42.5 \times 10^6 \Rightarrow \dfrac{k}{\sqrt{D}} = \dfrac{2.82 \times 10^7}{\sqrt{42.5 \times 10^6}} \Rightarrow \dfrac{k}{\sqrt{D}} \approx 4325.68$. Since $V = 2090$ and

$V < \dfrac{k}{\sqrt{D}}$, the trajectory is elliptic.

(b) For $V > \dfrac{k}{\sqrt{D}}, V > 4326$. the speed of Explorer IV should be 4326 meters per second or greater so its

trajectory is hyperbolic.

(c) If D is larger, then $\dfrac{k}{\sqrt{D}}$ is smaller, so smaller values for V satisfy $V > \dfrac{k}{\sqrt{D}}$.

Chapter 10 Review Exercises

1. The equation is $-x^2 = y$. See Figure 1.

3. The equation is $\dfrac{x^2}{25} + \dfrac{y^2}{49} = 1$. See Figure 3.

5. The equation is $\dfrac{y^2}{4} - \dfrac{x^2}{9} = 1$. See Figure 5.

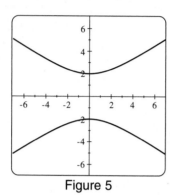

Figure 5

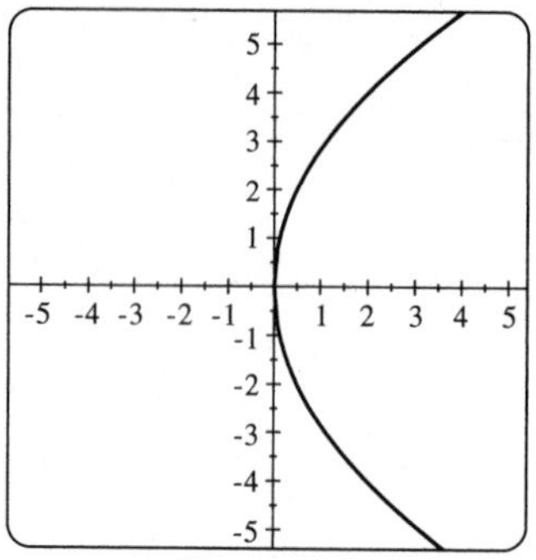

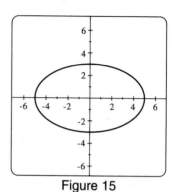

Figure 13

Figure 15

7. A parabola opening upwards; d

9. A circle; a

11. A hyperbola with horizontal transverse axis; e

13. The parabola opens to the right and $p = 2$. $y^2 = 4px$, so $y^2 = 8x$. See Figure 13.

15. The major axis is horizontal, $a = 5$ and $c = 4$; $b = \sqrt{a^2 - c^2} = \sqrt{25 - 16} = 3$.

   The equation is $\dfrac{x^2}{25} + \dfrac{y^2}{9} = 1$. See Figure 15.

17. The transverse axis is vertical, $c = 10$ and $b = 6$; $a = \sqrt{100 - 36} = 8$.

   The equation is $\dfrac{y^2}{64} - \dfrac{x^2}{36} = 1$. See Figure 17.

19. $x^2 = 4py \Rightarrow 4p = -4$, so $p = -1$. The vertex is at the origin and the focus is $(0, -1)$. See Figure 19.

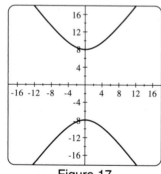

Figure 17

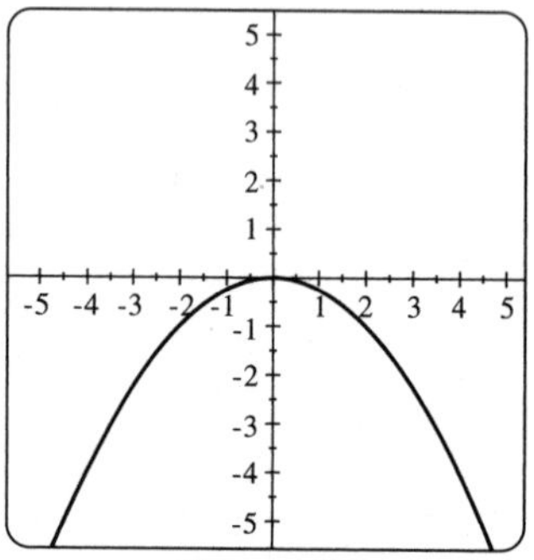

Figure 19

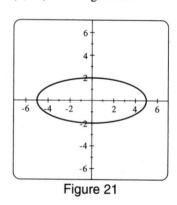

Figure 21

21. The major axis is horizontal, the center is at the origin, $a = 5$ and $b = 2$. $c^2 = 25 - 4 = 21$, so $c = \sqrt{21}$; the foci are $(\pm\sqrt{21}, 0)$. See Figure 21.

23. The center is the origin and the transverse axis is horizontal. $c^2 = a^2 + b^2 = 16 + 9 = 25$, so $c = 5$. The foci are $(\pm 5, 0)$. See Figure 23.

25. The equation represents a circle of radius 2, centered at the origin. If we think of the circle as an ellipse, both foci are at $(0, 0)$. See Figure 25.

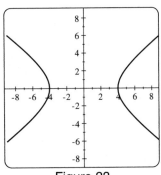

Figure 23

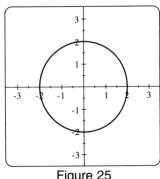

Figure 25

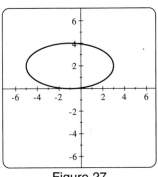

Figure 27

27. The equation $\dfrac{(y-2)^2}{4} + \dfrac{(x+1)^2}{16} = 1$ represents an ellipse with horizontal major axis and center $(-1, 2)$. See Figure 27.

29. The equation $\dfrac{(x-1)^2}{4} - \dfrac{(y+1)^2}{4} = 1$ represents a hyperbola with horizontal transverse axis and center $(1, -1)$. See Figure 29.

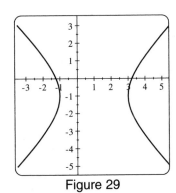

Figure 29

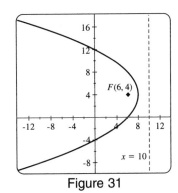

Figure 31

31. The equation $(y-4)^2 = -8(x-8)$ has the form $(y-k)^2 = 4p(x-h)$ with vertex (h, k) at $(8, 4)$, and $p = -2$. The focus is $(6, 4)$ and the directrix is $x = 10$. See Figure 31.

33. $y = \pm\sqrt{\dfrac{3}{4}x}$; See Figure 33.

[-6, 6, 1] by [-4, 4, 1]

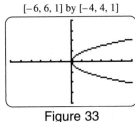

Figure 33

[-6, 6, 1] by [-4, 4, 1]

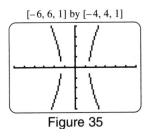

Figure 35

35. $y = \pm\sqrt{\dfrac{9x^2 - 17}{2}}$; See Figure 35. If breaks in the graph appear near the vertices, they should not be there. It is a result of the low resolution of the graphing calculator screen.

37. $-2x = y^2 + 8x + 14 \Rightarrow y^2 = -10x - 14 \Rightarrow (y-0)^2 = -10\left(x + \dfrac{7}{5}\right)$

39. $4x^2 + 8x + 25y^2 - 250y = -529 \Rightarrow 4(x^2 + 2x) + 25(y^2 - 10y) = -529 \Rightarrow$

$4(x^2 + 2x + 1) + 25(y^2 - 10y + 25) = -529 + 4 + 625 \Rightarrow 4(x + 1)^2 + 25(y - 5)^2 = 100 \Rightarrow$

$\dfrac{(x + 1)^2}{25} + \dfrac{(y - 5)^2}{4} = 1;$ The center is $(-1, 5)$. The vertices are $(-1 - 5, 5), (-1 + 5, 5)$ or $(-6, 5), (4, 5)$.

41. $x^2 + 4x - 4y^2 + 24y = 36 \Rightarrow (x^2 + 4x + 4) - 4(y^2 - 6y + 9) = 36 + 4 - 36 \Rightarrow$

$(x + 2)^2 - 4(y - 3)^2 = 4 \Rightarrow \dfrac{(x + 2)^2}{4} - \dfrac{(y - 3)^2}{1} = 1.$

The center is $(-2, 3)$. The vertices are $(-2 - 2, 3), (-2 + 2, 0)$ or $(-4, 3), (0, 3)$.

43. $x^2 + y^2 = 49$

45. Clear fractions: $x^2 + y^2 = 4 \Rightarrow y^2 = 4 - x^2$. Substituting in $x^2 + 4y^2 = 8$ gives $x^2 + 4(4 - x^2) = 8$ or

$3x^2 = 8$, so $x = \pm\sqrt{\dfrac{8}{3}}$. Substituting in $x^2 + y^2 = 4$ we find $y^2 = 4 - \dfrac{8}{3} = \dfrac{4}{3}$, so $y = \pm\sqrt{\dfrac{4}{3}}$. There are four

solutions: $\left(\pm\sqrt{\dfrac{8}{3}}, \pm\sqrt{\dfrac{4}{3}}\right)$.

47. The system is $\dfrac{x^2}{9} + \dfrac{y^2}{4} \le 1$ and $x + y \le 3$. See Figure 47.

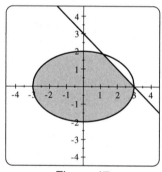

Figure 47

49. (a) $c = \sqrt{500^2 - 70^2} \approx 495.08$, and $a = 500$. The minimum distance is $a - c \approx 4.92$ million miles. The
maximum distance is $a + c \approx 995.08$ million miles.

(b) $2\pi\sqrt{\dfrac{500^2 + 70^2}{2}} \approx 2243$ million miles or 2.243 billion miles.

51. The equation of the ellipse is $\dfrac{x^2}{40^2} + \dfrac{y^2}{30^2} = 1$. Solving for y gives $y = 30\sqrt{1 - \dfrac{x^2}{40^2}}$. When $x = 10$,

$y = 30\sqrt{1 - \left(\dfrac{10}{40}\right)^2} \approx 29.05$ feet.

Chapter 11: Further Topics in Algebra

11.1: Sequences

Finding Terms of Sequences

1. $a_1 = 2(1) + 1 = 3;\ a_2 = 2(2) + 1 = 5;\ a_3 = 2(3) + 1 = 7;\ a_4 = 2(4) + 1 = 9.$

   The first four terms are 3, 5, 7, and 9.

3. $a_1 = 4(-2)^{1-1} = 4;\ a_2 = 4(-2)^{2-1} = -8;\ a_3 = 4(-2)^{3-1} = 16;\ a_4 = 4(-2)^{4-1} = -32.$

   The first four terms are 4, -8, 16, and -32.

5. $a_1 = \dfrac{1}{1^2 + 1} = \dfrac{1}{2};\ a_2 = \dfrac{2}{2^2 + 1} = \dfrac{2}{5};\ a_3 = \dfrac{3}{3^2 + 1} = \dfrac{3}{10};\ a_4 = \dfrac{4}{4^2 + 1} = \dfrac{4}{17}.$

   The first four terms are $\dfrac{1}{2}, \dfrac{2}{5}, \dfrac{3}{10}$, and $\dfrac{4}{17}$.

7. $a_1 = (-1)^1\left(\dfrac{1}{2}\right)^1 = -\dfrac{1}{2};\ a_2 = (-1)^2\left(\dfrac{1}{2}\right)^2 = \dfrac{1}{4};\ a_3 = (-1)^3\left(\dfrac{1}{2}\right)^3 = -\dfrac{1}{8};\ a_4 = (-1)^4\left(\dfrac{1}{2}\right)^4 = \dfrac{1}{16}.$

   The first four terms are $-\dfrac{1}{2}, \dfrac{1}{4}, -\dfrac{1}{8}$, and $\dfrac{1}{16}$.

9. $a_1 = (-1)^0\left(\dfrac{2}{1 + 2}\right) = \dfrac{2}{3};\ a_2 = (-1)^1\left(\dfrac{4}{1 + 4}\right) = -\dfrac{4}{5};\ a_3 = (-1)^2\left(\dfrac{8}{1 + 8}\right) = \dfrac{8}{9};$

   $a_4 = (-1)^3\left(\dfrac{16}{1 + 16}\right) = -\dfrac{16}{17}.$ The first four terms are $\dfrac{2}{3}, -\dfrac{4}{5}, \dfrac{8}{9}$, and $-\dfrac{16}{17}$.

11. $a_1 = 2 + 1^2 = 3;\ a_2 = 4 + 2^2 = 8;\ a_3 = 8 + 3^2 = 17;\ a_4 = 16 + 4^2 = 32.$

    The first four terms are 3, 8, 17, and 32.

13. The points (1, 2), (2, 4), (3, 3), (4, 5), (5, 3), (6, 6), (7, 4) lie on the graph. Therefore, the terms of the sequence

    are 2, 4, 3, 5, 3, 6, 4.

15. (a) $a_1 = 1;\ a_2 = 2a_1 = 2(1) = 2;\ a_3 = 2a_2 = 2(2) = 4;\ a_4 = 2a_3 = 2(4) = 8.$

    The first four terms are 1, 2, 4, and 8.

    (b) The graph of the points (1, 1), (2, 2), (3, 4), and (4, 8) is shown in Figure 15.

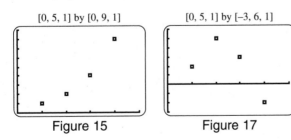

[0, 5, 1] by [0, 9, 1]   [0, 5, 1] by [−3, 6, 1]

Figure 15   Figure 17

17. (a) $a_1 = 2;\ a_2 = 5;\ a_3 = a_2 - a_1 = 5 - 2 = 3;\ a_4 = a_3 - a_2 = 3 - 5 = -2.$

    The first four terms are 2, 5, 3, and −2.

    (b) The graph of the points (1, 2), (2, 5), (3, 3), and (4, −2) is shown in Figure 17.

19. (a) $a_1 = 2$; $a_2 = a_1^2 = 2^2 = 4$; $a_3 = a_2^2 = 4^2 = 16$; $a_4 = a_3^2 = 16^2 = 256$.

   The first four terms are 2, 4, 16, and 256.

   (b) The graph of the points (1, 2), (2, 4), (3, 16), and (4, 256) is shown in Figure 19.

   (b) The graph of the points (1, 0), (2, 1), $\left(3, \dfrac{3}{2}\right)$, and $\left(4, \dfrac{43}{16}\right)$ is shown in Figure 20.

[0, 5, 1] by [0, 300, 50]   [0, 5, 1] by [0, 10, 1]   [0, 5, 1] by [0, 20, 2]

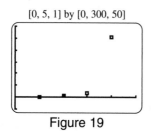   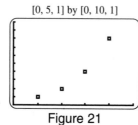   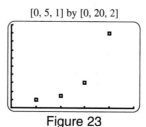

   Figure 19   Figure 21   Figure 23

21. (a) $a_1 = 1$; $a_2 = a_1 + 2 = 1 + 2 = 3$; $a_3 = a_2 + 3 = 3 + 3 = 6$; $a_4 = a_3 + 4 = 6 + 4 = 10$.

   The first four terms are 1, 3, 6, and 10.

   (b) The graph of the points (1, 1), (2, 3), (3, 6), and (4, 10) is shown in Figure 21.

23. (a) $a_1 = 2$; $a_2 = 3$; $a_3 = a_2 \cdot a_1 = 2 \cdot 3 = 6$; $a_4 = a_3 \cdot a_2 = 6 \cdot 3 = 18$

   The first four terms are 2, 3, 6, and 18.

   (b) The graph of the points (1, 2), (2, 3), (3, 6), and (4, 18) is shown in Figure 23.

Modeling Insect and Bacteria Populations

25. The insect population increases rapidly and then levels off at 5000 per acre.

27. (a) The initial density is 500. The population density each successive year is 0.8 of the previous year.

   Therefore $a_1 = 500$ and $a_n = 0.8a_{n-1}$.

   (b) $a_1 = 500$, $a_2 = 0.8(500) = 400$, $a_3 = 0.8(400) = 320$, $a_4 = 0.8(320) = 256$, $a_5 = 0.8(256) = 204.8$ and

   $a_6 = 0.8(204.8) = 163.84$. The population density is decreasing each year by 20%

   (c) The terms of the sequence 500, 400, 320, 256, ... are a geometric sequence with $a_1 = 500$ and $r = 0.8$.

   Therefore, the nth term is given by $a_n = 500(0.8)^{n-1}$.

29. (a) $a_1 = 8$, $a_2 = 2.9a_1 - 0.2a_1^2 = 2.9(8) - 0.2(8)^2 = 10.4$,

   $a_3 = 2.9a_2 - 0.2a_2^2 = 2.9(10.4) - 0.2(10.4)^2 = 8.528$.

   (b) Figures 29a & 29b show how to enter the sequence and a graph of the first twenty terms. The population

   density oscillates above and below 9.5 (approximately).

[0, 21, 1] by [0, 14, 1]

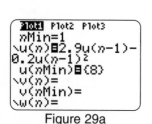   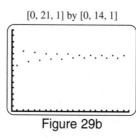

   Figure 29a   Figure 29b

Representations of Sequences

31. (a) Each term can be found by adding 2 to the previous term. A numerical representation for the first eight terms is shown in Figure 31a.

[0, 10, 1] by [0, 16, 1]

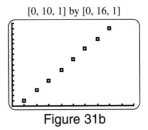

n	1	2	3	4	5	6	7	8
a_n	1	3	5	7	9	11	13	15

Figure 31a

Figure 31b

(b) A graphical representation is shown in Figure 31b and includes the points in Figure 31a. Notice that the points lie on a line with slope 2 because the sequence is arithmetic.

(c) To find the symbolic representation, we will use the formula $a_n = a_1 + (n - 1)d$, where $a_n = f(n)$. The common difference of this sequence is $d = 2$ and the first term is $a_1 = 1$. Therefore, a symbolic representation of the sequence is given by $a_n = 1 + (n - 1)2$ or $a_n = 2n - 1$.

33. (a) Each term can be found by subtracting 1.5 to the previous term. A numerical representation for the first eight terms is shown in Figure 33a.

[0, 12, 1] by [-4, 8, 1]

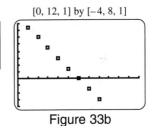

n	1	2	3	4	5	6	7	8
a_n	7.5	6	4.5	3	1.5	0	-1.5	-3

Figure 33a

Figure 33b

(b) A graphical representation is shown in Figure 33b and includes the points in Figure 33a. Notice that the points lie on a line with slope –1.5 because the sequence is arithmetic.

(c) To find the symbolic representation, we will use the formula $a_n = a_1 + (n - 1)d$, where $a_n = f(n)$. The common difference of this sequence is $d = -1.5$ and the first term is $a_1 = 7.5$. Therefore, a symbolic representation of the sequence is given by $a_n = 7.5 + (n - 1)(-1.5)$ or $a_n = -1.5n + 9$.

35. (a) Each term can be found by adding $\frac{3}{2}$ to the previous term. A numerical representation for the first eight terms is shown in Figure 35a.

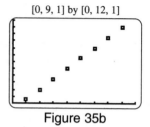

[0, 9, 1] by [0, 12, 1]

n	1	2	3	4	5	6	7	8
a_n	$\frac{1}{2}$	2	$\frac{7}{2}$	5	$\frac{13}{2}$	8	$\frac{19}{2}$	11

Figure 35a

Figure 35b

(b) A graphical representation is shown in Figure 35b and includes the points in Figure 35a. Notice that the points lie on a line with slope $\frac{3}{2}$ because the sequence is arithmetic.

(c) To find the symbolic representation, we will use the formula $a_n = a_1 + (n - 1)d$, where $a_n = f(n)$. The common difference of this sequence is $d = \frac{3}{2}$ and the first term is $a_1 = \frac{1}{2}$. Therefore, a symbolic representation of the sequence is given by $a_n = \frac{1}{2} + (n - 1)\left(\frac{3}{2}\right)$ or $a_n = \frac{3}{2}n - 1$.

37. (a) Each term can be found by multiplying the previous term by $\frac{1}{2}$. A numerical representation for the first eight terms is shown in Figure 37a.

(b) A graphical representation is shown in Figure 37b and includes the points in Figure 37a. Notice that the points lie on a curve that is decaying exponentially, because the sequence is geometric and the common ratio is less than one in absolute value.

(c) To find the symbolic representation, we will use the formula $a_n = a_1 r^{n-1}$, where $a_n = f(n)$. The common ratio of this sequence is $r = \frac{1}{2}$ and the first term is $a_1 = 8$. Therefore, a symbolic representation of the sequence is given by $a_n = 8\left(\frac{1}{2}\right)^{n-1}$.

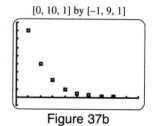

[0, 10, 1] by [−1, 9, 1]

n	1	2	3	4	5	6	7	8
a_n	8	4	2	1	$\frac{1}{2}$	$\frac{1}{4}$	$\frac{1}{8}$	$\frac{1}{16}$

Figure 37a

Figure 37b

39. (a) Each term can be found by multiplying the previous term by 2. A numerical representation for the first

eight terms is shown in Figure 39a.

[0, 10, 1] by [–10, 110, 10]

n	1	2	3	4	5	6	7	8
a_n	$\frac{3}{4}$	$\frac{3}{2}$	3	6	12	24	48	96

Figure 39a

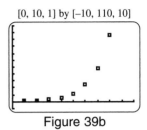

Figure 39b

(b) A graphical representation is shown in Figure 39b and includes the points in Figure 39a. Notice that the

points lie on a curve that is increasing exponentially, because the sequence is geometric and the common

ratio is greater than one in absolute value.

(c) To find the symbolic representation, we will use the formula $a_n = a_1 r^{n-1}$, where $a_n = f(n)$. The common

ratio of this sequence is $r = 2$ and the first term is $a_1 = \dfrac{3}{4}$. Therefore, a symbolic representation of the

sequence is given by $a_n = \dfrac{3}{4}(2)^{n-1}$.

41. (a) Each term can be found by multiplying the previous term by 2. A numerical representation for the first

eight terms is shown in Figure 41a.

[0, 9, 1] by [–20, 4, 2]

n	1	2	3	4	5	6	7	8
a_n	$-\frac{1}{4}$	$-\frac{1}{2}$	-1	-2	-4	-8	-16	-32

Figure 41a

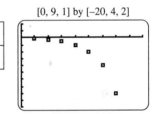

Figure 41b

(b) A graphical representation is shown in Figure 41b and includes the points in Figure 41a. Notice that the

points lie on a curve that is increasing exponentially, because the sequence is geometric and the common

ratio is greater than one in absolute value.

(c) To find the symbolic representation, we will use the formula $a_n = a_1 r^{n-1}$, where $a_n = f(n)$. The common

ratio of this sequence is $r = 2$ and the first term is $a_1 = -\dfrac{1}{4}$. Therefore, a symbolic representation of the

sequence is given by $a_n = -\dfrac{1}{4}(2)^{n-1}$.

43. Let $a_n = f(n)$, where $f(n) = dn + c$. Then, $f(n) = -2n + c$. Since $f(1) = -2(1) + c = 5 \Rightarrow c = 7$.

Thus, $a_n = f(n) = -2n + 7$. An alternate solution is to use the formula:

$a_n = a_1 + (n - 1)d = 5 + (n - 1)(-2) \Rightarrow a_n = -2n + 7$.

45. Let $a_n = f(n)$, where $f(n) = dn + c$. Then, $f(n) = 3n + c$. Since $f(3) = 3(3) + c = 1 \Rightarrow c = -8$.

Thus, $a_n = f(n) = 3n - 8$.

47. Let $a_n = f(n)$, where $f(n) = dn + c$. $a_2 = 5 \Rightarrow f(2) = 5$ and $a_6 = 13 \Rightarrow f(6) = 13$.

  Thus, $d = \dfrac{f(6) - f(2)}{6 - 2} = \dfrac{13 - 5}{4} = 2$. Then, $f(n) = 2n + c$. $f(2) = 2(2) + c = 5 \Rightarrow c = 1$.

  Thus, $a_n = f(n) = 2n + 1$.

49. Let $a_n = f(n)$, where $f(n) = dn + c$. $a_1 = 8 \Rightarrow f(1) = 8$ and $a_4 = 17 \Rightarrow f(4) = 17$.

  Thus, $d = \dfrac{f(4) - f(1)}{4 - 1} = \dfrac{17 - 8}{3} = 3$. Then, $f(n) = 3n + c$. $f(1) = 3(1) + c = 8 \Rightarrow c = 5$.

  Thus, $a_n = f(n) = 3n + 5$.

51. Let $a_n = f(n)$, where $f(n) = dn + c$. $a_5 = -4 \Rightarrow f(5) = -4$ and $a_8 = -2.5 \Rightarrow f(8) = -2.5$.

  Thus, $d = \dfrac{f(8) - f(5)}{8 - 5} = \dfrac{-2.5 - (-4)}{3} = 0.5$. Then, $f(n) = 0.5n + c$. $f(5) = 0.5(5) + c = -4 \Rightarrow c = -6.5$.

  Thus, $a_n = f(n) = 0.5n - 6.5$.

53. Let $a_n = f(n)$, where $f(n) = a_1 r^{n-1}$. Then, $a_n = f(n) = 2\left(\dfrac{1}{2}\right)^{n-1}$.

55. Let $a_n = f(n)$, where $f(n) = a_1 r^{n-1}$. Then, $a_n = f(n) = a_1\left(-\dfrac{1}{4}\right)^{n-1}$.

  Since $f(3) = a_1\left(-\dfrac{1}{4}\right)^{3-1} = \dfrac{1}{32} \Rightarrow a_1 = \dfrac{16}{32} = \dfrac{1}{2}$. Thus, $a_n = f(n) = \dfrac{1}{2}\left(-\dfrac{1}{4}\right)^{n-1}$.

57. Let $a_n = f(n)$, where $f(n) = a_1 r^{n-1}$. $a_3 = 2$ and $a_6 = \dfrac{1}{4} \Rightarrow \dfrac{1}{8} = \dfrac{\frac{1}{4}}{2} = \dfrac{a_6}{a_3} = \dfrac{a_1 r^{6-1}}{a_1 r^{3-1}} = \dfrac{r^5}{r^2} = r^3 \Rightarrow r = \dfrac{1}{2}$.

  Then, $f(n) = a_1\left(\dfrac{1}{2}\right)^{n-1}$. $f(3) = a_1\left(\dfrac{1}{2}\right)^{3-1} = 2 \Rightarrow a_1 = 8$. Thus, $a_n = f(n) = 8\left(\dfrac{1}{2}\right)^{n-1}$.

59. Let $a_n = f(n)$, where $f(n) = a_1 r^{n-1}$. $a_1 = 10$ and $a_2 = 2 \Rightarrow \dfrac{1}{5} = \dfrac{2}{10} = \dfrac{a_2}{a_1} = \dfrac{a_1 r^{2-1}}{a_1 r^{1-1}} = \dfrac{r^1}{r^0} = r^1 \Rightarrow r = \dfrac{1}{5}$.

  Thus, $a_n = f(n) = 10\left(\dfrac{1}{5}\right)^{n-1}$.

61. Let $a_n = f(n)$, where $f(n) = a_1 r^{n-1}$. $a_2 = -1$ and $a_7 = -32 \Rightarrow$

  $32 = \dfrac{-32}{-1} = \dfrac{a_7}{a_2} = \dfrac{a_1 r^{7-1}}{a_1 r^{2-1}} = \dfrac{r^6}{r^1} = r^5 \Rightarrow r = 2$.

  Then, $f(n) = a_1(2)^{n-1}$. $f(2) = a_1(2)^{2-1} = -1 \Rightarrow a_1 = -\dfrac{1}{2}$. Thus, $a_n = f(n) = -\dfrac{1}{2}(2)^{n-1}$.

Identifying Types of Sequences

63. Since $f(n) = 4 - 3n^3$ is not a linear function, it does not represent an arithmetic sequence.

65. Since $f(n) = 4n - (3 - n) = 5n - 3$ is a linear function, it represents an arithmetic sequence.

69. Since the common difference is -2, the table represents an arithmetic sequence.

71. Since $f(n) = 4(2)^{n-1}$ is written in the form $f(n) = cr^{n-1}$, it represents a geometric sequence.

73. Since $f(n) = -3(n)^2$ cannot be written in the form $f(n) = cr^{n-1}$, it does not represents a geometric sequence.

75. Since the plotted points appear to be collinear, the graph does not represent a geometric sequence.

77. Since there is no common ratio, the table does not represent a geometric sequence.

79. These terms represent an arithmetic sequence. Each term can be obtained by adding 7 to the previous term.

81. These terms represent a geometric sequence. Each term can be obtained by multiplying the previous term by 4.

83. These terms represent neither an arithmetic nor a geometric sequence. There is no common ratio or difference.

85. This sequence is arithmetic since the points lie on a line (and are evenly spaced). Since the sequence is decreasing the common difference must be negative. The slope of the line passing through these points is -1, so the common difference is $d = -1$.

87. The sequence is either geometric or arithmetic. This sequence must be geometric, since the point do not lie on a line. Since the terms alternate sign, the common ratio r is negative. The absolute value of the terms are dampening to a value of 0. Therefore, $|r| < 1$.

89. (a) A terms in this sequence can be found by adding the previous two terms. $a_1 = 1, a_2 = 1, a_3 = 2, a_4 = 3,$
    $a_5 = 5, a_6 = 8, a_7 = 13, a_8 = 21, a_9 = 34, a_{10} = 55, a_{11} = 89,$ and $a_{12} = 144.$

    (b) $\dfrac{a_2}{a_1} = \dfrac{1}{1} = 1, \dfrac{a_3}{a_2} = \dfrac{2}{1} = 2, \dfrac{a_4}{a_3} = \dfrac{3}{2} = 1.5, \dfrac{a_5}{a_4} = \dfrac{5}{3} \approx 1.6667, \dfrac{a_6}{a_5} = \dfrac{8}{5} = 1.6, \dfrac{a_7}{a_6} = \dfrac{13}{8} = 1.625,$

    $\dfrac{a_8}{a_7} = \dfrac{21}{13} \approx 1.6154, \dfrac{a_9}{a_8} = \dfrac{34}{21} = 1.6190, \dfrac{a_{10}}{a_9} = \dfrac{55}{34} \approx 1.6176, \dfrac{a_{11}}{a_{10}} = \dfrac{89}{55} = 1.6182,$ and

    $\dfrac{a_{12}}{a_{11}} = \dfrac{144}{89} \approx 1.6180.$ These ratios seem to be approaching a number near 1.618. This number is called the golden ratio.

    (c) $n = 2$: $a_1 \cdot a_3 - a_2^2 = (1)(2) - (1)^2 = 1 = (-1)^2$; $n = 3$: $a_2 \cdot a_4 - a_3^2 = (1)(3) - (2)^2 = -1 = (-1)^3$;
    $n = 4$: $a_3 \cdot a_5 - a_4^2 = (2)(5) - (3)^2 = 1 = (-1)^4$

91. (a) The salary of the first employee at the beginning of the nth year is given by:
    $$a_n = 30,000 + (n - 1)(2000) = 2000n + 28,000.$$
    This is a arithmetic sequence with $a_1 = 30,000$ and $d = 2000$.

    (b) The salary of the second employee at the beginning of the nth year is given by: $b_n = 30,000(1.05)^{n-1}$.
    This is a geometric sequence with $a_1 = 30,000$ and $d = 1.05$.

    (c) At the beginning of the 10th year each salary is:
    $$a_{10} = 2000(10) + 28,000 = \$48,000, b_{10} = 30,000(1.05)^{10-1} \approx \$46,540.$$

    At the beginning of the 20th year each salary is:
    $$a_{20} = 2000(20) + 28,000 = \$68,000, b_{20} = 30,000(1.05)^{20-1} \approx \$75,809.$$

    (d) The graph of each sequence is shown in Figure 91. With time the geometric sequence overtakes the arithmetic sequence since $r > 1$.

[0, 30, 10] by [0, 150,000, 50,000]

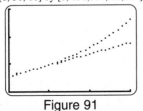

Figure 91

93. Let $a_1 = 2$. Then, $a_2 = \dfrac{1}{2}\left(a_1 + \dfrac{k}{a_1}\right) = \dfrac{1}{2}\left(2 + \dfrac{2}{2}\right) = 1.5$, $a_3 = \dfrac{1}{2}\left(a_2 + \dfrac{k}{a_2}\right) = \dfrac{1}{2}\left(1.5 + \dfrac{2}{1.5}\right) = 1.41\overline{6}$

In a similar manner, $a_4 \approx 1.414215686$, $a_5 \approx 1.414213562$, and $a_6 \approx 1.414213562$. Since $\sqrt{2} \approx 1.414213562$ this is a very accurate approximation.

95. Let $a_1 = 21$. Then, $a_2 = \dfrac{1}{2}\left(a_1 + \dfrac{k}{a_1}\right) = \dfrac{1}{2}\left(21 + \dfrac{21}{21}\right) = 11$, $a_3 = \dfrac{1}{2}\left(a_2 + \dfrac{k}{a_2}\right) = \dfrac{1}{2}\left(11 + \dfrac{21}{11}\right) = 6.\overline{45}$

In a similar manner, $a_4 \approx 4.854033291$, $a_5 \approx 4.59016621$, and $a_6 \approx 4.582581971$. Since $\sqrt{21} \approx 4.582575695$ this is an accurate approximation.

97. By definition $a_n = a_1 + (n - 1)d_1$ and $b_n = b_1(1 - 1)d_2$. Then

$c_n = a_n + b_n = [a_1 + (n - 1)d_1] + [b_1 + (n - 1)d_2] = (a_1 + b_1) + [(n - 1)d_1 + (n - 1)d_2] = (a_1 + b_1) + (n - 1)(d_1 + d_2) = c_1 + (n - 1)d$ where $c_1 = a_1 + b_1$ and $d = d_1 + d_2$.

11.2: Series

Finding Sums of Series

1. (a) $8518 + 9921 + 10{,}706 + 14{,}035 + 14{,}307 + 12{,}249$. The sum of the series is $69{,}736$, so there were $69{,}736$ escapes from state prisons between 1990 and 1995.

   (b) The series has six terms so it is a finite series.

3. $S_5 = 3(1) + 3(2) + 3(3) + 3(4) + 3(5) = 3 + 6 + 9 + 12 + 15 = 45$

5. $S_5 = (2(1) - 1) + (2(2) - 1) + (2(3) - 1) + (2(4) - 1) + (2(5) - 1) = 1 + 3 + 5 + 7 + 9 = 25$

7. $S_5 = (1^2 + 1) + (2^2 + 1) + (3^2 + 1) + (4^2 + 1) + (5^2 + 1) = 2 + 5 + 10 + 17 + 26 = 60$

9. $S_5 = \dfrac{1}{1 + 1} + \dfrac{2}{2 + 1} + \dfrac{3}{3 + 1} + \dfrac{4}{4 + 1} + \dfrac{5}{5 + 1} = \dfrac{1}{2} + \dfrac{2}{3} + \dfrac{3}{4} + \dfrac{4}{5} + \dfrac{5}{6} = \dfrac{71}{20}$

11. The first term is $a_1 = 3$ and the last term is $a_8 = 17$. To find the sum use:

$S_n = n\left(\dfrac{a_1 + a_n}{2}\right) \Rightarrow S_8 = 8\left(\dfrac{3 + 17}{2}\right) = 80$. The sum is 80.

13. The first term is $a_1 = 1$ and the last term is $a_{50} = 50$. To find the sum use:

$S_n = n\left(\dfrac{a_1 + a_n}{2}\right) \Rightarrow S_{50} = 50\left(\dfrac{1 + 50}{2}\right) = 1275$. The sum is 1275.

15. The first term is $a_1 = -7$ and $d = 3$. The last term is 101, but we must determine n. The term of 101 is

$\dfrac{101 - (-7)}{3} = 36$ terms after a_1. Thus, the last term is $a_{37} = 101$. To find the sum use the following:

$S_n = n\left(\dfrac{a_1 + a_n}{2}\right) \Rightarrow S_{37} = 37\left(\dfrac{-7 + 101}{2}\right) = 1739$. The sum is 1739.

17. The number of terms to be added is 40, so $n = 40$. The first term is $a_1 = 5(1) = 5$. The last term is

$a_{40} = 5(40) = 200$. $S_n = n\left(\dfrac{a_1 + a_n}{2}\right) \Rightarrow S_{40} = 40\left(\dfrac{5 + 200}{2}\right) = 4100$. The sum is 4100.

19. $S_{15} = 15\left(\dfrac{a_1 + a_{15}}{2}\right) \Rightarrow 255 = 15\left(\dfrac{3 + a_{15}}{2}\right) \Rightarrow 17 = \dfrac{3 + a_{15}}{2} \Rightarrow 34 = 3 + a_{15} \Rightarrow a_{15} = 31$

21. Since $a_1 = 4$ and $d = 2$, use the formula $S_n = \dfrac{n}{2}(2a_1 + (n-1)d)$.

$$S_{20} = \frac{20}{2}(2(4) + (20-1)2) = 10(8 + 19(2)) = 10(8 + 38) = 10(46) = 460$$

23. Since $a_1 = 10$ and $d = -\dfrac{1}{2}$, use the formula $S_n = \dfrac{n}{2}(2a_1 + (n-1)d)$.

$$S_{20} = \frac{20}{2}\left(2(10) + (20-1)\left(-\frac{1}{2}\right)\right) = 10\left(20 + 19\left(-\frac{1}{2}\right)\right) = 10\left(20 - \frac{19}{2}\right) = 10\left(\frac{21}{2}\right) = 105$$

25. Since $a_1 = 4$ and $a_{20} = 190.2$, use the formula $S_n = n\left(\dfrac{a_1 + a_n}{2}\right)$.

$$S_{20} = 20\left(\frac{4 + 190.2}{2}\right) = 20\left(\frac{194.2}{2}\right) = 20(97.1) = 1942$$

27. Since $a_1 = -2$ and $a_{11} = 50$, the common difference is $d = \dfrac{50 - (-2)}{11 - 1} = \dfrac{52}{10} = 5.2$.

Use the formula $S_n = \dfrac{n}{2}(2a_1 + (n-1)d)$.

$$S_{20} = \frac{20}{2}(2(-2) + (20-1)5.2) = 10(-4 + 19(5.2)) = 10(-4 + 98.8) = 10(94.8) = 948$$

29. Since $a_2 = 6$ and $a_{12} = 31$, the common difference is $d = \dfrac{31 - 6}{12 - 2} = \dfrac{25}{10} = 2.5$ and so $a_1 = 6 - 2.5 = 3.5$.

Use the formula $S_n = \dfrac{n}{2}(2a_1 + (n-1)d)$.

$$S_{20} = \frac{20}{2}(2(3.5) + (20-1)2.5) = 10(7 + 19(2.5)) = 10(7 + 47.5) = 10(54.5) = 545$$

31. The first term is $a_1 = 1$ and the common ratio is $r = 2$. Since there are 8 terms, the sum is

$$S_n = a_1\left(\frac{1 - r^n}{1 - r}\right) \Rightarrow S_8 = 1\left(\frac{1 - 2^8}{1 - 2}\right) = 255.$$

This sum can be verified by adding $1 + 2 + 4 + 8 + 16 + 32 + 64 + 128 = 255$.

33. The first term is $a_1 = 0.5$ and the common ratio is $r = 3$. Since there are 7 terms, the sum is

$$S_n = a_1\left(\frac{1 - r^n}{1 - r}\right) \Rightarrow S_7 = 0.5\left(\frac{1 - (3)^7}{1 - 3}\right) = 546.5.$$

This sum can be verified by adding $0.5 + 1.5 + 4.5 + 13.5 + 40.5 + 121.5 + 364.5 = 546.5$.

35. The first term is $a_1 = 3(2)^0 = 3$ and the common ratio is $r = 2$. Since there are 20 terms, the sum is

$$S_n = a_1\left(\frac{1 - r^n}{1 - r}\right) \Rightarrow S_{20} = 3\left(\frac{1 - 2^{20}}{1 - 2}\right) = 3{,}145{,}725.$$

37. Using the formula $S_n = a_1\left(\dfrac{1 - r^n}{1 - r}\right)$ with $a_1 = 1$ and $r = -\dfrac{1}{2}$,

$$S_4 = 1\left(\frac{1 - \left(-\frac{1}{2}\right)^4}{1 - \left(-\frac{1}{2}\right)}\right) = 0.625;\ \ S_7 = 1\left(\frac{1 - \left(-\frac{1}{2}\right)^7}{1 - \left(-\frac{1}{2}\right)}\right) = 0.671875;\ \ S_{10} = 1\left(\frac{1 - \left(-\frac{1}{2}\right)^{10}}{1 - \left(-\frac{1}{2}\right)}\right) = 0.666015625$$

39. Using the formula $S_n = a_1\left(\dfrac{1 - r^n}{1 - r}\right)$ with $a_1 = \dfrac{1}{3}$ and $r = 2$,

$$S_4 = \frac{1}{3}\left(\frac{1 - 2^4}{1 - 2}\right) = 5;\ \ S_7 = \frac{1}{3}\left(\frac{1 - 2^7}{1 - 2}\right) = 42.\overline{3};\ \ S_{10} = \frac{1}{3}\left(\frac{1 - 2^{10}}{1 - 2}\right) = 341$$

41. The first term is $a_1 = 1$ and the common ratio is $r = \dfrac{1}{3}$. The sum is $S = a_1\left(\dfrac{1}{1-r}\right) \Rightarrow S = 1\left(\dfrac{1}{1-\frac{1}{3}}\right) = \dfrac{3}{2}$.

43. The first term is $a_1 = 6$ and the common ratio is $r = -\dfrac{2}{3}$. The sum is

$$S = a_1\left(\dfrac{1}{1-r}\right) \Rightarrow S = 6\left(\dfrac{1}{1-\left(-\frac{2}{3}\right)}\right) = \dfrac{18}{5}.$$

45. The first term is $a_1 = 1$ and the common ratio is $r = -\dfrac{1}{10}$ or -0.1. The sum is

$$S = a_1\left(\dfrac{1}{1-r}\right) \Rightarrow S = 1\left(\dfrac{1}{1-(-0.1)}\right) = \dfrac{1}{1.1} = \dfrac{10}{11}.$$

47. The area of the largest square is 1, the area of the next largest square is $\dfrac{1}{2}$, and each successive square has an

area that is $\dfrac{1}{2}$ the area of the previous square. If there are an infinite number of squares, then the area is

represented by the geometric series $1 + \dfrac{1}{2} + \dfrac{1}{4} + \dfrac{1}{8} + \dfrac{1}{16} + \cdots$. This is an infinite geometric series with

$a_1 = 1$ and $r = \dfrac{1}{2}$, whose sum is $S = \dfrac{1}{1-r} = \dfrac{1}{1-\frac{1}{2}} = 2$. The area is 2.

49. Future value is given by $S_n = a_1\left(\dfrac{(1+i)^n - 1}{i}\right)$, where $A_0 = a_1$.

$$S_{20} = 2000\left(\dfrac{(1+0.08)^{20} - 1}{0.08}\right) \approx 91{,}523.93.$$ If $\$2000$ is deposited in an account at the end of each year for

20 years and the account pays 8% interest, the future value of this annuity will be $\$91{,}523.93$.

51. Future value is given by $S_n = A_0\left(\dfrac{(1+i)^n - 1}{i}\right)$, where $A_0 = a_1$.

$$S_5 = 10{,}000\left(\dfrac{(1+0.11)^5 - 1}{0.11}\right) \approx 62{,}278.01.$$ If $\$10{,}000$ is deposited in an account at the end of each year for

5 years and the account pays 11% interest, the future value of this annuity will be $\$62{,}278.01$.

Decimal Numbers and Geometric Series

53. $\dfrac{2}{3} = 0.6666666\ldots = 0.6 + 0.06 + 0.006 + 0.0006 + 0.00006 + \cdots$

55. $\dfrac{9}{11} = 0.81818181\ldots = 0.81 + 0.0081 + 0.000081 + 0.00000081 + \cdots$

57. $\dfrac{1}{7} = 0.142857142857\ldots = 0.142857 + 0.000000142857 + 0.000000000000142857 + \cdots$

59. The series $0.8 + 0.08 + 0.008 + 0.0008 + \cdots$ is an infinite geometric series with $a_1 = 0.8$ and $r = 0.1$.

$$S = \dfrac{a_1}{1-r} = \dfrac{0.8}{1-0.1} = \dfrac{8}{9}.$$

61. The series $0.45 + 0.0045 + 0.000045 + \cdots$ is an infinite geometric series with $a_1 = 0.45$ and $r = 0.01$.

$$S = \dfrac{a_1}{1-r} = \dfrac{0.45}{1-0.01} = \dfrac{45}{99} = \dfrac{5}{11}.$$

Summation Notation

63. $\displaystyle\sum_{k=1}^{4}(k+1) = (1+1)+(2+1)+(3+1)+(4+1) = 2+3+4+5 = 14$

65. $\displaystyle\sum_{k=1}^{8}4 = 4+4+4+4+4+4+4+4 = 32$

67. $\displaystyle\sum_{k=1}^{7}k^3 = 1^3+2^3+3^3+4^3+5^3+6^3+7^3 = 1+8+27+64+125+216+343 = 784$

69. $\displaystyle\sum_{k=4}^{5}(k^2-k) = (4^2-4)+(5^2-5) = 12+20 = 32$

71. $\displaystyle 1^4+2^4+3^4+4^4+5^4+6^4 = \sum_{k=1}^{6}k^4$

73. $\displaystyle 1+\frac{4}{3}+\frac{6}{4}+\frac{8}{5}+\frac{10}{6}+\frac{12}{7}+\frac{14}{8} = \sum_{k=1}^{7}\left(\frac{2k}{k+1}\right)$

75. $\displaystyle 1+\frac{1}{2^2}+\frac{1}{3^2}+\frac{1}{4^2}+\frac{1}{5^2}+\cdots = \sum_{k=1}^{\infty}\left(\frac{1}{k^2}\right)$

77. $\displaystyle\sum_{k=1}^{60}9 = 60(9) = 540$

79. $\displaystyle\sum_{k=1}^{15}5k = 5\sum_{k=1}^{15}k = 5\left[\frac{15(16)}{2}\right] = 600$

81. $\displaystyle\sum_{k=1}^{31}(3k-3) = 3\sum_{k=1}^{31}k - \sum_{k=1}^{31}3 = 3\left[\frac{31(32)}{2}\right]-31(3) = 1488-93 = 1395$

83. $\displaystyle\sum_{k=1}^{25}k^2 = \frac{(25)(26)(51)}{6} = 5525$

85. $\displaystyle\sum_{k=1}^{16}(k^2-k) = \sum_{k=1}^{16}k^2-\sum_{k=1}^{16}k = \frac{(16)(17)(33)}{6}-\frac{(16)(17)}{2} = 1496-136 = 1360$

87. $\displaystyle\sum_{k=5}^{24}k = \sum_{k=1}^{24}k-\sum_{k=1}^{4}k = \frac{(24)(25)}{2}-\frac{(4)(5)}{2} = 300-10 = 290$

89. $\displaystyle\sum_{k=1}^{n}k = 1+2+3+4+\cdots+n.$ This is an arithmetic series with $a_1 = 1$ and $a_n = n$. The sum is given by

$$S_n = n\left(\frac{a_1+a_n}{2}\right) = n\left(\frac{1+n}{2}\right) = \frac{n(n+1)}{2}.$$

91. The number of logs in the stack is given by $7+8+9+10+11+12+13+14+15$. This series is

arithmetic with first term $a_1 = 7$ and the last term $a_9 = 15$ so its sum is given by $S_9 = 9\left(\dfrac{7+15}{2}\right) = 99$. So,

there are 99 logs in the stack.

93. (a) Since each filter lets half of the impurities through, the series is can be written as

$$0.5(1)+0.5(0.5)+0.5(0.25)+\ldots = \sum_{k=1}^{n}0.5(0.5)^{k-1}.$$

(b) The series sums to 1 only if an infinite number of filters are used.

95. (a) With straight-line depreciation $\frac{1}{4}(10{,}000) = \$2500$ is depreciated each year. The sequence describing this

depreciation is the constant arithmetic sequence 2500, 2500, 2500, 2500. With the sum-of-the-years'-digits

method we note that $1 + 2 + 3 + 4 = 10$. Then there would be $\frac{4}{10}(10{,}000) = \$4000$ depreciated the first

year, $\frac{3}{10}(10{,}000) = \$3000$ the second year, $\frac{2}{10}(10{,}000) = \$2000$ the third year, and $\frac{1}{10}(10{,}000) = \$1000$

the last year. The sequence describing this depreciation is the arithmetic sequence 4000, 3000, 2000, 1000.

(b) The series for straight-line depreciation is $2500 + 2500 + 2500 + 2500$. The series for

sum-of-the-years'-digits depreciation is $4000 + 3000 + 2000 + 1000$.

97. The value of e^a is approximated by $e^a = 1 + a + \dfrac{a^2}{2!} + \dfrac{a^3}{3!} + \cdots + \dfrac{a^n}{n!}$. Apply the series by letting $a = 1$,

since $e^1 = e$. The first 8 terms sum to $e^1 \approx 1 + 1 + \dfrac{1}{2!} + \dfrac{1}{3!} + \dfrac{1}{4!} + \dfrac{1}{5!} + \dfrac{1}{6!} + \dfrac{1}{7!} =$

$1 + 1 + \dfrac{1}{2} + \dfrac{1}{6} + \dfrac{1}{24} + \dfrac{1}{120} + \dfrac{1}{720} + \dfrac{1}{5040} \approx 2.718254$. The actual value is $e \approx 2.718282$, so only 8 terms

of the series provides an approximation for e that is accurate to 5 decimal places.

Computing Partial Sums

99. One can either add the terms directly or apply the formula $S_n = a_1\left(\dfrac{1 - r^n}{1 - r}\right)$. We will apply the formula. In

this sequence $a_1 = 1$ and $r = \dfrac{1}{3}$. $S_2 = 1\left(\dfrac{1 - \left(\frac{1}{3}\right)^2}{1 - \left(\frac{1}{3}\right)}\right) = \dfrac{4}{3} \approx 1.3333$; $S_4 = 1\left(\dfrac{1 - \left(\frac{1}{3}\right)^4}{1 - \left(\frac{1}{3}\right)}\right) = \dfrac{40}{27} \approx 1.4815$;

$S_8 = 1\left(\dfrac{1 - \left(\frac{1}{3}\right)^8}{1 - \left(\frac{1}{3}\right)}\right) \approx 1.49977$; $S_{16} = 1\left(\dfrac{1 - \left(\frac{1}{3}\right)^{16}}{1 - \left(\frac{1}{3}\right)}\right) \approx 1.49999997$. The sum of the infinite series

$\displaystyle\sum_{n=1}^{\infty}\left(\dfrac{1}{3}\right)^{k-1}$ is given by $S = \dfrac{a_1}{1 - r} = \dfrac{1}{1 - \frac{1}{3}} = \dfrac{3}{2} = 1.5$. As n increases, the partial sums S_2, S_4, S_8, and S_{16}

become closer and closer to the value of 1.5.

101. One can either add the terms directly or apply the formula $S_n = a_1\left(\dfrac{1 - r^n}{1 - r}\right)$. We will apply the formula. In

this sequence $a_1 = 4$ and $r = -\dfrac{1}{10}$. $S_1 = 4\left(\dfrac{1 - \left(-\frac{1}{10}\right)^1}{1 - \left(-\frac{1}{10}\right)}\right) = 4$; $S_2 = 4\left(\dfrac{1 - \left(-\frac{1}{10}\right)^2}{1 - \left(-\frac{1}{10}\right)}\right) = 3.6$;

$S_3 = 4\left(\dfrac{1 - \left(-\frac{1}{10}\right)^3}{1 - \left(-\frac{1}{10}\right)}\right) = 3.64$; $S_4 = 4\left(\dfrac{1 - \left(-\frac{1}{10}\right)^4}{1 - \left(-\frac{1}{10}\right)}\right) = 3.636$; $S_5 = 4\left(\dfrac{1 - \left(-\frac{1}{10}\right)^5}{1 - \left(-\frac{1}{10}\right)}\right) = 3.6364$;

$S_6 = 4\left(\dfrac{1 - \left(-\frac{1}{10}\right)^6}{1 - \left(-\frac{1}{10}\right)}\right) = 3.63636$; The sum of the infinite series $\displaystyle\sum_{n=1}^{\infty} 4\left(-\dfrac{1}{10}\right)^{k-1}$ is given by

$S = \dfrac{a_1}{1 - r} = \dfrac{4}{1 - (-0.1)} = \dfrac{40}{11} = 3.\overline{63}$. As n increases, the partial sums become closer and closer to the

value of $3.\overline{63}$.

11.3: Counting

Counting

1. There are $2 \cdot 2 \cdot 2 \cdot 2 \cdot 2 \cdot 2 \cdot 2 \cdot 2 \cdot 2 \cdot 2 = 2^{10} = 1024$ different ways to answer the exam.

3. There are $2 \cdot 2 \cdot 2 \cdot 2 \cdot 2 \cdot 4 \cdot 4 \cdot 4 \cdot 4 \cdot 4 \cdot 4 \cdot 4 \cdot 4 \cdot 4 \cdot 4 = 2^5 \cdot 4^{10} = 33{,}554{,}432$ different ways.

5. There are 10 digits and 26 letters. There are $10 \cdot 10 \cdot 10 \cdot 26 \cdot 26 \cdot 26 = 17{,}576{,}000$ different license plates.

7. There are 36 digits and letters. There are $26 \cdot 26 \cdot 26 \cdot 36 \cdot 36 \cdot 36 = 820{,}025{,}856$ different license plates.

9. There are 3 choices for each of the 5 letters. There are $3 \cdot 3 \cdot 3 \cdot 3 \cdot 3 = 3^5 = 243$ different strings.

11. There are 5 choices for each of the 5 letters. There are $5 \cdot 5 \cdot 5 \cdot 5 \cdot 5 = 5^5 = 3125$ different strings.

13. Since there are 2 letters that can only be used once, each string must be 2 letters long. For the first position there are 2 choices and for the second position there is only 1 choice. There are $2 \cdot 1 = 2$ possible strings.

15. There are $4 \cdot 3 \cdot 2 \cdot 1 = 24$ possible strings.

17. Each combination can vary between 000 and 999. Thus, there are 1000 possibilities for each lock. That is, there are $1000 \cdot 1000 = 1{,}000{,}000$ different combinations in all.

19. There are 2 settings (on or off) for each of the 12 switches.

    There are $2 \cdot 2 \cdot 2 \cdot 2 \cdot 2 \cdot 2 \cdot 2 \cdot 2 \cdot 2 \cdot 2 \cdot 2 \cdot 2 = 2^{12} = 4096$ different codes for the garage door opener.

21. For the first letter there are 2 possibilities, whereas there are 26 possibilities for each of the last 3 positions.

    There are $2 \cdot 26 \cdot 26 \cdot 26 = 35{,}152$ different call letters possible. There is no shortage of call letters.

23. There are $2 \cdot 3 \cdot 4 = 24$ different packages that can be purchased.

25. There are $5 \cdot 10 \cdot 4 = 200$ different ways to order a salad, an entrée and a dessert.

27. $6! = 6 \cdot 5 \cdot 4 \cdot 3 \cdot 2 \cdot 1 = 720$

29. $10! = 10 \cdot 9 \cdot 8 \cdot 7 \cdot 6 \cdot 5 \cdot 4 \cdot 3 \cdot 2 \cdot 1 = 3{,}628{,}800$

Permutations

31. $P(n, r) = \dfrac{n!}{(n - r)!} \Rightarrow P(5, 3) = \dfrac{5!}{(5 - 3)!} = \dfrac{5!}{2!} = \dfrac{120}{2} = 60$

33. $P(n, r) = \dfrac{n!}{(n - r)!} \Rightarrow P(8, 1) = \dfrac{8!}{(8 - 1)!} = \dfrac{8!}{7!} = \dfrac{8 \cdot 7!}{7!} = 8$

35. $P(n, r) = \dfrac{n!}{(n - r)!} \Rightarrow P(7, 3) = \dfrac{7!}{(7 - 3)!} = \dfrac{7!}{4!} = \dfrac{7 \cdot 6 \cdot 5 \cdot 4!}{4!} = 7 \cdot 6 \cdot 5 = 210$

37. $P(n, r) = \dfrac{n!}{(n - r)!} \Rightarrow P(25, 2) = \dfrac{25!}{(25 - 2)!} = \dfrac{25!}{23!} = \dfrac{25 \cdot 24 \cdot 23!}{23!} = 25 \cdot 24 = 600$

39. $P(n, r) = \dfrac{n!}{(n - r)!} \Rightarrow P(10, 4) = \dfrac{10!}{(10 - 4)!} = \dfrac{10!}{6!} = \dfrac{10 \cdot 9 \cdot 8 \cdot 7 \cdot 6!}{6!} = 10 \cdot 9 \cdot 8 \cdot 7 = 5040$

41. $P(4, 4) = \dfrac{4!}{(4 - 4)!} = \dfrac{4!}{0!} = 4! = 24$

43. There are $5 \cdot 4 \cdot 3 \cdot 2 \cdot 1 = 120$ or $P(5, 5) = 5! = 120$ different orderings of the players.

45. There are $7 \cdot 6 \cdot 5 = 210$ different routes.

47. Initially any person can sit at the table. Then the remaining 6 people can sit in $6! = 720$ different ways. Since there is no difference between the people sitting clockwise or counterclockwise around the table, we must divide this result by 2 for the final answer. There are 360 different ways to seat the 7 people.

49. $P(9, 9) = \dfrac{9!}{(9 - 9)!} = \dfrac{9!}{0!} = 9! = 362,880$

51. Counting February 29th, $P(366, 5) = \dfrac{366!}{(366 - 5)!} = \dfrac{366!}{361!} = \dfrac{366 \cdot 365 \cdot 364 \cdot 363 \cdot 362!}{361!} \approx 6.39 \times 10^{12}$.

53. There are 8 choices for the first and fourth digits and 10 choices for each of the other digits.

$8 \cdot 10 \cdot 10 \cdot 8 \cdot 10 \cdot 10 \cdot 10 \cdot 10 \cdot 10 \cdot 10 = 6,400,000,000$

Combinations

55. $C(n, r) = \dfrac{n!}{(n - r)!\, r!} \Rightarrow C(3, 1) = \dfrac{3!}{(3 - 1)!\, 1!} = \dfrac{3!}{2!\, 1!} = \dfrac{6}{2} = 3$

57. $C(n, r) = \dfrac{n!}{(n - r)!\, r!} \Rightarrow C(6, 3) = \dfrac{6!}{(6 - 3)!\, 3!} = \dfrac{6!}{3!\, 3!} = \dfrac{720}{36} = 20$

59. $C(n, r) = \dfrac{n!}{(n - r)!\, r!} \Rightarrow C(5, 0) = \dfrac{5!}{(5 - 0)!\, 0!} = \dfrac{5!}{5!\, 0!} = \dfrac{5!}{5!} = 1$

61. $\dbinom{n}{r} = \dfrac{n!}{(n - r)!\, r!} \Rightarrow \dbinom{8}{2} = \dfrac{8!}{(8 - 2)!\, 2!} = \dfrac{8!}{6!\, 2!} = \dfrac{8 \cdot 7 \cdot 6!}{8!\, 2!} = \dfrac{56}{2} = 28$

63. $\dbinom{n}{r} = \dfrac{n!}{(n - r)!\, r!} \Rightarrow \dbinom{20}{18} = \dfrac{20!}{(20 - 18)!\, 18!} = \dfrac{20!}{2!\, 18!} = \dfrac{20 \cdot 19 \cdot 18!}{2!\, 18!} = \dfrac{20 \cdot 19}{2} = 190$

65. From 39 numbers a player picks 5 numbers. Since order is unimportant, there are $C(39, 5) = 575,757$ different ways of doing this.

67. Two women can be selected from 5 women in $C(5, 2)$ different ways. Two men can be selected from 3 men in $C(5, 2)$ different ways. The total number of committees is $C(5, 2) \cdot C(3, 2) = 10 \cdot 3 = 30$.

69. Three questions can be selected from 5 questions in $C(5, 3)$ different ways. Four questions can be selected from 5 questions in $C(5, 4)$ different ways. The total number of possibilities is $C(5, 3) \cdot C(5, 4) = 10 \cdot 5 = 50$.

71. Three red marbles can be drawn from 10 red marbles in $C(10, 3)$ different ways. Two blue marbles can be drawn from 12 blue marbles in $C(12, 2)$ different ways. There are $C(10, 3) \cdot C(12, 2) = 120 \cdot 66 = 7920$ ways.

73. Since order is not important, there are $C(24, 3) = 2024$ ways to do this.

75. $P(n, n - 1) = \dfrac{n!}{(n - (n - 1))!} = \dfrac{n!}{1} = n!$ and $P(n, n) = \dfrac{n!}{(n - n)!} = \dfrac{n!}{0!} = \dfrac{n!}{1} = n!$

For example $P(7, 6) = 5040 = P(7, 7)$.

11.4: The Binomial Theorem

Binomial Coefficients

1. $\dbinom{5}{4} = \dfrac{5!}{1!\, 4!} = \dfrac{5 \cdot 4!}{1 \cdot 4!} = 5$

3. $\dbinom{4}{0} = \dfrac{4!}{4!\,0!} = \dfrac{4!}{4! \cdot 1} = 1$

5. $\dbinom{6}{5} = \dfrac{6!}{1!\,5!} = \dfrac{6 \cdot 5!}{1 \cdot 5!} = \dfrac{6}{1} = 6$

7. $\dbinom{3}{3} = \dfrac{3!}{0!\,3!} = \dfrac{3!}{1 \cdot 3!} = 1$

Binomial Theorem

9. Three a's, two b's $\Rightarrow C(5, 2) = 10$ different strings.

11. Four a's, four b's $\Rightarrow C(8, 4) = 70$ different strings.

13. Five a's, zero b's $\Rightarrow C(5, 0) = 1$ string.

15. Four a's, one b $\Rightarrow C(5, 1) = 5$ different strings.

17. $(x + y)^2 = \dbinom{2}{0}x^2y^0 + \dbinom{2}{1}x^1y^1 + \dbinom{2}{2}x^0y^2 = x^2 + 2xy + y^2$

19. $(m + 2)^3 = \dbinom{3}{0}m^3(2)^0 + \dbinom{3}{1}m^2(2)^1 + \dbinom{3}{2}m^1(2)^2 + \dbinom{3}{3}m^0(2)^3 = m^3 + 6m^2 + 12m + 8$

21. $(2x - 3)^3 = \dbinom{3}{0}(2x)^3(-3)^0 + \dbinom{3}{1}(2x)^2(-3)^1 + \dbinom{3}{2}(2x)^1(-3)^2 + \dbinom{3}{3}(2x)^0(-3)^3 = 8x^3 - 36x^2 + 54x - 27$

23. $(p - q)^6 = \dbinom{6}{0}p^6(-q)^0 + \dbinom{6}{1}p^5(-q)^1 + \dbinom{6}{2}p^4(-q)^2 + \dbinom{6}{3}p^3(-q)^3 + \dbinom{6}{4}p^2(-q)^4 + \dbinom{6}{5}p^1(-q)^5 + \dbinom{6}{6}p^0(-q)^6$

    $= p^6 - 6p^5q + 15p^4q^2 - 20p^3q^3 + 15p^2q^4 - 6pq^5 + q^6$

25. $(2m - 3n)^3 = \dbinom{3}{0}(2m)^3(3n)^0 + \dbinom{3}{1}(2m)^2(3n)^1 + \dbinom{3}{2}(2m)^1(3n)^2 + \dbinom{3}{3}(2m)^0(3n)^3$

    $= 8m^3 + 36m^2n + 54mn^2 + 27n^3$

27. $(1 - x^2)^4 = \dbinom{4}{0}(1)^4(-x^2)^0 + \dbinom{4}{1}(1)^3(-x^2)^1 + \dbinom{4}{2}(1)^2(-x^2)^2 + \dbinom{4}{3}(1)^1(-x^2)^3 + \dbinom{4}{4}(1)^0(-x^2)^4$

    $= 1 - 4x^2 + 6x^4 - 4x^6 + x^8$

29. $(2p^3 - 3)^3 = \dbinom{3}{0}(2p^3)^3(-3)^0 + \dbinom{3}{1}(2p^3)^2(-3)^1 + \dbinom{3}{2}(2p^3)^1(-3)^2 + \dbinom{3}{3}(2p^3)^0(-3)^3$

    $= 8p^9 - 36p^6 + 54p^3 - 27$

Pascal's Triangle

31. Using Pascal's triangle, the coefficients are 1, 2, and 1.

    $(x + y)^2 = 1x^2y^0 + 2x^1y^1 + 1x^0y^2 = x^2 + 2xy + y^2$

33. Using Pascal's triangle, the coefficients are 1, 4, 6, 4, and 1.

    $(3x + 1)^4 = 1(3x)^4(1)^0 + 4(3x)^3(1)^1 + 6(3x)^2(1)^2 + 4(3x)^1(1)^3 + 1(3x)^0(1)^4$

    $= 81x^4 + 108x^3 + 54x^2 + 12x + 1$

35. Using Pascal's triangle, the coefficients are 1, 5, 10, 10, 5, and 1.

$$(2 - x)^5 = 1(2)^5(-x)^0 + 5(2)^4(-x)^1 + 10(2)^3(-x)^2 + 10(2)^2(-x)^3 + 5(2)^1(-x)^4 + 1(2)^0(-x)^5$$

$$= 32 - 80x + 80x^2 - 40x^3 + 10x^4 - x^5$$

37. Using Pascal's triangle, the coefficients are 1, 4, 6, 4, and 1.

$$(x^2 + 2)^4 = 1(x^2)^4(2)^0 + 4(x^2)^3(2)^1 + 6(x^2)^2(2)^2 + 4(x^2)^1(2)^3 + 1(x^2)^0(2)^4 = x^8 + 8x^6 + 24x^4 + 32x^2 + 16$$

39. Using Pascal's triangle, the coefficients are 1, 4, 6, 4, and 1.

$$(4x - 3y)^4 = 1(4x)^4(-3y)^0 + 4(4x)^3(-3y)^1 + 6(4x)^2(-3y)^2 + 4(4x)^1(-3y)^3 + 1(4x)^0(-3y)^4$$

$$= 256x^4 - 768x^3y + 864x^2y^2 - 432xy^3 + 81y^4$$

41. Using Pascal's triangle, the coefficients are 1, 6, 15, 20, 15, 6, and 1.

$$(m + n)^6 = m^6 + 6m^5n + 15m^4n^2 + 20m^3n^3 + 15m^2n^4 + 6mn^5 + n^6$$

43. Using Pascal's triangle, the coefficients are 1, 3, 3, and 1.

$$(2x^3 - y^2)^3 = 1(2x^3)^3(-y^2)^0 + 3(2x^3)^2(-y^2)^1 + 3(2x^3)^1(-y^2)^2 + 1(2x^3)^0(-y^2)^3 = 8x^9 - 12x^6y^2 + 6x^3y^4 - y^6$$

45. The fourth term of $(a + b)^9$ is $\binom{9}{9-3}a^{9-3}b^3 = \binom{9}{6}a^6b^3 = 84a^6b^3$.

47. The fifth term of $(x + y)^8$ is $\binom{8}{8-4}x^{8-4}y^4 = \binom{8}{4}x^4y^4 = 70x^4y^4$.

49. The fourth term of $(2x + y)^5$ is $\binom{5}{5-3}(2x)^{5-3}y^3 = \binom{5}{2}(2x)^2y^3 = 40x^2y^3$.

51. The sixth term of $(3x - 2y)^6$ is $\binom{6}{6-5}(3x)^{6-5}(-2y)^5 = \binom{6}{1}(3x)(-2y)^5 = -576xy^7$.

11.5: Mathematical Induction

1. $3 + 6 + 9 + \cdots + 3n = \dfrac{3n(n + 1)}{2}$

(i) Show that the statement is true for $n = 1$: $3(1) = \dfrac{3(1)(2)}{2} \Rightarrow 3 = 3$

(ii) Assume that S_k is true: $3 + 6 + 9 + \cdots + 3k = \dfrac{3k(k + 1)}{2}$

Show that S_{k+1} is true: $3 + 6 + \cdots + 3(k + 1) = \dfrac{3(k + 1)(k + 2)}{2}$

Add $3(k + 1)$ to each side of S_k: $3 + 6 + 9 + \cdots + 3k + 3(k + 1) = \dfrac{3k(k + 1)}{2} + 3(k + 1)$

$$= \dfrac{3k(k + 1) + 6(k + 1)}{2} = \dfrac{(k + 1)(3k + 6)}{2} = \dfrac{3(k + 1)(k + 2)}{2}$$

Since S_k implies S_{k+1}, the statement is true for every positive integer n.

3. $5 + 10 + 15 + \cdots + 5n = \dfrac{5n(n + 1)}{2}$

   (i) Show that the statement is true for $n = 1$: $5(1) = \dfrac{5(1)(2)}{2} \Rightarrow 5 = 5$

   (ii) Assume that S_k is true: $5 + 10 + 15 + \cdots + 5k = \dfrac{5k(k + 1)}{2}$

   Show that S_{k+1} is true: $5 + 10 + \cdots + 5(k + 1) = \dfrac{5(k + 1)(k + 2)}{2}$

   Add $5(k + 1)$ to each side of S_k: $5 + 10 + 15 + \cdots + 5k + 5(k + 1) = \dfrac{5k(k + 1)}{2} + 5(k + 1)$

   $= \dfrac{5k(k + 1) + 10(k + 1)}{2} = \dfrac{(k + 1)(5k + 10)}{2} = \dfrac{5(k + 1)(k + 2)}{2}$

   Since S_k implies S_{k+1}, the statement is true for every positive integer n.

5. $3 + 3^2 + 3^3 + \cdots + 3^n = \dfrac{3(3^n - 1)}{2}$

   (i) Show that the statement is true for $n = 1$: $3^1 = \dfrac{3(3^1 - 1)}{2} \Rightarrow 3 = 3$

   (ii) Assume that S_k is true: $3 + 3^2 + 3^3 + \cdots + 3^k = \dfrac{3(3^k - 1)}{2}$

   Show that S_{k+1} is true: $3 + 3^2 + 3^3 + \cdots + 3^{k+1} = \dfrac{3(3^{k+1} - 1)}{2}$

   Add 3^{k+1} to each side of S_k: $3 + 3^2 + 3^3 + \cdots + 3^k + 3^{k+1} = \dfrac{3(3^k - 1)}{2} + 3^{k+1} = \dfrac{3(3^k - 1) + 2(3^{k+1})}{2}$

   $= \dfrac{3^{k+1} - 3 + 2(3^{k+1})}{2} = \dfrac{3(3^{k+1}) - 3}{2} = \dfrac{3(3^{k+1} - 1)}{2}$

   Since S_k implies S_{k+1}, the statement is true for every positive integer n.

7. $1^3 + 2^3 + 3^3 + \cdots + n^3 = \dfrac{n^2(n + 1)^2}{4}$

   (i) Show that the statement is true for $n = 1$: $1^3 = \dfrac{1^2(1 + 1)^2}{4} \Rightarrow 1 = 1$

   (ii) Assume that S_k is true: $1^3 + 2^3 + 3^3 + \cdots + k^3 = \dfrac{k^2(k + 1)^2}{4}$

   Show that S_{k+1} is true: $1^3 + 2^3 + \cdots + (k + 1)^3 = \dfrac{(k + 1)^2(k + 2)^2}{4}$

   Add $(k + 1)^3$ to each side of S_k: $1^3 + 2^3 + 3^3 + \cdots + k^3 + (k + 1)^3 = \dfrac{k^2(k + 1)^2}{4} + (k + 1)^3$

   $= \dfrac{k^2(k + 1)^2 + 4(k + 1)^3}{4} = \dfrac{(k + 1)^2(k^2 + 4k + 4)}{4} = \dfrac{(k + 1)^2(k + 2)^2}{4}$

   Since S_k implies S_{k+1}, the statement is true for every positive integer n.

9. $\dfrac{1}{1\cdot 2}+\dfrac{1}{2\cdot 3}+\cdots+\dfrac{1}{n(n+1)}=\dfrac{n}{n+1}$

   (i) Show that the statement is true for $n=1$: $\dfrac{1}{1(1+1)}=\dfrac{1}{1+1}\Rightarrow\dfrac{1}{2}=\dfrac{1}{2}$

   (ii) Assume that S_k is true: $\dfrac{1}{1\cdot 2}+\dfrac{1}{2\cdot 3}+\cdots+\dfrac{1}{k(k+1)}=\dfrac{k}{k+1}$

      Show that S_{k+1} is true: $\dfrac{1}{1\cdot 2}+\dfrac{1}{2\cdot 3}+\cdots+\dfrac{1}{(k+1)(k+2)}=\dfrac{k+1}{k+2}$

      Add $\dfrac{1}{(k+1)(k+2)}$ to each side of S_k: $\dfrac{1}{1\cdot 2}+\dfrac{1}{2\cdot 3}+\cdots+\dfrac{1}{(k+1)(k+2)}=\dfrac{k}{k+1}+\dfrac{1}{(k+1)(k+2)}$

$$=\dfrac{k(k+2)+1}{(k+1)(k+2)}=\dfrac{k^2+2k+1}{(k+1)(k+2)}=\dfrac{(k+1)(k+1)}{(k+1)(k+2)}=\dfrac{k+1}{k+2}$$

Since S_k implies S_{k+1}, the statement is true for every positive integer n.

11. $\dfrac{4}{5}+\dfrac{4}{5^2}+\dfrac{4}{5^3}+\cdots+\dfrac{4}{5^n}=1-\dfrac{1}{5^n}$

   (i) Show that the statement is true for $n=1$: $\dfrac{4}{5^1}=1-\dfrac{1}{5^1}\Rightarrow\dfrac{4}{5}=\dfrac{4}{5}$

   (ii) Assume that S_k is true: $\dfrac{4}{5}+\dfrac{4}{5^2}+\dfrac{4}{5^3}+\cdots+\dfrac{4}{5^k}=1-\dfrac{1}{5^k}$

      Show that S_{k+1} is true: $\dfrac{4}{5}+\dfrac{4}{5^2}+\cdots+\dfrac{4}{5^{k+1}}=1-\dfrac{1}{5^{k+1}}$

      Add $\dfrac{4}{5^{k+1}}$ to each side of S_k: $\dfrac{4}{5}+\dfrac{4}{5^2}+\dfrac{4}{5^3}+\cdots+\dfrac{4}{5^k}+\dfrac{4}{5^{k+1}}=1-\dfrac{1}{5^k}+\dfrac{4}{5^{k+1}}$

$$=1-\dfrac{1}{5^k}\cdot\dfrac{5}{5}+\dfrac{4}{5^{k+1}}=1-\dfrac{5}{5^{k+1}}+\dfrac{4}{5^{k+1}}=1-\dfrac{1}{5^{k+1}}$$

Since S_k implies S_{k+1}, the statement is true for every positive integer n.

13. $\dfrac{1}{1\cdot 4}+\dfrac{1}{4\cdot 7}+\cdots+\dfrac{1}{(3n-2)(3n+1)}=\dfrac{n}{3n+1}$

   (i) Show that the statement is true for $n=1$: $\dfrac{1}{1\cdot 4}=\dfrac{1}{3(1)+1}\Rightarrow\dfrac{1}{4}=\dfrac{1}{4}$

   (ii) Assume that S_k is true: $\dfrac{1}{1\cdot 4}+\cdots+\dfrac{1}{(3k-2)(3k+1)}=\dfrac{k}{3k+1}$

      Show that S_{k+1} is true: $\dfrac{1}{1\cdot 4}+\cdots+\dfrac{1}{[3(k+1)-2][3(k+1)+1]}=\dfrac{k+1}{3(k+1)+1}$

      Add $\dfrac{1}{[3(k+1)-2][3(k+1)+1]}$ to each side of S_k: $\dfrac{1}{1\cdot 4}+\cdots+\dfrac{1}{[3(k+1)-2][3(k+1)+1]}$

$$=\dfrac{k}{3k+1}+\dfrac{1}{[3(k+1)-2][3(k+1)+1]}=\dfrac{k}{3k+1}+\dfrac{1}{(3k+1)(3k+4)}=\dfrac{k(3k+4)+1}{(3k+1)(3k+4)}$$

$$=\dfrac{3k^2+4k+1}{(3k+1)(3k+4)}=\dfrac{(3k+1)(k+1)}{(3k+1)(3k+4)}=\dfrac{k+1}{3k+4}=\dfrac{k+1}{3(k+1)+1}$$

Since S_k implies S_{k+1}, the statement is true for every positive integer n.

15. When $n=1$, $3^1<6(1)\Rightarrow 3<6$. When $n=2$, $3^2<6(2)\Rightarrow 9<12$.

    When $n=3$, $3^3<6(3)\Rightarrow 27>18$. For all $n\ge 3$, $3^n>6n$. The only values are 1 and 2.

17. When $n = 1, 2^1 > 1^2 \Rightarrow 2 > 1$. When $n = 2, 2^2 = 2^2 \Rightarrow 4 = 4$. When $n = 3, 2^3 < 3^2 \Rightarrow 8 < 9$.

    When $n = 4, 2^4 = 4^2 \Rightarrow 16 = 16$. For all $n \geq 5, 2^n > n^2$. The only values are 2, 3, and 4.

19. $(a^m)^n = a^{mn}$

    (i) Show that the statement is true for $n = 1$: $(a^m)^1 = a^{m \cdot 1} \Rightarrow a^m = a^m$

    (ii) Assume that S_k is true: $(a^m)^k = a^{mk}$

    Show that S_{k+1} is true: $(a^m)^{k+1} = a^{m(k+1)}$

    Multiply each side of S_k by a^m: $(a^m)^k \cdot (a^m)^1 = a^{mk} \cdot a^m \Rightarrow (a^m)^{k+1} = a^{mk+m} \Rightarrow (a^m)^{k+1} = a^{m(k+1)}$

    Since S_k implies S_{k+1}, the statement is true for every positive integer n.

21. $2^n > 2n$, if $n \geq 3$

    (i) Show that the statement is true for $n = 3$: $2^3 > 2(3) \Rightarrow 8 > 6$

    (ii) Assume that S_k is true: $2^k > 2k$

    Show that S_{k+1} is true: $2^{k+1} > 2(k + 1)$

    Multiply each side of S_k by 2: $2^k \cdot 2 > 2k \cdot 2 \Rightarrow 2^{k+1} > 2(k + 1)$

    Since S_k implies S_{k+1}, the statement is true for every positive integer $n \geq 3$.

23. $a^n > 1$, if $a > 1$

    (i) Show that the statement is true for $n = 1$: $a^1 > 1 \Rightarrow a > 1$, which is true by the given restriction.

    (ii) Assume that S_k is true: $a^k > 1$

    Show that S_{k+1} is true: $a^{k+1} > 1$

    Multiply each side of S_k by a: $a^k \cdot a > 1 \cdot a \Rightarrow a^{k+1} > a$

    Because $a > 1$, we may substitute 1 for a in the expression. That is $a^{k+1} > 1$

    Since S_k implies S_{k+1}, the statement is true for every positive integer n.

25. $a^n < a^{n-1}$, if $0 < a < 1$

    (i) Show that the statement is true for $n = 1$: $a^1 < a^0 \Rightarrow a < 1$, which is true by the given restriction.

    (ii) Assume that S_k is true: $a^k < a^{k-1}$

    Show that S_{k+1} is true: $a^{k+1} < a^k$

    Multiply each side of S_k by a: $a^k \cdot a < a^{k-1} \cdot a \Rightarrow a^{k+1} < a^k$

    Since S_k implies S_{k+1}, the statement is true for every positive integer n.

27. $n! > 2^n$, if $n \geq 4$

    (i) Show that the statement is true for $n = 4$: $4! > 2^4 \Rightarrow 24 > 16$

    (ii) Assume that S_k is true: $k! > 2^k$

    Show that S_{k+1} is true: $(k + 1)! > 2^{k+1}$

    Multiply each side of S_k by $k + 1$: $(k + 1)k! > 2^k(k + 1) \Rightarrow (k + 1)! > 2^k(k + 1)$

    Because $(k + 1) > 2$ for all $k \geq 4$, we may substitute 2 for $(k + 1)$ in the expression.

    That is $(k + 1)! > 2^k(2)$ or $(k + 1)! > 2^{k+1}$.

    Since S_k implies S_{k+1}, the statement is true for every positive integer $n \geq 4$.

29. The number of handshakes is $\dfrac{n^2 - n}{2}$ if $n \geq 2$.

    (i) Show that the statement is true for $n = 2$: The number of handshakes for 2 people is $\dfrac{2^2 - 2}{2} = \dfrac{2}{2} = 1$, which is true.

    (ii) Assume that S_k is true: The number of handshakes for k people is $\dfrac{k^2 - k}{2}$.

    Show that S_{k+1} is true: The number of handshakes for $k + 1$ people is

    $$\dfrac{(k+1)^2 - (k+1)}{2} = \dfrac{k^2 + 2k + 1 - k - 1}{2} = \dfrac{k^2 + k}{2}.$$

    When a person joins a group of k people, each person must shake hands with the new person.

    Since there are a total of k people that will shake hands with the new person, the total number of handshakes

    for $k + 1$ people is $\dfrac{k^2 - k}{2} + k = \dfrac{k^2 - k + 2k}{2} = \dfrac{k^2 + k}{2}$.

    Since S_k implies S_{k+1}, the statement is true for every positive integer $n \geq 2$.

31. The first figure has perimeter $P = 3$. When a new figure is generated, each side if the previous figure increases

    in length by a factor of $\dfrac{4}{3}$. Thus, the second figure has perimeter $P = 3\left(\dfrac{4}{3}\right)$, the third figure has perimeter

    $P = 3\left(\dfrac{4}{3}\right)^2$, and so on. In general, the nth figure has perimeter $P = 3\left(\dfrac{4}{3}\right)^{n-1}$.

33. With 1 ring, 1 move is required. With 2 rings, 3 moves are required. Note that $3 = 2 + 1$. With 3 rings, 7

    moves are required. Note that $7 = 2^2 + 2 + 1$.

    With n rings $2^{n-1} + 2^{n-2} + \cdots + 2^1 + 1 = 2^n - 1$ moves are required.

    (i) Show that the statement is true for $n = 1$: The number of moves for 1 ring is $2^1 - 1 = 1$, which is true.

    (ii) Assume that S_k is true: The number of moves for k rings is $2^k - 1$.

    Show that S_{k+1} is true: The number of moves for $k + 1$ rings is $2^{k+1} - 1$.

    Assume $k + 1$ rings are on the first peg. Since S_k is true, the top k rings can be moved to the second peg in

    $2^k - 1$ moves. Now move the bottom ring to the third peg. Since S_k is true, move the k rings from the

    second peg on top of the ring on the third peg in $2^k - 1$ moves. The total number of moves is

    $(2^k - 1) + 1 + (2^k - 1) = 2 \cdot 2^k - 1 = 2^{k+1} - 1$

    Since S_k implies S_{k+1}, the statement is true for every positive integer n.

11.6: Probability

Probability of an Event

1. Yes. The number is between 0 and 1.

3. No. The number is greater 1.

5. Yes. The number is between 0 and 1, inclusive.

7. No. This number is less than 0.

9. A head when tossing a fair coin $\Rightarrow \dfrac{1}{2}$.

11. Rolling a 2 with a fair die $\Rightarrow \dfrac{1}{6}$.

13. Guessing the correct answer for a true-false question $\Rightarrow \dfrac{1}{2}$.

15. Drawing an king from a standard deck of 52 cards $\Rightarrow \dfrac{4}{52} = \dfrac{1}{13}$.

17. The access code can vary between 0000 and 9999. This is 10,000 possibilities, so the probability of guessing

    the ATM code at random is $\dfrac{1}{10,000}$.

19. (a) The probability that their favorite is pepperoni is 0.43, so the probability it is not pepperoni is

    $1 - 0.43 = 0.57$ or 57%.

    (b) $19\% + 14\% = 33\%$ or 0.33

Probability of Compound Events

21. The probability of one tail is $\dfrac{1}{2}$. The events of tossing a coin are independent so the probability of tossing two

    tails is $\dfrac{1}{2} \cdot \dfrac{1}{2} = \dfrac{1}{4}$.

23. The probability of rolling either a 5 or a 6 is $\dfrac{2}{6} = \dfrac{1}{3}$. The probability of obtaining a 5 or 6 on three consecutive

    roles is $\dfrac{1}{3} \cdot \dfrac{1}{3} \cdot \dfrac{1}{3} = \dfrac{1}{27}$.

25. Refer to Table 8.10 in this section. There are 6 outcomes from a sample space of 36 outcomes that result in the

    event of the sum being 7. They are (1, 6), (2, 5), (3, 4), (4, 3), (5, 2), and (6, 1). Thus, the probability of

    rolling a sum of 7 is $\dfrac{6}{36} = \dfrac{1}{6}$.

27. The probability of rolling a die and not obtaining a 6 is $\dfrac{5}{6}$. The probability of rolling a die four times and not

    obtaining a 6 is $\dfrac{5}{6} \cdot \dfrac{5}{6} \cdot \dfrac{5}{6} \cdot \dfrac{5}{6} = \dfrac{625}{1296} \approx 0.482$.

29. The probability of drawing the first ace is $\dfrac{4}{52}$, the second ace $\dfrac{3}{51}$, the third ace $\dfrac{2}{50}$, and the fourth ace $\dfrac{1}{49}$. The

    probability of drawing four aces is $\dfrac{4}{52} \cdot \dfrac{3}{51} \cdot \dfrac{2}{50} \cdot \dfrac{1}{49} = \dfrac{24}{6,4978,400} = \dfrac{1}{270,725}$.

31. There are $\dbinom{13}{3}$ ways to draw 3 hearts, and there are $\dbinom{13}{2}$ ways to draw 2 diamonds. there are $\dbinom{52}{5}$ different

    poker hands. Thus, the probability of drawing 3 hearts and 2 diamonds is

    $$P(E) = \frac{n(E)}{n(S)} = \frac{\dbinom{13}{3} \cdot \dbinom{13}{2}}{\dbinom{52}{5}} = \frac{286 \cdot 78}{2,598,960} = 0.0086, \text{ or a } 0.86\% \text{ chance.}$$

33. There are 4 strings out of 20 that are defective. Therefore, there is a $\dfrac{4}{20} = 0.2$ probability or 20% chance of drawing a defective string and rejecting the box.

35. (a) Figure 35 shows a Venn diagram of the data.

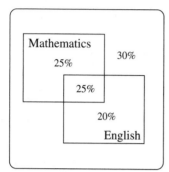

Figure 35

(b) The percentage of students needing help with mathematics, English, or both is $25 + 25 + 20 = 70\%$. The probability of needing help in mathematics, English, or both is 0.7.

(c) If M represents the event of a student needing help with mathematics, and E represents the event that a student needs help with English, then
$$P(M \text{ or } E) = P(M \cup E) = P(M) + P(E) - P(M \cap E) = 0.5 + 0.45 - 0.25 = 0.7.$$

37. (a) There were $4481 + 1614 = 6095$ books published in the areas of art and music out of a possible 119,923.

The probability of a new book or edition being in the area of art or music is $\dfrac{6095}{119{,}923} \approx 0.051$.

(b) The number of books published in science or religion was $7032 + 6659 = 13{,}691$. The probability of a new book or edition not being published in the areas of science or religion is
$$1 - \dfrac{13{,}691}{119{,}923} = \dfrac{106{,}232}{119{,}923} \approx 0.886.$$

39. The probability is $\dfrac{81}{100{,}000} \approx 0.00081$.

41. The probability of tossing a coin n times and obtaining n heads is $\left(\dfrac{1}{2}\right)^{n}$. As n increases the probability decreases. This agrees with intuition. The probability of tossing a long string of consecutive heads is small. The longer the string of heads, the less chance there is of it happening.

43. The possible outcomes for a sum of 5 or 6 are (1, 4), (2, 3), (3, 2), (4, 1), (1, 5), (2, 4), (3, 3), (4, 2), and (5, 1).

This is a total of 9 outcomes out of 36 possible outcomes. The probability is $\dfrac{9}{36} = \dfrac{1}{4}$.

45. There are four 2's, four 3's and four 4's in a standard deck. The probability is $\dfrac{12}{52} = \dfrac{3}{13}$.

47. (a) The probability of HT is $\dfrac{3}{4} \cdot \dfrac{1}{4} = \dfrac{3}{16}$.

    (b) The probability of HH is $\dfrac{3}{4} \cdot \dfrac{3}{4} = \dfrac{9}{16}$.

    (c) The probability of HHT is $\dfrac{3}{4} \cdot \dfrac{3}{4} \cdot \dfrac{1}{4} = \dfrac{9}{64}$.

    (d) The probability of THT is $\dfrac{1}{4} \cdot \dfrac{3}{4} \cdot \dfrac{1}{4} = \dfrac{3}{64}$.

49. (a) The only way to obtain a sum of 12 is for both dice to show a 6. Since these events are independent, the probability of two 6's is $(0.3)(0.3) = 0.09$ or 9%.

    (b) There are two ways to obtain a sum of 11. The red die shows a 5 and the blue die a 6, or vice versa. The roll of the two dice are independent. The probability of the red die showing a 5 is 0.2 and the probability of the blue die showing a 6 is 0.3. The probability of a sum of 11 is $(0.2)(0.3) = 0.06$. Similarly, the probability of the red die showing a 6 and the blue die showing a 5 is 0.06. Since these events are mutually exclusive, the probability of a sum of 11 is $0.06 + 0.06 = 0.12$ or 12%.

51. There are ten possibilities for each of the three digits. By the fundamental counting principle, there are $10 \cdot 10 \cdot 10 = 10^3 = 1000$ different ways to pick these numbers. There is only one winning number so the probability is $\dfrac{1}{1000}$.

53. (a) There are a total of $22 + 18 + 10 = 50$ marbles in the jar. Since 22 of them are red, there is a probability of $\dfrac{22}{50} = 0.44$ of drawing a red ball.

    (b) Since there is a 0.44 probability of drawing a red ball, there is a $1 - 0.44 = 0.56$ probability of not drawing a red ball.

    (c) Drawing a blue or green ball is equivalent to not drawing a red ball. The probability is the same as in part (b), which was 0.56.

55. (a) When the diet consists of 6% saccharin, there is approximately a 0.12 or 12% chance of developing bladder cancer over the lifetime of the animal.

    (b) There is little lifetime risk of bladder cancer until the percentage of saccharin in the diet exceeds 4%. As the amount of saccharin increases, the risk also increases. When the diet is 8% saccharin, there is approximately a 0.35 or 35% lifetime risk of developing bladder cancer.

Conditional Probability and Dependent Events

57. Since one card, a queen, has been drawn, there are 3 queens left in the set of 51 cards. So the probability of drawing a queen is $\dfrac{3}{51}$.

59. Since there are 4 kings in a total of 12 face cards, the probability of drawing a king is $\dfrac{4}{12} = \dfrac{1}{3}$.

61. Since 2 red out of 10 red marbles and 4 blue out of 23 blue marbles have already been drawn, 6 marbles out of 33 marbles have been removed from the jar. There are $23 - 4 = 19$ blue marbles left out of $33 - 6 = 27$ marble in the jar. The probability of drawing a blue marble next is $\dfrac{19}{27}$.

63. Let E_1 denote the event that it is cloudy and E_2 denote the event that it is windy. $P(E_1) = 0.30$ and

    $P(E_1 \text{ and } E_2) = P(E_1 \cap E_2) = 0.12$. $P(E_1 \cap E_2) = P(E_1) \cdot P(E_2, \text{ given that } E_1 \text{ has occured}) \Rightarrow$

    $0.12 = 0.30 \cdot P(E_2, \text{ given that } E_1 \text{ has occured}) \Rightarrow 0.4 = P(E_2, \text{ given that } E_1 \text{ has occured})$. The probability

    that it will be windy given that the day is cloudy is 40%.

65. The possibilities for the result that the first die is a 2 and the sum of the two dice is 7 or more, can be represent-

    ed by $\{(2, 5), (2, 6)\}$. The sample space can be represented by $\{(2, 1), (2, 2), (2, 3), (2, 4), (2, 5), \text{ and } (2, 6)\}$.

    Out of a total of 6, 2 outcomes satisfy the conditions, so the probability is $\dfrac{2}{6} = \dfrac{1}{3}$.

67. (a) Let D represent the event that part is defective. Since 18 of the 235 parts are defective, $P(D) = \dfrac{18}{235}$.

    (b) Let A represent the event that part is type A. Since 7 of the 18 parts are type A, $P(A, \text{ given } D) = \dfrac{7}{18}$.

    (c) $P(D \text{ and } A) = P(D \cap A) = P(D) \cdot P(A, \text{ given } D) = \dfrac{18}{235} \cdot \dfrac{7}{18} = \dfrac{7}{235}$.

69. (a) Let O represent the event that the number is odd. Since 8 of the 15 numbers are odd, $P(O) = \dfrac{8}{15}$.

    (b) Let E represent the event that the number is even. Since 7 of the 15 numbers are even, $P(E) = \dfrac{7}{15}$.

    (c) Let M represent the event that the number is prime. Since 6 of the 15 numbers are prime, $P(M) = \dfrac{6}{15} = \dfrac{2}{5}$.

    (d) Since 5 of the 6 prime numbers are odd, $P(M \cap O) = \dfrac{5}{15} = \dfrac{1}{3}$.

    (e) Since 1 of the 6 prime numbers is even, $P(M \cap O) = \dfrac{1}{15}$.

Chapter 11 Review Exercises

1. $a_1 = -3(1) + 2 = -1; a_2 = -3(2) + 2 = -4; a_3 = -3(3) + 2 = -7; a_4 = -3(4) + 2 = -10$

   The first four terms are $-1, -4, -7,$ and -10.

3. $a_1 = 0; a_2 = 2a_1 + 1 = 2(0) + 1 = 1; a_3 = 2a_2 + 1 = 2(1) + 1 = 3; a_4 = 2a_3 + 1 = 2(3) + 1 = 7$

   The first four terms are $0, 1, 3,$ and 7.

5. The points $(1, 5), (2, 3), (3, 1), (4, 2), (5, 4),$ and $(6, 6)$ lie on the graph. Therefore, the terms of this sequence

   are $5, 3, 1, 2, 4, 6$.

7. $a_1 = 0, a_2 = 3a_1 + 1 = 3(0) + 1 = 1, a_3 = 3a_2 + 1 = 3(1) + 1 = 4, a_4 = 3a_3 + 1 = 3(4) + 1 = 13,$

   $a_5 = 3a_4 + 1 = 3(13) + 1 = 40$. The first five terms are $0, 1, 4, 13,$ and 40.

9. (a) Each term can be found by adding 2 to the previous term, beginning with 2. It is an arithmetic sequence.

A numerical representation for the first eight terms is shown in Figure 9a.

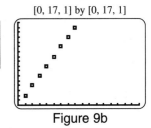

[0, 17, 1] by [0, 17, 1]

n	1	2	3	4	5	6	7	8
a_n	2	4	6	8	10	12	14	16

Figure 9a

Figure 9b

(b) A graphical representation for the first eight terms is shown in Figure 9b. Notice that the points lie on a line with slope 2 because the sequence is arithmetic with $d = 2$.

(c) The term of the sequence are the even, positive integers. This can be represented symbolically by $a_n = 2n$.

11. (a) Each term can be found by multiplying the previous term by $\frac{1}{3}$, beginning with 81. It is a geometric sequence.

A numerical representation for the first eight terms is shown in Figure 11a.

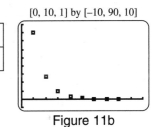

[0, 10, 1] by [−10, 90, 10]

n	1	2	3	4	5	6	7	8
a_n	81	27	9	3	1	$\frac{1}{3}$	$\frac{1}{9}$	$\frac{1}{27}$

Figure 11a

Figure 11b

(b) A graphical representation for the first eight terms is shown in Figure 11b. Notice that the points lie on a curve that is decaying exponentially because the common ratio is less than 1 in absolute value.

(c) To find the symbolic representation, we will use the formula $a_n = a_1 r^{n-1}$, where $a_n = f(n)$. The common ratio of this sequence is $r = \frac{1}{3}$ and the first term is $a_1 = 81$. Therefore a symbolic representation is given by $a_n = 81\left(\frac{1}{3}\right)^{n-1}$.

13. Let $a_n = f(n)$, where $f(n) = dn + c$. Then, $f(n) = 4n + c$. Since $f(3) = 4(3) + c = -3 \Rightarrow c = -15$.

Thus $a_n = f(n) = 4n - 15$.

15. Since f is a linear function, the sequence is arithmetic.

17. Since f is written in the form $f(n) = cr^{n-1}$, the sequence is geometric.

19. (a) The initial height is 4 feet. On the first rebound it reaches 90% of 4 or 3.6 feet. On the second rebound it attains a height of 90% of 3.6 or 3.24 feet. Each term in this sequence is found by multiplying the previous term by 0.9. Thus, the first five terms are 4, 3.6, 3.24, 2.916, 2.6244. This is a geometric sequence.

(b) Plot the points $(1, 4)$, $(2, 3.6)$, $(3, 3.24)$, $(4, 2.916)$, and $(5, 2.6244)$ as shown in Figure 19.

(c) The first term is 4 and the common ratio is 0.9. Thus, $a_n = 4(0.9)^{n-1}$.

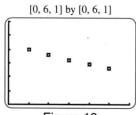

[0, 6, 1] by [0, 6, 1]

Figure 19

21. $S_5 = (4(1) + 1) + (4(2) + 1) + (4(3) + 1) + (4(4) + 1) + (4(5) + 1) = 5 + 9 + 13 + 17 + 21 = 65$

23. The first term is $a_1 = -2$ and the last term is $a_9 = 22$.

To find the sum use $S_n = n\left(\dfrac{a_1 + a_n}{2}\right) \Rightarrow S_9 = 9\left(\dfrac{-2 + 22}{2}\right) = 90$. The sum is 90.

25. The first term is $a_1 = 1$ and the common ratio is $r = 3$. Since there are 8 terms, the sum is

$$S_n = a_1\left(\frac{1 - r^n}{1 - r}\right) \Rightarrow S_8 = 1\left(\frac{1 - 3^8}{1 - 3}\right) = 3280.$$

27. The first term is $a_1 = 2$ and the common ratio is $r = \dfrac{1}{2}$. The sum is $S_n = a_1\left(\dfrac{1}{1 - r}\right) \Rightarrow S = 2\left(\dfrac{1}{1 - \frac{1}{2}}\right) = 4$.

29. The first term is $a_1 = 0.2$ and the common ratio is $r = \dfrac{1}{10}$.

The sum is $S_n = a_1\left(\dfrac{1}{1 - r}\right) \Rightarrow S = 0.2\left(\dfrac{1}{1 - \frac{1}{10}}\right) = \dfrac{2}{10} \cdot \dfrac{10}{9} = \dfrac{2}{9}$.

31. $\displaystyle\sum_{k=1}^{5}(5k + 1) = (5 + 1) + (10 + 1) + (15 + 1) + (20 + 1) + (25 + 1) = 6 + 11 + 16 + 21 + 26$

33. $1^3 + 2^3 + 3^3 + 4^3 + 5^3 + 6^3 = \displaystyle\sum_{k=1}^{6} k^3$

35. Since $a_1 = 5$ and $d = -3$, use the formula $S_n = \dfrac{n}{2}(2a_1 + (n - 1)d)$.

$$S_{30} = \frac{30}{2}(2(5) + (30 - 1)(-3)) = 15(10 + 29(-3)) = 15(-77) = -1155$$

37. $\dfrac{2}{11} = 0.18181818\ldots = 0.18 + 0.0018 + 0.000018 + 0.00000018 + \cdots$

39. There are $4^{20} \approx 1.1 \times 10^{12}$ different ways to answer the exam.

41. There are 50 choices for each number in the combination. So there are $50^4 = 6{,}250{,}000$ possible combinations.

43. $P(n, r) = \dfrac{n!}{(n - r)!} \Rightarrow P(6, 3) = \dfrac{6!}{(6 - 3)!} = \dfrac{6!}{3!} = \dfrac{6 \cdot 5 \cdot 4 \cdot 3!}{3!} = 6 \cdot 5 \cdot 4 = 120$

45. There are $P(5, 5) = 5! = 120$ different arrangements.

47. From a group of 6 people, we must select a set of 3 people. This can be done $C(6, 3) = 20$ different ways.

49. They can be selected in $C(10, 6) = 210$ different ways.

51. $1 + 3 + 5 + \cdots + (2n - 1) = n^2$

    (i) Show that the statement is true for $n = 1$: $2(1) - 1 = 1^2 \Rightarrow 1 = 1$

    (ii) Assume that S_k is true: $1 + 3 + 5 + \cdots + (2k - 1) = k^2$

        Show that S_{k+1} is true: $1 + 3 + \cdots + (2(k + 1) - 1) = (k + 1)^2$

        Add $2k + 1$ to each side of S_k: $1 + 3 + 5 + \cdots + (2k - 1) + (2k + 1) = k^2 + 2k + 1 = (k + 1)^2$

    Since S_k implies S_{k+1}, the statement is true for every positive integer n.

53. Since 1, 2, or 3 are three outcomes out of six possible outcomes, the probability is $\dfrac{3}{6} = \dfrac{1}{2}$.

55. There are $C(16, 2)$ different ways to select 2 batteries from a pack of 16. In order to avoid testing a defective

    battery, we must pick 2 batteries from the 14 good ones. This can be done in $C(14, 2)$ different ways. The

    probability that the box is not rejected is $\dfrac{C(14, 2)}{C(16, 2)} = \dfrac{91}{120} \approx 0.758$.

57. (a) The total number of marbles in the jar is $13 + 27 + 20 = 60$. Since 27 of the marbles are blue, there is a

        probability of $\dfrac{27}{60} = 0.45$ of drawing a blue marble.

    (b) Since there is a 0.45 probability of drawing a blue marble, there is a $1 - 0.45 = 0.55$ probability of drawing

        a marble that is not blue.

    (c) There is a $\dfrac{13}{60} \approx 0.217$ probability of drawing a red marble.

59. The sequence is graphed in Figure 59. Initially, the population density grows slowly, then it increases rapidly.

    After some time, it levels off near 4000 thousand (4,000,000) per acre.

[0, 16, 1] by [0, 5000, 1000]

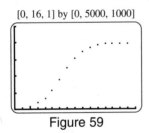

Figure 59

Chapters 1-11 Cumulative Review Exercises

1. $\dfrac{8863 - 8145}{8145} = \dfrac{717}{8145} \approx 0.088$ or about 8.8%

3. $\dfrac{5 - \sqrt[3]{4}}{\pi^2 - (\sqrt{3} + 1)} \approx 0.48$

5. $d = \sqrt{(1 - (-4))^2 + (-2 - 2)^2} = \sqrt{5^2 + (-4)^2} = \sqrt{25 + 16} = \sqrt{41}$

7.   For the graphs of parts (a) – (j), see Figures 7a – 7j.

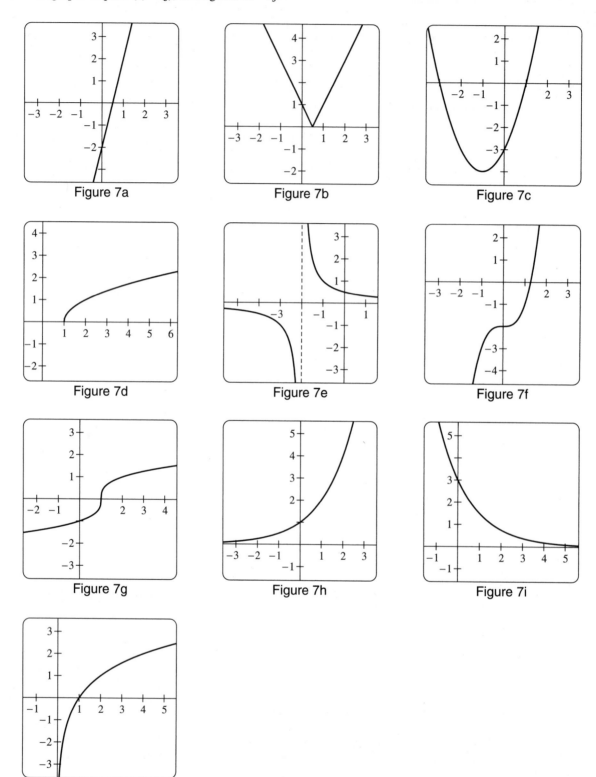

Figure 7a

Figure 7b

Figure 7c

Figure 7d

Figure 7e

Figure 7f

Figure 7g

Figure 7h

Figure 7i

Figure 7j

9. (a) $f(-3) = \dfrac{1}{(-3)^2 - 4} = \dfrac{1}{9 - 4} = \dfrac{1}{5}$; $f(a + 1) = \dfrac{1}{(a + 1)^2 - 4} = \dfrac{1}{a^2 + 2a + 1 - 4} = \dfrac{1}{a^2 + 2a - 3}$

(b) For f to be defined, $x^2 - 4 \neq 0$. Thus, the domain is $D = \{x \mid x \neq -2, x \neq 2\}$.

11. $\dfrac{f(-1) - f(-2)}{-1 - (-2)} = \dfrac{((-1)^3 - 4) - ((-2)^3 - 4)}{1} = -5 - (-12) = 7$

13. (a) The graph passes through the points $(0, -1)$ and $(4, 2)$. The slope is $m = \dfrac{2 - (-1)}{4 - 0} = \dfrac{3}{4}$.

The y-intercept is -1. The x-intercept can be found by noting that a 1-unit rise from the point $(0, -1)$ would

require a $\dfrac{4}{3}$-unit run to return to a line with slope $\dfrac{3}{4}$. That is, $m = \dfrac{\text{rise}}{\text{run}} = \dfrac{1}{\frac{4}{3}} = \dfrac{3}{4}$. The x-intercept is $\dfrac{4}{3}$.

(b) Since the slope is $\dfrac{3}{4}$ and the y-intercept is -1, the formula for f is $f(x) = \dfrac{3}{4}x - 1$.

(c) $\dfrac{3}{4}x - 1 = 0 \Rightarrow \dfrac{3}{4}x = 1 \Rightarrow x = \dfrac{4}{3}$

15. A line passing through the points $(2, -4)$ and $(-3, 2)$ has slope $m = \dfrac{2 - (-4)}{-3 - 2} = \dfrac{6}{-5} = -\dfrac{6}{5}$.

Using the point $(2, -4)$, the equation is $y = -\dfrac{6}{5}(x - 2) - 4 \Rightarrow y = -\dfrac{6}{5}x - \dfrac{8}{5}$.

17. A line that is parallel to the y-axis is vertical. A vertical line passing through $(-2, 4)$ has equation $x = -2$.

19. (a) $4(x - 2) + 1 = 3 - \dfrac{1}{2}(2x + 3) \Rightarrow 4x - 8 + 1 = 3 - x - \dfrac{3}{2} \Rightarrow 5x = \dfrac{17}{2} \Rightarrow x = \dfrac{17}{10}$

(b) $6x^2 = 13x + 5 \Rightarrow 6x^2 - 13x - 5 = 0 \Rightarrow (3x + 1)(2x - 5) = 0 \Rightarrow x = -\dfrac{1}{3}$ or $x = \dfrac{5}{2}$

(c) By the quadratic formula, $x = \dfrac{-b \pm \sqrt{b^2 - 4ac}}{2a} \Rightarrow x = \dfrac{-(-1) \pm \sqrt{(-1)^2 - 4(1)(-3)}}{2(1)} = \dfrac{1 \pm \sqrt{13}}{2}$

(d) $x^3 + x^2 = 4x + 4 \Rightarrow x^3 + x^2 - 4x - 4 = 0 \Rightarrow x^2(x + 1) - 4(x + 1) = 0 \Rightarrow (x^2 - 4)(x + 1) = 0 \Rightarrow$

$(x + 2)(x - 2)(x + 1) = 0 \Rightarrow x = -2, -1,$ or 2

(e) $x^4 - 4x^2 + 3 = 0 \Rightarrow (x^2 - 3)(x^2 - 1) = 0 \Rightarrow x^2 = 3$ or $x^2 = 1 \Rightarrow x = \pm\sqrt{3}$ or ± 1

(f) $\dfrac{1}{x - 3} = \dfrac{4}{x + 5} \Rightarrow (x - 3)(x + 5) \cdot \dfrac{1}{x - 3} = \dfrac{4}{x + 5} \cdot (x - 3)(x + 5) \Rightarrow x + 5 = 4(x - 3) \Rightarrow$

$x + 5 = 4x - 12 \Rightarrow -3x = -17 \Rightarrow x = \dfrac{17}{3}$

(g) $3e^{2x} - 5 = 23 \Rightarrow 3e^{2x} = 28 \Rightarrow e^{2x} = \dfrac{28}{3} \Rightarrow \ln e^{2x} = \ln\dfrac{28}{3} \Rightarrow 2x = \ln\dfrac{28}{3} \Rightarrow x = \dfrac{\ln\left(\frac{28}{3}\right)}{2} \approx 1.117$

(h) $2\log(x + 1) - 1 = 2 \Rightarrow 2\log(x + 1) = 3 \Rightarrow \log(x + 1) = \dfrac{3}{2} \Rightarrow 10^{\log(x+1)} = 10^{3/2} \Rightarrow$

$x + 1 = 10^{3/2} \Rightarrow x = 10^{3/2} - 1 \approx 30.623$

(i) $\sqrt{x + 3} + 4 = x + 1 \Rightarrow \sqrt{x + 3} = x - 3 \Rightarrow (\sqrt{x + 3})^2 = (x - 3)^2 \Rightarrow x + 3 = x^2 - 6x + 9 \Rightarrow$

$x^2 - 7x + 6 = 0 \Rightarrow (x - 1)(x - 6) = 0 \Rightarrow x = 1$ or $x = 6$. Note, 1 is extraneous. The only solution is 6.

(j) $|3x - 1| = 5 \Rightarrow 3x - 1 = -5$ or $3x - 1 = 5 \Rightarrow 3x = -4$ or $3x = 6 \Rightarrow x = -\dfrac{4}{3}$ or 2

21. (a) The solution to $f(x) = g(x)$ is the x-coordinate of the intersection point. The solution is 1.

(b) The graph of f is above the graph of g for all x-values to the left of 1. The solution set is $\{x \mid x < 1\}$.

(c) The graph of f is below the graph of g for all x-values to the right of 1. The solution set is $\{x \mid x \geq 1\}$.

23. Graph $Y_1 = -2.3X + 3.4$ and $Y_2 = \sqrt{(2)}X^2 - 1$ and find the intersection points as shown in Figure 23a and Figure 23b. The solutions to the equation are the x-coordinates of the intersection points, -2.8 and 1.1.

[–5, 5, 1] by [–5, 15, 1]

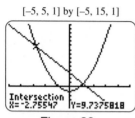

Figure 23a

[–5, 5, 1] by [–5, 15, 1]

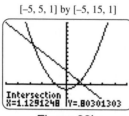

Figure 23b

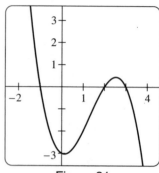

Figure 31

25. $f(x) = 3x^2 + 24x + 43 \Rightarrow f(x) = 3(x^2 + 8x) + 43 \Rightarrow f(x) = 3(x^2 + 8x + 16) + 43 - 48 \Rightarrow$
$f(x) = 3(x + 4)^2 - 5$

27. (a) Since the parabola opens downward, $a < 0$.

(b) The real zeros are the x-intercepts, -1 and 3.

(c) Since the quadratic function has two real zeros, the discriminant is positive.

(d) Using the zeros found above, the form is $f(x) = a(x - (-1))(x - 3)$ or $f(x) = (x + 1)(x - 3)$.

To find the value of a, note that $f(1) = 2$. That is $a(1 + 1)(1 - 3) = 2 \Rightarrow -4a = 2 \Rightarrow a = -\dfrac{1}{2}$.

The complete factored form is $f(x) = -\dfrac{1}{2}(x + 1)(x - 3)$.

29. (a) The graph of f is increasing on $(-\infty, -2] \cup [1, \infty)$ and decreasing on $[-2, 1]$.

(b) The zeros of f are the x-intercepts which approximately $-3.3, 0,$ and 1.8.

(c) The turning points are approximately $(-2, 2)$ and $(1, -0.7)$.

(d) There is a local minimum of -0.7 and a local maximum of 2.

31. See Figure 31.

33. $f(x) = 6(x - (-2))(x - (-1))(x - 1)(x - 2)$ or $f(x) = 6(x + 2)(x + 1)(x - 1)(x - 2)$

35. Since $3i$ is a zero its conjugate, $-3i$, is also a zero.

The complete factored form is $f(x) = 3(x - (-1))(x - (-3i))(x - 3i)$ or $f(x) = 3(x + 1)(x + 3i)(x - 3i)$.

To find the expanded form, first multiply $(x + 3i)(x - 3i) = x^2 + 9$. Continuing to multiply out gives

$f(x) = 3(x + 1)(x^2 + 9) = 3(x^3 + x^2 + 9x + 9)$ or $f(x) = 3x^3 + 3x^2 + 27x + 27$.

37. The zeros appear to be $-2, -1, 1,$ and 2. The complete factored function has the form

$f(x) = a(x - (-2))(x - (-1))(x - 1)(x - 2)$ or $f(x) = (x + 2)(x + 1)(x - 1)(x - 2)$.

To find a, note that $f(0) = -2$. That is $a(0 + 2)(0 + 1)(0 - 1)(0 - 2) = -2 \Rightarrow 4a = -2 \Rightarrow a = -\dfrac{1}{2}$.

The complete factored form is $f(x) = -\dfrac{1}{2}(x + 2)(x + 1)(x - 1)(x - 2)$.

39. $x = \dfrac{-b \pm \sqrt{b^2 - 4ac}}{2a} \Rightarrow x = \dfrac{-2 \pm \sqrt{2^2 - 4(1)(5)}}{2(1)} = \dfrac{-2 \pm \sqrt{-16}}{2} = \dfrac{-2 \pm 4i}{2} = -1 \pm 2i$

41. $\sqrt[5]{(x + 1)^3} = (x + 1)^{3/5}$. When $x = 31, (x + 1)^{3/5} = (31 + 1)^{3/5} = 32^{3/5} = (\sqrt[5]{32})^3 = 2^3 = 8$

43. (a) $(f + g)(2) = f(2) + g(2) = 3 + 0 = 3$

(b) $(fg)(0) = f(0) \cdot g(0) = -1(1) = -1$

(c) $(g \circ f)(1) = g(f(1)) = g(0) = 1$

(d) Note, for $g^{-1}(3)$, read the graph backward. $(g^{-1} \circ f)(-2) = g^{-1}(f(-2)) = g^{-1}(3) = -4$

45. Let $y = f(x)$: $y = \dfrac{x}{x + 1}$. Interchange x and y: $x = \dfrac{y}{y + 1}$. Then solve for y.

$x = \dfrac{y}{y + 1} \Rightarrow x(y + 1) = y \Rightarrow xy + x = y \Rightarrow x = y - xy \Rightarrow x = y(1 - x) \Rightarrow \dfrac{x}{1 - x} = y$

The inverse function is $f^{-1}(x) = \dfrac{x}{1 - x}$ or $f^{-1}(x) = -\dfrac{x}{x - 1}$.

47. Since each unit increase in x results in $f(x)$ increasing by a factor of 3, the function is exponential with base 3. Since $f(0) = 2$, the initial value is 2 and the function is $f(x) = 2(3^x)$.

49. Since $y = \dfrac{1}{2}$ when $x = 0, C = \dfrac{1}{2}$. Find the value of a by that $y = 1$ when $x = 1$. That is $\dfrac{1}{2}a^1 = 1 \Rightarrow a = 2$.

51. (a) $\log_2 \dfrac{1}{16} = \log_2 \dfrac{1}{2^4} = \log_2 2^{-4} = -4$

(b) $\log \sqrt{10} = \log 10^{1/2} = \dfrac{1}{2}$

(c) $\ln e^4 = 4$

(d) $\log_4 2 + \log_4 32 = \log_4 (2 \cdot 32) = \log_4 64 = \log_4 4^3 = 3$

53. $\log \sqrt{\dfrac{x + 1}{yz}} = \log \left(\dfrac{x + 1}{yz}\right)^{1/2} = \dfrac{1}{2}\log \left(\dfrac{x + 1}{yz}\right) = \dfrac{1}{2}[\log (x + 1) - \log yz]$

$= \dfrac{1}{2}[\log (x + 1) - (\log y + \log z)] = \dfrac{1}{2}[\log (x + 1) - \log y - \log z] = \dfrac{1}{2}\log (x + 1) - \dfrac{1}{2}\log y - \dfrac{1}{2}\log z$

55. $\log_4 52 = \dfrac{\ln 52}{\ln 4} \approx 2.850$

57. Since $\sec \theta$ is undefined for $x = \dfrac{\pi}{2} + \pi n$, where n is an integer,

the domain of $\sec 2x = \left\{x \mid x \neq \dfrac{\pi}{4} + \dfrac{\pi}{2}n, n \text{ is a integer}\right\}$

59. $\alpha = 90° - 54° \, 35' \, 12'' = 89° \, 59' \, 60'' - 54° \, 35' \, 12'' = 35° \, 24' \, 48''$

$\beta = 180° - 54° \, 35' \, 12'' = 179° \, 59' \, 60'' - 54° \, 35' \, 12'' = 125° \, 24' \, 48''$

61. $135° \cdot \dfrac{\pi}{180} = \dfrac{3\pi}{4}$

63. $\cot \dfrac{2\pi}{9} = \dfrac{\cos \frac{2\pi}{9}}{\sin \frac{2\pi}{9}} \approx 1.19$

65. The angle $\dfrac{5\pi}{6}$ intersects the unit circle at the point $\left(-\dfrac{\sqrt{3}}{2}, \dfrac{1}{2}\right)$.

$$\sin \theta = \dfrac{1}{2} \qquad\qquad \cos \theta = -\dfrac{\sqrt{3}}{2} \qquad\qquad \tan \theta = \dfrac{\frac{1}{2}}{-\frac{\sqrt{3}}{2}} = -\dfrac{1}{\sqrt{3}}$$

$$\csc \theta = \dfrac{1}{\frac{1}{2}} = 2 \qquad\qquad \sec \theta = \dfrac{1}{-\frac{\sqrt{3}}{2}} = -\dfrac{2}{\sqrt{3}} \qquad\qquad \cot \theta = \dfrac{-\frac{\sqrt{3}}{2}}{\frac{1}{2}} = -\sqrt{3}$$

67. $\cos^{-1}\left(\dfrac{1}{2}\right) = 60°$ or $\dfrac{\pi}{3}$

69. $(\sec t - 1)(\sec t + 1) = \sec^2 t - 1 = \tan^2 t$

71. Graph $Y_1 = 2 \sin (2X)$ and $Y_2 = \ln (X)$. Approximate answers are 1.47, 3.48 and 4.30 for $0 \le x \le 2\pi$.

See Figure 71.

[0, 2π, π/2] by [−3, 3, 1]

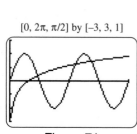

Figure 71

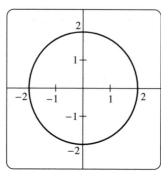

Figure 75

73. Area $= K = \dfrac{1}{2}(15)(20) \sin 30° = 75 \text{ ft}^2$

75. See Figure 75.

77. $-16i = 16(\cos 270° + i \sin 270°)$

$$w_0 = \sqrt{16}\left(\cos \dfrac{270° + 360° \cdot 0}{2} + i \sin \dfrac{270° + 360° \cdot 0}{2}\right) = 4(\cos 135° + i \sin 135°) = -2\sqrt{2} + 2\sqrt{2}i$$

$$w_1 = \sqrt{16}\left(\cos \dfrac{270° + 360° \cdot 1}{2} + i \sin \dfrac{270° + 360° \cdot 1}{2}\right) = 4(\cos 315° + i \sin 315°) = 2\sqrt{2} - 2\sqrt{2}i$$

The two square roots are $-2\sqrt{2} + 2\sqrt{2}i$ and $2\sqrt{2} - 2\sqrt{2}i$.

79. (a) Multiplying the first equation by -1 and adding the two equations will eliminate the variable x.

$$\begin{aligned} -2x - 3y &= -4 \\ 2x - 5y &= -12 \\ \hline -8y &= -16 \end{aligned}$$ Thus, $y = 2$. And so $2x + 3(2) = 4 \Rightarrow x = -1$. The solution is $(-1, 2)$.

(b) Multiplying the first equation by 2 and adding the two equations will eliminate both variables.

$$\begin{aligned} -4x + y &= 2 \\ 4x - y &= -2 \\ \hline 0 &= 0 \end{aligned}$$ This is an identity. The system is dependent with solutions $\{(x, y) | 4x - y = -2\}$.

(c) Adding the two equations will eliminate the variable y.

$$x^2 + y^2 = 16$$
$$2x^2 - y^2 = 11$$
$$\overline{ 3x^2 = 27}$$ Thus, $x = \pm 3$. And so $(\pm 3)^2 + y^2 = 16 \Rightarrow y^2 = 7 \Rightarrow y = \pm\sqrt{7}$.

There are four solutions, $(\pm 3, \pm\sqrt{7})$.

(d) Multiply the first equation by 2 and subtract the second equations to eliminate the variable x.

$$2x + 2y - 4z = -12$$
$$2x - y - 3z = -18$$
$$\overline{ 3y - z = 6}$$

Subtract the third equation from this *new* equation to eliminate both y and z.

$$3y - z = 6$$
$$3y - z = 6$$
$$\overline{ 0 = 0}$$

This is an identity which means there are an infinite number of solutions to the system.

Solving the third equation for y yields $3y - z = 6 \Rightarrow 3y = z + 6 \Rightarrow y = \dfrac{z + 6}{3}$. Substituting this result

for y in the first equation yields $x + \dfrac{z + 6}{3} - 2z = -6 \Rightarrow x = 2z - \dfrac{z + 6}{3} - 6 \Rightarrow x = \dfrac{5z - 24}{3}$.

The solutions set is $\left\{ (x, y, z) \mid x = \dfrac{5z - 24}{3},\, y = \dfrac{z + 6}{3},\, \text{and } z = z \right\}$.

81. For the graphs of parts (a) – (d), see Figures 81a – 81d.

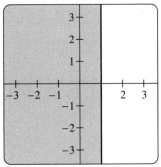

Figure 81a

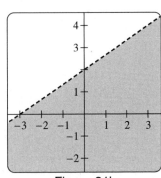

Figure 81b

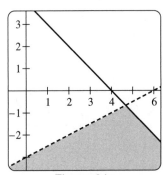

Figure 81c

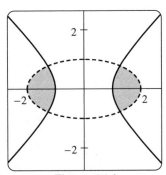

Figure 81d

83. $\begin{bmatrix} 1 & -2 & | & 1 & 0 \\ -3 & 4 & | & 0 & 1 \end{bmatrix} R_2 + 3R_1 \rightarrow \begin{bmatrix} 1 & -2 & | & 1 & 0 \\ 0 & -2 & | & 3 & 1 \end{bmatrix} R_1 - R_2 \rightarrow \begin{bmatrix} 1 & 0 & | & -2 & -1 \\ 0 & -2 & | & 3 & 1 \end{bmatrix} -\frac{1}{2}R_2 \rightarrow \begin{bmatrix} 1 & 0 & | & -2 & -1 \\ 0 & 1 & | & -\frac{3}{2} & -\frac{1}{2} \end{bmatrix}$

That is $A^{-1} = \begin{bmatrix} -2 & -1 \\ -1.5 & -0.5 \end{bmatrix}$.

85. $\det\left(\begin{bmatrix} -1 & 4 \\ 2 & 3 \end{bmatrix}\right) = -1(3) - 2(4) = -11$

$\det\left(\begin{bmatrix} 2 & 3 & -1 \\ 3 & -1 & 5 \\ 0 & 0 & -2 \end{bmatrix}\right) = 0(3(5) - (-1)(-1)) - 0(2(5) - 3(-1)) + -2(2(-1) - 3(3)) = -2(-11) = 22$

87. If the focus is at $\left(\dfrac{3}{4}, 0\right)$ and the vertex at $(0, 0)$, the parabola opens to the right and $p = \dfrac{3}{4}$. Substituting in

$(y - h)^2 = 4p(x - k)$, we get $(y - 0)^2 = 4\left(\dfrac{3}{4}\right)(x - 0)$ or $y^2 = 3x$.

89. Since the foci and the vertices lie on the y-axis, the hyperbola has a vertical transverse axis with an equation

of the form $\dfrac{y^2}{a^2} - \dfrac{x^2}{b^2} = 1$. $F(0, \pm 13) \Rightarrow c = 13$ and $V(0, \pm 5) \Rightarrow a = 5$. $c^2 = a^2 + b^2 \Rightarrow$

$b^2 = c^2 - a^2 = 169 - 25 = 144 \Rightarrow b = 12$. The equation of the hyperbola is $\dfrac{y^2}{25} - \dfrac{x^2}{144} = 1$.

91. Let $a_n = f(n)$, where $f(n) = dn + c$. $a_1 = 4 \Rightarrow f(1) = 4$ and $a_3 = 12 \Rightarrow f(3) = 12$.

Thus, $d = \dfrac{f(3) - f(1)}{3 - 1} = \dfrac{12 - 4}{2} = 4$. Then, $f(n) = 4n + c$. $f(1) = 4(1) + c = 4 \Rightarrow c = 0$.

Thus, $a_n = f(n) = 4n$.

93. (a) The first term is $a_1 = 2$ and the last term is $a_8 = 74$. There are 25 terms. To find the sum use:

$S_n = n\left(\dfrac{a_1 + a_n}{2}\right) \Rightarrow S_{25} = 25\left(\dfrac{2 + 74}{2}\right) = 950$. The sum is 950.

(b) The first term is $a_1 = 2$ and the common ratio is $r = 2$. Since there are 10 terms, the sum is

$S_n = a_1\left(\dfrac{1 - r^n}{1 - r}\right) \Rightarrow S_{10} = 2\left(\dfrac{1 - 2^{10}}{1 - 2}\right) = 2046$.

(c) The first term is $a_1 = 0.2$ and the common ratio is $r = \dfrac{1}{10}$ or 0.1. The sum is

$S = a_1\left(\dfrac{1}{1 - r}\right) \Rightarrow S = 0.2\left(\dfrac{1}{1 - (0.1)}\right) = \dfrac{0.2}{0.9} = \dfrac{2}{9}$.

95. There are $26 \cdot 26 \cdot 26 \cdot 10 \cdot 10 \cdot 10 \cdot 10 = 175{,}760{,}000$ such license plates.

97. $(2x - 1)^4 = \binom{4}{0}(2x)^4(-1)^0 + \binom{4}{1}(2x)^3(-1)^1 + \binom{4}{2}(2x)^2(-1)^2 + \binom{4}{3}(2x)^1(-1)^3 + \binom{4}{4}(2x)^0(-1)^4$

$= 16x^4 - 32x^3 + 24x^2 - 8x + 1$

99. The card must be one of the 16 cards (out of 52) that is either a heart or an ace. The probability is $\dfrac{16}{52} = \dfrac{4}{13}$.

101. There are 8 primes from 1 to 20 (2, 3, 5, 7, 11, 13, 17, 19). The probability is $\dfrac{8}{20} = \dfrac{2}{5}$.

Applications

103. $\frac{1}{3}\pi(1.3)^2 h = 10 \Rightarrow h = \frac{30}{1.3^2 \pi} \approx 5.7$ in.

105. From noon to 1:45 PM car A travelled $1.75(50) = 87.5$ miles and thus would be located $87.5 + 30 = 117.5$ miles north of the original location of car B. From noon to 1:45 PM car B travelled $1.75(50) = 87.5$ miles and thus would be located 87.5 miles east of its original location. The paths of the two cars form a right triangle with legs of length 117.5 and 87.5. The length of the hypotenuse is equal to the distance between the cars.
    $c^2 = 117.5^2 + 87.5^2 \Rightarrow c^2 = 21{,}462.5 \Rightarrow c \approx 146.5$ miles

107. Let x represent the time spent jogging at 7 miles per hour and let y represent the time spent jogging at 9 miles per hour. Then the system of equations to be solved is $x + y = 1.3$ and $7x + 9y = 10.5$. Solving the first equation for y gives $y = 1.3 - x$. Substituting this expression in the second equation and solving for x yields $7x + 9(1.3 - x) = 10.5 \Rightarrow -2x + 11.7 = 10.5 \Rightarrow -2x = -1.2 \Rightarrow x = 0.6$. From the first equation $0.6 + y = 1.3 \Rightarrow y = 0.7$. The jogger ran for 0.6 hours at 7 mph and 0.7 hours at 9 mph.

109. Let x represent the number of hours they work together. The the first person mows $\frac{x}{5}$ of the lawn in x hours and the second person mows $\frac{x}{4}$ of the lawn in x hours. Together they mow $\frac{x}{5} + \frac{x}{4}$ of the lawn in x hours. The job is complete when the fraction of the lawn mowed is 1. That is, we must solve $\frac{x}{5} + \frac{x}{4} = 1$.

     The LCD is 20. The first step is to multiply each side of the equation by the LCD.
     $$20 \cdot \left(\frac{x}{5} + \frac{x}{4}\right) = 1 \cdot 20 \Rightarrow 4x + 5x = 20 \Rightarrow 9x = 20 \Rightarrow x = \frac{20}{9} \approx 2.22 \text{ hours.}$$

111. (a) The slope of a line passing through the data points is $m = \dfrac{280 - 227}{2000 - 1980} = 2.65$. Using the data point $(1980, 227)$, the formula is $P(t) = 2.65(t - 1980) + 227$.

     (b) $2.65(t - 1980) + 227 = 350 \Rightarrow 2.65(t - 1980) = 123 \Rightarrow t - 1980 = \dfrac{123}{2.65} \Rightarrow t = \dfrac{123}{2.65} + 1980$
     This evaluates to about the year 2026.

113. When the data points are plotted they take on a parabolic shape with a vertex located at $(2, 6)$.
     Thus $h = 2$ and $k = 6$. Since $f(4) = 8$, $a(4 - 2)^2 + 6 = 8 \Rightarrow 4a + 6 = 8 \Rightarrow 4a = 2 \Rightarrow a = \dfrac{1}{2}$.
     The function is $f(x) = 0.5(x - 2)^2 + 6$.

115. $\tan 42° = \dfrac{105}{x} \Rightarrow x = 116.6$ feet

117. $\tan \theta = \dfrac{6}{7} \Rightarrow \tan^{-1}\left(\dfrac{6}{7}\right) \approx 40.6°$

119. From the graph of the region of feasible solutions (not shown), the vertices are $(1, 0)$, $(2, 0)$, and $(0, 3)$.
     The maximum value of P occurs at one of the vertices. For $(1, 0)$, $P = 2(1) + 3(0) = 2$.
     For $(2, 0)$, $P = 2(2) + 3(0) = 4$. For $(0, 3)$, $P = 2(0) + 3(3) = 9$. The maximum value is $P = 9$.

121. Let l represent the length of the rectangle and let w represent the width. Since the perimeter of a rectangle is given by $P = 2l + 2w$, and the given perimeter is 48, $2l + 2w = 48 \Rightarrow w = 24 - l$. Since the area of a rectangle is given by $A = lw$, and the given area is 143, $lw = 143 \Rightarrow l(24 - l) = 143 \Rightarrow 24l - l^2 = 143 \Rightarrow l^2 - 24l + 143 = 0 \Rightarrow (l - 13)(l - 11) = 0 \Rightarrow l = 13$ or $l = -11$. The length must be positive and so $l = 13$ and $w = 24 - 13 = 11$. The dimensions are 13 by 11 inches.

123. If the height attained by the ball is doubled for each bounce, the series sums to

$$2 \cdot 4\left(\frac{3}{4}\right)^0 + 2 \cdot 4\left(\frac{3}{4}\right) + 2 \cdot 4\left(\frac{3}{4}\right)^2 + 2 \cdot 4\left(\frac{3}{4}\right)^3 + \ldots = 8\left[\left(\frac{3}{4}\right)^0 + \left(\frac{3}{4}\right)^1 + \left(\frac{3}{4}\right)^2 + \left(\frac{3}{4}\right)^3 + \ldots\right] =$$

$8\sum_{k=1}^{\infty}\left(\frac{3}{4}\right)^{k-1} = 8\left(\frac{1}{1 - \frac{3}{4}}\right) = 8(4) = 32$. Since the ball was dropped from a height of 4 feet, it did not travel

upward for the first "bounce", so 4 must be subtracted. The total distance travelled is $32 - 4 = 28$ feet.

Chapter R: Basic Concepts from Algebra and Geometry

R.1: Formulas from Geometry

Rectangles

1. $A = LW = (15)(7) = 105 \text{ ft}^2$; $P = 2L + 2W = 2(15) + 2(7) = 30 + 14 = 44 \text{ ft}$

3. $A = LW = (100)(35) = 3500 \text{ m}^2$; $P = 2L + 2W = 2(100) + 2(35) = 200 + 70 = 270 \text{ m}$

5. $A = LW = (3x)(y) = 3xy$ square units; $P = 2L + 2W = 2(3x) + 2(y) = 6x + 2y$ units

7. Since the width is half of the length, $L = 2W$.

   $A = LW = (2W)(W) = 2W^2$ square units; $P = 2L + 2W = 2(2W) + 2(W) = 6W$ units

9. Since the length equals the width plus 5, $L = W + 5$.

   $A = LW = (W + 5)(W) = W(W + 5)$ square units; $P = 2L + 2W = 2(W + 5) + 2(W) = 4W + 10$ units

Triangles

11. $A = \frac{1}{2}bh = \left(\frac{1}{2}\right)(8)(5) = 20 \text{ cm}^2$

13. $A = \frac{1}{2}bh = \left(\frac{1}{2}\right)(5)(8) = 20 \text{ in.}^2$

15. $A = \frac{1}{2}bh = \left(\frac{1}{2}\right)(10.1)(730) = 3686.5 \text{ m}^2$

17. $A = \frac{1}{2}bh = \left(\frac{1}{2}\right)(2x)(6x) = 6x^2$ square units

19. $A = \frac{1}{2}bh = \left(\frac{1}{2}\right)(z)(5z) = \frac{5}{2}z^2$ square units

Circles

21. $C = 2\pi r = 2\pi(4) = 8\pi \approx 25.1 \text{ m}$; $A = \pi r^2 = \pi(4)^2 = 16\pi \approx 50.3 \text{ m}^2$

23. $C = 2\pi r = 2\pi(19) = 38\pi \approx 119.4 \text{ in.}$; $A = \pi r^2 = \pi(19)^2 = 361\pi \approx 1134.1 \text{ in.}^2$

25. $C = 2\pi r = 2\pi(2x) = 4\pi x$ units; $A = \pi r^2 = \pi(2x)^2 = 4\pi x^2$ square units

Pythagorean Theorem

27. $c^2 = a^2 + b^2 \Rightarrow c^2 = (60)^2 + (11)^2 \Rightarrow c^2 = 3721 \Rightarrow c = \sqrt{3721} \Rightarrow c = 61 \text{ ft}$

    $P = 60 + 11 + 61 = 132 \text{ ft}$

29. Since 21 feet is 7 yards, $c^2 = a^2 + b^2 \Rightarrow c^2 = (7)^2 + (11)^2 \Rightarrow c^2 = 170 \Rightarrow c = \sqrt{170} \Rightarrow c \approx 13.0 \text{ yd}$

    $P \approx 7 + 11 + 13.0 \approx 31.0 \text{ yd}$

31. $c^2 = a^2 + b^2 \Rightarrow b^2 = c^2 - a^2 \Rightarrow b^2 = (15)^2 - (6)^2 \Rightarrow b^2 = 189 \Rightarrow b = \sqrt{189} \Rightarrow b \approx 13.7 \text{ m}$

    $P \approx 6 + 13.7 + 15 \approx 34.7 \text{ m}$

33. $c^2 = a^2 + b^2 \Rightarrow a^2 = c^2 - b^2 \Rightarrow a^2 = (2)^2 - (1.2)^2 \Rightarrow a^2 = 2.56 \Rightarrow a = \sqrt{2.56} \Rightarrow a = 1.6$ mi

   $P = 1.6 + 1.2 + 2 = 4.8$ mi

35. $A = \dfrac{1}{2}bh = \left(\dfrac{1}{2}\right)(6)(3) = 9$ ft^2

37. $c^2 = a^2 + b^2 \Rightarrow b^2 = c^2 - a^2 \Rightarrow b^2 = (15)^2 - (11)^2 \Rightarrow b^2 = 104 \Rightarrow b = \sqrt{104} \Rightarrow b \approx 10.2$ in.

   $A = \dfrac{1}{2}bh = \left(\dfrac{1}{2}\right)(11)(\sqrt{104}) \approx 56.1$ in.^{2}

Rectangular Boxes

39. $V = LWH = (4)(3)(2) = 24$ ft^3; $S = 2LW + 2WH + 2LH = 2(4)(3) + 2(3)(2) + 2(4)(2) = 52$ ft^2

41. Since 1 foot is 12 inches, $V = LWH = (4.5)(4)(12) = 216$ in.^{3}

   $S = 2LW + 2WH + 2LH = 2(4.5)(4) + 2(4)(12) + 2(4.5)(12) = 240$ in.^{2}

43. $V = LWH = (3x)(2x)(x) = 6x^3$ cubic units

   $S = 2LW + 2WH + 2LH = 2(3x)(2x) + 2(2x)(x) + 2(3x)(x) = 22x^2$ square units

45. $V = LWH = (x)(2y)(3z) = 6xyz$ cubic units

   $S = 2LW + 2WH + 2LH = 2(x)(2y) + 2(2y)(3z) + 2(x)(3z) = 4xy + 12yz + 6xz$ square units

47. Since the length is twice W and the height is half W, $L = 2W$ and $H = 0.5W$

   $V = LWH = (2W)(W)(0.5W) = W^3$

Spheres

49. $V = \dfrac{4}{3}\pi r^3 \Rightarrow V = \dfrac{4}{3}\pi(3)^3 = 36\pi \approx 113.1$ ft^3; $S = 4\pi r^2 \Rightarrow S = 4\pi(3)^2 = 36\pi \approx 113.1$ ft^2

51. $V = \dfrac{4}{3}\pi r^3 \Rightarrow V = \dfrac{4}{3}\pi(5)^3 \approx 166.7\pi \approx 523.6$ cm^3; $S = 4\pi r^2 \Rightarrow S = 4\pi(5)^2 = 100\pi \approx 314.2$ cm^2

53. $V = \dfrac{4}{3}\pi r^3 \Rightarrow V = \dfrac{4}{3}\pi(5)^3 \approx 166.7\pi \approx 523.6$ mm^3; $S = 4\pi r^2 \Rightarrow S = 4\pi(5)^2 = 100\pi \approx 314.2$ mm^2

Cylinders

55. $V = \pi r^2 h \Rightarrow V = \pi(0.5)^2(2) = 0.5\pi \approx 1.6$ ft^3; $S_{side} = 2\pi rh \Rightarrow S_{side} = 2\pi(0.5)(2) = 2\pi \approx 6.3$ ft^2

   $S_{total} = 2\pi rh + 2\pi r^2 \Rightarrow S_{total} = 2\pi(0.5)(2) + 2\pi(0.5)^2 = 2.5\pi \approx 7.9$ ft^2

57. Since 1.5 feet is 18 inches, $V = \pi r^2 h \Rightarrow V = \pi(5)^2(18) = 450\pi \approx 1413.7$ in.^{3}

   $S_{side} = 2\pi rh \Rightarrow S_{side} = 2\pi(5)(18) = 180\pi \approx 565.5$ in.^{2}

   $S_{total} = 2\pi rh + 2\pi r^2 \Rightarrow S_{total} = 2\pi(5)(18) + 2\pi(5)^2 \approx 230\pi \approx 722.6$ in.^{2}

59. Since h is twice r, $h = 2(12) = 24$, thus $V = \pi r^2 h \Rightarrow V = \pi(12)^2(24) = 3456\pi \approx 10,857.3$ mm^3

   $S_{side} = 2\pi rh \Rightarrow S_{side} = 2\pi(12)(24) = 576\pi \approx 1809.6$ mm^2

   $S_{total} = 2\pi rh + 2\pi r^2 \Rightarrow S_{total} = 2\pi(12)(24) + 2\pi(12)^2 = 864\pi \approx 2714.3$ mm^2

61. Since the radius is half of the height, $h = 2r$, thus $V = \pi r^2 h \Rightarrow V = \pi(r)^2(2r) = 2\pi r^3$.

Cones

63. $V = \frac{1}{3}\pi r^2 h \Rightarrow V = \frac{1}{3}\pi(5)^2(6) \approx 157.1 \text{ cm}^3; \; S = \pi r \sqrt{r^2 + h^2} \Rightarrow S = \pi(5)\sqrt{(5)^2 + (6)^2} \approx 122.7 \text{ cm}^2$

65. Since 24 inches is 2 feet,

$$V = \frac{1}{3}\pi r^2 h \Rightarrow V = \frac{1}{3}\pi(2)^2(3) \approx 12.6 \text{ ft}^3; \; S = \pi r \sqrt{r^2 + h^2} \Rightarrow S = \pi(2)\sqrt{(2)^2 + (3)^2} \approx 22.7 \text{ ft}^2$$

67. Since h is three times r, $h = 3(2.4) = 7.2$.

$$V = \frac{1}{3}\pi r^2 h \Rightarrow V = \frac{1}{3}\pi(2.4)^2(7.2) \approx 43.4 \text{ ft}^3; \; S = \pi r \sqrt{r^2 + h^2} \Rightarrow S = \pi(2.4)\sqrt{(2.4)^2 + (7.2)^2} \approx 57.2 \text{ ft}^2$$

Similar Triangles

69. $\dfrac{x}{4} = \dfrac{5}{3} \Rightarrow 3x = 20 \Rightarrow x = \dfrac{20}{3} \approx 6.7$

71. $\dfrac{x}{7} = \dfrac{9}{6} \Rightarrow 6x = 63 \Rightarrow x = \dfrac{63}{6} = \dfrac{21}{2} = 10.5$

73. $\dfrac{x}{9} = \dfrac{10}{12} \Rightarrow 12x = 90 \Rightarrow x = \dfrac{90}{12} = \dfrac{15}{2} = 7.5$

R.2: Circles

Equations and Graphs of Circles

1. $x^2 + y^2 = 25 \Rightarrow (x - 0)^2 + (y - 0)^2 = 5^2 \Rightarrow$ Center: $(0, 0)$; Radius: 5

3. $x^2 + y^2 = 7 \Rightarrow (x - 0)^2 + (y - 0)^2 = (\sqrt{7})^2 \Rightarrow$ Center: $(0, 0)$; Radius: $\sqrt{7}$

5. $(x - 2)^2 + (y + 3)^2 = 9 \Rightarrow (x - 2)^2 + (y - (-3))^2 = 3^2 \Rightarrow$ Center: $(2, -3)$; Radius: 3

7. $x^2 + (y + 1)^2 = 100 \Rightarrow (x - 0)^2 + (y - (-1))^2 = 10^2 \Rightarrow$ Center: $(0, -1)$; Radius: 10

9. $5x^2 + 5y^2 = 20 \Rightarrow x^2 + y^2 = 4 \Rightarrow (x - 0)^2 + (y - 0)^2 = 2^2 \Rightarrow$ Center: $(0, 0)$; Radius: 2

11. $6x^2 + 6y^2 = 30 \Rightarrow x^2 + y^2 = 5 \Rightarrow (x - 0)^2 + (y - 0)^2 = (\sqrt{5})^2 \Rightarrow$ Center: $(0, 0)$; Radius: $\sqrt{5}$

13. $2(x - 4)^2 + 2(y + 5)^2 = 32 \Rightarrow (x - 4)^2 + (y - (-5))^2 = 4^2 \Rightarrow$ Center: $(4, -5)$; Radius: 4

15. Since the center is $(1, -2)$ and the radius is 1, the equation is $(x - 1)^2 + (y + 2)^2 = 1$.

17. Since the center is $(-2, 1)$ and the radius is 2, the equation is $(x + 2)^2 + (y - 1)^2 = 4$.

19. $(x - 3)^2 + (y - (-5))^2 = 8^2 \Rightarrow (x - 3)^2 + (y + 5)^2 = 64$

21. $(x - 3)^2 + (y - 0)^2 = 7^2 \Rightarrow (x - 3)^2 + y^2 = 49$

23. First find the radius using the distance formula: $r = \sqrt{(-3 - 0)^2 + (-1 - 0)^2} = \sqrt{10}$

$(x - 0)^2 + (y - 0)^2 = (\sqrt{10})^2 \Rightarrow x^2 + y^2 = 10$

25. First find the center using the midpoint formula: $C = \left(\dfrac{-5 + 1}{2}, \dfrac{-7 + 1}{2}\right) = (-2, -3)$

Then find the radius using the distance formula: $r = \sqrt{(-2 - 1)^2 + (-3 - 1)^2} = \sqrt{25} = 5$

$(x - (-2))^2 + (y - (-3))^2 = 5^2 \Rightarrow (x + 2)^2 + (y + 3)^2 = 25$

Finding the Center and Radius of a Circle

27. $x^2 + y^2 = 9 \Rightarrow (x - 0)^2 + (y - 0)^2 = 3^2 \Rightarrow$ Center: $(0, 0)$; Radius: 3. See Figure 27.

29. $x^2 + 2x + y^2 - 2y = 2 \Rightarrow (x^2 + 2x + 1) + (y^2 - 2y + 1) = 2 + 1 + 1 \Rightarrow (x + 1)^2 + (y - 1)^2 = 4 \Rightarrow$

$(x - (-1))^2 + (y - 1)^2 = 2^2 \Rightarrow$ Center: $(-1, 1)$; Radius: 2. See Figure 29.

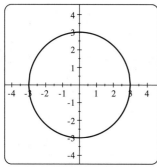

Figure 27

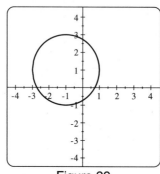

Figure 29

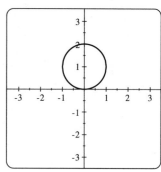

Figure 31

31. $x^2 + y^2 - 2y = 0 \Rightarrow x^2 + (y^2 - 2y + 1) = 0 + 1 \Rightarrow x^2 + (y - 1)^2 = 1 \Rightarrow$

$(x - 0)^2 + (y - 1)^2 = 1^2 \Rightarrow$ Center: $(0, 1)$; Radius: 1. See Figure 31.

33. $x^2 + 18x + y^2 + 10y = -97 \Rightarrow (x^2 + 18x + 81) + (y^2 + 10y + 25) = -97 + 81 + 25 \Rightarrow$

$(x + 9)^2 + (y + 5)^2 = 9 \Rightarrow (x - (-9))^2 + (y - (-5))^2 = 3^2 \Rightarrow$ Center: $(-9, -5)$; Radius: 3. See Figure 33.

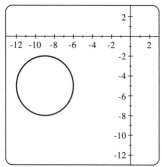

Figure 33

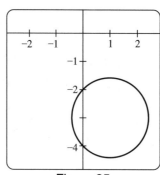

Figure 35

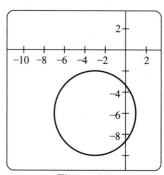

Figure 37

35. $2x^2 - 4x + 2y^2 + 12y + 16 = 0 \Rightarrow x^2 - 2x + y^2 + 6y + 8 = 0 \Rightarrow$

$(x^2 - 2x + 1) + (y^2 + 6y + 9) = -8 + 1 + 9 \Rightarrow (x - 1)^2 + (y + 3)^2 = 2 \Rightarrow$

$(x - 1)^2 + (y - (-3))^2 = (\sqrt{2})^2 \Rightarrow$ Center: $(1, -3)$; Radius: $\sqrt{2}$. See Figure 35.

37. $4x^2 + 24x + 4y^2 + 48y + 180 = 64 \Rightarrow x^2 + 6x + y^2 + 12y + 45 = 16 \Rightarrow$

$(x^2 + 6x + 9) + (y^2 + 12y + 36) = -45 + 16 + 9 + 36 \Rightarrow (x + 3)^2 + (y + 6)^2 = 16 \Rightarrow$

$(x - (-3))^2 + (y - (-6))^2 = 4^2 \Rightarrow$ Center: $(-3, -6)$; Radius: 4. See Figure 37.

R.3: Integer Exponents

Concepts

1. The base is 8 and the exponent is 3.

3. 7^3

5. No. $2^3 = 8$ and $3^2 = 9$

7. $\dfrac{1}{7^n}$

9. 5^{m-n}

11. 2^{mk}

13. 5000

Properties of Exponents

15. 2^3

17. 4^4

19. 3^0

21. $5^3 = 125$

23. $-2^4 = -16$

25. $5^0 = 1$

27. $\left(\dfrac{2}{3}\right)^3 = \dfrac{2^3}{3^3} = \dfrac{2 \cdot 2 \cdot 2}{3 \cdot 3 \cdot 3} = \dfrac{8}{27}$

29. $\left(-\dfrac{1}{2}\right)^4 = \left(-\dfrac{1}{2}\right) \cdot \left(-\dfrac{1}{2}\right) \cdot \left(-\dfrac{1}{2}\right) \cdot \left(-\dfrac{1}{2}\right) = \dfrac{1}{16}$

31. $4^{-3} = \dfrac{1}{4^3} = \dfrac{1}{64}$

33. $\dfrac{1}{2^{-4}} = 2^4 = 16$

35. $\left(\dfrac{3}{4}\right)^{-3} = \left(\dfrac{4}{3}\right)^3 = \left(\dfrac{4}{3}\right) \cdot \left(\dfrac{4}{3}\right) \cdot \left(\dfrac{4}{3}\right) = \dfrac{64}{27}$

37. $\dfrac{3^{-2}}{2^{-3}} = \dfrac{2^3}{3^2} = \dfrac{8}{9}$

39. $6^3 \cdot 6^{-4} = 6^{3+(-4)} = 6^{-1} = \dfrac{1}{6}$

41. $5^{-5} \cdot 5^2 = 5^{-5+2} = 5^{-3} = \dfrac{1}{5^3} = \dfrac{1}{125}$

43. $2x^2 \cdot 3x^{-3} \cdot x^4 = 6x^{2+(-3)+4} = 6x^3$

45. $10^0 \cdot 10^6 \cdot 10^2 = 10^{0+6+2} = 10^8 = 100{,}000{,}000$

47. $5^{-2} \cdot 5^3 \cdot 2^{-4} \cdot 2^3 = 5^{-2+3} \cdot 2^{-4+3} = 5^1 \cdot 2^{-1} = 5 \cdot \dfrac{1}{2} = \dfrac{5}{2}$

49. $2a^3 \cdot b^2 \cdot a^{-4} \cdot 4b^{-5} = 2a^{3+(-4)} \cdot 4b^{2+(-5)} = 8a^{-1}b^{-3} = \dfrac{8}{ab^3}$

51. $\dfrac{5^4}{5^2} = 5^{4-2} = 5^2 = 25$

53. $\dfrac{10^{-2} \cdot 10^3}{10^{-5}} = 10^{-2+3-(-5)} = 10^6 = 1{,}000{,}000$

55. $\dfrac{a^{-3}}{a^2 \cdot a} = a^{-3-(2+1)} = a^{-6} = \dfrac{1}{a^6}$

57. $\dfrac{24x^3}{6x} = \dfrac{24}{6}x^{3-1} = 4x^2$

59. $\dfrac{12a^2b^3}{18a^4b^2} = \dfrac{12}{18}a^{2-4} \cdot b^{3-2} = \dfrac{2}{3}a^{-2}b^1 = \dfrac{2b}{3a^2}$

61. $\dfrac{21x^{-3}y^4}{7x^4y^{-2}} = \dfrac{21}{7}x^{-3-4} \cdot y^{4-(-2)} = 3x^{-7}y^6 = \dfrac{3y^6}{x^7}$

63. $(5^3)^{-1} = 5^{3(-1)} = 5^{-3} = \dfrac{1}{5^3} = \dfrac{1}{125}$

65. $(y^4)^{-2} = y^{4(-2)} = y^{-8} = \dfrac{1}{y^8}$

67. $(4y^2)^3 = 4^3 \cdot y^{2 \cdot 3} = 64y^6$

69. $\left(\dfrac{4}{x}\right)^3 = \dfrac{4^3}{x^3} = \dfrac{64}{x^3}$

71. $\left(\dfrac{2x}{z^4}\right)^{-5} = \left(\dfrac{z^4}{2x}\right)^5 = \dfrac{(z^4)^5}{(2x)^5} = \dfrac{z^{4 \cdot 5}}{2^5 \cdot x^5} = \dfrac{z^{20}}{32x^5}$

73. $\dfrac{2}{(ab)^{-1}} = 2(ab)^1 = 2ab$

75. $\dfrac{2^{-3}}{2t^{-2}} = \dfrac{t^2}{2(2^3)} = \dfrac{t^2}{16}$

77. $\dfrac{6a^2b^{-3}}{4ab^{-2}} = \dfrac{6}{4} \cdot a^{2-1}b^{-3-(-2)} = \dfrac{3}{2} \cdot ab^{-1} = \dfrac{3a}{2b}$

79. $\dfrac{4m^2n^4}{m^{-2}n^{-3}} = 4m^{2-(-2)}n^{4-(-3)} = 4m^4n^7$

81. $\dfrac{5r^2st^{-3}}{25rs^{-2}t^2} = \dfrac{5}{25} \cdot r^{2-1}s^{1-(-2)}t^{-3-2} = \dfrac{1}{5} \cdot rs^3t^{-5} = \dfrac{rs^3}{5t^5}$

83. $(3x^2y^{-3})^{-2} = \dfrac{1}{(3x^2y^{-3})^2} = \dfrac{1}{3^2(x^2)^2(y^{-3})^2} = \dfrac{1}{9x^{2 \cdot 2}y^{-3 \cdot 2}} = \dfrac{1}{9x^4y^{-6}} = \dfrac{y^6}{9x^4}$

85. $\dfrac{(d^3)^{-2}}{(d^{-2})^3} = \dfrac{d^{3 \cdot (-2)}}{d^{-2 \cdot 3}} = \dfrac{d^{-6}}{d^{-6}} = d^{-6-(-6)} = d^0 = 1$

87. $\left(\dfrac{3t^2}{2t^{-1}}\right)^3 = \dfrac{(3t^2)^3}{(2t^{-1})^3} = \dfrac{3^3(t^2)^3}{2^3(t^{-1})^3} = \dfrac{27t^{2 \cdot 3}}{8t^{-1 \cdot 3}} = \dfrac{27t^6}{8t^{-3}} = \dfrac{27}{8} \cdot t^{6-(-3)} = \dfrac{27}{8} \cdot t^9 = \dfrac{27t^9}{8}$

89. $\dfrac{(-m^2n^{-1})^{-2}}{(mn)^{-1}} = \dfrac{(mn)^1}{(-m^2n^{-1})^2} = \dfrac{mn}{(-m^2)^2(n^{-1})^2} = \dfrac{mn}{m^{2 \cdot 2}n^{-1 \cdot 2}} = \dfrac{mn}{m^4n^{-2}} = m^{1-4}n^{1-(-2)} = m^{-3}n^3 = \dfrac{n^3}{m^3}$

91. $\left(\dfrac{2a^3}{6b}\right)^4 = \left(\dfrac{a^3}{3b}\right)^4 = \dfrac{(a^3)^4}{3^4b^4} = \dfrac{a^{3 \cdot 4}}{81b^4} = \dfrac{a^{12}}{81b^4}$

93. $\dfrac{8x^{-3}y^{-2}}{4x^{-2}y^{-4}} = \dfrac{8}{4}x^{-3-(-2)}y^{-2-(-4)} = 2x^{-1}y^2 = \dfrac{2y^2}{x}$

95. $\dfrac{(r^2t^2)^{-2}}{(r^3t)^{-1}} = \dfrac{r^3t}{(r^2t^2)^2} = \dfrac{r^3t}{(r^2)^2(t^2)^2} = \dfrac{r^3t}{r^{2 \cdot 2}t^{2 \cdot 2}} = \dfrac{r^3t}{r^4t^4} = r^{3-4}t^{1-4} = r^{-1}t^{-3} = \dfrac{1}{rt^3}$

97. $\dfrac{4x^{-2}y^3}{(2x^{-1}y)^2} = \dfrac{4x^{-2}y^3}{2^2(x^{-1})^2y^2} = \dfrac{4x^{-2}y^3}{4x^{-1\cdot2}y^2} = \dfrac{4x^{-2}y^3}{4x^{-2}y^2} = \left(\dfrac{4x^{-2}}{4x^{-2}}\right)y^{3-2} = 1y = y$

99. $\left(\dfrac{15r^2t}{3r^{-3}t^4}\right)^3 = \left(\dfrac{15}{3}r^{2-(-3)}t^{1-4}\right)^3 = (5r^5t^{-3})^3 = 5^3(r^5)^3(t^{-3})^3 = 125r^{5\cdot3}t^{-3\cdot3} = 125r^{15}t^{-9} = \dfrac{125r^{15}}{t^9}$

R.4: Polynomial Expressions

Concepts

1. $3x^2$; *Answers may vary.*

3. No, the exponents must match for each variable.

5. No, the opposite is $-x^2 - 1$.

7. $a^2 - b^2$

Monomials and Polynomials

9. $3x^3 + 5x^3 = 8x^3$

11. $5y^7 - 8y^7 = -3y^7$

13. $5x^2 + 8x + x^2 = 6x^2 + 8x$

15. $9x^2 - x + 4x - 6x^2 = 3x^2 + 3x$

17. $x^2 + 9x - 2 + 4x^2 + 4x = 5x^2 + 13x - 2$

19. $7y + 9x^2y - 5y + x^2y = 2y + 10x^2y = 10x^2y + 2y$

21. The degree is 2. The leading coefficient is 5.

23. The degree is 3. The leading coefficient is $-\dfrac{2}{5}$.

25. The degree is 5. The leading coefficient is 1.

27. $(5x + 6) + (-2x + 6) = 3x + 12$

29. $(2x^2 - x + 7) + (-2x^2 + 4x - 9) = 3x - 2$

31. $(4x) + (1 - 4.5x) = -0.5x + 1$

33. $(x^4 - 3x^2 - 4) + \left(-8x^4 + x^2 - \dfrac{1}{2}\right) = -7x^4 - 2x^2 - \dfrac{9}{2}$

35. $(2z^3 + 5z - 6) + (z^2 - 3z + 2) = 2z^3 + z^2 + 2z - 4$

37. $-(7x^3) = -7x^3$

39. $-(19z^5 - 5z^2 + 3z) = -19z^5 + 5z^2 - 3z$

41. $-(z^4 - z^2 - 9) = -z^4 + z^2 + 9$

43. $(5x - 3) - (2x + 4) = 5x - 3 - 2x - 4 = 3x - 7$

45. $(x^2 - 3x + 1) - (-5x^2 + 2x - 4) = x^2 - 3x + 1 + 5x^2 - 2x + 4 = 6x^2 - 5x + 5$

47. $(4x^4 + 2x^2 - 9) - (x^4 - 2x^2 - 5) = 4x^4 + 2x^2 - 9 - x^4 + 2x^2 + 5 = 3x^4 + 4x^2 - 4$

49. $(x^4 - 1) - (4x^4 + 3x + 7) = x^4 - 1 - 4x^4 - 3x - 7 = -3x^4 - 3x - 8$

51. $5x(x - 5) = 5x^2 - 25x$

53. $-5(3x + 1) = -15x - 5$

55. $5(y + 2) = 5y + 10$

57. $-2(5x + 9) = -10x - 18$

59. $(y - 3)6y = 6y^2 - 18y$

61. $-4(5x - y) = -20x + 4y$

63. $(y + 5)(y - 7) = y^2 - 7y + 5y - 35 = y^2 - 2y - 35$

65. $(7x - 3)(4 - 7x) = 28x - 49x^2 - 12 + 21x = -49x^2 + 49x - 12$

67. $(-2x + 3)(x - 2) = -2x^2 + 4x + 3x - 6 = -2x^2 + 7x - 6$

69. $\left(x - \dfrac{1}{2}\right)\left(x + \dfrac{1}{4}\right) = x^2 + \dfrac{1}{4}x - \dfrac{1}{2}x - \dfrac{1}{8} = x^2 - \dfrac{1}{4}x - \dfrac{1}{8}$

71. $(x^2 + 1)(2x^2 - 1) = 2x^4 - x^2 + 2x^2 - 1 = 2x^4 + x^2 - 1$

73. $(x + y)(x - 2y) = x^2 - 2xy + xy - 2y^2 = x^2 - xy - 2y^2$

75. $3x(2x^2 - x - 1) = 6x^3 - 3x^2 - 3x$

77. $-x(2x^4 - x^2 + 10) = -2x^5 + x^3 - 10x$

79. $(2x^2 - 4x + 1)(3x^2) = 6x^4 - 12x^3 + 3x^2$

81. $(x + 1)(x^2 + 2x - 3) = x^3 + 2x^2 - 3x + x^2 + 2x - 3 = x^3 + 3x^2 - x - 3$

83. $(2 - 3x)(5 - 2x) = 10 - 4x - 15x + 6x^2 = 10 - 19x + 6x^2$

85. $(x^2 + 2)(3x - 2) = 3x^3 - 2x^2 + 6x - 4$

87. $(x - 7)(x + 7) = x^2 - 7^2 = x^2 - 49$

89. $(3x - 4)(3x + 4) = (3x)^2 - 4^2 = 9x^2 - 16$

91. $(2x - 3y)(2x + 3y) = (2x)^2 - (3y)^2 = 4x^2 - 9y^2$

93. $(x + 4)^2 = x^2 + 2(4)x + 4^2 = x^2 + 8x + 16$

95. $(2x + 1)^2 = 4x^2 + 2(2x) + 1 = 4x^2 + 4x + 1$

97. $(x - 1)^2 = x^2 - 2(x) + 1 = x^2 - 2x + 1$

99. $(2 - 3x)^2 = 4 - 2(6x) + 9x^2 = 4 - 12x + 9x^2$

101. $3x(x + 1)(x - 1) = 3x(x^2 - 1) = 3x^3 - 3x$

103. $(2 - 5x^2)(2 + 5x^2) = 2^2 - (5x^2)^2 = 4 - 25x^4$

R.5: Factoring Polynomials

Greatest Common Factor

1. $10x - 15 = 5(2x - 3)$

3. $2x^3 - 5x = x(2x^2 - 5)$

5. $8x^3 - 4x^2 + 16x = 4x(2x^2 - x + 4)$

7. $5x^4 - 15x^3 + 15x^2 = 5x^2(x^2 - 3x + 3)$

9. $15x^3 + 10x^2 - 25x = 5x(3x^2 + 2x - 5)$

11. $6r^5 - 8r^4 + 12r^3 = 2r^3(3r^2 - 4r + 6)$

13. $8x^2y^2 - 24x^2y^3 = 8x^2y^2(1 - 3y)$

15. $18mn^2 - 12m^2n^3 = 6mn^2(3 - 2mn)$

17. $-4a^2 - 2ab + 6ab^2 = -2a(2a + b - 3b^2)$

Factoring by Grouping

19. $x^3 + 3x^2 + 2x + 6 = x^2(x + 3) + 2(x + 3) = (x + 3)(x^2 + 2)$

21. $6x^3 - 4x^2 + 9x - 6 = 2x^2(3x - 2) + 3(3x - 2) = (3x - 2)(2x^2 + 3)$

23. $z^3 - 5z^2 + z - 5 = z^2(z - 5) + 1(z - 5) = (z - 5)(z^2 + 1)$

25. $y^4 + 2y^3 - 5y^2 - 10y = y(y^3 + 2y^2 - 5y - 10) = y[y^2(y + 2) - 5(y + 2)] = y(y + 2)(y^2 - 5)$

27. $2x^3 - 3x^2 + 2x - 3 = x^2(2x - 3) + 1(2x - 3) = (x^2 + 1)(2x - 3)$

29. $2x^4 - x^3 + 4x - 2 = x^3(2x - 1) + 2(2x - 1) = (x^3 + 2)(2x - 1)$

31. $2ax - 6bx - ay + 3by = 2x(a - 3b) - y(a - 3b) = (2x - y)(a - 3b)$

Factoring Trinomials

33. $x^2 + 7x + 10 = (x + 2)(x + 5)$

35. $x^2 + 8x + 12 = (x + 2)(x + 6)$

37. $z^2 + z - 42 = (z - 6)(z + 7)$

39. $z^2 + 11z + 24 = (z + 3)(z + 8)$

41. $24x^2 + 14x - 3 = (4x + 3)(6x - 1)$

43. $6x^2 - x - 2 = (2x + 1)(3x - 2)$

45. $1 + x - 2x^2 = (1 - x)(1 + 2x)$

47. $20 + 7x - 6x^2 = (5 - 2x)(4 + 3x)$

49. $5x^3 + x^2 - 6x = x(5x^2 + x - 6) = x(x - 1)(5x + 6)$

51. $6x^3 + 21x^2 + 9x = 3x(2x^2 + 7x + 3) = 3x(x + 3)(2x + 1)$

53. $2x^2 - 14x + 20 = 2(x^2 - 7x + 10) = 2(x - 5)(x - 2)$

55. $60t^4 + 230t^3 - 40t^2 = 10t^2(6t^2 + 23t - 4) = 10t^2(t + 4)(6t - 1)$

57. $4m^3 + 10m^2 - 6m = 2m(2m^2 + 5m - 3) = 2m(m + 3)(2m - 1)$

Difference of Two Squares

59. $x^2 - 25 = (x - 5)(x + 5)$

61. $4x^2 - 25 = (2x - 5)(2x + 5)$

63. $36x^2 - 100 = 4(9x^2 - 25) = 4(3x - 5)(3x + 5)$

65. $64z^2 - 25z^4 = z^2(64 - 25z^2) = z^2(8 - 5z)(8 + 5z)$

67. $16x^4 - y^4 = (4x^2 - y^2)(4x^2 + y^2) = (2x - y)(2x + y)(4x^2 + y^2)$

69. The sum of two squares does not factor using real numbers.

71. $4 - r^2t^2 = (2 - rt)(2 + rt)$

73. $(x - 1)^2 - 16 = ((x - 1) - 4)((x - 1) + 4) = (x - 5)(x + 3)$

75. $4 - (z + 3)^2 = (2 - (z + 3))(2 + (z + 3)) = (-1 - z)(5 + z) = -(z + 1)(z + 5)$

Perfect Square Trinomials

77. $x^2 + 2x + 1 = (x + 1)^2$

79. $4x^2 + 20x + 25 = (2x + 5)^2$

81. $x^2 - 12x + 36 = (x - 6)^2$

83. $9z^3 - 6z^2 + z = z(9z^2 - 6z + 1) = z(3z - 1)^2$

85. $9y^3 + 30y^2 + 25y = y(9y^2 + 30y + 25) = y(3y + 5)^2$

87. $4x^2 - 12xy + 9y^2 = (2x - 3y)^2$

89. $9a^3b - 12a^2b + 4ab = ab(9a^2 - 12a + 4) = ab(3a - 2)^2$

Sum and Difference of Two Cubes

91. $x^3 - 1 = (x - 1)(x^2 + 1x + 1^2) = (x - 1)(x^2 + x + 1)$

93. $y^3 + z^3 = (y + z)(y^2 - yz + z^2)$

95. $8x^3 - 27 = (2x)^3 - 3^3 = (2x - 3)((2x)^2 + 2x(3) + 3^2) = (2x - 3)(4x^2 + 6x + 9)$

97. $x^4 + 125x = x(x^3 + 5^3) = x(x + 5)(x^2 - 5x + 5^2) = x(x + 5)(x^2 - 5x + 25)$

99. $8r^6 - t^3 = (2r^2 - t)((2r^2)^2 + 2r^2t + t^2) = (2r^2 - t)(4r^4 + 2r^2t + t^2)$

101. $10m^9 - 270n^6 = 10(m^9 - 27n^6) = 10(m^3 - 3n^2)((m^3)^2 + 3m^3n^2 + (3n^2)^2) =$
$10(m^3 - 3n^2)(m^6 + 3m^3n^2 + 9n^4)$

General Factoring

103. $16x^2 - 25 = (4x)^2 - 5^2 = (4x - 5)(4x + 5)$

105. $x^3 - 64 = x^3 - 4^3 = (x - 4)(x^2 + 4x + 4^2) = (x - 4)(x^2 + 4x + 16)$

107. $x^2 + 16x + 64 = (x + 8)(x + 8) = (x + 8)^2$

109. $5x^2 - 38x - 16 = (x - 8)(5x + 2)$

111. $x^4 + 8x = x(x^3 + 8) = x(x + 2)(x^2 - 2x + 4)$

113. $64x^3 + 8y^3 = 8(8x^3 + y^3) = 8(2x + y)(4x^2 - 2xy + y^2)$

115. $3x^2 - 5x - 8 = (x + 1)(3x - 8)$

117. $7a^3 + 20a^2 - 3a = a(7a^2 + 20a - 3) = a(a + 3)(7a - 1)$

119. $2x^3 - x^2 + 6x - 3 = x^2(2x - 1) + 3(2x - 1) = (x^2 + 3)(2x - 1)$

R.6: Rational Expressions

Simplifying Rational Expressions

1. $\dfrac{10x^3}{5x^2} = 2x$

3. $\dfrac{(x-5)(x+5)}{x-5} = x + 5$

5. $\dfrac{x^2-16}{x-4} = \dfrac{(x-4)(x+4)}{x-4} = x + 4$

7. $\dfrac{x+3}{2x^2+5x-3} = \dfrac{x+3}{(2x-1)(x+3)} = \dfrac{1}{2x-1}$

9. $-\dfrac{z+2}{4z+8} = -\dfrac{z+2}{4(z+2)} = -\dfrac{1}{4}$

11. $\dfrac{x^2+2x}{x^2+3x+2} = \dfrac{x(x+2)}{(x+1)(x+2)} = \dfrac{x}{x+1}$

13. $\dfrac{a^3+b^3}{a+b} = \dfrac{(a+b)(a^2-ab+b^2)}{a+b} = a^2 - ab + b^2$

Multiplication and Division of Rational Expressions

15. $\dfrac{1}{x^2} \cdot \dfrac{3x}{2} = \dfrac{3}{2x}$

17. $\dfrac{5x}{3} \div \dfrac{10x}{6} = \dfrac{5x}{3} \cdot \dfrac{6}{10x} = \dfrac{2}{2} = 1$

19. $\dfrac{x+1}{2x-5} \cdot \dfrac{x}{x+1} = \dfrac{x}{2x-5}$

21. $\dfrac{(x-5)(x+3)}{3x-1} \cdot \dfrac{x(3x-1)}{(x-5)} = \dfrac{x+3}{1} \cdot \dfrac{x}{1} = x(x+3)$

23. $\dfrac{x^2-2x-35}{2x^3-3x^2} \cdot \dfrac{x^3-x^2}{2x-14} = \dfrac{(x-7)(x+5)}{x^2(2x-3)} \cdot \dfrac{x^2(x-1)}{2(x-7)} = \dfrac{x+5}{2x-3} \cdot \dfrac{x-1}{2} = \dfrac{(x-1)(x+5)}{2(2x-3)}$

25. $\dfrac{6b}{b+2} \div \dfrac{3b^4}{2b+4} = \dfrac{6b}{b+2} \cdot \dfrac{2(b+2)}{3b^4} = \dfrac{2}{1} \cdot \dfrac{2}{b^3} = \dfrac{4}{b^3}$

27. $\dfrac{3a+1}{a^7} \div \dfrac{a+1}{3a^8} = \dfrac{3a+1}{a^7} \cdot \dfrac{3a^8}{a+1} = \dfrac{3a+1}{1} \cdot \dfrac{3a}{a+1} = \dfrac{3a(3a+1)}{a+1}$

29. $\dfrac{x+5}{x^3-x} \div \dfrac{x^2-25}{x^3} = \dfrac{x+5}{x(x^2-1)} \cdot \dfrac{x^3}{(x-5)(x+5)} = \dfrac{1}{x^2-1} \cdot \dfrac{x^2}{x-5} = \dfrac{x^2}{(x-5)(x^2-1)}$

Least Common Multiples

31. Since $12 = 2 \cdot 2 \cdot 3$ and $18 = 2 \cdot 3 \cdot 3$, the LCM is $2 \cdot 2 \cdot 3 \cdot 3 = 36$.

33. Since $5a^3 = 5 \cdot a \cdot a \cdot a$ and $10a = 2 \cdot 5 \cdot a$, the LCM is $2 \cdot 5 \cdot a \cdot a \cdot a = 10a^3$.

35. Since $z^2 - 4z = z(z-4)$ and $(z-4)^2 = (z-4)(z-4)$, the LCM is $z(z-4)(z-4) = z(z-4)^2$.

37. Since $x^2 - 6x + 9 = (x-3)(x-3)$ and $x^2 - 5x + 6 = (x-2)(x-3)$, the LCM is
$(x-2)(x-3)(x-3) = (x-2)(x-3)^2$.

Common Denominators

39. The factored denominators are $x + 1$ and 7. The LCD is $7(x + 1)$.

41. The factored denominators are $x + 4$ and $(x + 4)(x - 4)$. The LCD is $(x + 4)(x - 4)$.

43. The factored denominators are 2 and $2x + 1$ and $2(x - 2)$. The LCD is $2(2x + 1)(x - 2)$.

Addition and Subtraction of Rational Expressions

45. $\dfrac{4}{x + 1} + \dfrac{3}{x + 1} = \dfrac{7}{x + 1}$

47. $\dfrac{2}{x^2 - 1} - \dfrac{x + 1}{x^2 - 1} = \dfrac{2 - (x + 1)}{x^2 - 1} = \dfrac{2 - x - 1}{x^2 - 1} = \dfrac{-x + 1}{(x - 1)(x + 1)} = \dfrac{-(x - 1)}{(x + 1)(x - 1)} = -\dfrac{1}{x + 1}$

49. $\dfrac{x}{x + 4} - \dfrac{x + 1}{x(x + 4)} = \dfrac{x^2}{x(x + 4)} - \dfrac{x + 1}{x(x + 4)} = \dfrac{x^2 - (x + 1)}{x(x + 4)} = \dfrac{x^2 - x - 1}{x(x + 4)}$

51. $\dfrac{2}{x^2} - \dfrac{4x - 1}{x} = \dfrac{2}{x^2} - \dfrac{x(4x - 1)}{x^2} = \dfrac{2 - (4x^2 - x)}{x^2} = \dfrac{-4x^2 + x + 2}{x^2}$

53. $\dfrac{x + 3}{x - 5} + \dfrac{5}{x - 3} = \dfrac{(x + 3)(x - 3)}{(x - 5)(x - 3)} + \dfrac{5(x - 5)}{(x - 5)(x - 3)} = \dfrac{x^2 - 9 + 5x - 25}{(x - 5)(x - 3)} = \dfrac{x^2 + 5x - 34}{(x - 5)(x - 3)}$

55. $\dfrac{3}{x - 5} - \dfrac{1}{x - 3} - \dfrac{2x}{x - 5} = \dfrac{3(x - 3)}{(x - 5)(x - 3)} - \dfrac{x - 5}{(x - 5)(x - 3)} - \dfrac{2x(x - 3)}{(x - 5)(x - 3)} = \dfrac{-2(x^2 - 4x + 2)}{(x - 5)(x - 3)}$

57. $\dfrac{x}{x^2 - 9} + \dfrac{5x}{x - 3} = \dfrac{x}{(x - 3)(x + 3)} + \dfrac{5x(x + 3)}{(x - 3)(x + 3)} = \dfrac{x + 5x^2 + 15x}{(x - 3)(x + 3)} = \dfrac{x(5x + 16)}{(x - 3)(x + 3)}$

59. $\dfrac{b}{2b - 4} - \dfrac{b - 1}{b - 2} = \dfrac{b}{2(b - 2)} - \dfrac{2(b - 1)}{2(b - 2)} = \dfrac{b - (2b - 2)}{2(b - 2)} = \dfrac{-(b - 2)}{2(b - 2)} = -\dfrac{1}{2}$

61. $\dfrac{2x}{x - 5} + \dfrac{2x - 1}{3x^2 - 16x + 5} = \dfrac{2x(3x - 1)}{(x - 5)(3x - 1)} + \dfrac{2x - 1}{(x - 5)(3x - 1)} = \dfrac{6x^2 - 2x + 2x - 1}{(x - 5)(3x - 1)} = \dfrac{6x^2 - 1}{(x - 5)(3x - 1)}$

Clearing Fractions

63. The factored denominators are x and x^2. The LCD is x^2.

$$\frac{1}{x} + \frac{3}{x^2} = 0 \Rightarrow \frac{1(x^2)}{x} + \frac{3(x^2)}{x^2} = 0(x^2) \Rightarrow x + 3 = 0 \Rightarrow x = -3$$

65. The factored denominators are x and $2x - 1$. The LCD is $x(2x - 1)$.

$$\frac{1}{x} + \frac{3x}{2x - 1} = 0 \Rightarrow \frac{1(x)(2x - 1)}{x} + \frac{3x(x)(2x - 1)}{2x - 1} = 0(x)(2x - 1) \Rightarrow 2x - 1 + 3x^2 = 0 \Rightarrow$$

$$3x^2 + 2x - 1 = 0 \Rightarrow (3x - 1)(x + 1) = 0 \Rightarrow x = \frac{1}{3} \text{ or } x = -1$$

67. The factored denominators are $(3 - x)(3 + x)$ and $3 - x$. The LCD is $(3 - x)(3 + x)$.

$$\frac{2x}{9 - x^2} + \frac{1}{3 - x} = 0 \Rightarrow \frac{2x(3 - x)(3 + x)}{9 - x^2} + \frac{1(3 - x)(3 + x)}{3 - x} = 0(3 - x)(3 + x) \Rightarrow$$

$$2x + 3 + x = 0 \Rightarrow 3x + 3 = 0 \Rightarrow x = -1$$

69. The factored denominators are $2x$, $2x^2$ and x^3. The LCD is $2x^3$.

$$\frac{1}{2x} + \frac{1}{2x^2} - \frac{1}{x^3} = 0 \Rightarrow \frac{1(2x^3)}{2x} + \frac{1(2x^3)}{2x^2} - \frac{1(2x^3)}{x^3} = 0(2x^3) \Rightarrow x^2 + x - 2 = 0 \Rightarrow$$

$$(x + 2)(x - 1) = 0 \Rightarrow x = -2 \text{ or } x = 1$$

71. The factored denominators are x, $x + 5$ and $x - 5$. The LCD is $x(x + 5)(x - 5)$.

$$\frac{1}{x} - \frac{2}{x + 5} + \frac{1}{x - 5} = 0 \Rightarrow$$

$$\frac{1(x)(x + 5)(x - 5)}{x} - \frac{2(x)(x + 5)(x - 5)}{x + 5} + \frac{1(x)(x + 5)(x - 5)}{x - 5} = 0(x)(x + 5)(x - 5) \Rightarrow$$

$$(x + 5)(x - 5) - 2x(x - 5) + x(x + 5) = 0 \Rightarrow x^2 - 25 - 2x^2 + 10x + x^2 + 5x = 0 \Rightarrow$$

$$15x - 25 = 0 \Rightarrow x = \frac{25}{15} = \frac{5}{3}$$

Complex Fractions

73. $\dfrac{1 + \dfrac{1}{x}}{1 - \dfrac{1}{x}} = \dfrac{1 + \dfrac{1}{x}}{1 - \dfrac{1}{x}} \cdot \dfrac{x}{x} = \dfrac{x + 1}{x - 1}$

75. $\dfrac{\dfrac{1}{x - 5}}{\dfrac{4}{x} - \dfrac{1}{x - 5}} = \dfrac{\dfrac{1}{x - 5}}{\dfrac{4}{x} - \dfrac{1}{x - 5}} \cdot \dfrac{x(x - 5)}{x(x - 5)} = \dfrac{x}{4(x - 5) - x} = \dfrac{x}{4x - 20 - x} = \dfrac{x}{3x - 20}$

77. $\dfrac{\dfrac{1}{x} + \dfrac{2 - x}{x^2}}{\dfrac{3}{x^2} - \dfrac{1}{x}} = \dfrac{\dfrac{1}{x} + \dfrac{2 - x}{x^2}}{\dfrac{3}{x^2} - \dfrac{1}{x}} \cdot \dfrac{x^2}{x^2} = \dfrac{x + 2 - x}{3 - x} = \dfrac{2}{3 - x}$

79. $\dfrac{\dfrac{1}{x + 3} + \dfrac{2}{x - 3}}{2 - \dfrac{1}{x - 3}} = \dfrac{\dfrac{1}{x + 3} + \dfrac{2}{x - 3}}{2 - \dfrac{1}{x - 3}} \cdot \dfrac{(x - 3)(x + 3)}{(x - 3)(x + 3)} = \dfrac{x - 3 + 2(x + 3)}{2(x - 3)(x + 3) - (x + 3)} = \dfrac{3(x + 1)}{(x + 3)(2x - 7)}$

81. $\dfrac{\dfrac{4}{x - 5}}{\dfrac{1}{x + 5} + \dfrac{1}{x}} = \dfrac{\dfrac{4}{x - 5}}{\dfrac{1}{x + 5} + \dfrac{1}{x}} \cdot \dfrac{x(x - 5)(x + 5)}{x(x - 5)(x + 5)} = \dfrac{4x(x + 5)}{x(x - 5) + (x - 5)(x + 5)} = \dfrac{4x(x + 5)}{(x - 5)(2x + 5)}$

83. $\dfrac{\dfrac{1}{2a} - \dfrac{1}{2b}}{\dfrac{1}{a^2} - \dfrac{1}{b^2}} = \dfrac{\dfrac{1}{2a} - \dfrac{1}{2b}}{\dfrac{1}{a^2} - \dfrac{1}{b^2}} \cdot \dfrac{2a^2b^2}{2a^2b^2} = \dfrac{ab^2 - a^2b}{2b^2 - 2a^2} = \dfrac{ab(b - a)}{2(b + a)(b - a)} = \dfrac{ab}{2(a + b)}$

R.7: Radical Notation and Rational Exponents

Square Roots and Cube Roots

1. $-\sqrt{25} = -\sqrt{5^2} = -5$ and $\sqrt{25} = \sqrt{5^2} = 5$

3. $-\sqrt{121} = -\sqrt{11^2} = -11$ and $\sqrt{121} = \sqrt{11^2} = 11$

5. $-\sqrt{\dfrac{16}{25}} = -\sqrt{\left(\dfrac{4}{5}\right)^2} = -\dfrac{4}{5}$ and $\sqrt{\dfrac{16}{25}} = \sqrt{\left(\dfrac{4}{5}\right)^2} = \dfrac{4}{5}$

7. $-\sqrt{11} \approx -3.32$ and $\sqrt{11} \approx 3.32$

9. $\sqrt{144} = \sqrt{12^2} = 12$

11. $\sqrt{23} \approx 4.80$

13. $\sqrt{\dfrac{4}{49}} = \sqrt{\left(\dfrac{2}{7}\right)^2} = \dfrac{2}{7}$

15. Since $b < 0$, $\sqrt{b^2} = -b$

17. $\sqrt[3]{27} = \sqrt[3]{3^3} = 3$

19. $\sqrt[3]{-8} = \sqrt[3]{(-2)^3} = -2$

21. $\sqrt[3]{\dfrac{1}{27}} = \sqrt[3]{\left(\dfrac{1}{3}\right)^3} = \dfrac{1}{3}$

23. $\sqrt[3]{b^9} = \sqrt[3]{(b^3)^3} = b^3$

Radical Notation

25. $\sqrt{9} = 3$

27. $-\sqrt{5} \approx -2.24$

29. $\sqrt{z^2} = |z|$

31. $\sqrt[3]{27} = 3$

33. $\sqrt[3]{-64} = -4$

35. $\sqrt[3]{5} \approx 1.71$

37. $-\sqrt[3]{x^9} = -\sqrt[3]{(x^3)^3} = -x^3$

39. $\sqrt[3]{(2x)^6} = \sqrt[3]{((2x)^2)^3} = (2x)^2 = 4x^2$

41. $\sqrt[4]{81} = 3$

43. $\sqrt[5]{-7} \approx -1.48$

Rational Exponents

45. $6^{1/2} = \sqrt{6}$

47. $(xy)^{1/2} = \sqrt{xy}$

49. $y^{-1/5} = \dfrac{1}{\sqrt[5]{y}}$

51. The expression can be written $27^{2/3} = \sqrt[3]{27^2}$ or $(\sqrt[3]{27})^2$ and evaluated as $(\sqrt[3]{27})^2 = 3^2 = 9$.

53. The expression can be written $(-1)^{4/3} = \sqrt[3]{(-1)^4}$ or $(\sqrt[3]{-1})^4$ and evaluated as $(\sqrt[3]{-1})^4 = (-1)^4 = 1$.

55. The expression can be written $8^{-1/3} = \dfrac{1}{8^{1/3}} = \dfrac{1}{\sqrt[3]{8}}$ and evaluated as $\dfrac{1}{\sqrt[3]{8}} = \dfrac{1}{2}$.

57. The expression can be written $13^{-3/5} = \dfrac{1}{13^{3/5}} = \dfrac{1}{\sqrt[5]{13^3}}$ or $\dfrac{1}{(\sqrt[5]{13})^3}$. The result is not an integer.

59. $16^{1/2} = \sqrt{16} = 4$

61. $256^{1/4} = \sqrt[4]{256} = 4$

63. $32^{1/5} = \sqrt[5]{32} = 2$

65. $(-8)^{4/3} = (\sqrt[3]{-8})^4 = (-2)^4 = 16$

67. $2^{1/2} \cdot 2^{2/3} = 2^{1/2+2/3} = 2^{7/6} \approx 2.24$

69. $\left(\dfrac{4}{9}\right)^{1/2} = \dfrac{4^{1/2}}{9^{1/2}} = \dfrac{\sqrt{4}}{\sqrt{9}} = \dfrac{2}{3}$

71. $\dfrac{4^{2/3}}{4^{1/2}} = 4^{2/3-1/2} = 4^{1/6} \approx 1.26$

73. $4^{-1/2} = \dfrac{1}{4^{1/2}} = \dfrac{1}{\sqrt{4}} = \dfrac{1}{2}$

75. $(-8)^{-1/3} = \dfrac{1}{(-8)^{1/3}} = \dfrac{1}{\sqrt[3]{-8}} = \dfrac{1}{-2} = -\dfrac{1}{2}$

77. $\left(\dfrac{1}{16}\right)^{-1/4} = 16^{1/4} = \sqrt[4]{16} = 2$

79. $(2^{1/2})^3 = 2^{1/2\cdot3} = 2^{3/2} \approx 2.83$

81. $(x^2)^{3/2} = x^{2\cdot3/2} = x^3$

83. $(x^2y^8)^{1/2} = x^{2\cdot1/2} \cdot y^{8\cdot1/2} = xy^4$

85. $\sqrt[3]{x^3y^6} = (x^3y^6)^{1/3} = x^{3\cdot1/3} \cdot y^{6\cdot1/3} = xy^2$

87. $\sqrt{\dfrac{y^4}{x^2}} = \left(\dfrac{y^4}{x^2}\right)^{1/2} = \dfrac{y^{4\cdot1/2}}{x^{2\cdot1/2}} = \dfrac{y^2}{x}$

89. $\sqrt{y^3} \cdot \sqrt[3]{y^2} = (y^3)^{1/2} \cdot (y^2)^{1/3} = y^{3\cdot1/2} \cdot y^{2\cdot1/3} = y^{3/2} \cdot y^{2/3} = y^{3/2+2/3} = y^{13/6}$

91. $\left(\dfrac{x^6}{27}\right)^{2/3} = \dfrac{x^{6\cdot2/3}}{27^{2/3}} = \dfrac{x^4}{(\sqrt[3]{27})^2} = \dfrac{x^4}{3^2} = \dfrac{x^4}{9}$

93. $\left(\dfrac{x^2}{y^6}\right)^{-1/2} = \left(\dfrac{y^6}{x^2}\right)^{1/2} = \dfrac{y^{6\cdot1/2}}{x^{2\cdot1/2}} = \dfrac{y^3}{x}$

95. $\sqrt{\sqrt{y}} = (y^{1/2})^{1/2} = y^{1/2\cdot1/2} = y^{1/4}$

97. $(a^{-1/2})^{4/3} = a^{-1/2\cdot4/3} = a^{-2/3} = \dfrac{1}{a^{2/3}}$

99. $(a^3b^6)^{1/3} = a^{3\cdot1/3} \cdot b^{6\cdot1/3} = ab^2$

101. $\dfrac{(k^{1/2})^{-3}}{(k^2)^{1/4}} = \dfrac{k^{-3/2}}{k^{1/2}} = k^{-3/2-1/2} = k^{-4/2} = k^{-2} = \dfrac{1}{k^2}$

103. $\sqrt{b} \cdot \sqrt[4]{b} = b^{1/2} \cdot b^{1/4} = b^{1/2+1/4} = b^{3/4}$

105. $\sqrt{z} \cdot \sqrt[3]{z^2} \cdot \sqrt[4]{z^3} = z^{1/2} \cdot z^{2/3} \cdot z^{3/4} = z^{1/2+2/3+3/4} = z^{23/12}$

107. $p^{1/2}(p^{3/2} + p^{1/2}) = p^{1/2+3/2} + p^{1/2+1/2} = p^2 + p$

109. $\sqrt[3]{x}(\sqrt{x} - \sqrt[3]{x^2}) = x^{1/3}(x^{1/2} - x^{2/3}) = x^{1/3+1/2} - x^{1/3+2/3} = x^{5/6} - x$

111. $\sqrt{(-4)^2} = \sqrt{16} = 4$

113. $\sqrt{y^2} = |y|$

115. $\sqrt{(a + 3)^2} = |a + 3|$

117. $\sqrt{(x - 5)^2} = |x - 5|$

119. $\sqrt{x^2 - 2x + 1} = \sqrt{(x - 1)^2} = |x - 1|$

121. $\sqrt[4]{y^4} = (y^4)^{1/4} = y^{4 \cdot 1/4} = |y|$

123. $\sqrt[4]{x^{12}} = (x^{12})^{1/4} = x^{12 \cdot 1/4} = |x^3|$

125. $\sqrt[5]{x^5 y^{10}} = (x^5 y^{10})^{1/5} = x^{5 \cdot 1/5} y^{10 \cdot 1/5} = xy^2$

R.8: Radical Expressions

Multiplying and Dividing

1. $\sqrt{3} \cdot \sqrt{3} = \sqrt{3 \cdot 3} = \sqrt{9} = 3$

3. $\sqrt{2} \cdot \sqrt{50} = \sqrt{2 \cdot 50} = \sqrt{100} = 10$

5. $\sqrt[3]{4} \cdot \sqrt[3]{16} = \sqrt[3]{4 \cdot 16} = \sqrt[3]{64} = 4$

7. $\sqrt{\dfrac{9}{25}} = \dfrac{\sqrt{9}}{\sqrt{25}} = \dfrac{3}{5}$

9. $\sqrt{\dfrac{1}{2}} \cdot \sqrt{\dfrac{1}{8}} = \sqrt{\dfrac{1 \cdot 1}{2 \cdot 8}} = \sqrt{\dfrac{1}{16}} = \dfrac{\sqrt{1}}{\sqrt{16}} = \dfrac{1}{4}$

11. $\sqrt{\dfrac{x}{2}} \cdot \sqrt{\dfrac{x}{8}} = \sqrt{\dfrac{x \cdot x}{2 \cdot 8}} = \sqrt{\dfrac{x^2}{16}} = \dfrac{\sqrt{x^2}}{\sqrt{16}} = \dfrac{x}{4}$

13. $\dfrac{\sqrt{45}}{\sqrt{5}} = \sqrt{\dfrac{45}{5}} = \sqrt{9} = 3$

15. $\sqrt[4]{9} \cdot \sqrt[4]{9} = \sqrt[4]{9 \cdot 9} = \sqrt[4]{81} = 3$

17. $\dfrac{\sqrt[5]{64}}{\sqrt[5]{-2}} = \sqrt[5]{\dfrac{64}{-2}} = \sqrt[5]{-32} = -2$

19. $\dfrac{\sqrt{a^2 b}}{\sqrt{b}} = \sqrt{\dfrac{a^2 b}{b}} = \sqrt{a^2} = a$

21. $\sqrt[3]{\dfrac{x^3}{8}} = \dfrac{\sqrt[3]{x^3}}{\sqrt[3]{8}} = \dfrac{x}{2}$

23. $\sqrt{4x^4} = \sqrt{4} \cdot \sqrt{(x^2)^2} = 2x^2$

25. $\sqrt[4]{16x^4 y} = \sqrt[4]{16} \cdot \sqrt[4]{x^4} \cdot \sqrt[4]{y} = 2x\sqrt[4]{y}$

27. $\sqrt{3x} \cdot \sqrt{12x} = \sqrt{3 \cdot 12 \cdot x \cdot x} = \sqrt{36x^2} = \sqrt{36} \cdot \sqrt{x^2} = 6x$

29. $\sqrt[3]{8x^6 y^3 z^9} = \sqrt[3]{8} \cdot \sqrt[3]{(x^2)^3} \cdot \sqrt[3]{y^3} \cdot \sqrt[3]{(z^3)^3} = 2x^2 yz^3$

31. $\sqrt[4]{\dfrac{3}{4}} \cdot \sqrt[4]{\dfrac{27}{4}} = \sqrt[4]{\dfrac{3}{4} \cdot \dfrac{27}{4}} = \sqrt[4]{\dfrac{81}{16}} = = \dfrac{3}{2}$

33. $\sqrt[4]{25z} \cdot \sqrt[4]{25z} = \sqrt[4]{625z^2} = \sqrt[4]{625} \cdot \sqrt[4]{z^2} = 5\sqrt{z}$

35. $\sqrt[5]{\dfrac{7a}{b^2}} \cdot \sqrt[5]{\dfrac{b^2}{7a^6}} = \sqrt[5]{\dfrac{7ab^2}{7a^6b^2}} = \sqrt[5]{\dfrac{1}{a^5}} = \dfrac{1}{a}$

37. $\sqrt{200} = \sqrt{100 \cdot 2} = \sqrt{100} \cdot \sqrt{2} = 10\sqrt{2}$

39. $\sqrt[3]{81} = \sqrt[3]{27 \cdot 3} = \sqrt[3]{27} \cdot \sqrt[3]{3} = 3\sqrt[3]{3}$

41. $\sqrt[4]{64} = \sqrt[4]{16 \cdot 4} = \sqrt[4]{16} \cdot \sqrt[4]{4} = 2\sqrt[4]{4} = 2\sqrt[4]{2^2} = 2\sqrt{2}$

43. $\sqrt[5]{-64} = \sqrt[5]{-2^6} = \sqrt[5]{-2^5 \cdot 2} = \sqrt[5]{-2^5} \cdot \sqrt[5]{2} = -2\sqrt[5]{2}$

45. $\sqrt{8n^3} = \sqrt{(2n)^2 \cdot 2n} = \sqrt{(2n)^2} \cdot \sqrt{2n} = 2n\sqrt{2n}$

47. $\sqrt{12a^2b^5} = \sqrt{(2ab^2)^2 \cdot 3b} = \sqrt{(2ab^2)^2} \cdot \sqrt{3b} = 2ab^2\sqrt{3b}$

49. $\sqrt[3]{-125x^4y^5} = \sqrt[3]{(-5xy)^3 \cdot xy^2} = \sqrt[3]{(-5xy)^3} \cdot \sqrt[3]{xy^2} = -5xy\sqrt[3]{xy^2}$

51. $\sqrt[3]{5t} \cdot \sqrt[3]{125t} = \sqrt[3]{625t^2} = \sqrt[3]{5^4t^2} = \sqrt[3]{5^3 \cdot 5t^2} = \sqrt[3]{5^3} \cdot \sqrt[3]{5t^2} = 5\sqrt[3]{rt^2}$

53. $\sqrt[4]{\dfrac{9t^5}{r^8}} \cdot \sqrt[4]{\dfrac{9r}{5t}} = \sqrt[4]{\dfrac{81rt^5}{5r^8t}} = \sqrt[4]{\dfrac{81t^4}{5r^7}} = \dfrac{\sqrt[4]{(3t)^4}}{\sqrt[4]{r^4 \cdot 5r^3}} = \dfrac{3t}{r\sqrt[4]{5r^3}}$

55. $\sqrt{3} \cdot \sqrt[3]{3} = 3^{1/2} \cdot 3^{1/3} = 3^{1/2+1/3} = 3^{5/6} = \sqrt[6]{3^5}$

57. $\sqrt[4]{8} \cdot \sqrt[3]{4} = \sqrt[4]{2^3} \cdot \sqrt[3]{2^2} = 2^{3/4} \cdot 2^{2/3} = 2^{3/4+2/3} = 2^{17/12} = 2^{12/12+5/12} = 2 \cdot 2^{5/12} = 2\sqrt[12]{2^5}$

59. $\sqrt[4]{x^3} \cdot \sqrt[3]{x} = x^{3/4} \cdot x^{1/3} = x^{3/4+1/3} = x^{13/12} = x^{12/12} \cdot x^{1/12} = x\sqrt[12]{x}$

61. $\sqrt[4]{rt} \cdot \sqrt[3]{r^2t} = (rt)^{1/4} \cdot (r^2t)^{1/3} = r^{1/4}t^{1/4} \cdot r^{2/3}t^{1/3} = r^{1/4+2/3}t^{1/4+1/3} = r^{11/12}t^{7/12} = \sqrt[12]{r^{11}t^7}$

63. $2\sqrt{3} + 7\sqrt{3} = 9\sqrt{3}$

65. $\sqrt{x} + \sqrt{x} - \sqrt{y} = 2\sqrt{x} - \sqrt{y}$

67. $2\sqrt[3]{6} - 7\sqrt[3]{6} = -5\sqrt[3]{6}$

69. $3\sqrt{28} + 3\sqrt{7} = 3\sqrt{4 \cdot 7} + 3\sqrt{7} = 3 \cdot 2\sqrt{7} + 3\sqrt{7} = 9\sqrt{7}$

71. $\sqrt{44} - 4\sqrt{11} = \sqrt{4 \cdot 11} - 4\sqrt{11} = 2\sqrt{11} - 4\sqrt{11} = -2\sqrt{11}$

73. $2\sqrt[3]{16} + \sqrt[3]{2} - \sqrt{2} = 2\sqrt[3]{8 \cdot 2} + \sqrt[3]{2} - \sqrt{2} = 2 \cdot 2\sqrt[3]{2} + \sqrt[3]{2} - \sqrt{2} = 5\sqrt[3]{2} - \sqrt{2}$

75. $\sqrt[3]{xy} - 2\sqrt[3]{xy} = -\sqrt[3]{xy}$

77. $\sqrt{4x + 8} + \sqrt{x + 2} = \sqrt{4(x + 2)} + \sqrt{x + 2} = 2\sqrt{x + 2} + \sqrt{x + 2} = 3\sqrt{x + 2}$

79. $\dfrac{15\sqrt{8}}{4} - \dfrac{2\sqrt{2}}{5} = \dfrac{15 \cdot 2\sqrt{2}}{4} \cdot \dfrac{5}{5} - \dfrac{2\sqrt{2}}{5} \cdot \dfrac{4}{4} = \dfrac{150\sqrt{2}}{20} - \dfrac{8\sqrt{2}}{20} = \dfrac{150\sqrt{2} - 8\sqrt{2}}{20} = \dfrac{142\sqrt{2}}{20} = \dfrac{71\sqrt{2}}{10}$

81. $2\sqrt[4]{64} - \sqrt[4]{324} + \sqrt[4]{4} = 2\sqrt[4]{16 \cdot 4} - \sqrt[4]{81 \cdot 4} + \sqrt[4]{4} = 4\sqrt[4]{4} - 3\sqrt[4]{4} + \sqrt[4]{4} = 2\sqrt[4]{4} = 2\sqrt{2}$

83. $\sqrt{64x^3} - \sqrt{x} + 3\sqrt{x} = \sqrt{(8x)^2 \cdot x} - \sqrt{x} + 3\sqrt{x} = 8x\sqrt{x} - \sqrt{x} + 3\sqrt{x} = 2\sqrt{x}(4x + 1)$

85. $\sqrt[4]{81a^5b^5} - \sqrt[4]{ab} = \sqrt[4]{(3ab)^4 \cdot ab} - \sqrt[4]{ab} = 3ab\sqrt[4]{ab} - \sqrt[4]{ab} = (3ab - 1)\sqrt[4]{ab}$

87. $5\sqrt[3]{\dfrac{n^4}{125}} - 2\sqrt[3]{n} = 5\sqrt[3]{\dfrac{n^3}{125} \cdot n} - 2\sqrt[3]{n} = 5 \cdot \dfrac{n}{5}\sqrt[3]{n} - 2\sqrt[3]{n} = n\sqrt[3]{n} - 2\sqrt[3]{n} = (n - 2)\sqrt[3]{n}$

89. $(3 + \sqrt{7})(3 - \sqrt{7}) = 3^2 - (\sqrt{7})^2 = 9 - 7 = 2$

91. $(\sqrt{x} + 8)(\sqrt{x} - 8) = (\sqrt{x})^2 - 8^2 = x - 64$

93. $(\sqrt{ab} - \sqrt{c})(\sqrt{ab} + \sqrt{c}) = (\sqrt{ab})^2 - (\sqrt{c})^2 = ab - c$

95. $(\sqrt{x} - 7)(\sqrt{x} + 8) = (\sqrt{x})^2 + 8\sqrt{x} - 7\sqrt{x} - 56 = x + \sqrt{x} - 56$

97. $\dfrac{4}{\sqrt{3}} = \dfrac{4}{\sqrt{3}} \cdot \dfrac{\sqrt{3}}{\sqrt{3}} = \dfrac{4\sqrt{3}}{3}$

99. $\dfrac{5}{3\sqrt{5}} = \dfrac{5}{3\sqrt{5}} \cdot \dfrac{\sqrt{5}}{\sqrt{5}} = \dfrac{5\sqrt{5}}{3 \cdot 5} = \dfrac{5\sqrt{5}}{15} = \dfrac{\sqrt{5}}{3}$

101. $\sqrt{\dfrac{b}{12}} = \dfrac{\sqrt{b}}{\sqrt{12}} = \dfrac{\sqrt{b}}{\sqrt{12}} \cdot \dfrac{\sqrt{12}}{\sqrt{12}} = \dfrac{\sqrt{12b}}{12} = \dfrac{\sqrt{4 \cdot 3b}}{12} = \dfrac{2\sqrt{3b}}{12} = \dfrac{\sqrt{3b}}{6}$

103. $\dfrac{1}{3 - \sqrt{2}} = \dfrac{1}{3 - \sqrt{2}} \cdot \dfrac{3 + \sqrt{2}}{3 + \sqrt{2}} = \dfrac{3 + \sqrt{2}}{9 - 2} = \dfrac{3 + \sqrt{2}}{7}$

105. $\dfrac{\sqrt{2}}{\sqrt{5} + 2} = \dfrac{\sqrt{2}}{\sqrt{5} + 2} \cdot \dfrac{\sqrt{5} - 2}{\sqrt{5} - 2} = \dfrac{\sqrt{10} - 2\sqrt{2}}{5 - 4} = \dfrac{\sqrt{10} - 2\sqrt{2}}{1} = \sqrt{10} - 2\sqrt{2}$

107. $\dfrac{1}{\sqrt{7} - \sqrt{6}} = \dfrac{1}{\sqrt{7} - \sqrt{6}} \cdot \dfrac{\sqrt{7} + \sqrt{6}}{\sqrt{7} + \sqrt{6}} = \dfrac{\sqrt{7} + \sqrt{6}}{7 - 6} = \dfrac{\sqrt{7} + \sqrt{6}}{1} = \sqrt{7} + \sqrt{6}$

109. $\dfrac{\sqrt{z}}{\sqrt{z} - 3} = \dfrac{\sqrt{z}}{\sqrt{z} - 3} \cdot \dfrac{\sqrt{z} + 3}{\sqrt{z} + 3} = \dfrac{z + 3\sqrt{z}}{z - 9}$

111. $\dfrac{\sqrt{a} + \sqrt{b}}{\sqrt{a} - \sqrt{b}} = \dfrac{\sqrt{a} + \sqrt{b}}{\sqrt{a} - \sqrt{b}} \cdot \dfrac{\sqrt{a} + \sqrt{b}}{\sqrt{a} + \sqrt{b}} = \dfrac{a + 2\sqrt{ab} + b}{a - b}$

Appendix C: Partial Fractions

1. Multiply $\dfrac{5}{3x(2x + 1)} = \dfrac{A}{3x} + \dfrac{B}{2x + 1}$ by $3x(2x + 1) \Rightarrow 5 = A(2x + 1) + B(3x)$.

   Let $x = 0 \Rightarrow 5 = A(1) \Rightarrow A = 5$. Let $x = -\dfrac{1}{2} \Rightarrow 5 = B\left(-\dfrac{3}{2}\right) \Rightarrow B = -\dfrac{10}{3}$.

   The expression can be written $\dfrac{5}{3x} + \dfrac{-10}{3(2x + 1)}$.

3. Multiply $\dfrac{4x + 2}{(x + 2)(2x - 1)} = \dfrac{A}{x + 2} + \dfrac{B}{2x - 1}$ by $(x + 2)(2x - 1) \Rightarrow 4x + 2 = A(2x - 1) + B(x + 2)$.

   Let $x = -2 \Rightarrow -6 = A(-5) \Rightarrow A = \dfrac{6}{5}$. Let $x = \dfrac{1}{2} \Rightarrow 4 = B\left(\dfrac{5}{2}\right) \Rightarrow B = \dfrac{8}{5}$.

   The expression can be written $\dfrac{6}{5(x + 2)} + \dfrac{8}{5(2x - 1)}$.

5. Factoring $\dfrac{x}{x^2 + 4x - 5}$ results in $\dfrac{x}{(x + 5)(x - 1)}$.

   Multiply $\dfrac{x}{(x + 5)(x - 1)} = \dfrac{A}{x + 5} + \dfrac{B}{x - 1}$ by $(x + 5)(x - 1) \Rightarrow x = A(x - 1) + B(x + 5)$.

   Let $x = -5 \Rightarrow -5 = A(-6) \Rightarrow A = \dfrac{5}{6}$. Let $x = 1 \Rightarrow 1 = B(6) \Rightarrow B = \dfrac{1}{6}$.

   The expression can be written $\dfrac{5}{6(x + 5)} + \dfrac{1}{6(x - 1)}$.

7. Multiply $\dfrac{2x}{(x + 1)(x + 2)^2} = \dfrac{A}{x + 1} + \dfrac{B}{x + 2} + \dfrac{C}{(x + 2)^2}$ by $(x + 1)(x + 2)^2 \Rightarrow$

   $2x = A(x + 2)^2 + B(x + 1)(x + 2) + C(x + 1)$. Let $x = -1 \Rightarrow -2 = A(1) \Rightarrow A = -2$.

   Let $x = -2 \Rightarrow -4 = C(-1) \Rightarrow C = 4$. Let $x = 0$ with $A = -2$ and $C = 4 \Rightarrow 0 = -2(4) + B(2) + 4(1) \Rightarrow$

   $4 = 2B \Rightarrow B = 2$. The expression can be written $\dfrac{-2}{x + 1} + \dfrac{2}{x + 2} + \dfrac{4}{(x + 2)^2}$.

9. Multiply $\dfrac{4}{x(1 - x)} = \dfrac{A}{x} + \dfrac{B}{1 - x}$ by $x(1 - x) \Rightarrow 4 = A(1 - x) + B(x)$.

   Let $x = 0 \Rightarrow 4 = A(1) \Rightarrow A = 4$. Let $x = 1 \Rightarrow 4 = B(1) \Rightarrow B = 4$.

   The expression can be written $\dfrac{4}{x} + \dfrac{4}{1 - x}$.

11. Multiply $\dfrac{4x^2 - x - 15}{x(x + 1)(x - 1)} = \dfrac{A}{x} + \dfrac{B}{x + 1} + \dfrac{C}{x - 1}$ by $x(x + 1)(x - 1) \Rightarrow$

   $4x^2 - x - 15 = A(x - 1)(x + 1) + B(x)(x - 1) + C(x)(x + 1)$. Let $x = 0 \Rightarrow -15 = A(-1) \Rightarrow A = 15$.

   Let $x = 1 \Rightarrow -12 = C(1)(2) \Rightarrow C = -6$. Let $x = -1 \Rightarrow -10 = B(-1)(-2) \Rightarrow B = -5$.

   The expression can be written $\dfrac{15}{x} + \dfrac{-5}{x + 1} + \dfrac{-6}{x - 1}$.

13. By long division $\dfrac{x^2}{x^2 + 2x + 1} = 1 + \dfrac{-2x - 1}{(x + 1)^2}$.

Multiply $\dfrac{-2x - 1}{(x + 1)^2} = \dfrac{A}{x + 1} + \dfrac{B}{(x + 1)^2}$ by $(x + 1)^2 \Rightarrow -2x - 1 = A(x + 1) + B$.

Let $x = -1 \Rightarrow 1 = B$. Let $x = 0$ with $B = 1 \Rightarrow -1 = A + 1 \Rightarrow A = -2$.

The expression can be written $1 + \dfrac{-2}{x + 1} + \dfrac{1}{(x + 1)^2}$.

15. By long division $\dfrac{2x^5 + 3x^4 - 3x^3 - 2x^2 + x}{2x^2 + 5x + 2} = x^3 - x^2 + \dfrac{x}{2x^2 + 5x + 2} = x^3 - x^2 + \dfrac{x}{(2x + 1)(x + 2)}$.

Multiply $\dfrac{x}{(2x + 1)(x + 2)} = \dfrac{A}{2x + 1} + \dfrac{B}{x + 2}$ by $(2x + 1)(x + 2) \Rightarrow x = A(x + 2) + B(2x + 1)$.

Let $x = -\dfrac{1}{2} \Rightarrow -\dfrac{1}{2} = A\left(\dfrac{3}{2}\right) \Rightarrow A = -\dfrac{1}{3}$. Let $x = -2 \Rightarrow -2 = B(-3) \Rightarrow B = \dfrac{2}{3}$.

The expression can be written $x^3 - x^2 + \dfrac{-1}{3(2x + 1)} + \dfrac{2}{3(x + 2)}$.

17. By long division $\dfrac{x^3 + 4}{9x^3 - 4x} = \dfrac{1}{9} + \dfrac{\frac{4}{9}x + 4}{9x^3 - 4x} = \dfrac{1}{9} + \dfrac{\frac{4}{9}x + 4}{x(3x + 2)(3x - 2)}$.

Multiply $\dfrac{\frac{4}{9}x + 4}{x(3x + 2)(3x - 2)} = \dfrac{A}{x} + \dfrac{B}{3x + 2} + \dfrac{C}{3x - 2}$ by $x(3x + 2)(3x - 2) \Rightarrow$

$\dfrac{4}{9}x + 4 = A(3x + 2)(3x - 2) + B(x)(3x - 2) + C(x)(3x + 2)$. Let $x = 0 \Rightarrow 4 = A(-4) \Rightarrow A = -1$.

Let $x = -\dfrac{2}{3} \Rightarrow -\dfrac{8}{27} + 4 = B\left(-\dfrac{2}{3}\right)(-4) \Rightarrow \dfrac{100}{27} = \dfrac{8}{3}B \Rightarrow B = \dfrac{25}{18}$.

Let $x = \dfrac{2}{3} \Rightarrow \dfrac{8}{27} + 4 = C\left(\dfrac{2}{3}\right)(4) \Rightarrow \dfrac{116}{27} = \dfrac{8}{3}C \Rightarrow C = \dfrac{29}{18}$.

The expression can be written $\dfrac{1}{9} + \dfrac{-1}{x} + \dfrac{25}{18(3x + 2)} + \dfrac{29}{18(3x - 2)}$.

19. Multiply $\dfrac{-3}{x^2(x^2 + 5)} = \dfrac{A}{x} + \dfrac{B}{x^2} + \dfrac{Cx + D}{x^2 + 5}$ by $x^2(x^2 + 5) \Rightarrow$

$-3 = A(x)(x^2 + 5) + B(x^2 + 5) + (Cx + D)(x^2) \Rightarrow -3 = Ax^3 + 5Ax + Bx^2 + 5B + Cx^3 + Dx^2$.

Equate coefficients. For x^3: $0 = A + C$.

For x^2: $0 = B + D$. For x: $0 = 5A \Rightarrow A = 0$. For the constants: $-3 = 5B \Rightarrow B = -\dfrac{3}{5}$.

Substitute $A = 0$ in the first equation. $C = 0$. Substitute $B = -\dfrac{3}{5}$ in the second equation. $D = \dfrac{3}{5}$.

The expression can be written $\dfrac{-3}{5x^2} + \dfrac{3}{5(x^2 + 5)}$.

21. Multiply $\dfrac{3x - 2}{(x + 4)(3x^2 + 1)} = \dfrac{A}{x + 4} + \dfrac{Bx + C}{3x^2 + 1}$ by $(x + 4)(3x^2 + 1) \Rightarrow$

$3x - 2 = A(3x^2 + 1) + (Bx + C)(x + 4) \Rightarrow 3x - 2 = 3Ax^2 + A + Bx^2 + 4Bx + Cx + 4C.$

Let $x = -4 \Rightarrow -14 = 49A \Rightarrow A = -\dfrac{2}{7}.$ Equate coefficients.

For x^2: $0 = 3A + B \Rightarrow 0 = -\dfrac{6}{7} + B \Rightarrow B = \dfrac{6}{7}.$ For x: $3 = 4B + C \Rightarrow 2 = \dfrac{24}{7} + C \Rightarrow C = -\dfrac{3}{7}.$

The expression can be written $\dfrac{-2}{7(x + 4)} + \dfrac{6x - 3}{7(3x^2 + 1)}.$

23. Multiply $\dfrac{1}{x(2x + 1)(3x^2 + 4)} = \dfrac{A}{x} + \dfrac{B}{2x + 1} + \dfrac{Cx + D}{3x^2 + 4}$ by $x(2x + 1)(3x^2 + 4) \Rightarrow$

$1 = A(2x + 1)(3x^2 + 4) + B(x)(3x^2 + 4) + (Cx + D)(x)(2x + 1).$

Let $x = 0 \Rightarrow 1 = A(1)(4) \Rightarrow A = \dfrac{1}{4}.$ Let $x = -\dfrac{1}{2} \Rightarrow 1 = B\left(-\dfrac{1}{2}\right)\left(\dfrac{19}{4}\right) \Rightarrow B = -\dfrac{8}{19}.$

Multiply the right side out. $1 = A(6x^3 + 3x^2 + 8x + 4) + 3Bx^3 + 4Bx + 2Cx^3 + Cx^2 + 2Dx^2 + Dx \Rightarrow$

$1 = 6Ax^3 + 3Ax^2 + 8Ax + 4A + 3Bx^3 + 4Bx + 2Cx^3 + Cx^2 + 2Dx^2 + Dx.$ Equate coefficients.

For x^3: $0 = 6A + 3B + 2C \Rightarrow 0 = 6\left(\dfrac{1}{4}\right) + 3\left(-\dfrac{8}{19}\right) + 2C \Rightarrow 0 = \dfrac{9}{38} + 2C \Rightarrow C = -\dfrac{9}{76}.$

For x^2: $0 = 3A + C + 2D \Rightarrow 0 = \dfrac{3}{4} - \dfrac{9}{76} + 2D \Rightarrow 0 = \dfrac{48}{76} + 2D \Rightarrow D = -\dfrac{24}{76}.$

The expression can be written $\dfrac{1}{4x} + \dfrac{-8}{19(2x + 1)} + \dfrac{-9x - 24}{76(3x^2 + 4)}.$

25. Multiply $\dfrac{3x - 1}{x(2x^2 + 1)^2} = \dfrac{A}{x} + \dfrac{Bx + C}{2x^2 + 1} + \dfrac{Dx + E}{(2x^2 + 1)^2}$ by $x(2x^2 + 1)^2 \Rightarrow$

$3x - 1 = A(2x^2 + 1)^2 + (Bx + C)(x)(2x^2 + 1) + (Dx + E)(x).$

Let $x = 0 \Rightarrow -1 = A(1) \Rightarrow A = -1.$ Multiply the right side out.

$3x - 1 = A(4x^4 + 4x^2 + 1) + 2Bx^4 + Bx^2 + Cx + 2Cx^3 + Dx^2 + Ex \Rightarrow$

$3x - 1 = 4Ax^4 + 4Ax^2 + A + 2Bx^4 + Bx^2 + Cx + 2Cx^3 + Dx^2 + Ex.$

Equate coefficients. For x^4: $0 = 4A + 2B \Rightarrow 0 = -4 + 2B \Rightarrow B = 2.$ For x^3: $0 = 2C \Rightarrow C = 0.$

For x^2: $0 = 4A + B + D \Rightarrow 0 = -4 + 2 + D \Rightarrow D = 2.$ For x: $3 = C + E \Rightarrow 3 = 0 + E \Rightarrow E = 3$

The expression can be written $\dfrac{-1}{x} + \dfrac{2x}{2x^2 + 1} + \dfrac{2x + 3}{(2x^2 + 1)^2}.$

27. Multiply $\dfrac{-x^4 - 8x^2 + 3x - 10}{(x + 2)(x^2 + 4)^2} = \dfrac{A}{x + 2} + \dfrac{Bx + C}{x^2 + 4} + \dfrac{Dx + E}{(x^2 + 4)^2}$ by $(x + 2)(x^2 + 4)^2 \Rightarrow$

$-x^4 - 8x^2 + 3x - 10 = A(x^2 + 4)^2 + (Bx + C)(x + 2)(x^2 + 4) + (Dx + E)(x + 2)$.

Let $x = -2 \Rightarrow -64 = A(64) \Rightarrow A = -1$. Multiply the right side out.

$-x^4 - 8x^2 + 3x - 10 =$

$Ax^4 + 8Ax^2 + 16A + Bx^4 + 2Bx^3 + 4Bx^2 + 8Bx + Cx^3 + 2Cx^2 + 4Cx + 8C + Dx^2 + 2Dx + Ex + 2E$.

Equate coefficients.

For x^4: $-1 = A + B \Rightarrow -1 = -1 + B \Rightarrow B = 0$. For x^3: $0 = 2B + C \Rightarrow 0 = 0 + C \Rightarrow C = 0$.

For x^2: $-8 = 8A + 4B + 2C + D \Rightarrow -8 = -8 + 0 + 0 + D \Rightarrow D = 0$.

For x: $3 = 8B + 4C + 2D + E \Rightarrow 3 = 0 + 0 + 0 + E \Rightarrow E = 3$.

The expression can be written $\dfrac{-1}{x + 2} + \dfrac{3}{(x^2 + 4)^2}$.

29. By long division $\dfrac{5x^5 + 10x^4 - 15x^3 + 4x^2 + 13x - 9}{x^3 + 2x^2 - 3x} = 5x^2 + \dfrac{4x^2 + 13x - 9}{x^3 + 2x^2 - 3x} = 5x^2 + \dfrac{4x^2 + 13x - 9}{x(x + 3)(x - 1)}$.

Multiply $\dfrac{4x^2 + 13x - 9}{x(x + 3)(x - 1)} = \dfrac{A}{x} + \dfrac{B}{x + 3} + \dfrac{C}{x - 1}$ by $x(x + 3)(x - 1) \Rightarrow$

$4x^2 + 13x - 9 = A(x + 3)(x - 1) + B(x)(x - 1) + C(x)(x + 3)$ Let $x = 0 \Rightarrow -9 = A(-3) \Rightarrow A = 3$.

Let $x = -3 \Rightarrow -12 = B(-3)(-4) \Rightarrow B = -1$. Let $x = 1 \Rightarrow 8 = C(4) \Rightarrow C = 2$.

The expression can be written $5x^2 + \dfrac{3}{x} + \dfrac{-1}{x + 3} + \dfrac{2}{x - 1}$.

Appendix D: Rotation of Axes

1. $4x^2 + 3y^2 + 2xy - 5x = 8 \Rightarrow B^2 - 4AC = 2^2 - 4(4)(3) = 4 - 48 < 0$. Circle, ellipse, or a point.

3. $2x^2 + 3xy - 4y^2 = 0 \Rightarrow B^2 - 4AC = 3^2 - 4(2)(-4) = 9 + 32 > 0$. Hyperbola or 2 intersecting lines.

5. $4x^2 + 4xy + y^2 + 15 = 0 \Rightarrow B^2 - 4AC = 4^2 - 4(4)(1) = 16 - 16 = 0$. Parabola, one line, or 2 parallel lines.

7. $2x^2 + \sqrt{3}xy + y^2 + x = 5 \Rightarrow \cot 2\theta = \dfrac{A - C}{B} = \dfrac{2 - 1}{\sqrt{3}} \Rightarrow \cot 2\theta = \dfrac{1}{\sqrt{3}} \Rightarrow 2\theta = 60° \Rightarrow \theta = 30°$

9. $3x^2 + \sqrt{3}xy + 4y^2 + 2x - 3y = 12 \Rightarrow \cot 2\theta = \dfrac{A - C}{B} = \dfrac{3 - 4}{\sqrt{3}} \Rightarrow \cot 2\theta = -\dfrac{1}{\sqrt{3}} \Rightarrow 2\theta = 120° \Rightarrow$

   $\theta = 60°$

11. $x^2 - 4xy + 5y^2 = 18 \Rightarrow \cot 2\theta = \dfrac{A - C}{B} = \dfrac{1 - 5}{-4} = 1 \Rightarrow \cot 2\theta = 1 \Rightarrow 2\theta = 45° \Rightarrow \theta = 22.5°$

13. $x^2 - xy + y^2 = 6$ [1]; $\theta = 45°$; $x = x' \cos \theta - y' \sin \theta = \dfrac{\sqrt{2}}{2}x' - \dfrac{\sqrt{2}}{2}y'$ [2];

    $y = x' \sin \theta + y' \cos \theta = \dfrac{\sqrt{2}}{2}x' + \dfrac{\sqrt{2}}{2}y'$ [3]. Substitute [2] and [3] in [1].

    $\left(\dfrac{\sqrt{2}}{2}x' - \dfrac{\sqrt{2}}{2}y'\right)^2 - \left(\dfrac{\sqrt{2}}{2}x' - \dfrac{\sqrt{2}}{2}y'\right)\left(\dfrac{\sqrt{2}}{2}x' + \dfrac{\sqrt{2}}{2}y'\right) + \left(\dfrac{\sqrt{2}}{2}x' + \dfrac{\sqrt{2}}{2}y'\right)^2 = 6 \Rightarrow$

    $\dfrac{1}{2}x'^2 - x'y' + \dfrac{1}{2}y'^2 - \dfrac{1}{2}x'^2 + \dfrac{1}{2}y'^2 + \dfrac{1}{2}x'^2 + x'y' + \dfrac{1}{2}y'^2 = 6 \Rightarrow \dfrac{1}{2}x'^2 + \dfrac{3}{2}y'^2 = 6 \Rightarrow \dfrac{x'^2}{12} + \dfrac{y'^2}{4} = 1.$

    See Figure 13.

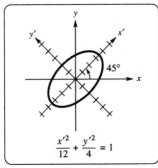

$\dfrac{x'^2}{12} + \dfrac{y'^2}{4} = 1$

Figure 13

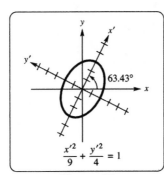

$\dfrac{x'^2}{9} + \dfrac{y'^2}{4} = 1$

Figure 15

15. $8x^2 - 4xy + 5y^2 = 36$ [1]; $\sin \theta = \dfrac{2}{\sqrt{5}}, y = 2, r = \sqrt{5}, x = \sqrt{5 - 4} = 1 \Rightarrow \cos \theta = \dfrac{1}{\sqrt{5}}$;

    $x = x' \cos \theta - y' \sin \theta = \dfrac{1}{\sqrt{5}}x' - \dfrac{2}{\sqrt{5}}y'$ [2]; $y = x' \sin \theta + y' \cos \theta = \dfrac{2}{\sqrt{5}}x' + \dfrac{1}{\sqrt{5}}y'$ [3].

    Substitute [2] and [3] in [1].

    $8\left(\dfrac{1}{\sqrt{5}}x' - \dfrac{2}{\sqrt{5}}y'\right)^2 - 4\left(\dfrac{1}{\sqrt{5}}x' - \dfrac{2}{\sqrt{5}}y'\right)\left(\dfrac{2}{\sqrt{5}}x' + \dfrac{1}{\sqrt{5}}y'\right) + 5\left(\dfrac{2}{\sqrt{5}}x' + \dfrac{1}{\sqrt{5}}y'\right)^2 = 36 \Rightarrow$

    $8\left(\dfrac{1}{5}x'^2 - \dfrac{4}{5}x'y' + \dfrac{4}{5}y'^2\right) - 4\left(\dfrac{2}{5}x'^2 - \dfrac{3}{5}x'y' - \dfrac{2}{5}y'^2\right) + 5\left(\dfrac{4}{5}x'^2 + \dfrac{4}{5}x'y' + \dfrac{1}{5}y'^2\right) = 36 \Rightarrow$

    $\dfrac{8}{5}x'^2 - \dfrac{32}{5}x'y' + \dfrac{32}{5}y'^2 - \dfrac{8}{5}x'^2 + \dfrac{12}{5}x'y' + \dfrac{8}{5}y'^2 + 4x'^2 + 4x'y' + y'^2 = 36 \Rightarrow 4x'^2 + 9y'^2 = 36 \Rightarrow$

    $\dfrac{x'^2}{9} + \dfrac{y'^2}{4} = 1.$ See Figure 15.

17. $3x^2 - 2xy + 3y^2 = 8$ [1]; $\cot 2\theta = \dfrac{A - C}{B} = \dfrac{3 - 3}{-2} = 0 \Rightarrow 2\theta = 90° \Rightarrow \theta = 45°$;

$x = x' \cos \theta - y' \sin \theta = \dfrac{\sqrt{2}}{2}x' - \dfrac{\sqrt{2}}{2}y'$ [2]; $y = x' \sin \theta + y' \cos \theta = \dfrac{\sqrt{2}}{2}x' + \dfrac{\sqrt{2}}{2}y'$ [3].

Substitute [2] and [3] in [1].

$$3\left(\dfrac{\sqrt{2}}{2}x' - \dfrac{\sqrt{2}}{2}y'\right)^2 - 2\left(\dfrac{\sqrt{2}}{2}x' - \dfrac{\sqrt{2}}{2}y'\right)\left(\dfrac{\sqrt{2}}{2}x' + \dfrac{\sqrt{2}}{2}y'\right) + 3\left(\dfrac{\sqrt{2}}{2}x' + \dfrac{\sqrt{2}}{2}y'\right)^2 = 8 \Rightarrow$$

$$3\left(\dfrac{1}{2}x'^2 - x'y' + \dfrac{1}{2}y'^2\right) - 2\left(\dfrac{1}{2}x'^2 - \dfrac{1}{2}y'^2\right) + 3\left(\dfrac{1}{2}x'^2 + x'y' + \dfrac{1}{2}y'^2\right) = 8 \Rightarrow$$

$$\dfrac{3}{2}x'^2 - 3x'y' + \dfrac{3}{2}y'^2 - x'^2 + y'^2 + \dfrac{3}{2}x'^2 + 3x'y' + \dfrac{3}{2}y'^2 = 8 \Rightarrow 2x'^2 + 4y'^2 = 8 \Rightarrow \dfrac{x'^2}{4} + \dfrac{y'^2}{2} = 1$$

See Figure 17.

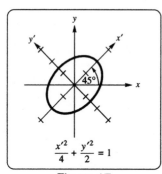

$$\dfrac{x'^2}{4} + \dfrac{y'^2}{2} = 1$$

Figure 17

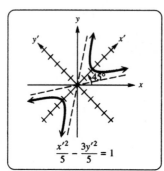

$$\dfrac{x'^2}{5} - \dfrac{3y'^2}{5} = 1$$

Figure 19

19. $x^2 - 4xy + y^2 = -5$ [1]; $\cot 2\theta = \dfrac{A - C}{B} = \dfrac{1 - 1}{-4} = 0 \Rightarrow 2\theta = 90° \Rightarrow \theta = 45°$;

$x = x' \cos \theta - y' \sin \theta = \dfrac{\sqrt{2}}{2}x' - \dfrac{\sqrt{2}}{2}y'$ [2]; $y = x' \sin \theta + y' \cos \theta = \dfrac{\sqrt{2}}{2}x' + \dfrac{\sqrt{2}}{2}y'$ [3].

Substitute [2] and [3] in [1].

$$\left(\dfrac{\sqrt{2}}{2}x' - \dfrac{\sqrt{2}}{2}y'\right)^2 - 4\left(\dfrac{\sqrt{2}}{2}x' - \dfrac{\sqrt{2}}{2}y'\right)\left(\dfrac{\sqrt{2}}{2}x' + \dfrac{\sqrt{2}}{2}y'\right) + \left(\dfrac{\sqrt{2}}{2}x' + \dfrac{\sqrt{2}}{2}y'\right)^2 = -5 \Rightarrow$$

$$\dfrac{1}{2}x'^2 - x'y' + \dfrac{1}{2}y'^2 - 4\left(\dfrac{1}{2}x'^2 - \dfrac{1}{2}y'^2\right) + \dfrac{1}{2}x'^2 + x'y' + \dfrac{1}{2}y'^2 = -5 \Rightarrow$$

$$\dfrac{1}{2}x'^2 - x'y' + \dfrac{1}{2}y'^2 - 2x'^2 + 2y'^2 + \dfrac{1}{2}x'^2 + x'y' + \dfrac{1}{2}y'^2 = -5 \Rightarrow -x'^2 + 3y'^2 = -5 \Rightarrow \dfrac{x'^2}{5} - \dfrac{3y'^2}{5} = 1.$$

See Figure 19.

21. $7x^2 + 6\sqrt{3}xy + 13y^2 = 64$ [1]; $\cot 2\theta = \dfrac{A - C}{B} = \dfrac{7 - 13}{6\sqrt{3}} = \dfrac{-6}{6\sqrt{3}} = -\dfrac{1}{\sqrt{3}} \Rightarrow 2\theta = 120° \Rightarrow \theta = 60°$;

$x = x'\cos\theta - y'\sin\theta = \dfrac{1}{2}x' - \dfrac{\sqrt{3}}{2}y'$ [2]; $y = x'\sin\theta + y'\cos\theta = \dfrac{\sqrt{3}}{2}x' + \dfrac{1}{2}y'$ [3].

Substitute [2] and [3] in [1].

$$7\left(\frac{1}{2}x' - \frac{\sqrt{3}}{2}y'\right)^2 + 6\sqrt{3}\left(\frac{1}{2}x' - \frac{\sqrt{3}}{2}y'\right)\left(\frac{\sqrt{3}}{2}x' + \frac{1}{2}y'\right) + 13\left(\frac{\sqrt{3}}{2}x' + \frac{1}{2}y'\right)^2 = 64 \Rightarrow$$

$$7\left(\frac{1}{4}x'^2 - \frac{\sqrt{3}}{2}x'y' + \frac{3}{4}y'^2\right) + 6\sqrt{3}\left(\frac{\sqrt{3}}{4}x'^2 - \frac{1}{2}x'y' - \frac{\sqrt{3}}{4}y'^2\right) + 13\left(\frac{3}{4}x'^2 + \frac{\sqrt{3}}{2}x'y' + \frac{1}{4}y'^2\right) = 64 \Rightarrow$$

$$\frac{7}{4}x'^2 - \frac{7\sqrt{3}}{2}x'y' + \frac{21}{4}y'^2 + \frac{18}{4}x'^2 - \frac{6\sqrt{3}}{2}x'y' - \frac{18}{4}y'^2 + \frac{39}{4}x'^2 + \frac{13\sqrt{3}}{2}x'y' + \frac{13}{4}y'^2 = 64 \Rightarrow$$

$16x'^2 + 4y'^2 = 64 \Rightarrow \dfrac{x'^2}{4} + \dfrac{y'^2}{16} = 1$. See Figure 21.

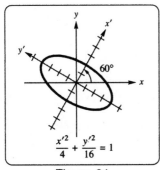

Figure 21

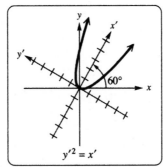

Figure 23

23. $3x^2 - 2\sqrt{3}xy + y^2 - 2x - 2\sqrt{3}y = 0$ [1]; $\cot 2\theta = \dfrac{A - C}{B} = \dfrac{3 - 1}{-2\sqrt{3}} = -\dfrac{1}{\sqrt{3}} \Rightarrow 2\theta = 120° \Rightarrow$

$\theta = 60°$; $x = x'\cos\theta - y'\sin\theta = \dfrac{1}{2}x' - \dfrac{\sqrt{3}}{2}y'$ [2]; $y = x'\sin\theta + y'\cos\theta = \dfrac{\sqrt{3}}{2}x' + \dfrac{1}{2}y'$ [3].

Substitute [2] and [3] in [1]. $3\left(\dfrac{1}{2}x' - \dfrac{\sqrt{3}}{2}y'\right)^2 - 2\sqrt{3}\left(\dfrac{1}{2}x' - \dfrac{\sqrt{3}}{2}y'\right)\left(\dfrac{\sqrt{3}}{2}x' + \dfrac{1}{2}y'\right) +$

$\left(\dfrac{\sqrt{3}}{2}x' + \dfrac{1}{2}y'\right)^2 - 2\left(\dfrac{1}{2}x' - \dfrac{\sqrt{3}}{2}y'\right) - 2\sqrt{3}\left(\dfrac{\sqrt{3}}{2}x' + \dfrac{1}{2}y'\right) = 0 \Rightarrow 3\left(\dfrac{1}{4}x'^2 - \dfrac{\sqrt{3}}{2}x'y' + \dfrac{3}{4}y'^2\right) -$

$2\sqrt{3}\left(\dfrac{\sqrt{3}}{4}x'^2 - \dfrac{1}{2}x'y' - \dfrac{\sqrt{3}}{4}y'^2\right) + \left(\dfrac{3}{4}x'^2 + \dfrac{\sqrt{3}}{2}x'y' + \dfrac{1}{4}y'^2\right) - x' + \sqrt{3}y' - 3x' - \sqrt{3}y' = 0 \Rightarrow$

$\dfrac{3}{4}x'^2 - \dfrac{3\sqrt{3}}{2}x'y' + \dfrac{9}{4}y'^2 - \dfrac{3}{2}x'^2 + \sqrt{3}x'y' + \dfrac{3}{2}y'^2 + \dfrac{3}{4}x'^2 + \dfrac{\sqrt{3}}{2}x'y' + \dfrac{1}{4}y'^2 - 4x' = 0 \Rightarrow$

$4y'^2 - 4x' = 0 \Rightarrow 4y'^2 = 4x' \Rightarrow y'^2 = x'$. See Figure 23.

25. $x^2 + 3xy + y^2 - 5\sqrt{2}y = 15$ [1]; $\cot 2\theta = \dfrac{A - C}{B} = \dfrac{1 - 1}{3} = 0 \Rightarrow 2\theta = 90° \Rightarrow \theta = 45°$;

$x = x' \cos\theta - y' \sin\theta = \dfrac{\sqrt{2}}{2}x' - \dfrac{\sqrt{2}}{2}y'$ [2]; $y = x' \sin\theta + y' \cos\theta = \dfrac{\sqrt{2}}{2}x' + \dfrac{\sqrt{2}}{2}y'$ [3].

Substitute [2] and [3] in [1]. $\left(\dfrac{\sqrt{2}}{2}x' - \dfrac{\sqrt{2}}{2}y'\right)^2 + 3\left(\dfrac{\sqrt{2}}{2}x' - \dfrac{\sqrt{2}}{2}y'\right)\left(\dfrac{\sqrt{2}}{2}x' + \dfrac{\sqrt{2}}{2}y'\right) +$

$\left(\dfrac{\sqrt{2}}{2}x' + \dfrac{\sqrt{2}}{2}y'\right)^2 - 5\sqrt{2}\left(\dfrac{1}{2}x' - \dfrac{\sqrt{2}}{2}y'\right) = 15 \Rightarrow$

$\dfrac{1}{2}x'^2 - x'y' + \dfrac{1}{2}y'^2 + 3\left(\dfrac{1}{2}x'^2 - \dfrac{1}{2}y'^2\right) + \dfrac{1}{2}x'^2 + x'y' + \dfrac{1}{2}y'^2 - 5x' - 5y' = 15 \Rightarrow$

$\dfrac{1}{2}x'^2 - x'y' + \dfrac{1}{2}y'^2 + \dfrac{3}{2}x'^2 - \dfrac{3}{2}y'^2 + \dfrac{1}{2}x'^2 + x'y' + \dfrac{1}{2}y'^2 - 5x' - 5y' = 15 \Rightarrow$

$\dfrac{5}{2}x'^2 - \dfrac{1}{2}y'^2 - 5x' - 5y' = 15 \Rightarrow 5x'^2 - 10x' - y'^2 - 10y' = 30 \Rightarrow$

$5(x'^2 - 2x' + 1) - (y'^2 + 10y' + 25) = 30 + 5 - 25 \Rightarrow 5(x' - 1)^2 - (y' + 5)^2 = 10 \Rightarrow$

$\dfrac{(x' - 1)^2}{2} - \dfrac{(y' + 5)^2}{10} = 1$. See Figure 25.

The graph of the equation is a hyperbola with its center at $(1, -5)$. By translating the axes of the $x'y'$-system

down 5 units and right 1 unit, we get an $x''y''$-coordinate system, in which the hyperbola is centered at the

origin. Thus $\dfrac{x''^2}{2} - \dfrac{y''^2}{10} = 1$.

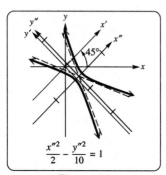

Figure 25

27. $4x^2 + 4xy + y^2 - 24x + 38y - 19 = 0$ [1]; $\cot 2\theta = \dfrac{A - C}{B} = \dfrac{4 - 1}{4} = \dfrac{3}{4} \Rightarrow 2\theta \approx 53.13° \Rightarrow$

$\theta \approx 26.57°$. For 2θ; $x = 3, y = 4, r = \sqrt{3^2 + 4^2} = 5 \Rightarrow \cos 2\theta = \dfrac{3}{5}$.

$\sin \theta = \sqrt{\dfrac{1 - \cos 2\theta}{2}} = \sqrt{\dfrac{1 - \frac{3}{5}}{2}} = \sqrt{\dfrac{2}{10}} = \dfrac{\sqrt{5}}{5}$; $\cos \theta = \sqrt{\dfrac{1 + \cos 2\theta}{2}} = \sqrt{\dfrac{1 + \frac{3}{5}}{2}} = \sqrt{\dfrac{8}{10}} = \dfrac{2\sqrt{5}}{5}$;

$x = x' \cos \theta - y' \sin \theta = \dfrac{2\sqrt{5}}{5}x' - \dfrac{\sqrt{5}}{5}y'$ [2]; $y = x' \sin \theta + y' \cos \theta = \dfrac{\sqrt{5}}{5}x' + \dfrac{2\sqrt{5}}{5}y'$ [3].

Substitute [2] and [3] in [1]. $4\left(\dfrac{2\sqrt{5}}{5}x' - \dfrac{\sqrt{5}}{5}y'\right)^2 + 4\left(\dfrac{2\sqrt{5}}{5}x' - \dfrac{\sqrt{5}}{5}y'\right)\left(\dfrac{\sqrt{5}}{5}x' + \dfrac{2\sqrt{5}}{5}y'\right) -$

$\left(\dfrac{\sqrt{5}}{5}x' + \dfrac{2\sqrt{5}}{5}y'\right)^2 - 24\left(\dfrac{2\sqrt{5}}{5}x' - \dfrac{\sqrt{5}}{5}y'\right) + 38\left(\dfrac{\sqrt{5}}{5}x' + \dfrac{2\sqrt{5}}{5}y'\right) = 19 \Rightarrow$

$4\left(\dfrac{4}{5}x'^2 - \dfrac{4}{5}x'y' + \dfrac{1}{5}y'^2\right) + 4\left(\dfrac{2}{5}x'^2 + \dfrac{3}{5}x'y' - \dfrac{2}{5}y'^2\right) + \left(\dfrac{1}{5}x'^2 + \dfrac{4}{5}x'y' + \dfrac{4}{5}y'^2\right) - \dfrac{48\sqrt{5}}{5}x' + \dfrac{25\sqrt{5}}{5}y' +$

$\dfrac{38\sqrt{5}}{5}x' + \dfrac{76\sqrt{5}}{5}y' = 19 \Rightarrow$

$\dfrac{16}{5}x'^2 - \dfrac{16}{5}x'y' + \dfrac{4}{5}y'^2 + \dfrac{8}{5}x'^2 + \dfrac{12}{5}x'y' - \dfrac{8}{5}y'^2 + \dfrac{1}{5}x'^2 + \dfrac{4}{5}x'y' + \dfrac{4}{5}y'^2 - \dfrac{48\sqrt{5}}{5}x' + \dfrac{24\sqrt{5}}{5}y' +$

$\dfrac{38\sqrt{5}}{5}x' + \dfrac{76\sqrt{5}}{5}y' = 19 \Rightarrow 5x'^2 - 2\sqrt{5}x' + 20\sqrt{5}y' = 19 \Rightarrow$

$5\left(x'^2 - \dfrac{2\sqrt{5}}{5}x' + \dfrac{1}{5}\right) + 20\sqrt{5}y' = 19 + 1 \Rightarrow 5\left(x' - \dfrac{\sqrt{5}}{5}\right)^2 = 20 - 20\sqrt{5}y \Rightarrow$

$\left(x' - \dfrac{\sqrt{5}}{5}\right)^2 = 4 - 4\sqrt{5}y' \Rightarrow \left(x' - \dfrac{\sqrt{5}}{5}\right)^2 = -4\sqrt{5}\left(y' - \dfrac{\sqrt{5}}{5}\right)$. See Figure 27.

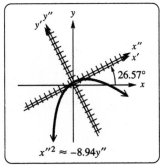

$x''^2 \approx -8.94y''$

Figure 27

29. $16x^2 + 24xy + 9y^2 - 130x + 90y = 0$ [1]; $\cot 2\theta = \dfrac{A - C}{B} = \dfrac{16 - 9}{24} = \dfrac{7}{24} \Rightarrow$

$2\theta \approx 73.74° \Rightarrow \theta \approx 36.87°$. For 2θ; $x = 7, y = 24, r = \sqrt{7^2 + 24^2} = 25 \Rightarrow \cos 2\theta = \dfrac{7}{25}$.

$\sin \theta = \sqrt{\dfrac{1 - \cos 2\theta}{2}} = \sqrt{\dfrac{1 - \frac{7}{25}}{2}} = \sqrt{\dfrac{18}{50}} = \dfrac{3}{5}$; $\cos \theta = \sqrt{\dfrac{1 + \cos 2\theta}{2}} = \sqrt{\dfrac{1 + \frac{7}{25}}{2}} = \sqrt{\dfrac{32}{50}} = \dfrac{4}{5}$;

$x = x' \cos \theta - y' \sin \theta = \dfrac{4}{5}x' - \dfrac{3}{5}y'$ [2]; $y = x' \sin \theta + y' \cos \theta = \dfrac{3}{5}x' + \dfrac{4}{5}y'$ [3].

Substitute [2] and [3] in [1]. $16\left(\dfrac{4}{5}x' - \dfrac{3}{5}y'\right)^2 + 24\left(\dfrac{4}{5}x' - \dfrac{3}{5}y'\right)\left(\dfrac{3}{5}x' + \dfrac{4}{5}y'\right) +$

$9\left(\dfrac{3}{5}x' + \dfrac{4}{5}y'\right)^2 - 130\left(\dfrac{4}{5}x' - \dfrac{3}{5}y'\right) + 90\left(\dfrac{3}{5}x' + \dfrac{4}{5}y'\right) = 0 \Rightarrow$

$16\left(\dfrac{16}{25}x'^2 - \dfrac{24}{25}x'y' + \dfrac{9}{25}y'^2\right) + 24\left(\dfrac{12}{25}x'^2 + \dfrac{7}{25}x'y' - \dfrac{12}{25}y'^2\right) + 9\left(\dfrac{9}{25}x'^2 + \dfrac{24}{25}x'y' + \dfrac{16}{25}y'^2\right) -$

$104x' + 78y' + 54x' + 72y' = 0 \Rightarrow \dfrac{256}{25}x'^2 - \dfrac{384}{25}x'y' + \dfrac{144}{25}y'^2 + \dfrac{288}{25}x'^2 + \dfrac{168}{25}x'y' - \dfrac{288}{25}y'^2 +$

$\dfrac{81}{25}x'^2 + \dfrac{216}{25}x'y' + \dfrac{144}{25}y'^2 - 50x' + 150y' = 0 \Rightarrow 25x'^2 - 50x' + 150y' = 0 \Rightarrow$

$25(x'^2 - 2x' + 1) = -150y' + 25 \Rightarrow 25(x' - 1)^2 = -150\left(y' - \dfrac{1}{6}\right) \Rightarrow (x' - 1)^2 = -6\left(y' - \dfrac{1}{6}\right)$.

See Figure 29. The graph of the equation is a parabola with its vertex at $\left(1, \dfrac{1}{6}\right)$. By translating the axes of the

$x'y'$-system up $\dfrac{1}{6}$ units and right 1 unit, we get an $x''y''$-coordinate system, in which the parabola is centered at

the origin. Thus $x''^2 = -6y''$.

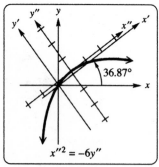

Figure 29